T0213211

Spencer J. Sherwin • David Moxey •
Joaquim Peiró • Peter E. Vincent • Christoph Schwab
Editors

Spectral and High Order Methods for Partial Differential Equations ICOSAHOM 2018

Selected Papers from the ICOSAHOM Conference, London, UK, July 9–13, 2018

 Springer

Editors
Spencer J. Sherwin
Department of Aeronautics
Imperial College
London, UK

David Moxey
College of Engineering, Mathematics &
Physical Sciences
University of Exeter
Exeter, UK

Joaquim Peiró
Department of Aeronautics
Imperial College
London, UK

Peter E. Vincent
Department of Aeronautics
Imperial College
London, UK

Christoph Schwab
Department of Mathematics
ETH Zürich
Zürich, Switzerland

ISSN 1439-7358   ISSN 2197-7100   (electronic)
Lecture Notes in Computational Science and Engineering
ISBN 978-3-030-39649-7   ISBN 978-3-030-39647-3   (eBook)
https://doi.org/10.1007/978-3-030-39647-3

Mathematics Subject Classification: 65M70, 65N35, 65N30, 74S25, 76M10, 76M22, 78M10, 78M22

This book is an open access publication.

Cover illustration: A 7th order accurate simulation of free stream turbulence passing over a turbine blade simulated using the Nektar++ package, courtesy of Andrea Cassinelli

This Springer imprint is published by the registered company Springer Nature Switzerland AG.
The registered company address is: Gewerbestrasse 11, 6330 Cham, Switzerland

Preface

This volume presents selected papers from the twelfth International Conference on Spectral and High-Order Methods (ICOSAHOM'18) that was held in London, United Kingdom, during the week of July 9–13th, 2018. These selected papers were refereed by members of the scientific committee of ICOSAHOM, as well as by other leading scientists.

The first ICOSAHOM conference was held in Como, Italy, in 1989 and marked the beginning of an international conference series in Montpellier, France (1992); Houston, TX, USA (1995); Tel Aviv, Israel (1998); Uppsala, Sweden (2001); Providence, RI, USA (2004); Beijing, China (2007); Trondheim, Norway (2009); Gammarth, Tunisia (2012); Salt Lake City, USA (2014); and Rio de Janeiro, Brazil (2016).

ICOSAHOM has established itself as the main meeting place for researchers with interests in the theoretical, applied, and computational aspects of high-order methods for the numerical solution of partial differential equations.

With over 360 attendees, ICOSAHOM '18 has been the largest edition of the conference series to date. The program consisted of eight invited speakers across the week from internationally renowned researchers, alongside 40 minisymposia (of around 300 presentations) dedicated to specialized topics in high-order methods, and approximately a further 90 contributed talks.

The content of these proceedings is organized as follows. First, contributions from the invited speakers are included. The remainder of the volume consists of refereed selected papers highlighting the broad spectrum of topics presented at ICOSAHOM '18.

The success of ICOSAHOM '18 was ensured through generous contributions and financial support of our sponsors: the Air Force Office of Scientific Research (AFSOR); the Platform for Research in Simulation Methods (PRISM) platform grant, funded by the Engineering and Physical Sciences Research Council (EPSRC); Rolls-Royce Ltd.; and, finally, the Department of Aeronautics at Imperial College London.

We would like to give special thanks to our local organizing committee for their efforts in organizing and promoting the event. In particular, we would also like to thank Mr. Andrea Cassinelli for his organizational efforts leading up to the conference, as well as the administrative staff of the Department of Aeronautics at Imperial College London for their help in coordinating the logistics of the event. We also thank the many student helpers for their advice, help, and support given to the delegates during the event itself, who all contributed to the smooth running of the event.

London, UK   Spencer J. Sherwin
Exeter, UK   David Moxey
London, UK   Joaquim Peiró
London, UK   Peter E. Vincent
Zürich, Switzerland   Christoph Schwab

Contents

Part I   Invited Papers

Stability of Wall Boundary Condition Procedures for Discontinuous Galerkin Spectral Element Approximations of the Compressible Euler Equations 3
Florian J. Hindenlang, Gregor J. Gassner, and David A. Kopriva

On the Order Reduction of Entropy Stable DGSEM for the Compressible Euler Equations 21
Florian J. Hindenlang and Gregor J. Gassner

A Review of Regular Decompositions of Vector Fields: Continuous, Discrete, and Structure-Preserving 45
Ralf Hiptmair and Clemens Pechstein

Model Reduction by Separation of Variables: A Comparison Between Hierarchical Model Reduction and Proper Generalized Decomposition 61
Simona Perotto, Michele Giuliano Carlino, and Francesco Ballarin

Recurrence Relations for a Family of Orthogonal Polynomials on a Triangle 79
Sheehan Olver, Alex Townsend, and Geoffrey M. Vasil

Part II   Contributed Papers

Greedy Kernel Methods for Center Manifold Approximation 95
Bernard Haasdonk, Boumediene Hamzi, Gabriele Santin, and Dominik Wittwar

An Adaptive Error Inhibiting Block One-Step Method for Ordinary Differential Equations 107
Jiaxi Gu and Jae-Hun Jung

Hermite Methods in Time 119
Rujie Gu and Thomas Hagstrom

**HPS Accelerated Spectral Solvers for Time Dependent Problems:
Part II, Numerical Experiments** 131
Tracy Babb, Per-Gunnar Martinsson, and Daniel Appelö

On the Use of Hermite Functions for the Vlasov–Poisson System 143
Lorella Fatone, Daniele Funaro, and Gianmarco Manzini

**HPS Accelerated Spectral Solvers for Time Dependent Problems:
Part I, Algorithms** 155
Tracy Babb, Per-Gunnar Martinsson, and Daniel Appelö

**High-Order Finite Element Methods for Interface Problems:
Theory and Implementations** 167
Yuanming Xiao, Fangman Zhai, Linbo Zhang, and Weiying Zheng

**Stabilised Hybrid Discontinuous Galerkin Methods for the Stokes
Problem with Non-standard Boundary Conditions** 179
Gabriel R. Barrenechea, Michał Bosy, and Victorita Dolean

RBF Based CWENO Method 191
Jan S. Hesthaven, Fabian Mönkeberg, and Sara Zaninelli

**Discrete Equivalence of Adjoint Neumann–Dirichlet div-grad
and grad-div Equations in Curvilinear 3D Domains** 203
Yi Zhang, Varun Jain, Artur Palha, and Marc Gerritsma

A Conservative Hybrid Method for Darcy Flow 215
Varun Jain, Joël Fisser, Artur Palha, and Marc Gerritsma

**High-Order Mesh Generation Based on Optimal Affine
Combinations of Nodal Positions** 229
Mike Stees and Suzanne M. Shontz

**Sparse Spectral-Element Methods for the Helically Reduced
Einstein Equations** 239
Stephen R. Lau

**Spectral Analysis of Isogeometric Discretizations of 2D Curl-Div
Problems with General Geometry** 251
Mariarosa Mazza, Carla Manni, and Hendrik Speleers

**Performance of Preconditioners for Large-Scale Simulations Using
Nek5000** 263
N. Offermans, A. Peplinski, O. Marin, E. Merzari, and P. Schlatter

**Two Decades Old Entropy Stable Method for the Euler Equations
Revisited** 273
Björn Sjögreen and H. C. Yee

A Mimetic Spectral Element Method for Free Surface Flows 285
L. Nielsen and B. Gervang

Spectral/hp Methodology Study for iLES-SVV on an Ahmed Body 297
Filipe F. Buscariolo, Spencer J. Sherwin, Gustavo R. S. Assi,
and Julio R. Meneghini

A High-Order Discontinuous Galerkin Solver for Multiphase Flows 313
Juan Manzanero, Carlos Redondo, Gonzalo Rubio, Esteban Ferrer,
Eusebio Valero, Susana Gómez-Álvarez, and Ángel Rivero-Jiménez

**High-Order Propagation of Jet Noise on a Tetrahedral Mesh Using
Large Eddy Simulation Sources** 325
M. A. Moratilla-Vega, V. Saini, H. Xia, and G. J. Page

Dynamical Degree Adaptivity for DG-LES Models 337
M. Tugnoli, A. Abbà, and L. Bonaventura

**A Novel Eighth-Order Diffusive Scheme for Unstructured
Polyhedral Grids Using the Weighted Least-Squares Method** 349
Duarte M. S. Albuquerque, Artur G. R. Vasconcelos, and Jose C. F. Pereira

An Explicit Mapped Tent Pitching Scheme for Maxwell Equations 359
Jay Gopalakrishnan, Matthias Hochsteger, Joachim Schöberl,
and Christoph Wintersteiger

**Viscous Diffusion Effects in the Eigenanalysis of (Hybridisable) DG
Methods** 371
Rodrigo C. Moura, Pablo Fernandez, Gianmarco Mengaldo,
and Spencer J. Sherwin

**Spectral Galerkin Method for Solving Helmholtz and Laplace
Dirichlet Problems on Multiple Open Arcs** 383
Carlos Jerez-Hanckes and José Pinto

**Explicit Polynomial Trefftz-DG Method for Space-Time
Elasto-Acoustics** 395
H. Barucq, H. Calandra, J. Diaz, and E. Shishenina

**An hp-Adaptive Iterative Linearization Discontinuous-Galerkin
FEM for Quasilinear Elliptic Boundary Value Problems** 407
Paul Houston and Thomas P. Wihler

Erosion Wear Evaluation Using Nektar++ 419
Manuel F. Mejía, Douglas Serson, Rodrigo C. Moura, Bruno S. Carmo,
Jorge Escobar-Vargas, and Andrés González-Mancera

**An Inexact Petrov-Galerkin Approximation for Gas Transport
in Pipeline Networks** 429
Herbert Egger, Thomas Kugler, and Vsevolod Shashkov

New Preconditioners for Semi-linear PDE-Constrained Optimal Control in Annular Geometries 441
Lasse Hjuler Christiansen and John Bagterp Jørgensen

DIRK Schemes with High Weak Stage Order 453
David I. Ketcheson, Benjamin Seibold, David Shirokoff, and Dong Zhou

Scheme for Evolutionary Navier-Stokes-Fourier System with Temperature Dependent Material Properties Based on Spectral/hp Elements 465
Jan Pech

Implicit Large Eddy Simulations for NACA0012 Airfoils Using Compressible and Incompressible Discontinuous Galerkin Solvers 477
Esteban Ferrer, Juan Manzanero, Andres M. Rueda-Ramirez, Gonzalo Rubio, and Eusebio Valero

SAV Method Applied to Fractional Allen-Cahn Equation 489
Xiaolan Zhou, Mejdi Azaiez, and Chuanju Xu

A First Meshless Approach to Simulation of the Elastic Behaviour of the Diaphragm 501
Nicola Cacciani, Elisabeth Larsson, Alberto Lauro, Marco Meggiolaro, Alessio Scatto, Igor Tominec, and Pierre-Frédéric Villard

An Explicit Hybridizable Discontinuous Galerkin Method for the 3D Time-Domain Maxwell Equations 513
Georges Nehmetallah, Stéphane Lanteri, Stéphane Descombes, and Alexandra Christophe

Entropy Conserving and Kinetic Energy Preserving Numerical Methods for the Euler Equations Using Summation-by-Parts Operators 525
Hendrik Ranocha

Multiwavelet Troubled-Cell Indication: A Comparison of Utilizing Theory Versus Outlier Detection 537
Mathea J. Vuik

An Anisotropic p-Adaptation Multigrid Scheme for Discontinuous Galerkin Methods 549
Andrés M. Rueda-Ramírez, Gonzalo Rubio, Esteban Ferrer, and Eusebio Valero

A Spectral Element Reduced Basis Method for Navier–Stokes Equations with Geometric Variations 561
Martin W. Hess, Annalisa Quaini, and Gianluigi Rozza

Iterative Spectral Mollification and Conjugation for Successive Edge Detection... 573
Robert E. Tuzun and Jae-Hun Jung

Small Trees for High Order Whitney Elements 587
Ana Alonso Rodríguez and Francesca Rapetti

Non-conforming Elements in Nek5000: Pressure Preconditioning and Parallel Performance 599
A. Peplinski, N. Offermans, P. F. Fischer, and P. Schlatter

Sparse Approximation of Multivariate Functions from Small Datasets Via Weighted Orthogonal Matching Pursuit 611
Ben Adcock and Simone Brugiapaglia

On the Convergence Rate of Hermite-Fejér Interpolation 623
Shuhuang Xiang and Guo He

Fifth-Order Finite-Volume WENO on Cylindrical Grids 637
Mohammad Afzal Shadab, Xing Ji, and Kun Xu

Part I
Invited Papers

Stability of Wall Boundary Condition Procedures for Discontinuous Galerkin Spectral Element Approximations of the Compressible Euler Equations

Florian J. Hindenlang, Gregor J. Gassner, and David A. Kopriva

1 Introduction

The ingredients for a reliable numerical method for the approximation of partial differential equations, e.g. one that will not blow up, include stable inter-element and physical boundary condition implementations. The recognition that the discontinuous Galerkin spectral element method (DGSEM) with Gauss-Lobatto quadratures satisfies a summation-by-parts (SBP) operators [4, 7] has allowed for the analysis of these schemes and to connect them with penalty collocation and SBP finite difference schemes. For instance, in [5], we showed that a split form approximation of the compressible Navier–Stokes equations was both linearly and entropy stable provided that the boundary conditions were properly imposed.

The importance of stable boundary condition procedures for hyperbolic equations has long been studied, especially in relation to finite difference methods, e.g. [3, 9, 10]. Only recently have they been studied for discontinuous Galerkin approximations. In [12], the authors showed that the reflection approach is stable when using an entropy conserving flux and an additional entropy stable dissipation

F. J. Hindenlang
Max Planck Institute for Plasma Physics, Garching, Germany
e-mail: florian.hindenlang@ipp.mpg.de

G. J. Gassner (✉)
Department for Mathematics and Computer Science, Center for Data and Simulation Science, University of Cologne, Cologne, Germany
e-mail: ggassner@math.uni-koeln.de

D. A. Kopriva
Florida State University, Tallahassee, FL, USA

San Diego State University, San Diego, CA, USA
e-mail: kopriva@math.fsu.edu

© The Author(s) 2020
S. J. Sherwin et al. (eds.), *Spectral and High Order Methods for Partial Differential Equations ICOSAHOM 2018*, Lecture Notes in Computational Science and Engineering 134, https://doi.org/10.1007/978-3-030-39647-3_1

term (EC-ES). In [2], the authors show that the reflection condition is stable if the numerical flux is either the Godunov or HLL flux.

In this paper, we analyze both the linear and entropy stability of two types of commonly used wall boundary condition procedures used with the DGSEM applied to the compressible Euler equations. In both cases, wall boundary conditions are implemented through a numerical flux. The boundary condition might be implemented through a special wall numerical flux that includes the boundary condition, or a fictitious external state applied to a Riemann solver approximation. We show how to construct special wall numerical fluxes that are stable, and study the behavior of the approximations. In particular, we show that the use of Riemann solvers at the boundaries introduce numerical dissipation in an amount that depends on the size of the normal Mach number at the wall.

2 The Compressible Euler Equations and the Wall Boundary Condition

We write the Euler equations as

$$\mathbf{u}_t + \sum_{i=1}^{3} \frac{\partial \mathbf{f}_i}{\partial x_i} = 0. \tag{1}$$

The state vector contains the conservative variables

$$\mathbf{u} = \begin{bmatrix} \varrho & \varrho\vec{v} & E \end{bmatrix}^T = \begin{bmatrix} \varrho & \varrho v_1 & \varrho v_2 & \varrho v_3 & E \end{bmatrix}^T. \tag{2}$$

In standard form, the components of the advective fluxes are

$$\mathbf{f}_1 = \begin{bmatrix} \varrho v_1 \\ \varrho v_1^2 + p \\ \varrho v_1 v_2 \\ \varrho v_1 v_3 \\ (E+p)v_1 \end{bmatrix} \quad \mathbf{f}_2 = \begin{bmatrix} \varrho v_2 \\ \varrho v_2 v_1 \\ \varrho v_2^2 + p \\ \varrho v_2 v_3 \\ (E+p)v_2 \end{bmatrix} \quad \mathbf{f}_3 = \begin{bmatrix} \varrho v_3 \\ \varrho v_3 v_1 \\ \varrho v_3 v_2 \\ \varrho v_3^2 + p \\ (E+p)v_3 \end{bmatrix}, \tag{3}$$

Here, ϱ, $\vec{v} = (v_1, v_2, v_3)^T$, p, E are the mass density, fluid velocities, pressure and total energy. We close the system with the ideal gas assumption, which relates the total energy and pressure

$$p = (\gamma - 1)\left(E - \frac{1}{2}\varrho \|\vec{v}\|^2\right), \tag{4}$$

where γ denotes the adiabatic coefficient. For a compact notation that simplifies the analysis, we define *block vectors* (with the double arrow)

$$\overset{\leftrightarrow}{\mathbf{f}} = \begin{bmatrix} \mathbf{f}_1 & \mathbf{f}_2 & \mathbf{f}_3 \end{bmatrix}^T, \tag{5}$$

so that the system of equations can be written in the compact form

$$\mathbf{u}_t + \vec{\nabla}_x \cdot \overset{\leftrightarrow}{\mathbf{f}} = 0. \tag{6}$$

The linear Euler equations are derived by linearizing about a constant mean state $(\bar{\varrho}, \bar{v}_1, \bar{v}_2, \bar{v}_3, \bar{p})$. We follow [11] for the symmetrization of the linearized equations, with the constants

$$a = \sqrt{\frac{\gamma - 1}{\gamma}} \bar{c}, \quad b = \frac{\bar{c}}{\sqrt{\gamma}}, \bar{c} = \sqrt{\frac{\gamma \bar{p}}{\bar{\varrho}}}, \tag{7}$$

where $\bar{c}$ is the sound speed of the constant mean state. The state variables become

$$\mathbf{u} = \begin{bmatrix} \varrho' & v_1 & v_2 & v_3 & p' \end{bmatrix}^T, \tag{8}$$

where $\vec{v}$ is the velocity perturbation from the mean state, and we introduce

$$\varrho' = b\frac{\tilde{\varrho}}{\bar{\varrho}}, \quad p' = \frac{1}{\bar{\varrho}a}\tilde{p} - \frac{1}{\sqrt{\gamma - 1}}\varrho', \tag{9}$$

which depend on the density and pressure perturbations $\tilde{\varrho}, \tilde{p}$. The flux vectors are

$$\mathbf{f}_i = \underline{A}_i \mathbf{u}, \quad \overset{\leftrightarrow}{\mathbf{f}} = \underline{\vec{A}}\mathbf{u} = \left(\underline{A}_1 \hat{x} + \underline{A}_2 \hat{y} + \underline{A}_3 \hat{z}\right)\mathbf{u}, \tag{10}$$

where [11]

$$\underline{A}_1 = \begin{bmatrix} \bar{v}_1 & b & 0 & 0 & 0 \\ b & \bar{v}_1 & 0 & 0 & a \\ 0 & 0 & \bar{v}_1 & 0 & 0 \\ 0 & 0 & 0 & \bar{v}_1 & 0 \\ 0 & a & 0 & 0 & \bar{v}_1 \end{bmatrix}, \quad \underline{A}_2 = \begin{bmatrix} \bar{v}_2 & 0 & b & 0 & 0 \\ 0 & \bar{v}_2 & 0 & 0 & 0 \\ b & 0 & \bar{v}_2 & 0 & a \\ 0 & 0 & 0 & \bar{v}_2 & 0 \\ 0 & 0 & a & 0 & \bar{v}_2 \end{bmatrix}, \quad \underline{A}_3 = \begin{bmatrix} \bar{v}_3 & 0 & 0 & b & 0 \\ 0 & \bar{v}_3 & 0 & 0 & 0 \\ 0 & 0 & \bar{v}_3 & 0 & 0 \\ b & 0 & 0 & \bar{v}_3 & a \\ 0 & 0 & 0 & a & \bar{v}_3 \end{bmatrix} \tag{11}$$

are constant symmetric matrices.

The linear equations have the property that the L^2 norm of the solution over a domain Ω is bounded by terms of the boundary data on $\partial\Omega$, only. Let

$$\langle \mathbf{v}, \mathbf{w} \rangle = \int_\Omega \mathbf{v}^T \mathbf{w} \, dx dy dz, \quad \left\langle \overleftrightarrow{\mathbf{f}}, \overleftrightarrow{\mathbf{g}} \right\rangle = \int_\Omega \sum_{i=1}^{3} \mathbf{f}_i^T \mathbf{g}_i \, dx dy dz. \tag{12}$$

represent the L^2 inner product of two state vectors $\mathbf{v}$ and $\mathbf{w}$ and two block vectors $\overleftrightarrow{\mathbf{f}}$ and $\overleftrightarrow{\mathbf{g}}$, respectively. Since the coefficient matrices are constant the product rule and symmetry of $\underline{\mathbf{A}}$ implies

$$\left\langle \vec{\nabla}_x \cdot \overleftrightarrow{\mathbf{f}}, \mathbf{u} \right\rangle = \left\langle \vec{\nabla}_x \cdot \left(\overleftrightarrow{\mathbf{A}} \mathbf{u} \right), \mathbf{u} \right\rangle = \left\langle \nabla_x \mathbf{u}, \overleftrightarrow{\mathbf{f}} \right\rangle. \tag{13}$$

Then it follows from Gauss' law (integration by parts) that

$$\left\langle \vec{\nabla}_x \cdot \overleftrightarrow{\mathbf{f}}, \mathbf{u} \right\rangle = \frac{1}{2} \int_{\partial\Omega} \mathbf{u}^T \overleftrightarrow{\mathbf{f}} \cdot \vec{n} \, dS, \tag{14}$$

where $\vec{n}$ is the outward normal to the surface of Ω. The norm of the solution therefore satisfies

$$\frac{d}{dt} ||\mathbf{u}||^2 = -\int_{\partial\Omega} \mathbf{u}^T \overleftrightarrow{\mathbf{f}} \cdot \vec{n} \, dS. \tag{15}$$

Replacing the boundary terms by boundary conditions leads to a bound on the solution in terms of the boundary data. The argument of the boundary integral on the right of (15) is

$$\mathbf{u}^T \overleftrightarrow{\mathbf{f}} \cdot \vec{n} = \mathbf{u}^T \left(\underline{\mathbf{A}} \cdot \vec{n} \right) \mathbf{u} = 2 \left(\varrho b + a p \right) v_n + (\vec{v} \cdot \vec{n})(\varrho^2 + |\vec{v}|^2 + p^2), \tag{16}$$

where v_n is the wall normal velocity, $v_n = \vec{v} \cdot \vec{n}$. Note that here, the mean flow must be chosen such that the normal flow vanishes at the wall boundary $\vec{v} \cdot \vec{n} = 0$, so that the boundary condition makes physical sense.

Therefore, with the no penetration wall condition $v_n = 0$ applied,

$$\frac{d}{dt} ||\mathbf{u}||^2 = 0, \tag{17}$$

and the (energy) norm of the solution is bounded for all time by its initial value.

The nonlinear equations, on the other hand, satisfy a bound on the entropy that depends only on the boundary data. For what follows, we assume that the solution is smooth so that we don't have to consider entropy generated at shock waves. We introduce the entropy density (scaled with $(\gamma - 1)$ for convenience) as

$$s(\mathbf{u}) = -\frac{\varrho\varsigma}{(\gamma - 1)}, \tag{18}$$

where $\varsigma = \ln(p) - \gamma \ln(\varrho)$ is the physical entropy. (The minus sign is conventional in the theory of hyperbolic conservation laws to ensure a decreasing entropy function.) The entropy flux for the Euler equations is

$$\vec{f}^{\varsigma}(\mathbf{u}) = \vec{v}\,s = -\frac{\varrho\varsigma\vec{v}}{(\gamma - 1)}. \tag{19}$$

Finally the entropy variables are

$$\mathbf{w} = \frac{\partial s(\mathbf{u})}{\partial \mathbf{u}} = \begin{bmatrix} \frac{\gamma - \varsigma}{\gamma - 1} - \beta\|\vec{v}\|^2, \\ 2\beta\vec{v} \\ -2\beta \end{bmatrix}, \qquad \beta = \frac{\varrho}{2p}. \tag{20}$$

The entropy pair contracts the solution and fluxes, meaning that it satisfies the relations

$$\mathbf{w}^T \mathbf{u}_t = \left(\frac{\partial s}{\partial \mathbf{u}}\right)^T \mathbf{u}_t = s_t(\mathbf{u}), \qquad \mathbf{w}^T \vec{\nabla}_x \cdot \vec{\mathbf{f}} = \vec{\nabla}_x \cdot \vec{f}^{\varsigma}. \tag{21}$$

When we multiply (6) with the entropy variables and integrate over the domain,

$$\langle \mathbf{w}(\mathbf{u}), \mathbf{u}_t \rangle + \langle \mathbf{w}(\mathbf{u}), \vec{\nabla}_x \cdot \vec{\mathbf{f}} \rangle = 0. \tag{22}$$

Next we use the properties of the entropy pair to contract (22) and use integration by parts to get

$$\langle s_t(\mathbf{u}), 1 \rangle = -\langle \vec{\nabla}_x \cdot \vec{f}^{\varsigma}, 1 \rangle = -\int_{\partial\Omega} \left(\vec{f}^{\varsigma} \cdot \vec{n}\right) dS \tag{23}$$

showing that, in the continuous case, the total entropy in the domain can only change via the boundary conditions.

In the case of a zero-mass flux boundary condition, with $v_n = \vec{v} \cdot \vec{n} = 0$, the entropy is not changed by the slip-wall boundary condition, since

$$-\vec{f}^{\varsigma} \cdot \vec{n} = \frac{\varrho\varsigma}{(\gamma - 1)} v_n = 0. \tag{24}$$

3 Stability Bounds for the DGSEM

The DGSEM is described in detail in [5] and elsewhere [1, 6]. We will only quickly summarize the approximation here. The domain, Ω is subdivided into non-overlapping, conforming, hexahedral elements. Each element is mapped to the reference element $E = [-1, 1]^3$. Associated with the transformation from the reference element is a set of contravariant coordinate vectors, $\vec{a}^i$, and transformation Jacobian, $\mathcal{J}$. Equation (6) transform to another conservation law on the reference element as

$$\mathcal{J}\mathbf{u}_t + \vec{\nabla}_\xi \cdot \overleftrightarrow{\tilde{\mathbf{f}}} = 0, \tag{25}$$

where $\overleftrightarrow{\tilde{\mathbf{f}}}$ is the contravariant flux vector with components $\tilde{\mathbf{f}}^i = \mathcal{J}\vec{a}^i \cdot \overleftrightarrow{\mathbf{f}}$.

The approximation of (25) proceeds as follows: A weak form is created by taking the inner product of the equation with a test function. The Gauss law is applied to the divergence term to separate the boundary from the interior contributions. The resulting weak form is then approximated: The solution vector is approximated by a polynomial of degree N interpolated at the Legendre–Gauss–Lobatto points. In the following, we will represent the true continuous solutions by lower case letter. Upper case letters will denote their polynomial approximations, except for the density, where the approximation is denoted by ρ. The volume fluxes are replaced by two-point numerical fluxes. In the linear case, the two point fluxes are immediately relatable to a split form of the equations. Integrals are replaced by Legendre–Gauss–Lobatto quadratures. Finally, the boundary fluxes are replaced by a numerical flux. See [5] and [8] for details.

The result is an approximation that is energy stable for the linearized equations if at every quadrature point along a physical boundary the numerical flux $\tilde{\mathbf{F}}^*$ satisfies the bound [5]

$$\mathbf{U}^T \left\{ \tilde{\mathbf{F}}^* - \frac{1}{2}\overleftrightarrow{\tilde{\mathbf{F}}} \cdot \hat{n} \right\} \geq 0, \tag{26}$$

where $\overleftrightarrow{\tilde{\mathbf{F}}}$ is the polynomial interpolation of the contravariant flux from the interior, $\hat{n}$ is the reference space outward normal direction, and $\mathbf{U}$ is the approximation of the state vector. Since the contravariant fluxes are proportional to the normal fluxes [6], we can change the condition (26) to

$$B_L \equiv \mathbf{U}^T \left\{ \mathbf{F}^* - \frac{1}{2}\overleftrightarrow{\mathbf{F}} \cdot \vec{n} \right\} \geq 0, \tag{27}$$

For entropy stability of the nonlinear equations, the boundary stability condition shown in [5] is proportional to

$$B_{NL} \equiv \mathbf{W}^T \left(\mathbf{F}^* - \left(\overleftrightarrow{\mathbf{F}} \cdot \vec{n} \right) \right) + \left(\vec{F}^{\,s} \cdot \vec{n} \right) \geq 0, \tag{28}$$

where $\vec{F}^{\,s}$ is the polynomial interpolation of the entropy flux, $\vec{f}^{\,s}$, and $\mathbf{W}$ is the interpolation of the entropy variables.

3.1 Linear Stability of Wall Boundary Condition Approximations

To find linearly stable implementations of the wall condition $v_n = 0$, one needs only find a numerical flux that satisfies it and the condition (27). For the linear equations, the approximation of the state vector is $\mathbf{U} = [\rho' \ \ \vec{V} \ \ P']^T$ and the normal contravariant flux is proportional to

$$\overleftrightarrow{\mathbf{F}} \cdot \vec{n} = \underline{\underline{\mathbf{A}}} \cdot \vec{n} \, \mathbf{U} = \begin{bmatrix} b V_n & n_1 Q & n_2 Q & n_3 Q & a V_n \end{bmatrix}^T, \tag{29}$$

where V_n is the approximation of the normal velocity at the wall computed from the interior, $Q = b\rho' + aP'$, and (n_1, n_1, n_3) are the three components of the physical space normal vector, $\vec{n}$. The numerical flux can be expressed as

$$\mathbf{F}^* = \underline{\underline{\mathbf{A}}} \cdot \vec{n} \, \mathbf{U}^* = \begin{bmatrix} b V_n^* & n_1 Q^* & n_2 Q^* & n_3 Q^* & a V_n^* \end{bmatrix}^T. \tag{30}$$

It then remains only to find Q^* so that (27) is satisfied when the normal wall condition $V_n^* = 0$ is applied. When we substitute the fluxes (29) and (30) into (27),

$$B_L = \frac{1}{2} \left\{ Q \left(2V_n^* - V_n \right) + V_n \left(2Q^* - Q \right) \right\} = \frac{1}{2} \left\{ 2Q V_n^* + 2V_n \left(Q^* - Q \right) \right\} \tag{31}$$

Substituting the wall boundary condition $V_n^* = 0$ yields the condition on Q^* for stability

$$V_n \left(Q^* - Q \right) \geq 0. \tag{32}$$

Neutral stability is thus ensured if ρ^* and P^* are computed from the interior, i.e. $\rho^* = \rho'$, $P^* = P'$ so that $Q^* = Q$.

In practice, the boundary condition is also implemented through the use of a Riemann solver and external state designed to imply the physical boundary condition to construct the numerical boundary flux. The exact upwind ($\varepsilon = 1$) normal Riemann flux and the central flux ($\varepsilon = 0$) for the linear system of equations is

$$\mathbf{F}^* \left(\mathbf{U}, \mathbf{U}^{\text{ext}} \right) = \frac{1}{2} \left\{ \vec{\overleftrightarrow{\mathbf{F}}} \left(\mathbf{U} \right) \cdot \vec{n} + \vec{\overleftrightarrow{\mathbf{F}}} \left(\mathbf{U}^{\text{ext}} \right) \cdot \vec{n} \right\} - \frac{\varepsilon}{2} \left| \underline{A}_n \right| \left(\mathbf{U}^{\text{ext}} - \mathbf{U} \right), \tag{33}$$

where $\underline{A}_n \equiv \vec{\underline{A}} \cdot \vec{n}$ is the normal coefficient matrix. The external state is set by using the interior values of the density and pressure and the negative of the value of the normal velocity,

$$\mathbf{U}^{\text{ext}} = \left[\rho' \quad \left(\vec{V} - 2V_n \vec{n} \right) \quad P' \right]^T. \tag{34}$$

For $\varepsilon = 0$, using the central (averaged) numerical flux, the interior flux contribution cancels and condition (27) reduces to

$$B_{L,0} = \frac{1}{2} \mathbf{U}^T \underline{A}_n \mathbf{U}^{\text{ext}} = \left[\rho' \quad \vec{V} \quad P' \right] \begin{bmatrix} 0 & n_1 b & n_2 b & n_3 b & 0 \\ n_1 b & 0 & 0 & 0 & n_1 a \\ n_2 b & 0 & 0 & 0 & n_2 a \\ n_3 b & 0 & 0 & 0 & n_3 a \\ 0 & n_1 a & n_2 a & n_3 a & 0 \end{bmatrix} \begin{bmatrix} \rho' \\ \vec{V} - 2V_n \vec{n} \\ P' \end{bmatrix}$$

$$= Q \left(-V \cdot \vec{n} \right) + \left(V \cdot \vec{n} \right) Q = 0, \tag{35}$$

which is neutrally stable, having no additional stabilizing dissipation. We note again, that the mean state for the linearization is chosen such that the normal mean velocity components are zero, resulting in the zeros on the diagonal of $\underline{A}_n$.

Substituting the exact upwind flux where $\varepsilon = 1$ into (27) and rearranging,

$$B_{L,1} = -\mathbf{U}^T \left| \underline{A}_n^- \right| \mathbf{U}^{\text{ext}} + \frac{1}{2} \mathbf{U}^T \left| \underline{A}_n \right| \mathbf{U}, \tag{36}$$

where $\underline{A}_n^- = \frac{1}{2} \left(\underline{A}_n - \left| \underline{A}_n \right| \right)$ is negative semidefinite. The second term is non-negative, depends only on the interior state, and adds stabilizing dissipation. From the matrix absolute value, the dissipation term is

$$\mathbf{U}^T \left| \underline{A}_n \right| \mathbf{U} = \frac{1}{\bar{c}} Q^2 + \bar{c}^3 \mathrm{Ma}_n^2, \tag{37}$$

where $\mathrm{Ma}_n = V_n / \bar{c}$ is the normal Mach number. Stability depends, then, on the value of the first term, which is where the boundary conditions are incorporated through the external state $\mathbf{U}^{ext}$ written in (34). Then

$$\mathbf{U}^T \left| \underline{A}_n^- \right| \mathbf{U}^{ext} = \frac{1}{2\bar{c}} Q^2 - \frac{\bar{c}^3}{2} \mathrm{Ma}^2. \tag{38}$$

Therefore, using the upwind numerical flux, (36) becomes

$$B_{L,1} = \bar{c}^3 \mathrm{Ma}_n^2 \geq 0, \tag{39}$$

as required. The amount of dissipation depends on how far the interior computed normal velocity deviates from zero.

The combination of the reflective state and local Lax-Friedrichs flux is also linearly stable. In that case the exact matrix absolute value is replaced by a diagonal matrix, $\left| \underline{A}_n \right| \approx |\lambda|_{\max} \underline{I}$. The jump term is added to the central (averaged) flux so

$$B_{L,LF} = -\frac{|\lambda|_{\max}}{2} \mathbf{U}^T \left(\mathbf{U}^{ext} - \mathbf{U} \right) = \bar{c}^2 |\lambda|_{\max} \mathrm{Ma}_n^2 \geq 0 \tag{40}$$

Finally, a dissipative version of the direct numerical flux (30) can be formed by looking at the reflective state approach. For instance, the equivalent to using the Lax-Friedrichs flux is to choose $\rho^* = \rho'$ and

$$P^* = P' + \frac{\bar{c}^3}{a} |\lambda|_{\max} \mathrm{Ma}_n. \tag{41}$$

Then $Q^* = Q + \bar{c}^3 |\lambda|_{\max} \mathrm{Ma}_n$ and

$$V_n \left(Q^* - Q \right) = \bar{c}^2 |\lambda|_{\max} \mathrm{Ma}_n^2 \geq 0. \tag{42}$$

A similar, though more complicated, modified P^* can be made to be equivalent to the exact upwind flux.

3.2 Entropy Stability of Wall Boundary Condition Approximations

As in the linear approximation, the wall boundary condition can be imposed for the nonlinear equations either by directly specifying the numerical flux or by computing it through a Riemann solver using a reflection external state that enforces the normal wall condition implicitly. Note that in this section, the discrete variables $(\rho, \vec{V}, P)$ describe the full nonlinear state.

For the nonlinear equations, we construct the numerical flux for a slip-wall as

$$\left(\overleftrightarrow{\mathbf{F}}\cdot\vec{n}\right)^* = \begin{bmatrix} 0 \\ P^*\vec{n} \\ 0 \end{bmatrix} \tag{43}$$

where we imposed $V_n = 0$ leading to a flux with no mass or energy transfer, and we introduce a wall pressure P^*, whose value will be chosen to ensure consistency and stability.

After some manipulations, the discrete entropy stability condition (28) becomes

$$-\rho V_n \left(\frac{\gamma - \varsigma}{(\gamma - 1)} - \beta||\vec{V}||^2\right)$$

$$+2\beta V_n \left(P^* - P - \rho||\vec{V}||^2\right) + 2\beta V_n \left(\rho E + P\right) - \frac{\rho\varsigma V_n}{(\gamma - 1)} =$$

$$-\rho V_n \left(\frac{\gamma}{(\gamma - 1)} - \beta||\vec{V}||^2\right) + 2\beta V_n \left(P^* + \rho E - \rho||\vec{V}||^2\right) = \tag{44}$$

$$\frac{\rho V_n}{P} \left(-\frac{\gamma}{(\gamma - 1)}P + \frac{1}{2}\rho||\vec{V}||^2 + P^* + \frac{P}{(\gamma - 1)} - \frac{1}{2}\rho||\vec{V}||^2\right) =$$

$$\rho V_n \left(\frac{P^*}{P} - 1\right) \geq 0$$

Therefore if we choose $P^* = P$, to be the internal pressure, the boundary flux does not contribute to the total entropy, independent of the inner normal velocity V_n. A value of P^* that leads to a dissipative boundary condition can be found either through exact solution of the Riemann problem at the boundary, or through the use of an external state and an approximate Riemann solver.

3.2.1 Exact Solution of the Riemann Problem

In [14] a symmetric 1D Riemann problem is exactly solved following Toro [13], to get the wall pressure P^*, accounting for the fact that V_n never vanishes discretely and therefore the wall pressure should be different from the interior pressure. The exact solution of the 1D Riemann problem reads as

$$\left(\frac{P^*}{P}\right)_{RP} = \begin{cases} 1 + \gamma \mathrm{Ma}_n \left(\frac{(\gamma+1)}{4}\mathrm{Ma}_n + \sqrt{\left(\frac{(\gamma+1)}{4}\mathrm{Ma}_n\right)^2 + 1}\right) > 1 & \text{for} \quad V_n > 0 \\ \left(1 + \frac{1}{2}(\gamma - 1)\mathrm{Ma}_n\right)^{\frac{2\gamma}{(\gamma-1)}} \leq 1 & \text{for} \quad V_n \leq 0 \end{cases} \tag{45}$$

with the normal Mach number, $\mathrm{Ma}_n = \frac{V_n}{c}$, and the sound speed $c = \sqrt{\gamma\frac{P}{\rho}}$.

As shown by Toro [13], the solution for the rarefaction has a limiting vacuum solution for $\mathrm{Ma}_n \leq -2(\gamma - 1)^{-1}$. We will restrict our analysis to normal Mach numbers yielding strictly positive pressure solutions only ($\mathrm{Ma}_n > -5$ for $\gamma = \frac{7}{5}$).

It is easy to see that using P^* from (45), the entropy inequality (44) is still satisfied for $|V_n| \neq 0$, and the added entropy scales with the discrete value of V_n at the boundary. Hence, for $h \to 0$, the discrete boundary condition converges to its physical counterpart, since $V_n \to 0$. The choice of P^* from (45) appears to stabilize under-resolved simulations, which can be now explained by the fact that the boundary flux always adds entropy for $|V_n| \neq 0$.

3.2.2 Using Approximate Riemann Solvers for the Boundary Flux

A well known strategy in finite volume methods is to mirror only the velocity of the internal state and solve an approximate Riemann problem to get the boundary flux, mostly just because of a simpler implementation, since an approximate Riemann solver is already available and used for the fluxes between the elements. For DG methods, see also, for example, [2] and [12] where reflection conditions are proved to be entropy stable.

The mirror state is set so that the mass and energy flux are zero. Let the inner state be labeled I, and the outer R. then the inner and outer states that satisfy the mirror condition are

$$\mathbf{U}^L = \begin{bmatrix} \rho & \rho \vec{V}_n & E \end{bmatrix}^T, \quad \mathbf{U}^R = \begin{bmatrix} \rho & \rho(\vec{V} - 2V_n \vec{n}) & E \end{bmatrix}^T \tag{46}$$

We show below under what conditions on the normal velocity V_n that the reflection condition is entropy stable for the Lax-Friedrichs, HLL and HLLC, Roe and EC-ES fluxes.

Lax-Friedrichs Flux

We start with the simplest approximate Riemann solver, the Lax-Friedrichs or Rusanov flux, which reads as

$$\left(\overset{\leftrightarrow}{\mathbf{F}} \cdot \vec{n} \right)^*_{\mathrm{LF}} = \frac{1}{2} \vec{n} \cdot \left(\overset{\leftrightarrow}{\mathbf{F}}(\mathbf{U}^L) + \overset{\leftrightarrow}{\mathbf{F}}(\mathbf{U}^R) \right) - \frac{|\lambda|_{\max}}{2} (\mathbf{U}^R - \mathbf{U}^L). \tag{47}$$

Inserting the states from (46), we get

$$\left(\overset{\leftrightarrow}{\mathbf{F}} \cdot \vec{n} \right)^*_{\mathrm{LF}} = \begin{bmatrix} 0 \\ (\rho V_n^2 + P)\vec{n} \\ 0 \end{bmatrix} - \frac{\lambda_{\max}}{2} \begin{bmatrix} 0 \\ -2\rho V_n \vec{n} \\ 0 \end{bmatrix} = \begin{bmatrix} 0 \\ (\rho V_n^2 + \rho V_n \lambda_{\max} + P)\vec{n} \\ 0 \end{bmatrix}.$$
$$\tag{48}$$

The maximum wave speed is normally approximated from the largest leftgoing and rightgoing wave speed,

$$\lambda_{\max} = \max(|V_n^L| + c^L, |V_n^R| + c^R) = |V_n| + c, \quad \text{since} \quad c^L = c^R = c,$$

$$V_n = V_n^L = -V_n^R \tag{49}$$

and thus gives a definition of P^*

$$\left(\frac{P^*}{P}\right)_{\text{LF}} = 1 + \gamma \text{Ma}_n \left(\text{Ma}_n + |\text{Ma}_n| + 1\right)$$

$$= \begin{cases} 1 + \gamma \text{Ma}_n (2\text{Ma}_n + 1) > 1 & \text{for} \quad V_n > 0 \\ 1 + \gamma \text{Ma}_n \qquad\qquad \leq 1 & \text{for} \quad V_n \leq 0 \end{cases}, \tag{50}$$

which shows that the Lax-Friedrichs flux satisfies the entropy inequality (44).

HLL and HLLC Flux

The HLL flux [13] is written as

$$\left(\overset{\leftrightarrow}{\mathbf{F}} \cdot \vec{n}\right)^*_{\text{HLL}} = \frac{1}{S^R - S^L} \left(\vec{n} \cdot \left(S^R \overset{\leftrightarrow}{\mathbf{F}}(\mathbf{U}^L) - S^L \overset{\leftrightarrow}{\mathbf{F}}(\mathbf{U}^R)\right) + S^L S^R \left(\mathbf{U}^R - \mathbf{U}^L\right)\right). \tag{51}$$

The leftgoing and rightgoing wave speeds are $S^L = V_n^L - c^L = -V_n^R - c^R = -S^R$ and the HLL flux reduces to

$$\left(\overset{\leftrightarrow}{\mathbf{F}} \cdot \vec{n}\right)^*_{\text{HLL}} = \frac{1}{2} \vec{n} \cdot \left(\overset{\leftrightarrow}{\mathbf{F}}(\mathbf{U}^L) + \overset{\leftrightarrow}{\mathbf{F}}(\mathbf{U}^R)\right) - \frac{S^R}{2} \left(\mathbf{U}^R - \mathbf{U}^L\right). \tag{52}$$

If we would choose S^R to be the maximum wave speed, the HLL flux would reduce to the Lax-Friedrichs flux. However, with $S^R = V_n^R + c^R = -V_n + c$, an even simpler relation for P^* is found, which also satisfies the entropy inequality

$$\left(\frac{P^*}{P}\right)_{\text{HLL}} = 1 + \gamma \text{Ma}_n \begin{cases} > 1 & \text{for} \quad V_n > 0 \\ \leq 1 & \text{for} \quad V_n \leq 0 \end{cases} \tag{53}$$

For the HLLC flux [13], one can show that since the Riemann problem is symmetric, the approximate wave speed of the contact discontinuity is $\lambda^* = 0$ and, choosing $S^R = -V_n + c$, HLLC reduces to the HLL flux.

$$\left(\frac{P^*}{P}\right)_{\text{HLLC}} = 1 + \gamma \text{Ma}_n = \left(\frac{P^*}{P}\right)_{\text{HLL}} \tag{54}$$

Roe Flux

For the original Roe method without entropy fix [13], the mean values are

$$\tilde{V}_n = \frac{\sqrt{\rho^L} V_n^L + \sqrt{\rho^R} V_n^R}{\sqrt{\rho^L} + \sqrt{\rho^R}} = 0, \quad \tilde{V}_{t_1} = V_{t_1}, \quad \tilde{V}_{t_2} = V_{t_2},$$

$$\tilde{c} = c\sqrt{1 + \frac{(\gamma - 1)}{2} \mathrm{Ma}_n^2}. \tag{55}$$

After some manipulations,

$$\left(\overset{\leftrightarrow}{\mathbf{F}} \cdot \vec{n}\right)^*_{\mathrm{Roe}} = \left(\overset{\leftrightarrow}{\mathbf{F}} \cdot \vec{n}\right) + \tilde{\lambda}_1 \tilde{\alpha}_1 \tilde{\mathbf{K}}^1 = \left(\overset{\leftrightarrow}{\mathbf{F}} \cdot \vec{n}\right) + (-\tilde{c})\frac{\rho V_n}{\tilde{c}} \begin{bmatrix} 1 \\ -\tilde{c} \\ V_{t_1} \\ V_{t_2} \\ \frac{1}{\rho}(\rho E + P) \end{bmatrix}$$

$$= \begin{bmatrix} 0 \\ (\rho V_n^2 + \rho V_n \tilde{c} + P)\vec{n} \\ 0 \end{bmatrix}. \tag{56}$$

with $\tilde{\lambda}_1 = \tilde{V}_n - \tilde{c} = -\tilde{c}$, $\alpha_1 = \rho V_n / \tilde{c}$ and $\tilde{\mathbf{K}}^1$ from [13]. This leads again to a definition of P^*

$$\left(\frac{P^*}{P}\right)_{\mathrm{Roe}} = 1 + \gamma \mathrm{Ma}_n \left(\mathrm{Ma}_n + \sqrt{1 + \frac{(\gamma - 1)}{2} \mathrm{Ma}_n^2}\right), \tag{57}$$

which fulfills the entropy inequality as long as

$$\mathrm{Ma}_n \geq -\sqrt{\frac{2}{3 - \gamma}}, \quad \text{for } \gamma = \frac{7}{5} \quad \mathrm{Ma}_n > -1.12. \tag{58}$$

Thus, the Roe flux is entropy stable for shocks, but not for supersonic rarefactions.

EC-ES Fluxes

We can also apply an entropy conservative (EC) flux that is used for interior element interfaces and add an entropy stable dissipation term (ES) to compute the boundary flux via the mirrored states (46). This is exactly the strategy proposed in Parsani et al. [12] to get the boundary flux. Such an EC-ES flux is presented in Winters et al. [15]

$$\left(\overset{\leftrightarrow}{\mathbf{F}} \cdot \vec{n}\right)^*_{\text{ES}} = \overset{\leftrightarrow}{\mathbf{F}}_{\text{EC}}\left(\mathbf{U}_L, \mathbf{U}_R\right) \cdot \vec{n} - \frac{1}{2}\underline{\underline{D}}\,\underline{\underline{H}}\left(\mathbf{w}^R - \mathbf{w}^L\right) \tag{59}$$

where $\underline{\underline{D}}$ is a dissipation matrix and the matrix $\underline{\underline{H}}\,[\![\mathbf{w}]\!] \simeq [\![\mathbf{u}]\!]$ is carefully derived from the left and right states. Details are given in [15], where two approaches for the dissipation are distinguished. One is a Lax-Friedrichs-type dissipation, scaling with the maximum eigenvalue $\lambda_{\max} = |V_n| + c$ (referred to as 'EC-LF'). The other is a Roe-type dissipation computed via the eigenstructure of the matrix $(\underline{\underline{D}}\,\underline{\underline{H}})$ (referred to as 'EC-Roe').

If we carefully insert the two mirrored boundary states into (59), we again get an equation for the modified pressure

$$\left(\frac{P^*}{P}\right)_{\text{EC-LF}} = 1 + \gamma \text{Ma}_n \left(|\text{Ma}_n| + 1\right) \tag{60}$$

for the Lax-Friedrichs-type dissipation and

$$\left(\frac{P^*}{P}\right)_{\text{EC-Roe}} = 1 + \gamma \text{Ma}_n \tag{61}$$

for the Roe-type dissipation. Both approaches lead to an entropy stable boundary flux when using a mirrored state. Note that the modified pressure of the EC-Roe flux (61) exactly matches the one of the HLL flux (53).

4 Discussion

In the previous section we have shown conditions under which a specified wall flux is stable. In the linear analysis, the central numerical flux adds no dissipation and is neutrally stable. In the nonlinear analysis, entropy is not generated if the numerical wall pressure is equal to the internal pressure, $P^* = P'$. For upwinded approximations, the amount of energy or entropy dissipation depends on the normal Mach number. Since the boundary condition is only imposed weakly through the numerical flux, the normal Mach number will not be exactly zero except in the convergence limit. In fact, flow computations (especially steady state ones) are usually initiated with an impulsive start, where the initial state is a uniform flow, and the normal Mach number is not zero. This has proved over time to be very robust in practice. The analysis above gives an explanation why.

In the linear analysis the dissipation due to imposing the boundary condition is proportional to the square of the normal Mach number. With an impulsive start initialization, this dissipation will be large. As the flow develops and the

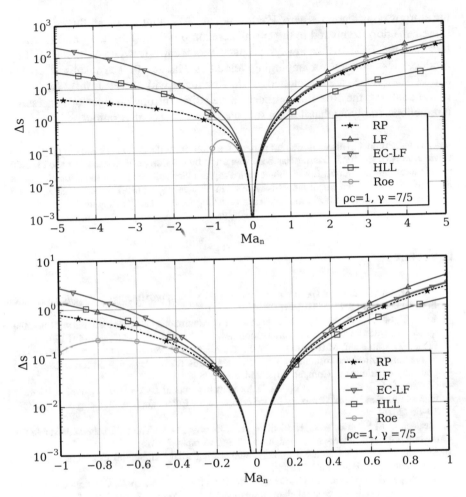

Fig. 1 Entropy contribution Δs (62) produced by the wall boundary flux. RP refers to the exact Riemann problem (45), LF to (50), EC-LF to (60), HLL to (53) and Roe to (57). Plotted over the normal Mach number ranges $|\mathrm{Ma}_n| \leq 5$ on the top and restricted to $|\mathrm{Ma}_n| \leq 1$ on the bottom

boundary condition is better enforced, the dissipation reduces, going away only as the approximate solution converges.

A similar effect is observed for the use of the different approximate Riemann solvers in the nonlinear analysis. In Fig. 1, we compare the entropy contribution

$$\Delta s = (\rho c)\mathrm{Ma}_n \left(\frac{P^*}{P} - 1 \right) \tag{62}$$

for the different wall boundary fluxes, over a range of normal Mach numbers for $(\rho c) = 1$ and $\gamma = 7/5$. When the boundary condition is exactly fulfilled ($\mathrm{Ma}_n = 0$), the entropy contribution is zero. For low normal Mach numbers, all fluxes have the same behavior. Compared to the exact Riemann problem (RP), the Lax-Friedrichs flux and the EC-LF flux always produce more entropy whereas the HLL flux produces less entropy for impinging velocities $Ma_n > 0$. The results of HLLC and EC-Roe fluxes are not plotted, as they coincide with the HLL flux. As shown in the analysis, the Roe flux produces a negative entropy change for supersonic rarefactions, implying that it is not suitable for all flow configurations.

Acknowledgements This work was supported by a grant from the Simons Foundation (#426393, David Kopriva). Gregor J. Gassner has been supported by the European Research Council (ERC) under the European Union's Eights Framework Program Horizon 2020 with the research project *Extreme*, ERC grant agreement no. 714487. Florian Hindenlang thanks Eric Sonnendrücker and the Max-Planck Institute for Plasma Physics in Garching for their constant support.

References

1. Black, K.: A conservative spectral element method for the approximation of compressible fluid flow. Kybernetika **35**(1), 133–146 (1999)
2. Chen, T., Shu, C.-W.: Entropy stable high order discontinuous Galerkin methods with suitable quadrature rules for hyperbolic conservation laws. J. Comput. Phys. **345**, 427–461 (2017)
3. Fisher, T., Carpenter, M.H., Nordström, J., Yamaleev, N.K., Swanson, C.: Discretely conservative finite-difference formulations for nonlinear conservation laws in split form: theory and boundary conditions. J. Comput. Phys. **234**, 353–375 (2013)
4. Gassner, G.J., Winters, A.R., Kopriva, D.A.: Split form nodal discontinuous Galerkin schemes with summation-by-parts property for the compressible Euler equations. J. Comput. Phys. **327**, 39–66 (2016)
5. Gassner, G.J., Winters, A.R., Hindenlang, F.J., Kopriva, D.A.: The BR1 scheme is stable for the compressible Navier–Stokes equations. J. Sci. Comput. (2018)
6. Kopriva, D.A.: Implementing Spectral Methods for Partial Differential Equations. Scientific Computation. Springer, Berlin (2009)
7. Kopriva, D.A., Gassner, G.: On the quadrature and weak form choices in collocation type discontinuous Galerkin spectral element methods. J. Sci. Comput. **44**(2), 136–155 (2010)
8. Kopriva, D.A., Gassner, G.J.: An energy stable discontinuous Galerkin spectral element discretization for variable coefficient advection problems. SIAM J. Sci. Comput. **36**(4), A2076–A2099 (2014)
9. Nordström, J.: Conservative finite difference formulations, variable coefficients, energy estimates and artificial dissipation. J. Sci. Comput. **29**(3), 375–404 (2006)
10. Nordström, J.: A roadmap to well posed and stable problems in computational physics. J. Sci. Comput. (2016). https://doi.org/10.1007/s10915-016-0303-9
11. Nordström, J., Svard, M.: Well-posed boundary conditions for the Navier–Stokes equations. SIAM J. Numer. Anal. **43**(3), 1231–1255 (2005)
12. Parsani, M., Carpenter, M.H., Nielsen, E.J.: Entropy stable wall boundary conditions for the three-dimensional compressible Navier–Stokes equations. J. Comput. Phys. **292**, 88–113 (2015)
13. Toro, E.F.: Riemann Solvers and Numerical Methods for Fluid Dynamics. Springer, Berlin (2009)

14. van der Vegt, J.J.W., van der Ven, H.: Slip flow boundary conditions in discontinuous Galerkin discretizations of the Euler equations of gas dynamics. In Mang, H.A., Rammenstorfer, F.G. (eds.) Proceedings of the 5th World Congress on Computational Mechanics (WCCM V), number NLR-TP in Technical Publications, pp. 1–16. National Aerospace Laboratory, NLR (2002)
15. Winters, A.R., Derigs, D., Gassner, G.J., Walch, S.: A uniquely defined entropy stable matrix dissipation operator for high Mach number ideal MHD and compressible Euler simulations. J. Comput. Phys. **332**, 274–289 (2017)

On the Order Reduction of Entropy Stable DGSEM for the Compressible Euler Equations

Florian J. Hindenlang and Gregor J. Gassner

1 Introduction

Discontinuous Galerkin spectral element collocation method (DGSEM) with either Legendre Gauss or Legendre-Gauss-Lobatto (LGL) nodes (see e.g. [14]) are among the most efficient variants in the class of element based high order methods, such as e.g. discontinuous Galerkin, flux reconstruction, or summation-by-parts (SBP) finite differences. In particular, the LGL variant, starting in [9], turned out to be similar to a SBP finite difference approximation with simultaneous-approximate-term technique (SAT). This relationship allowed to construct conservative skew-symmetric approximations, e.g. [9, 10, 21], and later enabled DGSEM-LGL approximations that are discretely entropy stable, e.g. [1, 3, 6, 8, 13, 17, 19, 20], and/or kinetic energy preserving [12]. These novel variants of nodal split form DG methods feature drastically increased non-linear robustness towards aliasing induced instabilities and favourable properties regarding the simulation of unresolved turbulence, e.g. [7, 23].

In addition to the very robust dissipative entropy stable versions, it is also possible to construct virtually dissipation free variants by choosing appropriate element interface numerical fluxes. These entropy conserving variants all show an odd-even behavior when experimentally testing the order of convergence, e.g. [9, 21], where the observed convergence order for even polynomial degrees

F. J. Hindenlang
Max Planck Institute for Plasma Physics, Garching, Germany
e-mail: florian.hindenlang@ipp.mpg.de

G. J. Gassner (✉)
Department for Mathematics and Computer Science, Center for Data and Simulation Science, University of Cologne, Cologne, Germany
e-mail: ggassner@math.uni-koeln.de

© The Author(s) 2020
S. J. Sherwin et al. (eds.), *Spectral and High Order Methods for Partial Differential Equations ICOSAHOM 2018*, Lecture Notes in Computational Science and Engineering 134, https://doi.org/10.1007/978-3-030-39647-3_2

N is N and for odd N is $N + 1$. Lately, a discussion emerged in the community, with interesting debates during the recent ICOSAHOM conference in London, where researchers reported non-optimal convergence behavior of the entropy stable DGSEM-LGL even with dissipative numerical surface fluxes, e.g. [6].

This paper contributes to this discussion and presents results of an experimental convergence order study for the compressible Euler equations with (1) the standard DGSEM with either Gauss and LGL nodes, (2) the entropy stable DGSEM with LGL nodes. For these nodal schemes, we test the convergence order with different numerical surface fluxes and report the results depending on the Mach number of the test case. The remainder of the paper is organized as follows: in the next section we describe the numerical model for our numerical experiments, in Sect. 3 we present our observed experimental convergence orders for different configurations and draw our conclusion in Sect. 4.

2 Numerical Model

We consider the compressible Euler equations defined in the domain $\Omega \subset \mathbb{R}^3$

$$\mathbf{u}_t + \sum_{i=1}^{3} \frac{\partial \mathbf{f}_i}{\partial x_i} = \mathbf{0}. \tag{1}$$

The state vector contains the conservative variables and the advective flux components are

$$\mathbf{u} = \begin{bmatrix} \varrho \\ \varrho \vec{v} \\ E \end{bmatrix} = \begin{bmatrix} \varrho \\ \varrho v_1 \\ \varrho v_2 \\ \varrho v_3 \\ E \end{bmatrix}, \ \mathbf{f}_1 = \begin{bmatrix} \varrho v_1 \\ \varrho v_1^2 + p \\ \varrho v_1 v_2 \\ \varrho v_1 v_3 \\ (E + p)v_1 \end{bmatrix}, \ \mathbf{f}_2 = \begin{bmatrix} \varrho v_2 \\ \varrho v_2 v_1 \\ \varrho v_2^2 + p \\ \varrho v_2 v_3 \\ (E + p)v_2 \end{bmatrix}, \ \mathbf{f}_3 = \begin{bmatrix} \varrho v_3 \\ \varrho v_3 v_1 \\ \varrho v_3 v_2 \\ \varrho v_3^2 + p \\ (E + p)v_3 \end{bmatrix}. \tag{2}$$

Here, ϱ, $\vec{v} = (v_1, v_2, v_3)^T$, p, E are the mass density, fluid velocities, pressure and total energy. We close the system with the ideal gas assumption, which relates the total energy and pressure

$$p = (\gamma - 1) \left(E - \frac{1}{2} \varrho \, \|\vec{v}\|^2 \right), \tag{3}$$

where γ denotes the adiabatic coefficient.

For our discretization, we subdivide the domain into non-overlapping hexahedral elements. For each element, we define a transfinite mapping to a unit reference space and use this mapping to transform the Eq. (1) from physical to reference space. A weak form is created by taking the inner product of the transformed equation with a test function. We use integration-by-parts for the flux term and approximate the resulting weak form as follows: the conservative variables are approximated by a polynomial in reference space with degree N, interpolated at the Gauss or LGL nodes. The volume fluxes are replaced by a standard interpolation of the non-linear flux function at the same Gauss/LGL nodes (standard DGSEM-Gauss or DGSEM-LGL), see e.g. [14]. For the LGL variant, we are also able to introduce the split form volume integral based on entropy conserving and kinetic energy preserving numerical volume fluxes (Split-DGSEM), e.g. [12] and [22], resulting in either the entropy conserving or entropy stable DGSEM variants, depending on the choice of numerical surface flux.

3 Convergence Results

In this section, we compare the convergence of the standard DGSEM and the entropy conservative and entropy stable discretization for different choices of the numerical flux and polynomial degrees $N = 2, 3, 4, 5$.

We choose the test case of a two-dimensional density wave, with a constant pressure and transported with a constant velocity, which was proposed for one-dimensional convergence tests in [4]. The density evolves as

$$\varrho(x_1, x_2, t) = 1 + 0.1 \sin \left(\pi \left((x_1 - v_1 t) + (x_2 - v_2 t) \right) \right) \tag{4}$$

with a prescribed velocity (v_1, v_2). The pressure is chosen as $p = 1/\gamma$ with $\gamma = 1.4$, so that the sound speed ranges between $c = 0.95 \ldots 1.05$. Thus, by changing the velocity, we change the Mach number of the flow $Ma = |\vec{v}|/c$. Three Mach numbers are chosen: $Ma \approx 0.2$ with $(v_1, v_2) = (0.1, 0.15)$, $Ma \approx 1.0$ with $(v_1, v_2) = (0.7, 0.65)$ and $Ma \approx 3.5$ with $(v_1, v_2) = (2.5, 2.4)$. The experimental order of convergence (EOC) is computed with the L_2 error of the density at $t = 1$.

The convergence study is performed with the open source, three-dimensional curvilinear split-form DG framework FLUXO (www.github.com/project-fluxo). As the test case is two-dimensional, we use fully periodic cartesian meshes of the domain $[-1, 1]^3$ with an equal number of elements in x- and y-directions and always one element in z-direction. Note that h_0 in the convergence tables refers to the coarsest mesh level, which is 4^2 elements for $N = 2, 3$ ($h_0 = 1/2$) and 2^2 elements for $N = 4, 5$ ($h_0 = 1$).

All simulation results are obtained with an explicit five stage, fourth order accurate low storage Runge–Kutta scheme [2], where a stable time step is computed according to the adjustable coefficient $CFL \in (0, 1]$ the local maximum wave speed, and the relative grid size, e.g. [11]. We made sure that the time integrator did not influence the spatial convergence order, by adjusting the CFL number accordingly.

3.1 Standard DGSEM

The convergence of the standard DGSEM with Gauss-Legendre nodes (DGSEM-Gauss) and with Legendre-Gauss-Lobatto (DGSEM-LGL) is shown in Tables 1 and 2, for the three Mach numbers and two choices of the numerical flux, namely the HLL (Harten, Lax, van Leer) flux and the Roe flux. The results of the LLF (local Lax-Friedrichs) flux and the HLLC flux (HLL variant with three waves, C for 'contact' wave) are reported in the Appendix, as the HLL results are similar to LLF, and HLLC behaves exactly the same as Roe, see Tables 4 and 5. Details on the properties and the implementation of the LLF, HLL, HLLC, and Roe fluxes are found in the book of Toro [18] and the references therein.

For the HLL flux and the low Mach number Ma = 0.2, we observe an odd-even behavior with an *order reduction* for even polynomial degrees $N = 2, 4$. Also for Ma = 1.0, the convergence for even degrees is slightly affected, whereas for the high Mach number, all fluxes converge with full order. Comparing the L_2 errors of the finest mesh for HLL and Roe for the low Mach number, HLL is less accurate for $N = 2, 4$ and more accurate for $N = 3, 5$.

All numerical fluxes are approximate Riemann solvers, but the LLF and HLL only use the maximum wave speeds, whereas the HLLC and Roe also take the contact wave into account, and therefore keep the full order of the scheme for all Mach numbers for this test case.

3.2 Entropy Conservative and Entropy Stable DGSEM

Now, we investigate the order reduction of the entropy conservative and entropy stable discretizations. Here, the standard DGSEM volume integral is replaced by split-form formulation (Split-DGSEM) using a two-point entropy conservative and kinetic energy preserving flux (ECKEP). If we choose the ECKEP flux at the surface, we get an entropy-conserving scheme. For entropy stability, we can use the LLF or HLL flux directly at the surface, or use the ECKEP flux and add a dissipation term, which must still satisfy the entropy inequality condition. In Winters et al. [22], such dissipation terms are carefully derived, using either only the maximum wave

Table 1 Experimental order of convergence of L_2 error to the exact density (4), using the standard DGSEM-Gauss with HLL and Roe fluxes

Volume disc. + Surface flux	Mach	Mesh level	N = 2		N = 3		N = 4		N = 5	
			$L_2(\varrho)$	EOC	$L_2(\varrho)$	EOC	$L_2(\varrho)$	EOC	$L_2(\varrho)$	EOC
DGSEM-Gauss + HLL	3.5	$h_0/2$	1.87e−04	3.34	8.57e−06	4.02	1.03e−05	5.02	6.76e−07	6.07
		$h_0/4$	2.27e−05	3.04	5.35e−07	4.00	3.30e−07	4.96	1.07e−08	5.99
		$h_0/8$	2.82e−06	3.00	3.34e−08	4.00	1.02e−08	5.01	1.66e−10	6.01
		$h_0/16$	3.53e−07	3.00 ■	2.09e−09	4.00 ■	3.22e−10	4.99 ■	2.60e−12	5.99 ■
	1.0	$h_0/2$	2.60e−04	2.55	5.92e−06	4.55	1.15e−05	4.39	6.74e−07	6.84
		$h_0/4$	3.74e−05	2.80	3.27e−07	4.18	4.94e−07	4.54	7.23e−09	6.54
		$h_0/8$	4.88e−06	2.94	1.95e−08	4.07	1.73e−08	4.84	9.98e−11	6.18
		$h_0/16$	6.18e−07	2.98 ■	1.20e−09	4.03 ■	5.60e−10	4.95 ■	1.52e−12	6.04 ■
	0.2	$h_0/2$	4.87e−04	1.70	4.36e−06	5.05	1.57e−05	2.56	9.80e−07	6.66
		$h_0/4$	1.08e−04	2.18	1.10e−07	5.31	9.90e−07	3.99	4.47e−09	7.77
		$h_0/8$	1.95e−05	2.46	5.84e−09	4.23	5.06e−08	4.29	3.96e−11	6.82
		$h_0/16$	3.06e−06	2.67 ⊠	2.34e−10	4.64 ■	2.26e−09	4.49 ⊠	4.57e−13	6.44 ■
DGSEM-Gauss + Roe	3.5	$h_0/2$	1.87e−04	3.34	8.57e−06	4.02	1.03e−05	5.02	6.76e−07	6.07
		$h_0/4$	2.27e−05	3.04	5.35e−07	4.00	3.30e−07	4.96	1.07e−08	5.99
		$h_0/8$	2.82e−06	3.00	3.34e−08	4.00	1.02e−08	5.01	1.66e−10	6.01
		$h_0/16$	3.53e−07	3.00 ■	2.09e−09	4.00 ■	3.22e−10	4.99 ■	2.60e−12	5.99 ■
	1.0	$h_0/2$	1.82e−04	3.07	8.76e−06	4.04	1.11e−05	5.07	6.95e−07	6.08
		$h_0/4$	2.26e−05	3.00	5.40e−07	4.02	3.44e−07	5.00	1.07e−08	6.01
		$h_0/8$	2.82e−06	3.00	3.35e−08	4.01	1.05e−08	5.04	1.67e−10	6.01
		$h_0/16$	3.53e−07	3.00 ■	2.09e−09	4.00 ■	3.25e−10	5.01 ■	2.61e−12	6.00 ■

(continued)

Table 1 (continued)

Volume disc. + Surface flux	Mach	Mesh level	$N = 2$		$N = 3$		$N = 4$		$N = 5$	
			$L_2(\varrho)$	EOC	$L_2(\varrho)$	EOC	$L_2(\varrho)$	EOC	$L_2(\varrho)$	EOC
	0.2	$h_0/2$	2.14e−04	2.65	1.04e−05	3.78	1.16e−05	4.53	7.76e−07	5.78
		$h_0/4$	2.22e−05	3.26	5.49e−07	4.25	3.49e−07	5.05	1.08e−08	6.17
		$h_0/8$	2.82e−06	2.98	3.76e−08	3.87	1.04e−08	5.07	1.70e−10	5.99
		$h_0/16$	3.53e−07	3.00 ■	2.07e−09	4.19 ■	3.38e−10	4.94 ■	2.64e−12	6.01 ■

Full order is marked with ■ ($\gtrsim N + 1$) and an order reduction with ⊠

Table 2 Experimental order of convergence of L_2 error to the exact density (4), using standard DGSEM-LGL with HLL and Roe fluxes

Volume disc. + Surface flux	Mach	Mesh level	N = 2		N = 3		N = 4		N = 5	
			$L_2(\varrho)$	EOC	$L_2(\varrho)$	EOC	$L_2(\varrho)$	EOC	$L_2(\varrho)$	EOC
DGSEM-LGL + HLL	3.5	$h_0/2$	1.22e−03	3.38	3.85e−05	4.03	4.34e−05	4.93	2.70e−06	5.93
		$h_0/4$	1.26e−04	3.27	2.41e−06	4.00	1.39e−06	4.96	4.33e−08	5.96
		$h_0/8$	1.48e−05	3.10	1.51e−07	4.00	4.34e−08	5.01	6.67e−10	6.02
		$h_0/16$	1.81e−06 ■	3.03 ■	9.42e−09 ■	4.00 ■	1.36e−09 ■	4.99 ■	1.05e−11 ■	5.98 ■
	1.0	$h_0/2$	1.04e−03	2.30	3.44e−05	4.44	4.11e−05	4.89	2.89e−06	6.33
		$h_0/4$	1.58e−04	2.72	1.88e−06	4.20	1.77e−06	4.54	3.85e−08	6.23
		$h_0/8$	2.11e−05	2.91	1.16e−07	4.02	6.25e−08	4.82	5.37e−10	6.16
		$h_0/16$	2.69e−06 ■	2.97 ■	7.18e−09 ■	4.01 ■	2.05e−09	4.93 ■	8.36e−12 ■	6.00 ■
	0.2	$h_0/2$	1.20e−03	1.96	4.86e−05	3.51	5.16e−05	3.91	3.85e−06	5.31
		$h_0/4$	2.72e−04	2.14	1.85e−06	4.71	2.99e−06	4.11	4.25e−08	6.50
		$h_0/8$	5.57e−05	2.29	1.21e−07	3.94	1.56e−07	4.26	5.45e−10	6.28
		$h_0/16$	1.01e−05 ⊠	2.47 ⊠	5.95e−09 ■	4.34 ■	7.19e−09 ■	4.44 ⊠	7.11e−12 ■	6.26 ■
DGSEM-LGL + Roe	3.5	$h_0/2$	1.22e−03	3.38	3.85e−05	4.03	4.34e−05	4.93	2.70e−06	5.93
		$h_0/4$	1.26e−04	3.27	2.41e−06	4.00	1.39e−06	4.96	4.33e−08	5.96
		$h_0/8$	1.48e−05	3.10	1.51e−07	4.00	4.34e−08	5.01	6.67e−10	6.02
		$h_0/16$	1.81e−06 ■	3.03 ■	9.42e−09 ■	4.00 ■	1.36e−09 ■	4.99 ■	1.05e−11 ■	5.98 ■
	1.0	$h_0/2$	9.17e−04	2.86	3.96e−05	3.94	4.41e−05	4.94	2.76e−06	6.02
		$h_0/4$	1.15e−04	2.99	2.41e−06	4.04	1.47e−06	4.90	4.44e−08	5.96
		$h_0/8$	1.44e−05	3.00	1.51e−07	4.00	4.38e−08	5.07	6.84e−10	6.02
		$h_0/16$	1.80e−06 ■	3.00 ■	9.42e−09 ■	4.00 ■	1.35e−09 ■	5.02 ■	1.08e−11 ■	5.99 ■

(continued)

Table 2 (continued)

Volume disc. + Surface flux	Mach	Mesh level	$N = 2$		$N = 3$		$N = 4$		$N = 5$	
			$L_2(\varrho)$	EOC	$L_2(\varrho)$	EOC	$L_2(\varrho)$	EOC	$L_2(\varrho)$	EOC
	0.2	$h_0/2$	9.26e−04	2.35	4.63e−05	3.45	4.26e−05	4.27	2.97e−06	5.37
		$h_0/4$	1.19e−04	2.96	2.40e−06	4.27	1.59e−06	4.74	4.48e−08	6.05
		$h_0/8$	1.43e−05	3.06	1.57e−07	3.93	4.34e−08	5.19	6.83e−10	6.04
		$h_0/16$	1.80e−06	2.99 ■	9.37e−09	4.07 ■	1.49e−09	4.86 ■	1.09e−11	5.97 ■

Full order is marked with ■ ($\gtrsim N + 1$) and an order reduction with ⊠

Table 3 Experimental order of convergence of L_2 error to the exact density (4), using entropy conservative ECKEP flux and entropy stable HLL and ECKEP-Roe fluxes

Volume disc. + Surface flux	Mach	Mesh level	N = 2		N = 3		N = 4		N = 5	
			$L_2(\varrho)$	EOC	$L_2(\varrho)$	EOC	$L_2(\varrho)$	EOC	$L_2(\varrho)$	EOC
Split-DGSEM + ECKEP	3.5	$h_0/2$	1.62e−03	4.23	8.45e−05	2.77	5.80e−05	5.88	4.63e−06	5.63
		$h_0/4$	1.30e−04	3.64	7.14e−06	3.56	1.60e−06	5.18	8.38e−08	5.79
		$h_0/8$	1.05e−05	3.62 ■	7.90e−07	3.18	4.56e−08	5.13	3.59e−09	4.54
		$h_0/16$	1.69e−06	2.64 ⊠	9.58e−08	3.04 ⊠	1.23e−09	5.21 ■	1.18e−10	4.93 ⊠
	1.0	$h_0/2$	1.41e−03	3.89	9.45e−05	2.34	7.71e−05	5.86	3.49e−06	6.00
		$h_0/4$	1.25e−04	3.49	1.26e−05	2.90	1.98e−06	5.28	5.00e−08	6.12
		$h_0/8$	1.48e−05	3.09	1.60e−06	2.98	4.06e−08	5.61	1.22e−09	5.36
		$h_0/16$	1.32e−06	3.48 ■	2.01e−07	3.00 ⊠	1.13e−09	5.17 ■	3.85e−11	4.98 ⊠
	0.2	$h_0/2$	1.13e−03	2.26	8.03e−05	2.98	5.95e−05	4.14	4.11e−06	5.14
		$h_0/4$	1.13e−04	3.32	1.02e−05	2.97	1.85e−06	5.01	1.33e−07	4.95
		$h_0/8$	1.12e−05	3.34 ■	1.29e−06	2.99	4.02e−08	5.52	4.21e−09	4.98
		$h_0/16$	1.98e−06	2.50 ⊠	1.61e−07	3.00 ⊠	1.43e−09	4.81 ■	1.32e−10	5.00 ⊠
Split-DGSEM + HLL	3.5	$h_0/2$	1.23e−03	3.36	3.88e−05	4.06	4.49e−05	4.94	3.01e−06	5.97
		$h_0/4$	1.27e−04	3.27	2.42e−06	4.00	1.43e−06	4.97	4.84e−08	5.96
		$h_0/8$	1.48e−05	3.10	1.51e−07	4.00	4.44e−08	5.01	7.48e−10	6.02
		$h_0/16$	1.81e−06	3.03 ■	9.46e−09	4.00 ■	1.40e−09	4.99 ■	1.18e−11	5.98 ■
	1.0	$h_0/2$	1.04e−03	2.30	3.48e−05	4.44	4.16e−05	4.90	3.48e−06	6.10
		$h_0/4$	1.58e−04	2.72	1.88e−06	4.21	1.79e−06	4.54	4.66e−08	6.23
		$h_0/8$	2.11e−05	2.90	1.16e−07	4.02	6.39e−08	4.81	6.08e−10	6.26
		$h_0/16$	2.69e−06	2.97 ■	7.20e−09	4.01 ■	2.11e−09	4.92 ■	9.28e−12	6.03 ■

(continued)

Table 3 (continued)

Volume disc. + Surface flux	Mach	Mesh level	$N = 2$		$N = 3$		$N = 4$		$N = 5$	
			$L_2(\varrho)$	EOC	$L_2(\varrho)$	EOC	$L_2(\varrho)$	EOC	$L_2(\varrho)$	EOC
	0.2	$h_0/2$	1.20e−03	1.96	4.90e−05	3.51	5.23e−05	3.92	4.47e−06	5.16
		$h_0/4$	2.72e−04	2.14	1.87e−06	4.71	3.02e−06	4.12	6.14e−08	6.19
		$h_0/8$	5.58e−05	2.29	1.21e−07	3.95	1.57e−07	4.26	6.18e−10	6.64
		$h_0/16$	1.01e−05	2.47 ⊠	5.98e−09	4.34 ■	7.28e−09	4.43 ⊠	8.20e−12	6.23 ■
Split-DGSEM + ECKEP-Roe	3.5	$h_0/2$	1.23e−03	3.36	3.88e−05	4.06	4.49e−05	4.94	3.01e−06	5.97
		$h_0/4$	1.27e−04	3.27	2.42e−06	4.00	1.43e−06	4.97	4.84e−08	5.96
		$h_0/8$	1.48e−05	3.10	1.51e−07	4.00	4.44e−08	5.01	7.48e−10	6.02
		$h_0/16$	1.81e−06	3.03 ■	9.46e−09	4.00 ■	1.40e−09	4.99 ■	1.18e−11	5.98 ■
	1.0	$h_0/2$	9.18e−04	2.86	3.98e−05	3.94	4.51e−05	4.92	3.08e−06	5.95
		$h_0/4$	1.15e−04	2.99	2.42e−06	4.04	1.51e−06	4.90	4.97e−08	5.95
		$h_0/8$	1.44e−05	3.00	1.51e−07	4.00	4.49e−08	5.07	7.66e−10	6.02
		$h_0/16$	1.80e−06	3.00 ■	9.46e−09	4.00 ■	1.38e−09	5.02 ■	1.21e−11	5.99 ■
	0.2	$h_0/2$	9.26e−04	2.35	4.65e−05	3.45	4.37e−05	4.26	3.26e−06	5.34
		$h_0/4$	1.19e−04	2.96	2.40e−06	4.27	1.63e−06	4.74	5.03e−08	6.02
		$h_0/8$	1.43e−05	3.06	1.58e−07	3.93	4.45e−08	5.19	7.66e−10	6.04
		$h_0/16$	1.80e−06	2.99 ■	9.40e−09	4.07 ■	1.53e−09	4.86 ■	1.22e−11	5.97 ■

Full order is marked with ■ ($\gtrsim N + 1$) and an order reduction with ⊠

speed (LLF-type) or incorporating all waves (Roe-type), which we will refer to as ECKEP-LLF and ECKEP-Roe fluxes.

In Table 3, we summarize the convergence of the dissipation-free ECKEP flux, the HLL and ECKEP-Roe flux. The results for LLF and ECKEP-LLF fluxes are found in the Appendix in Table 6, as they have the same convergence and error levels as the HLL flux. As expected, the dissipation-free surface flux (ECKEP) produces an order reduction for all Mach numbers for $N = 3, 5$, and for $N = 2$ full order is not kept in the last refinement step.

If we simply use the HLL flux, we have an entropy stable scheme, but an order reduction for $N = 2, 4$ can be observed for the low Mach number flow, analogously to the standard DGSEM-LGL scheme. Interestingly, the odd-even behavior switches between entropy conserving and entropy stable fluxes.

The ECKEP-Roe entropy stable flux accounts for all waves of the Riemann problem and adjusts the dissipation for each wave accordingly, which gives full order convergence for all Mach numbers.

4 Conclusions

In this work, we report the convergence of standard DGSEM Gauss and Gauss-Lobatto schemes to entropy conservative (EC) and entropy stable (ES) DGSEM schemes for the Euler equations, as there have been findings of order reduction for EC and ES schemes. We choose a simple density transport test case on a periodic domain and investigate the influence of the Mach number of the transport velocity.

The EC scheme is dissipation free and an order reduction is observed by the convergence study presented here, confirming many similar observations found in literature. We also confirm that the ES scheme can have an order reduction for low Mach numbers, but only if the entropy stable numerical flux relies on simple approximate Riemann solvers such as local Lax-Friedrichs or HLL. If all waves are accounted for in the dissipation term of the entropy stable flux as presented in [22], the full order is observed for all Mach numbers. In addition, we reproduce the same behavior for the standard DGSEM Gauss and Gauss-Lobatto schemes, where the LLF and HLL fluxes suffer from order reduction at low Mach number, and HLLC and Roe fluxes have full order for all Mach numbers.

We want to emphasize that the present convergence study should be seen merely as an observation, confirming that the numerical flux can have strong influence on the convergence order for both the standard DGSEM and the entropy stable DGSEM. Also, we stress that in our tests the order reduction is related to the form of the dissipation term in the numerical surface flux and is not related to the insufficient integration precision of the LGL-quadrature.

Based on the observations presented in this work, a possible explanation for the loss of convergence for the density transport at low Mach numbers when using LLF and HLL fluxes is the form of dissipation from the approximate Riemann solver. In the case of the density transport, the exact solution follows the characteristic with velocity $\vec{v}$. However, the approximate Riemann solver LLF and HLL consider only two waves with maximum velocity $\sim (|\vec{v}| + c)$ and do not consider the contact wave with velocity $\vec{v}$. Thus, the contact wave is dissipated proportional to $\sim (|\vec{v}| + c)$ and not to $|\vec{v}|$. For low Mach numbers, where $c > |\vec{v}|$, this causes over-upwinding. Over-upwinding was discussed in [5, 15]. It is not intuitive at first, but over-upwinding (over-penalization) can lead to a reduction of the in-built dissipation of the DG scheme, getting wave-propagation characteristics similar to a continuous Galerkin method [16]. This loss of in-built dissipation could be an explanation for the even-odd behavior we observed. However, it is still unclear why numerical surface fluxes with no in-built dissipation that are symmetric, e.g. EC flux, lead to an odd-even behavior in the convergence order and why numerical surface fluxes with over-upwinding, i.e. reduced dissipation due to over-penalization, cause an opposite even-odd behavior. What supports the explanation is the recovery of full convergence order for LLF and HLL when the difference in wave speed becomes smaller for higher Mach numbers, i.e. no over-upwinding. In contrast to LLF and HLL, the HLLC and Roe solvers take specifically the contact wave into account and adjust the dissipation accordingly and thus avoid strong over-upwinding by construction. In our tests, we always observe full convergence order for all Mach numbers for HLLC and Roe.

Lastly we note that a convergence study using a manufactured solution technique can be misleading, as full convergence order is found independent of the choice of numerical flux. Hence, the introduction of a source term to balance the prescribed solution overcomes possible deficiencies of the surface fluxes, showing the limit of the manufactured solution technique in this context. In the Appendix, the convergence results of a manufactured solution are reported.

Acknowledgements Gregor Gassner thanks the European Research Council for funding through the ERC Starting Grant "An Exascale aware and Un-crashable Space-Time-Adaptive Discontinuous Spectral Element Solver for Non-Linear Conservation Laws" (Extreme), ERC grant agreement no. 714487. Florian Hindenlang thanks Eric Sonnendrücker and the Max-Planck Institute for Plasma Physics in Garching for their constant support. We would also like to thank all participants of the ICOSAHOM 2018 for the valuable discussions on the topic of entropy stable schemes, which motivated this work.

Appendix

Additional Convergence Results

In this section, we present additional convergence results of the density wave test case for the DGSEM-Gauss and DGSEM-LGL with LLF and HLLC fluxes in Table 4 and Table 5, and also the entropy stable schemes with LLF and ECKEP-LLF fluxes in Table 6. The results for LLF-type fluxes behave like the HLL flux, and for the HLLC flux like the Roe-type fluxes presented in Table 3.

Table 4 Experimental order of convergence of L_2 error to the exact density (4), using DGSEM-Gauss with LLF and HLLC fluxes

Volume disc. + Surface flux	Mach	Mesh level	N = 2		N = 3		N = 4		N = 5	
			$L_2(\varrho)$	EOC	$L_2(\varrho)$	EOC	$L_2(\varrho)$	EOC	$L_2(\varrho)$	EOC
DGSEM-Gauss + LLF	3.5	$h_0/2$	2.42e−04	2.97	6.43e−06	4.41	1.05e−05	4.51	6.68e−07	6.60
		$h_0/4$	3.24e−05	2.90	3.71e−07	4.11	4.32e−07	4.60	8.04e−09	6.38
		$h_0/8$	4.15e−06	2.96	2.27e−08	4.03	1.47e−07	4.87	1.15e−10	6.13
		$h_0/16$	5.22e−07	2.99 ■	1.41e−09	4.01 ■	4.73e−10	4.96 ■	1.77e−12	6.02 ■
	1.0	$h_0/2$	3.13e−04	2.29	4.59e−06	4.84	1.18e−05	3.90	6.69e−07	7.40
		$h_0/4$	5.30e−05	2.56	2.25e−07	4.35	6.08e−07	4.28	5.56e−09	6.91
		$h_0/8$	7.43e−06	2.83	1.29e−08	4.12	2.47e−08	4.62	6.79e−11	6.35
		$h_0/16$	9.61e−07	2.95 ■	7.65e−10	4.07 ■	8.53e−10	4.85 ■	9.97e−13	6.09 ■
	0.2	$h_0/2$	4.95e−04	1.69	4.33e−06	5.07	1.58e−05	2.53	9.88e−07	6.68
		$h_0/4$	1.12e−04	2.15	1.06e−07	5.35	1.01e−06	3.97	4.46e−09	7.79
		$h_0/8$	2.06e−05	2.44	5.47e−09	4.28	5.23e−08	4.27	3.86e−11	6.85
		$h_0/16$	3.29e−06	2.65 ⊠	2.15e−10	4.67 ■	2.38e−09	4.46 ⊠	4.35e−13	6.47 ■
DGSEM-Gauss + HLLC	3.5	$h_0/2$	1.87e−04	3.34	8.57e−06	4.02	1.03e−05	5.02	6.76e−07	6.07
		$h_0/4$	2.27e−05	3.04	5.35e−07	4.00	3.30e−07	4.96	1.07e−08	5.99
		$h_0/8$	2.82e−06	3.00	3.34e−08	4.00	1.02e−08	5.01	1.66e−10	6.01
		$h_0/16$	3.53e−07	3.00 ■	2.09e−09	4.00 ■	3.22e−10	4.99 ■	2.60e−12	5.99 ■
	1.0	$h_0/2$	1.82e−04	3.07	8.76e−06	4.04	1.11e−05	5.07	6.95e−07	6.08
		$h_0/4$	2.26e−05	3.00	5.40e−07	4.02	3.44e−07	5.00	1.07e−08	6.01
		$h_0/8$	2.82e−06	3.00	3.35e−08	4.01	1.05e−08	5.04	1.67e−10	6.01
		$h_0/16$	3.53e−07	3.00 ■	2.09e−09	4.00 ■	3.25e−10	5.01 ■	2.61e−12	6.00 ■

(continued)

Table 4 (continued)

Volume disc. + Surface flux	Mach	Mesh level	$N = 2$		$N = 3$		$N = 4$		$N = 5$	
			$L_2(\varrho)$	EOC	$L_2(\varrho)$	EOC	$L_2(\varrho)$	EOC	$L_2(\varrho)$	EOC
	0.2	$h_0/2$	2.14e−04	2.65	1.04e−05	3.78	1.16e−05	4.53	7.76e−07	5.78
		$h_0/4$	2.22e−05	3.26	5.49e−07	4.25	3.49e−07	5.05	1.08e−08	6.17
		$h_0/8$	2.82e−06	2.98	3.76e−08	3.87	1.04e−08	5.07	1.70e−10	5.99
		$h_0/16$	3.53e−07	3.00 ■	2.07e−09	4.19 ■	3.38e−10	4.94 ■	2.64e−12	6.01 ■

Full order is marked with ■ ($\gtrsim N + 1$) and an order reduction with ⊠

Table 5 Experimental order of convergence of L_2 error to the exact density (4), using DGSEM-GL with LLF and HLLC fluxes

Volume disc. + Surface flux	Mach	Mesh level	$N = 2$		$N = 3$		$N = 4$		$N = 5$	
			$L_2(\varrho)$	EOC	$L_2(\varrho)$	EOC	$L_2(\varrho)$	EOC	$L_2(\varrho)$	EOC
DGSEM-LGL + LLF	3.5	$h_0/2$	1.23e−03	3.11	3.44e−05	4.36	4.02e−05	4.82	2.91e−06	6.14
		$h_0/4$	1.51e−04	3.02	2.00e−06	4.10	1.61e−06	4.64	3.80e−08	6.26
		$h_0/8$	1.89e−05	3.00	1.23e−07	4.02	5.55e−08	4.86	5.46e−10	6.12
		$h_0/16$	2.36e−06	3.00 ■	7.66e−09	4.01 ■	1.79e−09	4.95 ■	8.54e−12	6.00 ■
	1.0	$h_0/2$	1.10e−03	1.97	3.15e−05	4.83	4.02e−05	4.93	2.85e−06	6.73
		$h_0/4$	2.03e−04	2.43	1.68e−06	4.23	1.98e−06	4.34	3.68e−08	6.28
		$h_0/8$	2.99e−05	2.76	1.03e−07	4.02	8.35e−08	4.57	4.86e−10	6.24
		$h_0/16$	3.94e−06	2.93 ■	6.39e−09	4.01 ■	2.96e−09	4.82 ■	7.53e−12	6.01 ■
	0.2	$h_0/2$	1.21e−03	1.95	4.88e−05	3.51	5.20e−05	3.90	3.89e−06	5.31
		$h_0/4$	2.77e−04	2.12	1.86e−06	4.72	3.05e−06	4.09	4.27e−08	6.51
		$h_0/8$	5.78e−05	2.26	1.21e−07	3.94	1.61e−07	4.24	5.47e−10	6.29
		$h_0/16$	1.06e−05	2.44 ⊠	5.98e−09	4.34 ■	7.57e−09	4.41 ⊠	7.16e−12	6.26 ■
DGSEM-LGL + HLLC	3.5	$h_0/2$	1.22e−03	3.38	3.85e−05	4.03	4.34e−05	4.93	2.70e−06	5.93
		$h_0/4$	1.26e−04	3.27	2.41e−06	4.00	1.39e−06	4.96	4.33e−08	5.96
		$h_0/8$	1.48e−05	3.10	1.51e−07	4.00	4.34e−08	5.01	6.67e−10	6.02
		$h_0/16$	1.81e−06	3.03 ■	9.42e−09	4.00 ■	1.36e−09	4.99 ■	1.05e−11	5.98 ■
	1.0	$h_0/2$	9.17e−04	2.86	3.96e−05	3.94	4.41e−05	4.94	2.76e−06	6.02
		$h_0/4$	1.15e−04	2.99	2.41e−06	4.04	1.47e−06	4.90	4.44e−08	5.96
		$h_0/8$	1.44e−05	3.00	1.51e−07	4.00	4.38e−08	5.07	6.84e−10	6.02
		$h_0/16$	1.80e−06	3.00 ■	9.42e−09	4.00 ■	1.35e−09	5.02 ■	1.08e−11	5.99 ■

(continued)

Table 5 (continued)

Volume disc. + Surface flux	Mach	Mesh level	$N = 2$		$N = 3$		$N = 4$		$N = 5$	
			$L_2(\varrho)$	EOC	$L_2(\varrho)$	EOC	$L_2(\varrho)$	EOC	$L_2(\varrho)$	EOC
	0.2	$h_0/2$	9.26e−04	2.35	4.63e−05	3.45	4.26e−05	4.27	2.97e−06	5.37
		$h_0/4$	1.19e−04	2.96	2.40e−06	4.27	1.59e−06	4.74	4.48e−08	6.05
		$h_0/8$	1.43e−05	3.06	1.57e−07	3.93	4.34e−08	5.19	6.83e−10	6.04
		$h_0/16$	1.80e−06	2.99 ■	9.37e−09	4.07 ■	1.49e−09	4.86 ■	1.09e−11	5.97 ■

Full order is marked with ■ ($\gtrsim N + 1$) and an order reduction with ⊠

Table 6 Experimental order of convergence of L_2 error to the exact density (4), using entropy stable LLF and ECKEP-LLF flux

Volume disc. + Surface flux	Mach	Mesh level	N = 2		N = 3		N = 4		N = 5	
			$L_2(\varrho)$	EOC	$L_2(\varrho)$	EOC	$L_2(\varrho)$	EOC	$L_2(\varrho)$	EOC
Split-DGSEM + LLF	3.5	$h_0/2$	1.24e−03	3.11	3.49e−05	4.37	4.16e−05	4.85	3.37e−06	6.05
		$h_0/4$	1.52e−04	3.03	2.01e−06	4.12	1.64e−06	4.67	4.49e−08	6.23
		$h_0/8$	1.89e−05	3.00	1.23e−07	4.03	5.67e−08	4.85	6.21e−10	6.18
		$h_0/16$	2.36e−06	3.00 ■	7.68e−09	4.01 ■	1.84e−09	4.95 ■	9.58e−12	6.02 ■
	1.0	$h_0/2$	1.09e−03	1.97	3.21e−05	4.82	4.07e−05	4.94	3.77e−06	6.34
		$h_0/4$	2.03e−04	2.43	1.69e−06	4.25	1.99e−06	4.35	4.69e−08	6.33
		$h_0/8$	2.99e−05	2.76	1.04e−07	4.03	8.46e−08	4.56	5.57e−10	6.39
		$h_0/16$	3.94e−06	2.93 ■	6.41e−09	4.02 ■	3.03e−09	4.80 ■	8.38e−12	6.05 ■
	0.2	$h_0/2$	1.21e−03	1.95	4.92e−05	3.50	5.26e−05	3.91	4.52e−06	5.15
		$h_0/4$	2.77e−04	2.12	1.88e−06	4.71	3.07e−06	4.10	6.21e−08	6.19
		$h_0/8$	5.79e−05	2.26	1.22e−07	3.95	1.63e−07	4.24	6.21e−10	6.64
		$h_0/16$	1.06e−05	2.44 ⊠	6.01e−09	4.34 ■	7.66e−09	4.41 ⊠	8.26e−12	6.23 ■
Split-DGSEM + ECKEP-LLF	3.5	$h_0/2$	1.24e−03	3.10	3.49e−05	4.37	4.16e−05	4.84	3.38e−06	6.06
		$h_0/4$	1.52e−04	3.03	2.01e−06	4.12	1.64e−06	4.67	4.49e−08	6.23
		$h_0/8$	1.89e−05	3.01	1.23e−07	4.03	5.68e−08	4.85	6.21e−10	6.18
		$h_0/16$	2.36e−06	3.00 ■	7.68e−09	4.01 ■	1.84e−09	4.95 ■	9.58e−12	6.02 ■
	1.0	$h_0/2$	1.10e−03	1.97	3.20e−05	4.83	4.08e−05	4.94	3.77e−06	6.36
		$h_0/4$	2.04e−04	2.43	1.69e−06	4.25	2.00e−06	4.35	4.69e−08	6.33
		$h_0/8$	3.00e−05	2.76	1.04e−07	4.03	8.49e−08	4.56	5.57e−10	6.40
		$h_0/16$	3.95e−06	2.93 ■	6.41e−09	4.02 ■	3.04e−09	4.81 ■	8.38e−12	6.05 ■

(continued)

Table 6 (continued)

Volume disc. + Surface flux	Mach	Mesh level	$N = 2$		$N = 3$		$N = 4$		$N = 5$	
			$L_2(\varrho)$	EOC	$L_2(\varrho)$	EOC	$L_2(\varrho)$	EOC	$L_2(\varrho)$	EOC
	0.2	$h_0/2$	1.21e−03	1.95	4.93e−05	3.51	5.27e−05	3.91	4.53e−06	5.15
		$h_0/4$	2.78e−04	2.12	1.88e−06	4.71	3.08e−06	4.10	6.21e−08	6.19
		$h_0/8$	5.79e−05	2.26	1.22e−07	3.95	1.63e−07	4.24	6.21e−10	6.65
		$h_0/16$	1.07e−05	2.44 ⊠	6.01e−09	4.34 ■	7.67e−09	4.41 ⊠	8.26e−12	6.23 ■

Full order is marked with ■ ($\gtrsim N + 1$) and an order reduction with ⊠

Manufactured Solution with Source Term

Here, we run a convergence test with the method of manufactured solutions. To do so, we assume a two-dimensional solution of the form

$$\mathbf{u} = \left[\varrho, \varrho v_1, \varrho v_2, \varrho v_3, E\right]^T = \left[g, g, g, 0, g^2\right]^T \tag{5}$$

$$\text{with } g = g(x_1, x_2, t) = 0.5 \sin(2\pi(x_1 + x_2 - t)) + 2.$$

Note that the average Mach number in the domain is Ma $= 0.8$. Inserting (5) into the Euler equations, and using the fact that spatial and time derivatives are $g' = \partial_{x_1} g = \partial_{x_2} g = -\partial_t g$, we get an additional residual

$$\mathbf{u}_t + \sum_{i=1}^{3} \frac{\partial \mathbf{f}_i}{\partial x_i} = \begin{pmatrix} g' \\ (3\gamma - 2)g' + 2(\gamma - 1)gg' \\ (3\gamma - 2)g' + 2(\gamma - 1)gg' \\ 0 \\ (6\gamma - 2)g' + 2(2\gamma - 1)gg' \end{pmatrix} \tag{6}$$

To solve the inhomogeneous problem, we subtract the residual from the approximate solution in each Runge–Kutta step. Moreover, we run the test case up to the final time $t = 1.0$.

In the convergence results for the standard DGSEM Gauss and Gauss-Lobatto, we see that the LLF flux still leads to an order reduction for $N = 2, 4$, whereas full order is found for the HLL, HLLC and Roe fluxes, see Tables 7 and 8.

In Table 9 the entropy conservative scheme shows again an order reduction for $N = 3, 5$, and the LLF-Type dissipation too, for $N = 2, 4$, and for this test case, all entropy stable schemes exhibit full order.

Table 7 Experimental order of convergence of L_2 error of density for the manufactured solution (5), using DGSEM-Gauss with LLF, HLL, HLLC and Roe fluxes

Volume disc. + Surface flux	Mesh level	$N = 2$		$N = 3$		$N = 4$		$N = 5$	
		$L_2(\varrho)$	EOC	$L_2(\varrho)$	EOC	$L_2(\varrho)$	EOC	$L_2(\varrho)$	EOC
DGSEM-Gauss + LLF	$h_0/2$	2.30e−03	2.20	4.54e−05	5.34	1.13e−04	6.35	4.52e−05	4.38
	$h_0/4$	4.81e−04	2.25	1.99e−06	4.52	4.78e−06	4.56	2.37e−07	7.58
	$h_0/8$	9.48e−05	2.34	1.02e−07	4.28	2.88e−07	4.05	1.53e−09	7.28
	$h_0/16$	1.57e−05	2.60 ⊠	6.25e−09	4.03 ■	1.73e−08	4.05 ⊠	1.57e−11	6.61
DGSEM-Gauss + HLL	$h_0/2$	1.24e−03	2.84	5.46e−05	4.35	1.32e−04	5.22	1.47e−05	6.30
	$h_0/4$	1.17e−04	3.41	3.36e−06	4.02	2.89e−06	5.51	1.44e−07	6.67
	$h_0/8$	1.41e−05	3.04	1.85e−07	4.18	7.14e−08	5.34	1.74e−09	6.37
	$h_0/16$	1.76e−06	3.00 ■	1.07e−08	4.11 ■	2.15e−09	5.05 ■	2.24e−11	6.28 ■
DGSEM-Gauss + HLLC	$h_0/2$	1.24e−03	2.84	5.46e−05	4.35	1.32e−04	5.22	1.47e−05	6.30
	$h_0/4$	1.17e−04	3.41	3.36e−06	4.02	2.89e−06	5.51	1.44e−07	6.67
	$h_0/8$	1.41e−05	3.04	1.85e−07	4.18	7.14e−08	5.34	1.74e−09	6.37
	$h_0/16$	1.76e−06	3.00 ■	1.07e−08	4.11 ■	2.15e−09	5.05 ■	2.24e−11	6.28 ■
DGSEM-Gauss + Roe	$h_0/2$	1.24e−03	2.84	5.46e−05	4.35	1.32e−04	5.22	1.47e−05	6.30
	$h_0/4$	1.17e−04	3.41	3.36e−06	4.02	2.89e−06	5.51	1.44e−07	6.67
	$h_0/8$	1.41e−05	3.04	1.85e−07	4.18	7.14e−08	5.34	1.74e−09	6.37
	$h_0/16$	1.76e−06	3.00 ■	1.07e−08	4.11 ■	2.15e−09	5.05 ■	2.24e−11	6.28 ■

Full order is marked with ■ ($\gtrsim N + 1$) and an order reduction with ⊠

Table 8 Experimental order of convergence of L_2 error of density for the manufactured solution (5), using DGSEM-LGL with LLF, HLL, HLLC and Roe fluxes

Volume disc. + Surface flux	Mesh level	N = 2		N = 3		N = 4		N = 5	
		$L_2(\varrho)$	EOC	$L_2(\varrho)$	EOC	$L_2(\varrho)$	EOC	$L_2(\varrho)$	EOC
DGSEM-LGL + LLF	$h_0/2$	7.36e−03	2.85	3.15e−04	4.38	5.69e−04	5.78	9.33e−05	5.34
	$h_0/4$	1.33e−03	2.47	1.43e−05	4.46	2.04e−05	4.80	9.27e−07	6.65
	$h_0/8$	2.79e−04	2.25	7.99e−07	4.16	8.81e−07	4.54	9.34e−09	6.63
	$h_0/16$	5.31e−05	2.39 ⊠	4.72e−08	4.08 ■	5.94e−08	3.89 ⊠	1.37e−10	6.09 ■
DGSEM-LGL + HLL	$h_0/2$	5.32e−03	3.16	2.52e−04	4.00	3.84e−04	5.36	4.33e−05	6.24
	$h_0/4$	5.99e−04	3.15	1.38e−05	4.19	1.43e−05	4.75	4.58e−07	6.56
	$h_0/8$	7.25e−05	3.05	7.69e−07	4.17	2.92e−07	5.61	7.08e−09	6.02
	$h_0/16$	9.02e−06	3.01 ■	4.74e−08	4.02 ■	7.77e−09	5.23 ■	1.10e−10	6.01 ■
DGSEM-LGL + HLLC	$h_0/2$	5.32e−03	3.16	2.52e−04	4.00	3.84e−04	5.36	4.33e−05	6.24
	$h_0/4$	5.99e−04	3.15	1.38e−05	4.19	1.43e−05	4.75	4.58e−07	6.56
	$h_0/8$	7.25e−05	3.05	7.69e−07	4.17	2.92e−07	5.61	7.08e−09	6.02
	$h_0/16$	9.02e−06	3.01 ■	4.74e−08	4.02 ■	7.77e−09	5.23 ■	1.10e−10	6.01 ■
DGSEM-LGL + Roe	$h_0/2$	5.32e−03	3.16	2.52e−04	4.00	3.84e−04	5.36	4.33e−05	6.24
	$h_0/4$	5.99e−04	3.15	1.38e−05	4.19	1.43e−05	4.75	4.58e−07	6.56
	$h_0/8$	7.25e−05	3.05	7.69e−07	4.17	2.92e−07	5.61	7.08e−09	6.02
	$h_0/16$	9.02e−06	3.01 ■	4.74e−08	4.02 ■	7.77e−09	5.23 ■	1.10e−10	6.01 ■

Full order is marked with ■ ($\gtrsim N + 1$) and an order reduction with ⊠

Table 9 Experimental order of convergence of L_2 error of density for the manufactured solution (5), using entropy conservative and entropy stable schemes

Volume disc. + Surface flux	Mesh level	$N = 2$		$N = 3$		$N = 4$		$N = 5$	
		$L_2(\varrho)$	EOC	$L_2(\varrho)$	EOC	$L_2(\varrho)$	EOC	$L_2(\varrho)$	EOC
Split-DGSEM + ECKEP	$h_0/2$	1.31e−02	3.53	1.28e−03	2.88	5.62e−03	2.49	1.12e−03	4.05
	$h_0/4$	1.30e−03	3.34	1.13e−04	3.49	3.12e−04	4.17	5.97e−06	7.56
	$h_0/8$	1.23e−04	3.40	1.24e−05	3.20	2.00e−06	7.29	9.45e−08	5.98
	$h_0/16$	1.76e−05	2.80 ■	1.67e−06	2.89 ⊠	3.41e−08	5.87 ■	3.17e−09	4.90 ⊠
Split-DGSEM + LLF	$h_0/2$	7.60e−03	3.13	3.70e−04	3.91	6.56e−04	4.44	1.28e−04	4.75
	$h_0/4$	1.63e−03	2.22	1.90e−05	4.29	3.09e−05	4.41	2.01e−06	6.00
	$h_0/8$	3.40e−04	2.26	9.89e−07	4.26	1.83e−06	4.08	1.80e−08	6.80
	$h_0/16$	6.14e−05	2.47 ⊠	6.41e−08	3.95 ■	9.15e−08	4.32 ⊠	2.17e−10	6.37 ■
Split-DGSEM + ECKEP-LLF	$h_0/2$	7.66e−03	3.21	3.74e−04	3.98	6.79e−04	4.45	1.33e−04	4.75
	$h_0/4$	1.64e−03	2.23	1.90e−05	4.30	3.08e−05	4.46	2.06e−06	6.01
	$h_0/8$	3.41e−04	2.26	9.88e−07	4.26	1.83e−06	4.07	1.79e−08	6.85
	$h_0/16$	6.16e−05	2.47 ⊠	6.40e−08	3.95 ■	9.17e−08	4.32 ⊠	2.17e−10	6.37 ■
Split-DGSEM + HLL	$h_0/2$	5.82e−03	3.10	3.01e−04	3.85	5.11e−04	4.35	7.06e−05	5.27
	$h_0/4$	7.06e−04	3.04	2.04e−05	3.88	1.67e−05	4.94	1.08e−06	6.03
	$h_0/8$	8.63e−05	3.03	1.16e−06	4.14	5.08e−07	5.04	1.67e−08	6.02
	$h_0/16$	1.08e−05	3.00 ■	7.20e−08	4.01 ■	1.62e−08	4.97 ■	1.67e−08	6.02
Split-DGSEM + ECKEP-Roe	$h_0/2$	5.81e−03	3.11	3.01e−04	3.85	5.11e−04	4.35	2.64e−10	5.98 ■
	$h_0/4$	7.06e−04	3.04	2.04e−05	3.88	1.67e−05	4.94	7.06e−05	5.27
	$h_0/8$	8.63e−05	3.03	1.16e−06	4.14	5.08e−07	5.04	1.08e−06	6.03
	$h_0/16$	1.08e−05	3.00 ■	7.20e−08	4.01 ■	1.62e−08	4.97 ■	1.67e−08	6.02
								2.64e−10	5.98 ■

Full order is marked with ■ ($\gtrsim N + 1$) and an order reduction with ⊠

References

1. Bohm, M., Winters, A.R., Gassner, G.J., Derigs, D., Hindenlang, F., Saur, J.: An entropy stable nodal discontinuous Galerkin method for the resistive MHD equations. Part I: Theory and numerical verification. J. Comput. Phys. 108076 (2018). https://doi.org/10.1016/j.jcp.2018.06.027
2. Carpenter, M., Kennedy, C.: Fourth-order 2N-storage Runge-Kutta schemes. Technical Report NASA TM 109111 (1994)
3. Carpenter, M., Fisher, T., Nielsen, E., Frankel, S.: Entropy stable spectral collocation schemes for the Navier–Stokes equations: discontinuous interfaces. SIAM J. Sci. Comput. 36(5), B835–B867 (2014)
4. Chan, J.: On discretely entropy conservative and entropy stable discontinuous Galerkin methods. J. Comput. Phys. 362, 346–374 (2018)
5. Chan, J., Warburton, T.: On the penalty stabilization mechanism for upwind discontinuous Galerkin formulations of first order hyperbolic systems. Comput. Math. Appl. 74(12), 3099–3110 (2017)
6. Chen, T., Shu, C.-W.: Entropy stable high order discontinuous Galerkin methods with suitable quadrature rules for hyperbolic conservation laws. J. Comput. Phys. 345, 427–461 (2017)
7. Flad, D., Gassner, G.: On the use of kinetic energy preserving DG-schemes for large eddy simulation. J. Comput. Phys. 350, 782–795 (2017)
8. Friedrich, L., Winters, A.R., Del Rey Fernández, D.C., Gassner, G.J., Parsani, M., Carpenter, M.H.: An entropy stable h/p non-conforming discontinuous Galerkin method with the summation-by-parts property. J. Sci. Comput. 77, 689–725 (2018)
9. Gassner, G.: A skew-symmetric discontinuous Galerkin spectral element discretization and its relation to SBP-SAT finite difference methods. SIAM J. Sci. Comput. 35(3), A1233–A1253 (2013)
10. Gassner, G.J.: A kinetic energy preserving nodal discontinuous Galerkin spectral element method. Int. J. Numer. Methods Fluids 76(1), 28–50 (2014)
11. Gassner, G., Kopriva, D.A.: A comparison of the dispersion and dissipation errors of Gauss and Gauss–Lobatto discontinuous Galerkin spectral element methods. SIAM J. Sci. Comput. 33, 2560–2579 (2011)
12. Gassner, G.J., Winters, A.R., Kopriva, D.A.: Split form nodal discontinuous Galerkin schemes with summation-by-parts property for the compressible Euler equations. J. Comput. Phys. 327, 39–66 (2016)
13. Gassner, G.J., Winters, A.R., Hindenlang, F.J., Kopriva, D.A.: The BR1 scheme is stable for the compressible Navier–Stokes equations. J. Sci. Comput. 77, 154–200 (2018)
14. Kopriva, D.A.: Implementing Spectral Methods for Partial Differential Equations. Scientific Computation. Springer, Berlin (2009)
15. Manzanero, J., Ferrer, E., Rubio, G., Valero, E.: On the role of numerical dissipation in stabilising under-resolved turbulent simulations using discontinuous Galerkin methods. Preprint arXiv:1805.10519 (2018)
16. Moura, R., Sherwin, S., Peiró, J.: Eigensolution analysis of spectral/hp continuous Galerkin approximations to advection-diffusion problems: insights into spectral vanishing viscosity. J. Comput. Phys. 307, 401–422 (2016)
17. Parsani, M., Carpenter, M.H., Nielsen, E.J.: Entropy stable wall boundary conditions for the three-dimensional compressible Navier–Stokes equations. J. Comput. Phys. 292, 88–113 (2015)
18. Toro, E.: Riemann Solvers and Numerical Methods for Fluid Dynamics. Springer, Berlin (2009)
19. Wintermeyer, N., Winters, A.R., Gassner, G.J., Kopriva, D.A.: An entropy stable nodal discontinuous Galerkin method for the two dimensional shallow water equations on unstructured curvilinear meshes with discontinuous bathymetry. J. Comput. Phys. (2016, submitted). arXiv:1509.07096 [math.NA]

20. Wintermeyer, N., Winters, A.R., Gassner, G.J., Warburton, T.: An entropy stable discontinuous Galerkin method for the shallow water equations on curvilinear meshes with wet/dry fronts accelerated by GPUs. J. Comput. Phys. **375**, 447–480 (2018)
21. Winters, A.R., Gassner, G.J.: A comparison of two entropy stable discontinuous Galerkin spectral element approximations for the shallow water equations with non-constant topography. J. Comput. Phys. **301**, 357–376 (2015)
22. Winters, A.R., Derigs, D., Gassner, G.J., Walch, S.: A uniquely defined entropy stable matrix dissipation operator for high Mach number ideal MHD and compressible Euler simulations. J. Comput. Phys. **332**, 274–289 (2017)
23. Winters, A.R., Moura, R.C., Mengaldo, G., Gassner, G.J., Walch, S., Peiro, J., Sherwin, S.J.: A comparative study on polynomial dealiasing and split form discontinuous Galerkin schemes for under-resolved turbulence computations. J. Comput. Phys. **372**, 1–21 (2018)

A Review of Regular Decompositions of Vector Fields: Continuous, Discrete, and Structure-Preserving

Ralf Hiptmair and Clemens Pechstein

1 Introduction

For a bounded Lipschitz domain $\Omega \subset \mathbb{R}^3$ recall the classical L^2-orthogonal *Helmholtz decompositions*

$$\mathbf{L}^2(\Omega) = \nabla H_0^1(\Omega) \oplus \mathbf{H}(\operatorname{div} 0, \Omega) = \nabla H^1(\Omega) \oplus \mathbf{H}_0(\operatorname{div} 0, \Omega),$$

see, e.g., [9, Ch. XI, Sect. I]. They can be used to derive decompositions of (subspaces of) $\mathbf{H}(\mathbf{curl}, \Omega)$:

$$\mathbf{H}_0(\mathbf{curl}, \Omega) = \nabla H_0^1(\Omega) \oplus \mathbf{X}_N(\Omega), \quad \mathbf{X}_N(\Omega) := \mathbf{H}_0(\mathbf{curl}, \Omega) \cap \mathbf{H}(\operatorname{div} 0, \Omega),$$

$$\mathbf{H}(\mathbf{curl}, \Omega) = \nabla H^1(\Omega) \oplus \mathbf{X}_T(\Omega), \quad \mathbf{X}_T(\Omega) := \mathbf{H}(\mathbf{curl}, \Omega) \cap \mathbf{H}_0(\operatorname{div} 0, \Omega).$$

If the domain Ω is convex then the respective complementary space, $\mathbf{X}_N(\Omega)$ or $\mathbf{X}_T(\Omega)$, is continuously embedded in the space $\mathbf{H}^1(\Omega)$ of vector fields with Cartesian components in $H^1(\Omega)$, cf. [1]. Then one can, for instance, write any $\mathbf{u} \in \mathbf{H}(\mathbf{curl}, \Omega)$ as

$$\mathbf{u} = \nabla p + \mathbf{z}, \tag{1}$$

R. Hiptmair (✉)
Seminar for Applied Mathematics, ETH Zürich, Zürich, Switzerland
e-mail: hiptmair@sam.math.ethz.ch

C. Pechstein
Dassault Systèmes, Darmstadt, Germany
e-mail: clemens.pechstein@3ds.com

S. J. Sherwin et al. (eds.), *Spectral and High Order Methods for Partial Differential Equations ICOSAHOM 2018*, Lecture Notes in Computational Science and Engineering 134, https://doi.org/10.1007/978-3-030-39647-3_3

with $p \in H^1(\Omega)$ and $\mathbf{z} \in \mathbf{H}^1(\Omega)$. Since $\|\nabla p\|_{\mathbf{L}^2(\Omega)} \leq \|\mathbf{u}\|_{\mathbf{L}^2(\Omega)}$ one obtains (using the continuous embedding) the stability property[1]

$$\|\nabla p\|_{\mathbf{L}^2(\Omega)} + \|\mathbf{z}\|_{\mathbf{H}^1(\Omega)} \leq C \|\mathbf{z}\|_{\mathbf{H}(\mathbf{curl},\Omega)} . \tag{2}$$

A similar decomposition can be found for $\mathbf{u} \in \mathbf{H}_0(\mathbf{curl}, \Omega)$.

Generally, a decomposition of form (1) with the stability property (2) is called *regular decomposition*, even if $\mathbf{L}^2$-orthogonality does not hold. Actually, it turns out that (1)–(2) can be achieved even in cases where Ω is non-convex, in particular on non-smooth domains, or in cases where Ω or its boundary have non-trivial topology; only the $\mathbf{L}^2$-orthogonality has to be sacrificed, cf. [20].

Noting that $\nabla H^1(\Omega)$ is contained in the kernel of the **curl** operator and that— under mild smoothness assumptions on the domain—the whole kernel is spanned by $\nabla H^1(\Omega)$ plus a finite-dimensional *co-homology space* [15, Sect. 4] one can achieve a second decomposition,

$$\mathbf{u} = \mathbf{h} + \mathbf{z}, \tag{3}$$

with $\mathbf{h} \in \ker(\mathbf{curl}_{|\mathbf{H}(\mathbf{curl},\Omega)})$ and $\mathbf{z} \in \mathbf{H}^1(\Omega)$, where

$$\|\mathbf{h}\|_{\mathbf{L}^2(\Omega)} \leq C \|\mathbf{u}\|_{\mathbf{L}^2(\Omega)}, \qquad \|\mathbf{z}\|_{\mathbf{H}^1(\Omega)} \leq C \|\mathbf{curl}\ \mathbf{u}\|_{\mathbf{L}^2(\Omega)} . \tag{4}$$

The second stability estimate states that if $\mathbf{u}$ is already in the kernel of the **curl** operator, then $\mathbf{z}$ is zero. Hence, (1) the operator mapping $\mathbf{u}$ to $\mathbf{h}$ is a *projection* onto the kernel space and (2) the complement operator projects $\mathbf{u}$ to the function $\mathbf{z}$ of higher regularity $\mathbf{H}^1(\Omega)$. For trivial topology of Ω and $\partial\Omega$, the two decompositions (1)–(2) and (3)–(4) coincide.

As a few among many more [17, Sect. 1.5], we would like to highlight two important applications of these regular decompositions.

1. The second form (3)–(4), in the sequel called *rotation-bounded decomposition*, can be used to show that the operator underlying a certain boundary value problem for Maxwell's equations is a *Fredholm operator*. The key point is that the complement space of the kernel (from the view of the mentioned projections) is $\mathbf{H}^1(\Omega)$ which is *compactly* embedded in $\mathbf{L}^2(\Omega)$, see e.g., [14, 16] and references therein.
2. The first form (1)–(2), in the sequel called *gradient-based decomposition*, has been used to generate stable three-term splittings of a finite element subspace of $\mathbf{H}(\mathbf{curl}, \Omega)$, cf. [19–21, 23], which allows the construction of so-called *fictitious* or *auxiliary* space preconditioners for the ill-conditioned system matrix underlying the discretized Maxwell equations.

---

[1]Here and below C stands for a positive "generic constant" that may depend only on Ω, unless specified otherwise.

In both applications, it is desirable to obtain the decompositions for minimal smoothness of the domain, e.g., Lipschitz domains, which are not necessarily convex. Moreover, it is also desirable to go beyond decompositions of the entire space $\mathbf{H}(\mathbf{curl}, \Omega)$ and extend them to subspaces for which the appropriate trace vanishes on a "Dirichlet part" Γ_D of the boundary. In this case traces of the two summands should also vanish on Γ_D.

In the present paper, we provide regular decompositions of both types for subspaces of $\mathbf{H}(\mathbf{curl}, \Omega)$ (in Sect. 3) and $\mathbf{H}(\mathrm{div}, \Omega)$ (in Sect. 4) comprising functions with vanishing trace on a part Γ_D of the boundary $\partial\Omega$ for Lipschitz domains Ω of *arbitrary topology*. In particular, Ω is allowed to have handles, and $\partial\Omega$ and Γ_D may have several connected components. The Dirichlet boundary Γ_D must satisfy a certain smoothness assumption that we shall introduce in Sect. 2. In addition to the stability estimates (2) and (4), we show that the decompositions are stable even in $\mathbf{L}^2(\Omega)$.

In the final part of the manuscript, in Sect. 5, we establish regular decompositions of spaces of Whitney forms, which are lowest-order conforming finite element subspaces of $\mathbf{H}(\mathbf{curl}, \Omega)$ and $\mathbf{H}(\mathrm{div}, \Omega)$, respectively, built upon simplicial triangulations of Ω.

This note is based on [17] and is an abridged version of [18]. Please refer to this latter preprint for complete proofs of the results quoted below.

2 Preliminaries

Since subtle geometric arguments will play a major role for parts of the theory, we start with a precise characterization of the geometric setting: Let $\Omega \subset \mathbb{R}^3$ be an open, bounded, connected Lipschitz domain.[2] We write $\mathrm{d}(\Omega)$ for its diameter. Its boundary $\Gamma := \partial\Omega$ is partitioned according to $\Gamma = \Gamma_D \cup \Sigma \cup \Gamma_N$, with relatively open sets Γ_D and Γ_N. We assume that this provides a *piecewise C^1 dissection* of $\partial\Omega$ in the sense of [12, Definition 2.2]. Sloppily speaking, this means that Σ is the union of closed curves that are piecewise C^1.

Under the above assumptions on Ω and Γ_D, [12, Lemma 4.4] guarantees the existence of an open Lipschitz neighborhood Ω_Γ ("Lipschitz collar") of Γ and of a "bulge" $\Upsilon_D \subset \Omega_\Gamma \setminus \overline{\Omega}$. We recall the properties of bulge domains from [12, Sect. 2, Thm. 2.3], also stated in [17, Thm. 2.2]:

Theorem 1 (Bulge-Augmented Domain) *There exists a Lipschitz domain $\Upsilon_D \subset \mathbb{R}^3 \setminus \overline{\Omega}$, such that $\overline{\Upsilon}_D \cap \overline{\Omega} = \Gamma_D$, $\Omega^e := \Upsilon_D \cup \Gamma_D \cup \Omega$ is Lipschitz, $\mathrm{d}(\Omega^e) \leq 2\,\mathrm{d}(\Omega)$, and $\overline{\Upsilon}_D \subset \Omega_\Gamma$. Moreover, each connected component $\Gamma_{D,k}$ of Γ_D corresponds to a connected component $\Upsilon_{D,k}$ of Υ_D, and these have positive distance from each other.*

---

[2] Strongly Lipschitz, in the sense that the boundary is locally the graph of a Lipschitz continuous function.

Let

$$H^1_{\Gamma_D}(\Omega) := \{u \in H^1(\Omega) : (\gamma u)_{|\Gamma_D} = 0\},$$

$$\mathbf{H}_{\Gamma_D}(\mathbf{curl}, \Omega) := \{\mathbf{u} \in \mathbf{H}(\mathbf{curl}, \Omega) : (\boldsymbol{\gamma}_\tau \mathbf{u})_{|\Gamma_D} = 0\},$$

$$\mathbf{H}_{\Gamma_D}(\mathrm{div}, \Omega) := \{\mathbf{u} \in \mathbf{H}(\mathrm{div}, \Omega) : (\gamma_n \mathbf{u})_{|\Gamma_D} = 0\},$$

denote the standard Sobolev spaces where the distributional gradient, curl, or divergence is in L^2 and where the pointwise trace γu, the tangential trace $\boldsymbol{\gamma}_\tau \mathbf{u}$, or the normal trace $\gamma_n \mathbf{u}$, respectively, vanishes on the Dirichlet boundary Γ_D, see e.g. [3, 6, 26]. These space are linked via the *de Rham complex*,

$$\mathcal{K}_{\Gamma_D}(\Omega) \xrightarrow{\mathrm{id}} H^1_{\Gamma_D}(\Omega) \xrightarrow{\nabla} \mathbf{H}_{\Gamma_D}(\mathbf{curl}, \Omega) \xrightarrow{\mathbf{curl}} \mathbf{H}_{\Gamma_D}(\mathrm{div}, \Omega) \xrightarrow{\mathrm{div}} L^2(\Omega),$$
(5)

where

$$\mathcal{K}_{\Gamma_D}(\Omega) := \{v \in H^1_{\Gamma_D}(\Omega) : v = \mathrm{const}\} = \begin{cases} \mathrm{span}\{1\}, & \text{if } \Gamma_D = \emptyset, \\ \{0\}, & \text{otherwise.} \end{cases}$$

The range of each operator in (5) lies in the kernel space of the succeeding one, cf. [3, Lemma 2.2]. We define

$$\mathbf{H}_{\Gamma_D}(\mathbf{curl}\, 0, \Omega) := \{\mathbf{v} \in \mathbf{H}_{\Gamma_D}(\mathbf{curl}, \Omega) : \mathbf{curl}\, \mathbf{v} = 0\},$$

$$\mathbf{H}_{\Gamma_D}(\mathrm{div}\, 0, \Omega) := \{\mathbf{v} \in \mathbf{H}_{\Gamma_D}(\mathrm{div}, \Omega) : \mathrm{div}\, \mathbf{v} = 0\}.$$
(6)

Barring topological obstructions these kernels can be represented through potentials: Let $\beta_1(\Omega)$ denote the *first Betti number* of Ω (the number of "handles") and $\beta_2(\Omega)$ the *second Betti number* (the number of connected components of $\partial\Omega$ minus one). By the very definition of the Betti numbers as dimensions of co-homology spaces we have

$$\beta_1(\Omega) = 0 \implies \mathbf{H}(\mathbf{curl}\, 0, \Omega) = \nabla H^1(\Omega),$$
(7)

$$\beta_2(\Omega) = 0 \implies \mathbf{H}(\mathrm{div}\, 0, \Omega) = \mathbf{curl}\, \mathbf{H}(\mathbf{curl}, \Omega),$$
(8)

cf. [26]. We call Ω topologically trivial if $\beta_1(\Omega) = \beta_2(\Omega) = 0$.

3 Regular Decompositions and Potentials Related to H(curl)

Throughout we rely on the properties of Ω and Γ_D as introduced in Sect. 2 and use the notations from Theorem 1. We write C for positive "generic constants" and say that a constant "depends only on the shape of Ω and Γ_D", if it depends on the geometric setting alone, but is invariant with respect to similarity transformations. To achieve this the diameter of Ω will have to enter the estimates; we denote it by $d(\Omega)$.

3.1 Gradient-Based Regular Decomposition of H(curl)

The following theorem is essentially [17, Thm. 2.1].

Theorem 2 (Gradient-Based Regular Decomposition of H(curl)) *Let (Ω, Γ_D) satisfy the assumptions of Sect. 2. Then for each $\mathbf{u} \in \mathbf{H}_{\Gamma_D}(\mathbf{curl}, \Omega)$ there exist $\mathbf{z} \in \mathbf{H}_{\Gamma_D}^1(\Omega)$ and $p \in H_{\Gamma_D}^1(\Omega)$ depending linearly on $\mathbf{u}$ such that*

(i)
$$\mathbf{u} = \mathbf{z} + \nabla p,$$

(ii)
$$\|\mathbf{z}\|_{0,\Omega} + \|\nabla p\|_{0,\Omega} \leq C \|\mathbf{u}\|_{0,\Omega},$$

(iii)
$$\|\nabla \mathbf{z}\|_{0,\Omega} + \frac{1}{d(\Omega)}\|\mathbf{z}\|_{0,\Omega} \leq C\|\mathbf{curl}\,\mathbf{u}\|_{0,\Omega} + \frac{1}{d(\Omega)}\|\mathbf{u}\|_{0,\Omega},$$

with constants depending only on the shape of Ω and Γ_D, but not on $d(\Omega)$.

Remark 1 An early decomposition of a subspace of $\mathbf{H}(\mathbf{curl}, \Omega) \cap \mathbf{H}(\text{div}, \Omega)$ into a regular part in $\mathbf{H}^1(\Omega)$ and a singular part in $\nabla H^1(\Omega)$ can be found in [4] and in [5, Proposition 5.1], see also [7, Sect. 3] and references therein. Theorem 2 was proved in [14, Lemma 2.4] for the case of $\Gamma_D = \partial\Omega$ and without the $\mathbf{L}^2$-stability estimate, following [5, Proposition 5.1]. Pasciak and Zhao [28, Lemma 2.2] provided a version for simply connected Ω and the case $\Gamma_D = \partial\Omega$ *with* pure $\mathbf{L}^2$-stability, but p is only constant on each connected component of $\partial\Omega$ (see also Theorem 5 and Remark 3). This result was refined in [24, Thm. 3.1]. For the case $\Gamma_D = \emptyset$, [14, Lemma 2.4] gives a similar decomposition but ∇p must be replaced by an element from $\mathbf{H}(\mathbf{curl}\,0, \Omega)$ in general. Finally, Theorem 2 without the pure $\mathbf{L}^2$-stability was proved in [20, Thm. 5.2].[3]

Remark 2 The constant C in Theorem 2 depends mainly on the stability constants of key extension operators. If the bulge Υ_D has multiple components $\Upsilon_{D,k}$, the final estimate will depend on the relative distances between $\Upsilon_{D,k}$, $\Upsilon_{D,\ell}$, $k \neq \ell$ and the ratios $d(\Upsilon_{D,k})/d(\Omega)$.

---

[3] This reference contains a typo which is easily identified when inspecting the proof: In general, $\mathbf{z}$ cannot be estimated in terms of $\|\mathbf{curl}\,\mathbf{u}\|_{0,\Omega}$ but one must use the full $\mathbf{H}(\mathbf{curl})$ norm.

Remark 3 If $\Gamma_D = \partial\Omega$, one obtains only $p \in H^1(\Omega)$ being constant on each connected component of Γ_D but the improved bound

$$\| \nabla z \|_{0,\Omega} + d(\Omega)^{-1} \| z \|_{0,\Omega} \leq C \| \operatorname{\mathbf{curl}} u \|_{0,\Omega} .$$

Results on regular decompositions in this special case can be found in [24, 28].

3.2   Regular Potentials for Some Divergence-Free Functions

Let the domain Ω and the Dirichlet boundary part Γ_D be as introduced in Sect. 2 and let $\Gamma_i, i = 0, \ldots, \beta_2(\Omega)$, denote the connected components of $\partial\Omega$, where $\beta_2(\Omega)$ is the second Betti number of Ω.

We define the space[4]

$$\mathbf{H}_{\Gamma_D}(\operatorname{div} 00, \Omega) := \left\{ \mathbf{q} \in \mathbf{H}_{\Gamma_D}(\operatorname{div} 0, \Omega) : \langle \gamma_n \mathbf{q}, 1 \rangle_{\Gamma_i} = 0, \ i = 0, \ldots, \beta_2(\Omega) \right\}. \tag{9}$$

Above γ_n denotes the normal trace operator, and the duality pairing is that between $H^{-1/2}(\Gamma_i)$ and $H^{1/2}(\Gamma_i)$. If $\Gamma_D = \emptyset$ we simply drop the subscript Γ_D. Obviously,

$$\mathbf{H}_{\Gamma_D}(\operatorname{div} 00, \Omega) \subset \mathbf{H}(\operatorname{div} 00, \Omega) .$$

The next result identifies the above space as the range of the curl operator.

Theorem 3 (Regular Potential of Range(**curl**)) *Let* (Ω, Γ_D) *be as in Sect. 2 and assume in addition that each connected component* $\Upsilon_{D,k}$ *of the bulge has vanishing first Betti number,* $\beta_1(\Upsilon_{D,k}) = 0$. *Then*

$$\mathbf{H}_{\Gamma_D}(\operatorname{div} 00, \Omega) = \operatorname{\mathbf{curl}} \mathbf{H}_{\Gamma_D}(\operatorname{\mathbf{curl}}, \Omega) = \operatorname{\mathbf{curl}} \mathbf{H}^1_{\Gamma_D}(\Omega) ,$$

and for each $\mathbf{q} \in \mathbf{H}_{\Gamma_D}(\operatorname{div} 00, \Omega)$ *there exists* $\boldsymbol{\psi} \in \mathbf{H}^1_{\Gamma_D}(\Omega)$ *depending linearly on* $\mathbf{q}$ *such that*

$$\operatorname{\mathbf{curl}} \boldsymbol{\psi} = \mathbf{q} \quad and \quad \| \nabla \boldsymbol{\psi} \|_{0,\Omega} + \frac{1}{d(\Omega)} \| \boldsymbol{\psi} \|_{0,\Omega} \leq C \| \mathbf{q} \|_{0,\Omega} ,$$

where C *depends only on the shape of* Ω *and* Γ_D, *but not on* $d(\Omega)$

---

[4]Alternatively we can define $\mathbf{H}_{\Gamma_D}(\operatorname{div} 00, \Omega)$ as the functions in $\mathbf{H}_{\Gamma_D}(\operatorname{div} 0, \Omega)$ orthogonal to the *harmonic Dirichlet fields* $\mathbf{H}(\operatorname{div} 0, \Omega) \cap \mathbf{H}_0(\operatorname{\mathbf{curl}} 0, \Omega)$.

Remark 4 For the case that $\Gamma_D = \emptyset$, we reproduce the classical result

$$\mathbf{H}(\operatorname{div} 00, \Omega) = \operatorname{\mathbf{curl}} \mathbf{H}(\operatorname{\mathbf{curl}}, \Omega) = \operatorname{\mathbf{curl}} \mathbf{H}^1(\Omega),$$

see [11, Thm. 3.4]. In that case, Step 4 of the proof can be left out and $\boldsymbol{\psi} = \mathbf{w}_1$ which is why $\operatorname{div} \boldsymbol{\psi} = 0$ in Ω. This property, however, is lost in the general case.

3.3 Rotation-Bounded Regular Decomposition of H(curl)

We can now formulate another *new* variety of regular decompositions, for which the $\mathbf{H}^1$-component will vanish for curl-free fields.

Theorem 4 (Rotation-Bounded Regular Decomposition of H(curl) (I)) *Let* (Ω, Γ_D) *be as in Sect. 2 and assume, in addition, that each connected component* $\Upsilon_{D,k}$ *of the bulge has vanishing first Betti number,* $\beta_1(\Upsilon_{D,k}) = 0$. *Then, for each* $\mathbf{u} \in \mathbf{H}_{\Gamma_D}(\operatorname{\mathbf{curl}}, \Omega)$ *there exist* $\mathbf{z} \in \mathbf{H}^1_{\Gamma_D}(\Omega)$ *and a curl-free vector field* $\mathbf{h} \in \mathbf{H}_{\Gamma_D}(\operatorname{\mathbf{curl}} 0, \Omega)$, *depending linearly on* $\mathbf{u}$ *such that*

$$\mathbf{u} = \mathbf{z} + \mathbf{h},$$

$$\|\mathbf{h}\|_{0,\Omega} \leq \|\mathbf{u}\|_{0,\Omega} + C \, \mathrm{d}(\Omega) \, \|\operatorname{\mathbf{curl}} \mathbf{u}\|_{0,\Omega},$$

$$\|\nabla \mathbf{z}\|_{0,\Omega} + \frac{1}{\mathrm{d}(\Omega)}\|\mathbf{z}\|_{0,\Omega} \leq C \, \|\operatorname{\mathbf{curl}} \mathbf{u}\|_{0,\Omega},$$

where C *depends only on the shape of* Ω *and* Γ_D, *but not on* $\mathrm{d}(\Omega)$.

Remark 5 The constant C in Theorem 4 depends essentially on the stability constants of the divergence-free extension operator $E_{\Omega^e}^{\mathrm{div},0}$ and the (adapted) Stein extension operator $E_{\Upsilon_D}^{\nabla,\mathrm{Stein}}$.

Another *stronger* version of the rotation-bounded regular decomposition of **H(curl)** gets rid of the assumptions on the topology of the Dirichlet boundary and has improved stability properties (though with less explicit constants).

Theorem 5 (Rotation-Bounded Regular Decomposition of H(curl) (II)) *Let* (Ω, Γ_D) *be as in Sect. 2. Then for each* $\mathbf{u} \in \mathbf{H}_{\Gamma_D}(\operatorname{\mathbf{curl}}, \Omega)$ *there exist* $\mathbf{z} \in \mathbf{H}^1_{\Gamma_D}(\Omega)$ *and a curl-free* $\mathbf{h} \in \mathbf{H}_{\Gamma_D}(\operatorname{\mathbf{curl}} 0, \Omega)$ *depending linearly on* $\mathbf{u}$ *such that*

$$\mathbf{u} = \mathbf{z} + \mathbf{h},$$

$$\|\mathbf{z}\|_{0,\Omega} + \|\mathbf{h}\|_{0,\Omega} \leq C \, \|\mathbf{u}\|_{0,\Omega},$$

$$\|\nabla \mathbf{z}\|_{0,\Omega} + \mathrm{d}(\Omega)^{-1}\|\mathbf{z}\|_{0,\Omega} \leq C \, \|\operatorname{\mathbf{curl}} \mathbf{u}\|_{0,\Omega},$$

where C *depends only on the shape of* Ω *and* Γ_D, *but not on* $\mathrm{d}(\Omega)$.

Remark 6 For the case $\Gamma_D = \partial\Omega$ the result of the theorem is already proved by Remark 3 since we obtain $\mathbf{u} = \mathbf{z} + \nabla p$ with $\nabla p \in \nabla H^1_{0,\mathrm{const}}(\Omega) = \mathbf{H}_0(\mathbf{curl},\, \Omega)$.

Remark 7 We would like to emphasize that both in Theorems 2 and 5, the domain Ω may be non-convex, non-smooth, and may have non-trivial topology: It may have handles and its boundary may have multiple components. Also the Dirichlet boundary Γ_D may have multiple components, each of which with non-trivial topology. Moreover, we have the pure $\mathbf{L}^2(\Omega)$-stability in both theorems. In this sense, the results of Theorems 2 and 5 are superior to those found, e.g., in [7, Thm 3.4], [19] or the more recent ones in [8, Thm. 2.3], [22].

Remark 8 If Ω has vanishing first Betti number, $\beta_1(\Omega) = 0$, then $\mathbf{H}_{\Gamma_D}(\mathbf{curl}\,0,\, \Omega)$ $= \nabla H^1_{\Gamma_D,\mathrm{const}}(\Omega)$. Hence, we can split each $\mathbf{u} \in \mathbf{H}_{\Gamma_D}(\mathbf{curl},\, \Omega)$ into $\mathbf{z} \in \mathbf{H}^1_{\Gamma_D}(\Omega)$ and ∇p with $p \in H^1(\Omega)$ being constant on each connected component of Γ_D. If Γ_D is connected, then $p \in H^1_{\Gamma_D}(\Omega)$. Summarizing, if Ω has no handles and if Γ_D is connected, then we have the combined features of Theorems 2 and 5.

Finally, we mention that the regular decomposition theorems spawn projection operators that play a fundamental role in the analysis of weak formulations of Maxwell's equations in frequency domain [14, Sect. 5].

Corollary 1 *Let (Ω, Γ_D) be as in Sect. 2. Then there exist continuous projection operators* $\mathsf{R}\colon \mathbf{H}_{\Gamma_D}(\mathbf{curl},\, \Omega) \;\to\; \mathbf{H}^1_{\Gamma_D}(\Omega)$ *and* $\mathsf{N}\colon \mathbf{H}_{\Gamma_D}(\mathbf{curl},\, \Omega) \;\to\;$ $\mathbf{H}_{\Gamma_D}(\mathbf{curl}\,0,\, \Omega)$ *such that* $\mathsf{R} + \mathsf{N} = \mathrm{id}$ *and*

$$\|\mathsf{R}\mathbf{v}\|_{\mathbf{H}^1(\Omega)} + \|\mathsf{N}\mathbf{v}\|_{\mathbf{L}^2(\Omega)} \le C\, \|\mathbf{v}\|_{\mathbf{H}(\mathbf{curl},\Omega)} \qquad \forall \mathbf{v} \in \mathbf{H}(\mathbf{curl},\, \Omega),$$

where C is a constant independent of $\mathbf{v}$. Moreover, $\mathsf{F}\colon \mathbf{H}_{\Gamma_D}(\mathbf{curl},\, \Omega) \;\to\;$ $\mathbf{H}_{\Gamma_D}(\mathbf{curl},\, \Omega)$ *defined by* $\mathsf{F}\mathbf{v} := \mathsf{R}\mathbf{v} - \mathsf{N}\mathbf{v}$ *is an isomorphism.*

Remark 9 The $\mathbf{L}^2$-estimates from Theorem 4 then show that the corresponding operator R can be extended to a continuous operator mapping from $\mathbf{L}^2(\Omega)$ to $\mathbf{L}^2(\Omega)$.

4 Regular Decompositions and Potentials Related to H(div)

The developments of this section are largely parallel to those of Sect. 3 with some new aspects concerning extensions and topological considerations.

4.1 *Rotation-Based Regular Decomposition of* H(div)

The following theorem is the $\mathbf{H}(\mathbf{div})$-counterpart of Theorem 2.

Theorem 6 (Rotation-Based Regular Decomposition of H(div)) *Let* (Ω, Γ_D) *satisfy the assumptions made in Sect. 2. Then for each* $\mathbf{v} \in \mathbf{H}_{\Gamma_D}(\mathrm{div}, \Omega)$ *there exist* $\mathbf{z} \in \mathbf{H}^1_{\Gamma_D}(\Omega)$ *and* $\mathbf{q} \in \mathbf{H}^1_{\Gamma_D}(\Omega)$ *depending linearly on* $\mathbf{v}$ *such that*

$$\mathbf{v} = \mathbf{z} + \mathbf{curl}\, \mathbf{q},$$

$$\|\mathbf{z}\|_{0,\Omega} + \|\mathbf{curl}\, \mathbf{q}\|_{0,\Omega} + \frac{1}{d(\Omega)}\|\mathbf{q}\|_{0,\Omega} \le C\, \|\mathbf{v}\|_{0,\Omega}\,,$$

$$\|\nabla \mathbf{z}\|_{0,\Omega} + \frac{1}{d(\Omega)}\|\mathbf{z}\|_{0,\Omega} + \frac{1}{d(\Omega)}\|\nabla \mathbf{q}\|_{0,\Omega} \le C\big(\|\mathbf{curl}\, \mathbf{v}\|_{0,\Omega} + \frac{1}{d(\Omega)}\|\mathbf{v}\|_{0,\Omega}\big)\,,$$

with constant C *depending only on the shape of* Ω *and* Γ_D, *but not on* $d(\Omega)$.

4.2 Regular Potential with Prescribed Divergence

The next result carries Theorem 3 over to $\mathbf{H}(\mathrm{div})$.

Theorem 7 (Regular Potentials for the Image Space of div) *Let* (Ω, Γ_D) *be as in Sect. 2 and, in addition, assume that each connected component* $\Upsilon_{D,k}$ *of the bulge has a connected boundary, i.e.,* $\beta_2(\Upsilon_{D,k}) = 0$. *Then*

$$L^2(\Omega) = \mathrm{div}\, \mathbf{H}_{\Gamma_D}(\mathrm{div}, \Omega) = \mathrm{div}\, \mathbf{H}^1_{\Gamma_D}(\Omega).$$

Moreover, for each $v \in L^2(\Omega)$ *there exists* $\mathbf{q} \in \mathbf{H}^1_{\Gamma_D}(\Omega)$ *depending linearly on* v *such that, with a constant* C *depending on* Ω *and* Γ_D *but not on* $d(\Omega)$,

$$\mathrm{div}\, \mathbf{q} = v \quad and \quad \|\nabla \mathbf{q}\|_{0,\Omega} + \frac{1}{d(\Omega)}\|\mathbf{q}\|_{0,\Omega} \le C\, \|v\|_{0,\Omega}\,.$$

4.3 Divergence-Bounded Regular Decompositions of H(div)

We can now formulate other variants of regular decompositions of $\mathbf{H}(\mathrm{div})$ in analogy to what we did in Sect. 3.3.

Theorem 8 (Divergence-Bounded Regular Decomposition of H(div) (I)) *Let* (Ω, Γ_D) *be as in Sect. 2. In addition, assume that each connected component* $\Upsilon_{D,k}$ *of the bulge has a connected boundary, i.e.,* $\beta_2(\Upsilon_{D,k}) = 0$. *Then, for each* $\mathbf{v} \in \mathbf{H}_{\Gamma_D}(\mathrm{div}, \Omega)$ *there exists* $\mathbf{z} \in \mathbf{H}^1_{\Gamma_D}(\Omega)$ *and a divergence-free vector field*

$\mathbf{h} \in \mathbf{H}_{\Gamma_D}(\mathrm{div}\, 0, \Omega)$ *depending linearly on* $\mathbf{v}$ *such that*

$$\mathbf{v} = \mathbf{z} + \mathbf{h}, \tag{10}$$

$$\|\mathbf{h}\|_{0,\Omega} \leq \|\mathbf{v}\|_{0,\Omega} + C\, \mathrm{d}(\Omega) \|\, \mathrm{div}\, \mathbf{v}\|_{0,\Omega}, \tag{11}$$

$$\|\, \nabla\, \mathbf{z}\|_{0,\Omega} + \frac{1}{\mathrm{d}(\Omega)} \|\mathbf{z}\|_{0,\Omega} \leq C\, \|\, \mathrm{div}\, \mathbf{v}\|_{0,\Omega}, \tag{12}$$

where C depends only on the shape of Ω and Γ_D, but not on $\mathrm{d}(\Omega)$.

The last variant of $\mathbf{H}(\mathrm{div})$ regular decomposition of $\mathbf{H}(\mathrm{div})$ dispenses with the assumptions on the topology of the Dirichlet boundary and has better stability properties than the splitting from Theorem 8 (though with less explicit constants).

Theorem 9 (Divergence-Bounded Regular Decomposition of H(div) (II)) *Let* (Ω, Γ_D) *be as in Sect. 2. Then, for each* $\mathbf{v} \in \mathbf{H}_{\Gamma_D}(\mathrm{div}, \Omega)$ *there exists* $\mathbf{z} \in \mathbf{H}^1_{\Gamma_D}(\Omega)$ *and a divergence-free vector field* $\mathbf{h} \in \mathbf{H}_{\Gamma_D}(\mathrm{div}\, 0, \Omega)$ *depending linearly on* $\mathbf{v}$ *such that*

$$\mathbf{v} = \mathbf{z} + \mathbf{h}, \tag{13}$$

$$\|\mathbf{z}\|_{0,\Omega} + \|\mathbf{h}\|_{0,\Omega} \leq \|\mathbf{v}\|_{0,\Omega}, \tag{14}$$

$$\|\, \nabla\, \mathbf{z}\|_{0,\Omega} + \frac{1}{\mathrm{d}(\Omega)} \|\mathbf{z}\|_{0,\Omega} \leq C\, \|\, \mathrm{div}\, \mathbf{v}\|_{0,\Omega}, \tag{15}$$

where C depends only on the shape of Ω and Γ_D, but not on $\mathrm{d}(\Omega)$.

5 Discrete Counterparts of the Regular Decompositions

The discrete setting to which we want to extend the concept of regular decompositions is provided by finite element exterior calculus (FEEC, [2]) which introduces finite element subspaces of $\mathbf{H}(\mathbf{curl})$ and $\mathbf{H}(\mathrm{div})$ as special instances of spaces of discrete differential forms. In this section we confine ourselves to the lowest-order case of piecewise linear finite element functions.

Throughout, we assume that (Ω, Γ_D) is as in Sect. 2, and, additionally, that Ω is a polyhedron and that $\partial \Gamma_D$ consists of straight line segments. All considerations take for granted a shape-regular family of meshes $\{\mathcal{T}^h\}_h$ of Ω, consisting of tetrahedral elements, and resolving Γ_D in the sense that Γ_D is a union of faces of some of the tetrahedra.

The following finite element spaces will be relevant:

- the space $\mathcal{W}^0_{h,\Gamma_D}(\Omega)$ of $H^1_{\Gamma_D}(\Omega)$-conforming piecewise linear Lagrangian finite element functions,
- the space $\mathcal{W}^1_{h,\Gamma_D}(\Omega)$ of $\mathbf{H}_{\Gamma_D}(\mathbf{curl}, \Omega)$-conforming lowest order Nédélec elements, also known as *edge elements*,

- the space $\boldsymbol{W}^2_{h,\Gamma_D}(\Omega)$ of $\boldsymbol{H}_{\Gamma_D}(\mathrm{div},\Omega)$-conforming lowest order tetrahedral Raviart-Thomas finite elements, aka, *face elements*,
- the space $\boldsymbol{W}^0_{h,\Gamma_D}(\Omega) := [W^0_{h,\Gamma_D}(\Omega)]^3$ of piecewise linear globally continuous vector fields vanishing on Γ_D.

Functions in $\boldsymbol{W}^\ell_{h,\Gamma_D}(\Omega)$, $\ell = 1, 2, 3$, are so-called Whitney forms, lowest-order discrete differential forms of the first family as introduced in [13] and [2, Sect. 5].

5.1 Discrete Regular Decompositions for Edge Elements

Commuting projectors, also known as co-chain projectors, are the linchpin of FEEC theory [2, Sect. 7], and it is not different with our developments. Thus, let

$$R^0_{h,\Gamma_D} : H^1_{\Gamma_D}(\Omega) \to W^0_{h,\Gamma_D}(\Omega)$$

$$\text{and} \quad \mathbf{R}^1_{h,\Gamma_D} : \boldsymbol{H}_{\Gamma_D}(\mathbf{curl}, \Omega) \to \boldsymbol{W}^1_{h,\Gamma_D}(\Omega)$$

denote the *continuous, boundary-aware* cochain projectors from [17, Sect. 3.2.6], which extend the pioneering work [10] by Falk and Winther. These two linear operators are projectors onto their ranges, they fulfill the commuting property

$$\nabla(R^0_{h,\Gamma_D}\varphi) = \mathbf{R}^1_{h,\Gamma_D}(\nabla\varphi) \qquad \forall \varphi \in H^1_{\Gamma_D}(\Omega), \tag{16}$$

and local stability estimates

Theorem 10 ([17, Thm. 1.2]) *For each* $\mathbf{v}_h \in \boldsymbol{W}^1_{0,\Gamma_D}(\Omega)$ *there exists a continuous and piecewise linear vector field* $\mathbf{z}_h \in \boldsymbol{W}^0_{h,\Gamma_D}(\Omega)$, *a continuous and piecewise linear scalar function* $p_h \in W^0_{h,\Gamma_D}(\Omega)$, *and a remainder* $\tilde{\mathbf{v}}_h \in \boldsymbol{W}^1_{0,\Gamma_D}(\Omega)$, *all depending linearly on* $\mathbf{v}_h$, *providing the discrete regular decomposition*

$$\mathbf{v}_h = \mathbf{R}^1_{h,\Gamma_D}\mathbf{z}_h + \tilde{\mathbf{v}}_h + \nabla p_h$$

and satisfying the stability estimates

$$\|\mathbf{z}_h\|_{0,\Omega} + \|\nabla p_h\|_{0,\Omega} + \|\tilde{\mathbf{v}}_h\|_{0,\Omega} \le C \|\mathbf{v}_h\|_{0,\Omega}, \tag{17}$$

$$\|\nabla \mathbf{z}_h\|_{0,\Omega} + \|h^{-1}\tilde{\mathbf{v}}_h\|_{0,\Omega} \le C\big(\|\mathbf{curl}\,\mathbf{v}_h\|_{0,\Omega} + \tfrac{1}{d(\Omega)}\|\mathbf{v}_h\|_{0,\Omega}\big), \tag{18}$$

where C *is a generic constant that depends only on the shape of* (Ω, Γ_D), *but not on* $d(\Omega)$, *and on the shape regularity constant of* $\mathcal{T}^h(\Omega)$. *Above,* h^{-1} *is the piecewise constant function that is equal to* h_T^{-1} *on every element* T.

Obviously, this is a discrete counterpart of the regular decomposition of $\boldsymbol{H}(\mathbf{curl})$ from Theorem 2. The following theorem appears to be new and it corresponds to the rotation-bounded regular decomposition of Theorem 5. For the sake of brevity define the discrete nullspace of the curl operator

$$\mathcal{N}_h^1 := \{ \mathbf{v}_h \in \mathcal{W}_{h,\Gamma_D}^1(\Omega) : \mathbf{curl}\, \mathbf{v}_h = 0 \}. \tag{19}$$

If Ω and Γ_D have simple topology, $\mathcal{X}_h = \nabla\, \mathcal{W}_{h,\Gamma_D}^0(\Omega)$, but if the first Betti number of Ω is non-zero, or if Γ_D has multiple components, then a finite-dimensional co-homology space has to be added [2, Sect. 5.6].

Theorem 11 (Rotation-Bounded Discrete Regular Decomposition for Edge Elements) *For each* $\mathbf{v}_h \in \mathcal{W}_{0,\Gamma_D}^1(\Omega)$ *there exists a continuous and piecewise linear vector field* $\mathbf{z}_h \in \mathcal{W}_{h,\Gamma_D}^0(\Omega)$*, an* curl-free *edge element function* $\mathbf{h}_h \in \mathcal{N}_h^1$*, and a remainder* $\widetilde{\mathbf{v}}_h \in \mathcal{W}_{0,\Gamma_D}^1(\Omega)$*, all depending linearly on* $\mathbf{v}_h$*, providing the discrete regular decomposition*

$$\mathbf{v}_h = \mathbf{R}_{h,\Gamma_D}^1 \mathbf{z}_h + \widetilde{\mathbf{v}}_h + \mathbf{h}_h$$

and satisfying the stability bounds

$$\left. \begin{array}{l} \|\mathbf{z}_h\|_{0,\Omega} \\ \|\mathbf{h}_h\|_{0,\Omega} \\ \|\widetilde{\mathbf{v}}_h\|_{0,\Omega} \end{array} \right\} \leq C\,\|\mathbf{v}_h\|_{0,\Omega}\,, \qquad \left. \begin{array}{l} \|\nabla\, \mathbf{z}_h\|_{0,\Omega} \\ \|h^{-1}\widetilde{\mathbf{v}}_h\|_{0,\Omega} \end{array} \right\} \leq C\,\|\mathbf{curl}\, \mathbf{v}_h\|_{0,\Omega}\,,$$

where C *is a uniform constant that depends only on the shape of* (Ω, Γ_D)*, but not on* $\mathrm{d}(\Omega)$*, and on the shape regularity constant of* $\mathcal{T}^h(\Omega)$*.*

We stress that the statements of Theorems 10 and 11 do not hinge on *any* assumptions on the topological properties of Ω and Γ_D.

5.2 Discrete Regular Decompositions for Face Elements

For face elements, the construction of a boundary-aware co-chain projection operator

$$\mathbf{R}_{h,\Gamma_D}^2 : \mathbf{H}_{\Gamma_D}(\mathrm{div}, \Omega) \to \mathcal{W}_{h,\Gamma_D}^2(\Omega)$$

that commutes with $\mathbf{R}_{h,\Gamma_D}^1$ and the **curl**-operator has not yet been accomplished. Fortunately, in the case $\Gamma_D = \emptyset$, this operator is available from [10]. Thus, in the following, we treat only the case $\Gamma_D = \emptyset$ and just omit the subscript Γ_D. Then, from [10] we can borrow a linear operator $\mathbf{R}_h^2 : \mathbf{H}(\mathrm{div}, \Omega) \to \mathcal{W}_h^2(\Omega)$ such that

$$\mathbf{curl}\, \mathbf{R}_h^1 \mathbf{u} = \mathbf{R}_h^2\, \mathbf{curl}\, \mathbf{u} \qquad \forall \mathbf{u} \in \mathbf{H}(\mathbf{curl}, \Omega)\,. \tag{20}$$

The next result takes Theorem 6 to the discrete setting.

Theorem 12 (Discrete Regular Decomposition of $\mathcal{W}_h^2(\Omega)$) *For each vector field $\mathbf{v}_h$ in the lowest-order Raviart-Thomas space $\mathcal{W}_h^2(\Omega)$, there exists a continuous and piecewise linear vector field $\mathbf{z}_h \in \mathcal{W}_h^0(\Omega)$, a vector field $\mathbf{q}_h$ in the lowest-order Nédélec space $\mathcal{W}_h^1(\Omega)$, and a remainder $\widetilde{\mathbf{v}}_h \in \mathcal{W}_h^2(\Omega)$, all depending linearly on $\mathbf{v}_h$, providing the discrete regular decomposition*

$$\mathbf{v}_h = \mathbf{R}_h^2 \mathbf{z}_h + \widetilde{\mathbf{v}}_h + \mathbf{curl}\,\mathbf{q}_h\,,$$

and the stability estimates

$$\left.\begin{array}{c} \|\mathbf{z}_h\|_{0,\Omega} \\ \|\mathbf{curl}\,\mathbf{q}_h\|_{0,\Omega} + \frac{1}{d(\Omega)}\|\mathbf{q}_h\|_{0,\Omega} \\ \|\widetilde{\mathbf{v}}_h\|_{0,\Omega} \end{array}\right\} \leq C\,\|\mathbf{v}_h\|_{0,\Omega}\,,$$

$$\left.\begin{array}{c} \|\nabla\mathbf{z}_h\|_{0,\Omega} \\ \|h^{-1}\widetilde{\mathbf{v}}_h\|_{0,\Omega} \end{array}\right\} \leq C\|\operatorname{div}\mathbf{v}_h\|_{0,\Omega} + \frac{1}{d(\Omega)}\|\mathbf{v}_h\|_{0,\Omega}\,.$$

The constant C depends only on the shape of Ω, but not on $d(\Omega)$, and the shape-regularity of $\mathcal{T}^h(\Omega)$.

Finally, we present a counterpart to the divergence-bounded regular decomposition of Theorem 9. For convenience we introduce the space of divergence-free face element functions

$$\mathcal{N}_h^2 := \{\mathbf{q}_h \in \mathcal{W}_h^2(\Omega)\colon \operatorname{div}\mathbf{q}_h = 0\}\,. \tag{21}$$

Theorem 13 (Divergence-Bounded Discrete Regular Decomposition of $\mathcal{W}_h^2(\Omega)$) *For each vector field $\mathbf{v}_h$ in the lowest-order Raviart-Thomas space $\mathcal{W}_h^2(\Omega)$, there exists a continuous and piecewise linear vector field $\mathbf{z}_h \in \mathcal{W}_h^0(\Omega)$, an element $\mathbf{h}_h$ in the discrete divergence-free subspace $\mathcal{N}_h^2$, and a remainder $\widetilde{\mathbf{v}}_h \in \mathcal{W}_h^2(\Omega)$, all depending linearly on $\mathbf{v}_h$, providing the discrete regular decomposition*

$$\mathbf{v}_h = \mathbf{R}_h^2 \mathbf{z}_h + \widetilde{\mathbf{v}}_h + \mathbf{h}_h$$

and the stability estimates

$$\left.\begin{array}{c} \|\mathbf{z}_h\|_{0,\Omega} \\ \|\widetilde{\mathbf{v}}_h\|_{0,\Omega} \\ \|\mathbf{h}_h\|_{0,\Omega} \end{array}\right\} \leq C\,\|\mathbf{v}_h\|_{0,\Omega}\,, \qquad \left.\begin{array}{c} \|\nabla\mathbf{z}_h\|_{0,\Omega} \\ \|h^{-1}\widetilde{\mathbf{v}}_h\|_{0,\Omega} \end{array}\right\} \leq C\,\|\operatorname{div}\mathbf{v}_h\|_{0,\Omega}\,.$$

The constants C depend only on the shape of Ω, but not on $d(\Omega)$, and the shape regularity of $\mathcal{T}^h(\Omega)$.

Remark 10 The result of Theorem 13 can be viewed as an improvement of the decompositions in [25] which are elaborated for the case of essential boundary conditions on $\partial\Omega$.

Corollary 2 *If the second Betti number of Ω vanishes, that is, if $\partial\Omega$ is connected, then $\mathbf{h}_h$ in Theorem 13 can be chosen as $\mathbf{h}_h = \mathbf{curl}\,\mathbf{q}_h$ with $\mathbf{q}_h \in \mathcal{W}_h^1(\Omega)$ such that*

$$\mathbf{v}_h = \mathbf{R}_h^2 \mathbf{z} + \tilde{\mathbf{v}}_h + \mathbf{curl}\,\mathbf{q}_h \,,$$

with the bounds

$$\left.\begin{array}{r}\|\mathbf{z}_h\|_{0,\Omega} \\ \|\tilde{\mathbf{v}}_h\|_{0,\Omega} \\ \|\mathbf{curl}\,\mathbf{q}_h\|_{0,\Omega} \\ \mathrm{d}(\Omega)^{-1}\|\mathbf{q}_h\|_{0,\Omega}\end{array}\right\} \le C\,\|\mathbf{v}_h\|_{0,\Omega}\,, \qquad \left.\begin{array}{r}\|\nabla\,\mathbf{z}_h\|_{0,\Omega} \\ \|h^{-1}\tilde{\mathbf{v}}_h\|_{0,\Omega}\end{array}\right\} \le C\,\|\operatorname{div}\mathbf{v}_h\|_{0,\Omega}\,.$$

Remark 11 The result of Corollary 2 is an improvement of [19, Lemma 5.2] which assumes a domain Ω that is smooth enough to allow H^2-regularity of the Laplace problem (2-regular case, for details see [19, Sect. 3]). This lemma is used in [27] in a domain decomposition framework, where convex subdomains are assumed. With our improved version, this assumption can be weakened considerably.

Acknowledgements The second author would like to thank Dirk Pauly (Essen) for enlightening discussions.

References

1. Amrouche, C., Bernardi, C., Dauge, M., Girault, V.: Vector potentials in three-dimensional non-smooth domains. Math. Meth. Appl. Sci. **21**(9), 823–864 (1998)
2. Arnold, D., Falk, R., Winther, R.: Finite element exterior calculus, homological techniques, and applications. Acta Numer. **15**, 1–155 (2006)
3. Bauer, S., Pauly, D., Schomburg, M.: The Maxwell compactness property in bounded weak Lipschitz domains with mixed boundary conditions. SIAM J. Math. Anal. **48**(4), 2912–2943 (2016)
4. Birman, M.S., Solomyak, M.Z.: L^2-theory of the Maxwell operator in arbitrary domains. Usp. Mat. Nauk. **42**(6), 61–76 (1987). English Transliterated, Russian Math. Surveys **42**(6), 75–96 (1987)
5. Bonnet-Ben Dhia, A., Hazard, C., Lohrengel, S.: A singular field method for the solution of Maxwell's equations in polyhedral domains. SIAM J. Appl. Math. **59**(6), 2028–2044 (1999)
6. Buffa, A., Costabel, M., Sheen, D.: On traces for $\mathbf{H}(\mathbf{curl},\Omega)$ in Lipschitz domains. J. Math. Anal. Appl. **276**, 845–867 (2002)
7. Costabel, M., Dauge, M., Nicaise, S.: Singularities of Maxwell interface problems. Model. Math. Anal. Numer. **33**(2), 627–649 (1999)
8. Creusé, E., Nicaise, S., Tang, Z.: Helmholtz decomposition of vector fields with mixed boundary conditions and an application to a posteriori finite element analysis of the Maxwell system. Math. Methods Appl. Sci. **38**(4), 738–750 (2015)

9. Dautray, R., Lions, J.: Mathematical Analysis and Numerical Methods for Science and Technology. Springer, Berlin (2000). Original French edition published by Masson, S.A., Paris (1984)

10. Falk, R.S., Winther, R.: Local bounded cochain projections. Math. Comp. **83**(290), 2631–2656 (2014)

11. Girault, V., Raviart, P.A.: Finite Element Methods for Navier–Stokes Equations. Springer, New York (1986)

12. Gopalakrishnan, J., Qiu, W.: Partial expansion of a Lipschitz domain and some applications. Front. Math. China **7**(2), 249–272 (2012)

13. Hiptmair, R.: Canonical construction of finite elements. Math. Comp. **68**, 1325–1346 (1999)

14. Hiptmair, R.: Finite elements in computational electromagnetism. Acta Numer. **11**, 237–339 (2002)

15. Hiptmair, R.: Boundary element methods for eddy current computation. In: Schanz, M., Steinbach, O. (eds.) Boundary Element Analysis: Mathematical Aspects and Applications. Lecture Notes in Applied and Computational Mechanics, vol. 29, pp. 213–248. Springer, Heidelberg (2007)

16. Hiptmair, R.: Maxwell's equations: Continuous and discrete. In: Bermúdez de Castro, A., Valli, A. (eds.) Computational Electromagnetism. Lecture Notes in Mathematics, vol. 2148. Springer, Cham (2015)

17. Hiptmair, R., Pechstein, C.: Discrete regular decompositions of tetrahedral discrete 1-forms. In: Langer, U., Pauly, D., Repin, S. (eds.) Maxwell's Equations. Radon Series on Computational and Applied Mathematics, vol. 24, pp. 199–258. De Gruyter, Stuttgart (2019)

18. Hiptmair, R., Pechstein, C.: Regular decompositions of vector fields - continuous, discrete, and structure-preserving. Technical Report 2019-18, Seminar for Applied Mathematics, ETH Zürich (2019). https://www.sam.math.ethz.ch/sam_reports/reports_final/reports2019/2019-18.pdf

19. Hiptmair, R., Xu, J.: Nodal auxiliary space preconditioning in H(curl) and H(div) spaces. SIAM J. Numer. Anal. **45**, 2483–2509 (2007)

20. Hiptmair, R., Zheng, W.: Local multigrid in H(curl). J. Comput. Math. **27**(5), 573–603 (2009)

21. Hiptmair, R., Widmer, G., Zou, J.: Auxiliary space preconditioning in $H_0(\text{curl}; \Omega)$. Numer. Math. **103**, 435–459 (2006)

22. Hu, Q.: Convergence of HX preconditioner for Maxwell's equations with jump coefficients (i): various extensions of the regular Helmholtz decomposition. Technical Report arXiv:1708.05850v2 [math.AP] (2018)

23. Kolev, T.V., Pasciak, J.E., Vassilevski, P.S.: H(curl) auxiliary mesh preconditioning. Numer. Linear Algebr. Appl. **15**(5), 455–471 (2008)

24. Kolev, T.V., Vassilevski, P.S.: Parallel auxiliary space AMG for **H**(curl) problems. J. Comput. Math. **27**(5), 604–623 (2009)

25. Kolev, T.V., Vassilevski, P.S.: Parallel auxiliary space AMG for **H**(div) problems. SIAM J. Sci. Comput. **34**(6), A3079–A3098 (2012)

26. Monk, P.: Finite Element Methods for Maxwell's Equations. Numerical Mathematics and Scientific Computation. Oxford University Press, Oxford (2003)

27. Oh, D., Widlund, O.B., Zampini, S., Dohrmann, C.R.: BDDC algorithms with deluxe scaling and adaptive selection of primal constraints for Raviart–Thomas vector fields. Math. Comp. **87**, 659–692 (2018)

28. Pasciak, J.E., Zhao, J.: Overlapping Schwarz methods in H(curl) on polyhedral domains. Numer. Math. **10**(3), 211–234 (2002)

Model Reduction by Separation of Variables: A Comparison Between Hierarchical Model Reduction and Proper Generalized Decomposition

Simona Perotto, Michele Giuliano Carlino, and Francesco Ballarin

1 Introduction

This paper is meant as a first attempt to compare two procedures which share the idea of exploiting separation of variables to perform model reduction, albeit with different purposes. Proper Generalized Decomposition (PGD) is essentially employed as a powerful tool to deal with parametric problems in several fields of application [3, 14, 23]. Parametrized models characterize multi query contexts, such as parameter optimization, statistical analysis or inverse problems. Here, the computation of the solution for many different parameters demands, in general, a huge computational effort, and this justifies the development of model reduction techniques.

For this purpose, projection-based techniques, such as Proper Orthogonal Decomposition (POD) or Reduced Basis methods, are widely used in the literature [11]. The idea is to project the discrete operators onto a reduced space so that the problem can be solved rapidly in the lower dimensional space. PGD adopts a completely different way to deal with parameters. Here, parameters are considered

S. Perotto
MOX - Modeling and Scientific Computing, Dipartimento di Matematica, Politecnico di Milano, Milano, Italy
e-mail: simona.perotto@polimi.it

M. G. Carlino
Inria Bordeaux Sud-Ouest and Institut de Mathématiques de Bordeaux, University of Bordeaux, Talence, France
e-mail: michele-giuliano.carlino@inria.fr

F. Ballarin (✉)
MathLab, Mathematics Area, SISSA, Trieste, Italy
e-mail: francesco.ballarin@sissa.it

© The Author(s) 2020
S. J. Sherwin et al. (eds.), *Spectral and High Order Methods for Partial Differential Equations ICOSAHOM 2018*, Lecture Notes in Computational Science and Engineering 134, https://doi.org/10.1007/978-3-030-39647-3_4

61

as new independent variables of the problem, together with the standard space-time ones [5]. Although the dimensionality of the problem is inevitably increased, PGD transforms the computation of the solution for new values of the parameters into a plain evaluation of the reduced solution, with striking computational advantages.

Hierarchical-Model (HiMod) reduction has been proposed to improve one-dimensional (1D) partial differential equation (PDE) solvers for problems defined in domains with a geometrically dominant direction, like slabs or pipes [6, 20]. The main applicative field of interest is hemodynamics, in particular the modeling of blood flow in patient-specific geometries. Purely 1D hemodynamic models completely drop the transverse dynamics, which, however may be locally important (e.g., in the presence of a stenosis or an aneurism). HiMod aims at providing a numerical tool to incorporate the transverse components of the 3D solution into a conceptually 1D solver. To do this, the driving idea is to discretize main and transverse dynamics in a different way. The latter are generally of secondary importance and can be described by few degrees of freedom using a spectral approximation, in combination, for instance, with a finite element (FE) discretization of the mainstream.

The parametric version of HiMod (namely, HiPOD) is a more recent proposal [4, 13]. On the other hand, PGD is not so widely employed in a non-parametric setting, despite its original formulation [12]. Nevertheless, for the sake of comparison, in this paper we consider the non-parametric as well as the parametric versions of both the HiMod and PGD approaches. The goal is to begin a preliminary comparative analysis between the two methodologies, to highlight the respective weaknesses and strengths. The main limit of PGD remains its inability to deal with non-Cartesian geometries without losing the computational benefits arising from the separability of the spatial coordinates. HiMod turns out to be more flexible from a geometric viewpoint. On the other hand, PGD turns out to be extremely effective for parametric problems thanks to the explicit expression of the PGD solution in terms of the parameters, while HiPOD can be classified as a projection-based method with all the associated drawbacks. In perspective, the ultimate goal is to merge HiMod with PGD to emphasize the good features and mitigate the intrinsic limits of the two methods taken alone.

2 The HiMod Approach

Hierarchical Model reduction proved to be an efficient and reliable method to deal with phenomena characterized by dominant dynamics [10]. In general, the computational domain itself exhibits an intrinsic directionality. We assume $\Omega \subset \mathbb{R}^d$ ($d = 2, 3$) to coincide with a d-dimensional fiber bundle, $\Omega = \bigcup_{x \in \Omega_{1D}} \{x\} \times \gamma_x$, where $\Omega_{1D} \subset \mathbb{R}$ denotes the supporting fiber aligned with the main stream, while $\gamma_x \subset \mathbb{R}^{d-1}$ is the transverse fiber at $x \in \Omega_{1D}$, parallel to the transverse dynamics. For the sake of simplicity, we identify Ω_{1D} with a straight segment, (x_0, x_1). We refer to [15, 21] for the case where Ω_{1D} is curvilinear. From a computational viewpoint, the idea is to exploit a map, $\Psi : \Omega \rightarrow \hat{\Omega}$, transforming the physical domain, Ω, into a reference domain, $\hat{\Omega}$, and to make explicit computations in $\hat{\Omega}$ only. Typically, $\hat{\Omega}$ coincides with a rectangle in 2D, with a cylinder with circular section in 3D. To define Ψ, for each $x \in \Omega_{1D}$, we introduce the map, $\psi_x : \gamma_x \rightarrow \hat{\gamma}_{d-1}$, from fiber γ_x to the reference transverse fiber, $\hat{\gamma}_{d-1}$, so that the reference domain coincides with $\hat{\Omega} = \bigcup_{x \in \Omega_{1D}} \{x\} \times \hat{\gamma}_{d-1}$. The supporting fiber is preserved by map Ψ, which modifies the lateral boundaries only.

We consider now the (full) problem to be reduced. Due to the comparative purposes of the paper, we focus on a scalar elliptic equation, and, in particular, on the associated weak formulation,

$$\text{find } u \in V : a(u, v) = F(v) \quad \forall v \in V, \tag{1}$$

where $V \subseteq H^1(\Omega)$, $a(\cdot, \cdot) : V \times V \rightarrow \mathbb{R}$ is a continuous and coercive bilinear form and $F(\cdot) : V \rightarrow \mathbb{R}$ is a continuous linear functional. To provide the HiMod formulation for problem (1), we introduce the hierarchical reduced space

$$V_m = \left\{ v_m(x, \mathbf{y}) = \sum_{k=1}^{m} \tilde{v}_k(x) \psi_k(\psi_x(\mathbf{y})), \text{ with } \tilde{v}_k \in V_{1D}^h, x \in \Omega_{1D}, \mathbf{y} \in \gamma_x \right\} \tag{2}$$

for a modal index $m \in \mathbb{N}^+$, where $V_{1D}^h \subseteq H^1(\Omega_{1D})$ is a discrete space of dimension N_h associated with a partition $\mathcal{T}_h$ of Ω_{1D}, while $\{\varphi_k\}_{k=1}^{m}$ denotes a modal basis of functions orthogonal with respect to the $L^2(\hat{\gamma}_{d-1})$-scalar product. Index m sets the hierarchical level of the HiMod space, being $V_m \subset V_{m+1}$, for any m. Concerning V_{1D}^h, we adopt here a standard FE space, although any discrete space can be employed (see, e.g., [21], where an isogeometric discretization is used). Functions in V_{1D}^h have to include the boundary conditions on $\{x_0\} \times \gamma_{x_0}$ and $\{x_1\} \times \gamma_{x_1}$; analogously, the modal functions have to take into account the boundary data along the horizontal sides. In Sect. 4 further comments are provided about the selection of the modal basis and of the modal index m. The HiMod formulation for problem (1) thus reads

$$\text{find } u_m^{\text{HiMod}} \in V_m : a(u_m^{\text{HiMod}}, v_m) = F(v_m) \quad \forall v_m \in V_m. \tag{3}$$

To ensure the well-posedness of formulation (3) and the convergence of the HiMod approximation, u_m^{HiMod}, to the full solution, u, we endow the HiMod space with a conformity and a spectral approximability hypothesis, and we introduce a standard density assumption on the discrete space V_{1D}^h (see [20] for all the details).

The HiMod solution can be fully characterized by introducing a basis, $\{\theta_l\}_{l=1}^{N_h}$, for the space V_{1D}^h. Actually, each modal coefficient, $\tilde{u}_k$, of u_m^{HiMod} can be expanded in terms of such a basis, so that, we obtain the modal representation

$$u_m^{\text{HiMod}}(x, \mathbf{y}) = \sum_{k=1}^{m} \sum_{l=1}^{N_h} \tilde{u}_{k,l} \theta_l(x) \varphi_k(\psi_x(\mathbf{y})). \tag{4}$$

The actual unknowns of problem (3) become the mN_h coefficients $\{\tilde{u}_{k,l}\}_{k=1,l=1}^{m,N_h}$. With reference to the Poisson problem, $-\Delta u = f$, completed with full homogeneous Dirichlet boundary data, the corresponding HiMod formulation, after exploiting (4) in (3) and picking $v_m(x, \mathbf{y}) = \theta_i(x)\varphi_j(\psi_x(\mathbf{y}))$ with $i = 1, \ldots, N_h$ and $j = 1, \ldots, m$, reduces to the system of mN_h 1D equations in the mN_h unknowns $\{\tilde{u}_{k,l}\}_{k=1,l=1}^{m,N_h}$,

$$\sum_{k=1}^{m} \sum_{l=1}^{N_h} \tilde{u}_{k,l} \left[\int_{\Omega_{1D}} \left(\hat{r}_{jk}^{1,1}(x) \frac{d\theta_l}{dx}(x) \frac{d\theta_i}{dx}(x) + \hat{r}_{jk}^{1,0}(x) \frac{d\theta_l}{dx}(x)\theta_i(x) + \right. \right.$$
$$\left. \left. + \hat{r}_{jk}^{0,1}(x)\theta_l(x) \frac{d\theta_i}{dx}(x) + \hat{r}_{jk}^{0,0}(x)\theta_l(x)\theta_i(x) \right) \right] dx = \int_{\Omega_{1D}} \hat{f}_j(x)\theta_i(x)\, dx,$$

where $\hat{r}_{jk}^{a,b}(x) = \int_{\hat{\gamma}_{d-1}} r_{jk}^{a,b}(x, \hat{\mathbf{y}})|J|\, d\hat{\mathbf{y}}$ with $a, b = 0, 1$, $J = \det(\mathcal{D}_2^{-1}(x, \psi_x^{-1}(\hat{\mathbf{y}})))$ with $\mathcal{D}_2 = \mathcal{D}_2(x, \psi_x^{-1}(\hat{\mathbf{y}})) = \nabla_{\mathbf{y}}\psi_x$,

$$r_{jk}^{0,0}(x, \hat{\mathbf{y}}) = \varphi_k'(\hat{\mathbf{y}})\varphi_j'(\hat{\mathbf{y}})(\mathcal{D}_1^2 + \mathcal{D}_2^2), \quad r_{jk}^{0,1}(x, \hat{\mathbf{y}}) = \varphi_k'(\hat{\mathbf{y}})\varphi_j(\hat{\mathbf{y}})\mathcal{D}_1,$$
$$r_{jk}^{1,0}(x, \hat{\mathbf{y}}) = \varphi_k(\hat{\mathbf{y}})\varphi_j'(\hat{\mathbf{y}})\mathcal{D}_1, \qquad r_{jk}^{1,1}(x, \hat{\mathbf{y}}) = \varphi_k(\hat{\mathbf{y}})\varphi_j(\hat{\mathbf{y}}),$$

with $\mathcal{D}_1 = \mathcal{D}_1(x, \psi_x^{-1}(\hat{\mathbf{y}})) = \partial\psi_x/\partial x$, and $\hat{f}_j(x) = \int_{\hat{\gamma}_{d-1}} f(x, \psi_x^{-1}(\hat{\mathbf{y}}))\varphi_j(\hat{\mathbf{y}})$ $|J|\, d\hat{\mathbf{y}}$. Information associated with the transverse dynamics are lumped in the coefficients $\{\hat{r}_{jk}^{a,b}\}$, so that the HiMod system is solved on the supporting fiber, Ω_{1D}. Collecting the HiMod unknowns, by mode, in the vector $\mathbf{u}_m^{\text{HiMod}} \in \mathbb{R}^{mN_h}$, such that

$$\mathbf{u}_m^{\text{HiMod}} = [\tilde{u}_{1,1}, \tilde{u}_{1,2}, \ldots, \tilde{u}_{1,N_h}, \tilde{u}_{2,1}, \ldots, \tilde{u}_{m,1}, \ldots, \tilde{u}_{m,N_h}]^T, \tag{5}$$

we can rewrite the HiMod system in the compact form

$$A_m^{\text{HiMod}} \mathbf{u}_m^{\text{HiMod}} = \mathbf{f}_m^{\text{HiMod}}, \tag{6}$$

where $A_m^{\text{HiMod}} \in \mathbb{R}^{mN_h \times mN_h}$ and $\mathbf{f}_m^{\text{HiMod}} \in \mathbb{R}^{mN_h}$ are the HiMod stiffness matrix and right-hand side, respectively, with $[\mathbf{f}_m^{\text{HiMod}}]_{ji} = \int_{\Omega_{1D}} \hat{f}_j(x)\theta_i(x)dx$, and $[A_m^{\text{HiMod}}]_{ji,kl} = \sum_{a,b=0}^{1} \int_{\Omega_{1D}} \hat{r}_{jk}^{a,b}(x) \frac{d^a\theta_l}{dx}(x) \frac{d^b\theta_i}{dx}(x)dx$. According to (5), for each modal index j, between 1 and m, the nodal index, i, takes the values $1, \ldots, N_h$. Thus, HiMod reduction leads to solve a system of order mN_h, independently of the dimension of the full problem (1).

3   The PGD Approach

To perform PGD, we have to introduce on problem (1) a separability hypothesis with respect to both the spatial variables and the data [5, 22]. Thus, domain $\Omega \subset \mathbb{R}^d$ coincides with the rectangle $\Omega_x \times \Omega_y$ if $d = 2$, with the parallelepiped $\Omega_x \times \Omega_y \times \Omega_z$ (total separability) or with the cylinder $\Omega_x \times \Omega_y$ (partial separability) if $d = 3$, for $\Omega_x, \Omega_y, \Omega_z \subset \mathbb{R}$ and $\Omega_y \subset \mathbb{R}^2$, being $\mathbf{y} = (y, z)$. In the following, we focus on partial separability, since it is more suited to match HiMod reduction with PGD. Analogously, we assume that the generic problem data, $d = d(x, y, z)$, can be written as $d = d^x(x)d^{\mathbf{y}}(\mathbf{y})$. The separability is inherited by the PGD space

$$W_m = \left\{ w_m(x, \mathbf{y}) = \sum_{k=1}^m w_k^x(x)w_k^{\mathbf{y}}(\mathbf{y}), \text{ with } w_k^x \in W_h^x, \ w_k^{\mathbf{y}} \in W_h^{\mathbf{y}}, \ x \in \Omega_x, \ \mathbf{y} \in \Omega_{\mathbf{y}} \right\}, \tag{7}$$

where $W_h^x \subseteq H^1(\Omega_x)$ and $W_h^{\mathbf{y}} \subseteq H^1(\Omega_{\mathbf{y}}; \mathbb{R}^{d-1})$ are discrete spaces, with $\dim(W_h^x) = N_h^x$ and $\dim(W_h^{\mathbf{y}}) = N_h^{\mathbf{y}}$, associated with partitions, $\mathcal{T}_h^x$ and $\mathcal{T}_h^{\mathbf{y}}$, of Ω_x and $\Omega_{\mathbf{y}}$, respectively. In general, W_h^x and $W_h^{\mathbf{y}}$ are FE spaces, although, a priori, any discretization can be adopted. It turns out that W_m is a tensor function space, being $W_m = W_h^x \otimes W_h^{\mathbf{y}} \subseteq H^1(\Omega_x) \otimes H^1(\Omega_{\mathbf{y}}; \mathbb{R}^{d-1})$.

Index m plays the same role as in the HiMod reduction, setting the level of detail for the reduced solution (see Sect. 4 for possible criteria to choose m). PGD exploits the hierarchical structure in W_m to build the generic function $w_m \in W_m$. In particular, w_m is computed as

$$w_m(x, \mathbf{y}) = w_m^x(x)w_m^{\mathbf{y}}(\mathbf{y}) + \sum_{k=1}^{m-1} w_k^x(x)w_k^{\mathbf{y}}(\mathbf{y}), \tag{8}$$

where w_k^x and $w_k^{\mathbf{y}}$ are assumed known for $k = 1, \dots, m - 1$, so that the enrichment functions, w_m^x and $w_m^{\mathbf{y}}$, become the actual unknowns. To provide the PGD formulation for the Poisson problem considered in Sect. 2, we exploit representation (8) for the PGD approximation, u_m^{PGD}, and we pick the test function as $X(x)Y(\mathbf{y})$, with $X \in W_h^x$ and $Y \in W_h^{\mathbf{y}}$. The coupling between the unknowns, u_m^x and $u_m^{\mathbf{y}}$, leads to a nonlinear problem, which is tackled by means of the *Alternating Direction Strategy* (ADS) [5]. The idea is to look for u_m^x and $u_m^{\mathbf{y}}$, separately via a fixed point procedure. We introduce an auxiliary index to keep trace of the ADS iterations, so that, at the p-th ADS iteration we compute $u_m^{x,p}$ and $u_m^{\mathbf{y},p}$ starting from the previous approximations, $u_m^{x,g}$ and $u_m^{\mathbf{y},g}$ for $g = 1, \dots, p - 1$, following a two-step procedure. First, we compute $u_m^{x,p}$ by identifying $u_m^{\mathbf{y}}$ with $u_m^{\mathbf{y},p-1}$, and by selecting $Y(\mathbf{y}) = u_m^{\mathbf{y},p-1}$ in the test function. This yields, for any $X \in W_h^x$,

$$
\int_{\Omega_x} \left(u_m^{x,p}\right)' X' dx \int_{\Omega_y} \left[u_m^{\mathbf{y},p-1}\right]^2 d\mathbf{y} + \int_{\Omega_x} u_m^{x,p} X dx \int_{\Omega_y} \left[\left(u_m^{\mathbf{y},p-1}\right)'\right]^2 d\mathbf{y}
$$
$$
= \int_{\Omega_x} f^x X dx \int_{\Omega_y} f^{\mathbf{y}} u_m^{\mathbf{y},p-1} d\mathbf{y} - \sum_{k=1}^{m-1} \int_{\Omega_x} \left(u_k^x\right)' X' dx \int_{\Omega_y} u_k^{\mathbf{y}} u_m^{\mathbf{y},p-1} d\mathbf{y} \qquad (9)
$$
$$
- \sum_{k=1}^{m-1} \int_{\Omega_x} u_k^x X dx \int_{\Omega_y} \left(u_k^{\mathbf{y}}\right)' \left(u_m^{\mathbf{y},p-1}\right)' d\mathbf{y},
$$

where the separability of f is exploited (the dependence on the independent variables, x and $\mathbf{y}$, is omitted to simplify notation). Successively, we compute $u_m^{\mathbf{y},p}$, after setting u_m^x to $u_m^{x,p}$ and choosing function X as to $u_m^{x,p}$ in the test function, so that we obtain, for any $Y \in W_h^{\mathbf{y}}$,

$$
\int_{\Omega_x} \left[\left(u_m^{x,p}\right)'\right]^2 dx \int_{\Omega_y} u_m^{\mathbf{y},p} Y d\mathbf{y} + \int_{\Omega_x} \left[u_m^{x,p}\right]^2 dx \int_{\Omega_y} \left(u_m^{\mathbf{y},p}\right)' Y' d\mathbf{y}
$$
$$
= \int_{\Omega_x} f^x u_m^{x,p} dx \int_{\Omega_y} f^{\mathbf{y}} Y d\mathbf{y} - \sum_{k=1}^{m-1} \int_{\Omega_x} \left(u_k^x\right)' \left(u_m^{x,p}\right)' dx \int_{\Omega_y} u_k^{\mathbf{y}} Y d\mathbf{y} \qquad (10)
$$
$$
- \sum_{k=1}^{m-1} \int_{\Omega_x} u_k^x u_m^{x,p} dx \int_{\Omega_y} \left(u_k^{\mathbf{y}}\right)' Y' d\mathbf{y}.
$$

The algebraic counterpart of (9) and (10) is obtained by introducing a basis, $\mathcal{B}_x = \{\theta_\alpha^x\}_{\alpha=1}^{N_h^x}$ and $\mathcal{B}_{\mathbf{y}} = \{\theta_\beta^{\mathbf{y}}\}_{\beta=1}^{N_h^{\mathbf{y}}}$, for the space W_h^x and $W_h^{\mathbf{y}}$, respectively, so that $u_j^q(q) = \sum_{i=1}^{N_h^q} \tilde{u}_{ji}^q \theta_i^q(q)$, $u_m^{q,s}(q) = \sum_{i=1}^{N_h^q} \tilde{u}_{mi}^{q,s} \theta_i^q(q)$, with $q = x, \mathbf{y}, s = p, p-1, j = 1, \ldots, m-1$, and, likewise, $X(x) = \sum_{\alpha=1}^{N_h^x} \tilde{x}_\alpha \theta_\alpha^x(x)$ and $Y(\mathbf{y}) = \sum_{\beta=1}^{N_h^{\mathbf{y}}} \tilde{y}_\beta \theta_\beta^{\mathbf{y}}(\mathbf{y})$. Thanks to these expansions and to the arbitrariness of X and Y, we can rewrite (9) and (10) as

$$
\left\{\left[\left(\mathbf{u}_m^{\mathbf{y},p-1}\right)^T M^{\mathbf{y}} \mathbf{u}_m^{\mathbf{y},p-1}\right] K^x + \left[\left(\mathbf{u}_m^{\mathbf{y},p-1}\right)^T K^{\mathbf{y}} \mathbf{u}_m^{\mathbf{y},p-1}\right] M^x\right\} \mathbf{u}_m^{x,p} = \left[\left(\mathbf{u}_m^{\mathbf{y},p-1}\right)^T \mathbf{f}^{\mathbf{y}}\right] \mathbf{f}^x
$$
$$
- \sum_{k=1}^{m-1} \left\{\left[\left(\mathbf{u}_m^{\mathbf{y},p-1}\right)^T M^{\mathbf{y}} \mathbf{u}_k^{\mathbf{y}}\right] K^x + \left[\left(\mathbf{u}_m^{\mathbf{y},p-1}\right)^T K^{\mathbf{y}} \mathbf{u}_k^{\mathbf{y}}\right] M^x\right\} \mathbf{u}_k^x,
$$

$$(11)$$

and

$$
\left\{\left[\left(\mathbf{u}_m^{x,p}\right)^T K^x \mathbf{u}_m^{x,p}\right] M^{\mathbf{y}} + \left[\left(\mathbf{u}_m^{x,p}\right)^T M^x \mathbf{u}_m^{x,p}\right] K^{\mathbf{y}}\right\} \mathbf{u}_m^{\mathbf{y},p} = \left[\left(\mathbf{u}_m^{x,p}\right)^T \mathbf{f}^x\right] \mathbf{f}^{\mathbf{y}}
$$
$$
- \sum_{k=1}^{m-1} \left\{\left[\left(\mathbf{u}_m^{x,p}\right)^T K^x \mathbf{u}_k^x\right] M^{\mathbf{y}} + \left[\left(\mathbf{u}_m^{x,p}\right)^T M^x \mathbf{u}_k^x\right] K^{\mathbf{y}}\right\} \mathbf{u}_k^{\mathbf{y}},
$$

$$(12)$$

respectively, where vectors $\mathbf{u}_j^q, \mathbf{u}_m^{q,s} \in \mathbb{R}^{N_h^q}$ collect the PGD coefficients, being $\left[\mathbf{u}_j^q\right]_i = \tilde{u}_{ji}^q$, $\left[\mathbf{u}_m^{q,s}\right]_i = \tilde{u}_{mi}^{q,s}$ and $i = 1, \ldots, N_h^q$, $K^x, M^x \in \mathbb{R}^{N_h^x \times N_h^x}$ and $K^{\mathbf{y}}$, $M^{\mathbf{y}} \in \mathbb{R}^{N_h^{\mathbf{y}} \times N_h^{\mathbf{y}}}$ are the stiffness and mass matrices associated with x- and $\mathbf{y}$-variables, with $\left[K^x\right]_{\alpha l} = \int_{\Omega_x} \left(\theta_\alpha^x\right)' \left(\theta_l^x\right)' dx$, $\left[K^{\mathbf{y}}\right]_{\beta s} = \int_{\Omega_y} \left(\theta_\beta^{\mathbf{y}}\right)' \left(\theta_s^{\mathbf{y}}\right)' d\mathbf{y}$, $\left[M^x\right]_{\alpha l} =$

$\int_{\Omega_x} \theta_\alpha^x \theta_l^x \, dx$, $\left[M^y\right]_{\beta s} = \int_{\Omega_y} \theta_\beta^y \theta_s^y \, dy$, and where $\mathbf{f}^x \in \mathbb{R}^{N_h^x}$, $\mathbf{f}^y \in \mathbb{R}^{N_h^y}$, with $\left[\mathbf{f}^x\right]_l = \int_{\Omega_x} f^x \theta_l^x \, dx$, $\left[\mathbf{f}^y\right]_s = \int_{\Omega_y} f^y \theta_s^y \, dy$, for $\alpha, l = 1, \ldots, N_h^x$, $\beta, s = 1, \ldots, N_h^y$. Systems (11) and (12) are solved at each ADS iteration, so that the computational effort characterizing PGD is the one associated with the solution of two systems of order N_h^x and N_h^y, respectively, for each ADS iteration. When a certain stopping criterion is met (see the next section for more details), ADS procedure yields vectors $\mathbf{u}_m^x$ and $\mathbf{u}_m^y$ which identify the enrichment functions u_m^x and u_m^y.

4   HiMod Reduction Versus PGD

Both HiMod reduction and PGD exploit the separation of variables and, according to [5], belong to the a priori approaches, since they do not rely on any solution to the problem at hand. Nevertheless, we can easily itemize features which distinguish the two techniques. The most relevant ones concern the geometry of Ω, the selection of the transverse basis and of the modal index, and the numerical implementation of the two procedures. Pros and cons of the two methods are then here highlighted.

4.1   Domain Geometry

HiMod reduction and PGD advance precise hypotheses on the geometry of the computational domain.

According to the HiMod approach, Ω is expected to coincide with a fiber bundle and to be mapped into the reference domain, $\hat{\Omega}$, by a sufficiently regular transformation. Actually, map Ψ is assumed differentiable, while map ψ_x is required to be a C^1-diffeomorphism, for all $x \in \Omega_{1D}$ [20]. These hypotheses introduce some constraints, in particular, on the lateral boundary of Ω which, e.g., cannot exhibit kinks. Additionally, geometries of interest in many applications, such as bifurcations or, more in general, networks are ruled out from the demands on ψ_x and Ψ. An approach based on the domain decomposition technique is currently under investigation as a viable way to deal with such geometries. The isogeometric version of HiMod (i.e., the HIgaMod approach) will play a crucial role in view of HiMod simulations for the blood flow modeling in patient-specific geometries [21].

The constraints introduced by PGD on the geometry of Ω are more restrictive. The separability hypothesis leads to consider essentially only Cartesian domains. This considerably reduces the applicability of PGD to practical contexts. Some techniques are available in the literature to overcome this issue. For instance, in [9] a generic domain is embedded into a Cartesian geometry, while in [7] the authors introduce a parametrization map for quadrilateral domains.

Overall, HiMod reduction exhibits a higher geometric flexibility with respect to PGD, in its straightforward formulation. As discussed in Sect. 5, this limitation can be removed when considering a parametric setting.

4.2 Modeling of the Transverse Dynamics

In the HiMod expansion, y-components, $\varphi_k(\psi_x(\mathbf{y}))$, are selected before starting the model reduction. This choice, although coherent with an a priori approach, introduces a constraint on the dynamics that can be described, so that hints about the solution trend along the transverse direction can be helpful to select a representative modal basis. In the original proposal of the HiMod procedure, sinusoidal functions are employed according to a Fourier expansion [6, 20]. This turns out to be a reasonable choice when Dirichlet boundary conditions are assigned on the lateral surface, $\Gamma_{\text{lat}} = \{x\} \times \partial \gamma_x$, of Ω. Legendre polynomials, properly modified to include the homogeneous Dirichlet data and orthonormalized, are employed in [20] as an alternative to a trigonometric expansion. Nevertheless, Legendre polynomials require high-order quadrature rules to accurately compute coefficients $\{\hat{r}_{jk}^{a,b}\}$.

In [1], the concept of educated modal basis is introduced to impose generic boundary conditions on Γ_{lat}. The idea is to solve an auxiliary Sturm-Liouville eigenvalue problem on the transverse reference fiber $\hat{\gamma}_{d-1}$, to build a basis which automatically includes the boundary values on Γ_{lat}. The eigenfunctions of the Sturm-Liouville problem provide the modal basis. A first attempt to generalize the educated-HiMod reduction to three-dimensional (3D) cylindrical geometries is performed in [10], where the Navier-Stokes equations are hierarchically reduced to model the blood flow in pipes. This generalization is far from being straightforward due to the employment of polar coordinates. To overcome this issue, we are currently investigating the HIgaMod approach [21], which allows us to define the transverse basis as the Cartesian product of 1D modal functions, independently of the considered geometry.

Additionally, we remark that any modal basis can be precomputed on the transverse reference fiber before performing the HiMod reduction, thanks to the employment of map Ψ. This considerably simplifies computations.

When applying PGD, y-components are unknown as the ones associated with x. This leads to the nonlinear problems (9)–(10), thus loosing any advantage related to a precomputation of the HiMod modal basis. On the other hand, PGD does not constrain the transverse dynamic to follow a prescribed (e.g., sinusoidal) analytical shape as HiMod procedure does. The educated-Himod reduction clearly is out of this comparison, since the modal basis strictly depends on the problem at hand.

Finally, we observe that HiMod modes are orthonormal with respect to the $L^2(\hat{\gamma}_{d-1})$-norm. This property is not ensured by PGD.

Concerning the selection of the modal index m in (2) and (7), as a first attempt, both HiMod reduction and PGD resort to a trial-and-error approach, so that the modal index is gradually increased until a check on the accuracy of the reduced solution is satisfied. For instance, in [6, 20] a qualitative investigation of the contour plot of the HiMod approximation drives the choice of m. Concerning PGD, the check on the relative enrichment

$$\frac{\|u_m^x u_m^y\|_{L^2(\Omega)}}{\|u_1^x u_1^y\|_{L^2(\Omega)}} \leq \text{TOL}_E, \tag{13}$$

is usually employed, with TOL_E a user-defined tolerance [5]. An automatic selection of index m can yield a significant improvement. In [17, 19], an adaptive procedure is proposed for HiMod, based on an a posteriori modeling error analysis. In particular, the estimator in [17] is derived in a goal-oriented setting to control a quantity of interest, and exploits the hierarchical structure (i.e., the inclusion $V_m \subset V_{m+d}$, $\forall m, d \in \mathbb{N}^+$) typical of a HiMod reduction. A similar modeling error analysis is performed in [2] for PGD, although no adaptive algorithm is here set to automatically pick the reduced model. Paper [19] generalizes the a posteriori analysis in [17] to an unsteady setting, providing the tool to automatically select m together with the partition $\mathcal{T}_h$ along Ω_{1D} and the time step.

Finally, HiMod allows to tune the modal index along the domain Ω, according to the local complexity of the transverse dynamics. In particular, m can be varied in different areas of Ω or, in the presence of very localized dynamics, in correspondence with specific nodes of the partition $\mathcal{T}_h$. We refer to these two variants as to *piecewise* and *pointwise* HiMod reduction, in contrast to a *uniform* approach, where the same number of modes is adopted everywhere [16, 18]. This flexibility in the choice of m is currently not available for PGD. Adaptive strategies to select the modal index are available for the three variants of the HiMod procedure [17, 19].

4.3 Computational Aspects

From a computational viewpoint, HiMod reduction and PGD lead to completely different procedures. Indeed, for a fixed value of m, we have to solve the only system (6) of order mN_h when applying HiMod, in contrast to PGD which demands a multiple solution of systems (11)–(12) of order N_h^x and N_h^y, respectively because of the fixed point and the enrichment algorithms. Thus, the direct solution of a single system, in general of larger order, is replaced by an iterative solution of several and smaller systems. This heterogeneity makes a computational comparison between PGD and HiMod not so meaningful. We verify the reliability of the HiMod and PGD procedures on a common test case, by choosing in (1) $V = H_0^1(\Omega)$ with $\Omega = (0, 5) \times (0, 1)$, $a(u, v) = \int_\Omega [\mu \nabla u \cdot \nabla v + \mathbf{b} \cdot \nabla u] d\Omega$ for $\mu = 0.24$, $\mathbf{b} = [-5, 0]^T$, and $F(v) = \int_\Omega f v d\Omega$ with $f(x, y) = 50\{\exp[-((x - 2.85)/0.075)^2 - ((y - 0.5)/0.075)^2] + \exp[-((x - 3.75)/0.075)^2 - ((y - 0.5)/0.075)^2]\}$. For both the methods, we uniformly subdivide Ω_{1D} into 285 subintervals. We set the PGD discretization along y as well as the PGD and the HiMod index m in order

Fig. 1 Qualitative comparison between a HiMod (left) and a PGD (right) approximations

to ensure the same accuracy, TOL, on the reduced approximations with respect to a reference FE solution, computed on a 2500×500 structured mesh. In particular, for TOL $= 8 \cdot 10^{-3}$, we have to subdivide interval $(0, 1)$ into 20 uniform subintervals, and to set m to 6 and to 9 in the PGD and the HiMod discretization, respectively. Sinusoidal functions are chosen for the HiMod modal basis. The ADS iterations are controlled in terms of the relative increment, as

$$\frac{\|u_m^{x,p} u_m^{y,p} - u_m^{x,p-1} u_m^{y,p-1}\|_{L^2(\Omega)}}{\|u_m^{x,p} u_m^{y,p}\|_{L^2(\Omega)}} \leq \text{TOL}_{\text{FP}}, \tag{14}$$

with $\text{TOL}_{\text{FP}} = 10^{-2}$. Figure 1 shows the reduced approximations (which are fully comparable with the FE one, here omitted). The contourplots are very similar. The coarse PGD y-discretization justifies the slight roughness of the PGD contourlines.

Another distinguishing feature between HiMod and PGD is the domain discretization. Indeed, HiMod requires only the partition $\mathcal{T}_h$ along Ω_{1D}, independently of the dimension of Ω. No discretization is needed in the **y**-direction, although we have to carefully select the quadrature nodes to compute coefficients $\{\hat{r}_{jk}^{a,b}\}$. This task becomes particularly challenging when dealing with polar coordinates [10]. With PGD to benefit of the computational advantages associated with a 1D discretization, we are obliged to assume the full separability of Ω; actually, a partial separability demands a 1D partition for Ω_x, and a two-dimensional partition of $\Omega_\mathbf{y}$. As explained in Sect. 5, non-Cartesian domains require a 3D discretization of Ω.

Finally we analyze the interplay between the enrichment and the ADS iterations in the PGD reduction. We investigate the possible relationship between TOL_{FP} in (14) and TOL_{E} in (13), to verify if a small tolerance for the fixed point iteration improves the accuracy of the PGD approximation, thus reducing the number of enrichment steps. To do this, we adopt the same test case used above. Table 1 gathers the number of ADS iterations, #IT$_{\text{FP}}$, the number, m, of enrichment steps, and the CPU time[1] (in seconds) demanded by the PGD procedure, for two different values of TOL_{E} and three different choices of TOL_{FP}. In particular, in column #IT$_{\text{FP}}$ we specify the number of ADS iterations required by each enrichment step. As expected, there exists a link between the two tolerances, namely, when a higher accuracy constrains the fixed point iteration, a smaller number of enrichment steps is performed to ensure the accuracy TOL_{E}.

---

[1]The computations have been run on a Intel Core i5 Dual-Core CPU 2.7 GHz 8 GB RAM MacBook.

Table 1 Quantitative analysis for PGD in terms of fixed point iterations and enrichment steps

	$\text{TOL}_E = 2 \cdot 10^{-2}$			$\text{TOL}_E = 8 \cdot 10^{-3}$		
	#IT_{FP}	m	CPU [s]	#IT_{FP}	m	CPU [s]
$\text{TOL}_{\text{FP}} = 10^{-1}$	$\{2, 2, 2\}$	3	0.099640	$\{2, 2, 2, 2, 2\}$	5	0.337861
$\text{TOL}_{\text{FP}} = 10^{-2}$	$\{4, 3\}$	2	0.046756	$\{4, 3, 2, 2, 4\}$	5	0.358555
$\text{TOL}_{\text{FP}} = 10^{-3}$	$\{5, 5\}$	2	0.077958	$\{5, 5, 2, 7\}$	4	0.341748

5 HiMod Reduction and PGD for Parametrized Problems

The actual potential of PGD becomes more evident when considering a parametric setting, i.e., when problem (1) is replaced by the formulation

$$\text{find } u(\boldsymbol{\mu}) \in V : a(u(\boldsymbol{\mu}), v; \boldsymbol{\mu}) = F(v; \boldsymbol{\mu}) \quad \forall v \in V, \tag{15}$$

with $\boldsymbol{\mu}$ a parameter, which may represent any data of the problem, e.g., the coefficients of the considered PDE, the source term, a boundary value or the domain geometry.

The technique adopted by PGD to deal with the parametric dependence in (15) is very effective. Parameter $\boldsymbol{\mu}$ is considered as an additional independent variable which varies in a domain $\Omega_{\boldsymbol{\mu}}$ [5]. Thus, the PGD space (7) changes into the new one

$$W_m^{\mu} = \left\{ w_m(x, \mathbf{y}, \boldsymbol{\mu}) = \sum_{k=1}^{m} w_k^x(x) w_k^{\mathbf{y}}(\mathbf{y}) w_k^{\mu}(\boldsymbol{\mu}), \text{ with} \right.$$

$$\left. w_k^x \in W_h^x, \ w_k^{\mathbf{y}} \in W_h^{\mathbf{y}}, \ w_k^{\mu} \in W_h^{\mu}, x \in \Omega_x, \ \mathbf{y} \in \Omega_{\mathbf{y}}, \ \boldsymbol{\mu} \in \Omega_{\boldsymbol{\mu}} \right\}, \tag{16}$$

with W_h^{μ} a discretization of the space $L^2(\Omega_{\boldsymbol{\mu}}; \mathbb{R}^Q)$, being Q the length of vector $\boldsymbol{\mu}$. Generalizing the enrichment paradigm in (8), at the m-th step of the PGD approach applied to problem (15) we have to compute three unknown functions, u_m^x, $u_m^{\mathbf{y}}$ and u_m^{μ}, by picking the test function as $X(x)Y(\mathbf{y})Z(\boldsymbol{\mu})$, with $X \in W_h^x$, $Y \in W_h^{\mathbf{y}}$, $Z \in W_h^{\mu}$. Functions u_m^x, $u_m^{\mathbf{y}}$, u_m^{μ} are computed by ADS, which now coincides with a three-step procedure. Thus, with reference to the Poisson problem, $-\nabla \cdot (\mu \nabla u) = f$ completed with full homogeneous Dirichlet boundary conditions and for $\mu \equiv \mu$, we first compute $u_m^{x,p}$ by identifying $u_m^{\mathbf{y}}$ and u_m^{μ} with the previous approximations, $u_m^{\mathbf{y},p-1}$ and $u_m^{\mu,p-1}$, respectively and by selecting $Y(\mathbf{y})Z(\boldsymbol{\mu}) = u_m^{\mathbf{y},p-1} u_m^{\mu,p-1}$ in the test function. This leads to a linear system which generalizes (11), namely

$$
\begin{aligned}
&\left[\left(\mathbf{u}_m^{\mu,p-1}\right)^T M^\mu \mathbf{u}_m^{\mu,p-1}\right] \\
&\left\{\left[\left(\mathbf{u}_m^{\mathbf{y},p-1}\right)^T M^\mathbf{y} \mathbf{u}_m^{\mathbf{y},p-1}\right] K^x + \left[\left(\mathbf{u}_m^{\mathbf{y},p-1}\right)^T K^\mathbf{y} \mathbf{u}_m^{\mathbf{y},p-1}\right] M^x\right\} \mathbf{u}_m^{x,p} \\
&= \left[\left(\mathbf{u}_m^{\mathbf{y},p-1}\right)^T \mathbf{f}^\mathbf{y}\right]\left[\left(\mathbf{u}_m^{\mu,p-1}\right)^T \mathbf{f}^\mu\right] \mathbf{f}^x - \sum_{k=1}^{m-1}\left[\left(\mathbf{u}_m^{\mu,p-1}\right)^T M^\mu \mathbf{u}_k^\mu\right] \\
&\left\{\left[\left(\mathbf{u}_m^{\mathbf{y},p-1}\right)^T M^\mathbf{y} \mathbf{u}_k^\mathbf{y}\right] K^x + \left[\left(\mathbf{u}_m^{\mathbf{y},p-1}\right)^T K^\mathbf{y} \mathbf{u}_k^\mathbf{y}\right] M^x\right\} \mathbf{u}_k^x,
\end{aligned}
\tag{17}
$$

where $M^\mu \in \mathbb{R}^{N_h^\mu \times N_h^\mu}$ is the mass matrix associated with the parameter μ, with $\left[M^\mu\right]_{ij} = \int_{\Omega_\mu} \mu \theta_i^\mu \theta_j^\mu d\mu$ for $i, j = 1, \ldots, N_h^\mu$ and $\mathcal{B}_\mu = \{\theta_\gamma^\mu\}_{\gamma=1}^{N_h^\mu}$ a basis for the space W_h^μ, $\mathbf{f}^\mu \in \mathbb{R}^{N_h^\mu}$ with $\left[\mathbf{f}^\mu\right]_l = \int_{\Omega_\mu} f^\mu \theta_l^\mu d\mu$ for $l = 1, \ldots, N_h^\mu$ after assuming the separability $f = f^x f^\mathbf{y} f^\mu$ for the source term f, and where we employ the same notation as in (11)–(12) to denote vectors $\mathbf{u}_w^\mu$, $\mathbf{u}_m^{\mu,s}$, with $w = 1, \ldots, m - 1$, $s = p, p - 1$, collecting the PGD coefficients associated with the basis $\mathcal{B}_\mu$. Analogously, $u_m^{\mathbf{y},p}$ is computed by solving the generalization of the linear system (12) given by

$$
\begin{aligned}
&\left[\left(\mathbf{u}_m^{\mu,p-1}\right)^T M^\mu \mathbf{u}_m^{\mu,p-1}\right]\left\{\left[\left(\mathbf{u}_m^{x,p}\right)^T K^x \mathbf{u}_m^{x,p}\right] M^\mathbf{y} + \left[\left(\mathbf{u}_m^{x,p}\right)^T M^x \mathbf{u}_m^{x,p}\right] K^\mathbf{y}\right\} \mathbf{u}_m^{\mathbf{y},p} \\
&= \left[\left(\mathbf{u}_m^{x,p}\right)^T \mathbf{f}^x\right]\left[\left(\mathbf{u}_m^{\mu,p-1}\right)^T \mathbf{f}^\mu\right] \mathbf{f}^\mathbf{y} - \sum_{k=1}^{m-1}\left[\left(\mathbf{u}_m^{\mu,p-1}\right)^T M^\mu \mathbf{u}_k^\mu\right] \\
&\left\{\left[\left(\mathbf{u}_m^{x,p}\right)^T K^x \mathbf{u}_k^x\right] M^\mathbf{y} + \left[\left(\mathbf{u}_m^{x,p}\right)^T M^x \mathbf{u}_k^x\right] K^\mathbf{y}\right\} \mathbf{u}_k^\mathbf{y},
\end{aligned}
$$

after setting $u_m^x = u_m^{x,p}$, $u_m^\mu = u_m^{\mu,p-1}$ and $X(x)Z(\mu) = u_m^{x,p} u_m^{\mu,p-1}$ for the PGD test function. Finally, we have the additional linear system used to compute $u_m^{\mu,p}$,

$$
\begin{aligned}
&\left\{\left[\left(\mathbf{u}_m^{x,p}\right)^T K^x \mathbf{u}_m^{x,p}\right]\left[\left(\mathbf{u}_m^{\mathbf{y},p}\right)^T M^\mathbf{y} \mathbf{u}_m^{\mathbf{y},p}\right] + \left[\left(\mathbf{u}_m^{x,p}\right)^T M^x \mathbf{u}_m^{x,p}\right]\left[\left(\mathbf{u}_m^{\mathbf{y},p}\right)^T K^\mathbf{y} \mathbf{u}_m^{\mathbf{y},p}\right]\right\} \\
&M^\mu \mathbf{u}_m^{\mu,p} = \left[\left(\mathbf{u}_m^{x,p}\right)^T \mathbf{f}^x\right]\left[\left(\mathbf{u}_m^{\mathbf{y},p}\right)^T \mathbf{f}^\mathbf{y}\right] \mathbf{f}^\mu - \sum_{k=1}^{m-1}\left\{\left[\left(\mathbf{u}_m^{x,p}\right)^T K^x \mathbf{u}_k^x\right]\left[\left(\mathbf{u}_m^{\mathbf{y},p}\right)^T M^\mathbf{y} \mathbf{u}_k^\mathbf{y}\right]\right. \\
&\left.\left[\left(\mathbf{u}_m^{x,p}\right)^T M^x \mathbf{u}_k^x\right]\left[\left(\mathbf{u}_m^{\mathbf{y},p}\right)^T K^\mathbf{y} \mathbf{u}_k^\mathbf{y}\right]\right\} M^\mu \mathbf{u}_k^\mu,
\end{aligned}
$$

obtained for $u_m^x = u_m^{x,p}$, $u_m^\mathbf{y} = u_m^{\mathbf{y},p}$ and by selecting $X(x)Y(\mathbf{y}) = u_m^{x,p} u_m^{\mathbf{y},p}$ for the test function. From a computational viewpoint, at each ADS iteration, we have to solve now three linear systems of order N_h^x, $N_h^\mathbf{y}$, N_h^μ, respectively.

We investigate the reliability of PGD on problem (15), for $V = H_{\Gamma_{in} \cup \Gamma_{up} \cup \Gamma_{down}}^1$ (Ω) with $\Omega = (0, 3) \times (0, 1)$, $\Gamma_{in} = \{0\} \times (0, 1)$, $\Gamma_{up} = (0, 3) \times \{1\}$, $\Gamma_{down} = (0, 3) \times \{0\}$, $a(u, v) = \int_\Omega [\mu \nabla u \cdot \nabla v + \mathbf{b} \cdot \nabla u] d\Omega$ with $\mathbf{b} = [2.5, 0]^T$ and μ the parameter to be varied in $\Omega_\mu = [1, 5]$, $F(v) = \int_\Omega f v d\Omega$ with $f = 1$. The problem is completed with mixed boundary conditions, namely a homogeneous Dirichlet data on $\Gamma_{up} \cup \Gamma_{down}$, the non-homogeneous Dirichlet condition, $u = u_{in}$ with $u_{in} = y(1-y)$, on Γ_{in} and a homogeneous Neumann value on $\Gamma_{out} = \{3\} \times (0, 1)$. We apply the PGD reduction for $m = 2$, and we uniformly subdivide Ω_x, Ω_y, Ω_μ, being $N_h^x =$

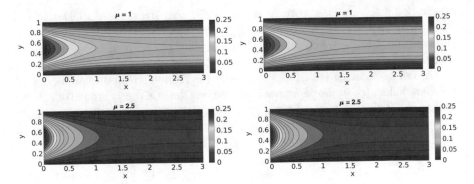

Fig. 2 Qualitative comparison between the reference (left) and the PGD (right) solutions, for $\mu = 1$ (top) and $\mu = 2.5$ (bottom)

150, $N_h^y = 50$, $N_h^\mu = 500$. The tolerance in (14) is set to 10^{-2}. Figure 2 compares the PGD approximation for $\mu = 1$ and $\mu = 2.5$ with a reference full solution coinciding with a linear FE approximation computed on a 300×100 structured mesh. The qualitative matching between the corresponding solutions is significant. From a quantitative viewpoint, the $L^2(\Omega)$-norm of the relative error associated with the PGD approximation does not vary significantly by increasing m, whereas a slight error reduction is detected by increasing μ.

The parametric counterpart of the HiMod reduction, known as HiPOD, merges HiMod with POD [4, 13]. HiPOD pursues a different goal with respect to PGD. Indeed, for a new value, μ^*, of the parameter, PGD provides an approximation for the full solution $u(\mu^*)$, while HiPOD approximates the HiMod solution associated with μ^*. The offline/online paradigm of POD is followed also by HiPOD. The peculiarity is that the offline step is now performed in the HiMod setting to contain the computational burden typical of this stage and by relying on the good properties of HiMod in terms of reliability-versus-accuracy balance. Thus, we choose P different values, $\mu = \mu_i$ with $i = 1, \ldots, P$, for parameter μ, and we collect the HiMod approximation for the corresponding problem (15) into the response matrix, $S = \left[\mathbf{u}_m^{\text{HiMod}}(\mu_1), \mathbf{u}_m^{\text{HiMod}}(\mu_2), \ldots, \mathbf{u}_m^{\text{HiMod}}(\mu_P) \right] \in \mathbb{R}^{mN_h \times P}$, according to representation (5). Successively, we define the null-average matrix

$$\mathcal{V} = S - \frac{1}{P} \sum_{i=1}^{P} \left[\mathbf{u}_m^{\text{HiMod}}(\mu_i), \mathbf{u}_m^{\text{HiMod}}(\mu_i), \ldots, \mathbf{u}_m^{\text{HiMod}}(\mu_i) \right] \in \mathbb{R}^{mN_h \times P},$$

and we apply the Singular Value Decomposition (SVD) to $\mathcal{V}$, so that $\mathcal{V} = \Phi \Sigma \Psi^T$, where $\Phi \in \mathbb{R}^{(mN_h) \times (mN_h)}$ and $\Psi \in \mathbb{R}^{P \times P}$ are the unitary matrices of the left- and of the right-singular vectors of $\mathcal{V}$, respectively while $\Sigma = \text{diag}(\sigma_1, \ldots, \sigma_P) \in \mathbb{R}^{(mN_h) \times P}$ denotes the pseudo-diagonal matrix of the singular values of $\mathcal{V}$, being $\sigma_1 \geq \sigma_2 \geq \cdots \geq \sigma_\rho \geq 0$ and $\rho = \min(mN_h, P)$ [8]. The POD basis is identified by the first l left singular vectors, ϕ_i, of $\mathcal{V}$, so that the reduced POD space is $V_{POD}^l = \text{span}\{\phi_1, \ldots, \phi_l\}$, with $\dim(V_{POD}^l) = l$ and $l \ll mN_h$. In the numerical assessment below, value l coincides with the smallest integer such that $\sigma_l^2 < \varepsilon$, with ε a prescribed tolerance.

The online phase of HiPOD approximates the HiMod solution to problem (15) for a new value, μ^*, of the parameter by exploiting the POD basis instead of solving system (6). This is performed via a projection step. After assembling the HiMod stiffness matrix and right-hand side, $A_m^{HiMod}(\mu^*)$ and $\mathbf{f}_m^{HiMod}(\mu^*)$, associated with the new value of the parameter, we solve the POD system of order l

$$A_{POD}(\mu^*)\mathbf{u}_{POD}(\mu^*) = \mathbf{f}_{POD}(\mu^*), \tag{18}$$

where $A_{POD}(\mu^*) = (\Phi_{POD}^l)^T A_m^{HiMod}(\mu^*) \Phi_{POD}^l$ and $\mathbf{f}_{POD}(\mu^*) = (\Phi_{POD}^l)^T \mathbf{f}_m^{HiMod}(\mu^*)$ denote the POD stiffness matrix and right-hand side, respectively with $\Phi_{POD}^l = [\phi_1, \ldots, \phi_l] \in \mathbb{R}^{(mN_h) \times l}$ the matrix collecting the POD basis vectors. The HiMod solution is thus approximated by vector $\Phi_{POD}^l \mathbf{u}_{POD}(\mu^*) \in \mathbb{R}^{mN_h}$, i.e., after solving a system of order l instead of mN_h. Overall, HiPOD requires to solve P linear systems of order mN_h during the offline phase, additionally to a system of order l in the online phase.

To check the performances of HiPOD, we adopt the test case used above for PGD, for the same values of the parameters, $\mu^* = 1$ and $\mu^* = 2.5$. The reference solution is the corresponding HiMod approximation computed by using $m = 15$ sinusoidal functions in the y-direction, and a linear FE discretization along the mainstream based on a uniform subdivision of Ω_{1D} into 50 subintervals. The same HiMod discretization is adopted to build the response matrix. Concerning the HiPOD approximation, we pick $P = 100$ by uniformly sampling the interval $[1, 5]$, and we select $\varepsilon = 2.5 \cdot 10^{-15}$. This choice sets the dimension of the POD space to $l = 8$, so that we have to solve a system of order 8 instead of 750. The contour plots in Fig. 3 qualitatively compare the HiMod solution with the HiPOD approximation for $l = 1$. The correspondence between the two approximations is good despite a single POD mode is employed (in such a case, system (18)

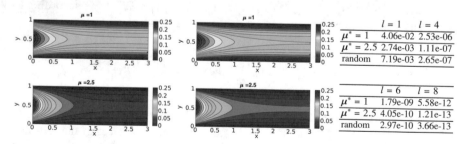

	$l = 1$	$l = 4$
$\mu^* = 1$	4.06e-02	2.53e-06
$\mu^* = 2.5$	2.74e-03	1.11e-07
random	7.19e-03	2.65e-07

	$l = 6$	$l = 8$
$\mu^* = 1$	1.79e-09	5.58e-12
$\mu^* = 2.5$	4.05e-10	1.21e-13
random	2.97e-10	3.66e-13

Fig. 3 Contour plots: comparison between the reference HiMod solution (left) and the HiPOD approximation with $l = 1$ (right), for $\mu^* = 1$ (top) and $\mu* = 2.5$ (bottom). Table: relative error between HiMod and HiPOD solutions with respect to the $L^2(\Omega)$-norm

reduces to a scalar equation). We do not provide the HiPOD approximations for $l = 8$ since they qualitatively coincide with the corresponding HiMod solution. The left panels can be additionally compared with the FE solutions in Fig. 2 to verify the reliability of the HiMod procedure. Finally, the table in Fig. 3 gathers the $L^2(\Omega)$-norm of the relative error between HiMod and HiPOD solutions, for four different POD bases and for three choices of the viscosity (1, 2.5 and the average over a sampling of 30 random values of μ). The error monotonically decreases for larger and larger values of l, independently of the choice for μ. If we compare the values for $\mu = 1$ and for $\mu = 2.5$ (one of the endpoints and the midpoint of the sampling interval, respectively), we notice a higher accuracy (of about one order of magnitude) for the latter choice. This is rather standard in projection-based reduced order modeling [11]. Concerning the computational saving in terms of CPU time, HiPOD method requires on average $O(10^{-3})[s]$ to be compared with $O(10)[s]$ demanded by HiMod, resulting in a speedup of 10^4.

Although PGD and HiPOD are not directly comparable due to the different purpose they pursue, we highlight the main pros and cons of the two methods. The explicit dependence of the approximation on the parameters makes PGD an ideal tool to efficiently deal with parametric problems. For any new parameter, a direct evaluation yields the corresponding PGD approximation. On the other hand, HiPOD suffers of the drawbacks typical of the projection-based methods. The main bottleneck is the assembling of the HiMod arrays involved in $A_{\text{POD}}(\mu^*)$ and $f_{\text{POD}}(\mu^*)$.

When PGD is applied to parametric problems, we recover the possibility to deal with any geometric domain. In such a case, a partial separability is applied to the problem, so that the space independent variables are kept together whereas parameters are separated. This approach clearly looses the computational advantages due to space separability. On the contrary, HiPOD inherits the geometric flexibility of the HiMod reduction, without giving up the spatial dimensional reduction of the problem.

Acknowledgements The authors thank Yves Antonio Brandes Costa Barbosa for his support in the HiMod simulations. This work has been partially funded by GNCS-INdAM 2018 project on "Tecniche di Riduzione di Modello per le Applicazioni Mediche". F. Ballarin also acknowledges the support by European Union Funding for Research and Innovation, Horizon 2020 Program, in the framework of European Research Council Executive Agency: H2020 ERC Consolidator Grant 2015 AROMA-CFD project 681447 "Advanced Reduced Order Methods with Applications in Computational Fluid Dynamics" (P. I. G. Rozza).

References

1. Aletti, M.C., Perotto, S., Veneziani, A.: HiMod reduction of advection-diffusion-reaction problems with general boundary conditions. J. Sci. Comput. **76**(1), 89–119 (2018)
2. Ammar, A., Chinesta, F., Diez, P., Huerta, A.: An error estimator for separated representations of highly multidimensional models. Comput. Methods Appl. Mech. Eng. **199**(25–28), 1872–1880 (2010)

3. Ammar, A., Cueto, E., Chinesta, F.: Reduction of the chemical master equation for gene regulatory networks using proper generalized decompositions. Int. J. Numer. Methods Biomed. Eng. **28**(9), 960–973 (2012)
4. Baroli, D., Cova, C.M., Perotto, S., Sala, L., Veneziani, A.: Hi-POD solution of parametrized fluid dynamics problems: Preliminary results. In: Model Reduction of Parametrized Systems. MS&A: Modeling, Simulation and Applications, vol.17, pp. 235–254. Springer, Cham (2017)
5. Chinesta, F., Keunings, R., Leygue, A.: The Proper Generalized Decomposition for Advanced Numerical Simulations: A Primer. SpringerBriefs in Applied Sciences and Technology. Springer International Publishing, Berlin (2014)
6. Ern, A., Perotto, S., Veneziani, A.: Hierarchical model reduction for advection-diffusion-reaction problems. In: Numerical Mathematics and Advanced Applications, pp. 703–710. Springer, Berlin (2008)
7. Ghnatios, C., Ammar, A., Cimetiere, A., Hamdouni, A., Leygue, A., Chinesta, F.: First steps in the space separated representation of models defined in complex domains. In: 11th Biennial Conference on Engineering Systems Design and Analysis, pp. 37–42. Nantes (2012)
8. Golub, G.H., Van Loan, C.F.: Matrix Computations. Johns Hopkins Studies in the Mathematical Sciences, 4th edn. Johns Hopkins University Press, Baltimore (2013)
9. González, D., Ammar, A., Chinesta, F., Cueto, E.: Recent advances on the use of separated representations. Internat. J. Numer. Methods Eng. **81**(5), 637–659 (2010)
10. Guzzetti, S., Perotto, S., Veneziani, A.: Hierarchical model reduction for incompressible fluids in pipes. Internat. J. Numer. Methods Eng. **114**(5), 469–500 (2018)
11. Hesthaven, J.S., Rozza, G., Stamm, B.: Certified Reduced Basis Methods for Parametrized Partial Differential Equations. SpringerBriefs in Mathematics. Springer, Cham; BCAM Basque Center for Applied Mathematics, Bilbao (2016)
12. Ladevèze, P., Passieux, J.-C., Néron, D.: The LATIN multiscale computational method and the proper generalized decomposition. Comput. Methods Appl. Mech. Eng. **199**(21–22), 1287–1296 (2010)
13. Lupo Pasini, M., Perotto, S., Veneziani, A.: HiPOD: Hierarchical model reduction driven by a Proper Orthogonal Decomposition for parametrized advection-diffusion-reaction problems. In preparation
14. Niroomandi, S., González, D., Alfaro, I., Bordeu, F., Leygue, A., Cueto, E., Chinesta, F.: Real-time simulation of biological soft tissues: a PGD approach. Int. J. Numer. Methods Biomed. Eng. **29**(5), 586–600 (2013)
15. Perotto, S.: Hierarchical model (Hi-Mod) reduction in non-rectilinear domains. In: Domain Decomposition Methods in Science and Engineering XXI. Lecture Notes in Computational Science and Engineering, vol. 98, pp. 477–485. Springer, Cham (2014)
16. Perotto, S.: A survey of hierarchical model (Hi-Mod) reduction methods for elliptic problems. In: Numerical Simulations of Coupled Problems in Engineering. Computational Methods in Applied Sciences, vol. 33, pp. 217–241. Springer, Cham (2014)
17. Perotto, S., Veneziani, A.; Coupled model and grid adaptivity in hierarchical reduction of elliptic problems. J. Sci. Comput. **60**(3), 505–536 (2014)
18. Perotto, S., Zilio, A.: Hierarchical model reduction: three different approaches. In: Numerical Mathematics and Advanced Applications 2011, pp. 851–859. Springer, Heidelberg (2013)
19. Perotto, S., Zilio, A.: Space-time adaptive hierarchical model reduction for parabolic equations. Adv. Model. Simul. Eng. Sci. **2**, 25 (2015)
20. Perotto, S., Ern, A., Veneziani, A.: Hierarchical local model reduction for elliptic problems: a domain decomposition approach. Multiscale Model. Simul. **8**(4), 1102–1127 (2010)
21. Perotto, S., Reali, A., Rusconi, P., Veneziani, A.: HIGAMod: a hierarchical isogeometric approach for model reduction in curved pipes. Comput. Fluids **142**, 21–29 (2017)

22. Pruliere, E., Chinesta, F., Ammar, A.: On the deterministic solution of multidimensional parametric models using the proper generalized decomposition. Math. Comput. Simul. **81**(4), 791–810 (2010)
23. Signorini, M., Zlotnik, S., Díez, P.: Proper generalized decomposition solution of the parameterized Helmholtz problem: application to inverse geophysical problems. Int. J. Numer. Methods Eng. **109**(8), 1085–1102 (2017)

Recurrence Relations for a Family of Orthogonal Polynomials on a Triangle

Sheehan Olver, Alex Townsend, and Geoffrey M. Vasil

1 Introduction

In 1975, Koornwinder described a general procedure for constructing multivariate orthogonal polynomials from univariate ones [4, §3.7.2]. The procedure allows for the construction of seven classes of bivariate orthogonal polynomials from Jacobi polynomials, some of which were previously known [9]. In this paper, we consider a four-parameter variant of Koornwinder's class IV polynomials (the four-parameter variant was not constructed by Koornwinder) defined as [2]

$$
\begin{aligned}
P_{n,k}^{(a,b,c,d)}(x, y) &= P_{n-k}^{(2k+b+c+d+1,a)}(2x - 1)(1 - x)^k P_k^{(c,b)}\left(-1 + \frac{2y}{1-x}\right) \\
&= \tilde{P}_{n-k}^{(2k+b+c+d+1,a)}(x)(1 - x)^k \tilde{P}_k^{(c,b)}\left(\frac{y}{1-x}\right),
\end{aligned}
\tag{1}
$$

where $a, b, c > -1$, n and k are integers such that $n \geq k \geq 0$, $P_k^{(a,b)}(x)$ is the Jacobi polynomial of degree k [7, Table 18.3.1], and $\tilde{P}_k^{(a,b)}$ is the Jacobi polynomial of degree k shifted to have support on $(0, 1)$. Koornwinder's construction derives the polynomials with $d = 0$, which we denote by $P_{n,k}^{(a,b,c)}$. The polynomials in (1)

S. Olver (✉)
Department of Mathematics, Imperial College, London, UK
e-mail: s.olver@imperial.ac.uk

A. Townsend
Department of Mathematics, Cornell University, Ithaca, NY, USA
e-mail: townsend@cornell.edu

G. M. Vasil
School of Mathematics and Statistics, University of Sydney, Sydney, NSW, Australia
e-mail: geoffrey.vasil@sydney.edu.au

© The Author(s) 2020
S. J. Sherwin et al. (eds.), *Spectral and High Order Methods for Partial Differential Equations ICOSAHOM 2018*, Lecture Notes in Computational Science and Engineering 134, https://doi.org/10.1007/978-3-030-39647-3_5

are orthogonal on the right-angled triangle $\{(x, y) : 0 < x < 1, 0 < y < 1 - x\}$ with respect to the weight function $w_{a,b,c,d}(x, y) = x^a y^b (1 - x - y)^c (1 - x)^d$.

The basis $P_{n,k}^{(a,b,c)}$ has been used extensively by the spectral element community, see the overview in [3]. The recurrence relations we derive can be employed to reduce partial differential operators to sparse matrices, enabling efficient solution of linear partial differential equations defined on triangles, which will be the topic of a future paper. This is analogous to the ultraspherical spectral method for solving ordinary differential equations on bounded intervals [8]. A similar idea using a hierarchy of Zernike polynomials, which are bivariate orthogonal polynomials on the unit disk, is used in [13] to develop a sparse spectral method for solving partial differential equations defined on the disk [13]. On the disk, polar coordinates allow for radially symmetric partial differential operators to be reduced to ordinary differential operators acting on Jacobi polynomials [13]. This simplification does not translate to non-radially symmetric partial differential operators on the disk, nor partial differential operators on the triangle.

Several of the formulae in this paper have already be derived by directly employing recurrence relations satisfied by Jacobi polynomials [14]. Our approach via ladder operators is a more systematic study that derives previously unreported recurrence relations for $P_{n,k}^{(a,b,c)}$. We also hope to use ladder operators to derive sparse recurrence relations for multivariate orthogonal polynomials built from Jacobi polynomials on higher-dimensional simplices.

Throughout this paper, the recurrence relations hold for choices of the parameters n, k, a, b, c, and d that make the Jacobi polynomials well-defined. Moreover, we take $P_{-1}^{(a,b)}(x) = 0$. Also, note that orthogonal polynomials remain orthogonal after an affine transformation so the recurrence relations in this paper for (1) on a right-angled triangle can be extended to any triangle, including triangles with the corners permuted.

The paper is structured as follows. In the next section, we give 12 ladder operators for Jacobi polynomials and use them to derive sparse recurrence relations for $P_n^{(a,b)}$. In Sect. 3 we give 24 ladder operators for (1) and write down the corresponding sparse recurrence relations for $P_{n,k}^{(a,b,c,d)}$. In Sect. 4, we use the ladder operators to derive a collection of sparse recurrence relations for differentiation, conversion, and multiplication that are satisfied by $P_{n,k}^{(a,b,c)}$. Section 5 applies these sparse recurrence relations to efficiently calculating Laplacians of functions on the triangle.

2 Ladder Operators for Jacobi Polynomials

We give 12 ordinary differential operators that increment or decrement the parameters and degree of Jacobi polynomials by zero or one. Each ladder operator maps $P_n^{(a,b)}(x)$ to $P_{\tilde{n}}^{(\tilde{a},\tilde{b})}(x)$, where $|\tilde{n} - n| \le 1$, $|\tilde{a} - a| \le 1$, and $|\tilde{b} - b| \le 1$.

Definition 1 The following operators are ladder operators for Jacobi polynomials:

$$\mathcal{L}_1 u = \frac{du}{dx}$$

$$\mathcal{L}_1^{\dagger} u = ((1+x)a - (1-x)b)u - (1-x^2)\frac{du}{dx}$$

$$\mathcal{L}_2 u = (a+b+n+1)u + (1+x)\frac{du}{dx}$$

$$\mathcal{L}_2^{\dagger} u = (2a + (1-x)n)u - (1-x^2)\frac{du}{dx}$$

$$\mathcal{L}_3 u = (a+b+n+1)u - (1-x)\frac{du}{dx}$$

$$\mathcal{L}_3^{\dagger} u = (2b + (1+x)n)u + (1-x^2)\frac{du}{dx}$$

$$\mathcal{L}_4 u = ((1+x)a - (1-x)(b+n+1))u - (1-x^2)\frac{du}{dx} \qquad \mathcal{L}_4^{\dagger} u = -nu + (1+x)\frac{du}{dx}$$

$$\mathcal{L}_5 u = ((1+x)(a+n+1) - (1-x)b)u - (1-x^2)\frac{du}{dx} \qquad \mathcal{L}_5^{\dagger} u = nu + (1-x)\frac{du}{dx}$$

$$\mathcal{L}_6 u = bu + (1+x)\frac{du}{dx} \qquad\qquad\qquad \mathcal{L}_6^{\dagger} u = au - (1-x)\frac{du}{dx}.$$

The notation for the ladder operators is chosen so that $\mathcal{L}_s^{\dagger} \mathcal{L}_s P_n^{(a,b)}$ and $\mathcal{L}_s \mathcal{L}_s^{\dagger} P_n^{(a,b)}$ are scalar multiples of $P_n^{(a,b)}$ for $1 \leq s \leq 6$. These ladder operators are carefully constructed to give rise to sparse recurrence relations for Jacobi polynomials.

Lemma 1 *The ladder operators give sparse recurrence relations for Jacobi polynomials:*

$$\mathcal{L}_1 P_n^{(a,b)} = \tfrac{1}{2}(n+a+b+1)P_{n-1}^{(a+1,b+1)} \qquad \mathcal{L}_1^{\dagger} P_n^{(a,b)} = 2(n+1)P_{n+1}^{(a-1,b-1)}$$

$$\mathcal{L}_2 P_n^{(a,b)} = (n+a+b+1)P_n^{(a+1,b)} \qquad \mathcal{L}_2^{\dagger} P_n^{(a,b)} = 2(n+a)P_n^{(a-1,b)}$$

$$\mathcal{L}_3 P_n^{(a,b)} = (n+a+b+1)P_n^{(a,b+1)} \qquad \mathcal{L}_3^{\dagger} P_n^{(a,b)} = 2(n+b)P_n^{(a,b-1)}$$

$$\mathcal{L}_4 P_n^{(a,b)} = 2(n+1)P_{n+1}^{(a-1,b)} \qquad \mathcal{L}_4^{\dagger} P_n^{(a,b)} = (n+b)P_{n-1}^{(a+1,b)}$$

$$\mathcal{L}_5 P_n^{(a,b)} = 2(n+1)P_{n+1}^{(a,b-1)} \qquad \mathcal{L}_5^{\dagger} P_n^{(a,b)} = (n+a)P_{n-1}^{(a,b+1)}$$

$$\mathcal{L}_6 P_n^{(a,b)} = (n+b)P_n^{(a+1,b-1)} \qquad \mathcal{L}_6^{\dagger} P_n^{(a,b)} = (n+a)P_n^{(a-1,b+1)}.$$

Proof The relationship for $\mathcal{L}_1$ is a formula for the derivative of $P_n^{(a,b)}(x)$ [7, 18.9.15] and relationship $\mathcal{L}_1^{\dagger}$ is equivalent to [7, 18.9.16]. Six more follow from expressing the left- and right-hand sides in terms of $_2F_1$ functions using [7, 18.5.7] and the reflection formula $P_n^{(a,b)}(x) = (-1)^n P_n^{(b,a)}(-x)$: $\mathcal{L}_4^{\dagger}$ and $\mathcal{L}_4^{\dagger}$ are equivalent to [7, 15.5.3], $\mathcal{L}_6^{\dagger}$ is equivalent to [7, 15.5.4], $\mathcal{L}_4$ and $\mathcal{L}_5$ are equivalent to [7, 15.5.5], and $\mathcal{L}_6$ is equivalent to [7, 15.5.6].

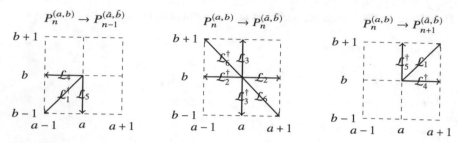

Fig. 1 Illustration of the 12 ladder operators for Jacobi polynomials in Definition 1

The relationship for $\mathcal{L}_2$ follows from combining [7, 18.9.5] and [7, 18.9.6]. The relationships for $\mathcal{L}_2^\dagger$ follows by writing

$$\mathcal{L}_2^\dagger = \mathcal{L}_1^\dagger + (n + a + b)(1 - x)$$

and then using [7, 18.9.5] and [7, 18.9.6]. Finally, $\mathcal{L}_3$ and $\mathcal{L}_3^\dagger$ follow just as $\mathcal{L}_2$ and $\mathcal{L}_2^\dagger$, using the reflection formula.

Remark 1 We note that the first-order differential operators occurring in Lemma 1 form together an action of the Lie algebra sl(4) [1, 5, 6].

Figure 1 illustrates the ladder operators and how they increment or decrement the parameters associated to a Jacobi polynomial.

The ladder operators can be easily adapted to the shifted Jacobi polynomials, denoted by $\tilde{P}_k^{(a,b)}$, which are supported on $(0, 1)$, with x in place of $1 + x$, and no factors of 2. The corresponding recurrence relations for $\tilde{P}_k^{(a,b)}$ are the same as in Lemma 1, except the multiplicative factors of $\frac{1}{2}$ and 2 are replaced by 1.

3 Ladder Operators for $P_{n,k}^{(a,b,c,d)}$

The 12 ladder operators for the Jacobi polynomials in Sect. 2 allow us to derive 24 ladder operators for $P_{n,k}^{(a,b,c,d)}$. The ladder operators are carefully defined so that they map $P_{n,k}^{(a,b,c,d)}$ to a scalar multiple of $P_{\tilde{n},\tilde{k}}^{(\tilde{a},\tilde{b},\tilde{c},\tilde{d})}$, where the new parameters in $P_{\tilde{n},\tilde{k}}^{(\tilde{a},\tilde{b},\tilde{c},\tilde{d})}$ are n, k, a, b, c or d, respectively, incremented or decremented by 0 or 1.

To highlight the symmetries of the right-angled triangle and make the recurrences more convenient to write down, we define

$$z := 1 - x - y \qquad \text{and} \qquad \frac{\partial}{\partial z} := \frac{\partial}{\partial y} - \frac{\partial}{\partial x},$$

as in [14]. Now, the variables x, y, and z have the convenient property that any affine transformation that maps the triangle onto itself has the effect of exchanging the roles of x, y, and z.

Definition 2 The following operators are ladder operators for $P_{n,k}^{(a,b,c,d)}$. The first set of 12 are:

$$\mathcal{M}_{0,1}u = \frac{\partial u}{\partial y} \qquad\qquad \mathcal{M}_{0,1}^{\dagger}u = (yc - zb)u - yz\frac{\partial u}{\partial y}$$

$$\mathcal{M}_{0,2}u = (k+b+c+1)u + y\frac{\partial u}{\partial y} \qquad \mathcal{M}_{0,2}^{\dagger}u = \left(c+k-\frac{yk}{1-x}\right)u - \frac{y}{1-x}z\frac{\partial u}{\partial y}$$

$$\mathcal{M}_{0,3}u = (k+b+c+1)u - x\frac{\partial u}{\partial y} \qquad \mathcal{M}_{0,3}^{\dagger}u = \left(b+\frac{ky}{1-x}\right)u + \frac{y}{1-x}z\frac{\partial u}{\partial y}$$

$$\mathcal{M}_{0,4}u = (yc - z(b+k+1))u - yz\frac{\partial u}{\partial y} \qquad \mathcal{M}_{0,4}^{\dagger}u = -\frac{k}{1-x}u + \frac{y}{1-x}\frac{\partial u}{\partial y}$$

$$\mathcal{M}_{0,5}u = (y(c+k+1))u - zb - yz\frac{\partial u}{\partial y} \qquad \mathcal{M}_{0,5}^{\dagger}u = \frac{k}{1-x}u + \left(1-\frac{y}{1-x}\right)\frac{\partial u}{\partial y}$$

$$\mathcal{M}_{0,6}u = cu - z\frac{\partial u}{\partial y} \qquad\qquad \mathcal{M}_{0,6}^{\dagger}u = bu + y\frac{\partial u}{\partial y}.$$

The second set of 12 are:

$$\mathcal{M}_{1,0}u = \frac{k}{1-x}u + \frac{\partial u}{\partial x} - \frac{y}{1-x}\frac{\partial u}{\partial y},$$

$$\mathcal{M}_{1,0}^{\dagger}u = (x(k+a+b+c+d+1)-a)u - x(1-x)\frac{\partial u}{\partial x} + xy\frac{\partial u}{\partial y}$$

$$\mathcal{M}_{2,0}u = (n+k+a+b+c+d+2)u + \frac{xk}{1-x}u + x\frac{\partial u}{\partial x} - \frac{xy}{1-x}\frac{\partial u}{\partial y}$$

$$\mathcal{M}_{2,0}^{\dagger}u = (n+k+b+c+d+1-xn)u - x(1-x)\frac{\partial u}{\partial x} + xy\frac{\partial u}{\partial y}$$

$$\mathcal{M}_{3,0}u = (n+a+b+c+d+2)u - (1-x)\frac{\partial u}{\partial x} + y\frac{\partial u}{\partial y}$$

$$\mathcal{M}_{3,0}^{\dagger}u = (a+xn)u + x(1-x)\frac{\partial u}{\partial x} - xy\frac{\partial u}{\partial y}$$

$$\mathcal{M}_{4,0}u = (x(n+a+b+c+d+2)-a-n+k-1)u - x(1-x)\frac{\partial u}{\partial x} + xy\frac{\partial u}{\partial y}$$

$$\mathcal{M}_{4,0}^{\dagger}u = \frac{k}{1-x}u - nu + x\frac{\partial u}{\partial x} - \frac{xy}{1-x}\frac{\partial u}{\partial y}, \qquad \mathcal{M}_{5,0}u = nu + (1-x)\frac{\partial u}{\partial x} - y\frac{\partial u}{\partial y}$$

$$\mathcal{M}_{5,0}^{\dagger}u = x(n+a+b+c+d+2)u - au - x(1-x)\frac{\partial u}{\partial x} + xy\frac{\partial u}{\partial y}$$

$$\mathcal{M}_{6,0}u = au + \frac{xk}{1-x}u + x\frac{\partial u}{\partial x} - \frac{xy}{1-x}\frac{\partial u}{\partial y},$$

$$\mathcal{M}_{6,0}^{\dagger}u = (k+b+c+d+1)u - (1-x)\frac{\partial u}{\partial x} + y\frac{\partial u}{\partial y}.$$

The notation for the ladder operators is chosen so that the recurrence relations in Theorem 1 are derived for $\mathcal{M}_{s,0}$ (resp. $\mathcal{M}_{0,s}$) by applying $\mathcal{L}_s$ or $\mathcal{L}_s^{\dagger}$ to the first (resp. second) Jacobi polynomial in $P_{n,k}^{(a,b,c,d)}(x,y)$ for $1 \le s \le 6$. Moreover, we know that $\mathcal{M}_{s,0}^{\dagger}\mathcal{M}_{s,0}P_{n,k}^{(a,b,c,d)}$, $\mathcal{M}_{0,s}^{\dagger}\mathcal{M}_{0,s}P_{n,k}^{(a,b,c,d)}$, $\mathcal{M}_{s,0}\mathcal{M}_{s,0}^{\dagger}P_{n,k}^{(a,b,c,d)}$, and $\mathcal{M}_{0,s}\mathcal{M}_{0,s}^{\dagger}P_{n,k}^{(a,b,c,d)}$ are scalar multiples of $P_{n,k}^{(a,b,c,d)}$ for $1 \le s \le 6$.

The ladder operators in Definition 2 correspond to 24 sparse recurrence relations for $P_{n,k}^{(a,b,c,d)}$. To derive these recurrences, we first express the partial derivatives of $P_{n,k}^{(a,b,c,d)}$ as derivatives of shifted Jacobi polynomials.

Proposition 1 *The following relationships hold:*

$$\tilde{P}_{n-k}^{(2k+b+c+d+1,a)}(x)(1-x)^k \left[\tilde{P}_k^{(c,b)}\right]' \left(\tfrac{y}{1-x}\right) = (1-x)\tfrac{\partial}{\partial y} P_{n,k}^{(a,b,c,d)}(x,y), \tag{2}$$

$$\left[\tilde{P}_{n-k}^{(2k+b+c+d+1,a)}\right]'(x)(1-x)^{k+1}\tilde{P}_k^{(c,b)}\left(\tfrac{y}{1-x}\right) = \left(k+(1-x)\tfrac{\partial}{\partial x} - y\tfrac{\partial}{\partial y}\right) P_{n,k}^{(a,b,c,d)}(x,y). \tag{3}$$

Proof The first relationship is immediate. The second relationship follows from the chain-rule:

$$(1-x)\tfrac{\partial}{\partial x}\left(f(x)(1-x)^k g\left(\tfrac{y}{1-x}\right)\right) = f'(x)(1-x)^{k+1} g\left(\tfrac{y}{1-x}\right)$$

$$- kf(x)(1-x)^k g(x) + yf(x)(1-x)^{k-1} g'\left(\tfrac{y}{1-x}\right)$$

and an application of (2) to simplify the last term.

The 24 sparse recurrence relations for $P_{n,k}^{(a,b,c,d)}$ are given in the following theorem.

Theorem 1 *Let $t = a+b+c+d$. The first set of 12 are:*

$$\mathcal{M}_{0,1} P_{n,k}^{(a,b,c,d)} = (k+b+c+1)P_{n-1,k-1}^{(a,b+1,c+1,d)} \qquad \mathcal{M}_{0,1}^\dagger P_{n,k}^{(a,b,c,d)} = (k+1)P_{n+1,k+1}^{(a,b-1,c-1,d)}$$

$$\mathcal{M}_{0,2} P_{n,k}^{(a,b,c,d)} = (k+b+c+1)P_{n,k}^{(a,b,c+1,d-1)} \qquad \mathcal{M}_{0,2}^\dagger P_{n,k}^{(a,b,c,d)} = (k+c)P_{n,k}^{(a,b,c-1,d+1)}$$

$$\mathcal{M}_{0,3} P_{n,k}^{(a,b,c,d)} = (k+b+c+1)P_{n,k}^{(a,b+1,c,d-1)} \qquad \mathcal{M}_{0,3}^\dagger P_{n,k}^{(a,b,c,d)} = (k+b)P_{n,k}^{(a,b-1,c,d+1)}$$

$$\mathcal{M}_{0,4} P_{n,k}^{(a,b,c,d)} = (k+1)P_{n+1,k+1}^{(a,b,c-1,d-1)} \qquad \mathcal{M}_{0,4}^\dagger P_{n,k}^{(a,b,c,d)} = (k+b)P_{n-1,k-1}^{(a,b,c+1,d+1)}$$

$$\mathcal{M}_{0,5} P_{n,k}^{(a,b,c,d)} = (k+1)P_{n+1,k+1}^{(a,b-1,c,d-1)} \qquad \mathcal{M}_{0,5}^\dagger P_{n,k}^{(a,b,c,d)} = (k+c)P_{n-1,k-1}^{(a,b+1,c,d+1)}$$

$$\mathcal{M}_{0,6} P_{n,k}^{(a,b,c,d)} = (k+c)P_{n,k}^{(a,b+1,c-1,d)} \qquad \mathcal{M}_{0,6}^\dagger P_{n,k}^{(a,b,c,d)} = (k+b)P_{n,k}^{(a,b-1,c+1,d)}.$$

The second set of 12 are:

$$\mathcal{M}_{1,0} P_{n,k}^{(a,b,c,d)} = (n+k+t+2)P_{n-1,k}^{(a+1,b,c,d+1)}$$

$$\mathcal{M}_{1,0}^\dagger P_{n,k}^{(a,b,c,d)} = (n-k+1)P_{n+1,k}^{(a-1,b,c,d-1)}$$

$$\mathcal{M}_{2,0} P_{n,k}^{(a,b,c,d)} = (n+k+t+2)P_{n,k}^{(a,b,c,d+1)}$$

$$\mathcal{M}_{2,0}^\dagger P_{n,k}^{(a,b,c,d)} = (n+k+t-a+1)P_{n,k}^{(a,b,c,d-1)}$$

$$\mathcal{M}_{3,0} P_{n,k}^{(a,b,c,d)} = (n + k + t + 2) P_{n,k}^{(a+1,b,c,d)}$$

$$\mathcal{M}_{3,0}^{\dagger} P_{n,k}^{(a,b,c,d)} = (n - k + a) P_{n,k}^{(a-1,b,c,d)}$$

$$\mathcal{M}_{4,0} P_{n,k}^{(a,b,c,d)} = (n - k + 1) P_{n+1,k}^{(a,b,c,d-1)}$$

$$\mathcal{M}_{4,0}^{\dagger} P_{n,k}^{(a,b,c,d)} = (n - k + a) P_{n-1,k}^{(a,b,c,d+1)}$$

$$\mathcal{M}_{5,0} P_{n,k}^{(a,b,c,d)} = (n + k + t - a + 1) P_{n-1,k}^{(a+1,b,c,d)}$$

$$\mathcal{M}_{5,0}^{\dagger} P_{n,k}^{(a,b,c,d)} = (n - k + 1) P_{n+1,k}^{(a-1,b,c,d)}$$

$$\mathcal{M}_{6,0} P_{n,k}^{(a,b,c,d)} = (n - k + a) P_{n,k}^{(a-1,b,c,d+1)}$$

$$\mathcal{M}_{6,0}^{\dagger} P_{n,k}^{(a,b,c,d)} = (n + k + t - a + 1) P_{n,k}^{(a+1,b,c,d-1)}.$$

Proof We present the proof of $\mathcal{M}_{0,1} P_{n,k}^{(a,b,c,d)} = (k + b + c + 1) P_{n-1,k-1}^{(a,b+1,c+1,d)}$. By the definition of $P_{n,k}^{(a,b,c,d)}$ in (1), the chain rule, and the relationship in (2), we have

$$\frac{\partial}{\partial y} P_{n,k}^{(a,b,c,d)}(x, y) = \tilde{P}_{n-k}^{(2k+b+c+d+1,a)}(x)(1 - x)^{k-1} [\tilde{P}_k^{(c,b)}]' \left(\frac{y}{1-x} \right)$$

$$= (k + c + b + 1) \tilde{P}_{n-k}^{(2k+b+c+d+1,a)}(x)(1 - x)^{k-1} \tilde{P}_{k-1}^{(c+1,b+1)} \left(\frac{y}{1-x} \right), \tag{4}$$

where the last equality comes from applying $\mathcal{L}_1$ in Definition 1. The final expression in (4) is equivalent to $(k + b + c + 1) P_{n-1,k-1}^{(a,b+1,c+1,d)}$. The manipulations for the remaining recurrence relations are similar, except with different choices of the operators $\mathcal{L}_s$ or $\mathcal{L}_s^{\dagger}$ and combinations of (2) and (3).

4 Sparse Recurrence Relations for $P_{n,k}^{(a,b,c)}$

We can combine the ladder operators in Sect. 3 to derive sparse recurrence relations between $P_{n,k}^{(a,b,c)}$ polynomials with different parameters and their partial derivatives. These recurrence relations are analogous to many of the sparse recurrence relations for Jacobi polynomials [7, §18.9].

4.1 Differentiation

The partial derivatives of $P_{n,k}^{(a,b,c)}$ can be written in terms of Jacobi polynomials on the triangle with incremented parameters, which is analogous to a recurrence relation for the derivative of a Jacobi polynomial [7, 18.9.15]. A similar recurrence for $P_{n,k}^{(a,b,c)}$ can be found in [14, Prop. 4.6, 4.7, & 4.8].

Corollary 1 *The following recurrence relations hold:*

$$(2k+b+c+1)\frac{\partial}{\partial x}P_{n,k}^{(a,b,c)} = (n+k+a+b+c+2)(k+b+c+1)P_{n-1,k}^{(a+1,b,c+1)}$$

$$+ (k+b)(n+k+b+c+1)P_{n-1,k-1}^{(a+1,b,c+1)},$$

$$\tag{5}$$

$$\frac{\partial}{\partial y}P_{n,k}^{(a,b,c)} = (k+b+c+1)P_{n-1,k-1}^{(a,b+1,c+1)}, \tag{6}$$

$$(2k+b+c+1)\frac{\partial}{\partial z}P_{n,k}^{(a,b,c)} = -(n+k+a+b+c+2)(k+b+c+1)P_{n-1,k}^{(a+1,b+1,c)}$$

$$+ (k+c)(n+k+b+c+1)P_{n-1,k-1}^{(a+1,b+1,c)}. $$

$$\tag{7}$$

Proof The recurrence (5) follows from the fact that $(\mathcal{M}_{1,0}\mathcal{M}_{0,2} + \mathcal{M}_{0,4}^{\dagger}\mathcal{M}_{6,0}^{\dagger})u = (2k+b+c+1)\frac{\partial u}{\partial x}$ when $d = 0$. The relationship (6) is equivalent to the relation given by $\mathcal{M}_{0,1}$ in Theorem 1 when $d = 0$. Finally, (7) follows from the fact that $(\mathcal{M}_{1,0}\mathcal{M}_{0,3} - \mathcal{M}_{0,5}^{\dagger}\mathcal{M}_{6,0}^{\dagger})u = -(2k+b+c+1)\frac{\partial u}{\partial z}$.

The derivatives of weighted versions of $P_{n,k}^{(a,b,c)}$ also satisfy sparse recurrence relations, which are analogous to an expression for the derivative of a weighted Jacobi polynomial [7, 18.9.16].

Corollary 2 *The following recurrence relations hold:*

$$-(2k+b+c+1)\frac{\partial}{\partial x}\left(x^a y^b z^c P_{n,k}^{(a,b,c)}\right) = x^{a-1}y^b z^{c-1}\Big((k+c)(n-k+1)P_{n+1,k}^{(a-1,b,c-1)}$$

$$+ (k+1)(n-k+a)P_{n+1,k+1}^{(a-1,b,c-1)}\Big),$$

$$\frac{\partial}{\partial y}\left(x^a y^b z^c P_{n,k}^{(a,b,c)}\right) = -(k+1)x^a y^{b-1}z^{c-1}P_{n+1,k+1}^{(a,b-1,c-1)},$$

$$(2k+b+c+1)\frac{\partial}{\partial z}\left(x^a y^b z^c P_{n,k}^{(a,b,c)}\right) = x^{a-1}y^{b-1}z^c\Big((k+b)(n-k+1)P_{n+1,k}^{(a-1,b-1,c)}$$

$$- (k+1)(n-k+a)P_{n+1,k+1}^{(a-1,b-1,c)}\Big).$$

Proof The first recurrence follows from

$$(\mathcal{M}_{0,2}^{\dagger}\mathcal{M}_{1,0}^{\dagger} + \mathcal{M}_{0,4}\mathcal{M}_{6,0})u = (2k+b+c+1)(cx - az - xz\frac{\partial}{\partial x})u$$

$$= -(2k+b+c+1)x^{1-a}z^{1-c}\frac{\partial}{\partial x}(x^a z^c u).$$

The second recurrence holds since

$$\mathcal{M}_{0,1}^{\dagger}u = (cy - bz - yz\frac{\partial}{\partial y})u = -y^{1-b}z^{1-c}\frac{\partial}{\partial x}(y^b z^c u).$$

The third recurrence is derived from the fact that

$$(M_{0,3}^\dagger M_{1,0}^\dagger - M_{0,5} M_{6,0})u = (2k+b+c+1)(bx - ay + xy(\tfrac{\partial}{\partial y} - \tfrac{\partial}{\partial x}))u$$

$$= (2k+b+c+1)x^{1-a}y^{1-b}z^{-c}\tfrac{\partial}{\partial z}(x^a y^b z^c u).$$

4.2 Conversion

Recurrence relations for conversion allow us to express $P_{n,k}^{(a,b,c)}$ in terms of Jacobi polynomials on the triangle with different parameters. Here, we give the recurrence relations that increment the parameters, which are analogues of [7, 18.9.3]. Similar relations can be found in [14, Prop. 4.4].

Corollary 3 *The following recurrence relations hold:*

$$(2n+a+b+c+2)P_{n,k}^{(a,b,c)} = (n+k+a+b+c+2)P_{n,k}^{(a+1,b,c)}$$

$$+ (n+k+b+c+1)P_{n-1,k}^{(a+1,b,c)}, \tag{8}$$

$$(2n+a+b+c+2)(2k+b+c+1)P_{n,k}^{(a,b,c)}$$

$$= (n+k+a+b+c+2)(k+b+c+1)P_{n,k}^{(a,b+1,c)}$$

$$- (n-k+a)(k+b+c+1)P_{n-1,k}^{(a,b+1,c)} + (k+c)(n+k+b+c+1)P_{n-1,k-1}^{(a,b+1,c)}$$

$$- (k+c)(n-k+1)P_{n,k-1}^{(a,b+1,c)}, \tag{9}$$

$$(2n+a+b+c+2)(2k+b+c+1)P_{n,k}^{(a,b,c)}$$

$$= (n+k+a+b+c+2)(k+b+c+1)P_{n,k}^{(a,b,c+1)}$$

$$- (n-k+a)(k+b+c+1)P_{n-1,k}^{(a,b,c+1)} - (k+b)(n+k+b+c+1)P_{n-1,k-1}^{(a,b,c+1)}$$

$$+ (k+b)(n-k+1)P_{n,k-1}^{(a,b,c+1)}. \tag{10}$$

Proof The recurrence relation in (8) follows from the fact that $(M_{30} + M_{50})u = (2n+a+b+c+2)u$ when $d = 0$. Since $(M_{2,0} - M_{4,0}^\dagger)u = (2n+a+b+c+d+2)u$ and $(M_{2,0}^\dagger - M_{4,0})u = (2n+a+b+c+d+2)(1-x)u$, we obtain

$$(M_{0,2}M_{2,0} - M_{0,2}M_{4,0}^\dagger - M_{0,4}^\dagger M_{2,0}^\dagger + M_{0,4}^\dagger M_{4,0})u = (2k+b+c+1)(2n+a+b+c+d+2)u,$$

The recurrence relation (10) immediately follows. Similarly, (9) holds since

$$(M_{0,3}M_{2,0} - M_{0,3}M_{4,0}^\dagger + M_{0,5}^\dagger M_{2,0}^\dagger - M_{0,5}^\dagger M_{4,0})u = (2k+b+c+1)(2n+a+b+c+d+2)u.$$

4.3 Multiplication

Recurrence relations for multiplication allow one to express $x P_{n,k}^{(a,b,c)}$, $y P_{n,k}^{(a,b,c)}$, and $z P_{n,k}^{(a,b,c)}$ in terms of a sum of Jacobi polynomials on the triangle with potentially different parameters. The recurrences in Corollary 4 are analogous to the recurrence relations for $P_n^{(a,b)}$ found in [7, 18.9.6].

Corollary 4 *The following recurrence relations hold:*

$$(2n + a + b + c + 2)x P_{n,k}^{(a,b,c)} = (n - k + a)P_{n,k}^{(a-1,b,c)} + (n - k + 1)P_{n+1,k}^{(a-1,b,c)},$$

(11)

$$(2k + b + c + 1)(2n + a + b + c + 2)y P_{n,k}^{(a,b,c)} = (k + b)(n + k + b + c + 1)P_{n,k}^{(a,b-1,c)}$$

$$- (k + 1)(n - k + a)P_{n,k+1}^{(a,b-1,c)} - (k + b)(n - k + 1)P_{n+1,k}^{(a,b-1,c)}$$

$$+ (k + 1)(n + k + a + b + c + 2)P_{n+1,k+1}^{(a,b-1,c)},$$

(12)

$$(2k + b + c + 1)(2n + a + b + c + 2)z P_{n,k}^{(a,b,c)} = (k + c)(n + k + b + c + 1)P_{n,k}^{(a,b,c-1)}$$

$$+ (k + 1)(n - k + a)P_{n,k+1}^{(a,b,c-1)} - (k + c)(n - k + 1)P_{n+1,k}^{(a,b,c-1)}$$

$$- (k + 1)(n + k + a + b + c + 2)P_{n+1,k+1}^{(a,b,c-1)}.$$

(13)

Proof The recurrence relation in (11) follows from the fact that $(M_{3,0}^{\dagger} + M_{5,0}^{\dagger})u = (2n + a + b + c + d + 2)xu$. Since

$$(M_{03}^{\dagger}M_{20}^{\dagger} - M_{03}^{\dagger}M_{40} + M_{05}M_{20} - M_{05}M_{40}^{\dagger})u = (2k+b+c+1)(2n+a+b+c+d+2)yu$$

holds, we find that (12) is satisfied. Finally, (13) follows from

$$(M_{02}^{\dagger}M_{20}^{\dagger} - M_{02}^{\dagger}M_{40} + M_{04}M_{40}^{\dagger} - M_{04}M_{20})u = (2k+b+c+1)(2n+a+b+c+d+2)zu.$$

Combining the recurrence relations in Corollaries 3 and 4, we can derive expressions for $x P_{n,k}^{(a,b,c)}$, $y P_{n,k}^{(a,b,c)}$, and $z P_{n,k}^{(a,b,c)}$ in terms of a sum of Jacobi polynomials on the triangle with parameters (a, b, c). These are analogous to the three-term recurrence relation for Jacobi polynomials [7, 18.9.2]. Since these recurrence relations are long, we refer the reader to [2, pp. 80–81].

4.4 Differential Eigenvalue Problems

The polynomials $P_{n,k}^{(a,b,c)}$ are eigenfunctions for second-order differential operators (see [2, (5.3.4)] and [14, Prop. 4.11]), and the ladder operators in Sect. 3 make it easy to derive this fact.

Theorem 2 *The polynomial $P_{n,k}^{(a,b,c)}$ satisfies two second-order differential eigenproblems:*

$$zy \frac{\partial^2}{\partial y^2} P_{n,k}^{(a,b,c)} + ((1+b)(1-x)-(2+b+c)y)\frac{\partial}{\partial y} P_{n,k}^{(a,b,c)} = -k(k+b+c+1)P_{n,k}^{(a,b,c)}$$

and

$$x(1-x)\frac{\partial^2}{\partial x^2} P_{n,k}^{(a,b,c)} - 2xy\frac{\partial^2}{\partial x \partial y} P_{n,k}^{(a,b,c)} + y(1-y)\frac{\partial^2}{\partial y^2} P_{n,k}^{(a,b,c)}$$
$$+ (a+1-(a+b+c+3)x)\frac{\partial}{\partial x} P_{n,k}^{(a,b,c)} + (b+1-(a+b+c+3)y)\frac{\partial}{\partial y} P_{n,k}^{(a,b,c)}$$
$$= -n(n+a+b+c+d+2)P_{n,k}^{(a,b,c,d)}(x,y).$$

Proof The first equation follows from

$$\mathcal{M}_{0,1}\mathcal{M}_{0,1}^{\dagger} P_{n,k}^{(a,b,c,d)}(x,y) = k(k+b+c+1)P_{n,k}^{(a,b,c,d)}(x,y)$$

and the second from

$$\left[\frac{\mathcal{M}_{0,1}\mathcal{M}_{0,1}^{\dagger} - k(k+b+c+1)}{1-x} + \mathcal{M}_{3,0}\mathcal{M}_{3,0}^{\dagger} + \mathcal{M}_{4,0}\mathcal{M}_{4,0}^{\dagger} \right] P_{n,k}^{(a,b,c,d)}(x,y)$$
$$= (1+a-k+n)(3+t+2n)P_{n,k}^{(a,b,c,d)}(x,y).$$

5 Application: Calculating Laplacians

We can use the recurrence relationships in this paper to calculate partial derivatives too. Slevinsky's fast triangle transform [11] (which builds on his fast spherical harmonic transform [10]) as implemented in the FastTransform multithreaded C code [12] gives an efficient and stable routine for calculating the expansion coefficients on the triangle in $O(d^2 \log^2 d)$ operations, where d is the polynomial degree. The partial derivative recurrences (see Corollary 1) show us how to calculate the expansion coefficients of $\frac{\partial p}{\partial x}$ and $\frac{\partial p}{\partial y}$ with coefficients associated to the parameters $(a+1, b, c+1)$ and $(a, b+1, c+1)$, respectively. Moreover, Corollary 3 informs us how to convert from expansions with parameters (a, b, c) to $(a+1, b, c)$, $(a, b+1, c)$ and $(a, b, c+1)$. We can combine these various recurrences relations to compute

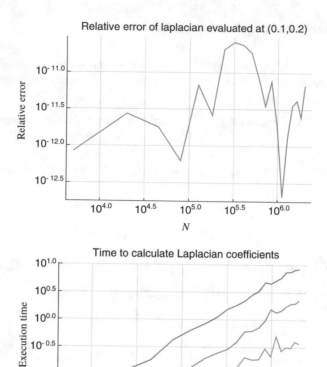

Fig. 2 Top: Error when evaluating the Laplacian of $f(x, y) = \cos(nxy/40)$ at $(x, y) = (0.1, 0.2)$ by expanding in degree $N = (n + 1)(n + 2)/2$ Jacobi polynomials on the triangle with $(a, b, c) = (0, 0, 0)$ and using the recurrences. Bottom: Execution times of (1) the fast transform, (2) constructing the recurrences as 8 banded-block-banded matrices, and (3) applying the matrices

the coefficients of the Laplacian in the basis $P_{n,k}^{(2,2,2)}$ in an optimal complexity of $O(N) = O(d^2)$ operations, where d is the polynomial degree and $N = d(d + 1)/2$ is the total number of degrees of freedom.

For example, the Laplacian of $f(x, y) = \cos(nxy/40)$ can be computed by first approximating f on the unit right-angled triangle to within machine precision by a polynomial, and then employing various recurrence relationships to calculate its Laplacian (Fig. 2). To do this efficiently, we store the recurrence relations as banded-block-banded matrices to take advantage of fast banded matrix-vector multiplication.

6 Conclusion

We introduce ladder operators for systematically deriving sparse recurrence relations for differentiation, conversion, and multiplication of Jacobi and orthogonal polynomials on the triangle. We use these recurrences to efficiently apply partial differential operators, in particular for calculating Laplacians. The importance of these relationships is that they allow general linear partial differential operators with polynomial coefficients to be represented as sparse operators acting on orthogonal polynomial expansions. This application will be the topic of a subsequent paper.

References

1. Derezinski, J., Majewski, P.: From conformal group to symmetries of hypergeometric type equations. Symmetry Integrability Geometry: Methods Appl. **12**, 108 (2016)
2. Dunkl, C.F., Xu, Y.: Orthogonal Polynomials of Several Variables, 2nd edn. Cambridge University Press, Cambridge (2014)
3. Karniadakis, G., Sherwin, S.: Spectral/hp Element Methods for Computational Fluid Dynamics. Oxford University Press, Oxford (2013)
4. Koornwinder, T.: Two-variable analogues of the classical orthogonal polynomials. In: Theory and Application of Special Functions (Proceedings of an Advanced Seminar Sponsored by the Mathematics Research Center, the University of Wisconsin, Madison, 1975), pp. 435–495
5. Miller, Jr., W.: Lie theory and generalizations of the hypergeometric functions. SIAM J. Appl. Math. **25**, 226–235 (1973)
6. Miller, Jr., W.: Symmetries of differential equations. The hypergeometric and Euler–Darboux equations. SIAM J. Math. Anal. **4**, 314–328 (1973)
7. Olver, F.W.J., Lozier, D.W., Boisvert, R.F, Clark, C.W. (eds): NIST Handbook of Mathematical Functions. Cambridge University Press, Cambridge (2010)
8. Olver, S., Townsend, A.: A fast and well-conditioned spectral method. SIAM Rev. **55**, 462–489 (2013)
9. Proriol, J.: Sur une famille de polynômes á deux variables orthogonaux dans un triangle. C. R. Acad. Sci. Paris **257**, 2459–2461 (1957)
10. Slevinsky, R.M.: Fast and backward stable transforms between spherical harmonic expansions and bivariate Fourier series. Appl. Comp. Harm. Anal. **47**(3), 585–606 (2019)
11. Slevinsky, R.M.: Conquering the pre-computation in two-dimensional harmonic polynomial transforms. Preprint arXiv:1711.07866 (2017)
12. Slevinsky, R.M.: FastTransforms. v0.1. https://github.com/MikaelSlevinsky/FastTransforms
13. Vasil, G.M., Burns, K.J., Lecoanet, D., Olver, S., Brown, B.P., Oishi, J.S.: Tensor calculus in polar coordinates using Jacobi polynomials. J. Comp. Phys. **325**, 53–73 (2016)
14. Xu, Y.: Approximation and orthogonality in Sobolev spaces on a triangle. Constr. Approx. **46**, 349–434 (2017)

Part II
Contributed Papers

Greedy Kernel Methods for Center Manifold Approximation

Bernard Haasdonk, Boumediene Hamzi, Gabriele Santin, and Dominik Wittwar

1   Introduction

Center manifold theory plays an important role in the study of the stability of dynamical systems when the equilibrium point is not hyperbolic. It isolates the complicated asymptotic behavior by locating the center manifold which is an invariant manifold tangent to the subspace spanned by the eigenspace of eigenvalues on the imaginary axis. Then, the dynamics of the original system will be essentially determined by the restriction of this dynamics on the center manifold since the local dynamic behavior "transverse" to this invariant manifold is relatively simple as it corresponds to the flows in the local stable (and unstable) manifolds. In practice, one does not compute the center manifold and its dynamics exactly since this requires the resolution of a quasilinear partial differential equation which is not easily solvable. In most cases of interest, an approximation of degree two or three of the solution is sufficient. Then, the reduced dynamics on the center manifold can be determined, its stability can be studied and then conclusions about the stability of the original system can be obtained [1, 3, 4, 6, 8].

In this article, we use greedy kernel methods to construct a data-based approximation of the center manifold. The present work is a preliminary study that is intended to introduce our concept and algorithm, and to test it on some examples.

B. Haasdonk · G. Santin · D. Wittwar
Inst. of Applied Analysis and Numerical Simulation, University of Stuttgart, Stuttgart, Germany
e-mail: bernard.haasdonk@mathematik.uni-stuttgart.de;
gabriele.santin@mathematik.uni-stuttgart.de; dominik.wittwar@mathematik.uni-stuttgart.de

B. Hamzi (✉)
Department of Mathematics, Imperial College London, London, UK

S. J. Sherwin et al. (eds.), *Spectral and High Order Methods for Partial Differential Equations ICOSAHOM 2018*, Lecture Notes in Computational Science and Engineering 134, https://doi.org/10.1007/978-3-030-39647-3_6

95

2 Background

We consider a large dimensional dynamical system

$$\dot{x} = f(x), \quad x \in D, \tag{1}$$

where $f : D \to \mathbb{R}^n$ is a continuously differentiable function over the domain $D \subset \mathbb{R}^n$ such that $0 \in D$. We are interested in the study of the behavior of the system around an equilibrium point $x \in D$, i.e., $f(x) = 0$, possibly analyzing a smaller dimensional system.

Without loss of generality, we may assume that the equilibrium is $x = 0$, and, letting $L = \frac{\partial f}{\partial x}(x)|_{x=0}$, we can rewrite (1) as

$$\dot{x} = f(x) = Lx + N(x),$$

with a suitable nonlinear component N, and denote as $\sigma_{\mathbb{R}}(L)$ the set of real parts of the eigenvalues of L. A classical result relates the stability of the equilibrium with the spectrum of L, and in particular it is known that if L has all its eigenvalues with negative real parts, i.e., $\sigma_{\mathbb{R}}(L) \subset \mathbb{R}_{<0}$, then the origin is asymptotically stable, and if L has some eigenvalues with positive real parts, then the origin is unstable. If instead $\sigma_{\mathbb{R}}(L) \subset \mathbb{R}_{\leq 0}$, the linearization fails to determine the stability properties of the origin, and thus the analysis of this situation requires to employ additional tools.

In this case, we can first use a linear change of coordinates to separate the zero and the negative eigenvalues, i.e., we can rewrite (1) as

$$\dot{x} = L_1 x + N_1(x, y)$$
$$\dot{y} = L_2 y + N_2(x, y) \tag{2}$$

where $L_1 \in \mathbb{R}^{d \times d}$ is such that $\sigma_{\mathbb{R}}(L_1) = \{0\}$ and $L_2 \in \mathbb{R}^{m \times m}$ with $m := n - d$ is such that $\sigma_{\mathbb{R}}(L_2) \subset \mathbb{R}_{<0}$. The nonlinear functions $N_1 : \mathbb{R}^d \times \mathbb{R}^m \to \mathbb{R}^d$ and $N_2 : \mathbb{R}^d \times \mathbb{R}^m \to \mathbb{R}^m$ are continuously differentiable. Intuitively, we expect the stability of the equilibrium to only depend on the nonlinear term $N_1(x, y)$. This intuition turns out to be correct, and indeed it can be properly formalized by means of the center manifold theorem.

We start by recalling a sufficient condition for the existence of a center manifold.

Theorem 1 ([1]) *If N_1 and N_2 are twice continuously differentiable and are such that*

$$N_i(0, 0) = 0, \quad \frac{\partial N_i}{\partial x}(0, 0) = 0, \quad \frac{N_i}{\partial y}(0, 0) = 0, \quad i = 1, 2,$$

and if the eigenvalues of L_1 have zero real parts, and all the eigenvalues of L_2 have negative real parts, then there exists a neighbourhood $\Omega \subset \mathbb{R}^d$ of the origin $0 \in \mathbb{R}^d$

and a center manifold $h : \Omega \to \mathbb{R}^m$ for (2), i.e., $y = h(x)$ is an invariant manifold for (2),[1] h is smooth, and

$$h(0) = 0, \quad Dh(0) = 0. \tag{3}$$

Under the assumptions of this theorem, using (2) we deduce that h satisfies the PDE

$$L_2 h(x) + N_2(x, h(x)) = Dh(x) \left(L_1 x + N_1(x, h(x)) \right), \tag{4}$$

and the following center manifold theorem ensures that there are smooth solutions to this PDE. Moreover, it also allows to deduce the stability of the origin of the full order system (2) from the stability of the origin of a reduced order system called the *center dynamics.*

Theorem 2 (Center Manifold Theorem [1]) *The equilibria $x = 0$, $y = 0$ of the original dynamics is locally asymptotically stable (resp. unstable) if and only if the equilibrium $x = 0$ of the center dynamics (dynamics on the center manifold)*

$$\dot{x} = L_1 x + N_1(x, h(x)), \tag{5}$$

is locally asymptotically stable (resp. unstable).

In particular, this result guarantees that, after solving the PDE (4), the problem of analyzing the stability properties of the system (2) reduces to analyzing the nonlinear stability of the lower dimensional system (5). This second problem is of smaller dimension and thus, provided the knowledge of h, the approach is attractive to obtain information on the system (1) via a reduced model.

Moreover, we remark that an exact knowledge of h is not required for this purpose, i.e., it is sufficient to have an approximate solution of the PDE (4). Indeed, it is frequently sufficient to compute only the low degree terms of the Taylor series expansion of h around $x = 0$, i.e., if $(\cdot)^{[k]}$ is the degree k part of the Taylor series of h, the approximation

$$h(x) \approx h^{[1]}x + h^{[2]}(x) + h^{[3]}(x) + \ldots + h^{[d-1]}(x) \tag{6}$$

is sufficient to obtain an approximation of the dynamics of order ε^d as $\|x\| \le \varepsilon$. The approximation (6) can be obtained by coefficient comparison, thus rewriting

---

[1] A differentiable manifold $\mathcal{M}$ is said to be invariant under the flow of a vector field X if for $x \in \mathcal{M}$, $F_t(x) \in \mathcal{M}$ for small $t > 0$, where $F_t(x)$ is the flow of X.

the PDE (4) as a set of algebraic equations as

$$L_2 h^{[1]} = h^{[1]} L_1$$

$$L_2 h^{[2]}(x) + N_2^{[2]}(x, h^{[1]}(x)) = \frac{\partial h^{[2]}}{\partial x}(x) \left(L_1 x_1 + N_1^{[2]}(x, h^{[1]}(x)) \right)$$

$$L_2 h^{[3]}(x) + \left(N_2(x, h^{[2]}(x)) \right)^{[2]} = \frac{\partial h^{[2]}}{\partial x}(x) \left(L_1 x + \left(N_1(x, h^{[2]}(x)) \right)^{[2]} \right)$$

. . . .

We remark that this methodology is valid for parameterized dynamical systems and is used to study the stability of dynamical systems with bifurcations.

Nevertheless, even this approximated knowledge of h can be difficult to obtain in practice for a general ODE. To overcome this limitation, and since an approximated knowledge of the manifold is sufficient, our goal in this paper is to find a data-based approximation of the center manifold. This approximation is based solely on the knowledge of the splitting (1) and on the numerical computation of a set of trajectories of the system, and it provides an approximation of h which can be used to study the system stability.

3 Kernel Approximation

We want to build a surrogate model $s_h : \Omega \to \mathbb{R}^m$ which approximates the center manifold h on a suitable set $\Omega \subset \mathbb{R}^d$, in the sense that $s_h(x) \approx h(x)$ for all $x \in \Omega$. This model is constructed in a data-based way, i.e., we assume some knowledge of the map h on a finite set of input parameters, or training data. In practice, such values are computed from high-fidelity numerical approximations, which will be discussed in detail in the following.

The surrogate is based on kernel approximation, which allows the use of scattered data, i.e., we do not require any grid structure on the set of training data. Moreover, since the unknown function h is vector-valued, we employ here matrix-valued kernels. Details on kernel-based approximation can be found e.g. in [9], and the extension to the vectorial case is detailed e.g. in [5, 10]. We recall here only that a positive definite matrix-valued kernel on Ω is a function $K : \Omega \times \Omega \to \mathbb{R}^{m \times m}$ such that $K(x, y) = K(y, x)^T$ for all $x, y \in \Omega$ and $[K(x_i, x_j)]_{i,j=1}^N \in \mathbb{R}^{mN \times mN}$ is positive semidefinite for any set $\{x_1, \ldots, x_N\} \subset \Omega$ of pairwise distinct points, for all $N \in \mathbb{N}$. Associated to a positive definite kernel there is a unique Hilbert space $\mathcal{H}$ of functions $\Omega \to \mathbb{R}^m$, named native space, where the kernel is reproducing, meaning that $K(\cdot, x)\alpha$ is the Riesz representer of the directional point evaluation $\delta_x^\alpha(f) := \alpha^T f(x)$, for all $\alpha \in \mathbb{R}^m, x \in \Omega$.

We consider here a twice continuously differentiable matrix-valued kernel k on Ω, and we use a specific functional formulation for our approximation and a specific cost function, in order to construct a surrogate that is well suited for the particular approximation task.

In detail, the approximant takes the form

$$s_h(x) = \sum_{i=1}^{n_1} K(x, x_i^{(1)})\alpha_i + \sum_{j=1}^{n_2} \sum_{i=1}^{m} \partial_i^{(2)} K(x, x_j^{(2)})\beta_{i,j},$$

with centers $x_i^{(1)} \in X^{(1)} = \left\{ x_1^{(1)}, \ldots, x_{n_1}^{(1)} \right\}$, $x_j^{(2)} \in X^{(2)} = \left\{ x_1^{(2)}, \ldots, x_{n_2}^{(2)} \right\}$ and coefficient vectors $\alpha_i, \beta_{i,j} \in \mathbb{R}^m$. Here the superscript $\partial^{(2)}$ denotes that the derivative with regards to the second kernel component is taken.

Subsequently, we assume to have a sufficient amount of data $X_{N^*} = \{x_1, \ldots, x_{N^*}\}$ and $Y_{N^*} = \{y_1, \ldots, y_{N^*}\}$ which, for example, is generated by running a numerical scheme to compute discrete trajectories for different initial values (x_0, y_0). For this step, we need to assume that the variable splitting (2) is known in advance. Note that this is not a severe restriction, as for a general ODE (1) the required state transformation can be determined by eigenvalue decomposition of L.

Observe that we do not know if a data pair (x_i, y_i) lies on the center manifold, i.e. if $y_i = h(x_i)$ holds. We only know that the data converges asymptotically to the center manifold as $x_i \to 0$. Thus, an interpolation-based surrogate which merely interpolates the data on a given subset $X \subset X_{N^*}$ seems ill-suited for our purposes. Instead we consider another set of conditions to define the approximant. First, we still require the conditions in (3) to be satisfied by our approximation. Moreover, for the given subsets $X = \{x_1, \ldots, x_N\}$ and $Y = \{y_1, \ldots, y_N\}$, we compute our approximant by minimizing the following functional $J : \mathscr{H} \to \mathbb{R}$ under the constraint $s(0) = 0$, $Ds(0) = 0$:

$$J(s) := \|s\|_{\mathscr{H}}^2 + \sum_{i=1}^{N} (s(x_i) - y_i)^T \omega_i (s(x_i) - y_i). \tag{7}$$

Here $\omega_i \in \mathbb{R}^{m \times m}$ is a positive definite weight matrix. It can be shown that (7) has a unique minimizer s_h (see [11]). In particular s_h and its derivative Ds_h have the form

$$s_h(x) = \sum_{i=1}^{N+1} K(x, x_i)\alpha_i + \sum_{i=1}^{m} \partial_i^{(2)} K(x, 0)\beta_i, \tag{8}$$

$$Ds_h(x) = \sum_{i=1}^{N+1} D^{(1)} K(x, x_i)\alpha_i + \sum_{i=1}^{m} D^{(1)} \partial_i^{(2)} K(x, 0)\beta_i,$$

where we set $x_{N+1} := 0$. The coefficient vectors α_i, β_i can be computed by solving the system

$$\begin{pmatrix} A + W & B \\ B^T & C \end{pmatrix} \begin{pmatrix} \alpha \\ \beta \end{pmatrix} = \begin{pmatrix} Y \\ Z \end{pmatrix}, \tag{9}$$

with

$$A := \left(K(x_i, x_j) \right)_{i,j} \in \mathbb{R}^{m(N+1) \times m(N+1)},$$

$$W := \text{diag} \left(\omega_1^{-1}, \ldots, \omega_N^{-1}, 0 \right) \in \mathbb{R}^{m(N+1) \times m(N+1)},$$

$$B := \left(\partial_j^{(2)} K(x_i, 0) \right)_{i,j} \in \mathbb{R}^{m(N+1) \times m^2},$$

$$C := \left(\partial_i^{(1)} \partial_j^{(2)} k(0, 0) \right)_{i,j} \in \mathbb{R}^{m^2 \times m^2},$$

$$Y := (y_1^T, \ldots, y_n^T, 0)^T \in \mathbb{R}^{m(N+1)},$$

$$Z := 0 \in \mathbb{R}^{m^2 \times m}.$$

The weight matrices ω_i can either be chosen manually, or a regularizing function $r : \Omega \to \mathbb{R}^{m \times m}$ can be prescribed such that $\omega_i = r(x_i)$ is symmetric and positive definite. In our numerical examples in Sect. 4 we chose a constant regularization function, i.e.

$$\omega_i = r(x_i) = \lambda I_m$$

for some $\lambda > 0$. However, one might consider a more general approach, where the weight increases as the data tends to the origin, i.e. $\omega_i \succeq \omega_j$ if $\|x_i\| \leq \|x_j\|$.

3.1 Greedy Approximation

If the technique of the previous section is used as it is, the surrogate (8) is given by an expansion with N^* terms, where N^* is the number of points in the training set. Therefore, the model evaluation might not be efficient enough if the model is built using a too large dataset. Furthermore, the computation of the coefficients in (8) requires the solution of the linear system (9), whose size again scales with the size of the training set, and which can be severely ill-conditioned for non well-placed points.

To mitigate both problems, we employ an algorithm that aims at selecting small subsets X_N, Y_N of points such that the surrogate computed with these sets is a sufficiently good approximation of the one which uses the full sets. The algorithm selects the points in a greedy way, i.e., one point at a time is selected and added to the current training set. In this way, it is possible to identify a good set without the need to solve a nearly infeasible combinatorial problem.

The selection is performed using the P-greedy method of [2] applied to the kernel K, such that the set of points is selected before the computation of the surrogate. The number of points, and therefore the expansion size and evaluation speed, is depending on a prescribed target accuracy $\varepsilon_{tol} > 0$. For details on the method implementation and its convergence properties we refer to [7].

4 Numerical Examples

We test now our method on three different examples. In each of them, we specify the setting and the parameters used to build the surrogate and visualize our approximation to the center manifold. Additionally, we compute the pointwise residual

$$r(x) = Ds_h(x)\left(L_1 x + N_1(x, s_h(x))\right) - \left(L_2 s_h(x) + N_2(x, s_h(x))\right),$$

which measures how well the surrogate s_h satisfies the ODE (4).

In all the three examples, the greedy algorithm is used to select a suitable subset of the points, and in all cases the procedure is stopped with a prescribed ε_{tol}. In the first two examples we set $\varepsilon_{tol} := 10^{-15}$, while $\varepsilon_{tol} := 10^{-10}$ is used in the last one.

4.1 Example 1

We consider the 2-dimensional system

$$\dot{x} = L_1 x + N_1(x, y) = 0 + xy$$
$$\dot{y} = L_2 y + N_2(x, y) = -y + x^2. \tag{10}$$

We generate the training data by solving (10) with an implicit Euler scheme for initial time $t_0 = 0$, final time $T = 1000$ and with the time step $\Delta t = 0.1$. We initiate the numerical procedure with initial values $(x_0, y_0) \in \{\pm 0.8\} \times \{\pm 0.8\}$ and store the resulting data pairs in X and Y after discarding all data whose x-values are not contained in the neighborhood $[-0.1, 0.1]$ which results in $N^* = 38{,}248$ data pairs.

We run the greedy algorithm for the kernels $k_1(x, y) := \left(1 + xy/2\right)^4$ and $k_2(x, y) = e^{-(x-y)^2/2}$. This results in the sets X_1 and X_2 which contain 14 and 6 points, respectively. The corresponding approximations s_1 and s_2 for the constant

Fig. 1 Approximations s_1 and s_2 of the center manifold

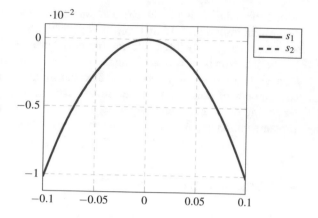

Fig. 2 Residuals r_1 and r_2 of the center manifold

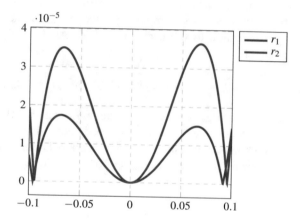

regularization function $r \equiv 10^{-10}$ are plotted in Fig. 1 over the domain $[-0.1, 0.1]$. The pointwise residual is depicted in Fig. 2.

4.2 Example 2

We consider the 2-dimensional system

$$\dot{x} = L_1 x + N_1(x, y) = 0 - xy$$
$$\dot{y} = L_2 y + N_2(x, y) = -y + x^2 - 2y^2. \tag{11}$$

The training data is generated the same way as in Example 1. We again use the kernels k_1 and k_2. The greedy algorithm gives sets X_1 and X_2 of size 12 and 6, respectively. The evaluation of the approximations s_1 and s_2 over the neighborhood

Fig. 3 Approximations s_1 and s_2 of the center manifold

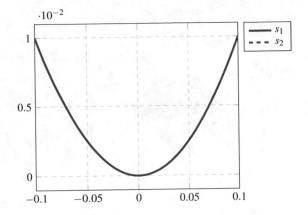

Fig. 4 Residuals r_1 and r_2 of the center manifold

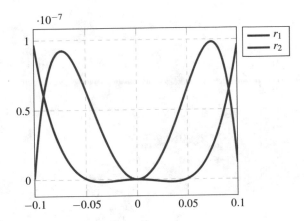

$[-0.1, 0.1]$ can be seen in Fig. 3, while the respective pointwise residuals are plotted in Fig. 4.

4.3 Example 3

We consider the $(2 + 1)$-dimensional system

$$\dot{x} = L_1 x + N_1(x, y) = \begin{pmatrix} 0 & -1 \\ 1 & 0 \end{pmatrix} \begin{pmatrix} x_1 \\ x_2 \end{pmatrix} + \begin{pmatrix} x_1 y \\ x_2 y \end{pmatrix} \tag{12}$$

$$\dot{y} = L_2 y + N_2(x, y) = -y - x_1^2 - x_2^2 + y^2.$$

We generate the training data in a similar fashion as before. We again use the implicit Euler scheme with start time $t = 0$, final time $T = 1000$ and with time step

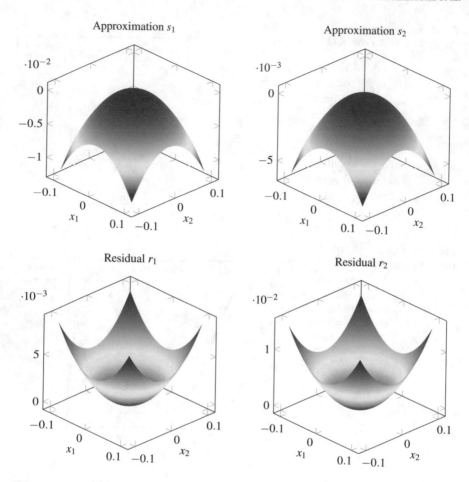

Fig. 5 Approximations s_1 and s_2 of the center manifold and corresponding residuals r_1 and r_2

$\Delta t = 0.1$. The Euler method is performed for initial data $(x_0, y_0) \in \{\pm 0.8\}^3$ and the resulting trajectories are stored in X and Y, where only data with $x \in [-0.1, 0.1]^2$ was considered; this leads to $N^* = 78,796$ data pairs. We use the kernels $k_1(x, y) = (1 + x^T y/2)^4$ and $k_2(x, y) = e^{-\|x-y\|_2^2/2}$, and the greedy-selected sets have the size 21 (for k_1) and 25 (for k_2), respectively. The approximations s_1, s_2 and their corresponding residuals r_1 and r_2 computed over the domain $[-0.1, 0.1]^2$. The results can be seen in Fig. 5.

We remark that in all the three experiments both kernels give comparable results in terms of error magnitude, and they both provide a good approximation of the manifold.

5 Conclusions

In this paper we introduced a novel algorithm to approximate the center manifold of a given ODE using a data-based surrogate.

This algorithm computes an approximation of the manifold from a set of numerical trajectories with different initial data. It is based on kernel methods, which allow the use of the scattered data generated by these simulations as training points. Moreover, an application-specific ansatz and cost function have been employed in order to enforce suitable properties on the surrogate.

Several numerical experiments suggested that the present method can reach a significant accuracy, and that it has the potential to be used as an effective model reduction technique. It seems promising to apply this approach to high dimensional systems as the approximation technique straightforwardly can be extended and is less prone to the curse of dimensionality than grid-based approximation techniques. An interesting extension would consist of determining the decomposition (2) in a data-based fashion by suitable processing of the trajectory data.

Acknowledgements The first, third, and fourth authors would like to thank the German Research Foundation (DFG) for support within the Cluster of Excellence in Simulation Technology (EXC 310/2) at the University of Stuttgart. The second author thanks the European Commission for financial support received through Marie Curie Fellowships.

References

1. Carr, J.: Applications of Centre Manifold Theory. Applied Mathematical Sciences, vol. 35. Springer, Berlin (1981)
2. De Marchi, S., Schaback, R., Wendland, H.: Near-optimal data-independent point locations for radial basis function interpolation. Adv. Comput. Math. **23**(3), 317–330 (2005)
3. Henry, D.: Geometric Theory of Semilinear Parabolic Equations. Lecture Notes in Mathematics, vol. 840. Springer, Berlin (1981)
4. Kelley, A.: The stable, center-stable, center, center-unstable, unstable manifolds. J. Differ. Equ. **3**, 546–570 (1967)
5. Micchelli, C.A., Pontil, M.: On learning vector-valued functions. Neural Comput. **17**(1), 177–204 (2005)
6. Pliss, V.A.: A reduction principle in the theory of stability of motion. Izv. Akad. Nauk SSSR Ser. Mat. **28**, 1297–1324 (1964)
7. Santin, G., Haasdonk, B.: Convergence rate of the data-independent P-greedy algorithm in kernel-based approximation. Dolomites Res. Notes Approx. **10**, 68–78 (2017)
8. Shoshitaishvili, A.N.: Bifurcations of topological type of singular points of vector fields that depend on parameters. Funkcional. Anal. i Priložen. **6**(2), 97–98 (1972)
9. Wendland, H.: Scattered Data Approximation. Cambridge Monographs on Applied and Computational Mathematics, vol. 17. Cambridge University Press, Cambridge (2005)
10. Wittwar, D., Santin, G., Haasdonk, B.: Interpolation with uncoupled separable matrix-valued kernels. Dolomites Res. Notes Approx. **11**, 23–29 (2018)
11. Wittwar, D., Santin, G., Haasdonk, B.: Weighted regularized interpolation with matrix valued kernels. Technical Report, University of Stuttgart, (2019, in preparation)

An Adaptive Error Inhibiting Block One-Step Method for Ordinary Differential Equations

Jiaxi Gu and Jae-Hun Jung

1 Introduction

General linear methods have been extensively studied for solving ODEs. Among the large family of general linear methods the diagonally implicit multistage integration methods (DIMSIMs) in [1] are the special cases, which exhibit considerable potential for efficient implementation, providing the global error of the same order as the local truncation error. In [2], it was demonstrated that finite difference methods for PDEs can be constructed such that their convergence rates, or the order of their global errors, are higher than the order of the truncation errors. Following this idea, Ditkowski and Gottlieb devised the error inhibiting strategy in [3] by inhibiting the lowest order term in the truncation error from accumulating over time and thus showed that the global error of the scheme is one order higher than the local truncation error. The form of the error inhibiting scheme is inspired by the work of [7], where a block of s new step values is obtained at each step. The key idea of this method is to construct a coefficient matrix that has the null space where the local truncation error resides.

In this work, we further improved the original error inhibiting method by introducing an additional free parameter used in the radial basis function

J. Gu
Department of Mathematics, University at Buffalo, The State University of New York, Buffalo, NY, USA
e-mail: jiaxigu@buffalo.edu

J.-H. Jung (✉)
Department of Mathematics, University at Buffalo, The State University of New York, Buffalo, NY, USA

Department of Data Science, Ajou University, Suwon, South Korea
e-mail: jaehun@buffalo.edu

© The Author(s) 2020
S. J. Sherwin et al. (eds.), *Spectral and High Order Methods for Partial Differential Equations ICOSAHOM 2018*, Lecture Notes in Computational Science and Engineering 134, https://doi.org/10.1007/978-3-030-39647-3_7

(RBF) approximations. The main idea of the proposed method is to adopt the free parameter in the reconstruction of the error inhibiting method and to control it for further possible error cancellations. This results in a higher order of convergence than the original method. One advantage is that the proposed method does not need any additional conditions, so it is efficient to implement.

The next section will review the explicit error inhibiting block one-step method. In Sect. 3, we will explain the RBF interpolation. In Sect. 4, we show how the new method can be derived followed by Sect. 5 where numerical results are provided verifying that the convergence rate of the proposed method is increased by one order. A brief conclusion and an outline of our future research are presented in Sect. 6.

2 Error Inhibiting Block One-Step Method

Consider the initial value problem for the first-order ODE below

$$u'(t) = f(t, u(t)), \; t \geqslant a$$
$$u(a) = u_a \tag{1}$$

where we assume $f(t, u)$ is uniformly Lipschitz continuous in u and continuous in t. We choose a value h for the step size and set $t_n = a + nh$ a discrete sequence in the time domain. Denote the numerical approximation of the solution $u(t_n)$ by v_n.

Define the solution vector U_n by

$$U_n = \left[u_{n+\frac{s-1}{s}}, \cdots, u_{n+\frac{1}{s}}, u_n \right]^T,$$

where $u_{n+\frac{j}{s}} = u(t_{n+jh/s})$ is the exact solution at $t = t_n + \frac{jh}{s}$ for $j = 0, \cdots, s - 1$. The corresponding approximation vector V_n is defined as

$$V_n = \left[v_{n+\frac{s-1}{s}}, \cdots, v_{n+\frac{1}{s}}, v_n \right]^T.$$

In [3], the scheme is formulated as

$$V_{n+1} = Q V_n \tag{2}$$

where the operator Q is represented by the following

$$Q = A + hBf$$

and $A, B \in \mathbb{R}^{s \times s}$. There are 4 sufficient conditions imposed on the matrices A and B in order to be error inhibiting:

1. **rank**$(A) = 1$.
2. The only non-zero eigenvalue of A is 1 and its corresponding eigenvector is

$$[1, \cdots, 1]^T.$$

3. A can be diagonalized.
4. The matrices A and B are constructed such that when the local truncation error is multiplied by the discrete solution operator, we have

$$||Q\tau_v|| \leqslant O(h) \cdot ||\tau_v||.$$

This is accomplished by requiring that the leading order term of the local truncation error is in the eigenspace of A associated with the zero eigenvalue.

We derive those matrices of A and B with symbolic computation. As an example of the derivation of the error inhibiting method, we consider the construction of the scheme with $s = 2$. The solution vector is then

$$U_n = [u_{n+1/2}, u_n]^T,$$

and the corresponding approximation vector is given by

$$V_n = [v_{n+1/2}, v_n]^T.$$

In order to satisfy those conditions listed above we first select

$$A = \begin{bmatrix} 1 - v & v \\ 1 - v & v \end{bmatrix}, \tag{3}$$

which can be diagonalized as

$$A = \begin{bmatrix} 1 - v & v \\ 1 - v & v \end{bmatrix} = \begin{bmatrix} v - 1 & v \\ v - 1 & v - 1 \end{bmatrix} \begin{bmatrix} 1 & 0 \\ 0 & 0 \end{bmatrix} \begin{bmatrix} -1 & \frac{v}{v-1} \\ 1 & -1 \end{bmatrix}. \tag{4}$$

Then conditions 1, 2 and 3 are satisfied. Further suppose that

$$B = \begin{bmatrix} b_{11} & b_{12} \\ b_{21} & b_{22} \end{bmatrix}. \tag{5}$$

Then

$$V_{n+1} = \begin{bmatrix} 1-\upsilon & \upsilon \\ 1-\upsilon & \upsilon \end{bmatrix} V_n + h \begin{bmatrix} b_{11} & b_{12} \\ b_{21} & b_{22} \end{bmatrix} \begin{bmatrix} f_{n+1/2} \\ f_n \end{bmatrix} \tag{6}$$

where $f_{n+1/2} = f(t_{n+1/2}, \upsilon_{n+1/2})$ and $f_n = f(t_n, \upsilon_n)$. The components of V_{n+1} are

$$\upsilon_{n+3/2} = (1-\upsilon)\upsilon_{n+1/2} + \upsilon\upsilon_n + h(b_{11} f_{n+1/2} + b_{12} f_n),$$

$$\upsilon_{n+1} = (1-\upsilon)\upsilon_{n+1/2} + \upsilon\upsilon_n + h(b_{21} f_{n+1/2} + b_{22} f_n).$$

We write each difference equation in the form of error normalized by the step size and then insert the exact solutions to the ODE into the difference equation. Expanding $u_{n+3/2}$, u_{n+1} and $u_{n+1/2}$ around $t = t_n$ in Taylor series gives the local truncation error

$$\tau_n = (\tau_{n+1/2}, \tau_n)^T,$$

where

$$\tau_{n+1/2} = \frac{1}{2}(2 - 2b_{11} - 2b_{12} + \upsilon)u_n' + \frac{1}{8}(8 - 4b_{11} + \upsilon)u_n'' h$$

$$+ \frac{1}{48}(26 - 6b_{11} + \upsilon)u_n^{(3)} h^2 + O(h^3), \tag{7}$$

$$\tau_n = \frac{1}{2}(1 - 2b_{21} - 2b_{22} + \upsilon)u_n' + \frac{1}{8}(3 - 4b_{21} + \upsilon)u_n'' h$$

$$+ \frac{1}{48}(7 - 6b_{21} + \upsilon)u_n^{(3)} h^2 + O(h^3). \tag{8}$$

Vanishing the coefficients of the constant term and the term h in (7) and (8), and equating the quotient of the coefficient of the terms h^2 in (7) and (8) to $\frac{\upsilon}{\upsilon-1}$, the condition 4 is satisfied.

Finally we have the desired scheme as in [3]

$$V_{n+1} = \frac{1}{6} \begin{bmatrix} -1 & 7 \\ -1 & 7 \end{bmatrix} V_n + \frac{h}{24} \begin{bmatrix} 55 & -17 \\ 25 & 1 \end{bmatrix} \begin{bmatrix} f_{n+1/2} \\ f_n \end{bmatrix}, \tag{9}$$

and correspondingly the local truncation error is 2nd order convergent as expected

$$\tau_n = \frac{23}{576} \begin{bmatrix} 7 \\ 1 \end{bmatrix} u_n^{(3)} h^2 + O(h^3). \tag{10}$$

3 RBF Interpolation

Now we briefly explain the RBF interpolation in one dimension. Suppose that for a domain $\Omega \subset \mathbb{R}$, a data set $\{(x_i, u_i)\}_{i=0}^N$ is given where u_i is the value of the unknown function $u(x)$ at $x = x_i \in \Omega$. We use the RBFs $\phi : \Omega \to \mathbb{R}$ defined by $\phi_i(x) = \phi(|x - x_i|, \epsilon_i)$, where $|x - x_i|$ is the distance between x and x_i and ϵ_i is a free parameter. The reconstruction of a function, $u(x)$, is then made by a linear combination of RBFs

$$I_N^R u(x) = \sum_{i=0}^N \lambda_i \phi(|x - x_i|, \epsilon_i), \tag{11}$$

where λ_i are the expansion coefficients to be determined. Using the interpolation condition $I_N^R u(x_i) = u_i$, $i = 0, \cdots, N$, we could find the expansion coefficients λ_i by solving the linear system

$$\begin{bmatrix} \phi(|x_0 - x_0|, \epsilon_0) & \phi(|x_0 - x_1|, \epsilon_1) & \cdots & \phi(|x_0 - x_N|, \epsilon_N) \\ \phi(|x_1 - x_0|, \epsilon_0) & \phi(|x_1 - x_1|, \epsilon_1) & \cdots & \phi(|x_1 - x_N|, \epsilon_N) \\ \vdots & \vdots & & \vdots \\ \phi(|x_N - x_0|, \epsilon_0) & \phi(|x_N - x_1|, \epsilon_1) & \cdots & \phi(|x_N - x_N|, \epsilon_N) \end{bmatrix} \cdot \begin{bmatrix} \lambda_0 \\ \lambda_1 \\ \vdots \\ \lambda_N \end{bmatrix} = \begin{bmatrix} u_0 \\ u_1 \\ \vdots \\ u_N \end{bmatrix}. \tag{12}$$

If we choose the multiquadric RBF with all the free parameters equal, then the interpolation matrix, A, becomes a symmetric matrix with all diagonal entries 1,

$$A = \begin{bmatrix} 1 & \sqrt{1 + \epsilon^2(x_0 - x_1)^2} & \cdots & \sqrt{1 + \epsilon^2(x_0 - x_N)^2} \\ \sqrt{1 + \epsilon^2(x_1 - x_0)^2} & 1 & \cdots & \sqrt{1 + \epsilon^2(x_1 - x_N)^2} \\ \vdots & \vdots & & \vdots \\ \sqrt{1 + \epsilon^2(x_N - x_0)^2} & \sqrt{1 + \epsilon^2(x_N - x_1)^2} & \cdots & 1 \end{bmatrix}. \tag{13}$$

Consider the case of three equally spaced nodes x_0, x_1, x_2 with $x_0 < x_1 < x_2$. Let h be the grid spacing. Then the linear system becomes

$$\begin{bmatrix} 1 & \sqrt{1 + \epsilon^2 h^2} & \sqrt{1 + 4\epsilon^2 h^2} \\ \sqrt{1 + \epsilon^2 h^2} & 1 & \sqrt{1 + \epsilon^2 h^2} \\ \sqrt{1 + 4\epsilon^2 h^2} & \sqrt{1 + \epsilon^2 h^2} & 1 \end{bmatrix} \cdot \begin{bmatrix} \lambda_0 \\ \lambda_1 \\ \lambda_2 \end{bmatrix} = \begin{bmatrix} u_0 \\ u_1 \\ u_2 \end{bmatrix}. \tag{14}$$

By the closed-form expression for the RBF interpolant in [4],

$$I_2^R u(x) = \sum_{i=0}^{2} \frac{u_i}{\det(A)} \det(A_i(x)). \tag{15}$$

where $A_i(x)$, a 3×3 matrix, is obtained by replacing the ith row of A with the row vector

$$\left[\sqrt{1 + \epsilon^2 (x - x_0)^2} \quad \sqrt{1 + \epsilon^2 (x - x_1)^2} \quad \sqrt{1 + \epsilon^2 (x - x_2)^2} \right].$$

Differentiating the interpolant, we obtain the first-order derivative

$$\frac{d}{dx} I_2^R u(x) = \sum_{i=0}^{2} \frac{u_i}{\det(A)} \cdot \frac{d}{dx} \det(A_i(x)). \tag{16}$$

We then estimate the derivative of u at $x = x_1$ as we do in polynomial interpolation for the central difference formula:

$$\frac{d}{dx} I_2^R u(x_1) = \frac{\sqrt{1 + 4\epsilon^2 h^2} + 1}{4h\sqrt{1 + \epsilon^2 h^2}} (u_2 - u_0). \tag{17}$$

By employing the Taylor expansion of the quotient on the right-hand side of (17), we have

$$\frac{d}{dx} I_2^R u(x_1) = \left[\frac{1}{2h} + \epsilon^2 \frac{h}{4} + O(h^3) \right] (u_2 - u_0). \tag{18}$$

The main feature of the RBF method is that it contains a free parameter, ϵ, which we could make use of to further inhibit the errors. In the following section, we will show that using the parameter ϵ coupled with h^p terms, where $p \geqslant 2$, we can increase the order of local truncation error and further promote the order of global error by adopting the error inhibiting scheme.

4   Construction of the Adaptive Error Inhibiting Scheme

Following the main feature of the RBF method explained in the preceding section, we try to establish a similar explicit block one-step scheme that provides a higher order of convergence by adding one more block of the free parameters ϵ_1 and ϵ_2

coupled with h^p term. With $p = 3$, we have

$$V_{n+1} = \begin{bmatrix} 1 - \upsilon & \upsilon \\ 1 - \upsilon & \upsilon \end{bmatrix} V_n + h \begin{bmatrix} b_{11} & b_{12} \\ b_{21} & b_{22} \end{bmatrix} \begin{bmatrix} f_{n+1/2} \\ f_n \end{bmatrix} + h^3 \begin{bmatrix} 0 & \epsilon_1 \\ 0 & \epsilon_2 \end{bmatrix} \begin{bmatrix} f_{n+1/2} \\ f_n \end{bmatrix}. \tag{19}$$

We measure the one-step error normalized by the step size as in Sect. 2. Expanding $u_{n+3/2}$, u_{n+1} and $u_{n+1/2}$ around $t = t_n$ in Taylor series again yields the local truncation error

$$\tau_n = [\tau_{n+1/2}, \tau_n]^T,$$

where

$$\tau_{n+1/2} = \frac{1}{2}(2 - 2b_{11} - 2b_{12} + \upsilon)u_n' + \frac{1}{8}(8 - 4b_{11} + \upsilon)u_n''h +$$

$$\left(-\epsilon_1 u_n' + \frac{1}{48}(26 - 6b_{11} + \upsilon)u_n^{(3)} \right) h^2 + \frac{1}{384}(80 - 8b_{11} + \upsilon)u_n^{(4)}h^3 + O(h^4), \tag{20}$$

$$\tau_n = \frac{1}{2}(1 - 2b_{21} - 2b_{22} + \upsilon)u_n' + \frac{1}{8}(3 - 4b_{21} + \upsilon)u_n''h +$$

$$\left(-\epsilon_2 u_n' + \frac{1}{48}(7 - 6b_{21} + \upsilon)u_n^{(3)} \right) h^2 + \frac{1}{384}(15 - 8b_{21} + \upsilon)u_n^{(4)}h^3 + O(h^4). \tag{21}$$

Annihilating the first two terms in (20) and (21), and equating the quotient of the coefficient of the terms h^3 in (20) and (21) to $\frac{\upsilon}{\upsilon-1}$, we have the scheme

$$V_{n+1} = \frac{1}{7} \begin{bmatrix} -1 & 8 \\ -1 & 8 \end{bmatrix} V_n + \frac{h}{28} \begin{bmatrix} 64 & -20 \\ 29 & 1 \end{bmatrix} \begin{bmatrix} f_{n+1/2} \\ f_n \end{bmatrix} + h^3 \begin{bmatrix} 0 & \epsilon_1 \\ 0 & \epsilon_2 \end{bmatrix} \begin{bmatrix} f_{n+1/2} \\ f_n \end{bmatrix}. \tag{22}$$

We can easily check that the scheme (22) satisfies those four conditions in Sect. 2. Further annihilating the coefficients of the term h^2, we get the optimal values of ϵ_1 and ϵ_2:

$$\epsilon_1 = \frac{47u_n^{(3)}}{168u_n'}, \tag{23}$$

$$\epsilon_2 = \frac{9u_n^{(3)}}{224u_n'}. \tag{24}$$

Our new scheme has the truncation error

$$\tau_n = \frac{55}{2688} \begin{bmatrix} 8 \\ 1 \end{bmatrix} u_n^{(4)} h^3 + O(h^4). \tag{25}$$

Note that in our new scheme, we need the value of $u_n^{(3)}$ at each step. This higher order derivative can be computed by repeated differentiation of the function f on the right-hand side of (1) twice. However, we choose to estimate the third-order derivative. For u_n', we use the given condition from (1), i.e. $u'(t) = f(t, u(t))$. For the third-order derivative $u_n^{(3)}$, we employ the second-order central difference formula for $f''(t, u(t))$ at $t = t_n$ as

$$u_n^{(3)} = f''(t_n, u_n) \approx \frac{4(f_{n+1/2} + 2f_n - f_{n-1/2})}{h^2}, \tag{26}$$

where $f_{n+1/2}$, f_n and $f_{n-1/2}$ are given values. For this computation, no additional conditions are necessary. The truncation error is still third order accurate, $O(h^3)$, as in (25), so by the error inhibiting strategy we end up with a global error that is $O(h^4)$, which will soon be confirmed in the following section.

We conclude this section with a comparison of three methods. For DIMSIM of type 3,

$$\begin{bmatrix} v_{n+2} \\ v_{n+1} \end{bmatrix} = \frac{1}{4} \begin{bmatrix} 7 & -3 \\ 7 & -3 \end{bmatrix} \begin{bmatrix} v_{n+1} \\ v_n \end{bmatrix} + \frac{h}{8} \begin{bmatrix} 9 & -7 \\ -3 & -3 \end{bmatrix} \begin{bmatrix} f_{n+1} \\ f_n \end{bmatrix},$$

two steps v_n and v_{n+1} are employed to update the step v_{n+1} and obtain the step v_{n+2}.
For error inhibiting scheme,

$$\begin{bmatrix} v_{n+3/2} \\ v_{n+1} \end{bmatrix} = \frac{1}{6} \begin{bmatrix} -1 & 7 \\ -1 & 7 \end{bmatrix} \begin{bmatrix} v_{n+1/2} \\ v_n \end{bmatrix} + \frac{h}{24} \begin{bmatrix} 55 & -17 \\ 25 & 1 \end{bmatrix} \begin{bmatrix} f_{n+1/2} \\ f_n \end{bmatrix},$$

two steps v_n and $v_{n+1/2}$ are involved to generate the next two steps v_{n+1} and $v_{n+3/2}$. For our method (if we utilize (26) and substitute (23), (24) for respective ϵ_1 and ϵ_2 in (22) to avoid the zero denominator),

$$\begin{bmatrix} v_{n+3/2} \\ v_{n+1} \\ v_{n+1/2} \end{bmatrix} = \frac{1}{7} \begin{bmatrix} -1 & 8 & 0 \\ -1 & 8 & 0 \\ 1 & 0 & 0 \end{bmatrix} \begin{bmatrix} v_{n+1/2} \\ v_n \\ v_{n-1/2} \end{bmatrix} + \frac{h}{168} \begin{bmatrix} 572 & -496 & 188 \\ 201 & -48 & 27 \\ 0 & 0 & 0 \end{bmatrix} \begin{bmatrix} f_{n+1/2} \\ f_n \\ f_{n-1/2} \end{bmatrix},$$

we use previous three steps $v_{n-1/2}$, v_n and $v_{n+1/2}$ to evolve the next two steps v_{n+1} and $v_{n+3/2}$. In [5] the stability analysis has been done for the adaptive radial basis function methods for IVPs and it has been shown that some adaptive methods have a better stability condition than the original ones. However, it seems that the adaptive error inhibiting method is more computationally expensive than the original one when the approximation of (26) is used.

5 Numerical Results

We start with the nonlinear first-order differential equation used in [3]

$$u' = -u^2, \ t \geqslant 0 \tag{27}$$
$$u(0) = 1.$$

The exact solution of the example is $u(t) = 1/(t+1)$. The left figure of Fig. 1 shows the global errors at the time $t = 1$ versus N, the number of steps, in logarithmic scale for the type-3 DIMSIM (blue), the original error inhibiting scheme (red) and our proposed method (green). As seen in the figure, our proposed method is the most accurate among those three methods and yields high order convergence which is 4th order. Table 1 shows the convergence with N for (27). The type-3 DIMSIM yields the 2nd order accuracy, the original error inhibiting scheme yields the 3rd order accuracy and our proposed method yields the 4th order accuracy.

Next we consider the following problem used in [6] where the solution changes rapidly between $[-2, 2]$

$$u' = -4t^3 u^2, \ t \geqslant -10 \tag{28}$$
$$u(-10) = 1/10001.$$

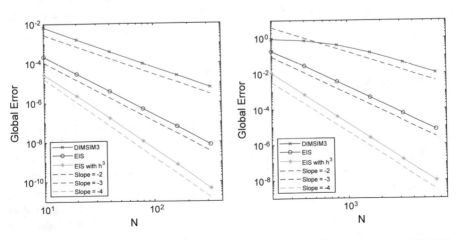

Fig. 1 Global error versus N in logarithmic scale. Left: (27). Right: (28). Blue: DIMSIM (DIMSIM3) 2nd order. Red: error inhibiting scheme (EIS) 3rd order. Green: our proposed method (EIS with h^3) 4th order

Table 1 Global error and order of convergence for $u' = -u^2$ with $u(0) = 1$

Method	N	Global error	order
DIMSIM type-3	10	6.60E−3	
	20	1.60E−3	2.0702
	40	3.82E−4	2.0402
	80	9.41E−5	2.0208
	160	2.34E−5	2.0105
	320	5.82E−6	2.0053
Error inhibiting scheme	10	2.17E−4	
	20	2.89E−5	2.9118
	40	3.73E−6	2.9536
	80	4.74E−7	2.9763
	160	5.97E−8	2.9880
	320	7.50E−9	2.9940
Error inhibiting scheme with h^3 term	10	2.71E−5	
	20	2.24E−6	3.5935
	40	1.64E−7	3.7781
	80	1.11E−8	3.8833
	160	7.22E−10	3.9400
	320	4.61E−11	3.9698

The exact solution is $u(t) = 1/(t^4 + 1)$. The right figure of Fig. 1 shows the global errors at $t = 0$ versus N in logarithmic scale for the type-3 DIMSIM (blue), the original error inhibiting method (red) and our proposed method (green). We verify again that our proposed method is indeed the most accurate and yields the highest order of convergence. Table 2 shows the convergence with N for (28). Although the type-3 DIMSIM does not reveal the 2nd order accuracy in the beginning, it eventually exhibits the order of accuracy as expected. The original error inhibiting scheme is 3rd order accurate and our proposed method 4th order accurate.

6 Conclusions

In this note, we modified and improved the original error inhibiting block one-step method proposed in [3] by introducing a free parameter. By exploiting the parameter, the local truncation error is further reduced resulting in higher order of the global error. It is numerically demonstrated that, with the proposed method, the local truncation error is of the 3rd order and the global error of the 4th order. As mentioned in Sect. 4, we will investigate the stability of the error inhibiting method and our proposed method as well as relaxing the fourth constraint in error inhibiting method in our future research.

Table 2 Global error and order of convergence for $u' = -4t^3u^2$ with $u(-10) = 1/10001$

Method	N	Global error	Order
DIMSIM type-3	200	9.05E−1	
	400	7.24E−1	0.3221
	800	4.07E−1	0.8293
	1600	1.49E−1	1.4476
	3200	4.24E−2	1.8158
	6400	1.10E−2	1.9475
Error inhibiting scheme	200	1.86E−1	
	400	2.80E−2	2.7294
	800	3.60E−3	2.9639
	1600	4.50E−4	2.9965
	3200	5.63E−5	3.0002
	6400	7.03E−6	3.0005
Error inhibiting scheme with h^3 term	200	1.14E−2	
	400	6.57E−4	4.1132
	800	3.94E−5	4.0620
	1600	2.41E−6	4.0307
	3200	1.49E−7	4.0123
	6400	9.91E−9	3.9122

Acknowledgements The authors thank Adi Ditkowski for introducing the error inhibiting method to us and communicating with us on the subject. The research is partially supported by Ajou University.

References

1. Butcher, J.C.: Diagonally-implicit multi-stage integration method. Appl. Numer. Math. **11**, 347–363 (1993)
2. Ditkowski, A.: High order finite difference schemes for the heat equation whose convergence rates are higher than their truncation errors. In: R.M. Kirby et al. (eds.), Spectral and High Order Methods for Partial Differential Equations ICOSAHOM 2014, pp. 167–178. Springer, Switzerland (2015)
3. Ditkowski, A., Gottlieb, S.: Error inhibiting block one-step schemes for ordinary differential equations. J. Sci. Comput. **73**, 691–711 (2017)
4. Fornberg, B., Weight, G., Larsson, E.: Some observations regarding interpolants in the limit of flat radial basis functions. Comput. Math. Appl. **47**, 37–55 (2004)
5. Gu, J. Jung, J.-H.: Adaptive radial basis function methods for initial value problems. J. Sci. Comput. **82**, 47 (2020). https://doi.org/10.1007/s10915-020-01140-0
6. Sauer, T.: Numerical Analysis, 2nd edn. Pearson, New York (2012)
7. Shampine, L.F., Watts, H.A.: Block implicit one-step methods. Math. Comp. **23**, 731–740 (1969)

Hermite Methods in Time

Rujie Gu and Thomas Hagstrom

1 Introduction

Over the past decade a number of works have appeared which exploit the unique properties of Hermite-Birkhoff interpolation in space to construct arbitrary-order discretization methods for hyperbolic [1, 2, 4–7, 10, 14, 16–19] as well as Schrödinger [3] equations. The precise form of the interpolant in a single cell, which here we write in one dimension labelled t, is

$$u(t) \approx \mathscr{I}u(t) \in \Pi^{2m+1}, \quad t \in (t_{j-1}, t_j), \tag{1}$$

$$\frac{d^k}{dt^k} \mathscr{I}u(t_\ell) = \frac{d^k u}{dt^k}(t_\ell); \quad k = 0, \ldots, m, \quad \ell = j-1, j, \tag{2}$$

where Π^{2m+1} denotes the polynomials of degree $2m + 1$. (In higher dimensions one uses a tensor-product cell interpolant based on vertex data consisting of mixed derivatives of order through m in each Cartesian coordinate.)

In contrast, there has been little work on analogous methods for time discretization. A recent exception is the manuscript by Liu et al. [15]. They develop methods for second-order semilinear hyperbolic equations using interpolants of the form (1)–(2) combined with a reformulation of the evolution problem using exact solutions of the linear part. They demonstrate excellent long-time performance.

The outline of the paper is as follows. In Sect. 2 we list a few properties of piecewise Hermite-Birkhoff interpolation. In Sect. 3 we construct the time-stepping

R. Gu · T. Hagstrom (✉)
Southern Methodist University, Dallas, TX, USA
e-mail: rujieg@smu.edu; thagstrom@smu.edu

© The Author(s) 2020
S. J. Sherwin et al. (eds.), *Spectral and High Order Methods for Partial Differential Equations ICOSAHOM 2018*, Lecture Notes in Computational Science and Engineering 134, https://doi.org/10.1007/978-3-030-39647-3_8

schemes and establish some basic results, with a few numerical experiments described in Sect. 4.

2 Basic Properties of Hermite-Birkhoff Interpolation

Hermite interpolants have a number of interesting properties which make them very attractive for the solution of differential equations; see, e.g., [2]. Here we will mainly use the simplest. Precisely, for $t \in (t_{j-1}, t_j)$, the Peano representation of the local error can be easily derived by noting that $e = u - \mathcal{I}u$ solves the two point boundary value problem

$$\frac{d^{2m+2}e}{dt^{2m+2}} = \frac{d^{2m+2}u}{dt^{2m+2}}, \quad \frac{d^k e}{dt^k} = 0, \quad t = t_{j-1}, t_j, \quad k = 0, \ldots m. \tag{3}$$

Thus

$$e(t) = \int_{t_{j-1}}^{t_j} K_j(t, s) \frac{d^{2m+2}u}{dt^{2m+2}}(s)ds, \tag{4}$$

where the kernel K_j is the Green's function for (3). Simple scaling arguments combined with the transformation $t = t_{j-1} + zh_j$ then show that $e = O(h_j^{2m+2})$ where $h_j = t_j - t_{j-1}$ is the time step. A fundamental feature of piecewise Hermite interpolation is the following orthogonality property. For **any** functions $v(t), w(t)$

$$\int_{t_{j-1}}^{t_j} \frac{d^{m+1}\mathcal{I}v}{dt^{m+1}}(t) \cdot \frac{d^{m+1}(w - \mathcal{I}w)}{dt^{m+1}}(t)dt = 0, \tag{5}$$

which in particular implies that interpolation reduces the H^{m+1} seminorm.

3 Time-Stepping Methods

We begin by considering the initial value problem for a first-order system ordinary differential equations:

$$\frac{du}{dt} = f(u, t), \quad u(t_0) = u_0, \quad u(t) \in \mathbb{R}^d. \tag{6}$$

Given a discrete time sequence $t_j > t_{j-1}$, $j = 1, \ldots, N$, with time steps $h_j = t_j - t_{j-1}$ we write down the Picard integral formulation of the time evolution over a single step

$$u(t_j) = u(t_{j-1}) + \int_{t_{j-1}}^{t_j} f(u(s), s)ds. \qquad (7)$$

The construction of our time integration formula proceeds in three steps. We denote by v_j the approximation to $u(t_j)$.

1. Given v_{j-1} and assuming for the moment that v_j is known, use the differential equation to compute m scaled derivatives of its solution, $V_\ell(t)$, satisfying $V_\ell(t_\ell) = v_\ell$, $\ell = j, j-1$. Setting

$$F_\ell(t) = f(V_\ell(t), t), \qquad (8)$$

these are recursively defined by the formula

$$\frac{d^k V_\ell}{dt^k}(t_\ell) = \frac{d^{k-1} F_\ell}{dt^{k-1}}(t_\ell), \quad k = 0, \ldots m. \qquad (9)$$

2. Construct the Hermite-Birkhoff interpolant of this data; that is the polynomial $P_{j-1/2}(t; v_{j-1}, v_j)$ of degree $2m + 1$, satisfying

$$\frac{d^k P_{j-1/2}}{dt^k}(t_\ell; v_{j-1}, v_j) = \frac{d^k V_\ell}{dt^k}(t_\ell); \quad \ell = j - 1, j, \quad k = 0, \ldots, m.. \qquad (10)$$

3. Approximate (7) by replacing $u(t_\ell)$ by v_ℓ and replacing the integral by a $q + 1$-point quadrature rule with f evaluated at the Hermite interpolant:

$$v_j = v_{j-1} + h_j \sum_{k=0}^{q} w_k f(P_{j-1/2}(t_{j,k}; v_{j-1}, v_j)). \qquad (11)$$

4. Solve (11) for v_j. Note that this is a system of d nonlinear equations for any m; that is, unlike standard implicit Runge–Kutta methods, the size of the nonlinear system is independent of the order.

We remark that we have not studied in detail the unique solvability of (11) in the stiff case. In our numerical experiments we used the solution at the current time step as an initial approximation for Newton iterations and simply accepted the solution to which the iterates converged.

To emphasize the ideas we write down some specific examples of methods with $m = 1$ and $m = 2$ making the simplifying assumption of autonomy; that is $f = f(u)$. The derivation of methods of arbitrary order is straightforward and the formulas can be trivially obtained using software capable of symbolic computations. To apply them at higher order one must evaluate higher derivatives of f, which is also possible using automatic differentiation tools [11].

Example (m = 1) Set

$$\tau = \frac{t - t_{j-1}}{h_j}.$$

Now the interpolant $P_{j-1/2}(t; v_{j-1}, v_j)$ is given by:

$$P_{j-1/2}(t; v_{j-1}, v_j) = \sum_{k=0}^{3} a_k \tau^k, \tag{12}$$

where

$$a_0 = v_{j-1}, \quad a_1 = h_j f(v_{j-1}),$$

$$a_2 = 3\left(v_j - v_{j-1}\right) - h_j \left(2f(v_{j-1}) + f(v_j)\right), \tag{13}$$

$$a_3 = -2\left(v_j - v_{j-1}\right) + h_j \left(f(v_{j-1}) + f(v_j)\right).$$

We next introduce a quadrature rule which is exact for polynomials of degree 3. Possible choices include the 2-point Gauss-Legendre (14) rules, or the 3-point Gauss-Radau (15) or Gauss-Lobatto rules. Note that by using two different rules we obtain a possible error indicator. Here are the two different methods used below. Note that the methods are identical if f is linear.

$$v_j = v_{j-1} + \frac{h_j}{2} \left(f\left(P_{j-1/2}(\alpha_-; v_{j-1}, v_j)\right) + f\left(P_{j-1/2}(\alpha_+; v_{j-1}, v_j)\right) \right), \tag{14}$$

$$v_j = v_{j-1} + \frac{h_j}{36} \left(\beta_+ f\left(P_{j-1/2}(\gamma_-; v_{j-1}, v_j)\right) + \beta_- f\left(P_{j-1/2}(\gamma_+; v_{j-1}, v_j)\right) \right.$$
$$\left. + 4f\left(v_j\right) \right), \tag{15}$$

$$\alpha_\pm = \left(1 \pm \frac{1}{\sqrt{3}}\right), \quad \gamma_\pm = \frac{4 \pm \sqrt{6}}{10}, \quad \beta_\pm = 16 \pm \sqrt{6}. \tag{16}$$

A time step is executed by solving the nonlinear system, (14) or (15), for v_j.

Example (m = 2) Now we also need the second time derivative of u,

$$\frac{d^2 u}{dt^2} = \frac{d}{dt} f(u) = J(u) f(u), \tag{17}$$

where $J(u)$ is the Jacobian derivative. The Hermite interpolant can now be written:

$$P_{j-1/2}(t; v_{j-1}, v_j) = \sum_{k=0}^{5} a_k \tau^k, \tag{18}$$

where

$$a_0 = v_{j-1}, \quad a_1 = h_j f(v_{j-1}), \quad a_2 = \frac{h_j^2}{2} J(v_{j-1}) f(v_{j-1}), \tag{19}$$

$$a_3 = 10 \left(v_j - v_{j-1} \right) - h_j \left(6f(v_{j-1}) + 4f(v_j) \right) + \frac{h_j^2}{2} \left(-3J(v_{j-1}) f(v_{j-1}) + J(v_j) f(v_j) \right),$$

$$a_4 = -15 \left(v_j - v_{j-1} \right) + h_j \left(8f(v_{j-1}) + 7f(v_j) \right) + \frac{h_j^2}{2} \left(3J(v_{j-1}) f(v_{j-1}) - J(v_j) f(v_j) \right),$$

$$a_5 = 6 \left(v_j - v_{j-1} \right) - 3h_j \left(f(v_{j-1}) + f(v_i) \right) + \frac{h_j^2}{2} \left(-J(v_{j-1}) f(v_{j-1}) + J(v_j) f(v_j) \right).$$

Again we can now use, for example, the 3-point Gauss-Legendre or 4-point Gauss-Radau quadrature rules to produce the equation we must solve for v_j.

3.1 Stability and Consistency

The consistency of the method is a straightforward consequence of its construction, and its linear stability properties can also be established.

Theorem 1 *Assume that the quadrature rule has positive weights and is exact for polynomials of degree $2m + 1$. Then the implicit Hermite method is A-stable and accurate of order $2m + 2$.*

Proof Assume that f is smooth and that $u(t) \in C^{2m+2}(0, T)$. Using (4), standard estimates for quadrature errors, and the Picard formula (7) we find for the truncation error

$$\tau_j = \frac{u(t_j) - u(t_{j-1})}{h_j} - \sum_k w_k f(P_{j-1/2}(t_{j,k}; u(t_{j-1}), u(t_j))), \tag{20}$$

$$|\tau_j| = \frac{1}{h_j} \left| \int_{t_{j-1}}^{t_j} f(u(s)) ds - h_j \sum_k w_k f(P_{j-1/2}(t_{j,k}; u(t_{j-1}), u(t_j))) \right|$$

$$\leq \frac{1}{h_j} \left| \int_{t_{j-1}}^{t_j} f(u(s)) ds - h_j \sum_k w_k f(u(t_{j,k})) \right|$$

$$+ \left| \sum_k w_k \left(f(u(t_{j,k})) - f(P_{j-1/2}(t_{j,k}; u(t_{j-1}), u(t_j))) \right) \right|$$

$$\leq C h_j^{2m+2}. \tag{21}$$

Now consider the Dahlquist test problem, $f(u) = \lambda u$. In this case all quadrature rules which are exact for the Hermite interpolant produce the same method. As interpolation is linear, we have that the coefficients of the interpolant are linear combinations $h_j^k \lambda^k v_{j-1}$ and $h_j^k \lambda^k v_j$, $k = 0, \dots m$. The Picard integral then increases the powers of $h_j \lambda$ by one so that the implicit system (11) can be rearranged to:

$$Q_+(h_j\lambda)v_j = Q_-(h_j\lambda)v_{j-1} \Rightarrow v_j = \frac{Q_-(h_j\lambda)}{Q_+(h_j\lambda)} v_{j-1}, \tag{22}$$

where $Q_\pm(h_j\lambda)$ are polynomials of degree $m + 1$. Consistency implies

$$e^{h_j\lambda} = \frac{Q_-(h_j\lambda)}{Q_+(h_j\lambda)} + O\left((h_j\lambda)^{2m+3}\right). \tag{23}$$

The only rational function of the given degree with this accuracy is the diagonal Padé approximant. We thus conclude that our methods are A-stable [12].◊

4 Numerical Experiments

Our first experiments treat standard problems from the ode literature and are restricted to the fourth and sixth order methods described above with either Gauss-Legendre or Gauss-Radau quadrature. Our practical implementations employ the classical Aitken algorithm adapted to Hermite interpolation to directly evaluate $P_{j-1/2}(t_{j,k}, v_{j-1}, v_j)$ and solve (11) using Newton's method with the Jacobian of the implicit system approximated by finite differences. For adaptive computations we

1. Compute v_j using the Gauss-Radau-based formulas,
2. Compute a residual, ρ_j, by substituting v_j into the Gauss-Legendre-based formulas.

We then adjust the time step by the simple rule

$$h_{j+1} = \left(\frac{\text{tol}}{\rho_j}\right)^{1/(2m+3)} h_j, \tag{24}$$

while also imposing a minimum time step.

Our final experiment examines the use of the method for evolving spectral discretizations of initial-boundary value problems for the Schrödinger equation.

4.1 Arentsorf Orbit

We first consider the problem of computing a periodic solutions of the restricted three-body problem which we reformulate as a first-order system of four variables:

$$\frac{d^2 y_1}{dt^2} = y_1 + 2\frac{dy_2}{dt} - (1-\mu)\frac{y_1 + \mu}{\left((y_1 + \mu)^2 + y_2^2\right)^{3/2}} - \mu\frac{y_1 - (1-\mu)}{\left((y_1 - (1-\mu))^2 + y_2^2\right)^{3/2}},$$

$$\frac{d^2 y_2}{dt^2} = y_2 - 2\frac{dy_1}{dt} - (1-\mu)\frac{y_2}{\left((y_1 + \mu)^2 + y_2^2\right)^{3/2}} - \mu\frac{y_2}{\left((y_1 - (1-\mu))^2 + y_2^2\right)^{3/2}},$$

$$\mu = 0.012277471, \quad y_1(0) = .994, \quad \frac{dy_1}{dt}(0) = y_2(0) = 0,$$

$$\frac{dy_2}{dt}(0) = -2.01585106379082\ldots, \quad T = 17.06521656015796\ldots.$$

(For graphs of the solution see [13, Ch. II].)

We note that this problem is not considered to be stiff. The main difficulty is a need for very small time steps when the orbits approach the singularities of f. However, we use it to verify convergence at the design order when (woefully inefficient) uniform time steps are employed and to test the utility of our naive time step adaptivity algorithm.

Results for fixed (small) time steps are displayed in Table 1. We observe that convergence is at design order and that the results for the two quadrature formulas are comparable, though the fourth order Radau method is somewhat more accurate than Gauss-Legendre with roles reversed at sixth order. The sixth order methods are more accurate with larger time steps. The error is simply $\sqrt{(y_1(T) - y_1(0))^2 + (y_2(T) - y_2(0))^2}$.

Results for adaptive computations with $m = 2$ are shown in Table 2. Obviously, the adaptive methods lead to a very significant reduction in the number of time steps; an accuracy of 10^{-7} is achieved with 264 steps of the adaptive method

Table 1 Convergence with fixed time steps for the Arentsorf orbit problem

Gauss-Legendre $m = 1$			Gauss-Radau $m = 1$			Gauss-Legendre $m = 2$			Gauss-Radau $m = 2$		
h	Error	Rate	h	Error	Rate	h	Error	Rate	h	Error	Rate
5.69(−4)	2.30(−4)		5.69(−4)	9.65(−5)		8.53(−4)	2.98(−6)		8.53(−4)	4.46(−6)	
4.88(−4)	1.24(−4)	4.01	4.88(−4)	5.21(−5)	4.00	6.83(−4)	7.82(−7)	6.00	6.83(−4)	1.16(−6)	6.03
4.27(−4)	7.26(−5)	4.01	4.27(−4)	3.05(−5)	4.00	5.69(−4)	2.62(−7)	6.00	5.69(−4)	3.88(−7)	6.02
3.79(−4)	4.53(−5)	4.01	3.79(−4)	1.91(−5)	4.00	4.88(−4)	1.04(−7)	6.00	4.88(−4)	1.53(−7)	6.01
3.41(−4)	2.97(−5)	4.01	3.41(−4)	1.25(−5)	4.00	4.27(−4)	4.66(−8)	6.00	4.27(−4)	6.87(−8)	6.01
3.10(−4)	2.03(−5)	4.00	3.10(−4)	8.54(−6)	4.00	3.79(−4)	2.30(−8)	6.00	3.79(−4)	3.39(−8)	6.01
2.84(−4)	1.43(−5)	4.00	2.84(−4)	6.03(−6)	4.00	3.41(−4)	1.22(−8)	6.00	3.41(−4)	1.80(−8)	6.00

Table 2 Time steps and error as a function of tolerance for adaptive solutions of the Arentsorf problem with $m = 2$

Tol	Error	Steps
10^{-6}	8.21(−3)	65
10^{-8}	1.85(−5)	136
10^{-10}	1.15(−8)	264

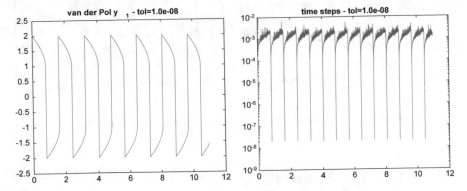

Fig. 1 Solution and time step history for the van der Pol oscillator with tolerance 10^{-8}

while 35,000 uniform steps are required. Due to the sensitivity of the problem, the global error is much larger than the error tolerance, but is reduced in proportion to it.

4.2 Van der Pol Oscillator

Our second example is the van der Pol oscillator problem, which again we rewrite as a first order system:

$$\frac{d^2 y}{dt^2} = \epsilon^{-1}\left((1 - y^2)\frac{dy}{dt} - y\right),\tag{25}$$

$$\epsilon = 10^{-6}, \quad y(0) = 2, \quad \frac{dy}{dt}(0) = 0.$$

We solve up to $T = 11$ using the adaptive method with $m = 2$. We plot the solution and the time step histories for a tolerance of 10^{-10} in Fig. 1. Note that very small steps are needed to resolve the fast transitions, while the problem is quite stiff in the regions where y is nearly constant. Plots for the other tolerances tested, 10^{-6} and 10^{-10}, are similar though the number of time steps required varies.

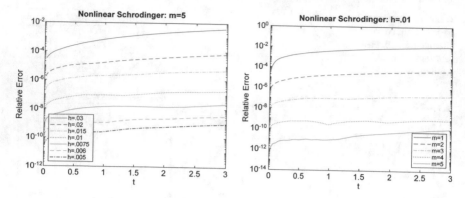

Fig. 2 Left: Relative errors for NLS with various time steps and $m = 3$. Right: Relative errors for NLS with $h = .01$ and varying m

4.3 Schrödinger Equation

Lastly, we apply the method to evolve a Fourier pseudospectral discretization of the nonlinear Schrödinger equation. Precisely we consider the real problem

$$\frac{\partial v}{\partial t} = -\frac{\partial^2 w}{\partial x^2} - \left(v^2 + w^2\right) w, \qquad \frac{\partial w}{\partial t} = \frac{\partial^2 v}{\partial x^2} + \left(v^2 + w^2\right) v, \qquad (26)$$

for $x \in (-8, 8)$, $t \in (0, 3)$ with periodic boundary conditions $\frac{\partial v}{\partial x} = \frac{\partial w}{\partial x} = 0$ at $x = \pm 1$. We approximate the periodization of the exact solitary wave solution

$$v(x, t) = \sqrt{50} \cos(rx - st) \cdot \text{sech}(5(x - ct)),$$
$$w(x, t) = \sqrt{50} \sin(rx - st) \cdot \text{sech}(5(x - ct)), \qquad (27)$$

with $c = 2\pi$, $r = \pi$, $s = \pi^2 - 25$. We note that the amplitude of the solitary wave is reduced by about 17 digits at a distance of 8 from its peak so that the interaction with periodic copies is negligible over the simulation time. We use 512 Fourier modes in the computation of the derivatives and experiments show that this is sufficient to represent the solitary wave to machine precision. The implicit system was solved using Newton iterations each time step. In Fig. 2 we present results for $m = 3$ (8th order) with varying time step and for m varying from 1 to 5 (order 4 through 12) with $h = 10^{-2}$. In both cases we observe rapid convergence. We also tabulate the errors at the final time and calculate the convergence rates when $m = 3$ in Table 3. The results are clearly consistent with the design order.

Table 3 Relative errors for the Fourier pseudospectral discretization of the NLS (26) with solitary wave solution (27)

$m = 3$			$h = 0.01$	
h	Error	Rate	m	Error
3.0(−2)	4.0(−3)		1	1.2(−2)
2.0(−2)	6.7(−5)	10.1	2	5.0(−5)
1.5(−2)	4.9(−6)	9.1	3	1.8(−7)
1.0(−2)	1.8(−7)	8.1	4	8.6(−10)
7.5(−3)	2.2(−8)	7.3	5	1.2(−10)
6.0(−3)	3.2(−9)	8.7		
5.0(−3)	9.4(−10)	6.8		

5   Conclusions and Future Work

In conclusion, we have demonstrated that Hermite-Birkhoff interpolation can be used to develop singly-implicit A-stable timestepping methods of arbitrary order. A number of possible generalizations and improvements to the method are possible. These include

1. Stability analysis for variable coefficient or nonlinear problems using the projection properties (5);
2. Improved time step/order adaptivity;
3. Preconditioning of the implicit system for applications to partial differential equations such as spectral/pseudospectral discretizations of equations of Schrödinger type (e.g. integration preconditioners [8, 9]);
4. Development of IMEX schemes combining Hermite and Taylor polynomials.

Acknowledgements This work was supported in part by NSF Grant DMS-1418871. Any opinions, findings, and conclusions or recommendations expressed in this material are those of the author and do not necessarily reflect the views of the National Science Foundation.

References

1. Appelö, D., Hagstrom, T.: On advection by Hermite methods. Pac. J. Appl. Math. **4**, 125–139 (2012)
2. Appelö, D., Hagstrom, T.: Solving PDEs with Hermite Interpolation. Lecture Notes in Computational Science, pp. 31–49. Springer, Berlin (2015)
3. Appelö, D., Kreiss, G., Wang, S.: An explicit Hermite-Taylor method for the Schrödinger equation. Commun. Comput. Phys **21**, 1207–1230 (2017)
4. Appelö, D., Hagstrom, T., Vargas, A.: Hermite methods for the scalar wave equation. SIAM J. Sci. Comput. **40**, A3902–A3927 (2018)
5. Chen, R., Hagstrom, T.: P-adaptive Hermite methods for initial value problems. ESAIM: Math. Model. Numer. Anal. **46**, 545–557 (2012)
6. Chen, X., Appelö, D., Hagstrom, T.: A hybrid Hermite-discontinuous Galerkin method for hyperbolic systems with application to Maxwell's equations. J. Comput. Phys. **257**, 501–520 (2014)

7. Chidwagyai, P., Nave, J.-C., Rosales, R., Seibold, B.: A comparative study of the efficiency of jet schemes. Int. J. Numer. Anal. Model.-B **3**, 297–306 (2012)
8. Coutsias, E., Hagstrom, T., Hesthaven, J., Torres, D.: Integration preconditioners for differential operators in spectral τ methods. Houst. J. Math. **1995**, 21–38 (1996). Special issue: ICOSAHOM
9. Coutsias, E., Hagstrom, T., Torres, D.: An efficient spectral method for ordinary differential equations with rational function coefficients. Math. Comp. **65**, 611–635 (1996)
10. Goodrich, J., Hagstrom, T., Lorenz, J.: Hermite methods for hyperbolic initial-boundary value problems. Math. Comp. **75**, 595–630 (2006)
11. Griewank, A.: Evaluating Derivatives: Principles and Techniques of Algorithmic Differentiation. SIAM, Philadelphia (2000)
12. Hairer, E., Wanner, G.: Solving Ordinary Differential Equations II, Stiff and Differential-Algebraic Problems. Springer, New York (1996)
13. Hairer, E., Norsett, S., Wanner, G.: Solving Ordinary Differential Equations I, Nonstiff Problems. Springer, New York (1992)
14. Kornelus, A., Appelö, D.: Flux-conservative Hermite methods for simulation of nonlinear conservation laws. J. Sci. Comput. **76**, 24–47 (2018)
15. Liu, C., Iserles, A., Wu, W.: Symmetric and arbitrarily high-order Birkhoff-Hermite time integrators and their long-time behaviour for solving nonlinear Klein-Gordon equations. J. Comput. Phys. **356**, 1–30 (2018)
16. Seibold, B., Rosales, R., Nave, J.-C.: Jet schemes for advection problems. Discrete Contin. Dyn. Syst. Ser. B **17**, 1229–1259 (2012)
17. Vargas, A., Chan, J., Hagstrom, T., Warburton, T.: GPU Acceleration of Hermite Methods for Simulation of Wave Propagation. Lecture Notes in Computational Science, pp. 357–368. Springer, Berlin (2017)
18. Vargas, A., Chan, J., Hagstrom, T., Warburton, T.: Variations on Hermite methods for wave propagation. Commun. Comput. Phys. **22**, 303–337 (2017)
19. Vargas, A., Hagstrom, T., Chan, J., Warburton, T.: Leapfrog time-stepping for Hermite methods. J. Sci. Comput. **80**, 289–314 (2019)

HPS Accelerated Spectral Solvers for Time Dependent Problems: Part II, Numerical Experiments

Tracy Babb, Per-Gunnar Martinsson, and Daniel Appelö

1 Introduction

In this chapter describes a highly computationally efficient solver for equations of the form

$$\kappa \frac{\partial u}{\partial t} = \mathscr{L}u(\boldsymbol{x}, t) + h(u, \boldsymbol{x}, t), \quad \boldsymbol{x} \in \Omega, t > 0, \tag{1}$$

with initial data $u(\boldsymbol{x}, 0) = u_0(\boldsymbol{x})$. Here $\mathscr{L}$ is an elliptic operator acting on a fixed domain Ω and h is lower order, possibly nonlinear terms. We take κ to be real or imaginary, allowing for parabolic and Schrödinger type equations. We desire the benefits that can be gained from an implicit solver, such as L-stability and stiff accuracy, which means that the computational bottleneck will be the solution of a sequence of elliptic equations set on Ω. In situations where the elliptic equation to be solved is the same in each time-step, it is highly advantageous to use a *direct* (as opposed to *iterative*) solver. In a direct solver, an approximate solution operator to the elliptic equation is built once. The cost to build it is typically higher than the cost required for a single elliptic solve using an iterative method such as multigrid, but the upside is that after it has been built, each subsequent solve is very fast. In this chapter, we argue that a particularly efficient direct solver to use in this context is a method obtained by combining a multidomain spectral collocation discretization (a

T. Babb · D. Appelö (✉)
University of Colorado, Boulder, CO, USA
e-mail: tracy.babb@colorado.edu; daniel.appelo@colorado.edu

P.-G. Martinsson
University of Texas, Austin, TX, USA
e-mail: pgm@ices.utexas.edu

© The Author(s) 2020
S. J. Sherwin et al. (eds.), *Spectral and High Order Methods for Partial Differential Equations ICOSAHOM 2018*, Lecture Notes in Computational Science and Engineering 134, https://doi.org/10.1007/978-3-030-39647-3_9

131

so-called "patching method", see e.g. Ch. 5.13 in [3]) with a nested dissection type solver. It has recently been demonstrated [1, 7, 12] that this combined scheme, which we refer to as a "Hierarchial Poincaré–Steklov (HPS)" solver, can be used with very high local discretization orders (up to $p = 20$ or higher) without jeopardizing either speed or stability, as compared to lower order methods.

In this chapter, we investigate the stability and accuracy that is obtained when combining high-order time-stepping schemes with the HPS method for solving elliptic equations. We restrict attention to relatively simple geometries (mostly rectangles). The method can without substantial difficulty be generalized to domains that can naturally be expressed as a union of rectangles, possibly mapped via curvilinear smooth parameter maps.

A longer version of this chapter with additional details is available at [2]. Also note that the conclusions are deferred to Part II of this paper (same issue).

2　The Hierarchical Poincaré–Steklov Method

In this section, we describe a computationally efficient and highly accurate technique for solving an elliptic PDE of the form

$$[Au](x) = g(x), \qquad x \in \Omega,$$
$$u(x) = f(x), \qquad x \in \Gamma, \tag{2}$$

where Ω is a domain with boundary Γ, and where A is a variable coefficient elliptic differential operator

$$[Au](x) = -c_{11}(x)[\partial_1^2 u](x) - 2c_{12}(x)[\partial_1 \partial_2 u](x) - c_{22}(x)[\partial_2^2 u](x)$$
$$+ c_1(x)[\partial_1 u](x) + c_2(x)[\partial_2 u](x) + c(x) u(x)$$

with smooth coefficients. In the present context, (2) represents an elliptic solve that is required in an implicit time-descretization technique of a parabolic PDE, as discussed in Sect. 1. For simplicity, let us temporarily suppose that the domain Ω is rectangular; the extension to more general domains is discussed in Remark 1.

Our ambition here is merely to provide a high level description of the method; for implementation details, we refer to [1, 2, 7–9, 12, 13].

2.1　Discretization

We split the domain Ω into $n_1 \times n_2$ boxes, each of size $h \times h$. Then on each box, we place a $p \times p$ tensor product grid of Chebyshev nodes, as shown in Fig. 1. We use collocation to discretize the PDE (2). With $\{x_i\}_{i=1}^N$ denoting the collocation points,

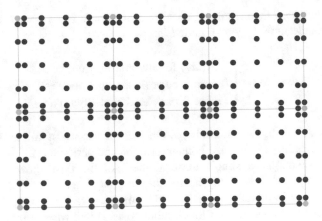

Fig. 1 The domain Ω is split into $n_1 \times n_2$ squares, each of size $h \times h$. In the figure, $n_1 = 3$ and $n_2 = 2$. Then on each box, a $p \times p$ tensor product grid of Chebyshev nodes is placed, shown for $p = 7$. At red nodes, the PDE (2) is enforced via collocation of the spectral differentiation matrix. At the blue nodes, we enforce continuity of the normal fluxes. Observe that the corner nodes (gray) are excluded from consideration

the vector u that represents our approximation to the solution u of (2) is given simply by $\mathsf{u}(i) \approx u(x_i)$. We then discretize (2) as follows:

1. For each collocation node that is *internal* to a box (red nodes in Fig. 1), we enforce (2) by directly collocating the spectral differential operator supported on the box, as described in, e.g., Trefethen [15].

2. For each collocation node on an *edge* between two boxes (blue nodes in Fig. 1), we enforce that the normal fluxes across the edge be continuous. For instance, for a node x_i on a vertical line, we enforce that $\partial u / \partial x_1$ is continuous across the edge by equating the values for $\partial u / \partial x_1$ obtained by spectral differentiation of the boxes to the left and to the right of the edge. For an edge node that lies on the external boundary Γ, simply evaluate the normal derivative at the node, as obtained by spectral differentiation in the box that holds the node.

3. All corner nodes (gray in Fig. 1) are dropped from consideration. For an elliptic operator of the form (2) with $c_{12} = 0$, it turns out that these values do not contribute to any of the spectral derivatives on the interior nodes, which means that the method without corner nodes is mathematically equivalent to the method with corner nodes, see [5, Sec. 2.1] for details. When $c_{12} \neq 0$, one must in order to drop the corner nodes include an extrapolation operator when evaluating the terms involving the spectral representation of the mixed derivative $\partial^2 u / \partial x_1 \partial x_2$. This may lead to a slight drop in the order of convergence, but the difference is hardly noticeable in practice, and the exclusion of corner nodes greatly simplifies the implementation of the method.

Since we exclude the corner nodes from consideration, the total number of nodes in the grid equals $N = (p - 2)(p\, n_1 n_2 + n_1 + n_2) \approx p^2 n_1 n_2$. The discretization procedure described then results in an $N \times N$ matrix A. For a node i, the value of $\mathsf{A}(i, :)\mathsf{u}$ depends on what type of node i is:

$$\mathsf{A}(i, :)\mathsf{u} \approx \begin{cases} [Au](x_i) \text{ for any interior (red) node,} \\ 0 \text{ for any edge node (blue) not on } \Gamma, \\ \partial u/\partial n \text{ for any edge node (blue) on } \Gamma. \end{cases}$$

This matrix A can be used to solve BVPs with a variety of different boundary conditions, including Dirichlet, Neumann, Robin, and periodic [12].

In many situations, a simple uniform mesh of the type shown in Fig. 1 is not optimal, since the regularity in the solution may vary greatly, due to corner singularities, localized loads, etc. The HPS method can easily be adapted to handle local refinement. The essential difficulty that arises is that when boxes of different sizes are joined, the collocation nodes along the joint boundary will not align. It is demonstrated in [1, 5] that this difficulty can stably and efficiently be handled by incorporating local interpolation operators.

2.2 A Hierarchical Direct Solver

A key observation in previous work on the HPS method is that the sparse linear system that results from the discretization technique described in Sect. 2.1 is particularly well suited for direct solvers, such as the well-known multifrontal solvers that compute an LU-factorization of a sparse matrix. The key is to minimize fill-in by using a so called nested dissection ordering [4, 6]. Such direct solvers are very powerful in a situation where a sequence of linear systems with the same coefficient matrix needs to be solved, since each solve is very fast once the coefficient matrix has been factorized. This is precisely the environment under consideration here. The particular advantage of combining the multidomain spectral collocation discretization described in Sect. 2.1 is that the time required for factorizing the matrix is *independent* of the local discretization order. As we will see in the numerical experiments, this enables us to attain both very high accuracy, and very high computational efficiency.

Remark 1 (General Domains) For simplicity we restrict attention to rectangular domains in this chapter. The extension to domains that can be mapped to a union of rectangles via smooth coordinate maps is relatively straight-forward, since the method can handle variable coefficient operators [12, Sec. 6.4]. Some care must be exercised since singularities may arise at intersections of parameter maps, which may require local refinement to maintain high accuracy.

The direct solver described exactly mimics the classical nested dissection method, and has the same asymptotic complexity of $O(N^{1.5})$ for the "build" (or "factorization") stage, and then $O(N \log N)$ cost for solving a system once the coefficient matrix has been factorized. Storage requirements are also $O(N \log N)$. A more precise analysis of the complexity that takes into account the dependence on the order p of the local discretization shows [1] that $T_{\text{build}} \sim N p^4 + N^{1.5}$, and $T_{\text{solve}} \sim N p^2 + N \log N$.

3 Time-Stepping Methods

For high-order time-stepping of (1), we use the so called Explicit, Singly Diagonally Implicit Runge–Kutta (ESDIRK) methods. These methods have a Butcher diagram with a constant diagonal γ and are of the form

0	0					
2γ	γ	γ				
c_3	$a_{3,1}$	$a_{3,2}$	γ			
$\vdots$	$\vdots$	$\vdots$	$\ddots$	$\ddots$		
c_{s-1}	$a_{s-1,1}$	$a_{s-1,2}$	$a_{s-1,3}$	$\cdots$	γ	
1	b_1	b_2	b_3	$\cdots$	b_{s-1}	γ
	b_1	b_2	b_3	$\cdots$	b_{s-1}	γ

ESDIRK methods offer the advantages of stiff accuracy and L-stability. They are particularly attractive when used in conjunction with direct solvers since the elliptic solve required in each stage involves the same coefficient matrix $(I - h\gamma \mathscr{L})$, where h is the time-step.

In general we split the right hand side of (1) into a stiff part, $F^{[1]}$, that will be treated implicitly using ESDIRK methods, and a part, $F^{[2]}$, that will be treated explicitly (with a Butcher table denoted $\hat{c}$, $\hat{A}$, and $\hat{b}$). Precisely we will use the Additive Runge–Kutta (ARK) methods by Carpenter and Kennedy [11], of order 3, 4 and 5.

We may choose to formulate the Runge–Kutta method in terms of either solving for slopes or solving for stage solutions. We denote these the k_i formulation and the u_i formulation, respectively. When solving for slopes the stage computation is

$$k_i^n = F^{[1]}(t_n + c_i \Delta t, u^n + \Delta t \sum_{j=1}^{s} a_{ij} k_j^n + \Delta t \sum_{j=1}^{s} \hat{a}_{ij} l_j^n), \quad i = 1, \ldots, s, \quad (3)$$

$$l_i^n = F^{[2]}(t_n + c_i \Delta t, u^n + \Delta t \sum_{j=1}^{s} a_{ij} k_j^n + \Delta t \sum_{j=1}^{s} \hat{a}_{ij} l_j^n), \quad i = 1, \ldots, s. \quad (4)$$

Note that the explicit nature of (4) is encoded in the fact that the elements on the diagonal and above in $\hat{A}$ are zero. Once the slopes have been computed the solution at the next time-step is assembled as

$$u^{n+1} = u^n + \Delta t \sum_{j=1}^{s} b_j k_j^n + \Delta t \sum_{j=1}^{s} \hat{b}_j l_j^n. \tag{5}$$

If the method is instead formulated in terms of solving for the stage solutions the implicit solves take the form

$$u_i^n = u^n + \Delta t \sum_{j=1}^{s} \left(a_{ij} F^{[1]}(t_n + c_j \Delta t, u_j^n) + \hat{a}_{ij} F^{[2]}(t_n + c_j \Delta t, u_j^n) \right),$$

and the explicit update for u^{n+1} is given by

$$u^{n+1} = u^n + \Delta t \sum_{j=1}^{s} b_j (F^{[1]}(t_n + c_j \Delta t, u_j^n) + F^{[2]}(t_n + c_j \Delta t, u_j^n)).$$

The two formulations are algebraically equivalent but offer different advantages. For example, when working with the slopes we do not observe (see experiments presented in the second part of this paper) any order reduction due to time-dependent boundary conditions (see e.g. the analysis by Rosales et al. [14]). On the other hand and as discussed in some detail below, in solving for the slopes the HPS framework requires an additional step to enforce continuity.

We note that it is generally preferred to solve for the slopes when implementing implicit Runge–Kutta methods, particularly when solving very stiff problems where the influence of roundoff (or solver tolerance) errors can be magnified by the Lipschitz constant when solving for the stages directly.

Remark 2 The HPS method for elliptic solves was previously used in [10], which considered a linear hyperbolic equation

$$\frac{\partial u}{\partial t} = \mathscr{L}u(x, t), \quad x \in \Omega, t > 0,$$

where $\mathscr{L}$ is a skew-Hermitian operator. The evolution of the numerical solution can be performed by approximating the propagator $\exp(\tau \mathscr{L}) : L^2(\Omega) \to L^2(\Omega)$ via a rational approximation

$$\exp(\tau \mathscr{L}) \approx \sum_{m=-M}^{M} b_m (\tau \mathscr{L} - \alpha_m)^{-1}.$$

If application of $(\tau \mathscr{L} - \alpha_m)^{-1}$ to the current solution can be reduced to the solution of an elliptic-type PDE it is straightforward to apply the HPS scheme to each term in the approximation. A drawback with this approach is that multiple operators must be formed and it is also slightly more convenient to time step non-linear equations using the Runge–Kutta methods we use here.

There are two modifications to the HPS algorithm that are necessitated by the use of ARK time integrators, we discuss these in the next two subsections.

3.1 Neumann Data Correction in the Slope Formulation

In the HPS algorithm the PDE is enforced on interior nodes and continuity of the normal derivative is enforced on the leaf boundary. Now, due to the structure of the update formula (5), if at some time u^n has an error component in the null space of the operator that is used to solve for a slope k_i, then this will remain throughout the solution process. Although this does not affect the stability of the method it may result in loss of relative accuracy as the solution evolves. As a concrete example consider the heat equation

$$u_t = u_{xx}, \quad x \in [0, 2], t > 0, \tag{6}$$

with the initial data $u(x, 0) = 1 - |x - 1|$, and with homogenous Dirichlet boundary conditions. We discretize this on two leaves which we denote by α and β.

Now in the k_i formulation, we solve several PDEs for the k_i values and update the solution as

$$u^{n+1} = u^n + \Delta t \sum_{j=1}^{s} b_j k_j^n.$$

Here, even though the individual slopes have continuous derivatives the kink in u^n will be propagated to u^{n+1}. In this particular example we would end up with the incorrect steady state solution $u(x, t) = 1 - |x - 1|$.

Fortunately, this can easily be mitigated by adding a consistent penalization of the jump in the derivative of the solution during the merging of two leaves (for details see Section 4 in [1]). That is, if we denote the jump by $[[\cdot]]$ we replace the condition $0 = [[Tk + h^k]]$ where Tk is the derivative from the homogenous part and h^k is the derivative for the particular solution (of the slope) by the condition $[[Tk + h^k - \Delta t^{-1} h^u]] = 0$. In comparison to [1] we get the slightly modified merge formula

$$k_{i,3} = \left(T_{3,3}^{\alpha} - T_{3,3}^{\beta}\right)^{-1} \left(T_{3,2}^{\beta} k_{i,2} - T_{3,1}^{\alpha} k_{i,1} + h_3^{k,\beta} - h_3^{k,\alpha} - \frac{1}{\Delta t}(h_3^{u,\alpha} - h_3^{u,\beta})\right),$$

along with the modified equation for the fluxes of the particular solution on the parent box

$$
\begin{bmatrix} v_1 \\ v_2 \end{bmatrix} = \left(\begin{bmatrix} T^\alpha_{1,1} & 0 \\ 0 & T^\beta_{2,2} \end{bmatrix} + \begin{bmatrix} T^\alpha_{1,3} \\ T^\beta_{2,3} \end{bmatrix} (T^\alpha_{3,3} - T^\beta_{3,3})^{-1} [- T^\alpha_{3,1} \mid T^\beta_{3,2}] \right) \begin{bmatrix} k_{i,1} \\ k_{i,2} \end{bmatrix} +
$$

$$
\begin{bmatrix} h^{k,\alpha}_1 \\ h^{k,\beta}_2 \end{bmatrix} + \begin{bmatrix} T^\alpha_{1,3} \\ T^\beta_{2,3} \end{bmatrix} (T^\alpha_{3,3} - T^\beta_{3,3})^{-1} \left(h^\beta_3 - h^\alpha_3 - \frac{1}{\Delta t}(h^{u,\alpha}_3 - h^{u,\beta}_3) \right).
$$

Due to space we must refer to [1] for a detailed discussion of these equations. Briefly, $h^{k,\alpha}$ and $h^{k,\beta}$ above denote the spectral derivative on each child's boundary for the particular solution to the PDE for k_i and are already present in [1]. However, $h^{u,\alpha}$ and $h^{u,\beta}$, which denote the spectral derivative of u^n on the boundary from each child box, are new additions.

The above initial data is of course extreme but we note that the problem persists for any non-polynomial initial data with the size of the (stationary) error depending on resolution of the simulation. We further note that the described penalization removes this problem without affecting the accuracy or performance of the overall algorithm.

Remark 3 Although for linear constant coefficient PDE it may be possible to project the initial data in a way so that interior affine functions do not cause the difficulty above, for greater generality, we have chosen to enforce the extra penalization throughout the time evolution.

Remark 4 When utilizing the u_i formulation in a purely implicit problem we do not encounter the difficulty described above. This is because we enforce continuity of the derivative in u^n_s when solving

$$
(I - \Delta t \gamma \mathscr{L}) u^n_s = u^n + \Delta t \mathscr{L} \left(\sum_{j=1}^{s-1} a_{sj} u^n_j \right) + \Delta t \sum_{j=1}^{s-1} a_{sj} g(x, t_n + c_j \Delta t),
$$

followed by the update $u^{n+1} = u^n_s$.

3.2 Enforcing Continuity in the Explicit Stage

The second modification is to the first explicit stage in the k_i formulation. Solving a problem with no forcing this stage is simply

$$
k^n_1 = \mathscr{L}(u_n).
$$

When, for example, $\mathscr{L}$ is the Laplacian, we must evaluate it on all nodes on the interior of the physical domain. This includes the nodes on the boundary between two leafs where the spectral approximation to the Laplacian can be different if we use values from different leaves. The seemingly obvious choice, replacing the Laplacian on the leaf boundary by the average, leads to instability. However, stability can be restored if we enforce $k_1^n = \mathscr{L}(u_n)$ on the interior of each leaf and continuity of the derivative across each leaf boundary. Algorithmically, this is straightforward as these are the same conditions that are enforced in the regular HPS algorithm, except in this case we simply have an identity equation for k_1 on the interior nodes instead of a full PDE.

Although it is convenient to enforce continuity of the derivative using the regular HPS algorithm it can be done in a more efficient fashion by forming a separate system of equations involving only data on the leaf boundary nodes. In a single dimension on a discretization with n leafs this reduces the work associated with enforcing continuity of the derivative across leaf boundary nodes from solving $n \times (p-1) - 1$ equations for $n \times (p-1) - 1$ unknowns to solving a tridiagonal system of equations $n - 1$ equations for $n - 1$ unknowns.

In two dimensions the system is slightly different, but if we have $n \times n$ leafs with $p \times p$ Chebyshev nodes on each leaf then eliminating the explicit equations for the interior nodes reduces the system to $(p-2) \times 2n$ independent tridiagonal systems of $n - 1$ equations with $n - 1$ unknowns for a total of $(p-2) \times 2n \times (n-1)$ equations with $(p-2) \times 2n \times (n-1)$ unknowns.

When the u_i formulation is used for a fully implicit problem the intermediate stage values still requires us to evaluate $\mathscr{L}u^n$, but this quantity only enters through the body load in the intermediate stage PDEs. The explicit first stage in this formulation is simply $u_1^n = u^n$. Furthermore, while we must calculate

$$u^{n+1} = u^n + \Delta t \mathscr{L}\left(\sum_{j=1}^{s} a_{sj} u_j^n\right),$$

this is equivalent to u_s^n since $b_j = a_{sj}$ and we simply take $u^{n+1} = u_s^n$.

When both explicit and implicit terms are present, we proceed differently. Now, the values of u_i^n look almost identical to the implicit case and we still avoid the problem of an explicit "solve" in u_1^n, but we also have

$$u^{n+1} = u^n + \Delta t \sum_{j=1}^{s} b_j (F^{[1]}(t_n + c_j \Delta t, u_j^n) + F^{[2]}(t_n + c_j \Delta t, u_j^n))$$

The ESDIRK method has the property that $b_j = a_{sj}$, but for the explicit Runge–Kutta method we have $b_j \neq \hat{a}_{sj}$. When the explicit operator $F^{[2]}$ does not contain partial derivatives we need not enforce continuity of the derivative and can simply reformulate the method as

$$u^{n+1} = u_s^n + \Delta t \sum_{j=1}^{s} (a_{sj} - \hat{a}_{sj}) F^{[2]}(t_n + c_j \Delta t, u_j^n)$$

4 Boundary Conditions

The above description for Runge–Kutta methods does not address how to impose boundary conditions for a system of ODEs resulting from a discretization of a PDE. In particular, the different formulations incorporate boundary conditions in slightly different ways.

In this work we consider Dirichlet, Neumann, and periodic boundary conditions. For periodic boundary conditions the intermediate stage boundary conditions are enforced to be periodic for both formulations. As the k_i stage values are approximations to the time derivative of u, the imposed Dirichlet boundary conditions for $x \in \Gamma$ are $k_i^n = u_t(x, t_n + c_i \Delta t)$. When solving for u_i one may attempt to enforce boundary conditions using $u_i = u(x, t + c_i \Delta t), x \in \Gamma$. However, as demonstrated in part two of this series and discussed in detail in [14], this results in order reduction for time dependent boundary conditions.

In the HPS algorithm, Neumann or Robin boundary conditions are mapped to Dirichlet boundary conditions using the linear Dirichlet to Neumann operator as discussed for example in [1].

References

1. Babb, T., Gillman, A., Hao, S., Martinsson, P.: An accelerated Poisson solver based on multidomain spectral discretization. BIT Numer. Math. **58**, 851–879 (2018)
2. Babb, T., Martinsson, P.-G., Appelö, D.: HPS accelerated spectral solvers for time dependent problems (2018). arXiv:1811.04555
3. Canuto, C., Hussaini, M.Y., Quarteroni, A., Zang, T.A.: Spectral Methods: Evolution to Complex Geometries and Applications to Fluid Dynamics. Springer, Berlin (2007)
4. Duff, I., Erisman, A., Reid, J.: Direct Methods for Sparse Matrices. Oxford University Press, Oxford (1989)
5. Geldermans, P., Gillman, A.: An adaptive high order direct solution technique for elliptic boundary value problems. SIAM J. Sci. Comput. **41**(1), A292–A315 (2019). arXiv:1710.08787
6. George, A.: Nested dissection of a regular finite element mesh. SIAM J. Numer. Anal. **10**, 345–363 (1973)
7. Gillman, A., Martinsson, P.: A direct solver with $\mathcal{O}(N)$ complexity for variable coefficient elliptic PDEs discretized via a high-order composite spectral collocation method. SIAM J. Sci. Comput. **36**, A2023–A2046 (2014). arXiv:1307.2665
8. Gillman, A., Barnett, A., Martinsson, P.-G.: A spectrally accurate direct solution technique for frequency-domain scattering problems with variable media. BIT Numer. Math. **55**, 141–170 (2015)

9. Hao, S., Martinsson, P.: A direct solver for elliptic PDEs in three dimensions based on hierarchical merging of Poincaré-Steklov operators. J. Comput. Appl. Math. **308**, 419–434 (2016)
10. Haut, T., Babb, T., Martinsson, P., Wingate, B.: A high-order scheme for solving wave propagation problems via the direct construction of an approximate time-evolution operator. IMA J. Numer. Anal. **36**, 688–716 (2016)
11. Kennedy, C., Carpenter, M.: Additive Runge–Kutta schemes for convection-diffusion-reaction equations. Appl. Numer. Math. 44, 139–181 (2003)
12. Martinsson, P.: A direct solver for variable coefficient elliptic PDEs discretized via a composite spectral collocation method. J. Comput. Phys. **242**, 460–479 (2013)
13. Martinsson, P.: The hierarchical Poincaré-Steklov (HPS) solver for elliptic PDEs: a tutorial (2015). arXiv:1506.01308
14. Rosales, R., Seibold, B., Shirokoff, D., Zhou, D.: Order reduction in high-order Runge–Kutta methods for initial boundary value problems (2017). arXiv:1712.00897
15. Trefethen, L.: Spectral Methods in Matlab. SIAM, Philadelphia (2000)

On the Use of Hermite Functions for the Vlasov–Poisson System

Lorella Fatone, Daniele Funaro, and Gianmarco Manzini

1 Introduction

A semi-Lagrangian spectral method has been proposed in [8] for the numerical approximation of the nonrelativistic Vlasov–Poisson equations, which describe the dynamics of a collisionless plasma of charged particles, coupled under the effect of their own electric field. We assume for simplicity that the development of the plasma is only due to electrons. Moreover, we just treat the case of a 1D-1V distribution function, defined in a phase space consisting of the two one-dimensional independent variables x (space) and v (velocity). The approximation introduced in [8] has been initially developed and tested on Fourier-Fourier periodic discretizations, for both variables in the phase space. In the successive paper [9], the approximation in the variable v has been approached with the help of Hermite functions, i.e., Hermite polynomials multiplied by the Gaussian weight $\exp(-v^2)$.

Semi-Lagrangian methods for plasma physics calculations were originally proposed in [5, 18] and more recently in [6, 15, 16]. By this approach, at different times, the solution is approximated at the nodes of a Cartesian grid covering the space-

L. Fatone
Dipartimento di Matematica, Università degli Studi di Camerino, Camerino, Italy
e-mail: lorella.fatone@unicam.it

D. Funaro
Dipartimento di Scienze Chimiche e Geologiche, Università degli Studi di Modena e Reggio Emilia, Modena, Italy
e-mail: daniele.funaro@unimore.it

G. Manzini (✉)
Group T-5, Applied Mathematics and Plasma Physics, Theoretical Division, Los Alamos National Laboratory, Los Alamos, NM, USA
e-mail: gmanzini@lanl.gov

© The Author(s) 2020
S. J. Sherwin et al. (eds.), *Spectral and High Order Methods for Partial Differential Equations ICOSAHOM 2018*, Lecture Notes in Computational Science and Engineering 134, https://doi.org/10.1007/978-3-030-39647-3_10

143

velocity domain. The solution at each space-velocity node is traced back along the characteristic curve originating backward from that node. In [8] a high-order Taylor expansion of the characteristic curves is used to trace back the solution in time, which is then approximated by spectral interpolation. Such a method guarantees the conservation of the main physical quantities (charge, mass, and momentum).

The first attempt in using Hermite polynomials to solve the Vlasov equation dates back to the work [10], where the Hermite basis is used in the velocity variable to describe a plasma in a physical state near the thermodynamic equilibrium. Within this approach, exact discrete conservation laws can be constructed [7, 13, 14, 20, 21]. The weight function of the Hermite basis can be generalized by introducing a parameter α in such a way that it becomes $\exp(-\alpha^2 v^2)$. A proper choice of this parameter can significantly improve the convergence [2, 3, 19]. This fact was also confirmed in earlier works on plasmas physics based on Hermite spectral methods (see [11, 17] and more recently [4]).

The paper is organized as follows. In Sect. 2, we present the continuous model, i.e., the 1D-1V Vlasov equation. In Sect. 3, we introduce the spectral approximation in the phase space. In Sect. 4, we present the semi-Lagrangian schemes based on an approximation of the characteristic curves coupled with a second-order backward differentiation formula (BDF). In Sect. 5, we numerically assess the performance of the method for a standard test case, and we show how the solution's behavior can be affected by the choice of a certain parameter β, acting on the location of Hermite weight function.

2 The Continuous Model

We deal with the 1D-1V Vlasov equation defined in the domain $\Omega = \Omega_x \times \mathbb{R}$, with $\Omega_x \subseteq \mathbb{R}$. The unknown $f = f(t, x, v)$ denotes the probability of finding negative charged particles at the location x with velocity v. This is solution of the problem

$$\frac{\partial f}{\partial t} + v \frac{\partial f}{\partial x} - E(t, x) \frac{\partial f}{\partial v} = 0, \qquad t \in (0, T], \ x \in \Omega_x, \ v \in \mathbb{R}. \tag{1}$$

At time $t = 0$ we have the initial distribution $f(0, x, v) = \bar{f}(x, v)$. The problem is nonlinear, since the electric field E is coupled with f. Indeed, we set

$$\frac{\partial E}{\partial x}(t, x) = 1 - \rho(t, x) = 1 - \int_{\mathbb{R}} f(t, x, v) dv, \tag{2}$$

where ρ denotes the electron charge density. System (1)–(2) in the unknowns f and E is a simplification of the Vlasov–Poisson equations in two or three dimensional space domains. Uniqueness of the solution is ensured by imposing that

$$\int_{\Omega_x} E(t, x) dx = 0, \qquad \text{which implies that} \qquad \int_{\Omega_x} \rho(t, x) dx = |\Omega_x|, \tag{3}$$

where $|\Omega_x|$ is the size of Ω_x. We assume periodic boundary conditions in the variable x and a suitable exponential decay at infinity for the variable v. After integration and by using the boundary constraints, we obtain the conservation of mass

$$\frac{d}{dt} \int_\Omega f(t, x, v)\, dx\, dv = 0. \tag{4}$$

When f and E are smooth enough, for a sufficiently small $\delta > 0$, the local system of characteristics associated with (1) is given by the curves $(X(\tau), V(\tau))$ solving

$$\frac{dX}{d\tau} = -V(\tau), \qquad \frac{dV}{d\tau} = E(\tau, X(\tau)), \qquad \tau \in]t - \delta, t + \delta[, \tag{5}$$

with the condition that $(X(t), V(t)) = (x, v)$ when $\tau = t$. With this setting we have in mind that for $\tau > 0$ we proceed backward. Under suitable regularity assumptions, there exists a unique solution of the Vlasov–Poisson problem (1)–(2) which is formally obtained by propagating the initial condition along the characteristic curves described by (5), i.e. we have

$$f(t, x, v) = \bar{f}(X(t), V(t)), \tag{6}$$

where we recall that $\bar{f}$ is the initial datum. By using the first-order approximation

$$X(\tau) = x - v(\tau - t), \qquad V(\tau) = v + E(t, x)(\tau - t), \tag{7}$$

the Vlasov equation is satisfied up to an error decaying as $|\tau - t|$, for τ tending to t.

3   Phase-Space Discretization

We briefly recall the construction of the approximation method proposed in [8]. At each point of a given grid, the new value of the discrete solution is set up to be equal to the value obtained by going backward, by a suitably small amount, along the local characteristic lines. The algorithm follows from a Taylor expansion of arbitrary order, where the derivatives in the variable x and v are carried out with spectral accuracy. In particular, for the variable x we consider the domain $\Omega_x = [0, 2\pi[$. Given the positive integer N, we have the equispaced nodes $x_i = 2\pi i/N$, $i = 0, 1, \ldots, N - 1$. Regarding the direction v, when M is a given positive integer, the nodes v_j, $j = 0, 1, \ldots, M - 1$, are the zeros of H_M, which is the Hermite polynomial of degree M.

We introduce the polynomial Lagrangian basis functions for the x and v variables, that are $B_i^{(N)}(x_n) = \delta_{in}$ and $B_j^{(M)}(v_m) = \delta_{jm}$, where δ_{ij} is the usual Kronecker symbol. We recall that Hermite functions are obtained from Hermite polynomials after multiplication by the weight $\omega(v) = e^{-v^2}$. We also define the discrete spaces

$$\mathbf{X}_N = \text{span}\left\{ B_i^{(N)} \right\}_{i=0,1,\ldots,N-1}, \quad \mathbf{Y}_{N,M} = \text{span}\left\{ B_i^{(N)} B_j^{(M)} \omega \right\}_{\substack{i=0,1,\ldots,N-1 \\ j=0,1,\ldots,M-1}}. \quad (8)$$

Any function $f_{N,M}$ that belongs to $\mathbf{Y}_{N,M}$ can be represented as

$$f_{N,M}(x, v) = \sum_{i=0}^{N-1} \sum_{j=0}^{M-1} c_{ij} B_i^{(N)}(x) B_j^{(M)}(v) \omega(v), \quad (9)$$

where the coefficients of such an expansion are given by $c_{ij} = f_{N,M}(x_i, v_j)$.

In the following, the matrices $d_{ni}^{(N,s)}$ and $d_{mj}^{(M,s)}$ denote the s-th derivative of $B_i^{(N)}$ evaluated at point x_n and $(B_j^{(M)} \omega)$ evaluated at point v_m

$$d_{ni}^{(N,s)} = \frac{d^s B_i^{(N)}}{dx^s}(x_n) \quad \text{and} \quad d_{mj}^{(M,s)} = \frac{d^s \left(B_j^{(M)} \omega \right)}{dv^s}(v_m). \quad (10)$$

As a special case, we set $d_{ni}^{(N,0)} = \delta_{ni}$, $d_{mj}^{(M,0)} = \delta_{mj}$.

Now, let us assume that the one-dimensional function $E_N \in \mathbf{X}_N$ is known. Given $\Delta t > 0$, by taking $\tau = t - \Delta t$ in formula (7), we define the new set of points $\tilde{x}_{nm} = x_n - v_m \Delta t$ and $\tilde{v}_{nm} = v_m + E_N(x_n) \Delta t$. To evaluate a function $f_{N,M} \in \mathbf{Y}_{N,M}$ at the new points $(\tilde{x}_{nm}, \tilde{v}_{nm})$ through the coefficients c_{ij}, we use a Taylor expansion in time. By omitting the terms in Δt of order higher than one, we get

$$B_i^{(N)}(\tilde{x}_{nm}) \left(B_j^{(M)} \omega \right)(\tilde{v}_{nm}) \approx$$

$$\delta_{in} \delta_{jm} \omega(v_m) - v_m \Delta t \, \delta_{jm} d_{ni}^{(N,1)} \omega(v_m) + E_N(x_n) \Delta t \, \delta_{in} d_{mj}^{(M,1)}. \quad (11)$$

By substituting (11) in (9), we obtain the approximation

$$f_{N,M}(\tilde{x}_{nm}, \tilde{v}_{nm}) = \sum_{i=0}^{N-1} \sum_{j=0}^{M-1} c_{ij} B_i^{(N)}(\tilde{x}_{nm}) B_j^{(M)}(\tilde{v}_{nm}) \omega(\tilde{v}_{nm})$$

$$\approx c_{nm} \omega(v_m) - v_m \omega(v_m) \Delta t \sum_{i=0}^{N-1} d_{ni}^{(N,1)} c_{im} + E_N(x_n) \Delta t \sum_{j=0}^{M-1} d_{mj}^{(M,1)} c_{nj}, \quad (12)$$

which is the main building block for more advanced schemes.

4 Discretization of the Vlasov Equation

Given the time instants $t^k = k\Delta t = kT/K$ for any integer $k = 0, 1, \ldots, K$, we consider the approximation of the unknowns f and E of problem (1)–(2), given by

$$\left(f_{N,M}^{(k)}(x, v), \; E_N^{(k)}(x) \right) \simeq \left(f(t^k, x, v), \; E(t^k, x) \right), \qquad x \in \Omega_x, \; v \in \mathbb{R}, \qquad (13)$$

where the function $f_{N,M}^{(k)}$ belongs to $\mathbf{Y}_{N,M}$ and the function $E_N^{(k)}$ belongs to $\mathbf{X}_N$. Concerning the density function, we define

$$\rho_N^{(k)}(x) = \int_{\Omega_v} f_{N,M}^{(k)}(x, v)\, dv \; \simeq \; \rho(t^k, x). \qquad (14)$$

Hence, at any time step k, we express $f_{N,M}^{(k)}$ in the following way

$$f_{N,M}^{(k)}(x, v) = \sum_{i=0}^{N-1} \sum_{j=0}^{M-1} c_{ij}^{(k)}\, B_i^{(N)}(x)\, B_j^{(M)}(v)\omega(v), \qquad (15)$$

where $c_{ij}^{(k)} = f_{N,M}^{(k)}(x_i, v_j)$. At time $t = 0$, we use the initial condition $c_{ij}^{(0)} = f(0, x_i, v_j) = \bar{f}(x_i, v_j)$.

Suppose that $E_N^{(k)}$ is given at step k. According to [8], we write

$$E_N^{(k)}(x) = -\sum_{n=1}^{N/2} \frac{1}{n} \left[\hat{a}_n^{(k)} \sin(nx) - \hat{b}_n^{(k)} \cos(nx) \right], \qquad (16)$$

where the discrete Fourier coefficients $\hat{a}_n^{(k)}$ and $\hat{b}_n^{(k)}$, $n = 1, 2, \ldots, N/2$, are suitably related to those of $\rho_N^{(k)}$.

By taking $\tau = t - \Delta t$ in (7), we define $\tilde{x}_{nm} = x_n - v_m \Delta t$ and $\tilde{v}_{nm} = v_m + E_N^{(k)}(x_n)\Delta t$. The distribution function f is expected to remain constant along the characteristics. The most straightforward discretization method is obtained by advancing the coefficients according to the approximation

$$f_{N,M}^{(k+1)}(x_n, v_m) \approx f_{N,M}^{(k)}(\tilde{x}_{nm}, \tilde{v}_{nm}). \qquad (17)$$

This states that the value of $f_{N,M}^{(k+1)}$, at the grid points and time step $(k+1)\Delta t$, is assumed to correspond to the previous value at time $k\Delta t$, recovered by going backwards along the characteristics. To compute $\tilde{v}_{nm}$, we should use $E_N^{(k+1)}(x_n)$ instead of $E_N^{(k)}(x_n)$. However, the distance between these two quantities is of the order of Δt, so that the replacement has no practical effects on the accuracy of first-order methods. Between each step k and the successive one, we need to update the electric field. This can be done by using the Gaussian quadrature formula in (14), so obtaining

$$\rho_N^{(k)}(x_i) = \sum_{j=0}^{M-1} \frac{1}{\omega(v_j)} f_{N,M}^{(k)}(x_i, v_j) \, w_j = \sum_{j=0}^{M-1} \frac{1}{\omega(v_j)} c_{ij}^{(k)} \, w_j, \tag{18}$$

where w_j, for $j = 1, \ldots, M - 1$, are the quadrature weights. Afterwards, in order to compute the new point-values $E_N^{(k+1)}(x_n)$ of the electric field, it is necessary to integrate $\rho_N^{(k)}$. By using approximation (12) in (17), we end up with the first-order explicit scheme of Euler type:

$$c_{nm}^{(k+1)} = c_{nm}^{(k)} + \Delta t \, \Phi_{nm}^{(k)}, \tag{19}$$

where

$$\Phi_{nm}^{(k)} = -v_m \sum_{i=0}^{N-1} d_{ni}^{(N,1)} c_{im}^{(k)} + E_N^{(k)}(x_n) \sum_{j=0}^{M-1} d_{mj}^{(M,1)} c_{nj}^{(k)} \frac{1}{\omega(v_m)}. \tag{20}$$

The parameter Δt must satisfy a suitable CFL condition, which is obtained by requiring that the point $(\tilde{x}_{nm}, \tilde{v}_{nm})$ falls inside the box $]x_{n-1}, x_{n+1}[\times]v_{m-1}, v_{m+1}[$. A straightforward way to increase the time accuracy is to use a multistep discretization scheme as the second-order accurate two-step BDF scheme. We have

$$f_{N,M}^{(k+1)}(x_n, v_m) \approx \frac{4}{3} f_{N,M}^{(k)}(\tilde{x}_{nm}, \tilde{v}_{nm}) - \frac{1}{3} f_{N,M}^{(k-1)}(\tilde{\tilde{x}}_{nm}, \tilde{\tilde{v}}_{nm}), \tag{21}$$

where $(\tilde{x}_{nm}, \tilde{v}_{nm})$ is the point obtained from (x_n, v_m) going back of one step Δt along the characteristic lines. Similarly, the point $(\tilde{\tilde{x}}_{nm}, \tilde{\tilde{v}}_{nm})$ is obtained by going two steps back along the characteristic lines, i.e., by using $2\Delta t$ instead of Δt when computing $\tilde{x}_{nm}$ and $\tilde{v}_{nm}$. Despite the fact that a BDF scheme is commonly presented as an implicit technique, in our context (f constant along the characteristics) it assumes the form of an explicit method. In terms of the coefficients, we end up with the scheme

$$c_{nm}^{(k+1)} = \frac{4}{3} \left(c_{nm}^{(k)} + \Delta t \, \Phi_{nm}^{(k)} \right) - \frac{1}{3} \left(c_{nm}^{(k-1)} + 2\Delta t \, \Phi_{nm}^{(k-1)} \right)$$

$$= \frac{4}{3} c_{nm}^{(k)} - \frac{1}{3} c_{nm}^{(k-1)} + \frac{2}{3} \Delta t \left[-v_m \sum_{i=0}^{N-1} d_{ni}^{(N,1)} (2c_{im}^{(k)} - c_{im}^{(k-1)}) \right.$$

$$\left. + E_N^{(k)}(x_n) \sum_{j=0}^{M-1} d_{mj}^{(M,1)} (2c_{nj}^{(k)} - c_{nj}^{(k-1)}) \frac{1}{\omega(v_m)} \right]. \tag{22}$$

From theoretical considerations and the experiments in [8], it turns out that the above method is actually second-order accurate in Δt. Higher order schemes can be obtained with similar principles. All the above schemes guarantee mass conservation (see (4) for the continuous case), which is a crucial physical property.

For practical purposes, it is advisable to make the change of variable $f(t, x, v) = p(t, x, v) \exp(-v^2)$ in the Vlasov equation, so obtaining

$$\frac{\partial p}{\partial t} + v \frac{\partial p}{\partial x} - E(t, x) \left[\frac{\partial p}{\partial v} - 2vp \right] = 0, \ t \in (0, T], \ x \in \Omega_x, \ v \in \mathbb{R}. \tag{23}$$

At time step k, the function $p(t^k, x, v)$ is approximated by a function $p_{N,M}^{(k)}(x, v)$ in such a way that $p_{N,M}^{(k)} e^{-v^2}$ belongs to the finite dimensional space $\mathbf{Y}_{N,M}$.

A generalization consists in introducing a real parameter α and assuming that the weight function is $\omega(v) = \exp(-\alpha^2 v^2)$. The approximation scheme can be easily adjusted by modifying nodes and weights of the Gaussian formula, through a multiplication by suitable constants. The difficulty in the implementation is practically the same, but, as observed in [9], the results are quite sensitive to the variation of α.

5 Numerical Experiments

The numerical scheme here proposed is validated in the standard two-stream instability benchmark test. We consider the Vlasov–Poisson problem (1)–(2) where we set $\Omega_x = [0, 4\pi[, \Omega_v = [-5, 5]$. The initial solution is given by

$$\bar{f}(x, v) = \frac{1}{2a\sqrt{2\pi}} \Big[G_R(v) + G_L(v) \Big] (1 + \epsilon \cos(\kappa x)), \tag{24}$$

where $G_R(v) = e^{-\alpha^2(v-\beta)^2}$ and $G_R(v) = e^{-\alpha^2(v+\beta)^2}$ are two Gaussians centered symmetrically at the points $v = \pm\beta$. The parameters for (24) are: $a = 1/\sqrt{8}$, $\epsilon = 10^{-3}, \kappa = 0.5, \alpha = \bar{\alpha} = 2, \beta = \bar{\beta} = 1$.

In all the experiments that follow, we integrate up to time $T = 30$ using the second-order BDF scheme with a suitably small time step, in order to guarantee stability and a good accuracy. In this way we can concentrate our attention to the spectral approximation in the variable x and v. A study of the convergence rate in time of the proposed numerical scheme can be found in [8]. First of all, in Fig. 1 we show the results at time $T = 30$ of the solution recovered by the Fourier-Fourier method, by choosing $N = 2^5$, $M = 2^6$ and time step equal to $\Delta t = 0.00125$. This will be the referring figure for the successive comparisons. Besides we show the corresponding time evolution of $|\hat{a}_1^{(k)}|$, the first Fourier mode of the electric field $E_N^{(k)}$ in (16). The behavior of this last quantity is predicted by theoretical

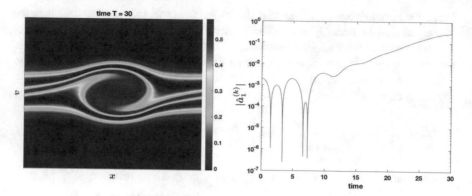

Fig. 1 Two-stream instability test: approximated distribution function at time $T = 30$ obtained by using the Fourier-Fourier method with $N = 2^5$, $M = 2^6$, $\Delta t = 0.00125$, and the corresponding time evolution of the first Fourier mode of the electric field $E_N^{(k)}$, i.e. $|\hat{a}_1^{(k)}|$ in (16)

considerations, and the slope of the "segment" starting at $T = 15$ agrees with the expectancy [1, Chapter 5].

As done in [9], we perform a series of experiments using less degrees of freedom than those actually necessary to resolve accurately the equation. In practice, we set $N = M = 2^4$. In this way, we could for instance detect what happens by varying the parameters α and β. Of course, if we increase the number of degrees of freedom, the numerical solution improves and cannot be distinguished from the referring one shown in Fig. 1. The purpose in [9] was to check what happens by varying the parameter α in the Hermite weight $\exp(-\alpha^2 v^2)$. The conclusions are that the approximate solution is very sensitive to the choice of α and that there are values of α that perform better than others. In general these values are those belonging to a neighbourhood of $\alpha = 1$. Moreover, in [9], we note that keeping α constantly equal to the value that better fits the initial datum (i.e. $\alpha = \bar{\alpha} = 2$ for (24)) may create instability as time increases. For such motivations, since at the moment a practical algorithm able to vary α in a dynamical way during the computations is not available, in the numerical experiments that follow we fix $\alpha = 1$, while play with β.

Due to the particular initial condition, we adopt a two-species decomposition of the Vlasov equation, where the distribution function is given by the sum of two electron distribution functions, i.e., $f = f_R + f_L$. These distribution functions refer to the two initial electron distributions, so that $f_R = p_R G_R$ and $f_L = p_L G_L$, where p_L and p_R are given polynomials. We consider the two systems of electrons described by the distribution functions f_L and f_R at the initial time as distinct plasma species that maintain their diversity throughout the whole numerical simulation. Therefore, we can split the Vlasov equation into two equations that are still of Vlasov type and are solvable independently, although they are coupled through the same electric field, which depends on the total charge density. This amounts to approximate two independent equations of the same type of that given

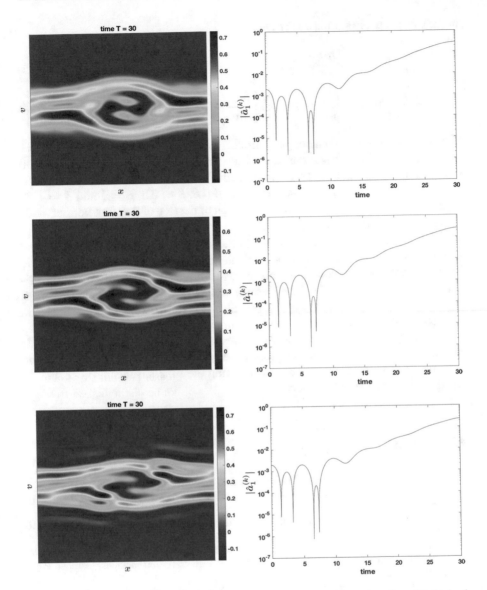

Fig. 2 Two-stream instability test: approximated distribution function at time $T = 30$ obtained by using the Fourier–Hermite method with $N = M = 2^4$, $\Delta t = 0.01$, $\alpha = 1$ (left panel) and the corresponding time evolution of the first Fourier mode of the electric field $E_N^{(k)}$, i.e. $|\hat{a}_1^{(k)}|$ in (16) (right panel) when $\beta = 0.5$ (top), $\beta = 1$ (center) and $\beta = 1.5$ (bottom)

in (23), respectively shifted by $\pm\beta$, i.e.

$$\frac{\partial p_R}{\partial t} + (v - \beta)\frac{\partial p_R}{\partial x} - E(t, x)\left[\frac{\partial p_R}{\partial v} - 2\alpha^2(v - \beta)p_R\right] = 0, \tag{25}$$

$$\frac{\partial p_L}{\partial t} + (v + \beta)\frac{\partial p_L}{\partial x} - E(t, x)\left[\frac{\partial p_L}{\partial v} - 2\alpha^2(v + \beta)p_L\right] = 0. \tag{26}$$

The two unknowns are then coupled through the density function as in (2).

The plots of Fig. 2 show the numerical distribution function at time $T = 30$ obtained by using the Fourier–Hermite method with $N = M = 2^4$, $\Delta t = 0.01$, $\alpha = 1$ and different values of the parameter β (i.e. $\beta = 0.5$, $\beta = 1$ and $\beta = 1.5$), together with the corresponding time evolution of the (log of the) first Fourier mode of the electric field $E_N^{(k)}$, i.e. $|\hat{a}_1^{(k)}|$ in (16).

The distribution functions presented in the left column of Fig. 2 are visibly and significantly different depending on β, while the first Fourier mode of the electric field shown in the right column seems to be less affected. These differences practically confirm that the choice of the Hermite weight functions $\omega(v) = \exp(-\alpha^2(v \pm \beta)^2)$ is a crucial aspect of the method (see also [11, 12, 17, 22]). This conclusion is heuristic. Unfortunately, there is no space enough for a deeper quantitative analysis in these pages. The question deserves however further investigation. Moreover, it would be advisable to develop appropriate algorithms allowing for the automatic adjustment of both parameters α and β during the time advancing procedure, in order to optimize the performance.

References

1. Bittencourt, J.A.: Fundamentals of Plasma Physics. Springer, New York (2004)
2. Boyd, J.P.: The rate of convergence of Hermite function series. Math. Comput. **35**, 1309–1316 (1980)
3. Boyd, J.P.: Asymptotic coefficients of Hermite function series. J. Comput. Phys. **54**, 382–410 (1984)
4. Camporeale, E., et al. On the velocity space discretization for the Vlasov–Poisson system: comparison between implicit Hermite spectral and Particle-in-Cell methods. Comput. Phys. Commun. **198**, 47–58 (2016)
5. Cheng, C.Z., Knorr, G.: The integration of the Vlasov equation in configuration space. J. Comput. Phys. **22**(3), 330–351 (1976)
6. Crouseilles, N., Respaud, T., Sonnendrücker, E.: A forward semi-Lagrangian method for the numerical solution of the Vlasov equation. Comput. Phys. Commun. **180**(10), 1730–1745 (2009)
7. Delzanno, G.L.: Multi-dimensional, fully-implicit, spectral method for the Vlasov–Maxwell equations with exact conservation laws in discrete form. J. Comput. Phys. **301**, 338–356 (2015)
8. Fatone, D., Funaro, L., Manzini, G.: Arbitrary-order time-accurate semi-Lagrangian spectral approximations of the Vlasov–Poisson system. J. Comput. Phys. **384**, 349–375 (2019)

9. Fatone, D., Funaro, L., Manzini, G.: A semi-Lagrangian spectral method for the Vlasov–Poisson system based on Fourier, Legendre and Hermite polynomials. Commun. Appl. Math. Comput. **1**, 333–360 (2019)
10. Grad, H.: On the kinetic theory of rarefied gases. Commun. Pure Appl. Math. **2**(4), 331–407 (1949)
11. Holloway, J.P.: Spectral velocity discretizations for the Vlasov-Maxwell equations. Transp. Theory Stat. Phys. **25**(1), 1–32 (1996)
12. Ma, H., Sun, W., Tang, T.: Hermite spectral methods with a time-dependent scaling for parabolic equations in unbounded domains. SIAM J. Numer. Anal. **43**, 58–75 (2005)
13. Manzini, G., et al. A Legendre-Fourier spectral method with exact conservation laws for the Vlasov–Poisson system. J. Comput. Phys. **317**, 82–107 (2016)
14. Manzini, G., Funaro, D., Delzanno, G.L.: Convergence of spectral discretizations of the Vlasov–Poisson system. SIAM J. Numer. Anal. **55**(5), 2312–2335 (2017)
15. Qiu, J.-M., Christlieb, A.: A conservative high order semi-Lagrangian WENO method for the Vlasov equation. J. Comput. Phys. **229**(4), 1130–1149 (2010)
16. Qiu, J.-M., Russo, G.: A high order multidimensional characteristic tracing strategy for the Vlasov–Poisson system. J. Sci. Comput. **71**, 414–434 (2017)
17. Schumer, J.W., Holloway, J.P.: Vlasov simulations using velocity-scaled Hermite representations. J. Comput. Phys. **144**(2), 626–661 (1998)
18. Sonnendrücker, E., et al.: The semi-Lagrangian method for the numerical resolution of the Vlasov equation. J. Comput. Phys. **149**(2), 201–220 (1999)
19. Tang, T.: The Hermite spectral method for Gaussian-type functions. SIAM J. Sci. Comput. **14**(3), 594–606 (1993)
20. Vencels, J., et al.: Spectral solver for multi-scale plasma physics simulations with dynamically adaptive number of moments. Proc. Comput. Sci. **51**, 1148–1157 (2015)
21. Vencels, J., et al.: SpectralPlasmaSolver: a spectral code for multiscale simulations of collisionless, magnetized plasmas. J. Phys. Conf. Series **719**(1), 012022 (2016)
22. Xiang, X.-M., Wang, Z.-Q.: Generalized Hermite approximations and spectral method for partial differential equations in multiple dimensions. J. Sci. Comput. **57**, 229–253 (2013)

HPS Accelerated Spectral Solvers for Time Dependent Problems: Part I, Algorithms

Tracy Babb, Per-Gunnar Martinsson, and Daniel Appelö

1 Introduction

In this chapter, part two in a two part series, describes a sequence of numerical experiments demonstrating the performance of a highly computationally efficient solver for equations of the form

$$\kappa \frac{\partial u}{\partial t} = \mathscr{L}u(x, t) + g(u, x, t), \quad x \in \Omega, t > 0, \tag{1}$$

with initial data $u(x, 0) = u_0(x)$. Here $\mathscr{L}$ is an elliptic operator acting on a fixed domain Ω and f is lower order, possibly nonlinear terms. We take κ to be real or imaginary, allowing for parabolic and Schrödinger type equations.

The "Hierarchial Poincaré–Steklov (HPS)" solver has already been demonstrated to be a highly competitive spectrally accurate solver for elliptic problems [1, 4, 7] and has also been used together with a class of exponential integrators [5], to evolve solutions to hyperbolic differential equations. As just mentioned, the focus here is on differential equations in the form (1) whose discretization leads to stiff system of ODE that can beneficially be advanced in time using Explicit, Singly Diagonally Implicit Runge–Kutta (ESDIRK) methods. ESDIRK methods offer the advantages of stiff accuracy and L-stability and are well suited for the HPS algorithm as they only require a single matrix factorization. They are also easily combined with

T. Babb · D. Appelö (✉)
University of Colorado, Boulder, CO, USA
e-mail: tracy.babb@colorado.edu; daniel.appelo@colorado.edu

P.-G. Martinsson
University of Texas, Austin, TX, USA
e-mail: pgm@ices.utexas.edu

© The Author(s) 2020
S. J. Sherwin et al. (eds.), *Spectral and High Order Methods for Partial Differential Equations ICOSAHOM 2018*, Lecture Notes in Computational Science and Engineering 134, https://doi.org/10.1007/978-3-030-39647-3_11

explicit Runge–Kutta method leading to so called Additive Runge–Kutta (ARK) methods [6].

To this end we investigate the stability and accuracy that is obtained when combining high-order time-stepping schemes with the HPS method for solving elliptic equations. We restrict attention to relatively simple geometries (rectangles) but note that the method can without difficulty be generalized to domains that can be expressed as a union of rectangles, possibly mapped via curvilinear smooth parameter maps.

The rest of this chapter is organized as follows. In Sect. 2 we present results illustrating that the order reduction phenomena for DIRK methods observed in [8] can be circumvented when formulating the time stepping in terms of slopes (with boundary conditions differentiated in time) rather than formulating it in terms of stage solutions. In Sect. 3 we present numerical results for Schrödingers equation in two dimensions and in Sect. 4 we present numerical results for a nonlinear problem, viscous Burgers' equation in two dimensions. Finally, in Sect. 5 we summarize and conclude. For a longer description of the method we refer to thee first part of this paper and to [2].

2   Time Dependent Boundary Conditions

This section discusses time-dependent boundary conditions within the two different Runge–Kutta formulations. In particular, we investigate the order reduction that has been documented in [8] for implicit Runge–Kutta methods and earlier in [3] for explicit Runge–Kutta methods.

In this first experiment, introduced in [8], we solve the heat equation in one dimension

$$u_t = u_{xx} + f(t), \qquad x \in [0, 2], \quad t > 0. \tag{2}$$

We set the initial data, Dirichlet boundary conditions and the forcing $f(t)$ so that exact solution is $u(x, t) = \cos(t)$. This example is designed to eliminate the effect of the spatial discretization, with the solution being constant in space and allows for the study of possible order reduction near the boundaries.

We use the HPS scheme in space and use 32 leafs with $p = 32$ Chebyshev nodes per leaf. We apply the third, fourth, and fifth order ESDIRK methods from [6]. We consider solving for the intermediate solutions, or as we refer to it below "the u_i formulation" with the boundary condition enforced as $u_i^n = \cos(t_n + c_i \Delta t)$. We also consider solving for the stages, which we refer to as "the k_i formulation" with boundary conditions imposed as $k_i^n = -\sin(x, t_n + c_i \Delta t)$.

Error reduction for time dependent boundary conditions has been studied both in the context of explicit Runge–Kutta methods in e.g. [3] and more recently for implicit Runge–Kutta methods in [8]. In [8] the authors report observed orders of accuracy equal to two (for the solution u) for DIRK methods of order 2, 3, and 4 for

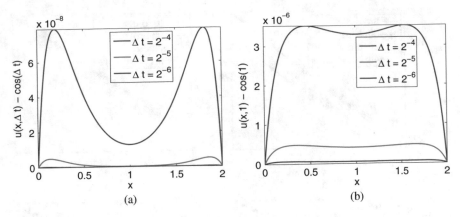

Fig. 1 The error in solving (2). Results are for a third order ESDIRK. (**a**) Displays the single step error which converges with fourth order of accuracy. (**b**) Displays the global error at $t = 1$ converging at third order. Both errors converge at one order higher than what is expected from the analysis in [8]

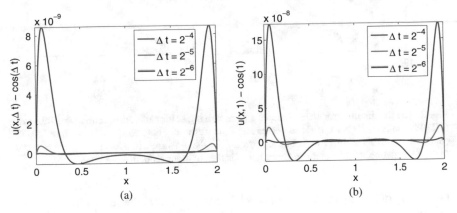

Fig. 2 The error in solving (2). Results are for a fifth order ESDIRK. (**a**) Displays the single step error which converges with fourth order of accuracy. (**b**) Displays the global error at $t = 1$ converging at third order. Both errors converge at one order higher than what is expected from the analysis in [8] but still lower than expected

the problem (2) discretized with a finite difference method on a fine grid (the spatial errors are zero) using the u_i formulation.

Figures 1 and 2 show the error for the third and fifth order ESDIRK methods, respectively, as a function of x for a single step and at the final time $t = 1$. Figure 3 shows the maximum error for the third, fourth, and fifth order methods as a function of time step Δt after a single step and at the final time $t = 1$.

In general, for a method of order p we expect that the single step error decreases as Δt^{p+1} while the global error decreases as Δt^p. However, with time dependent

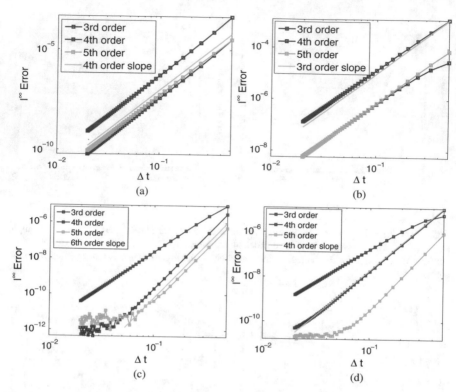

Fig. 3 The maximum error (here denoted l^∞) in solving (2) for the third, fourth, and fifth order ESDIRK methods for a sequence of decreasing time steps. (**a, c**) are errors after one time step and (**b, d**) are the errors at time $t = 1$. The top row are for the u_i formulation and the bottom row is for the k_i formulation. Note that the k_i formulation is free of order reduction

boundary conditions implemented as $u_i^n = \cos(t_n + c_i \Delta t)$ the results in [8] indicate that the rate of convergence will not exceed two for the single step or global error.

The results for the third order method ($p = 3$) displayed in Fig. 1 show that the single step error decreases as Δt^{p+1} while the global error decreases as Δt^p, which is better than the results documented in [8]. However, we still see that a boundary layer appears to be forming, but it is of the same order as the error away from the boundary. The results for the fifth order method ($p = 5$) displayed in Fig. 2 show that the single step error decreases as Δt^4 while the global error decreases as Δt^3, which is still better than the results documented in [8]. However, the boundary layer is giving order reduction from Δt^{p+1} for the single step error and Δt^p for the global error. We note that our observations differ from those in [8] but that this possibly can be attributed to the use of a ESDIRK method rather than a DIRK method.

We repeat the experiment but now we use the k_i formulation for Runge–Kutta methods and for the boundary condition we enforce $k_i^n = -\sin(t_n + c_i \Delta t)$. The intuition here is that k_i^n is an approximation to u_t at time $t_n + c_i \Delta t$ and we use the value of u_t for the boundary condition of k_i^n. Intuitively we expect that the fact that we reduce the index of the system of differential algebraic equation in the u_i formulation by differentiating the boundary conditions can restore the design order of accuracy.

In the previous examples the Runge–Kutta method introduced an error on the interior while the solution on the boundary was exact. If the error on the boundary is on the same order of magnitude as the error on the interior then the error in u_{xx} is of the correct order, but when the value of u is exact on the boundary it introduces a larger error in u_{xx}. In the k_i formulation, for each intermediate stage we find $u_{xx} = 0$ and then $k_i^n = -\sin(t_n + c_i \Delta t)$ on the interior and on the boundary. So at a fixed time the solution is constant in x and a boundary layer does not form. Additionally, the error is constant in x at any fixed time and for a method of order p we obtain the expected behavior where the single step error decreases as Δt^{p+1} and the global error decreases as Δt^p.

Figure 3 shows the maximum error for the third, fourth, and fifth order methods as a function of time step Δt after a single step and at the final time $t = 1$. The results show that the methods behave exactly as we expect. The single step error behaves as Δt^{p+1} for the third and fifth order methods and Δt^{p+2} for the fourth order method. The fourth order method gives sixth order error in a single step because the exact solution is $u(x, t) = \cos(t)$, which has every other derivative equal to zero at $t = 0$ and for a single step we start at $t = 0$. The global error behaves as Δt^p for each method.

3   Schrödinger Equation

Next we consider the Schrödinger equation for $u = u(x, y, t)$

$$i\hbar u_t = -\frac{\hbar^2}{2M}\Delta u + V(x, y)u, \quad t > 0, \quad (x, y) \in [x_l, x_r] \times [y_b, y_t], \tag{3}$$

$$u(x, y, 0) = u_0(x, y).$$

Here we nondimensionalize in a way equivalent to setting $M = 1, \hbar = 1$ in the above equation. We choose the potential to be the harmonic potential

$$V(x, y) = \frac{1}{2}\left(x^2 + y^2\right).$$

This leads to an exact solution

$$u(x, y, t) = Ae^{-it}e^{-\frac{(x^2+y^2)}{2}}, \tag{4}$$

where we set $A = 1/\sqrt{\sqrt{\pi}}$ and solve until $t = 2\pi$ on the domain $(x, y) \in [-8, 8]^2$.

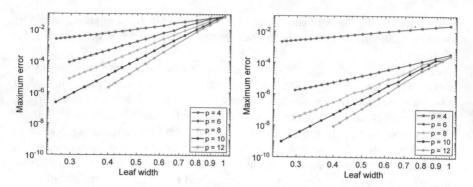

Fig. 4 Error in the Schrödinger equation as a function of leaf size. The exact solution is given in Eq. (4)

Table 1 Estimated rates of convergence for different Runge–Kutta methods and different orders of approximation

p	4	6	8	10	12
ESDIRK3	2.59	5.73	7.72	9.69	11.47
ESDIRK4	1.89	6.47	7.82	9.76	11.69
ESDIRK5	1.84	4.42	7.69	9.71	11.48

The computational domain is subdivided into $n_x \times n_y$ panels with $p \times p$ points on each panel. To begin, we study the order of accuracy with respect to leaf size. To eliminate the effect of time-stepping errors we scale $\Delta t = h^{p/q_{RK}}$, where q_{RK} is the order of the Runge–Kutta method. In Fig. 4 we display the errors as a function of the leaf size for $p = 4, 6, 8, 10, 12, 16$ and for the third and fifth order Runge–Kutta methods ($q_{RK} = 3, 5$). The rates of convergence are found for all three Runge–Kutta methods and summarized in Table 1. As can be seen from the table, $p = 4$ appears to converge at second order, while for higher p we generally observe a rate of convergence approaching to p.

In this problem the efficiency of the method is limited by the order of the Runge–Kutta methods. However, as our methods are unconditionally stable we may enhance the efficiency by using Richardson extrapolation to achieve a highly accurate solution in time. We solve the same problem, but now we fix $p = 12$ and take $5 \cdot 2^n$ time steps, with $n = 0, 1, \ldots, 5$. For the third order ESDIRK method we use 60×60 leaf boxes. For the fourth order ESDIRK method we use 90×90 leaf boxes. For the fifth order ESDIRK method we use 120×120 leaf boxes. Table 2 shows that we can easily achieve much higher accuracy by using Richardson extrapolation.

Finally, we solve a problem without an analytic solution. In this problem the initial data

$$u(x, y, t) = 3 \sin(x) \sin(y) e^{-(x^2+y^2)},$$

Table 2 Estimated errors at the final time after Richardson extrapolation

q_{RK}/extrapolations	0	1	2	3	4	5	6
3		1.32 (−1)	1.01 (−2)	1.27 (−4)	1.17 (−5)	6.98 (−8)	8.62 (−10)	7.40 (−6)
4		2.70 (−4)	6.46 (−6)	1.23 (−7)	2.95 (−10)	1.59 (−11)	3.70 (−14)	1.20 (−11)
5		1.28 (−3)	9.67 (−6)	6.30 (−8)	1.86 (−10)	4.11 (−13)	9.27 (−14)	5.08 (−11)

The notation $d(-p)$ means $d \cdot 10^{-p}$

Table 3 Errors computed against a p and h refined solution

p/panels	2	4	8	16	32
8	1.11 (0)	1.39 (−1)	8.74 (−3)	1.50 (−4)	2.45 (−6)
Rate	*	3.00	3.99	5.87	5.92
10	5.87 (−1)	3.16 (−2)	4.62 (−4)	6.17 (−6)	5.21 (−8)
Rate	*	4.21	6.10	6.22	6.89

The errors are maximum errors at the final time $t = 4$. The notation $d(-p)$ means $d \cdot 10^{-p}$

interacts with the weak and slightly non-symmetric potential

$$V(x, y) = 1 - e^{-(x+0.9y)^4},$$

allowing the solution to reach the boundary where we impose homogenous Dirichlet conditions.

We evolve the solution until time $t = 4$ using $p = 8$ and 10 and 2, 4, 8, 16 and 32 leaf boxes in each direction of a domain of size 12×12. The errors computed against a reference solution with $p = 12$ and with 32 leaf boxes can be found in Table 3.

In Fig. 5 we display snapshots of the magnitude of the solution at the initial time $t = 0$, the intermediate times $t \approx 1.07$, $t \approx 1.68$ and at the final time $t = 4.0$.

4 Burgers' Equation in Two Dimensions

As a first step towards a full blown flow solver we solve Burgers' equation in two dimensions using the additive Runge–Kutta methods described in the first part of this paper. Precisely, we solve the system

$$\mathbf{u}_t + \mathbf{u} \cdot \nabla \mathbf{u} = \varepsilon \Delta \mathbf{u}, \quad \mathbf{x} \in [-\pi, \pi]^2, \quad t > 0, \tag{5}$$

where $\mathbf{u} = [u(x, y, t), v(x, y, t)]^T$ is the vector containing the velocities in the x and y directions.

The first problem we solve uses the initial condition $\mathbf{u} = 5[-y, x]^T \exp(-3r^2)$ and the boundary conditions are taken to be no-slip boundary conditions on all sides.

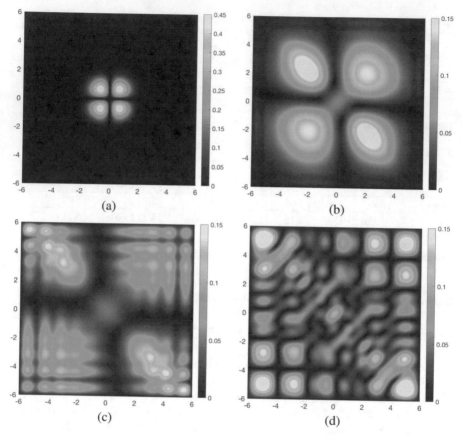

Fig. 5 Snapshots of the magnitude of the solution at the initial time (**a**) $t = 0$, the intermediate times (**b**) $t \approx 1.07$, (**c**) $t \approx 1.68$ and at the final time (**d**) $t = 4.0$

We solve the problem using 24×24 leafs, $p = 24$, $\varepsilon = 0.005$, and the fifth order ARK method found in [6]. We use a time step of $k = 1/80$ and solve until time $t_{max} = 5$. The low viscosity combined with the initial condition produces a rotating flow resembling a vortex that steepens up over time.

In Fig. 6 we can see the velocities at times $t = 0.5$ and $t = 1$. The fluid rotates and expands out and eventually forms a shock like transition. This creates a sharp flow region with large gradients resulting in a flow that may be difficult to resolve with a low order accurate method. These sharp gradients can be seen in the two vorticity plots in Fig. 6 along with the speed and vorticity plots in Fig. 7.

In our second experiment we consider a cross stream of orthogonal flows. We use an initial condition of

$$\mathbf{u} = [8y\, e^{-36\left(\frac{y}{2}\right)^8}, -8x e^{-36\left(\frac{x}{2}\right)^8}]^T, \tag{6}$$

and time independent boundary conditions that are compatible with the initial data.

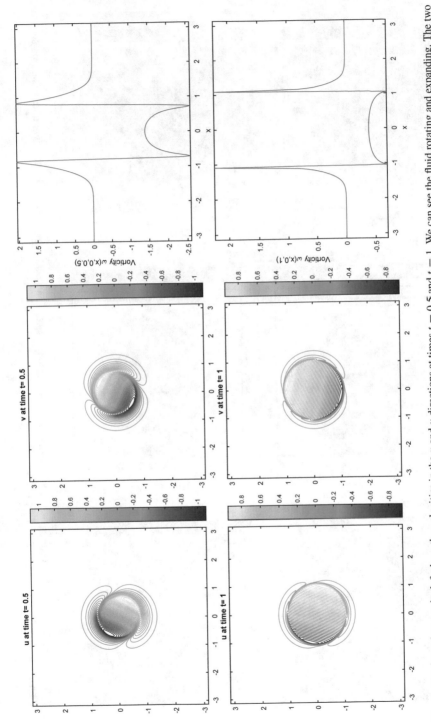

Fig. 6 The four plots on the left show the velocities in the x and y directions at times $t = 0.5$ and $t = 1$. We can see the fluid rotating and expanding. The two plots in the third column show the vorticity at these times. We see sharp gradients near the edge of the rotating fluid

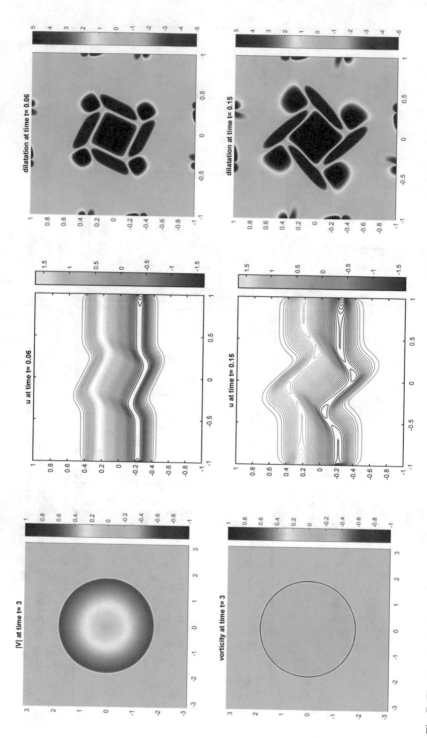

Fig. 7 The plots in the first column correspond to the rotating shock problem. The second and third columns show the velocity in the x direction and the dilatation at times $t = 0.06$ and $t = 0.15$ for the cross flow problem

This initial horizontal velocity drops to zero quickly as we approach $|y| = 0.5$. For $|y| < 0.5$ the exponential term approaches exp(0) and the velocity behaves like $u = 8y$. The flow has changed slightly by $t = 0.06$, but we can see in Fig. 7 the flow is moving to the right for $y > 0$ and the flow is moving the left for $y < 0$ and all significant behavior is in $|y| < 0.5$. A plot of the velocity v would show similar behavior. We also use 24×24 leafs, $p = 24$, $\epsilon = 0.025$, $k = 1/200$, and $t_{max} = 0.75$. We show plots of the horizontal velocity u and the dilatation at time $t = 0.06$ and $t = 0.15$. We only show plots before time $t = 0.15$ when the fluid is hardest to resolve and we observe that after $t = 0.15$ the cross streams begin to dissipate. This problem contains sharp interfaces inside $\mathbf{x} \in [-0.5, 0.5]^2$.

5 Conclusion

In this two part series we have demonstrated that the spectrally accurate Hierarchial Poincaré–Steklov solver can be easily extended to handle time dependent PDE problems with a parabolic principal part by using ESDIRK methods. We have outlined the advantages of the two possible ways to formulate implicit Runge–Kutta methods within the HPS scheme and demonstrated the capabilities on both linear and non-linear examples.

There are many avenues for future work, for example:

- Extension of the solvers to compressible and incompressible flows.
- Application of the current solvers to inverse and optimal design problems, in particular for problems where changes in parameters do not require new factorizations.

References

1. Babb, T., Gillman, A., Hao, S., Martinsson, P.: An accelerated Poisson solver based on multidomain spectral discretization. BIT Numer. Math. **58**, 851–879 (2018)
2. Babb, T., Martinsson, P.-G., Appelö, D.: HPS accelerated spectral solvers for time dependent problems (2018). arXiv:1811.04555
3. Carpenter, M., Gottlieb, D., Abarbanel, S., Don, W.-S.: The theoretical accuracy of Runge–Kutta time discretizations for the initial boundary value problem: a study of the boundary error. SIAM J. Sci. Comput. **16**, 1241–1252 (1995)
4. Gillman, A., Martinsson, P.: A direct solver with $\mathcal{O}(N)$ complexity for variable coefficient elliptic PDEs discretized via a high-order composite spectral collocation method. SIAM J. Sci. Comput. **36**, A2023–A2046 (2014). arXiv:1307.2665
5. Haut, T., Babb, T., Martinsson, P., Wingate, B.: A high-order scheme for solving wave propagation problems via the direct construction of an approximate time-evolution operator. IMA J. Numer. Anal. **36**, 688–716 (2016)
6. Kennedy, C., Carpenter, M.: Additive Runge–Kutta schemes for convection-diffusion-reaction equations. Appl. Numer. Math. **44**, 139–181 (2003)

7. Martinsson, P.: A direct solver for variable coefficient elliptic PDEs discretized via a composite spectral collocation method. J. Comp. Phys. **242**, 460–479 (2013)
8. Rosales, R., Seibold, B., Shirokoff, D., Zhou, D.: Order reduction in high-order Runge–Kutta methods for initial boundary value problems (2017). arXiv:1712.00897

High-Order Finite Element Methods for Interface Problems: Theory and Implementations

Yuanming Xiao, Fangman Zhai, Linbo Zhang, and Weiying Zheng

1 Introduction

The interface problems which involve partial differential equations having discontinuous coefficients across certain interfaces are often encountered in fluid dynamics, electromagnetics and materials science. Because of the low global regularity and the irregular geometry of the interface, the standard numerical methods which are efficient for smooth solutions usually lead to loss in accuracy across the interface.

For arbitrarily shaped interface Γ, it is known that optimal or nearly optimal convergence rate can be recovered if body-fitted finite element meshes are used, see e.g. [6, 8, 20, 29]. Here, by "body-fitted meshes" we mean an element of the underlying mesh is required to intersect with the interface only through its boundaries (Fig. 1). Unfortunately, when the geometry is complex, this usually leads to a nontrivial interface meshing problem. Therefore, numerous modified finite difference methods based only on simple Cartesian grids have been proposed in the literature. We refer to the immersed boundary method [24], the immersed interface method [17, 18], the ghost fluid method [21], and the references therein. In the

Y. Xiao (✉)
Department of Mathematics, Nanjing University, Nanjing, China
e-mail: xym@nju.edu.cn

F. Zhai
Department of Applied Mathematics, Nanjing Forestry University, Nanjing, China

L. Zhang · W. Zheng
State Key Laboratory of Scientific and Engineering Computing, Academy of Mathematics and Systems Science, Chinese Academy of Sciences, Beijing, China

School of Mathematical Sciences, University of Chinese Academy of Sciences, Beijing, China
e-mail: zlb@lsec.cc.ac.cn; zwy@lsec.cc.ac.cn

© The Author(s) 2020
S. J. Sherwin et al. (eds.), *Spectral and High Order Methods for Partial Differential Equations ICOSAHOM 2018*, Lecture Notes in Computational Science and Engineering 134, https://doi.org/10.1007/978-3-030-39647-3_12

Fig. 1 A body-fitted, shape regular mesh

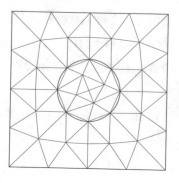

finite element setting, we refer to the work of the immersed finite element method [7, 11, 19], the multiscale finite element method [9], the penalty finite element method [1].

In the past decade, a combination of the extended finite element method (XFEM) with the Nitsche scheme has become a popular discretization method. As the first attempt, an *unfitted finite element method* was proposed in [13] which can be viewed as a linear and consistent modification of [1]. This approach has motivated a number of works, e.g., the unfitted finite element method [4, 5, 12], the Ghost penalty method [2, 3], the unfitted discontinuous Galerkin methods [22]. Although significant progresses in the error analyses of some methods have been made, the development of high-order accurate unfitted FEMs with rigorous error analysis is still challenging. We refer to the work of [14–16, 22, 27, 28] which claim high order approximations. In [22], an *hp-unfitted discontinuous Galerkin method* for Problem (1) was considered, and optimal h-convergence for arbitrary p was shown for the two-dimensional case in the energy norm and in the L^2-norm. With an extra flux penalty term applied on the interface, [27] gave better hp a priori error estimates in both two and three dimensions. In [15, 16], an isoparametric finite element method with a high order geometrical approximation of level set domains was presented. The analysis reveals optimal order error bounds with respect to h for the geometry approximation and for the finite element approximation. In [14, 28], various issues related to unfitted methods was addressed, including the dependence of error estimates on the diffusion coefficients, the condition number of the discrete system, and the choice of stabilization parameters.

The Nitsche-XFEM can be interpreted as applying interior penalty (IP) methods on the interface, and our method falls into this category. The major step in our variant is an appropriate choice of the mesh and geometry dependent weights in the average (see (6)), which lead to trace and inverse inequalities for possibly degenerated sub-elements (see (9)). We note that in our approach, the penalization is applied only to the jump of the solution values across the interface (compared with the bilinear form in [27]). The optimal h-convergence rate for arbitrary high-order discretization in the energy and L_2-norm are proved regardless of the dimension. We refer to [14–16] for the similar estimates with respect to h and [27] for a refined version with respect to both h and p.

Efficient implementations of this method are then discussed in two aspects. We first consider an optimal multigrid solver for the generated linear system. We use the continuous FE space as a "background" subspace, with some smoothing operations added near the interface, to formulate a nested geometrical multigrid method. We prove the optimality of this special multiplicative multigrid method, which means the method converges uniformly with respect to the mesh size, and is independent of the location of the interface relative to the meshes. Since the assembling of the stiffness matrix will require integration over curved surfaces and volumes, we then implement a robust and arbitrarily high order numerical quadrature algorithm by transforming surface and volume integrals into multiple 1-D integrals. The code for the algorithm is freely available in the open source finite element toolbox Parallel Hierarchical Grid (PHG) [26]. We also refer to [23, 25] for different approaches to compute integrals on curved sub-elements and their curved boundaries.

The layout of this paper is as follows. In Sect. 2 we introduce the XFE spaces and reformulate the interface problem (1) in DG schemes. The H^1- and L^2- error estimates of both schemes—which attain the optimal order of the convergence rate in respect to mesh size h—are given. In Sect. 3, we give an optimal multigrid method for the aforementioned DG-XFE schemes. Numerical examples for both two and three dimensions are reported in Sect. 4, to illustrate the high accuracy of the algorithm.

2 XFE and DG Schemes for Interface Problems

We consider the following elliptic interface problem for u: Let $\Omega = \Omega_1 \cup \Gamma \cup \Omega_2$ be a bounded and convex polygonal or polyhedral domain in $\mathbb{R}^d$, $d = 2$ or 3, where Ω_1 and Ω_2 are two subdomains of Ω and are separated by a C^2-smooth interface Γ (see Fig. 2 for an illustration of a unit square that contains a circle as an interface),

$$
\begin{cases}
-\nabla \cdot (\alpha(\mathbf{x})\nabla u) = f, & \text{in } \Omega_1 \cup \Omega_2, \\
[\alpha(\mathbf{x})\nabla u] = g_N, & \text{on } \Gamma, \\
[u] = g_D, & \text{on } \Gamma, \\
u = 0, & \text{on } \partial\Omega.
\end{cases}
\tag{1}
$$

Here $\alpha(\mathbf{x}) = \alpha_i$, $i = 1, 2$, is a piecewise constant function on the partition $\Omega_1 \cup \Omega_2$.

Denote by $\{\mathcal{T}_h\}$, a family of conforming, quasi-uniform, and regular partitions of Ω into triangles and parallelograms/tetrahedrons and parallelepipeds. As K is of regular shape, there is a constant γ_0 such that

$$
h_K^d \leq \gamma_0 |K|, \quad \forall K \in \mathcal{T}_h.
\tag{2}
$$

We define the set of all elements intersected by Γ as $\mathcal{T}_h^\Gamma = \{K \in \mathcal{T}_h : |K \cap \Gamma| \neq 0\}$. Each $\mathcal{T}_h^\Gamma$ induces a partition of interface Γ, which we denote by $\mathcal{E}_h^\Gamma = \{e_K : e_K = $

Fig. 2 Domain
$\Omega = \Omega_1 \cup \Gamma \cup \Omega_2$ with an
unfitted mesh

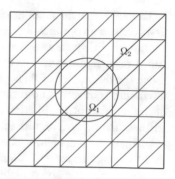

$K \cap \Gamma, K \in \mathcal{T}_h^\Gamma\}$. For any $K \in \mathcal{T}_h^\Gamma$, let $K_i = K \cap \Omega_i$ denote the part of K in Ω_i and $\boldsymbol{n}_i$ be the unit outward normal vector on ∂K_i with $i = 1, 2$. As Γ is of class C^2, it is easy to prove that (cf.[6, 31]) each interface segment/patch e_K is contained in a strip of width δ and satisfies

$$\delta \le \gamma_1 h_K^2 \quad \text{and} \quad |\boldsymbol{n}_i(\mathbf{x}) - \boldsymbol{n}_i(\mathbf{y})| \le \gamma_2 h_K, \quad \forall \mathbf{x}, \mathbf{y} \in e_K. \tag{3}$$

We define the weighted average $\{\cdot\}$ and the jump $[\cdot]$ on $e \in \mathcal{E}_h^\Gamma$ by

$$\{v\} = \kappa_1 v_1 + \kappa_2 v_2, \qquad\qquad [v] = v_1 \boldsymbol{n}_1 + v_2 \boldsymbol{n}_2, \tag{4}$$

$$\{\boldsymbol{q}\} = \kappa_1 \boldsymbol{q}_1 + \kappa_2 \boldsymbol{q}_2, \qquad\qquad [\boldsymbol{q}] = \boldsymbol{q}_1 \cdot \boldsymbol{n}_1 + \boldsymbol{q}_2 \cdot \boldsymbol{n}_2. \tag{5}$$

For the stability analysis of our schemes, we define (κ_1, κ_2) on each element as follows:

$$\kappa_i = \begin{cases} 0, & \text{if } \frac{|K_i|}{|K|} < c_0 h_K, \\ 1, & \text{if } \frac{|K_i|}{|K|} > 1 - c_0 h_K, \\ \frac{|K_i|}{|K|}, & \text{otherwise}. \end{cases} \tag{6}$$

Clearly, $0 \le \kappa_i \le 1$ and $\kappa_1 + \kappa_2 = 1$ so that $\{\cdot\}$ is a convex combination along Γ. Roughly speaking, we adopt the weight $\kappa_i = \frac{|K_i|}{|K|}$ suggested in [13] for general sub-elements and we set $\kappa_i = 0$ for $|K_i| < c h_K^{d+1}$. Here, the user-defined constant $c_0 \ge 2\gamma_0 \gamma_1$ and γ_0, γ_1 are constants defined in (2) and (3), respectively. The dependence of c_0 on these generic constants is elaborated in Lemma 1.

Let χ_i be the characteristic function on Ω_i with $i = 1, 2$. Given a mesh $\mathcal{T}_h$, let V_h be the continuous piecewise polynomial function space of degree $p \ge 1$ on the mesh. Let $V_h^0 := V_h \cap H_0^1(\Omega)$, $V_h^1 := V_h^0 \cdot \chi_1$ and $V_h^2 := V_h^0 \cdot \chi_2$. We define the XFE space as $V_h^\Gamma = V_h^1 + V_h^2$.

Then, the DG-XFE method for the interface problem is: Find $u_h \in V_h^\Gamma$ such that

$$B_h(u_h, v_h) = F_h(v_h), \quad \forall v_h \in V_h^\Gamma, \tag{7}$$

where

$$B_h(w, v) := \int_{\Omega_1 \cup \Omega_2} \alpha(x) \nabla w \cdot \nabla v - \int_{\Gamma} \{\alpha(x) \nabla w\} \cdot [v]$$

$$- \beta \int_{\Gamma} [w] \cdot \{\alpha(x) \nabla v\} + \sum_{K \in \mathcal{T}_h^{\Gamma}} \frac{\eta_\beta}{h_K} \int_{K \cap \Gamma} [w] \cdot [v],$$

$$F_h(v) := \int_{\Omega} f v + \int_{\Gamma} g_N(\kappa_1 v_2 + \kappa_2 v_1)$$

$$- \beta \int_{\Gamma} g_D \cdot \{\alpha(x) \nabla v\} + \sum_{K \in \mathcal{T}_h^{\Gamma}} \frac{\eta_\beta}{h_K} \int_{K \cap \Gamma} g_D \cdot [v],$$

For η_β sufficiently large, the norm corresponding to the bilinear form $B_h(\cdot, \cdot)$ is uniformly equivalent to $\| \cdot \|_{B_h}$, which is defined by

$$\|v\|_{B_h}^2 = |v|_{1,\Omega_1 \cup \Omega_2}^2 + \sum_{K \in \mathcal{T}_h^{\Gamma}} \eta_\beta h_K^{-1} \|[v]\|_{L^2(e_K)}^2 + \sum_{K \in \mathcal{T}_h^{\Gamma}} \eta_\beta^{-1} h_K \|\{\alpha(x) \nabla v\}\|_{L^2(e_K)}^2. \quad (8)$$

The crucial component in regard to establishing this equivalence result and also the stability of bilinear forms is the control on the weighted normal derivatives, which is stated as a trace and inverse inequality in Lemma 1.

Lemma 1 ([27, 28]) *Let γ_0 and γ_1 be constants defined in (2) and (3), respectively. If we choose $c_0 \geq 2\gamma_0\gamma_1$ in the definition (6) of κ, there exists a positive constant h_0 such that for all $h \in (0, h_0]$ and any interface segment/patch $e_K = K \cap \Gamma \in \mathcal{E}_h^{\Gamma}$, the following estimates hold on both sub-elements of K:*

$$\|\kappa_i^{1/2} v_i\|_{L^2(e_K)} \leq \frac{C}{h_K^{1/2}} \|v_i\|_{L^2(K_i)}, \quad v_i \in \mathcal{P}_p(K_i), \ i = 1, 2. \quad (9)$$

The coercivity and boundedness of $B_h(\cdot, \cdot)$ in its norm $\| \cdot \|_{B_h}^2$ is then a direct consequence of the Cauchy–Schwarz inequality.

Lemma 2 *Let $V = H^2(\Omega_1 \cup \Omega_2)$ and $V(h) = V_h^{\Gamma} + V$, we have*

$$B_h(w, v) \leq C_b \|w\|_{B_h} \|v\|_{B_h}, \quad \forall w, v \in V(h), \quad (10)$$

and

$$B_h(v, v) \geq C_s \|v\|_{B_h}^2, \quad \forall v \in V_h^{\Gamma}, \quad (11)$$

provided the penalty parameter η_β is chosen sufficiently large.

The XFE space has optimal approximation quality for piecewise smooth functions in $H^p(\Omega_1 \cup \Omega_2)$. The following theorem is proved in [28] as an analogue of Cea's lemma.

Theorem 1 *Assume that the interface Γ is C^2 smooth and that the solution of the elliptic interface problem (1) satisfies $u \in H^s(\Omega_1 \cup \Omega_2)$, where $s \geq 2$ is an integer. Let $\mu = \min\{p + 1, s\}$. The following error estimates hold for any $h \in (0, h_0]$: If η_β is chosen sufficiently large (see (11)) and u_h is the solution to the first scheme of (7), then*

$$\|u - u_h\|_{B_h} \lesssim h^{\mu-1}\|u\|_{H^s(\Omega_1 \cup \Omega_2)}, \quad \forall 0 < h \leq h_0. \tag{12}$$

The hidden constants in the above estimates are dependent on the angle condition of the mesh $\mathcal{T}_h$, the degree of the polynomials, the parameter in the scheme, and $\alpha(\mathbf{x})$, but are independent of the location of the interface relative to the mesh. Here, the constant h_0 is from Lemma 1.

3 An Optimal Multigrid Method for (7)

In this section, we propose a two-level geometric multigrid solver of the finite element problem (7). It is well known that the element K with a "small" cut (i.e. $|K \cap \Omega_i|/|K| \ll 1$) would have adverse effect on the conditioning of the resulting stiffness matrices (see e.g. [3]). Our approach is based on the general theory of the successive subspace correction (SSC) method of solving on a linear vector space $\tilde{V} = \sum_{i=0}^{J} V_i$ with inner product $(\cdot, \cdot)$ the equation $(Au, v) = (f, v)$, where $A : \tilde{V} \to \tilde{V}$ is a symmetric positive definite operator.

We apply SSC for a relatively simple case of two subspaces (i.e. $J = 2$), that is, $\tilde{V} = V_h^\Gamma = V_1 \oplus V_2$, with $V_1 = V_h^0$ and $V_2 = \tilde{V}_h^\Gamma$, where $\tilde{V}_h^\Gamma \subset V_h^\Gamma$ is the space of nodal basis functions that vanish on $\tilde{\mathcal{N}}_h := \{x_j : |\text{supp}(\psi_j) \cap \Gamma| = 0\}$. With a slight abuse of notation, the DG-XFE scheme induces a symmetric positive definite operator B_h for $\beta = 1$. Let $\tilde{B}_h$ and $\tilde{B}_h^\Gamma$ be the restrictions of B_h on V_h^0 and $\tilde{V}_h^\Gamma$, respectively. Let $\tilde{R}_h : V_h^0 \to V_h^0$ be approximately an inverse of $\tilde{B}_h$. We have this two-level successive subspace correction method (Algorithm 1). The similar idea has been employed in a special linear case in [32] and analyzed using the framework given in [30, 33].

Algorithm 1 The multigrid method for (7)

Implement this iterative procedure until converge:

1. do subspace correction on V_h^0 with an inexact solver $\tilde{R}_h$;
2. do subspace correction on $\tilde{V}_h^\Gamma$ with an exact solver $(\tilde{B}_h^\Gamma)^{-1}$.

Obviously, Algorithm 1 defines an iterative method for solving $B_h u_h = f_h$. Denote by A_h the iterator of the method, then the error contract property is summarized as the following theorem.

Theorem 2 ([28]) *Assume that* $\|I - \tilde{R}_h \tilde{B}_h\|_{\tilde{B}_h} \leqslant \rho < 1$. *Then Algorithm 1 is uniformly convergent with respect to the mesh size with*

$$\|I - A_h B_h\|_{B_h}^2 \leqslant \frac{\Lambda}{1 - \rho^2 + \Lambda},$$

where Λ *is a constant independent of* h.

When $\mathcal{T}_h$ is a shape-regular grid with a geometrical multilevel structure, then a geometric multigrid process can be implemented on V_h^0, and the approximate inverse $\tilde{R}_h$ of $\tilde{B}_h$ can be chosen to be the iterator of V-cycle multigrid method.

4 Numerical Tests

In this section, we present some initial results to demonstrate the high-order accuracy and robustness of our method. A 2-D example was implemented in MATLAB. The numerical experiment for a 3-D case was carried out in the open source finite element toolbox PHG [26].

4.1 High-Order Numerical Quadratures on "cut" Elements

Assembling the local stiffness matrix and the corresponding RHS for $K \in \mathcal{T}_h^\Gamma$ requires integration over irregularly shaped manifolds:

$$I = \int_{K \cap \Omega_i} u(\mathbf{x}) d\mathbf{x} \quad \text{and} \quad I = \int_{K \cap \Gamma} u(\mathbf{x}) d\Gamma, \tag{13}$$

where Γ is defined by the zero level set of a piecewise smooth function.

Our implementation of (13) relies on a general-purpose and arbitrarily high order numerical quadrature algorithm proposed in [10]. The basic idea is to choose a local coordinate system with three orthogonal directions, decompose integrals in (13) into multiple 1-D integrals along these directions, and use 1-D Gaussian quadratures to compute these integrals. For 1-D Gaussian quadratures to work, the local coordinate system should be suitably chosen according to properties of K and Γ to prevent essential singularities from appearing in the 1-D integrands, and the integration intervals are divided into subintervals at the non essential singularities of the integrands. We note that the proposed algorithm only requires finding roots of univariate nonlinear functions in given intervals and evaluating the integrand, the level set function, and the gradient of the level set function at given points. It can achieve arbitrarily high order by increasing the orders of Gaussian quadratures, and does not need extra a priori knowledge about the integrand and the level set function.

This algorithm has been implemented in the file `src/quad-interface.c` and `include/phg/quad-interface.h` in PHG [26]. Extensive $h-$ and $p-$convergence tests have been performed in [10] and included in a sample code `test/quad_test2.c`.

4.2 2-D Numerical Examples

Let domain Ω be the unit square $(0, 1)^2$ and interface Γ be the zero level set of the function $\varphi(x) = (x_1 - 0.5)^2 + (x_2 - 0.5)^2 - 1/7$. The subdomain Ω_1 is characterized by $\varphi(x) < 0$ and Ω_2 by $\varphi(x) > 0$. The domain Ω is partitioned into grids of squares with the same size h. The exact solution is chosen as

$$u(x_1, x_2) = \begin{cases} 1/\alpha_1 \exp(x_1 x_2), & (x_1, x_2) \in \Omega_1, \\ 1/\alpha_2 \sin(\pi x_1) \sin(\pi x_2), & (x_1, x_2) \in \Omega_2. \end{cases}$$

The right-hand side can be computed accordingly.

We implement Algorithm 1, with V-cycle geometric multigrid based on the unfitted grid $\mathcal{T}_h$ playing as the coarse grid corrector. In each pre- and post-smoothing stage of V-cycle iterator, we perform Gauss-Seidel for two times. We record the numerical results in Table 1. In these examples, the initial guess is $\mathbf{0}$, and the stopping criterion is

$$\|f_h - B_h u_h^{(k)}\|_\infty / \|f_h - B_h u_h^{(0)}\|_\infty < 10^{-10}.$$

From Table 1, we can see that the multigrid method converges uniformly with respect to the mesh size, which confirms our theoretical results.

Table 1 Numerical performance of Algorithm 1 (2-D example)

		h	2^{-2}	2^{-3}	2^{-4}	2^{-5}	2^{-6}
$\alpha_1 : \alpha_2 = 1 : 10$	$p = 1$	#iter	7	10	10	11	12
	$p = 2$	#iter	13	10	12	13	14
$\alpha_1 : \alpha_2 = 10 : 1$	$p = 1$	#iter	24	30	29	27	25
	$p = 2$	#iter	24	23	22	21	20

4.3 3-D Numerical Examples

The settings of this numerical experiment are as follows. The domain $\Omega = (0, 1)^3$. The interfaces are two touched spheres of radius 0.1 centered at $(0.4, 0.5, 0.5)$ and $(0.6, 0.5, 0.5)$. The exact solution is given by

$$u(x_1, x_2, x_3) = \begin{cases} \exp(x_1 + x_2 + x_3), & (x_1, x_2, x_3) \in \Omega_1, \\ \sin(x_1)\sin(x_2)\sin(x_3), & (x_1, x_2, x_3) \in \Omega_2. \end{cases}$$

The discontinuous coefficient function is defined such that $\alpha_1 = 1$ and $\alpha_2 = 100$.

A convergence study is performed on a series of meshes generated by uniform refinements of an initial mesh consisting of 6 congruent tetrahedra. Relative errors and convergence rates of numerical solutions for P_p elements for $p = 1, 2, 3$ and 4 are listed in Table 2, with the quadrature order $q = 2p + 3$. The convergence rates are optimal for both $H^1(\Omega)$-errors (order p) and $L^2(\Omega)$-errors (order $p + 1$). For the

Table 2 Errors and convergence orders of the numerical solutions (3-D example)

Number of elements	Degrees of freedom	Relative H^1 error		Relative L^2 error	
		Error	Order	Error	Order
P_1 element($p = 1, q = 2p + 3 = 5$)					
768	189	1.690e−01	–	1.686e−02	–
6144	1241	7.510e−02	1.17	3.403e−03	2.31
49,152	9009	3.514e−02	1.10	9.618e−04	1.82
393,216	68,705	1.658e−02	1.08	2.272e−04	2.08
3,145,728	536,769	8.145e−03	1.03	4.869e−05	2.22
P_2 element($p = 2, q = 2p + 3 = 7$)					
768	1241	9.041e−03	–	4.150e−04	–
6144	9009	2.026e−03	2.16	4.323e−05	3.26
49,152	68,705	4.973e−04	2.03	5.171e−06	3.06
393,216	536,769	1.234e−04	2.01	6.413e−07	3.01
3,145,728	4,243,841	3.070e−05	2.01	7.965e−08	3.01
P_3 element ($p = 3, q = 2p + 3 = 9$)					
768	3925	2.175e−03	–	8.394e−05	–
6144	29,449	3.793e−05	5.84	5.864e−07	7.16
49,152	228,241	4.743e−06	3.00	3.683e−08	3.99
393,216	1,797,409	5.932e−07	3.00	2.321e−09	3.99
3,145,728	14,266,945	7.414e−08	3.00	1.456e−10	3.99
P_4 element ($p = 4, q = 2p + 3 = 11$)					
768	9009	2.971e−03	–	9.606e−05	–
6144	68,705	6.042e−07	12.26	7.560e−09	13.63
49,152	536,769	3.778e−08	4.00	2.380e−10	4.99
393,216	4,243,841	2.362e−09	4.00	7.481e−12	4.99

time being, however, the design of multigrid solver for 3-D case is still on-going. The computations for P_1, P_2 and P_3 elements were done using the 64-bit double precision and the linear systems were solved using MUMPS, but for P_4 element, to eliminate influences of roundoff errors, the computations were done using the 80-bit extended double precision and the linear systems were solved using the GMRES method with MUMPS in double precision as its preconditioner. The performance of Algorithm 1 will be reported in a future work.

Acknowledgements The authors would like to acknowledge the funding support of this research by the National Key Research and Development Program of China under grant number 2017YFC0209804, National Natural Science Foundation of China under grant number 11101208, and National Center for Mathematics and Interdisciplinary Sciences of Chinese Academy of Sciences. The authors are grateful to Jinchao Xu and Haijun Wu for the fruitful discussions and suggestions, to Dr. Huaqing Liu for his valuable help on preparing the numerical examples.

References

1. Babuška, I.: The finite element method for elliptic equations with discontinuous coefficients. Computing **5**, 207–213 (1970)
2. Becker, R., Burman, E., Hansbo, P.: A Nitsche extended finite element method for incompressible elasticity with discontinuous modulus of elasticity. Comput. Methods Appl. Mech. Eng. **198**, 3352–3360 (2009)
3. Burman, E.: Ghost penalty. C.R. Math. **348**, 1217–1220 (2010)
4. Burman, E., Hansbo, P.: Fictitious domain finite element methods using cut elements: II. A stabilized Nitsche method. Appl. Num. Math. **62**, 328–341 (2012)
5. Burman, E., Guzman, J., Sanchez, M.A., Sarkis, M.: Robust flux error estimation of an unfitted Nitsche method for high-contrast interface problems. IMA J. Numer. Anal. **38**, 646–668 (2018)
6. Chen, Z., Zou, J.: Finite element methods and their convergence for elliptic and parabolic interface problems. Numer. Math. **79**, 175–202 (1998)
7. Chen, Z., Xiao, Y., Zhang, L.: The adaptive immersed interface finite element method for elliptic and Maxwell interface problems. J. Comput. Phys. **228**, 5000–5019 (2009)
8. Chen, Z., Wu, Z., Xiao, Y.: An adaptive immersed finite element method with arbitrary Lagrangian-Eulerian scheme for parabolic equations in time variable domains. Int. J. Numer. Anal. Model. **12**, 567–591 (2015)
9. Chu, C.-C., Graham, I.G., Hou, T.Y.: A new multiscale finite element method for high-contrast elliptic problems. Math. Comput. **79**, 1915–1955 (2010)
10. Cui, T., Leng, W., Liu, H., Zhang, L., Zheng, W.: High-order numerical quadratures in a tetrahedron with an implicitly defined curved interface, ACM Transactions on Mathematical Software, in press.
11. Gong, Y., Li, B., Li, Z.: Immersed-interface finite-element methods for elliptic interface problems with non-homogeneous jump conditions. SIAM J. Numer. Anal. **46**, 472–495 (2008)
12. Hansbo, P.: Nitsche's method for interface problems in computational mechanics. GAMM-Mitt. **47**, 183–206 (2005)
13. Hansbo, A., Hansbo, P.: An unfitted finite element method, based on Nitsche's method for elliptic interface problems. Comput. Methods Appl. Mech. Eng. **191**, 5537–5552 (2002)
14. Huang, P., Wu, H., Xiao, Y.: An unfitted interface penalty finite element method for elliptic interface problems. Comput. Methods Appl. Mech. Eng. **323**, 439–460 (2017)
15. Lehrenfeld, C., Reusken, A.: L^2-error analysis of an isoparametric unfitted finite element method for elliptic interface problems (2016). Preprint. arXiv:1604.04529

16. Lehrenfeld, C., Reusken, A.: Analysis of a high-order unfitted finite element method for elliptic interface problems. IMA J. Numer. Anal. **38**, 1351–1387 (2018)
17. LeVeque, R., Li, Z.: The immersed interface method for elliptic equations with discontinuous coefficients and singular sources. SIAM J. Numer. Anal. **31**, 1019–1044 (1994)
18. Li, Z., Ito, K.: The Immersed Interface Method: Numerical Solutions of PDEs Involving Interfaces and Irregular Domains. SIAM, Philadephia (2006)
19. Li, Z., Lin, T., Wu, X.: New Cartesian grid methods for interface problems using the finite element formulation. Numer. Math. **96**, 61–98 (2003)
20. Li, J., Melenk, J.M., Wohlmuth, B., Zou, J.: Optimal a priori estimates for higher order finite elements for elliptic interface problems. Appl. Numer. Math. **60**, 19–37 (2010)
21. Liu, X., Fedkiw, R.P., Kang, M.: A boundary condition capturing method for Poisson's equation on irregular domains. J. Comput. Phys. **160**, 151–178 (2000)
22. Massjung, R.: An unfitted discontinuous Galerkin method applied to elliptic interface problems. SIAM J. Numer. Anal. **50**(6), 3134–3162 (2012)
23. Müller, B., Kummer, F., Oberlack, M.: Highly accurate surface and volume integration on implicit domains by means of moment-fitting. Int. J. Numer. Methods Eng. **96**, 512–528 (2013)
24. Peskin, C.S.: Numerical analysis of blood flow in the heart. J. Comput. Phys. **25**, 220–252 (1977)
25. Saye, R.I.: High-order quadrature methods for implicitly defined surfaces and volumes in hyperrectangles. SIAM J. Sci. Comput. **37**, A993–A1019 (2015)
26. The toolbox Parallel Hierarchical Grid (PHG). http://lsec.cc.ac.cn/phg
27. Wu, H., Xiao, Y.: An unfitted hp-interface penalty finite element method for elliptic interface problems. J. Comput. Math. **37**, 316–339 (2019). http://arXiv.org/abs/1007.2893
28. Xiao, Y., Xu, J., Wang, F.: High-order extended finite element methods for solving interface problems. Comput. Methods Appl. Mech. Eng. **364**, 112964 (2020). https://doi.org/10.1016/j.cma.2020.112964
29. Xu, J.: Estimate of the convergence rate of finite element solutions to elliptic equations of second order with discontinuous coefficients (in Chinese). Nat. Sci. J. Xiangtan Univ. **1**, 1–5 (1982)
30. Xu, J.: Iterative methods by space decomposition and subspace correction. SIAM Rev. **34**, 581–613 (1992)
31. Xu, J.: Estimate of the convergence rate of finite element solutions to elliptic equations of second order with discontinuous coefficients (2013, preprint). arXiv:1311.4178
32. Xu, J., Zhang, S.: Optimal finite element methods for interface problems. In: Dickopf, T., et al. (eds.), Domain Decomposition Methods in Science and Engineering XXII. Lecture Notes in Computational Science and Engineering, vol. 104, pp. 77–91. Springer, Cham (2016)
33. Xu, J., Zikatanov, L.: The method of alternating projections and the method of subspace corrections in Hilbert space. J. Am. Math. Soc. **15**, 573–597 (2002)

Stabilised Hybrid Discontinuous Galerkin Methods for the Stokes Problem with Non-standard Boundary Conditions

Gabriel R. Barrenechea, Michał Bosy, and Victorita Dolean

1 Introduction

The interest of this paper is to discretise the Stokes problem with non-standard boundary conditions. In [1], a hybrid discontinuous Galerkin (hdG) method was proposed and analysed for this problem. The finite element method used was the combination of BDM elements of order k for the velocity, and discontinuous elements of order $k - 1$ for the pressure. In this paper we increase the order of the pressure space to k, while keeping the order for the velocity space fixed as k. Since this pair does not satisfy the inf-sup condition, a stabilisation term needs to be added.

The stabilisation term referred to above can be built using a diversity of approaches, but, roughly speaking, the stabilisation can be residual or non-residual. In [8] the authors added a mesh-dependent term penalising the gradient of the pressure to the formulation. Later, in [14] this method was restricted and reinterpreted

G. R. Barrenechea
Department of Mathematics and Statistics, University of Strathclyde, Glasgow, UK
e-mail: gabriel.barrenechea@strath.ac.uk

M. Bosy
Department of Mathematics and Statistics, University of Strathclyde, Glasgow, UK

Dipartimento di Matematica "F. Casorati", Universitá degli Studi di Pavia, Pavia, Italy
e-mail: michal.bosy@unipv.it

V. Dolean (✉)
Department of Mathematics and Statistics, University of Strathclyde, Glasgow, UK

University Côte d'Azur, CNRS, LJAD, Nice Cedex, France
e-mail: work@victoritadolean.com

© The Author(s) 2020
S. J. Sherwin et al. (eds.), *Spectral and High Order Methods for Partial Differential Equations ICOSAHOM 2018*, Lecture Notes in Computational Science and Engineering 134, https://doi.org/10.1007/978-3-030-39647-3_13

as a Petrov–Galerkin scheme leading to the first consistent stabilised method, and further developments were presented in the works [7] and [13]. For a review of different residual stabilised finite element methods for the Stokes problem, see the review paper [2].

Now, due to their nature, residual methods include unphysical couplings to the formulation, and modify all the entries of the stiffness matrix. Hence, non-residual methods where only a positive semi-definite term penalising the pressure is added have also being proposed. Examples of this type of methods are the pressure gradient projection [9] and local pressure gradient stabilisation [3]. The methods just mentioned typically use two nested meshes in order to build the method. Thus, to avoid this complication, the local pressure gradient stabilisation has been also presented on the same mesh in [12]. Additionally, methods that use fluctuations of the pressure gradient are not effective when the finite element space for pressure is the piecewise constant space. The usual way to overcome this is to add pressure jumps to the formulation, as it has been done, e.g., in [16]. These have been shown to be very effective, but they do somehow temper with the data structure of the code. To avoid this, the authors in [10] present an approach that is based on polynomial-pressure-projection. This method works for low order of polynomials as was shown in [4], and preserves symmetry of the original equation.

In the light of the discussion of the previous paragraphs, in this work we propose a stabilised hdG method for the Stokes problem with non-standard boundary conditions. The method is reminiscent of the Dorhmann–Bochev method (from [10]), but uses the same velocity space used in the hdG method from [1].

1.1 Notations and Model Problem

Let Ω be an open polygonal domain in $\mathbb{R}^2$ with Lipschitz boundary $\Gamma := \partial\Omega$. We use boldface font for tensor or vector variables e.g. u is a velocity vector field. The scalar variables will be italic e.g. p denotes pressure scalar value. We define the stress tensor $\sigma := \nu\nabla u - p I$ (where $\nu > 0$ is the fluid viscosity and I is the identity matrix) and the flux as $\sigma_n := \sigma\, n$. In addition, we denote normal and tangential components as follows $u_n := u \cdot n$, $u_t := u \cdot t$, $\sigma_{nn} := \sigma_n \cdot n$, where n is the outward unit normal vector to the boundary Γ and t is a vector tangential to Γ such that $n \cdot t = 0$.

For $D \subset \Omega$, we use the standard $L^2(D)$ space with the following norm

$$\|f\|_D^2 := \int_D f^2 \, dx \text{ for all } f \in L^2(D).$$

Let us define, for $m \in \mathbb{N}$, the following Sobolev spaces

$$H^m(D) := \left\{ v \in L^2(D) : \forall |\alpha| \leq m \; \partial^\alpha v \in L^2(D) \right\},$$

$$H(div, D) := \left\{ v \in [L^2(D)]^2 : \nabla \cdot v \in L^2(D) \right\},$$

where, for $\alpha = (\alpha_1, \alpha_2) \in \mathbb{N}^2$, $|\alpha| = \alpha_1 + \alpha_2$, and $\partial^\alpha = \frac{\partial^{|\alpha|}}{\partial x_1^{\alpha_1} \partial x_2^{\alpha_2}}$. In addition, we will use the standard semi-norm and norm for the Sobolev space $H^m(D)$

$$|f|^2_{H^m(D)} := \sum_{|\alpha|=m} \|\partial^\alpha f\|^2_D, \quad \|f\|^2_{H^m(D)} := \sum_{k=0}^{m} |f|^2_{H^k(D)} \; \forall f \in H^m(D).$$

In this work, we consider the two dimensional Stokes problem with tangential-velocity and normal-flux (TVNF) boundary conditions

$$\begin{cases} -\nu \Delta u + \nabla p = f & \text{in } \Omega, \\ \nabla \cdot u = 0 & \text{in } \Omega, \\ \sigma_{nn} = g & \text{on } \Gamma, \\ u_t = 0 & \text{on } \Gamma, \end{cases} \tag{1}$$

where $u : \bar{\Omega} \to \mathbb{R}^2$ is the unknown velocity field, $p : \bar{\Omega} \to \mathbb{R}$ the pressure, $\nu > 0$ the viscosity, which is considered to be constant, and $f \in [L^2(\Omega)]^2$, $g \in L^2(\Gamma)$ are given functions. The restriction to homogeneous Dirichlet conditions on u_t is made only to simplify the presentation.

Let $\{\mathcal{T}_h\}_{h>0}$ be a regular family of triangulations of $\bar{\Omega}$ made of triangles. For each triangulation $\mathcal{T}_h$, $\mathcal{E}_h$ denotes the set of its edges. In addition, for each of element $K \in \mathcal{T}_h$, $h_K := \text{diam}(K)$, and we denote $h := \max_{K \in \mathcal{T}_h} h_K$. We define following Sobolev spaces on the triangulation $\mathcal{T}_h$ and the set of all edges in $\mathcal{E}_h$

$$L^2(\mathcal{E}_h) := \left\{ v : v|_E \in L^2(E) \; \forall E \in \mathcal{E}_h \right\},$$

$$H^m(\mathcal{T}_h) := \left\{ v \in L^2(\Omega) : v|_K \in H^m(K) \; \forall K \in \mathcal{T}_h \right\} \text{ for } m \in \mathbb{N},$$

with the corresponding broken norms.

Now we will introduce the finite element spaces that discretise the above spaces. Let $k \geq 1$. We start by introducing the velocity and pressure spaces. To discretise the velocity u we use the Brezzi–Douglas–Marini space (see [5, Section 2.3.1]) of order $k \geq 1$ defined by

$$BDM_h^k := \left\{ v_h \in H(div, \Omega) : v_h|_K \in \left[\mathbb{P}_k(K) \right]^2 \; \forall K \in \mathcal{T}_h \right\}.$$

Associated to this space, we introduce the BDM projection $\Pi^k : [H^1(\Omega)]^2 \to BDM_h^k$ defined in [5, Section 2.5]. The pressure is discretised using the following space

$$Q_h^k := \left\{ q_h \in L^2(\Omega) : q_h|_K \in \mathbb{P}_k(K) \ \forall K \in \mathcal{T}_h \right\}.$$

Associated to this space we define the local $L^2(K)$-projection $\Psi_K^k : L^2(K) \to \mathbb{P}_k(K)$ for each $K \in \mathcal{T}_h$ defined as follows. For every $w \in L^2(K)$, $\Psi_K^k(w)$ is the unique element of $\mathbb{P}_k(K)$ satisfying $\int_K \Psi_K^k(w)v_h dx = \int_K wv_h dx \ \forall v_h \in \mathbb{P}_k(K)$, and we define the continuous projection $\Psi^k|_K = \Psi_K^k$ for all $K \in \mathcal{T}_h$.

The last ingredient needed in the method described below is a finite element space associated to a family of Lagrange multipliers associated to the edges of the triangulation. These multipliers will be denoted by $\tilde{u}$ and are meant to approximate the tangential trace of the velocity u on the edges of the triangulation. For this, and in order to propose a discretisation with fewer degrees of freedom, we discretise the Lagrange multiplier $\tilde{u}$ using the space

$$M_{h,0}^{k-1} := \left\{ \tilde{v}_h \in L^2(\mathcal{E}_h) : \tilde{v}_h|_E \in \mathbb{P}_{k-1}(E) \ \forall E \in \mathcal{E}_h, \ \tilde{v}_h = 0 \text{ on } \Gamma \right\}.$$

Furthermore, we introduce for all $E \in \mathcal{E}_h$ the $L^2(E)$-projection $\Phi_E^{k-1} : L^2(E) \to \mathbb{P}_{k-1}(E)$ defined as follows. For every $\tilde{w} \in L^2(E)$, $\Phi_E^{k-1}(\tilde{w})$ is the unique element of $\mathbb{P}_{k-1}(E)$ satisfying $\int_E \Phi_E^{k-1}(\tilde{w})\tilde{v}_h \, ds = \int_E \tilde{w}\tilde{v}_h \, ds \ \forall \tilde{v}_h \in \mathbb{P}_{k-1}(E)$, and we denote $\Phi^{k-1} : L^2(\mathcal{E}_h) \to M_h^{k-1}$ defined as $\Phi^{k-1}|_E := \Phi_E^{k-1}$ for all $E \in \mathcal{E}_h$.

2 The Stabilised Method

Our approach is to write the discrete problem with the same degree of polynomials for velocity and pressure spaces. In other words, denoting $V_h := BDM_h^k \times M_{h,0}^{k-1}$, we want to use the space $V_h \times Q_h^k$, instead of $V_h \times Q_h^{k-1}$ as it was done in [1]. To do this, we need the proper stabilisation term, because this choice of spaces does not guarantee inf-sup stability.

The first ingredient in the definition of the stabilised method for (1) we use the same bilinear forms as in [1], this is

$$a\left((w_h, \tilde{w}_h), (v_h, \tilde{v}_h)\right) := \sum_{K \in \mathcal{T}_h} \left(\int_K \nu \nabla w_h : \nabla v_h \, dx \right.$$

$$\left. - \int_{\partial K} \nu (\partial_n w_h)_t \left((v_h)_t - \tilde{v}_h\right) ds + \varepsilon \int_{\partial K} \nu \left((w_h)_t - \tilde{w}_h\right) (\partial_n v_h)_t \, ds \right.$$

$$+ \nu \frac{\tau}{h_K} \int_{\partial K} \Phi^{k-1}\big((\boldsymbol{w_h})_t - \tilde{w}_h\big) \Phi^{k-1}\big((\boldsymbol{v_h})_t - \tilde{v}_h\big)\, ds\Bigg)$$

$$b\big((\boldsymbol{v_h}, \tilde{v}_h), q_h\big) := - \sum_{K \in \mathcal{T}_h} \int_K q_h \nabla \cdot \boldsymbol{v_h}\, dx,$$

where $\varepsilon \in \{-1, 1\}$ and $\tau > 0$ is a stabilisation parameter. In addition, to compensate for the non-inf-sup stability of the finite element spaces we have chosen, we introduce the bilinear form

$$s\left(p_h, q_h\right) := \frac{1}{\nu} \int_\Omega \left(p_h - \Psi^{k-1} p_h\right)\left(q_h - \Psi^{k-1} q_h\right)\, dx.$$

With these ingredients we can now present the finite element method analysed in this work: Find $(\boldsymbol{u_h}, \tilde{u}_h, p_h) \in \boldsymbol{V_h} \times Q_h^k$ such that for all $(\boldsymbol{v_h}, \tilde{v}_h, q_h) \in \boldsymbol{V_h} \times Q_h^k$

$$A\big((\boldsymbol{u_h}, \tilde{u}_h, p_h), (\boldsymbol{v_h}, \tilde{v}_h, q_h)\big) = \int_\Omega \boldsymbol{f} \boldsymbol{v_h}\, dx + \int_\Gamma g(\boldsymbol{v_h})_n\, ds, \tag{2}$$

where

$$A\big((\boldsymbol{u_h}, \tilde{u}_h, p_h), (\boldsymbol{v_h}, \tilde{v}_h, q_h)\big) := a\big((\boldsymbol{u_h}, \tilde{u}_h), (\boldsymbol{v_h}, \tilde{v}_h)\big) + b\big((\boldsymbol{v_h}, \tilde{v}_h), p_h\big)$$

$$+ b\big((\boldsymbol{u_h}, \tilde{u}_h), q_h\big) - s\left(p_h, q_h\right).$$

2.1 Well-Posedness of the Discrete Problem

Let us consider the following norm on $\boldsymbol{V_h}$ (see [1, Lemma 3.2] for a proof that this is actually a norm in $\boldsymbol{V_h}$)

$$||| (\boldsymbol{w_h}, \tilde{w}_h) |||^2 := \nu \sum_{K \in \mathcal{T}_h} \left(|\boldsymbol{w_h}|^2_{H^1(K)} + h_K \|\partial_n \boldsymbol{w_h}\|^2_{\partial K} + \frac{\tau}{h_K} \left\|\Phi^{k-1}\big((\boldsymbol{w_h})_t - \tilde{w}_h\big)\right\|^2_{\partial K}\right).$$

The first step towards proving the stability of Method (2) is the following weak inf-sup condition for b.

Lemma 1 *There exist constants* $C_1, C_2 > 0$, *independent of* h_K *and* ν, *such that*

$$\sup_{(\boldsymbol{v_h}, \tilde{v}_h) \in \boldsymbol{V_h}} \frac{b\big((\boldsymbol{v_h}, \tilde{v}_h), q_h\big)}{||| (\boldsymbol{v_h}, \tilde{v}_h) |||} \geq C_1 \|q_h\|_\Omega - C_2 \left\|q_h - \Psi^{k-1} q_h\right\|_\Omega \quad \forall q_h \in Q_h^k.$$
$$\tag{3}$$

Proof We consider an arbitrary $q_h \in Q_h^k$. Let $\tilde{\Omega}$ be a convex, open, Lipschitz set such that $\Omega \subset \tilde{\Omega}$, and let us consider following extension

$$\hat{q}_h := \begin{cases} q_h \text{ in } & \Omega \\ 0 \text{ in } \tilde{\Omega} \setminus \Omega \end{cases}.$$

Let now ϕ be the unique weak solution of the problem

$$\begin{cases} -\Delta\phi = \hat{q}_h \text{ in } \tilde{\Omega} \\ \phi = 0 \text{ on } \partial\Omega \end{cases}.$$

Since $\tilde{\Omega}$ is convex, then $\phi \in H^2(\tilde{\Omega})$. Then $w := \nabla\phi|_{\Omega}$ belongs to $[H^1(\Omega)]^2$, and for $\tilde{w} := w_t$,

$$b\left((w, \tilde{w}), q_h\right) = \|q_h\|_{\Omega}^2 \quad \forall q_h \in Q_h^k. \tag{4}$$

In addition, applying standard regularity results, see [5, Section 1.2], we get

$$\|w\|_{H^1(\Omega)} \leq \|\nabla\phi\|_{H^1(\tilde{\Omega})} \leq c_1\|q_h\|_{\Omega}. \tag{5}$$

In [1, Lemma 3.5] it is shown that there exists a Fortin operator $\mathbf{\Pi} : \left[H^1(\Omega)\right]^2 \to V_h$ satisfying the following condition: for all $v \in [H^1(\Omega)]^2$ the following holds

$$b\left((v, \tilde{v}), q_h\right) = b\left(\mathbf{\Pi}(v), q_h\right) \quad \forall q_h \in Q_h^{k-1}, \tag{6}$$

$$|||\mathbf{\Pi}(v)||| \leq C\sqrt{v}\|v\|_{H^1(\Omega)}. \tag{7}$$

Let $(w_h, \tilde{w}_h) := \mathbf{\Pi}(w)$, then thanks to (6), (4) and the continuity of b (see [1, Lemma 3.3])

$$b\left((w_h, \tilde{w}_h), q_h\right) = b\left((w, \tilde{w}), q_h\right) - b\left((w - w_h, \tilde{w} - \tilde{w}_h), q_h - \Psi^{k-1}q_h\right)$$

$$\geq \|q_h\|_{\Omega}^2 - c_2\sqrt{\sum_{K \in \mathcal{T}_h} |w_h - w|_{H^1(K)}^2} \left\|q_h - \Psi^{k-1}q_h\right\|_{\Omega}.$$

Using the approximation properties of the BDM interpolation operator (see [5, Preposition 2.5.1]) and (5)

$$b\left((w_h, \tilde{w}_h), q_h\right) \geq \left(\frac{1}{c_1}\|q_h\|_{\Omega} - c_2 c_3 \left\|q_h - \Psi^{k-1}q_h\right\|_{\Omega}\right) |w|_{H^1(\Omega)}$$

$$\geq \left(C_1\|q_h\|_{\Omega} - C_2 \left\|q_h - \Psi^{k-1}q_h\right\|_{\Omega}\right) |||(w_h, \tilde{w}_h)|||,$$

where, in the last estimate we have used the stability of the Fortin operator Π in the $||| \cdot |||$ norm (7). This proves the result with $C_1 = \frac{1}{C\sqrt{\nu c_1}}$ and $C_2 = \frac{c_2 c_3}{C\sqrt{\nu}}$. $\qquad\square$

Before showing an inf-sup condition, we prove the continuity of bilinear form A.

Lemma 2 *There exists a constant $C > 0$ such that, for all $\left(w_h, \tilde{w}_h\right), \left(v_h, \tilde{v}_h\right) \in V_h$ and $r_h, q_h \in Q_h^k$, we have*

$$\left| A\left(\left(w_h, \tilde{w}_h, r_h\right), \left(v_h, \tilde{v}_h, q_h\right)\right) \right| \leq C ||| \left(w_h, \tilde{w}_h, r_h\right) |||_h ||| \left(v_h, \tilde{v}_h, q_h\right) |||_h.$$

$$(8)$$

Proof We use the continuity of the bilinear forms (see [1, Lemma 3.3]) and the fact that the projection is a bounded operator. $\qquad\square$

The final step towards stability is proving the inf-sup condition for bilinear form A.

Lemma 3 *There exists $\beta > 0$ independent of h_K such that for all $\left(w_h, \tilde{w}_h, r_h\right) \in V_h \times Q_h^k$ the following holds*

$$\sup_{\left(v_h, \tilde{v}_h, q_h\right) \in V_h \times Q_h^k} \frac{A\left(\left(w_h, \tilde{w}_h, r_h\right), \left(v_h, \tilde{v}_h, q_h\right)\right)}{||| \left(v_h, \tilde{v}_h, q_h\right) |||_h} \geq \beta ||| \left(w_h, \tilde{w}_h, r_h\right) |||_h. \quad (9)$$

As a consequence, Problem (2) is well-posed.

Proof Let $\left(w_h, \tilde{w}_h, r_h\right) \in V_h \times Q_h^k$. The idea of the proof is to construct an appropriate $\left(v_h, \tilde{v}_h, q_h\right)$ such that

$$A\left(\left(w_h, \tilde{w}_h, r_h\right), \left(v_h, \tilde{v}_h, q_h\right)\right) \geq c ||| \left(w_h, \tilde{w}_h, r_h\right) |||_h ||| \left(v_h, \tilde{v}_h, q_h\right) |||_h.$$

To achieve that we use coercivity of a (see [1, Lemma 3.4]), continuity of a (see [1, Lemma 3.3]) and Lemma 2. For details see [6]. $\qquad\square$

2.2 Error Analysis

In this section we present the error estimates for the method. The addition of the stabilising bilinear form $s(\cdot, \cdot)$ introduced a consistency error. However according to [4], this should not be viewed as a serious flaw, as this consistency error can be bounded in an optimal way. The following result is the first step towards that goal.

Lemma 4 *Let $(u, p) \in \left[H^1(\Omega) \cap H^2(\mathcal{T}_h) \right]^2 \times L^2(\Omega)$ be the solution of the problem (1) and $\tilde{u} = u_t$ on all edges of $\mathcal{E}_h$. If $\left(u_h, \tilde{u}_h, p_h\right) \in V_h \times Q_h^k$ solves (2),* *then for all* $\left(\boldsymbol{v_h}, \tilde{v}_h, q_h\right) \in \boldsymbol{V_h} \times Q_h^k$ *the following holds*

$$A\left(\left(\boldsymbol{u} - \boldsymbol{u_h}, \tilde{u} - \tilde{u}_h, p - p_h\right), \left(\boldsymbol{v_h}, \tilde{v}_h, q_h\right)\right) = s\left(p, q_h\right). \tag{10}$$

Next, we introduce the following norm

$$|||(\boldsymbol{u}, \tilde{u}, p)|||_h := |||(\boldsymbol{u}, \tilde{u})||| + \frac{1}{\sqrt{v}} \|p\|_\Omega, \tag{11}$$

and prove the following variant of Cea's lemma [11, Lemma 2.28] for this stabilised Stokes problem.

Lemma 5 *Let* $\left(\boldsymbol{u}, p\right) \in \left[H^1\left(\Omega\right) \cap H^2\left(\mathcal{T}_h\right)\right]^2 \times L^2\left(\Omega\right)$ *be the solution of the problem* (1) *and* $\tilde{u} = u_t$ *on all edges of* $\mathcal{E}_h$. *If* $\left(\boldsymbol{u_h}, \tilde{u}_h, p_h\right) \in \boldsymbol{V_h} \times Q_h^k$ *solves* (2), *then there exists* $C > 0$, *independent of h and* v, *such that*

$$||| \left(\boldsymbol{u} - \boldsymbol{u_h}, \tilde{u} - \tilde{u}_h, p - p_h\right) |||_h \leq C \inf_{\left(\boldsymbol{v_h}, \tilde{v}_h, q_h\right) \in \boldsymbol{V_h} \times Q_h^k} ||| \left(\boldsymbol{u} - \boldsymbol{v_h}, \tilde{u} - \tilde{v}_h, p - q_h\right) |||_h$$

$$+ \frac{C}{\sqrt{v}} \left\| p - \Psi^{k-1} p \right\|_\Omega. \tag{12}$$

Proof It is a combination of Lemmas 1, 2 and 3. For details see [6].   □

Lemma 6 *Let* $\left(\boldsymbol{u}, p\right) \in \left[H^1\left(\Omega\right) \cap H^2\left(\mathcal{T}_h\right)\right]^2 \times L^2\left(\Omega\right)$ *be the solution of the problem* (1) *and* $\tilde{u} = u_t$ *on all edges of* $\mathcal{E}_h$. *If* $\left(\boldsymbol{u_h}, \tilde{u}_h, p_h\right) \in \boldsymbol{V_h} \times Q_h^k$ *solves* (2), *then there exists* $C > 0$, *independent of h and* v, *such that*

$$||| \left(\boldsymbol{u} - \boldsymbol{u_h}, \tilde{u} - \tilde{u}_h, p - p_h\right) |||_h \leq C h^k \left(\sqrt{v} \|\boldsymbol{u}\|_{H^{k+1}(\mathcal{T}_h)} + \frac{1}{\sqrt{v}} \|p\|_{H^k(\mathcal{T}_h)}\right).$$

Proof It is a combination of [1, Lemmas 3.8] and Lemma 5 with the local L^2-projection approximation [11, Theorem 1.103].   □

3   Numerical Experiments

The computational domain is the unit square $\Omega = (0, 1)^2$. We present the results for $k = 1$, that is the discrete space is given by $\boldsymbol{BDM}_h^1 \times M_{h,0}^0 \times Q_h^1$. We test both the symmetric method ($\varepsilon = -1$) and the non-symmetric method ($\varepsilon = 1$). We have followed the recommendation given in [15, Section 2.5.2] and taken $\tau = 6$.

We choose the right hand side f and the boundary condition g such that the exact solution is given by

$$u = \mathrm{curl}\left[\left(1 - \cos((1-x)^2)\right)\sin(x^2)\sin(y^2)\left(1 - \cos((1-y)^2)\right)\right], \quad p = \tan(xy).$$

In Fig. 1a and b we depict the errors for both the symmetric and non-symmetric cases, respectively. We can see that they not only validate the theory from Sect. 2.2, but also perform an optimal h^2 convergence rate for $\|u - u_h\|_\Omega$. Furthermore, we observe an increased order of convergence for $\|p - p_h\|_\Omega$. In fact, the error seems to decrease with $O(h^{3/2})$, rather than the $O(h)$ predicted by the theory.

To stress the last point made in the previous paragraph, in Table 1 we compare the L^2 error of the pressure ($\|p - p_h\|_\Omega$) for hdG method introduced in [1] and stabilised hdG method from Sect. 2. Columns $p_h \in Q_h^0$ are associated with hdG method and $p_h \in Q_h^1$ with stabilised hdG ones. There, we confirm that the pressure

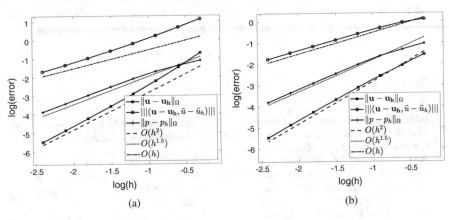

Fig. 1 Convergence the stabilised method with $k = 1$. (**a**) Symmetric bilinear form ($\varepsilon = -1$). (**b**) Non-symmetric bilinear form ($\varepsilon = 1$)

Table 1 Comparison of the error of the pressure $\|p - p_h\|_\Omega$

h	Symmetric bilinear form ($\varepsilon = -1$)		Non-symmetric bilinear form ($\varepsilon = 1$)	
	$p_h \in Q_h^0$	$p_h \in Q_h^1$	$p_h \in Q_h^0$	$p_h \in Q_h^1$
2^{-1}	0.152296	0.077228	0.159019	0.090624
2^{-2}	0.082775	0.041790	0.084875	0.047488
2^{-3}	0.042620	0.020500	0.043313	0.009449
2^{-4}	0.021357	0.008338	0.021513	0.003516
2^{-5}	0.010676	0.003083	0.010707	0.001269
2^{-6}	0.005340	0.001105	0.005346	0.002171
2^{-7}	0.002671	0.000392	0.002672	0.000453
2^{-8}	0.001336	0.000139	0.001336	0.000161

error for the stabilised version is much smaller than the one for the inf-sup stable case, in addition to having an increased order of convergence.

4 Conclusion

In this work we have applied the idea introduced in [10] to stabilise the hdG method proposed in [1] for the Stokes problem with TVNF boundary conditions. The method adds a simple, symmetric, term to the formulation, and allowed us to use a higher order pressure space, which, in turn, improved the pressure convergence (although a proof of this fact is, in general, not available). This approach was also applied to NVTF boundary conditions (see [6]) and can be used for other discontinuous Galerkin methods that deal with Stokes or nearly incompressible elasticity problems.

Future testing using higher order discretisations is needed to assess whether this approach provides an increase of the convergence rate for the pressure. Thus, the numerical tests with higher order of polynomials for discontinuous finite methods is interest for further research to look for the improvement of the convergence.

References

1. Barrenechea, G.R., Bosy, M., Dolean, V., Nataf, F., Tournier, P.-H.: Hybrid discontinuous Galerkin discretisation and domain decomposition preconditioners for the Stokes problem. Comput. Methods Appl. Math. **19**(4), 703–722 (2019)
2. Barth, T., Bochev, P.B., Gunzburger, M., Shadid, J.: A taxonomy of consistently stabilized finite element methods for the Stokes problem. SIAM J. Sci. Comput. **25**(5), 1585–1607 (2004)
3. Becker, R., Braack, M.: A finite element pressure gradient stabilization for the Stokes equations based on local projections. Calcolo **38**(4), 173–199 (2001)
4. Bochev, P.B., Dohrmann, C.R., Gunzburger, M.D.: Stabilization of low-order mixed finite elements for the Stokes equations. SIAM J. Numer. Anal. **44**(1), 82–101 (2006)
5. Boffi, D., Brezzi, F., Fortin, M.: Mixed Finite Element Methods and Applications. Springer Series in Computational Mathematics, vol. 44. Springer, Heidelberg (2013)
6. Bosy, M.: Efficient discretisation and domain decomposition preconditioners for incompressible fluid mechanics. Ph.D. Thesis, University of Strathclyde (2017). https://doi.org/10.13140/RG.2.2.24947.17444
7. Brezzi, F., Douglas, J., Jr.: Stabilized mixed methods for the Stokes problem. Numer. Math. **53**(1–2), 225–235 (1988)
8. Brezzi, F., Pitkäranta, J.: On the stabilization of finite element approximations of the Stokes equations. In: Efficient Solutions of Elliptic Systems (Kiel, 1984). Notes on Numerical Fluid Mechanics, vol. 10, pp. 11–19. Friedr. Vieweg, Braunschweig (1984)
9. Codina, R., Blasco, J.: Analysis of a pressure-stabilized finite element approximation of the stationary Navier-Stokes equations. Numer. Math. **87**(1), 59–81 (2000)
10. Dohrmann, C.R., Bochev, P.B.: A stabilized finite element method for the Stokes problem based on polynomial pressure projections. Int. J. Numer. Methods Fluids **46**(2), 183–201 (2004)
11. Ern, A., Guermond, J.L.: Theory and Practice of Finite Elements. Applied Mathematical Sciences, vol. 159. Springer, New York (2004)

12. Ganesan, S., Matthies, G., Tobiska, L.: Local projection stabilization of equal order interpolation applied to the Stokes problem. Math. Comput. **77**(264), 2039–2060 (2008)
13. Hughes, T.J.R., Franca, L.P.: A new finite element formulation for computational fluid dynamics. VII. The Stokes problem with various well-posed boundary conditions: symmetric formulations that converge for all velocity/pressure spaces. Comput. Methods Appl. Mech. Eng. **65**(1), 85–96 (1987)
14. Hughes, T.J.R., Franca, L.P., Balestra, M.: A new finite element formulation for computational fluid dynamics. V. Circumventing the Babuška-Brezzi condition: a stable Petrov-Galerkin formulation of the Stokes problem accommodating equal-order interpolations. Comput. Methods Appl. Mech. Eng. **59**(1), 85–99 (1986)
15. Lehrenfeld, C.: Hybrid discontinuous Galerkin methods for solving incompressible flow problems. Dissertation, Rheinisch-Westfälischen Technischen Hochschule Aachen (2010)
16. Silvester, D.J., Kechkar, N.: Stabilised bilinear-constant velocity-pressure finite elements for the conjugate gradient solution of the Stokes problem. Comput. Methods Appl. Mech. Eng. **79**(1), 71–86 (1990)

RBF Based CWENO Method

Jan S. Hesthaven, Fabian Mönkeberg, and Sara Zaninelli

1 Introduction

A broad range of physical phenomena can be described by hyperbolic conservation laws of the form

$$u_t + f(u)_x = 0, \quad (x, t) \in \mathbb{R} \times \mathbb{R}_+,$$
$$u(0) = u_0, \tag{1}$$

with the conserved variables $u : \mathbb{R} \times \mathbb{R}_+ \to \mathbb{R}^N$ and the flux function $f : \mathbb{R}^N \to \mathbb{R}^N$. The nonlinear behavior of f can lead to complex solutions, most notably shocks. It is well-known that high-order methods give good results for smooth data, but for discontinuous ones spurious oscillations are introduced. A popular class of methods to solve (1) is the finite volume method, which is based on a discretization in space $\ldots < x_{i-1/2} < x_{i+1/2} < \ldots$ and the average values $\bar{u}_i$ of its cells $C_i = [x_{i-1/2}, x_{i+1/2}]$. It is defined by the semi-discrete scheme

$$\frac{d\bar{u}_i}{dt} = -\frac{F_{i+1/2} - F_{i-1/2}}{\Delta x}, \tag{2}$$

where the numerical flux term $F_{i+1/2}$ depends on the values $\{\bar{u}_{i-k}, \ldots, \bar{u}_{i+p-k}\}$ with $0 \leq k \leq p - 1$. For more details we refer the reader to [15, 20, 22].

The class of essentially nonoscillatory (ENO) methods, introduced by Harten et al. [14], reduces spurious oscillations to a minimum. They are based on a monotone numerical flux function $F(u, v)$ and high-order accurate reconstruction $s_i(x)$ for

J. S. Hesthaven · F. Mönkeberg (✉) · S. Zaninelli
École Polytechnique Fédérale de Lausanne (EPFL), Lausanne, Switzerland
e-mail: jan.hesthaven@epfl.ch; fabian.monkeberg@epfl.ch; sara.zaninelli@epfl.ch

© The Author(s) 2020
S. J. Sherwin et al. (eds.), *Spectral and High Order Methods for Partial Differential Equations ICOSAHOM 2018*, Lecture Notes in Computational Science and Engineering 134, https://doi.org/10.1007/978-3-030-39647-3_14

each cell i. The central idea is to choose the least oscillating interpolation function s_i and define the numerical flux $F_{i+1/2} = F(u^+_{i+1/2}, u^-_{i+1/2})$ with $u^\pm_{i+1/2}$ being the evaluation of s_{i+1} and s_i at the interface $x_{i+1/2}$. Based on the ENO method, Jiang and Shu [19] introduced the weighted ENO (WENO) method which considers different interpolation polynomials, based on different stencils, and combines them in a nonoscillatory manner to maximize the attainable accuracy. Further results on ENO and WENO methods can be found in [10, 11, 16].

2   CWENO

The CWENO method is based on the WENO method and was introduced by Levy et al. [23] as a third order method. Further analysis and generalization to higher orders on general grids can be found in [6, 7].

Let us consider the standard semi-discrete formulation (2) with a monotone flux function $F(u, v)$. The goal is to construct a reconstruction $P_{rec,i}$ for each cell C_i based on the stencil $\{C_{i-k}, \ldots, C_{i+k}\}$ for $k \in \mathbb{N}$. In the smooth regions the algorithm should choose a polynomial of degree $2k$ which interpolates the central stencil $\bar{u}_{i-k}, \ldots, \bar{u}_{i+k}$ in the mean value sense. In case of a non-smooth solution it chooses a polynomial of degree k on one stencil $\{C_{i-k+l}, \ldots, C_{i+l}\}$ that avoids the discontinuity. Given the reconstruction, the high-order numerical flux is $F_{i+1/2} = F(P_{rec,i+1}(x_{i+1/2}), P_{rec,i}(x_{i+1/2}))$.

Specifically, let us consider P_{opt} as the polynomial of degree $2k$ that interpolates all data in the $2k + 1$ stencil and the polynomials P_l of degree k that interpolate the data on the stencil $\{C_{i-k+l-1}, \ldots, C_{i+l-1}\}$ for $l = 1, \ldots, k + 1$. Furthermore, the reconstruction depends on the choice of the positive real coefficients $d_0, \ldots, d_{k+1} \in [0, 1]$ such that $\sum_{l=0}^{k+1} d_l = 1, d_0 \neq 0$. Then, the reconstruction polynomial of degree $2k$ is

$$P_{rec}(x) = \sum_{l=0}^{k+1} \omega_l P_l(x), \tag{3}$$

with

$$P_0(x) = \frac{1}{d_0}\left(P_{opt}(x) - \sum_{l=1}^{k+1} d_l P_l(x)\right), \tag{4}$$

and the nonlinear coefficients ω_l that are defined as

$$\omega_l = \frac{\alpha_l}{\sum_{i=0}^{k+1} \alpha_i}, \qquad \alpha_l = \frac{d_l}{(I[P_l] + \bar{\epsilon})^t}, \tag{5}$$

where $I[P_l]$ indicates the smoothness of P_l, $1 \gg \bar{\epsilon} > 0$ and $t \geq 2$. A classical indicator of smoothness in the cell C for a polynomial is the Jiang–Shu indicator [19]

$$I[P] = \sum_{l>0} \text{diam}(C)^{2l-1} \int_C \left(\frac{d^l}{dx^l} P(x)\right)^2 dx. \tag{6}$$

The choice of $\bar{\epsilon}$ is of importance: if it is too small, it might affect the order of convergence. On the other hand if it is too big, spurious oscillations may occur. Cravero et al. [7] show that the choice $\bar{\epsilon} = \hat{\epsilon} h^p$ for $p = 1, 2$ leads to the maximal order of convergence. As proposed in [7] we define the coefficients d_j over the temporary weights

$$\hat{d}_j = \hat{d}_{k+2-j} = j, \qquad 1 \leq j \leq \frac{k+2}{2}, \tag{7}$$

and we choose $d_0 \in (0, 1)$ for the high-order polynomial. This gives us a possible choice for the coefficients

$$d_j = \frac{\hat{d}_j}{\sum_{i>0} \hat{d}_i} (1 - d_0). \tag{8}$$

The main difference with respect to the classical WENO method is that for the smooth case we are not constructing P_{opt} out of the polynomials P_l, but we build it independently by resolving an additional system of equations. This method has the advantage that it is easier to generalize on general grids in high dimensions, while maintaining high-order accuracy.

3 Radial Basis Functions

An alternative to the classical polynomial interpolation is the interpolation with radial basis functions (RBF). RBFs were proposed in the seminal work by Hardy [13]. They have been successfully applied in scattered data interpolation [4, 9, 17, 24, 27] and as a basis for a generalized finite difference method (RBF-FD) [5, 12]. The advantage is its flexibility in high dimensions and the possibility to reduce the risk of ill-conditioned point constellations. Its disadvantage is the ill-conditioning of the interpolation matrix for small grid sizes [8, 21, 26].

The RBF interpolation is based on a basis $\mathscr{B}$, obtained from a univariate continuous function $\phi : \mathbb{R}^d \to \mathbb{R}$, composed with the Euclidean norm centered at the data points

$$\phi(x - x_j) := \phi(\varepsilon \|x - x_j\|), \tag{9}$$

Table 1 Commonly used RBFs with $\mathbb{N} \not\ni \nu > 0, k \in \mathbb{N}$ and $\varepsilon > 0$

RBF	$\phi(r)$	Order
Infinitely smooth RBFs		
Multiquadratics	$(1 + (\varepsilon r)^2)^\nu$	$\lceil \nu \rceil$
Inverse multiquadratics	$(1 + (\varepsilon r)^2)^{-\nu}$	0
Gaussians	$\exp(-(\varepsilon r)^2)$	0
Piecewise smooth RBFs		
Polyharmonic splines	r^{2k-d}	k
	$r^{2k-d} \log(r)$	k

with the shape parameter ε. Some common RBFs can be found in Table 1. Thus, for given scattered data points $X = (x_1, \ldots, x_n)^T$ with $x_j \in \mathbb{R}^d$ and corresponding values $f_1, \ldots, f_n \in \mathbb{R}$ we look for

$$s(x) = \sum_{j=1}^n a_j \phi(x - x_j) + p(x), \tag{10}$$

with a polynomial $p \in \Pi_{m-1}(\mathbb{R}^d)$, $m \in \mathbb{N}$, the interpolation condition $s(x_j) = f_j$ and the additional constraints

$$\sum_{j=1}^n a_j q(x_j) = 0, \qquad \text{for all } q \in \Pi_{m-1}(\mathbb{R}^d), \tag{11}$$

with the coefficients $a_j \in \mathbb{R}$ for all $j = 1, \ldots, n$.

The same concept can be applied in the case of cell-averages. We seek functions

$$s(x) = \sum_{j=1}^n a_j \lambda_{C_j}^\xi \phi(x - \xi) + p(x), \qquad p \in \Pi_{m-1}(\mathbb{R}^d), \tag{12}$$

such that

$$\lambda_{C_j} s = \bar{u}_j, \qquad \text{for all } j = 1, \ldots, n, \tag{13a}$$

$$\sum_{j=1}^n a_j \lambda_C(p) = 0, \qquad \text{for all } C \in \{C_1, \ldots, C_n\}, \tag{13b}$$

with the averaging operator $\lambda_C^x f(x) = \frac{1}{|C|} \int_C f(x) dx$. A well-known problem with RBFs is the high condition number of the interpolation matrix for small grid sizes or small shape parameters [8, 21, 26]. This problem can be resolved by using the vector-valued rational approximation method [28].

4 RBF-CWENO

Methods combining RBFs and essentially nonoscillatory methods have been proposed, e.g. RBFs with ENO [18, 25], RBFs with WENO [1–3]. The advantage of the CWENO method over the WENO method is its flexibility on general grids and its independence of the construction of a high-order interpolation function out of lower order ones. This facilitates the use of the whole grid in smooth regions and is important for non-polynomial interpolation functions which cannot be combined to an higher order function.

We propose the RBF-CWENO method which works as the classical CWENO method with the reconstruction function (3) and the weights (5), but as interpolation function we use RBFs instead of polynomials. Since the problem of the ill-conditioning can be solved by using the vector-valued rational approximation method [28], the main challenge for RBF methods is the choice of the smoothness indicator. For polyharmonic splines, Aboyar et al. [1] use the semi-norm of the Beppo-Levi space and Bigoni et al. [3] use a modified version of the Jiang-Shu indicator (6).

4.1 Smoothness Indicator

The smoothness indicator is the heart of the essentially nonoscillatory methods. We consider one based on the one introduced by Bigoni and Hesthaven [3]

$$
I_i[s] = \sum_{l=1}^{g+1} \Delta x_i^{2l-1} \int_{C_i} \left(\frac{\partial^l p(x)}{\partial x^l} \right)^2 dx
$$

$$
+ \Delta x_i^{2g+1} \int_{C_i} \left(\frac{\partial^{g+1}}{\partial x^{g+1}} \Big[\sum_{j=1}^{g+1} a_j \lambda_{C_j}^\xi \phi(\|x - \xi\|) \Big] dx \right)^2 , \tag{14}
$$

where the first part is the sum of the derivatives of the polynomial part and the second term expresses the highest derivative of the RBF-part. The original Jiang-Shu indicator applied to (12) would include the lower derivatives of the RBF-part plus all mixed terms, but we find this to be less efficient. For simplicity the integrals can be approximated with a simple mid-point rule.

We face again the problem of ill-conditioning when recovering the coefficients a_i. Numerical examples indicate that small shape parameter improve the accuracy, but they do not affect the choice of the stencil using this smoothness indicator. Thus, we use a bigger shape parameter ε_R, that is smaller than the smallest distance to a singularity

$$\varepsilon_R = 0.95(\max_{i,j \leq N} \|x_i - x_j\|)^{-1}, \tag{15}$$

which ensures the solvability of the system of equations [28].

5 Numerical Results

We now discuss the numerical results of the RBF-CWENO method and compare it with the RBF-WENO method [3] and the classical ENO method [14]. All methods are using the Lax-Friedrichs numerical flux and integration in time is done using the SSPRK-5 method [15] with time step $dt = CFL \cdot \Delta x / \lambda_{max}$ and the maximal eigenvalue λ_{max} of $\nabla_u F$. Furthermore, we use the vector-valued rational approximation approach [28] to circumvent ill-conditioning of the interpolation matrix and a shape parameter $\varepsilon = 0.1$. For the nonlinear weights (5) we choose $\bar{\epsilon} = \hat{\epsilon} h^2$ with $\hat{\epsilon} = 0.1$.

5.1 Linear Advection Equation

Let us consider the linear advection equation

$$u_t + a u_x = 0, \qquad x \in [0, 1], \tag{16}$$

with wave speed $a = 1$, initial condition $u_0(x) = sin(2\pi x)$ and periodic boundary conditions [22]. Note that for $k = 3$ we expect the order of convergence to be 7, therefore we use the reduced time step $dt = CFL \cdot \Delta x^{7/5} / \lambda_{max}$ to recover the right order of convergence. The correct order of convergence of the RBF-CWENO method is shown in Table 2 and it seems to be more accurate than the RBF-WENO method.

5.2 Burger's Equation

Considering the Burger's equation

$$u_t + \frac{1}{2}(u^2)_x = 0, \qquad x \in [0, 1], \tag{17}$$

Table 2 Convergence rates of RBF-CWENO using multiquadratics for the linear advection equation at time $t = 0.05$

| | | RBF-CWENO | | | | | | RBF-WENO | |
| | | L_h^1 | | L_h^2 | | L_h^∞ | | L_h^2 | |
k	N	Error	Rate	Error	Rate	Error	Rate	Error	Rate
1	16	5.6409e−04	–	2.1702e−04	–	1.5903e−04	–	1.5754e−02	–
	32	7.6612e−05	2.75	2.4817e−05	2.99	1.6221e−05	3.15	4.8924e−03	1.69
	64	1.0082e−05	2.79	2.5297e−06	3.15	1.3561e−06	3.42	1.2608e−03	1.96
	128	1.3812e−06	2.74	2.4032e−07	3.24	9.6982e−08	3.63	9.2931e−05	3.76
	256	2.1322e−07	2.57	2.3289e−08	3.21	6.5703e−09	3.71	2.3008e−06	5.34
2	16	2.3796e−05	–	7.3671e−06	–	4.1241e−06	–	5.4401e−04	–
	32	3.5783e−06	2.61	8.3093e−07	3.01	3.9675e−07	3.22	4.4938e−05	3.60
	64	2.8691e−07	3.48	5.9366e−08	3.63	3.6940e−08	3.27	3.4787e−06	3.69
	128	1.4563e−08	4.11	2.5775e−09	4.32	1.3965e−09	4.51	2.5956e−07	3.74
	256	6.8835e−10	4.20	9.6168e−11	4.53	4.4249e−11	4.75	1.9221e−08	3.76
3	16	3.8815e−05	–	1.3319e−05	–	7.7293e−06	–	2.2578e−04	–
	32	4.3423e−07	6.48	1.3452e−07	6.63	8.1494e−08	6.57	7.3483e−06	4.94
	64	5.1821e−09	6.39	1.4750e−09	6.51	8.8273e−10	6.54	1.4075e−07	5.71
	128	7.6636e−11	6.08	1.6792e−11	6.46	7.8655e−12	6.81	1.4510e−09	6.60
	256	1.1554e−12	6.05	1.5855e−13	6.73	6.9487e−14	6.82	2.0120e−11	6.17

We use shape parameter $\varepsilon - 0.1$, CFL $= 0.01$

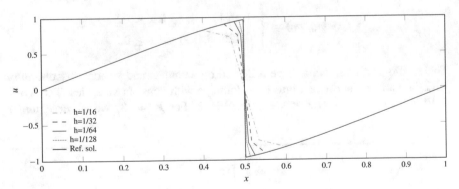

Fig. 1 Burger's equation at $t = 0.3$ with $u_0 = \sin(2\pi x)$ solved by using RBF-CWENO method with MQ interpolants of order $k = 3$

we analyze its robustness with respect to discontinuities. In Fig. 1 we report the results performed with $CFL = 0.5$ at $t = 0.3$. We observe no oscillations around the discontinuity at $x = 0.5$ and as expected an increasing accuracy for increasing number of elements.

5.3 Euler Equations

The one-dimensional Euler equations express conservation of mass, momentum and the total energy. They can be described by the density ρ, the mass flow m, the energy per unit volume E and the pressure p through

$$
\begin{pmatrix} \rho \\ m \\ E \end{pmatrix}_t + \begin{pmatrix} m \\ \frac{m^2}{\rho} + p \\ \frac{m}{\rho}(E + p) \end{pmatrix}_x = 0, \tag{18}
$$

with $p = \mathscr{R}\rho T = (\gamma - 1)(E - \frac{1}{2}\frac{m^2}{\rho})$ for an ideal gas with the ratio of specific heat $\gamma = 1.4$ [15]. For $k = 3$ we need to change the nonlinear weights (5) by using $\bar{\epsilon} = \hat{\epsilon}h^2$ with $\hat{\epsilon} = 10^{-6}$ to avoid oscillations.

5.3.1 Sod's Shock Tube Problem

The Sod's shock tube problem describes two colliding gases in [0, 1] with different densities given by the initial conditions

$$
(\rho_0, m_0, p_0) = \begin{cases} (1, 0, 1) & \text{if } x < 0.5 \\ (0.125, 0, 0.1) & \text{if } x \geq 0.5 \end{cases}. \tag{19}
$$

This results in a rarefaction wave followed by a contact and a shock discontinuity which separates the domain into four domains with constant variables. The RBF-CWENO method resolves it well, see Fig. 2. For $k = 3$, we observe minor

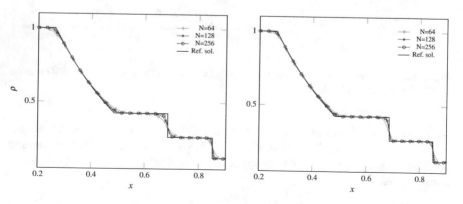

Fig. 2 Results for the Sod shock tube problem at $t = 0.2$ solved by using RBF-CWENO with MQ interpolants of order $k = 2, 3$ on characteristic variables (left: $k = 2$, right: $k = 3$)

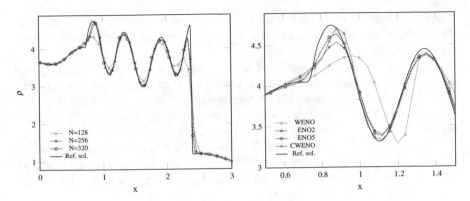

Fig. 3 Results for the Euler shock entropy problem at $t = 1.8$ solved by using RBF-CWENO with MQ interpolants of order $k = 2$ on characteristic variables (Left) and a comparison with WENO, ENO2 and ENO5 for $N = 256$ cells (Right)

oscillations, but their amplitude decreases for increasing number of elements. Furthermore, we observe the increasing accuracy for $k = 3$ compared to $k = 2$.

5.3.2 Shu–Osher Shock-Entropy Wave Interaction Problem

The Shu–Osher problem describes the interaction of a discontinuity with a low frequency wave which introduces some high frequent waves. Its initial conditions are

$$(\rho_0, m_0, p_0) = \begin{cases} (3.857143, 2.629369, 10.33333) & \text{if } x < -4 \\ (1 + 0.2\sin(5x), 0, 1) & \text{if } x \geq -4 \end{cases}. \qquad (20)$$

In Fig. 3, we observe on the left side the increasing accuracy for increasing number of elements for $k = 2$. On the right side we see its good approximative behaviour compared to the existing methods ENO2, ENO5 and the corresponding WENO. In particular we observe that the performance of the RBF-CWENO ($k = 2$) is comparable to ENO5 and superior to WENO ($k = 2$).

6 Conclusion

In this work, we introduce the RBF-CWENO method that relies on the CWENO method [23] and the use of radial basis functions for the interpolation. We develop a smoothness indicator that is based on RBFs but works similarly to the one for polynomials. Furthermore, we tackle the problem about the choice of the weight

$1 \gg \bar{\epsilon} > 0$. For $\bar{\epsilon} = \hat{\epsilon}h^2$ with $\hat{\epsilon} = 0.1$ we get the right order of convergence, but for the 7th order method ($k = 3$) we choose $\hat{\epsilon} = 10^{-6}$ to reduce spurious oscillations for the Euler equations.

Moreover, we should point out that the choice of the linear weight d_0 can influence the result; indeed if it is too close to 1 then the reconstruction almost coincides with P_{opt}, which can lead to spurious oscillations in case of discontinuous solutions. We present multiple numerical examples to show the robustness of the method.

We can conclude that the RBF-CWENO method works comparable to the existing RBF-WENO and ENO methods in one dimension. The advantage of RBFs is clearer when considering unstructured grids in higher dimensions where polynomial reconstruction is complex.

References

1. Aboiyar, T., Georgoulis, E.H., Iske, A.: High order WENO finite volume schemes using poly-harmonic spline reconstruction. In: Proceedings of the International Conference on Numerical Analysis and Approximation Theory NAAT2006, Cluj-Napoca (Romania). Department of Mathematics, University of Leicester (2006)
2. Aboiyar, T., Georgoulis, E.H., Iske, A.: Adaptive ADER methods using kernel-based polyharmonic spline WENO reconstruction. SIAM J. Sci. Comput. **32**(6), 3251–3277 (2010)
3. Bigoni, C., Hesthaven, J.S.: Adaptive WENO methods based on radial basis function reconstruction. J. Sci. Comput. **72**(3), 986–1020 (2017)
4. Buhmann, M.D.: Radial Basis Functions: Theory and Implementations. Cambridge University Press, Cambridge (2003)
5. Chandhini, G., Sanyasiraju, Y.: Local RBF-ŘFD solutions for steady convection–diffusion problems. Int. J. Numer. Methods Eng. **72**(3), 352–378 (2007)
6. Cravero, I., Semplice, M.: On the accuracy of WENO and CWENO reconstructions of third order on nonuniform meshes. J. Sci. Comput. **67**(3), 1219–1246 (2016)
7. Cravero, I., Puppo, G., Semplice, M., Visconti, G.: CWENO: uniformly accurate reconstructions for balance laws (2016). Preprint. arXiv:1607.07319
8. Driscoll, T.A., Fornberg, B.: Interpolation in the limit of increasingly flat radial basis functions. Comput. Math. Appl. **43**(3), 413–422 (2002)
9. Duchon, J.: Splines Minimizing Rotation-Invariant Semi-Norms in Sobolev Spaces, pp. 85–100. Springer, Berlin (1977)
10. Fjordholm, U.S., Ray, D.: A sign preserving WENO reconstruction method. J. Sci. Comput. 1–22 (2016)
11. Fjordholm, U.S., Mishra, S., Tadmor, E.: Arbitrarily high-order accurate entropy stable essentially nonoscillatory schemes for systems of conservation laws. SIAM J. Numer. Anal. **50**(2), 544–573 (2012)
12. Fornberg, B., Lehto, E., Powell, C.: Stable calculation of gaussian-based RBF-FD stencils. Comput. Math. Appl. **65**(4), 627–637 (2013)
13. Hardy, R.L.: Multiquadric equations of topography and other irregular surfaces. J. Geophys. Res. **76**(8), 1905–1915 (1971)
14. Harten, A., Engquist, B., Osher, S., Chakravarthy, S.R.: Uniformly high order accurate essentially non-oscillatory schemes, iii. J. Comput. Phys. **71**(2), 231–303 (1987)
15. Hesthaven, J.S.: Numerical Methods for Conservation Laws: From Analysis to Algorithms. Society for Industrial and Applied Mathematics, Philadelphia (2017)

16. Hu, C., Shu, C.-W.: Weighted essentially non-oscillatory schemes on triangular meshes. J. Comput. Phys. **150**(1), 97–127 (1999)
17. Iske, A.: Multiresolution Methods in Scattered Data Modelling, vol. 37. Springer, Berlin (2004)
18. Iske, A., Sonar, T.: On the structure of function spaces in optimal recovery of point functionals for ENO-schemes by radial basis functions. Numer. Math. **74**(2), 177–201 (1996)
19. Jiang, G.-S., Shu, C.-W.: Efficient implementation of weighted ENO schemes. J. Comput. Phys. **126**(1), 202–228 (1996)
20. Kröner, D.: Numerical Schemes for Conservation Laws. Wiley, Chichester (1997)
21. Larsson, E., Fornberg, B.: Theoretical and computational aspects of multivariate interpolation with increasingly flat radial basis functions. Comput. Math. Appl. **49**(1), 103–130 (2005)
22. LeVeque, R.J.: Numerical Methods for Conservation Laws. Springer Science & Business Media, New York (1992)
23. Levy, D., Puppo, G., Russo, G.: Central WENO schemes for hyperbolic systems of conservation laws. ESAIM: Math. Model. Numer. Anal. **33**(3), 547–571 (1999)
24. Micchelli, C.A.: Interpolation of scattered data: distance matrices and conditionally positive definite functions. Constr. Approx. **2**(1), 11–22 (1986)
25. Mönkeberg, F., Hesthaven, J.S.: Entropy stable essentially nonoscillatory methods based on RBF reconstruction. ESAIM: Math. Model. Numer. Anal. **53**, 925–958 (2019)
26. Schaback, R.: Multivariate interpolation by polynomials and radial basis functions. Constr. Approx. **21**(3), 293–317 (2005)
27. Wendland, H.: Scattered Data Approximation. Cambridge University Press, Cambridge (2004)
28. Wright, G.B., Fornberg, B.: Stable computations with flat radial basis functions using vector-valued rational approximations. J. Comput. Phys. **331**, 137–156 (2017)

Discrete Equivalence of Adjoint Neumann–Dirichlet div-grad and grad-div Equations in Curvilinear 3D Domains

Yi Zhang, Varun Jain, Artur Palha, and Marc Gerritsma

1 Introduction

In $\mathbb{R}^d$, given a bounded domain Ω with Lipschitz boundary $\partial\Omega$ and $\hat{\sigma}_n \in H^{-1/2}(\partial\Omega) = \operatorname{tr} H(\operatorname{div}, \Omega)$, $\omega \in H^1(\Omega)$ solves the Neumann problem,

$$\begin{cases} \dfrac{\partial\omega}{\partial\boldsymbol{n}} = \hat{\sigma}_n & \text{on } \partial\Omega \\ -\operatorname{div}\left(\operatorname{grad}\omega\right) + \omega = 0 & \text{in } \Omega \end{cases}, \tag{1}$$

if and only if $\sigma \in H(\operatorname{div}, \Omega)$ which solves the Dirichlet problem,

$$\begin{cases} \sigma \cdot \boldsymbol{n} = \hat{\sigma}_n & \text{on } \partial\Omega \\ -\operatorname{grad}\left(\operatorname{div}\sigma\right) + \sigma = 0 & \text{in } \Omega \end{cases}, \tag{2}$$

satisfies $\sigma = \operatorname{grad}\omega$ [3]. This is obvious at the continuous level. The question is whether we can find a set of finite dimensional function spaces such that $\sigma^h = \operatorname{grad}\omega^h$ holds if ω^h and σ^h solve the discrete Neumann and Dirichlet problems respectively. The answer is yes.

Y. Zhang (✉) · V. Jain · M. Gerritsma
Delft University of Technology, Delft, Netherlands
e-mail: y.zhang-14@tudelft.nl; v.jain@tudelft.nl; m.i.gerritsma@tudelft.nl

A. Palha
Eindhoven University of Technology, Eindhoven, Netherlands
e-mail: a.palha@tue.nl

© The Author(s) 2020
S. J. Sherwin et al. (eds.), *Spectral and High Order Methods for Partial Differential Equations ICOSAHOM 2018*, Lecture Notes in Computational Science and Engineering 134, https://doi.org/10.1007/978-3-030-39647-3_15

203

Throughout this paper, we restrict ourselves to $\mathbb{R}^3$. We will first construct the primal polynomial spaces and their algebraic dual representations, and then use them to discretize problems (1) and (2) such that the identity $\sigma^h = \text{grad } \omega^h$ holds at the discrete level in any curvilinear domain for any polynomial approximation degree. This work extends [7, 9], where similar dual Neumann–Dirichlet problems are considered, to 3-dimensional space. These primal spaces and their algebraic dual representations can be ideal for the so-called mimetic or structure-preserving discretizations [1, 4, 8, 11, 12]. Together with their trace spaces, they can be used for the hybrid finite element methods which first decompose the domains into discontinuous elements then connect them with Lagrange multipliers living in the trace spaces [2, 13, 14].

The outline of this paper is as follows: In Sect. 2, we introduce the construction of polynomial spaces and their algebraic dual representations. The discrete formulations of the Neumann–Dirichlet problems and the proof of their equivalence at the discrete level follow in Sect. 3. A 3-dimensional numerical test case is then presented in Sect. 4. Finally, conclusions are drawn in Sect. 5.

2 Function Spaces

2.1 Primal Polynomial Spaces

Let $-1 = \xi_0^i < \xi_1^i < \cdots < \xi_{I^i}^i = 1, i = 1, 2, 3$, being three partitionings of $[-1, 1]$. The associated Lagrange polynomials are

$$h_j(\xi^i) = \prod_{m=0, m \neq j}^{I^i} \frac{\xi^i - \xi_m^i}{\xi_j^i - \xi_m^i}, \quad j = 0, 1, \cdots, I^i.$$

They are polynomials of degree I^i which satisfy the Kronecker delta property, $h_j(\xi_k^i) = \delta_{jk}$. The associated edge functions can be derived as [6],

$$e_j(\xi^i) = -\sum_{k=0}^{j-1} \frac{dh_k(\xi^i)}{d\xi^i}, \quad j = 1, 2, \cdots, I^i,$$

which are polynomials of degree $I^i - 1$. Edge functions also satisfy the Kronecker delta property, but in the integral sense,

$$\int_{\xi_{k-1}^i}^{\xi_k^i} e_j(\xi^i) \, d\xi^i = \delta_{jk}.$$

Consider a reference domain $\Omega_{\text{ref}}|_{\xi^1,\xi^2,\xi^3} := [-1, 1]^3$. With the tensor product, we can construct finite dimensional scalar function space $\mathcal{P}^{I^1,I^2,I^3}$ spanned by polynomial basis functions

$$\left\{ h_i(\xi^1) h_j(\xi^2) h_k(\xi^3) \right\},$$

and vector-valued function space $\mathcal{L}^{I^1,I^2,I^3}$ spanned by polynomial basis functions

$$\left\{ e_i(\xi^1) h_j(\xi^2) h_k(\xi^3), \ h_i(\xi^1) e_j(\xi^2) h_k(\xi^3), \ h_i(\xi^1) h_j(\xi^2) e_k(\xi^3) \right\}.$$

Let $\omega^h \in \mathcal{P}^{I^1,I^2,I^3}$ be

$$\omega^h = \sum_{i=0}^{I^1} \sum_{j=0}^{I^2} \sum_{k=0}^{I^3} w_{i,j,k} h_i(\xi^1) h_j(\xi^2) h_k(\xi^3). \tag{3}$$

Due to the way of constructing the edge functions, we can easy derive $\rho^h = \text{grad } \omega^h \in \mathcal{L}^{I^1,I^2,I^3}$,

$$\rho^h = \text{grad } \omega^h = (\rho_1, \ \rho_2, \ \rho_3)^{\mathsf{T}},$$

where [6],

$$\rho_1 = \sum_{i=1}^{I^1} \sum_{j=0}^{I^2} \sum_{k=0}^{I^3} \left(w_{i,j,k} - w_{i-1,j,k} \right) e_i(\xi^1) h_j(\xi^2) h_k(\xi^3),$$

$$\rho_2 = \sum_{i=0}^{I^1} \sum_{j=1}^{I^2} \sum_{k=0}^{I^3} \left(w_{i,j,k} - w_{i,j-1,k} \right) h_i(\xi^1) e_j(\xi^2) h_k(\xi^3),$$

$$\rho_3 = \sum_{i=0}^{I^1} \sum_{j=0}^{I^2} \sum_{k=1}^{I^3} \left(w_{i,j,k} - w_{i,j,k-1} \right) h_i(\xi^1) h_j(\xi^2) e_k(\xi^3).$$

Let $\underline{\omega}, \underline{\rho}$ be the vectors of expansion coefficients of ω^h, ρ^h. We can obtain

$$\underline{\rho} = \mathbb{E} \, \underline{\omega}, \tag{4}$$

where $\mathbb{E}$ is called the incidence matrix. The incidence matrix is very sparse, only consists of ± 1 as non-zero entries. If we squeeze, stretch or distort the domain, of course, the polynomial basis functions change, but the incidence matrix will remain the same. It only depends on the topology of the mesh and the numbering of the

degrees of freedom. And it is exact. In other words, it introduces no extra error.
All these features make it an excellent discrete counterpart of the grad operator.
Examples of incidence matrices can be found in [8, 10–12].

For a comprehensive explanation of these polynomial basis functions, we refer
to [6]. In isogeometric analysis, tensor-product B-splines with similar properties
have been developed, see, for example [5]. For tetrahedral elements, an analogue
development can be found in [15].

From (3), we can derive the trace of ω^h, for example, on the back boundary of
Ω_{ref}, $\Gamma_b = \left\{ \xi^1 = -1, \, \xi^2, \, \xi^3 \in [-1, 1] \right\}$,

$$\text{tr}_b \, \omega^h = \sum_{j=0}^{I^2} \sum_{k=0}^{I^3} w_{0,j,k} h_0(-1) h_j(\xi^2) h_k(\xi^3).$$

Let $\underline{\omega}_b$ be the vector of expansion coefficients of $\text{tr}_b \, \omega^h$. Clearly, there exists a linear
operator $\mathbb{N}_b$ such that

$$\underline{\omega}_b = \mathbb{N}_b \, \underline{\omega}.$$

The same processes can be done for other boundaries. If we collect the traces
of ω^h on all boundaries and combine their vectors of expansion coefficients and
corresponding linear operators, we can eventually obtain

$$\underline{\omega}_{\text{tr}} = \mathbb{N} \, \underline{\omega},$$

where the matrix $\mathbb{N}$, like $\mathbb{E}$, is sparse and only depends on the topology of the mesh
and the numbering of the degrees of freedom. Furthermore, it contains only 1 as
non-zero entries. An example of $\mathbb{N}$ can be found in [7]. Now, we can conclude that
the trace space, $P^{I^1, I^2, I^3} = \text{tr} \, \mathcal{P}^{I^1, I^2, I^3}$, is given as

$$P^{I^1, I^2, I^3} := P_{-1}^{I^2, I^3} \cup P_1^{I^2, I^3} \cup P_{-1}^{I^1, I^3} \cup P_1^{I^1, I^3} \cup P_{-1}^{I^1, I^2} \cup P_1^{I^1, I^2},$$

where $P_{-1}^{I^2, I^3}$ is the space spanned by $\left\{ h_0(-1) h_j(\xi^2) h_k(\xi^3) \right\}$, $P_1^{I^2, I^3}$ is the
space spanned by $\left\{ h_{I^1}(1) h_j(\xi^2) h_k(\xi^3) \right\}$ and so on. Notice that the polynomial
basis functions in $\left\{ h_0(-1) h_j(\xi^2) h_k(\xi^3) \right\}$ are exactly the same as those in
$\left\{ h_{I^1}(1) h_j(\xi^2) h_k(\xi^3) \right\}$ because $h_0(-1) = h_{I^1}(1) = 1$. But here we still distinguish
them because they represent basis functions at different boundaries.

2.2 Algebraic Dual Polynomial Spaces

We first consider the space $\mathcal{P}^{I^1,I^2,I^3}$. Let $\mathbb{M}_\mathcal{P}$ be the symmetric mass matrix, for example,

$$\mathbb{M}_{\mathcal{P}i+j(I^1+1)+k(I^1+1)(I^2+1),\, l+m(I^1+1)+n(I^1+1)(I^2+1)} :=$$

$$\iiint_{\Omega_{\text{ref}}} h_i(\xi^1)h_j(\xi^2)h_k(\xi^3)h_l(\xi^1)h_m(\xi^2)h_n(\xi^3)\, \mathrm{d}\xi^1\mathrm{d}\xi^2\mathrm{d}\xi^3.$$

The associated algebraic dual polynomial representations, or simply dual polynomials, are linear combinations of the polynomial basis functions, or simply primal polynomials, defined in the previous section,

$$\left[\widetilde{h_{0,0,0}}(\xi^1,\xi^2,\xi^3),\ \cdots,\ \widetilde{h_{I^1,I^2,I^3}}(\xi^1,\xi^2,\xi^3)\right]$$

$$:= \left[h_0(\xi^1)h_0(\xi^2)h_0(\xi^3),\ \cdots,\ h_{I^1}(\xi^1)h_{I^2}(\xi^2)h_{I^3}(\xi^3)\right]\mathbb{M}_\mathcal{P}^{-1}.$$

These dual polynomials are always well-defined. This is because the primal polynomials are linearly independent. So the mass matrix $\mathbb{M}_\mathcal{P}$ is injective and surjective, therefore invertible. Let the finite dimensional space spanned by $\left\{\widetilde{h_{i,j,k}}(\xi^1,\xi^2,\xi^3)\right\}$ be denoted by $\widetilde{\mathcal{P}}^{I^1,I^2,I^3}$. We say $\widetilde{\mathcal{P}}^{I^1,I^2,I^3}$ is the algebraic dual space of the primal space $\mathcal{P}^{I^1,I^2,I^3}$. Note that $\mathcal{P}^{I^1,I^2,I^3}$ and $\widetilde{\mathcal{P}}^{I^1,I^2,I^3}$ actually represent the same space. The change of basis functions only leads to a different representation. Therefore, we also call the algebraic dual space a dual representation. Let $\widetilde{\mathbb{M}}_\mathcal{P}$ be the mass matrix of $\widetilde{\mathcal{P}}^{I^1,I^2,I^3}$, we can easily see that

$$\widetilde{\mathbb{M}}_\mathcal{P}\mathbb{M}_\mathcal{P} = I, \tag{5}$$

where I is the identity matrix. Similarly, we can derive the algebraic dual space $\widetilde{\mathcal{L}}^{I^1,I^2,I^3}$ of the primal space $\mathcal{L}^{I^1,I^2,I^3}$. Let $\widetilde{\mathbb{M}}_\mathcal{L}$ and $\mathbb{M}_\mathcal{L}$ be their mass matrices, we have

$$\widetilde{\mathbb{M}}_\mathcal{L}\mathbb{M}_\mathcal{L} = I. \tag{6}$$

If $\rho^h \in \mathcal{L}^{I^1,I^2,I^3}$, σ^h, whose vector of expansion coefficients $\underline{\sigma}$ satisfies

$$\underline{\sigma} = \mathbb{M}_\mathcal{L}\,\underline{\rho}, \tag{7}$$

will be the representation of ρ^h in the algebraic dual space $\widetilde{\mathcal{L}}^{I^1,I^2,I^3}$.

To explain how the algebraic dual space of the trace space P^{I^1,I^2,I^3} is derived, we take P^{I^2,I^3}_{-1} as example. We already know that P^{I^2,I^3}_{-1} is a space spanned by primal polynomials $\left\{ h_0(-1)h_j(\xi^2)h_k(\xi^3) \right\}$. With these primal polynomials, we can compute its mass matrix, denoted by $\mathbb{M}_b$. The dual polynomials are then computed by

$$\left[\widetilde{h_{0,0,0}}(-1,\xi^2,\xi^3), \; \cdots, \; \widetilde{h_{0,I^2,I^3}}(-1,\xi^2,\xi^3) \right]$$

$$= \left[h_0(-1)h_1(\xi^2)h_1(\xi^3), \; \cdots, \; h_0(-1)h_{I^2}(\xi^2)h_{I^3}(\xi^3) \right] \mathbb{M}_b^{-1}.$$

The algebraic dual space $\widetilde{P}^{I^2,I^3}_{-1}$ is spanned by dual polynomials $\left\{ \widetilde{h_{0,j,k}}(-1,\xi^2,\xi^3) \right\}$. The algebraic dual space of the trace space P^{I^1,I^2,I^3} eventually can be written as

$$\widetilde{P}^{I^1,I^2,I^3} = \widetilde{P}^{I^2,I^3}_{-1} \cup \widetilde{P}^{I^2,I^3}_{1} \cup \widetilde{P}^{I^1,I^3}_{-1} \cup \widetilde{P}^{I^1,I^3}_{1} \cup \widetilde{P}^{I^1,I^2}_{-1} \cup \widetilde{P}^{I^1,I^2}_{1}.$$

The divergence of $\sigma^h \in \widetilde{\mathcal{L}}^{I^1,I^2,I^3}$ can be done with the help of the boundary value $\hat{\sigma}^h \in \widetilde{P}^{I^1,I^2,I^3}$. With vector proxies, it can be written as

$$\underline{\text{div } \sigma^h} = \mathbb{N}^\mathsf{T} \underline{\hat{\sigma}^h} - \mathbb{E}^\mathsf{T} \underline{\sigma^h}. \tag{8}$$

A detailed introduction of algebraic dual polynomial spaces is given in [9].

2.3   Function Spaces in Curvilinear Domains

So far, all polynomial spaces are defined only in the reference domain $\Omega_{\text{ref}}|_{\xi^1,\xi^2,\xi^3} = [-1,1]^3$. Consider an arbitrary domain Ω and a C^1 diffeomorphism $\Phi : \Omega_{\text{ref}}|_{\xi^1,\xi^2,\xi^3} \to \Omega|_{x^1,x^2,x^3}$. In Ω, the primal polynomials change. Therefore, the mass matrices will also change. But the process of constructing dual polynomials does not change. And as we mentioned before, the metric-independent incidence matrix $\mathbb{E}$ and the matrix $\mathbb{N}$ remain the same. The way of converting polynomials in Cartesian domain into those in curvilinear domains follows the general coordinate transformation process, for example, see [16].

From now on, notations mentioned in this section not only refer to the reference domain Ω_{ref}, but also refer to the physical domain Ω.

3 Weak Formulations

3.1 Discrete Neumann Problem

With integration by parts, we can derive the weak formulation of the Neumann problem, (1), written as: For given $\hat{\sigma} \in H^{-1/2}(\partial\Omega)$, find $\omega \in H^1(\Omega)$ such that

$$\left(\operatorname{grad} \omega,\ \operatorname{grad} \bar{\omega}\right)_{L^2} + (\omega,\ \bar{\omega})_{L^2} = \langle \operatorname{tr} \bar{\omega},\ \hat{\sigma} \rangle, \quad \forall \bar{\omega} \in H^1(\Omega). \tag{9}$$

Note that on the right hand side, we use $\langle \cdot, \cdot \rangle$ to represent the duality pairing between $\operatorname{tr} \bar{\omega} \in H^{1/2}(\partial\Omega)$ and $\hat{\sigma} \in H^{-1/2}(\partial\Omega)$. We use finite dimensional space $\mathcal{P}^{I^1, I^2, I^3}$ to approximate the space $H^1(\Omega)$ and use the algebraic dual trace space $\widetilde{P}^{I^1, I^2, I^3}$ to approximate the space $H^{-1/2}(\partial\Omega)$. Then we obtain

$$\left(\operatorname{grad} \omega^h,\ \operatorname{grad} \bar{\omega}^h\right)_{L^2} = \underline{\bar{\omega}}^{h,\mathsf{T}}\, \mathbb{E}^{\mathsf{T}} \mathsf{M}_{\mathcal{L}} \mathbb{E}\, \underline{\omega}^h,$$

$$\left(\omega^h,\ \bar{\omega}^h\right)_{L^2} = \underline{\bar{\omega}}^{h,\mathsf{T}}\, \mathsf{M}_{\mathcal{P}}\, \underline{\omega}^h,$$

and

$$\int_{\partial\Omega} \operatorname{tr} \bar{\omega}^h\, \hat{\sigma}^h\, \mathrm{d}\Gamma = \underline{\bar{\omega}}^{h,\mathsf{T}}\, \mathbb{N}^{\mathsf{T}}\, \underline{\hat{\sigma}}^h,$$

which eventually leads to the discrete formulation of (9),

$$\mathbb{E}^{\mathsf{T}} \mathsf{M}_{\mathcal{L}} \mathbb{E}\, \underline{\omega}^h + \mathsf{M}_{\mathcal{P}}\, \underline{\omega}^h = \mathbb{N}^{\mathsf{T}}\, \underline{\hat{\sigma}}^h. \tag{10}$$

3.2 Discrete Dirichlet Problem

For the Dirichlet problem, (2), the weak formulation is given as: For given $\hat{\sigma} \in H^{-1/2}(\partial\Omega)$, find $\sigma \in H(\operatorname{div}, \Omega)$, $\operatorname{tr} \sigma = \hat{\sigma}$ such that

$$(\operatorname{div} \sigma,\ \operatorname{div} \bar{\sigma})_{L^2} + (\sigma,\ \bar{\sigma})_{L^2} = 0, \quad \forall \bar{\sigma} \in H_0(\operatorname{div}, \Omega). \tag{11}$$

We use algebraic dual space $\widetilde{\mathcal{L}}^{I^1, I^2, I^3}$ to approximate $H(\operatorname{div}, \Omega)$. With $\hat{\sigma}^h \in \widetilde{P}^{I^1, I^2, I^3}$ given and (8), we obtain

$$\left(\operatorname{div} \sigma^h,\ \operatorname{div} \bar{\sigma}^h\right)_{L^2} = -\underline{\bar{\sigma}}^{h,\mathsf{T}}\, \mathbb{E}\widetilde{\mathsf{M}}_{\mathcal{P}} \left(\mathbb{N}^{\mathsf{T}}\, \underline{\hat{\sigma}}^h - \mathbb{E}^{\mathsf{T}}\, \underline{\sigma}^h\right),$$

and

$$\left(\sigma^h,\ \bar{\sigma}^h\right)_{L^2} = \underline{\bar{\sigma}}^{h,\mathsf{T}}\, \widetilde{\mathsf{M}}_{\mathcal{L}}\, \underline{\sigma}^h.$$

Therefore, the discrete formulation of (11) is written as

$$\mathrm{E}\widetilde{\mathrm{M}}_{\mathcal{P}}\mathrm{E}^{\mathsf{T}}\,\underline{\sigma}^h + \widetilde{\mathrm{M}}_{\mathcal{L}}\,\underline{\sigma}^h = \mathrm{E}\widetilde{\mathrm{M}}_{\mathcal{P}}\mathrm{N}^{\mathsf{T}}\,\underline{\hat{\sigma}}^h. \tag{12}$$

3.3 *Equivalence Between Discrete Formulations*

Now it is time to check if the equivalence between (1) and (2) holds at the discrete level. In other words, it is time to check if the statement that $\underline{\omega}^h$ solves (10) if and only if $\underline{\sigma}^h = \mathrm{grad}\,\underline{\omega}^h$ solves (12) is correct.

From (4) and (7), we know that $\underline{\sigma}^h$,

$$\underline{\sigma}^h = \mathrm{M}_{\mathcal{L}}\mathrm{E}\,\underline{\omega}^h, \tag{13}$$

is the vector representation of $\mathrm{grad}\,\omega^h$ in the dual space. If we insert (13) into (12), we obtain

$$\mathrm{E}\widetilde{\mathrm{M}}_{\mathcal{P}}\mathrm{E}^{\mathsf{T}}\mathrm{M}_{\mathcal{L}}\mathrm{E}\,\underline{\omega}^h + \widetilde{\mathrm{M}}_{\mathcal{L}}\mathrm{M}_{\mathcal{L}}\mathrm{E}\,\underline{\omega}^h = \mathrm{E}\widetilde{\mathrm{M}}_{\mathcal{P}}\mathrm{N}^{\mathsf{T}}\,\underline{\hat{\sigma}}^h. \tag{14}$$

From (10), we know that

$$\mathrm{E}^{\mathsf{T}}\mathrm{M}_{\mathcal{L}}\mathrm{E}\,\underline{\omega}^h = -\mathrm{M}_{\mathcal{P}}\,\underline{\omega}^h + \mathrm{N}^{\mathsf{T}}\,\underline{\hat{\sigma}}^h. \tag{15}$$

By inserting (15) into (14), we get

$$\mathrm{E}\widetilde{\mathrm{M}}_{\mathcal{P}}\left(-\mathrm{M}_{\mathcal{P}}\,\underline{\omega}^h + \mathrm{N}^{\mathsf{T}}\,\underline{\hat{\sigma}}^h\right) + \widetilde{\mathrm{M}}_{\mathcal{L}}\mathrm{M}_{\mathcal{L}}\mathrm{E}\,\underline{\omega}^h = \mathrm{E}\widetilde{\mathrm{M}}_{\mathcal{P}}\mathrm{N}^{\mathsf{T}}\,\underline{\hat{\sigma}}^h. \tag{16}$$

From (5) and (6), we know that (16) holds, which proves the equivalence.

If the equivalence holds, relation $\left\|\omega^h\right\|_{H^1(\Omega)} = \left\|\sigma^h\right\|_{H(\mathrm{div},\Omega)}$ should also be satisfied. To prove this, we have

$$\left\|\sigma^h\right\|^2_{H(\mathrm{div},\Omega)} \overset{(8)}{=} \underline{\sigma}^{h,\mathsf{T}}\,\widetilde{\mathrm{M}}_{\mathcal{L}}\,\underline{\sigma}^h + \left(\mathrm{N}^{\mathsf{T}}\,\underline{\hat{\sigma}}^h - \mathrm{E}^{\mathsf{T}}\underline{\sigma}^h\right)^{\mathsf{T}}\widetilde{\mathrm{M}}_{\mathcal{P}}\left(\mathrm{N}^{\mathsf{T}}\,\underline{\hat{\sigma}}^h - \mathrm{E}^{\mathsf{T}}\underline{\sigma}^h\right)$$

$$\overset{(13)}{=} \left(\mathrm{M}_{\mathcal{L}}\mathrm{E}\,\underline{\omega}^h\right)^{\mathsf{T}}\widetilde{\mathrm{M}}_{\mathcal{L}}\left(\mathrm{M}_{\mathcal{L}}\mathrm{E}\,\underline{\omega}^h\right)$$

$$+ \left[\mathrm{N}^{\mathsf{T}}\,\underline{\hat{\sigma}}^h - \mathrm{E}^{\mathsf{T}}\left(\mathrm{M}_{\mathcal{L}}\mathrm{E}\,\underline{\omega}^h\right)\right]^{\mathsf{T}}\widetilde{\mathrm{M}}_{\mathcal{P}}\left[\mathrm{N}^{\mathsf{T}}\,\underline{\hat{\sigma}}^h - \mathrm{E}^{\mathsf{T}}\left(\mathrm{M}_{\mathcal{L}}\mathrm{E}\,\underline{\omega}^h\right)\right]$$

$$\overset{(10)}{=} \underline{\omega}^{h,\mathsf{T}}\,\mathrm{E}^{\mathsf{T}}\mathrm{M}_{\mathcal{L}}\mathrm{E}\,\underline{\omega}^h + \underline{\omega}^{h,\mathsf{T}}\,\mathrm{M}_{\mathcal{P}}\widetilde{\mathrm{M}}_{\mathcal{P}}\mathrm{M}_{\mathcal{P}}\,\underline{\omega}^h$$

$$= \left\|\omega^h\right\|^2_{H^1(\Omega)},$$

where we constantly use (5) and (6) and the fact that mass matrices are symmetric.

4 Numerical Test

Consider the mapping Φ which maps the Cartesian reference domain $\Omega_{\text{ref}}|_{\xi^1,\xi^2,\xi^3} :=$ $[-1, 1]^3$ into the physical domain $\Omega|_{x^1,x^2,x^3} = [0, 1]^3$ by

$$x^i = \frac{1}{2} + \frac{1}{2}\left(\xi^i + c\prod_j \sin(\pi\xi^j)\right), \quad i = 1, 2, 3.$$

When the deformation coefficient $c = 0$, the domain Ω is Cartesian. Otherwise the domain is curvilinear, meaning that a curvilinear coordinate system parametrizes Ω. Examples of such curvilinear domains in $\mathbb{R}^2$ are shown in Fig. 1.

A manufactured solution of the Neumann problem, (1), is

$$\omega_{\text{exact}} = e^{x^1} + e^{x^2} + e^{x^3}.$$

Clearly, $\sigma_{\text{exact}} = \text{grad}\,\omega_{\text{exact}} = \left(e^{x^1}, e^{x^2}, e^{x^3}\right)^{\mathsf{T}}$ solves the Dirichlet problem, (2).

In the domains of different deformation coefficient c, with the boundary condition $\hat{\sigma} = \text{tr}\,\sigma_{\text{exact}}$ imposed, we solve the discrete formulations (10) and (12) using Gauss–Lobatto–Legendre (GLL) polynomial spaces of degree $I^1 = I^2 = I^3 = N$.

The results of the L^2-error of $\left(\sigma^h - \text{grad}\,\omega^h\right)$ are shown in Fig. 2 (Left) where we can see that the relation $\sigma^h = \text{grad}\,\omega^h$ is preserved up to the machine precision. With the growth of the polynomial degree, the error increases slowly because of the accumulation of the machine error as the amount of degrees of freedom grows significantly.

In Table 1, the results of the H^1-norm of ω^h and $H(\text{div})$-norm of σ^h are presented. It is shown that the relation $\left\|\omega^h\right\|_{H^1(\Omega)} = \left\|\sigma^h\right\|_{H(\text{div},\Omega)}$ holds for all polynomial degrees irrespective of whether we use the Cartesian domain, $c = 0$, or

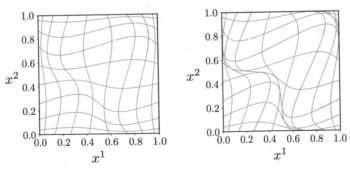

Fig. 1 Curvilinear domains for $c = 0.15$ (Left) and $c = 0.3$ (Right) in $\mathbb{R}^2$. The gray lines illustrate the coordinate lines

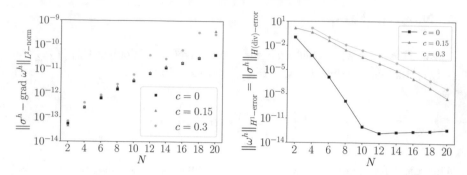

Fig. 2 The L^2-error of $\left(\sigma^h - \text{grad } \omega^h\right)$ (Left) and the p-convergence of the H^1-error of ω^h (Right) for $N = 2, 4, \cdots, 20$ and $c = 0, 0.15, 0.3$

Table 1 The H^1-norm of ω^h and $H(\text{div})$-norm of σ^h for polynomial degree $N = 2, 4, \cdots, 20$ and deformation coefficient $c = 0, 0.15, 0.3$

	$c = 0$		$c = 0.15$		$c = 0.3$	
N	$\left\|\omega^h\right\|_{H^1}$	$\left\|\sigma^h\right\|_{H(\text{div})}$	$\left\|\omega^h\right\|_{H^1}$	$\left\|\sigma^h\right\|_{H(\text{div})}$	$\left\|\omega^h\right\|_{H^1}$	$\left\|\sigma^h\right\|_{H(\text{div})}$
2	6.0720702909	6.0720702909	5.8899445673	5.8899445673	6.7381947027	6.7381947027
4	6.0730653395	6.0730653395	6.0567452129	6.0567452129	5.8849807780	5.8849807780
6	6.0730653668	6.0730653668	6.0729332275	6.0729332275	6.0721137212	6.0721137212
8	6.0730653668	6.0730653668	6.0730647051	6.0730647051	6.0730525346	6.0730525346
10	6.0730653668	6.0730653668	6.0730653557	6.0730653557	6.0730648440	6.0730648440
12	6.0730653668	6.0730653668	6.0730653665	6.0730653665	6.0730653428	6.0730653428
14	6.0730653668	6.0730653668	6.0730653667	6.0730653667	6.0730653663	6.0730653663
16	6.0730653668	6.0730653668	6.0730653668	6.0730653668	6.0730653667	6.0730653667
18	6.0730653668	6.0730653668	6.0730653668	6.0730653668	6.0730653668	6.0730653668
20	6.0730653668	6.0730653668	6.0730653668	6.0730653668	6.0730653668	6.0730653668

curvilinear domains, $c = 0.15, 0.3$. It is also seen that the results always converge to the analytical value $\|\omega_{\text{exact}}\|_{H^1} = \left\|\sigma^h\right\|_{H(\text{div})} = 6.0730653668$. The p-convergence for the H^1-error of ω^h, therefore also for the $H(\text{div})$-error of σ^h, is shown in Fig. 2 (Right), which shows the exponential convergence of the method.

5 Conclusions

By constructing and using primal polynomial spaces and their algebraic dual representations both in the domain and on the boundary, we successfully preserve the equivalence of the div-grad Neumann problem and the grad-div Dirichlet problem at the discrete level in 3-dimensional curvilinear domains. This suggests the further usage of these spaces to structure-preserving methods and hybrid methods.

References

1. Bochev, P.B., Hyman, J.M.: Principles of mimetic discretizations of differential operators. In: Compatible Spatial Discretizations, pp. 89–119. Springer, New York (2006)
2. Brezzi, F., Fortin, M.: Mixed and Hybrid Finite Element Methods, vol. 15. Springer Science & Business Media, New York (2012)
3. Carstensen, C., Demkowicz, L., Gopalakrishnan, J.: Breaking spaces and forms for the DPG method and applications including Maxwell equations. Comput. Math. Appl. **72**(3), 494–522 (2016)
4. Castillo, J.E., Miranda, G.F.: Mimetic Discretization Methods. Chapman and Hall/CRC, London (2013)
5. Evans, J.A., Scott, M.A., Shepherd, K.M., Thomas, D.C., Vázquez Hernández, R.: Hierarchical B-spline complexes of discrete differential forms. IMA J. Numer. Anal. **40**(1), 422–473 (2020)
6. Gerritsma, M.: Edge functions for spectral element methods. In: Spectral and High Order Methods for Partial Differential Equations, pp. 199–207. Springer, Berlin (2011)
7. Gerritsma, M., Jain, V., Zhang, Y., Palha, A.: Algebraic dual polynomials for the equivalence of curl-curl problems (2018). arXiv:1805.00114
8. Gerritsma, M., Palha A., Jain, V., Zhang, Y.: Mimetic spectral element method for anisotropic diffusion. In: Numerical Methods for PDEs. Springer SEMA SIMAI Series, vol. 15, pp. 31–74. Springer, Berlin (2018)
9. Jain, V., Zhang, Y., Palha, A., Gerritsma, M.: Construction and application of algebraic dual polynomial representations for finite element methods (2017). arXiv:1712.09472
10. Jain, V., Zhang, Y., Fisser J., Palha, A., Gerritsma, M.: A conservative hybrid method for Darcy flow (ICOSAHOM 2018, accepted)
11. Kreeft, J., Gerritsma, M.: Mixed mimetic spectral element method for Stokes flow: a pointwise divergence-free solution. J. Comput. Phys. **240**, 284–309 (2013)
12. Palha, A., Rebelo, P.P., Hiemstra, R., Kreeft, J., Gerritsma, M.: Physics-compatible discretization techniques on single and dual grids, with application to the Poisson equation of volume forms. J. Comput. Phys. **257**, 1394–1422 (2014)
13. Pian, T.H.: Derivation of element stiffness matrices by assumed stress distributions. AIAA J. **2**(7), 1333–1336 (1964)
14. Pian, T.H., Tong, P.: Basis of finite element methods for solid continua. Int. J. Numer. Methods Eng. **1**(1), 3–28 (1969)
15. Rapetti, F.: High order edge elements on simplicial meshes. ESAIM: Math. Model. Numer. Anal. **41**(6), 1001–1020 (2007)
16. Steinberg, S.: Fundamentals of Grid Generation. CRC Press, Boca Raton (1993)

A Conservative Hybrid Method for Darcy Flow

Varun Jain, Joël Fisser, Artur Palha, and Marc Gerritsma

1 Introduction

Hybrid formulations [1, 3, 10] are classical domain decomposition methods which reduce the problem of solving one global system to many small local systems. The local systems can then be efficiently solved independently of each other in parallel.

In this work we present a hybrid mimetic spectral element formulation to solve Darcy flow. We follow [8] which render the constraints on divergence of mass flux, the pressure gradient and the inter-element continuity metric free. The resulting system is extremely sparse and shows a reduced growth in condition number as compared to a non-hybrid system.

This document is structured as follows: In Sect. 2 we define the weak formulation for Darcy flow. The basis functions are introduced in Sect. 3. The evaluation of weighted inner product and duality pairings are discussed in Sect. 4. In Sect. 5 we discuss the formulation of discrete algebraic system. In Sect. 6 we present results for a test case taken from [7].

V. Jain (✉) · J. Fisser · A. Palha · M. Gerritsma
Faculty of Aerospace Engineering, TU Delft, Delft, The Netherlands
e-mail: V.Jain@tudelft.nl; A.PalhaDaSilvaClerigo@tudelft.nl; M.I.Gerritsma@tudelft.nl

© The Author(s) 2020
S. J. Sherwin et al. (eds.), *Spectral and High Order Methods for Partial Differential Equations ICOSAHOM 2018*, Lecture Notes in Computational Science and Engineering 134, https://doi.org/10.1007/978-3-030-39647-3_16

2 Darcy Flow Formulation

For $\Omega \in \mathbb{R}^d$, where d is the dimension of the domain, the governing equations for Darcy flow, are given by,

$$
\begin{cases} u + \mathbb{A}\,\nabla p = 0 \\ \nabla \cdot u = f \end{cases} \quad \text{in } \Omega \quad \text{and} \quad \begin{cases} \partial\Omega = \Gamma_D \cup \Gamma_N \\ p = \hat{p} & \text{on } \Gamma_D \\ u \cdot n = \hat{u}_n & \text{on } \Gamma_N \end{cases},
$$

where, u is the velocity, p is the pressure, f the prescribed RHS term, $\mathbb{A}$ is a $d \times d$ symmetric positive definite matrix, $\hat{p}$ and $\hat{u}_n$ are the prescribed pressure and flux boundary conditions, respectively.

2.1 Notations

For $f, g \in L^2(\Omega)$, $(f, g)_\Omega$ denotes the usual L^2-inner product.

For vector-valued functions in L^2 we define the weighted inner product by,

$$
(u, v)_{\mathbb{A}^{-1}, \Omega} = \int_\Omega \left(u, \mathbb{A}^{-1} v \right) d\Omega, \tag{1}
$$

where $(\cdot\,,\cdot)$ denotes the pointwise inner product.

Duality pairing, denoted by $\langle\cdot,\cdot\rangle_\Omega$, is the outcome of a linear functional on $L^2(\Omega)$ acting on elements from $L^2(\Omega)$.

Let Ω_K be a disjoint partitioning of Ω with total number of elements K, and K_i is any element in Ω_K, such that, $K_i \in \Omega_K$. We define the following broken Sobolev spaces [2], $H\left(\text{div}; \Omega_K\right) = \prod_i H\left(\text{div}; K_i\right)$, and $H^{1/2}(\partial\Omega_K) = \prod_i H^{1/2}(\partial K_i)$.

2.2 Weak Formulation

The Lagrange functional for Darcy flow is defined as,

$$
\mathcal{L}\left(u, p, \lambda; f\right) = \tfrac{1}{2} \int_{\Omega_K} u^T \mathbb{A}^{-1} u \; d\Omega_K - \int_{\Omega_K} p \left(\nabla \cdot u - f\right) d\Omega_K
$$
$$
+ \int_{\partial\Omega_K \backslash \Gamma_D} \lambda \left(u \cdot n\right) d\Gamma + \int_{\Gamma_D} \hat{p} \left(u \cdot n\right) d\Gamma - \int_{\Gamma_N} \lambda \left(\hat{u}_n\right) d\Gamma
$$

The variational problem is then given by: For given $f \in L^2(\Omega_K)$, $\hat{p} \in H^{1/2}(\Gamma_D)$ and $\hat{u}_n \in H^{-1/2}(\Gamma_N)$ find $u \in H(\text{div}; \Omega_K)$, $p \in L^2(\Omega_K)$, $\lambda \in H^{\frac{1}{2}}(\partial\Omega_K)$, such that,

$$
\begin{cases}
(v, u)_{\mathbb{A}^{-1}, \Omega_K} - \langle \nabla \cdot v, p \rangle_{\Omega_K} + \langle (v \cdot n), \lambda \rangle_{\partial\Omega_K \setminus \Gamma_D} = -\langle v \cdot n, \hat{p} \rangle_{\Gamma_D} & \forall v \in H(\text{div}; \Omega_K) \\
-\langle q, \nabla \cdot u \rangle_{\Omega_K} = -\langle q, f \rangle_{\Omega_K} & \forall q \in L^2(\Omega_K) \\
\langle \mu, (u \cdot n) \rangle_{\partial\Omega_K \setminus \Gamma_D} = \langle \mu, \hat{u}_n \rangle_{\Gamma_N} & \forall \mu \in H^{\frac{1}{2}}(\partial\Omega_K)
\end{cases}
$$

$$(2)$$

3   Basis Functions

3.1   Primal and Dual Nodal Degrees of Freedom

Let ξ_j, $j = 0, 1, \ldots, N$, be the $N + 1$ Gauss–Lobatto–Legendre (GLL) points in $I \in [-1, 1]$. The Lagrange polynomials $h_i(\xi)$ through ξ_j, of degree N, given by,

$$
h_i(\xi) = \frac{\left(\xi^2 - 1\right) L'_N(\xi)}{N(N+1) L_N(\xi_i)(\xi - \xi_i)},
$$

form the 1D primal nodal polynomials which satisfy, $h_i(\xi_j) = \delta_{ij}$.

Let a^h and b^h be two polynomials expanded in terms of $h_i(\xi)$. The L^2—inner product is then given by,

$$
\left(a^h, b^h\right)_I = \mathbf{a}^T \mathbb{M}^{(0)} \mathbf{b}, \quad \text{where} \quad \mathbb{M}^{(0)}_{i,j} = \int_{-1}^{1} h_i(\xi) \, h_j(\xi) \, d\xi,
$$

and, $\mathbf{a} = \lceil a_0 \, a_1 \, \ldots \, a_N \rceil$ and $\mathbf{b} = \begin{bmatrix} b_0 \, b_1 \, \ldots \, b_N \end{bmatrix}$ are the nodal degrees of freedom. We define the algebraic *dual* degrees of freedom, $\tilde{\mathbf{a}}$, such that the duality pairing is simply the vector dot product between primal and dual degrees of freedom,

$$
\left\langle a^h, b^h \right\rangle_I = \tilde{\mathbf{a}}^T \mathbf{b} := \mathbf{a}^T \mathbb{M}^{(0)} \mathbf{b} \quad \Rightarrow \quad \tilde{\mathbf{a}} = \mathbb{M}^{(0)} \mathbf{a}.
$$

Thus, the dual degrees of freedom are linear functionals of primal degrees of freedom.

3.2 Primal and Dual Edge Degrees of Freedom

The edge polynomials, for the N edges between $N + 1$ GLL points (ξ_{j-1}, ξ_j), of polynomial degree $N - 1$, are defined as [4],

$$e_j(\xi) = -\sum_{k=0}^{j-1} \frac{dh_k}{d\xi}(\xi) , \quad \text{such that} \quad \int_{\xi_{j-1}}^{\xi_j} e_i(\xi) = \delta_{ij} .$$

Let p^h and q^h be two polynomials expanded in edge basis functions. The inner product in L^2 space is given by,

$$\left(p^h, q^h\right)_I = \mathbf{p}^T \mathbb{M}^{(1)} \mathbf{q} , \quad \text{where} \quad \mathbb{M}^{(1)}_{i,j} = \int_{-1}^1 e_i(\xi) \, e_j(\xi) \, d\xi ,$$

and, $\mathbf{p} = [p_1 \, p_2 \, \ldots \, p_N]$ and $\mathbf{q} = [q_1 \, q_2 \, \ldots \, q_N]$ are the edge degrees of freedom. As before, we define the *dual* degrees of freedom such that,

$$\left\langle p^h, q^h\right\rangle_I = \tilde{\mathbf{p}}^T \mathbf{q} := \mathbf{p}^T \mathbb{M}^{(1)} \mathbf{q} \quad \Rightarrow \quad \tilde{\mathbf{p}} = \mathbb{M}^{(1)} \mathbf{p} .$$

A similar construction can be used for dual degrees of freedom in higher dimensions. For construction of the dual degrees of freedom in 2D see [8] and for 3D see [9].

3.3 Differentiation of Nodal Polynomial Representation

Let $a^h(\xi)$ be expanded in Lagrange polynomials, then

$$\frac{d}{d\xi} a^h(\xi) = \frac{d}{d\xi} \sum_{i=0}^N a_i h_i(\xi) = \sum_{i=1}^N (a_i - a_{i-1}) e_i(\xi) . \tag{3}$$

Therefore, taking the derivative of a polynomial involves two steps: First, take the difference of degrees of freedom; second, change of basis from nodal to edge [4].

4 Discrete Inner Product and Duality Pairing

For 2D domains, the higher dimensional primal basis are constructed using the tensor product of the 1D basis.

For the weak formulation (2) we expand the velocity u^h in primal edge basis as,

$$u^h (\xi, \eta) = \sum_{i=0}^{N} \sum_{j=1}^{N} u_{xi,j} \, h_i(\xi) \, e_j(\eta) \, \hat{\imath} + \sum_{i=1}^{N} \sum_{j=0}^{N} u_{yi,j} \, e_i(\xi) \, h_j(\eta) \, \hat{\jmath} \,, \qquad (4)$$

where $u_{xi,j}$ denotes the flux, $\int u \cdot n$, over the vertical edges and $u_{yi,j}$ the flux over the horizontal edges, see Fig. 1.

4.1 Weighted Inner Product

Using (1) and the expansions in (4), the weighted inner product is evaluated as,

$$\left(v^h, u^h\right)_{\mathbb{A}^{-1}, \Omega_K} = \sum_{K_i} v_{K_i}^T \, \mathbb{M}^{(1)}_{\mathbb{A}^{-1}, K_i} \, u_{K_i} \,,$$

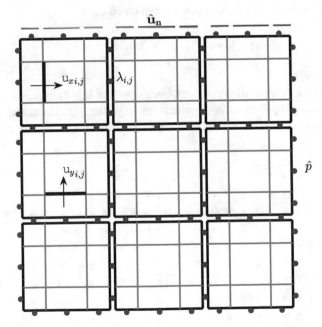

Fig. 1 Discretized domain for $K = 3 \times 3$, $N = 3$. The blue dots represent the pressure boundary condition $\hat{p}$, and the blue edges represent the velocity boundary condition $\hat{u}_n$

where, $\mathbf{u}_{K_i}$ are the degrees of freedom in element K_i, and

$$M^{(1)}_{\mathbb{A}^{-1}, K_i} = \int_{K_i} \begin{pmatrix} h_i(\xi)\, e_j(\eta) \\ e_i(\xi)\, h_j(\eta) \end{pmatrix}^T \mathbb{A}^{-1}(\xi, \eta) \begin{pmatrix} h_i(\xi)\, e_j(\eta) \\ e_i(\xi)\, h_j(\eta) \end{pmatrix} dK_i \, .$$

For mapping of elements please refer to [6].

4.2 Divergence of Velocity

Divergence of velocity, $\nabla \cdot u^h$, is evaluated using (3), but now for 2D,

$$\begin{aligned}
\nabla \cdot u^h &= \frac{\partial}{\partial x} \sum_{i=0}^{N} \sum_{j=1}^{N} \mathsf{u}_{x_{i,j}}\, h_i(\xi) e_j(\eta) + \frac{\partial}{\partial y} \sum_{i=1}^{N} \sum_{j=0}^{N} \mathsf{u}_{y_{i,j}}\, e_i(\xi) h_j(\eta) \\
&= \sum_{i,j=1}^{N} \left(\mathsf{u}_{x_{i,j}} - \mathsf{u}_{x_{i-1,j}} + \mathsf{u}_{y_{i,j}} - \mathsf{u}_{y_{i,j-1}} \right) e_i(\xi)\, e_j(\eta)
\end{aligned}$$

(5)

For pressure we will use dual degrees of freedom. Therefore the weak constraint on divergence of velocity is a duality pairing evaluated as,

$$\left\langle q^h, \nabla \cdot u^h \right\rangle_{\Omega_K} = \sum_{K_i} \tilde{\mathbf{q}}_{K_i}^T\, \mathbb{E}^{2,1}\, \mathbf{u}_{K_i} \, ,$$

where $\mathbb{E}^{2,1}$ represents the discrete divergence operator. It is an incidence matrix that is metric-free and topological, and remains the same for each element in Ω_K. For an extensive discussion on the incidence matrix, see for instance [6]. For an element of degree $N = 3$,

$$\mathbb{E}^{2,1} = \begin{bmatrix}
-1 & 0 & 0 & 1 & 0 & 0 & 0 & 0 & 0 & 0 & 0 & -1 & 0 & 0 & 1 & 0 & 0 & 0 & 0 & 0 & 0 & 0 \\
0 & 0 & 0 & -1 & 0 & 0 & 1 & 0 & 0 & 0 & 0 & 0 & -1 & 0 & 0 & 1 & 0 & 0 & 0 & 0 & 0 & 0 \\
0 & 0 & 0 & 0 & 0 & 0 & -1 & 0 & 0 & 1 & 0 & 0 & 0 & -1 & 0 & 0 & 1 & 0 & 0 & 0 & 0 & 0 \\
0 & -1 & 0 & 0 & 1 & 0 & 0 & 0 & 0 & 0 & 0 & 0 & 0 & 0 & -1 & 0 & 0 & 1 & 0 & 0 & 0 & 0 \\
0 & 0 & 0 & 0 & -1 & 0 & 0 & 1 & 0 & 0 & 0 & 0 & 0 & 0 & 0 & -1 & 0 & 0 & 1 & 0 & 0 & 0 \\
0 & 0 & 0 & 0 & 0 & 0 & -1 & 0 & 0 & 1 & 0 & 0 & 0 & 0 & 0 & 0 & -1 & 0 & 0 & 1 & 0 & 0 \\
0 & 0 & -1 & 0 & 0 & 1 & 0 & 0 & 0 & 0 & 0 & 0 & 0 & 0 & 0 & 0 & -1 & 0 & 0 & 1 & 0 & 0 \\
0 & 0 & 0 & 0 & 0 & -1 & 0 & 0 & 1 & 0 & 0 & 0 & 0 & 0 & 0 & 0 & 0 & -1 & 0 & 0 & 1 & 0 \\
0 & 0 & 0 & 0 & 0 & 0 & 0 & 0 & -1 & 0 & 0 & 1 & 0 & 0 & 0 & 0 & 0 & 0 & 0 & -1 & 0 & 0 & 1
\end{bmatrix} .$$

4.3 Connectivity Matrix

The connectivity matrix ensures continuity of the velocity flux across the elements. λ is the interface variable defined between the elements, shown as red dots in Fig. 1. λ acts as Lagrange multiplier that imposes the continuity constraint given by,

$$\left\langle \mu^h, u^h \cdot n \right\rangle_{\partial\Omega_K \backslash \Gamma_D} = \sum_K \tilde{\mu}_{K_i}^T \, \mathbb{N} \, \mathbf{u}_{K_i} = \tilde{\mu}^T \, \mathbb{E}_{\mathbb{N}} \, \mathbf{u} \ ,$$

where $\mathbb{N}$ is the discrete trace operator. It is a sparse matrix that consists of $1, -1$ and 0 only. For construction of $\mathbb{N}$ please refer to [5]. $\mathbb{E}_{\mathbb{N}}$ is the assembled $\mathbb{N}$ for all elements. For, $K = 2 \times 2$, $N = 2$, $\mathbb{E}_{\mathbb{N}}$ is shown in (6). The matrix size of $\mathbb{E}_{\mathbb{N}}$ is 8×64, but it has only 16 non-zero entities. It is an extremely sparse matrix that is metric free and the location of $\pm$ valued entries depend only on the connection between different elements.

$$\mathbb{E}_{\mathbb{N}} = \begin{bmatrix} & 1 & & & & -1 & & & & & & & & & & \\ & 1 & & & & & -1 & & & & & & & & & \\ & & & & & & & 1 & & & & -1 & & & & \\ & & & & & & & & 1 & & & & -1 & & & \\ & & 1 & & & & & & & & -1 & & & & & \\ & & 1 & & & & & & & -1 & & & & & & \\ & & & & & 1 & & & & & & & & -1 & & \\ & & & & & & 1 & & & & & & & & -1 & \end{bmatrix} .$$

$$(6)$$

5 Discrete Formulation

Using the weighted inner product and duality pairings discussed in Sect. 4, we can write the discrete form of weak formulation in (2) as,

$$\begin{bmatrix} \mathbb{B} & \mathbb{E}_{\mathbb{N}}^T \\ \mathbb{E}_{\mathbb{N}} & 0 \end{bmatrix} \begin{bmatrix} \mathbf{X} \\ \lambda \end{bmatrix} = \begin{bmatrix} \mathbf{F} \\ 0 \end{bmatrix} ,$$

$$(7)$$

where, $\mathbb{B}$ is an invertible block diagonal matrix given by,

$$
\mathbb{B} =
\begin{bmatrix}
\mathbb{M}^{(1)}_{\mathbb{A}^{-1}, K_1} & \mathbb{E}^{2,1^T} & & & & & \\
\mathbb{E}^{2,1} & 0 & & & & & \\
& & \mathbb{M}^{(1)}_{\mathbb{A}^{-1}, K_2} & \mathbb{E}^{2,1^T} & & & \\
& & \mathbb{E}^{2,1} & 0 & & & \\
& & & & \ddots & \ddots & \\
& & & & & \ddots & \ddots \\
& & & & & & \mathbb{M}^{(1)}_{\mathbb{A}^{-1}, K_K} & \mathbb{E}^{2,1^T} \\
& & & & & & \mathbb{E}^{2,1} & 0
\end{bmatrix},
\tag{8}
$$

$\mathbb{E}_{\mathbb{N}}$ is as given in (6), $\mathbf{X} = \sum_i \begin{bmatrix} \mathbf{u} \\ \mathbf{p} \end{bmatrix}_{K_i}$, and $\mathbf{F} = \sum_i \begin{bmatrix} \hat{\mathbf{p}} \\ \mathbf{f} \end{bmatrix}_{K_i}$, where $\mathbf{f}$ are the expansion coefficients of $f^h(x, y) = \sum_{i,j}^{N} \mathbf{f}_{ij}\, e_i(x)\, e_j(y)$.

In (8), the mass matrix $\mathbb{M}^{(1)}_{\mathbb{A}^{-1}, K_i}$ is the only dense matrix and also the only matrix that changes with each local element, K_i. $\mathbb{E}_{\mathbb{N}}$ is a sparse incidence matrix for the global system and $\mathbb{E}^{2,1}$ is a sparse incidence matrix for the local systems that remains the same for each element.

Using the Schur complement method, the global system (7) can be reduced to solve for $\boldsymbol{\lambda}$, [1],

$$
\boldsymbol{\lambda} = \left(\mathbb{E}_{\mathbb{N}}\mathbb{B}^{-1}\,\mathbb{E}_{\mathbb{N}}{}^{T}\right)^{-1} \cdot \left(\mathbb{E}_{\mathbb{N}}\mathbb{B}^{-1}\mathbf{F}\right).
\tag{9}
$$

To evaluate $\boldsymbol{\lambda}$ in (9) we need $\mathbb{B}^{-1}$ that can be calculated efficiently by taking inverse of each block of $\mathbb{B}$ separately. This part is trivially parallelized. Once the $\boldsymbol{\lambda}$ is determined the solution in each element, K_i, can be evaluated independent of each other.

The system (9) solves for interface degrees of freedom between the elements and will always be smaller than the full global system. For a comparison of the size of $\boldsymbol{\lambda}$ system with full system see Table 1 (for 2D), and Table 2 (for 3D). On the left of Tables 1 and 2 we see that, for constant K, increasing the order of polynomial basis the growth in size of $\boldsymbol{\lambda}$ system is less than the growth in size of full system. Thus, hybrid formulations are beneficial for high order methods where local degrees of freedom of an element are much higher than interface degrees of freedom.

Table 1 For 2D

N	Full system	λ only	λ/full	K	Full system	λ only	λ/full
5	825	60	0.07	400	15,480	2280	0.15
10	3000	120	0.04	1600	62,160	9360	0.15
15	6525	180	0.03	3600	140,040	21,240	0.15
20	11,400	240	0.02	6400	249,120	37,920	0.15
25	17,625	300	0.02	10,000	389,400	59,400	0.15

Left: Number of total unknowns as a function of N, for $K = 3 \times 3$. Right: Number of total unknowns as a function of the number of elements K, for $N = 3$

Table 2 For 3D

N	Full system	λ only	λ/full	K	Full system	λ only	λ/full
5	16,875	1350	0.08	8000	1,285,200	205,200	0.16
10	121,500	5400	0.04	64,000	10,324,800	1,684,800	0.16
15	394,875	12,150	0.03	216,000	34,894,800	5,734,800	0.16
20	918,000	21,600	0.02	512,000	82,771,200	13,651,200	0.16
25	1,771,875	33,750	0.02	1,000,000	161,730,000	26,730,000	0.17

Left: Number of total unknowns as a function of N, for $K = 3 \times 3 \times 3$. Right: Number of total unknowns as a function of the number of elements K, for $N = 3$

On the right of Tables 1 and 2 we see that, for constant N, the λ system is smaller than the full system, although the growth ratio of the size of λ and full systems do not change significantly.

6   Results

In this section we present the results for a test problem from [7] by solving system (7). The domain of the test problem is, $\Omega \in [0, 1]^2$. The RHS term is defined as,

$$f_{ex} = \nabla \cdot (-\mathbb{A}\nabla p_{ex}) , \quad \text{where} ,$$

$$\mathbb{A} = \frac{1}{x^2+y^2+\alpha} \begin{pmatrix} 10^{-3}x^2 + y^2 + \alpha & \left(10^{-3} - 1\right)xy \\ \left(10^{-3} - 1\right)xy & x^2 + 10^{-3}y^2 + \alpha \end{pmatrix} ; \quad \alpha = 0.1 ,$$

$$p_{ex} = \sin\left(2\pi x\right)\sin\left(2\pi y\right)$$

and Dirichlet boundary conditions are imposed along the entire boundary, $\Gamma_D = \partial\Omega$ and $\Gamma_N = \emptyset$. We solve this problem on an orthogonal and a curved mesh, see Fig. 2.

The same problem was earlier addressed in [6], but for a method with continuous elements and *primal* basis functions only. For the configuration $K = 3 \times 3$, $N = 6$, we compare the sparsity structure of the two approaches in Fig. 3. On the left we see

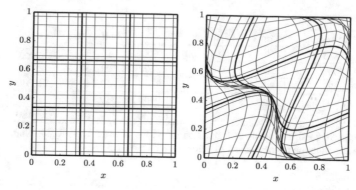

Fig. 2 Mesh configuration: $K = 3 \times 3$, $N = 6$. Left: orthogonal. Right: curved

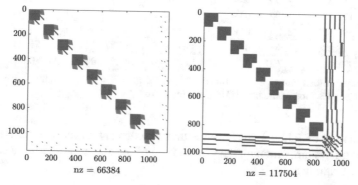

Fig. 3 Sparsity plots $K = 3 \times 3$, $N = 6$. Left: hybrid elements method. Right: continuous element method

the hybrid formulation, and on the right we see the continuous elements formulation [6]. The number of non zero entries are almost half in the hybrid formulation, 66,384, as compared to the continuous element formulation, 117,504. Here, the sparsity is due to use of algebraic dual degrees of freedom and is not because of hybridization of the scheme.

In Fig. 4, on the left we compare the growth in condition number, for the λ system (9) with full continuous element system, for $N = 7$ on the curved mesh, with increasing number of elements, K. We observe similar growth rates for hybrid and continuous formulation, however the condition number for continuous elements formulation is almost $O\left(10^2\right)$ higher. On the right we see the growth in condition number with increasing polynomial degree for $K = 9 \times 9$ on the curved mesh. A reduced growth rate in condition number for hybrid formulation is observed. Thus hybrid formulations are beneficial for high order methods.

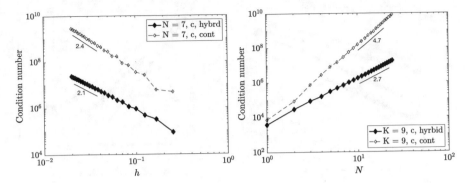

Fig. 4 Growth in condition number for hybrid elements in dark line, and continuous elements in dotted line. Left: h-refinement; Right: N-refinement. 'c' refers to the curved mesh

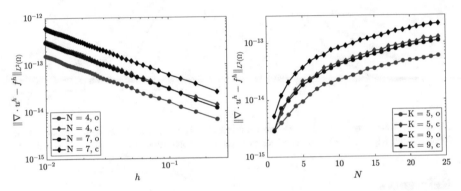

Fig. 5 L^2-error in divergence of velocity: Left: h-refinement; Right: N-refinement. 'o' refers to the orthogonal mesh and 'c' to the curved mesh

In Fig. 5 we show the L^2-error for $\|\nabla \cdot \boldsymbol{u}^h - f^h\|$. On the left side as a function of element size, $h = 1/\sqrt{K}$, and on the right side as a function of polynomial degree of the basis functions. In both cases the maximum error observed is of $O\left(10^{-12}\right)$.

In Fig. 6, on the top two figures we show the error in the $H\left(\text{div}; \Omega\right)$ norm for the velocity; and at the bottom two figures we show the error in $L^2\left(\Omega\right)$ norm for the pressure. On the left we have h-convergence plots, and on the right we have N-convergence plots. In all the figures, for the same number of elements, K, and polynomial degree, N, the error is higher for the curved mesh.

On the left we see that the error decreases with the element size. The slope of error rate of convergence is N, which is optimal for both curved and orthogonal meshes. On the right we see exponential convergence of the error with increasing polynomial degree of basis for both orthogonal and curved meshes.

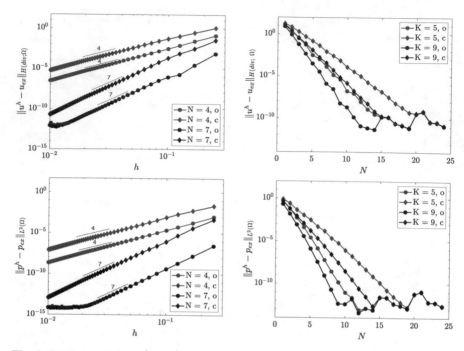

Fig. 6 Top row: error in $H\left(\mathrm{div};\,\Omega\right)$ norm for velocity; Bottom row: L^2-error in pressure. Left: h-refinement; Right: N-refinement. 'o' refers to the orthogonal mesh and 'c' to the curved mesh

References

1. Boffi, D., Brezzi, F., Fortin, M.: Mixed Finite Elements Methods and Applications. Springer Series in Computational Mechanics. Springer, Berlin (2010)
2. Carstensen, C., Demkowicz, L., Gopalakrishnan, J.: Breaking spaces and forms for the DPG method and applications including Maxwell equations. Comput. Math. Appl. **72**, 494–522 (2016)
3. Cockburn, B.: Static Condensation, Hybridization, and the Devising of the HDG Methods. Lecture Notes in Computational Science and Engineering, vol. 114. Springer, Berlin (2015)
4. Gerritsma, M.: Edge functions for spectral element methods. In: Spectral and High Order Methods for Partial Differential Equations, pp. 199–208. Springer, Berlin (2011)
5. Gerritsma, M., Jain, V., Zhang, Y., Palha, A.: Algebraic dual polynomials for the equivalence of curl-curl problems (2018). arXiv:1805.00114
6. Gerritsma, M., Palha, A., Jain, V., Zhang, Y.: Mimetic spectral element method for anisotropic diffusion. In: Numerical Methods for PDEs, pp. 31–74. Springer, Berlin (2018)
7. Herbin, R., Hubert, F.: Benchmark on Discretization Schemes for Anisotropic Diffusion Problems on General Grids. ISTE, Finite Volumes for Complex Applications V, pp. 659–692. Wiley, London (2008)
8. V. Jain, Y. Zhang, A. Palha, M. Gerritsma, Construction and application of algebraic dual polynomial representations for finite element methods (2017). arXiv:1712.09472

9. Zhang, Y., Jain, V., Palha, A., Gerritsma, M.: Discrete equivalence of adjoint Neumann-Dirichlet div-grad and grad-div equations in curvilinear 3D domains. In: Spectral and High Order Methods for Partial Differential Equations ICOSAHOM 2018. Springer, Cham (2020). https://doi.org/10.1007/978-3-030-39647-3_3
10. Zhang, Y., Jain, V., Palha, A., Gerritsma, M.: The discrete Steklov-Poincaré operator using algebraic dual polynomials. Comput. Methods Appl. Math. **19**(3), 645–661. https://doi.org/10.1515/cmam-2018-0208

High-Order Mesh Generation Based on Optimal Affine Combinations of Nodal Positions

Mike Stees and Suzanne M. Shontz

1 Introduction

The advantage of high-order numerical methods for solving partial differential equations is their higher degree of accuracy compared to low-order numerical methods. A major hurdle in the usage of these methods in the presence of complex geometries is the absence of robust high-order mesh generation methods [23]. In other words, these methods need a high-order mesh that accurately captures the features of the geometry to achieve their full potential [1, 10].

The typical approach for generating high-order meshes is to transform a coarse linear mesh [2–6, 9, 11, 12, 14, 16, 17, 19–22, 24]. At a high-level, these transformations usually consist of the following three steps: (1) the low-order mesh is enriched with additional nodes; (2) the new nodes that lie along the boundary of the mesh are moved to the true boundary; (3) the interior nodes are moved based on the boundary deformation. The main challenge of these methods arises from step (2). In particular, the curving of the elements along the boundary can result in invalid mesh elements. With that in mind, these high-order mesh generation methods use different approaches in step (3) in an effort to obtain a valid high-order mesh. Methods for transforming the linear mesh usually fall into two groups. The first

M. Stees (✉)
Department of Electrical Engineering and Computer Science, Information and Telecommunication Technology Center, University of Kansas, Lawrence, KS, USA
e-mail: mstees@ku.edu

S. M. Shontz
Department of Electrical Engineering and Computer Science, Bioengineering Program, Information and Telecommunication Technology Center, University of Kansas, Lawrence, KS, USA
e-mail: shontz@ku.edu

© The Author(s) 2020
S. J. Sherwin et al. (eds.), *Spectral and High Order Methods for Partial Differential Equations ICOSAHOM 2018*, Lecture Notes in Computational Science and Engineering 134, https://doi.org/10.1007/978-3-030-39647-3_17

229

group of methods transform the mesh based on the solution to a partial differential equation [3, 12, 14, 24]. The second group of methods are based on optimization of an objective function [2, 4, 5, 9, 16–21].

In this paper, we describe an optimization-based approach for generating high-order meshes based on affine combinations of nodal positions. The remainder of this paper is organized as follows. In Sect. 2, we present our new method for high-order mesh generation. In Sect. 3, we demonstrate the performance of our proposed method on several aerospace engineering geometries. Finally, in Sect. 4, we offer some concluding remarks and possible directions for our future work.

2 High-Order Mesh Generation Based on Affine Combinations of Nodal Positions

In this section, we present our optimization-based method for high-order mesh generation. Our proposed method uses affine combinations of nodal positions to determine the movement of the interior nodes after deforming the boundary. Our method consists of three steps. First, for each interior node in the high-order straight-sided mesh, an optimization problem is solved to calculate a set of weights that relates the interior node to its neighbors. Second, the boundary nodes are moved to the true boundary. Third, the new positions of the interior nodes are calculated by solving a linear system of equations using the weights and the new boundary positions. In spirit, this method is similar to the weight-based method that we proposed in [19] with two major differences. The first difference is that we propose an affine combination of nodal positions in this work, as opposed to a convex combination. This change allows us to remove the inequality constraint and log-barrier term, leaving only the equality constraints. We also propose an alternative objective function that when combined with the equality constraints allows us to directly solve the optimization problem via a QR factorization. This change results in simplified computational complexity and faster execution time.

To frame our discussion of the method, we introduce the following notation for the 2D formulation of the problem; the 3D formulation is similar. Let the x- and y-coordinates of the ith interior node be represented as (x_i, y_i). In addition, define the x- and y-coordinates of the vertices adjacent to node i as $\{(x_j, y_j) : j \in N_i\}$, where N_i is the set of neighbors of node i. For each interior node i, this information can be represented as the following linear system, where w_{ij} are the weights:

$$\sum_{j \in N_i} w_{ij} x_j = x_i$$

$$\sum_{j \in N_i} w_{ij} y_j = y_i,$$

where

$$N_i = \{\text{high-order nodes of the patch to which } i \text{ belongs}\}.$$

There are several potential choices for the local neighboring set based on use of the low-order nodes, high-order nodes, or both. We include only the high-order nodes as neighbors, as only the high-order boundary nodes move during the boundary deformations. Using either the low-order nodes or both the low- and high-order nodes would dampen the effect the boundary deformation has on the interior nodes, which might lead to tangling near the boundary. Including additional nodes as neighbors would also result in a less sparse matrix when solving (7). Another important consideration is that while the weight calculation is based on only the local neighbors, the position of an interior node is indirectly affected by the deformation of all the interior nodes through the solution of (7).

Adding the additional constraint that the weights sum to one results in the following linear system $Aw = b$ for finding an affine combination of the x- and y-coordinates of the vertices adjacent to node i:

$$\begin{bmatrix} x_1 \; x_2 \; \ldots \; x_n \\ y_1 \; y_2 \; \ldots \; y_n \\ 1 \; 1 \; \ldots \; 1 \end{bmatrix} \begin{bmatrix} w_{i1} \\ w_{i2} \\ \vdots \\ w_{in} \end{bmatrix} = \begin{bmatrix} x_i \\ y_i \\ 1 \end{bmatrix},$$

where $n = |N_i|$. Based on the set of neighbors, this linear system will be underdetermined (i.e., $A = m \times n$ with $m < n$) in general. If we assume that A has full rank, we can find one particular solution to our problem by requiring that w has the smallest norm of any solution. This results in the following optimization problem:

$$\min_w ||w||_2^2 \tag{1}$$

$$\text{subject to } Aw = b. \tag{2}$$

From the Karush–Kuhn–Tucker (KKT) theory [13], we know that the following conditions must hold for a solution (w^*, λ^*) to our problem to be optimal:

$$\nabla_w \mathcal{L}(w^*, \lambda^*) = 0 \tag{3}$$

$$Aw^* - b = 0 \tag{4}$$

$$\lambda^*(Aw^* - b) = 0. \tag{5}$$

The Lagrangian of our problem is given by:

$$\mathcal{L}(w, \lambda) = w^T w - \lambda^T (Aw - b),$$

where λ are the Lagrange multipliers.

Using (3)–(5), we can find the following solution pair (w^*, λ^*) as follows:

$$\nabla_w \mathcal{L}(w, \lambda) = 2w - A^T \lambda.$$

$$\nabla_w \mathcal{L}(w^*, \lambda^*) = 0 \Rightarrow w^* = \frac{1}{2} A^T \lambda^*.$$

$$Aw^* - b = 0 \Rightarrow A(\frac{1}{2} A^T \lambda^*) - b = 0.$$

$$\lambda^* = 2(AA^T)^{-1} b$$

$$w^* = \frac{1}{2} A^T \lambda^* = \frac{1}{2} A^T 2(AA^T)^{-1} b = A^T (AA^T)^{-1} b.$$

Although we have verified that (w^*, λ^*) is a stationary point, we cannot yet claim that it is a minimum. To do so, we must investigate $\nabla_w^2 \mathcal{L}(w^*, \lambda^*)$:

$$\nabla_w^2 \mathcal{L}(w, \lambda) = 2I_{|N_i| \times |N_i|}$$

$$\nabla_w^2 \mathcal{L}(w^*, \lambda^*) = 2I_{|N_i| \times |N_i|}.$$

From the second-order sufficient conditions, if w^* satisfies (3)–(5) and the following condition is satisfied:

$$z^T \nabla_w^2 \mathcal{L}(w^*, \lambda^*) z > 0, \text{ for all } z \in C(w^*, \lambda^*), z \neq 0, \tag{6}$$

where $C(w^*, \lambda^*) = \{z \mid \nabla_w c(w^*)^T z = 0\}$ is the critical cone and $c(w) = Aw - b$, then our solution is a minimum. Since $\nabla_w^2 \mathcal{L}(w^*, \lambda^*)$ is symmetric positive definite, the inequality in (6) is satisfied for any choice of z. Thus we can conclude that our solution w^* is a minimum of (1)–(2).

Now that we have established that w^* is our solution, we will discuss calculating it via a reduced QR factorization. Suppose that $A^T = QR$, where $Q_{n \times m}$, $R_{m \times m}$ is upper triangular, and $Q^T Q = I_{m \times m}$. Substituting in the QR factorization of A^T into w^*, we get the following:

$$w^* = A^+ b = A^T (AA^T)^{-1} b = QR(R^T Q^T QR)^{-1} b = QR(R^T R)^{-1} b$$

$$= QRR^{-1} R^{-T} b = QR^{-T} b.$$

Rearranging this into linear system form, we have:

$$R^T Q^T w^* = b.$$

If we let $t = Q^T w^*$, then $R^T t = b$ and $w^* = Qt$. Thus calculating w^* involves a QR decomposition of A^T, solving the lower triangular system $R^T t = b$ by forward substitution, and calculating the matrix-vector product Qt.

After calculating the weights, a boundary deformation is applied. The final step is to solve for the new locations of the interior nodes $[\hat{x}_I, \hat{y}_I]$ by solving the following global linear system:

$$A_I[\hat{x}_I, \hat{y}_I] = -A_B[\hat{x}_B, \hat{y}_B], \tag{7}$$

where $\hat{x}_B$ and $\hat{y}_B$ are the new x- and y-coordinates for the boundary nodes, and A_I and A_B contain the weights for the interior nodes and boundary nodes, respectively. In this global linear system, each row of the weight matrix corresponds to an interior node with nonzero entries for the node's neighbors and zero entries for the remainder of the row. The resulting global weight matrix is very sparse with irregular structure. In an effort to shift the nonzero entries closer to the diagonal, we apply the sparse reverse Cuthill–McKee ordering provided in Matlab. In Fig. 1, we show the matrix sparsity plots for the natural node ordering and the updated node ordering for the first two examples in Sect. 3. After applying the matrix reordering, the linear system is solved using a sparse LU factorization.

3   Numerical Experiments

In this section, we demonstrate the results from applying our method to generate several high-order meshes. We use Gmsh [7, 8, 15, 21] to generate the initial straight-sided high-order meshes. Our method then uses this mesh to calculate the weights (step 1). Next, we curve the boundary nodes (step 2) using Gmsh. The positions of the interior nodes in the resulting curved high-order mesh are then updated (step 3) by our method. For each example, we show the mesh which results from our method (with high-order nodes visible), the mesh element distortion for the curved boundary mesh generated using Gmsh, and the distortion for the mesh resulting from our method. When reporting the mesh distortion, we list the minimum distortion, maximum distortion, average distortion computed over all elements (referred to as Avg1 in figures), and average distortion computed over curved elements (referred to as Avg2 in figures). The ideal distortion value is 1, indicating that the element is straight. We also list the execution times needed for steps 1 and 3 of our method (excluding I/O) in Table 1. The code was run using Matlab R2018a, and the wall-clock execution times were measured on a machine with 16GB of RAM and a Ryzen 7 1700 CPU. All mesh visualizations and distortion evaluations were done using Gmsh. Our first example is a third-order mesh of a square region around

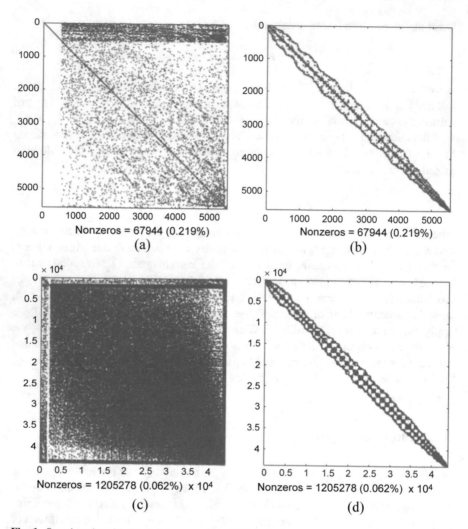

Fig. 1 Sparsity plots for the first two examples in Sect. 3: (**a** and **c**) the nonzero entries using the original node ordering; (**b** and **d**) the nonzero entries after applying the sparse reverse Cuthill–McKee ordering

a NACA0012 airfoil. In Fig. 2a, b, we show the mesh resulting from our method and a table of the mesh quality values as measured by the distortion metric. In this example, our method increased the minimum distortion from 0.744 to 0.799, while causing only minor changes in the average distortion.

In our second example, we extrude the NACA0012 airfoil and create a third-order mesh of the resulting region. In Fig. 3a, b, we show the mesh resulting from our method, and a table of the mesh quality values. For this example, our method improved the minimum distortion by 0.125, increasing it from 0.317 to 0.442.

Table 1 The number of elements and the wall clock times for steps 1 and 3 of our method (excluding I/O) for each example

Example	Number of elements	Execution time (s)	
		Original ordering	New reordering
NACA0012 airfoil	1312	11.51	3.61
Extruded NACA0012 airfoil	13,895	4826.09	958.14
Airbus A319	50,400	13,956.43	693.11

Fig. 2 NACA0012 airfoil example: (**a**) the mesh resulting from our method and (**b**) the mesh quality as measured by the element distortion metric

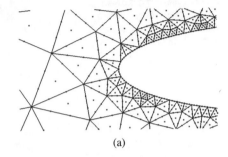

(a)

Example	Distortion			
	Min	Max	Avg1	Avg2
original mesh	0.744	1.000	0.999	0.994
resulting mesh	0.799	1.000	0.997	0.997

(b)

Fig. 3 Extruded NACA0012 airfoil example: (**a**) the mesh resulting from our method and (**b**) the mesh quality as measured by the element distortion metric

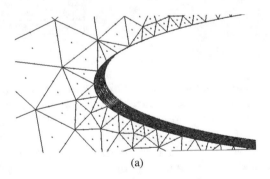

(a)

Example	Distortion			
	Min	Max	Avg1	Avg2
original mesh	0.317	1.000	0.997	0.994
resulting mesh	0.442	1.000	0.995	0.995

(b)

Our third and final example is a second-order mesh of an Airbus A319 aircraft. Unlike our previous examples, this geometry resulted in tangled elements after curving the boundary. Although our method still increased the minimum quality, it was not able to untangle the mesh. To address this, we applied the high-order

Fig. 4 Airbus A319
example: (**a**) surfaces of the
mesh resulting from our
method; (**b**) a view of a cut
through the interior volume
mesh, and (**c**) the quality of
the mesh with only boundary
curving, the quality of the
mesh resulting from our
method, and the quality of
both meshes after applying
the regularization scheme
available in Gmsh

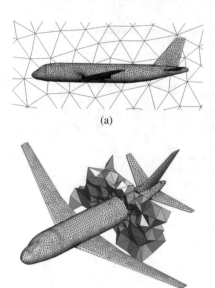

(a)

(b)

	Distortion			
Example	**Min**	**Max**	**Avg1**	**Avg2**
original mesh	−0.878	0.975	0.943	1.000
resulting mesh	−0.449	0.970	0.970	1.000
original mesh after untangling	0.206	0.975	0.945	1.000
resulting mesh after untangling	0.211	0.971	0.970	1.000

(c)

regularization scheme available in Gmsh as a post-processing step. Aside from changing the target Jacobian range to 0.3–2 and fixing the boundary nodes, all other parameters were left at their default values. The untangling for the original mesh took 14.14 s, while untangling the mesh resulting from our method required only 1.64 s. In Fig. 4a–c, we show the surfaces of the mesh resulting from our method, a view of a cut through the interior volume mesh, and a table of the mesh quality values. In Fig. 4c, we list the distortion for the mesh after curving the boundary, the distortion for the mesh resulting from our method, and the distortions of both meshes after applying the regularization scheme in Gmsh.

Aside from the third test case, all of these examples were relatively straightforward. In each case, our method increased the minimum distortion when compared to only curving the nodes along the boundary. While additional testing is necessary to confirm this, our results for the third example seem to indicate that our method could be used to reduce the severity of the mesh tangling and thus simplify the work for an untangling method during post-processing.

4 Concluding Remarks and Future Work

We have presented a new optimization-based method for generating high-order meshes. Our examples have shown that the proposed method based on affine combinations of nodal positions tends to improve the quality of the most distorted elements, while causing minor changes to the least distorted elements. While our approach is optimization-based, we have demonstrated that the optimization problem can be solved directly using a QR factorization as opposed to the typical iterative optimization approach. This change results in lessened computational complexity and reduced execution time.

As part of our future work, we will consider other definitions for the set of neighbors of an interior node. We will also investigate other aspects of the linear system including other node reordering schemes and solvers. Finally, we will apply the untangling method that we proposed in [20] after extending it to 3D.

Acknowledgements The work of the first author was funded in part by the Madison and Lila Self Graduate Fellowship and NSF CCF grant 1717894. The work of the second author was supported in part by NSF grants CCF 1717894 and OAC 1808553. The authors wish to thank the anonymous referee for his or her careful reading of the paper and for the helpful suggestions which strengthened it.

References

1. Bassi, F., Rebay, S.: High-order accurate discontinuous finite element solution of the 2D Euler equations. J. Comput. Phys. **138**(2), 251–285 (1997)
2. Dey, S., Shephard, M.S.: Curvilinear mesh generation in 3D. In: Proceedings of the 8th International Meshing Roundtable (1999)
3. Fortunato, M., Persson, P.-O.: High-order unstructured curved mesh generation using the Winslow equations. J. Comput. Phys. **307**(2016), 1–14 (2016)
4. Gargallo-Peiró, A., Roca, X., Peraire, J., Sarrate, J.: Distortion and quality measures for validating and generating high-order tetrahedral meshes. Eng. Comput. **31**(3), 423–437 (2015)
5. Gargallo-Peiró, A., Roca, X., Peraire, J., Sarrate, J.: Optimization of a regularized distortion measure to generate curved high-order unstructured tetrahedral meshes. Int. J. Numer. Methods Eng. **103**(5), 342–363 (2015)
6. George, P.L., Borouchaki., H.: Construction of tetrahedral meshes of degree two. Int. J. Numer. Methods Eng. **90**(9), 1156–1182 (2012)
7. Geuzaine, C., Remacle, J.-F.: Gmsh: a 3-D finite element mesh generator with built-in pre-and post-processing facilities. Int. J. Numer. Methods Eng. **79**(11), 1309–1331 (2009)
8. Johnen, A., Remacle, J.-F., Geuzaine, C.: Geometrical validity of high-order triangular finite elements. Eng. Comput. **30**(3), 375–382 (2014)
9. Karman, S.L., Erwin, J.T., Glasby, R.S., Stefanski, D.: High-order mesh curving using WCN mesh optimization. In: 46th AIAA Fluid Dynamics Conference, p. 3178 (2016)
10. Luo, X., Shephard, M.S., Remacle, J.-F.: The influence of geometric approximation on the accuracy of high order methods. Rensselaer SCOREC report, 1, 2001
11. Moxey, D., Green, M., Sherwin, S., Peiró, J.: An isoparametric approach to high-order curvilinear boundary-layer meshing. Comput. Methods Appl. Mech. Eng. **283**, 636–650 (2015)

12. Moxey, D., Ekelschot, D., Keskin, Ü., Sherwin, S.J., Peiró, J.: High-order curvilinear meshing using a thermo-elastic analogy. Comput. Aided Des. **72**, 130–139 (2016)
13. Nocedal, J., Wright, S.: Numerical Optimization, 2nd edn. Springer Series in Operations Research and Financial Engineering. Springer, Berlin (2006)
14. Persson, P.-O., Peraire, J.: Curved mesh generation and mesh refinement using Lagrangian solid mechanics. In: Proceedings of the 47th AIAA Aerospace Sciences Meeting Including the New Horizons Forum and Aerospace Exposition, p. 949 (2009)
15. Remacle, J.-F., Chevaugeon, N., Marchandise, E., Geuzaine, C.: Efficient visualization of high-order finite elements. Int. J. Numer. Methods Eng. **69**(4), 750–771 (2007)
16. Roca, X., Gargallo-Peiró, A., Sarrate, J.: Defining quality measures for high-order planar triangles and curved mesh generation. In: Proceedings of the 20th International Meshing Roundtable, pp. 365–383. Springer, Berlin (2012)
17. Ruiz-Gironés, E., Sarrate, J., Roca, X.: Generation of curved high-order meshes with optimal quality and geometric accuracy. In: Proceedings of the 25th International Meshing Roundtable, vol. 163, pp. 315–327. Procedia Engineering (2016)
18. Sherwin, S.J., Peiró, J.: Mesh generation in curvilinear domains using high-order elements. Int. J. Numer. Methods Eng. **53**(1), 207–223 (2001)
19. Stees, M., Shontz, S.M.: A high-order log barrier-based mesh generation and warping method. In: Proceedings of the 26th International Meshing Roundtable, vol. 203, pp. 180–192. Procedia Engineering (2017)
20. Stees, M., Shontz, S.M.: An angular approach to untangling high-order curvilinear triangular meshes. In: Proceedings of the 27th International Meshing Roundtable (2018)
21. Toulorge, T., Geuzaine, C., Remacle, J.-F., Lambrechts, J.: Robust untangling of curvilinear meshes. J. Comput. Phys. **254**, 8–26 (2013)
22. Turner, M., Moxey, D., Peiró, J., Gammon, M., Pollard, C.R., Bucklow, H.: A framework for the generation of high-order curvilinear hybrid meshes for CFD simulations. In: Proceedings of the 26th International Meshing Roundtable, vol. 203, pp. 206–218. Procedia Engineering (2017)
23. Wang, Z.J., Fidkowski, K., Abgrall, R., Bassi, F., Caraeni, D., Cary, A., Deconinck, H., Hartmann, R., Hillewaert, K., Huynh, H.T., et al. High-order CFD methods: current status and perspective. Int. J. Numer. Methods Fluids **72**(8), 811–845 (2013)
24. Xie, Z.Q., Sevilla, R., Hassan, O., Morgan, K.: The generation of arbitrary order curved meshes for 3D finite element analysis. Comput. Mech. **51**(3), 361–374 (2013)

Sparse Spectral-Element Methods for the Helically Reduced Einstein Equations

Stephen R. Lau

1 Introduction

To model the inspiral and merger of binary objects (blackholes or neutron stars), many researchers have been solving the Einstein equations numerically. Such simulation involves both the construction of gravitational initial data at time t_0 and its subsequent evolution to a final time $t_F \gg t_0$. Interpretation of experimental detections of gravitational waves relics on numerical simulation. Moreover, detection of weak signals is facilitated by statistical techniques alongside "template banks" of numerically generated signals. We consider a nonstandard problem, solution of the Einstein equations reduced by helical symmetry, as described by Beetle, Bromley, Hernández, and Price (BBHP) [1, 2]. Heuristically, helical reduction is a data+evolution synthesis. Although solutions to the BBHP equations are ultimately unphysical, they may approximate the early phase of inspiral and serve as reduced order models. Moreover, they may provide excellent "trial data" (the starting point for the construction of $t = t_0$ initial data). Finally, they would address bewitching mathematical issues concerning exact helical symmetry in general relativity.

We consider a spectral element approach [3–8] for solving the BBHP equations. Although the equations involve a mixed-typed operator L, we solve them via relaxation using a Broyden-Krylov approach. The computational domain which surrounds the compact objects is split into 11+ subdomains (blocks, spherical shells, and cylindrical shells with classical spectral expansions thereon). To rapidly solve the linear systems arising in our scheme, we have developed sparse modal methods based on the application of spectral integration matrices, extending ideas originally described in the 1990s by Coutsias, Hagstrom, Hesthaven, and Torres.

S. R. Lau (✉)
Mathematics and Statistics, University of New Mexico, Albuquerque, NM, USA
e-mail: lau@math.unm.edu

© The Author(s) 2020
S. J. Sherwin et al. (eds.), *Spectral and High Order Methods for Partial Differential Equations ICOSAHOM 2018*, Lecture Notes in Computational Science and Engineering 134, https://doi.org/10.1007/978-3-030-39647-3_18

We use preconditioned GMRES to solve these systems, with standard domain decomposition methods. In addition, we have developed fast methods for inversion of subdomain approximations of L, either via modal-based preconditioning or direct schemes.

2 Background

This section first describes the helically reduced wave equation (HRWE), a model for the helically reduced Einstein equations. Our HRWE description fixes ideas.

2.1 Helically Reduced Wave Equation

The wave equation is $\Box\psi = 0$, where $\Box = -\partial_t^2 + \Delta_\mathbf{x}$. Assume that ψ rotates rigidly with rate Ω. With the z-axis as the rotation axis, ψ then depends on time t only through $\varphi = \phi - \Omega t$, where ϕ is the azimuthal angle. Via the $\partial_t \rightarrow -\Omega\partial_\varphi$ replacement, $\Box\psi = 0$ becomes the HRWE $L\psi = 0$, where $L = \Delta_{\tilde{\mathbf{x}}} - \Omega^2(\tilde{x}\partial_{\tilde{y}} - \tilde{y}\partial_{\tilde{x}})^2$ in terms of co-rotating coordinates $(\tilde{t}, \tilde{x}, \tilde{y}, \tilde{z}) = (t, x\cos\Omega t + y\sin\Omega t, y\cos\Omega t - x\sin\Omega t, z)$.

We adopt a "2-center domain" $\mathscr{D}$, a 3d ball with two smaller 3d balls excised from it; see Fig. 1. Its boundary $\partial\mathscr{D} = \partial S_I^- \cup \partial S_{II}^- \cup \partial S_{\text{out}}^+$ is the union of two inner spheres (the $-$'s) and one outer sphere (the $+$). We consider the mixed-type problem

$$L\psi = 0, \qquad \psi = f^- \text{ on } \partial S_I^- \cup \partial S_{II}^-, \qquad \left(\partial_r - \Omega\partial_\varphi + r^{-1}\right)\psi = f^+ \text{ on } \partial S_{\text{out}}^+, \tag{1}$$

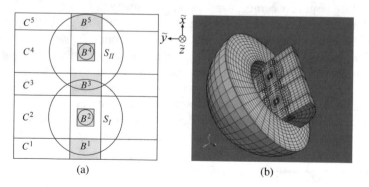

(a)   (b)

Fig. 1 2-center domain decomposition. The left panel depicts the z-cross section of the inner subdomains with S_{out} suppressed. The right panel depicts all subdomains, although for visualization the outer radius for S_{out} has been chosen rather small. In this work the blocks B^2 and B^4 are absent. (a) Inner domain decomposition. (b) Double cross section

with inner Dirichlet conditions and an outer radiation boundary condition on ψ. For simplicity here, we have put on $\partial S_{\text{out}}^+$ a simple Sommerfeld condition; in practice, f^+ is a nonlocal function of ψ enforcing an exact Dirichlet-to-Neumann map [4].

A class of solutions to (1) stems from Liénard-Wiechert potentials. Indeed, consider $L\psi = -16\pi\gamma M\delta(\tilde{\mathbf{x}} - \tilde{\mathbf{x}}_p)$. The "particle" location $\tilde{\mathbf{x}}_p = (\pm a, 0, 0)$ is the center of either ∂S_{II}^- or ∂S_I^-; whence the equation is homogeneous on $\mathcal{D}$. The mass M and relativistic factor $\gamma = (1 - v^2)^{-1/2}$ are constants, with $v = a\Omega < 1$ so the particle moves subluminally in the (t, x, y, z) frame. The retarded solution to this problem is

$$\psi(\tilde{x}, \tilde{y}, z) = \psi(\rho\cos\varphi, \rho\sin\varphi, z) = \frac{4\gamma M}{\lambda \mp v\rho\sin(\varphi + \Omega\lambda)} =: \gamma^2 \frac{4M}{\mathcal{R}}. \tag{2}$$

Evaluation of this expression involves a numerical component: solution of the fixed point equation $\lambda = \left[z^2 + \rho^2 + a^2 \mp 2a\rho\cos(\varphi + \Omega\lambda)\right]^{1/2}$ for (the retarded time) λ.

2.2 Helically Reduced Einstein Equations

We consider the vacuum Einstein equations in Landau-Lifshitz form. Write the (densitized contravariant) metric tensor as $\mathfrak{g}^{\mu\nu} = \eta^{\mu\nu} - h^{\mu\nu}$, where $\eta^{\mu\nu} = \text{diag}(-1, 1, 1, 1)$ is the flat metric and $h^{\mu\nu}$ is the *metric perturbation*. Assume the *harmonic gauge condition* $\partial_\nu h^{\mu\nu} = 0$. Then the vacuum Einstein equations can be expressed as

$$\Box h^{\mu\nu} = S_{\tau\phi\gamma\alpha}^{\mu\nu\kappa\beta}(\mathfrak{g})\frac{\partial h^{\tau\phi}}{\partial x^\kappa}\frac{\partial h^{\gamma\alpha}}{\partial x^\beta} + h^{\alpha\beta}\frac{\partial^2 h^{\mu\nu}}{\partial x^\alpha \partial x^\beta}, \tag{3}$$

where $S_{\tau\phi\gamma\alpha}^{\mu\nu\kappa\beta}(\mathfrak{g})$ depends on $\mathfrak{g}^{\mu\nu}$ and its inverse $\mathfrak{g}_{\mu\nu}$ (but not on derivatives of either). Einsteins equations are then a constrained system of 10 nonlinear wave equations.

The BBHP reduction of the Einstein equations is similar to the one outlined for the wave equation. Technically, it assumes the existence of a *Killing vector field*, but we give a brief and heuristic description of their approach. The challenge is that the perturbations $h^{\mu\nu}$ themselves are not "helical scalars"; helical reduction is therefore not tantamount to the replacement $\Box h^{\mu\nu} \to L h^{\mu\nu}$. However, BBHP have introduced helical scalars ψ^A through which the reduction can be carried out. These are

$$\psi^{(nn)} = h^{tt}, \qquad \psi^{(n0)} = \sqrt{2}h^{tz}, \quad \psi^{(00)} = \sqrt{\tfrac{1}{3}(h^{xx} + h^{yy} + h^{zz})}$$

$$\psi^{(20)} = -\sqrt{\tfrac{1}{6}(h^{xx} + h^{yy} - 2h^{zz})}, \quad \psi^{(n1)} = e^{i\Omega t}(-h^{tx} + ih^{ty}) \tag{4}$$

$$\psi^{(21)} = e^{i\Omega t}(-h^{xz} + ih^{yz}), \qquad \psi^{(22)} = e^{2i\Omega t}\left[\tfrac{1}{2}(h^{xx} - h^{yy}) - ih^{xy}\right].$$

These 4 real and 3 complex quantities contain the 10 degrees of freedom in the $h^{\mu\nu}$. We express this transformation as $\psi^A = e^{i\mu(A)\Omega t} M^A{}_{\mu\nu} h^{\mu\nu}$, where A runs over the (tensor-spherical-harmonic) labels, and $\mu(A)$ is 0,1, or 2. Contraction of $M^A{}_{\mu\nu}$ on (3), with subsequent helical reduction based on the action of $\Box$ on ψ^A, yields

$$L\psi^A - 2i\mu(A)\Omega^2\partial_\varphi\psi^A + \mu^2(A)\Omega^2\psi^A =$$

$$h^{\alpha\beta}\psi^A_{,\alpha\beta} - 2i\mu(A)\Omega h^{t\beta}\psi^A_{,\beta} - \mu^2(A)\Omega^2 h^{tt}\psi^A + M^A{}_{\mu\nu}S^{\mu\nu\kappa\beta}_{\tau\phi\gamma\alpha}(\mathfrak{g})h^{\tau\phi}{}_{,\kappa}h^{\gamma\alpha}{}_{,\beta}.$$
$$(5)$$

Here L is the operator appearing in the HRWE. Similar to the boundary conditions appearing in (1), the boundary conditions we adopt for (5) are

$$\psi^A = (f^A)^- \text{ on } \partial S^-_I \cup \partial S^-_{II}, \qquad \left(\partial_r - \Omega\partial_\varphi + r^{-1}\right)e^{i\mu(A)\varphi}\psi^A = (f^A)^+ \text{ on } \partial S^+_{out}.$$
$$(6)$$

Again for simplicity, here we have a Sommerfeld condition on ∂S^+_{out}, but in practice use a nonlocal outgoing condition based on an exact Dirichlet-to-Neumann map.

Price has written down the analog of (2) for linearized gravity, i.e. (5) when the right-hand side of the equation is set to zero. This solution may be viewed as an exact solution to $\Box h^{\mu\nu} = -16\pi T^{\mu\nu}$, where the stress energy tensor $T^{\mu\nu}$ corresponds to a massive point particle in an eternal circular orbit (as discussed, such a point source is excised from our domain $\mathscr{D}$). Price's solution is analogous to the electromagnetic solution given by G. A. Schott, and it is given by the leading $1/\mathcal{R}$ terms in the appendix expressions (12). Unfortunately, $\partial_\nu h^{\mu\nu} \neq 0$ for this solution.

3 Sparse Spectral Element Methods

This section summarizes our numerical methods. It is necessarily impressionistic, as even an incomplete presentation of details would take too much space.

3.1 Overview

We split $\mathscr{D}$ into subdomains, here with the minimal configuration of 11 subdomains: blocks $B^{1,3,5}$; cylinders $C^{1,2,3,4,5}$; an inner shell S_I around "particle" I; an inner shell S_{II} around "particle" II; and an outer shell S_{out}. This corresponds to a "binary blackhole" (BBH) domain with two excised inner balls. For a "binary neutron star" (BNS) domain, we further split both S_I and S_{II} into two overlapping concentric spheres (a stellar surface then resides in each overlap [8]), and fill in the excised regions with two extra blocks $B^{2,4}$. Figure 1 depicts a BNS domain. Pioneered by

Pfeiffer et al. [9], such decompositions are used in the `EllipticSolver` of SpEC [10].

The unknowns in our approach are the modal expansion coefficients associated with subdomain expansions in terms of classical (Chebyshev, Fourier, and spherical harmonic) basis functions. As described below, we make extensive use of integration matrices to achieve sparse representations of the relevant operators. Before presenting details, we first address how we handle the nonlinearities in (5). Let $\widetilde{\boldsymbol{\psi}}^A$ (the vector of unknowns) be a concatenation of the modal coefficients from all 11 subdomains. Then, as sketched below, upon approximation (5) becomes

$$\mathscr{B}\mathscr{L}\widetilde{\boldsymbol{\psi}}^A = \mathscr{B}\widetilde{\boldsymbol{g}}^A, \tag{7}$$

where $\mathscr{B}\mathscr{L}$ approximates the operator on the left-hand side of (5), with $\mathscr{B}$ representing the action of "integration preconditioning" (see below) on each subdomain. For linearized gravity, with the right-hand side of (5) set to zero, the vector $\widetilde{\boldsymbol{g}}^A$ is zero, save for select entries related to the inner Dirichlet values $(f^A)^-$ of ψ^A on $\partial S_I^- \cup \partial S_{II}^-$. For the full Einstein equations $\widetilde{\boldsymbol{g}}^A$ depends on $\widetilde{\boldsymbol{\psi}}^A$, and its evaluation relies on spectral analysis/synthesis (forward/backward transform) and numerical differentiation on each subdomain. We then view

$$\widetilde{\boldsymbol{\psi}}_k^A = (\mathscr{B}\mathscr{L})^{-1}\mathscr{B}\widetilde{\boldsymbol{g}}^A(\widetilde{\boldsymbol{\psi}}_{k-1}^B) \tag{8}$$

as a fixed-point equation, accelerating its convergence with the Broyden algorithm.

This approach relies on approximation and inversion of the operator appearing on the left-hand side of (5). Reference [4] is a detailed account of the case $\mu(A) = 0$, i.e. the HRWE. For $\mu(A) = 1$ or 2, the operator mixes the U^A and V^A in $\psi^A = U^A + iV^A$. We have not yet described our treatment of this scenario, but note that it relies on Schur-complement techniques. Here we describe only the $\mu(A) = 0$ case.

3.2 Integration Preconditioning and Other Key Aspects

We use "integration preconditioning" [11] in order to achieve sparse linear systems. Especially for the cylinders and inner shells, the details are formidable. We convey the basic ideas with the Laplacian Δ, rather than L, on a cube (suppressing tildes on the co-rotating coordinates). Let $u(x, y, z) \approx \sum_{i=0}^{N_x}\sum_{j=0}^{N_y}\sum_{k=0}^{N_z} \widetilde{u}_{ijk}T_i(x)T_j(y)T_k(z)$ obey $\Delta u = g$ on $C = [-1, 1]^3$. An approximation of the Poisson equation is

$$\left(D_x^2 \otimes I_y \otimes I_z + I_x \otimes D_y^2 \otimes I_z + I_x \otimes I_y \otimes D_z^2\right)\widetilde{\mathbf{u}} = \widetilde{\mathbf{g}}, \tag{9}$$

with $\widetilde{\mathbf{u}}$ the vector of $\widetilde{u}_{ijk}$ with appropriate ordering, and D^2 representing double differentiation in the modal Chebyshev basis. Let $B_{[2]}^2$ represent double integration in the modal basis, where the [2] indicates that the first two rows have all zero entries. To (9) we apply the "preconditioner" $\mathcal{B} = B_{x[2]}^2 \otimes B_{y[2]}^2 \otimes B_{z[2]}^2$, thereby reaching

$$\left(I_{x[2]} \otimes B_{y[2]}^2 \otimes B_{z[2]}^2 + B_{x[2]}^2 \otimes I_{y[2]} \otimes B_{z[2]}^2 + B_{x[2]}^2 \otimes B_{y[2]}^2 \otimes I_{z[2]}\right)\widetilde{\mathbf{u}} = \mathcal{B}\widetilde{\mathbf{g}}. \quad (10)$$

The system (10) is sparse, and the number of *empty rows* (all 0's) equals the number of *tau-conditions* to be enforced, auxiliary equations enforcing, say, $u|_{\partial C} = f$.

"Integration preconditioning" of (9) results in the sparse system (10), but the issue of condition number is subtle. Indeed, passage from (9) to (10) arguably *worsens conditioning*. For this reason, integration preconditioning has been viewed as bad for PDE; we call the technique *integration sparsification*. Conditioning issues are then surmounted either by further *genuine* preconditioning on top of the sparsification or fast direct solves. Our use of 2-center domains with sparse modal-tau methods features: (a) sparse representation of L on subdomains; (b) "gluing" of conforming and overlapping subdomains; (c) modal-based preconditioning of subdomain solves; (d) "fast" direct-solves on blocks and S_{out}; (e) standard global preconditioning; (f) low-rank treatment of stellar surfaces; (g) nonlocal domain reduction.

3.3 Current Complexity Estimates

We aim to solve the linear systems (say approximating the HRWE or the Helmholtz equation posed on the 2-center domain $\mathcal{D}$) arising in our problems at *demonstrably* sub-quadratic complexity; indeed, we believe that an order-$\frac{5}{3}$ complexity is achievable. This is the complexity associated with matrix-vector multiplication; despite our sparse representations, the "gluing" of overlapping subdomains in our decomposition of $\mathcal{D}$ prevents realization of a linear-complexity matrix-vector product.

To document progress towards our goal, we summarize our current complexity estimates associated with solution of the HRWE on each subdomain type. These solves serve as part of our preconditioner for the global GMRES solution of the HRWE on $\mathcal{D}$. For this discussion, let N represent the total number of modes on a given subdomain; e.g. $N = (N_x + 1)(N_y + 1)(N_z + 1)$ for the block considered above.

For S_{out} let $\mathcal{M}$ be the matrix which represents $r^2 L$ (the r^2 factor here is explained in [4]) and includes inserted tau-vectors to enforce boundary conditions. Ignoring tau-vectors, $\mathcal{M}$ is block diagonal in the spectral space of spherical harmonics indexed by (ℓ, m). View its elements as $\mathcal{M}_{\ell mk, \ell'm'k'}$, where ℓmk is a "clumped index' and k, k' are Chebyshev indices. Then, apart from tau-vectors, $\mathcal{M}_{\ell mk, \ell'm'k'} = 0$, unless $\ell = \ell'$ and $m = m'$. Moreover, each $(N_r + 1) \times (N_r + 1)$ block $\mathcal{M}_{\ell mk, \ell mk'}$ is itself sparse and banded. While these desirable structures are somewhat spoiled by the tau-conditions, through the use of the Woodbury formula and band solvers, for S_{out} we are able to directly invert M (i.e. solve the HRWE on S_{out}) at $O(N)$ cost.

For the inner shells, S_I and S_{II}, our representation M of r^2L is only block banded. Indeed, the centers of these shells lie off the rotation axis, and r^2L mixes spherical harmonic modes. Its spectral representation [4] is remarkably complicated, and relies on identities for spherical harmonics found in the treatise [12]. We solve the HRWE on inner shells via preconditioned GMRES, with a modal block-Jacobi preconditioner defined by inversion of the diagonal blocks $M_{\ell mk, \ell mk'}$. Apart from tau-conditions, these blocks are again sparse and banded, and therefore amenable to the fast methods alluded to in the last bullet. Construction and reuse of this block-Jacobi preconditioner therefore has $O(N)$ cost. Moreover, we have empirically observed (see [4]) that such preconditioning yields low and essentially *resolution independent* iteration counts. While more analysis is needed, from a practical standpoint solution of the HRWE on an inner shell has an $O(N)$ cost.

The situation on blocks is worse. Part of our global preconditioner for solving the HRWE on $\mathscr{D}$ involves inversion of the Poisson problem on blocks as an approximation to the HRWE (and inversion of the Helmholtz equation when the spin index $\mu(A) \neq 0$). This works extremely well, likely due to the fact that $\Omega \ll 1$ and the blocks are close to the rotation center. In any case, we solve the Poisson/Helmholtz problem on a block via a direct approach [13] based on a rank-augmenting generalization of the Woodbury formula. This direct method is empirically well-conditioned and low-memory. Moreover, it has an $O(N^2)$ set-up cost with a small constant, followed by an $O(N^{4/3})$ cost for subsequent solves. If possible, we hope to reduce the set-up cost to an $O(N^{5/3})$ complexity.

The situation on cylinders is worst of all, although to date we have not focused much attention on these subdomains. We solve the HRWE on cylinders (or the collection of cylinders) via GMRES, with modal-based preconditioners that empirically yield resolution-independent iteration counts. Application of the preconditioner currently involves an $O(N^{7/3})$ set-up cost, followed by an $O(N^{5/3})$ cost for subsequent solves. Here, we believe improvement is possible.

4 Numerical Tests

Our decomposition of $\mathscr{D}$ is from Table IV of [4], except here ∂S_{out} has $r_{out} = 15$. For that table ∂S_I^- has radius $r_{I,min} = 0.4$ and center $(\tilde{x}_I = -0.9, 0, 0)$, and ∂S_{II}^- has radius $r_{II,min} = 0.3$ and center $(\tilde{x}_{II} = 1.0, 0, 0)$. The coordinates $\tilde{X}, \tilde{Y}, \tilde{Z}$ in [4] are $\tilde{y}, \tilde{z}, \tilde{x}$ here. The subdomain truncations (number of modes) adopted here are nearly the same as those in [4]; however, here we only record the total number of modes over all subdomains. Unless otherwise stated, $\Omega = 0.075$, $M_I = 0.05$, and $M_{II} = 0.1$.

4.1 Comparison with Exact Solutions

Our first test uses `PriceLG`, the aforementioned solution for linearized gravity due to Price. Here the solution is a superposition of two point-sources with the above masses. Each source point's contribution to the helical scalars is defined by the leading $1/\mathcal{R}$ term in (12), and these expressions seed inner Dirichlet conditions on $\partial S_I^- \cup \partial S_{II}^-$. The outer boundary conditions are nonlocal conditions which are exact for this solution. Table 1 lists errors for $\psi^{(nn)}$; errors for the other scalars are similar. These are relative L_2-errors (against the exact solution) computed on both B^3 and S_{out} via interpolation onto uniform reference grids. Since the problem is linear, each line of table corresponds to a single GMRES solve performed in parallel on 10 nodes. For the table middle about 0.4% of the matrix entries are nonzero. The last table line has $\|\partial_\nu h^{\mu\nu}\|_{\text{rms}} \simeq (3.9420\text{e-}12, 1.7310\text{e-}03, 8.9752\text{e-}05, 9.0076\text{e-}16)$, with the rms calculation taken over the (relatively coarse) dual-nodal subdomain grids. The first and last components of $\partial_\nu h^{\mu\nu}$ vanish for the exact `PriceLG` solution.

Our next test is `SchwarzH`, the Schwarzschild metric in harmonic coordinates:

$$h^{tt} = -1 + r^{-2}(r+M)^3/(r-M), \quad h^{jk} = r^{-2}M^2 v^j v^k. \tag{11}$$

Here the radius r and the direction cosines $v = (\sin\theta\cos\phi, \sin\theta\sin\phi, \cos\theta)$ are chosen relative to a point $(x_0, y_0, z_0) = (-0.9 + 1.37\text{e-}3, -1.6854\text{e-}4, 2.9985\text{e-}3)$ which is off-center but close to the center of ∂S_I^-. The mass is $M = 0.05$, and for this choice the horizon of the blackhole lies inside of (but is not concentric with) ∂S_I^-. For this test $\Omega = 0$, and the exact solution (11) seeds inner Dirichlet boundary conditions on both ∂S_I^- and ∂S_{II}^-. On $\partial S_{\text{out}}^+$ rather than radiation conditions, we adopt an inhomogeneous Neumann condition based on the exact solution. Table 2 lists

Table 1 `PriceLG` solution

Truncation	Shell error	Block error	tGMRES	iGMRES
23,114	1.1003e-05	4.6050e-06	0.50e-05	7
93,067	5.4950e-08	8.0179e-08	0.50e-07	5
271,197	4.7606e-10	2.7153e-10	0.50e-09	4
553,149	1.3446e-12	1.3198e-12	0.50e-11	4

Here relative L_2 errors are listed only for $\psi^{(nn)}$. The lowest resolution run has zero initial iterate; afterwards the initial iterate stems from the previous solution. Respectively, tGMRES and iGMRES are the tolerance and iteration number for the GMRES solve

Table 2 `SchwarzH` solution

Truncation	Shell error	Block error	tBROY	iBROY
23,114	6.5414e-05	1.5238e-05	5.0e-05	6
93,067	1.7162e-06	3.3810e-07	5.0e-07	4
271,197	9.5944e-09	1.6022e-09	5.0e-09	4
553,149	1.1266e-11	3.6899e-12	5.0e-11	4

As for the `PriceLG` test, only errors for $\psi^{(nn)}$ are listed

Table 3 Full GR with the same boundary conditions as `PriceLG`

Truncation	Shell error (GR)	Block error (GR)	Residual	iBROY
23,114	1.5303e-05	7.6669e-06	3.7429e-07	7
93,067	1.3194e-07	1.4670e-07	8.3651e-10	6
271,197	1.2266e-10	2.4074e-10	2.9149e-11	4
553,149	-----	-----	4.0790e-13	4

errors with the same meanings as before. Now $\|\partial_\nu h^{\mu\nu}\|_{rms}$ for each μ is comparable to the corresponding table errors; i.e. the gauge constraint converges to zero. Table 2 also lists the number iBROY of iterations performed by the Broyden solver to achieve the tolerance tBROY. Each Broyden iteration itself involves a linear solve via GMRES. Each GMRES tolerance is tBROY/10, the same as the corresponding line in Table 1.

4.2 Gauge Constraint Tests

Our next two tests explore to what extent the gauge constraint $\partial_\nu h^{\mu\nu} = 0$ is satisfied for the Einstein problem (5) and (6) in a binary scenario. For the first test we redo the `PriceLG` test, except now with the Einstein equations. The inner and outer boundary conditions are exactly as before. Table 3 again lists errors for $\psi^{(nn)}$ with the same meanings as before. Errors are computed against the finest-resolution numerical solution. The table also lists the L_2-norm of the nonlinear residual. The last table line has $\|\partial_\nu h^{\mu\nu}\|_{rms} \simeq$ (3.2031e-03 5.1309e-03 4.5881e-03 4.6711e-03). That is, the gauge constraint does not converge to zero. Since the harmonic gauge is not satisfied, we cannot view these as solutions to the Einstein equations!

Presumably, the violation of the harmonic gauge in the preceding example stems from the fact that $\partial_\nu h^{\mu\nu} \neq 0$ for the Price solution used to fix the inner boundary conditions. Beetle, Bromley, and, Hernández, and Price have given a refined set of inner boundary conditions, based on the near-field asymptotics of a moving Schwarzschild blackhole and meant to improve on the point-particle boundary conditions. The appendix lists (our understanding of) these conditions. With the hope that these refined boundary will result in lower gauge errors, we have performed the previous test with them. Convergence of the numerical solution is similar, but $\|\partial_\nu h^{\mu\nu}\|_{rms}$ has comparable size and still does not converge to zero.

5 Conclusions and Acknowledgments

Our tentative conclusion for helically symmetric BBH models is that violation of the gauge constraint stems from imperfect inner boundary conditions. We have also found a persistent gauge error in our BNS models, despite being several orders of magnitude smaller in that context. The BNS model, with stars in place of excised regions, involves no inner boundary conditions. We believe that in this context it is the outer boundary conditions which give rise to the constraint violation. Likely, at both the inner and outer boundaries some of the helical scalars need to be fixed using the gauge constraint itself (similar to "constraint preserving boundary conditions" used in evolution codes). The author is grateful for correspondence with Richard H. Price, and for assistance from UNM's Center for Advanced Research Computing.

Appendix: Beetle, Bromley, Hernández, and Price Inner Boundary Conditions

This appendix lists expansions for the helical scalars which somewhat generalize the ones in [1]. We need two expansions, one for a "particle" at $(-a_I, 0, 0)$, and one at $(a_{II}, 0, 0)$. Each has its own mass $M_{I,II}$. The top choice of $\pm$ or $\mp$ refers to II and the bottom to I. For both a_I and a_{II}, we define $v = a\Omega$, $\gamma = (1 - v^2)^{-1/2}$, $\mathcal{R} = \gamma\lambda \mp v\gamma\rho\sin(\varphi + \Omega\lambda)$, $K^R = -\rho\cos\varphi \pm a\cos(\Omega\lambda) \pm v\gamma\mathcal{R}\sin(\Omega\lambda)$, and $K^I = \rho\sin\varphi \pm a\sin(\Omega\lambda) \mp v\gamma\mathcal{R}\cos(\Omega\lambda)$. For λ see after (2). Assuming $M/\mathcal{R} \ll 1$, we have

$$\psi^{(nn)} \sim \gamma^2\left(\frac{4M}{\mathcal{R}} + \frac{7M^2}{\mathcal{R}^2}\right) + \frac{M^2(\lambda - \gamma\mathcal{R})^2}{\mathcal{R}^4} \tag{12a}$$

$$\psi^{(n0)} \sim 0 \cdot \left(\frac{4M}{\mathcal{R}} + \frac{7M^2}{\mathcal{R}^2}\right) + \sqrt{2}\frac{M^2(\lambda - \gamma\mathcal{R})z}{\mathcal{R}^4} \tag{12b}$$

$$\psi^{(00)} \sim \frac{v^2\gamma^2}{\sqrt{3}}\left(\frac{4M}{\mathcal{R}} + \frac{7M^2}{\mathcal{R}^2}\right) + \frac{M^2[\lambda^2 - 2\gamma\lambda\mathcal{R} + (2 + v^2\gamma^2)\mathcal{R}^2]}{\sqrt{3}\mathcal{R}^4} \tag{12c}$$

$$\psi^{(20)} \sim -\frac{v^2\gamma^2}{\sqrt{6}}\left(\frac{4M}{\mathcal{R}} + \frac{7M^2}{\mathcal{R}^2}\right) - \frac{M^2[\lambda^2 - 3z^2 - 2\gamma\lambda\mathcal{R} + (2 + v^2\gamma^2)\mathcal{R}^2]}{\sqrt{6}\mathcal{R}^4} \tag{12d}$$

$$\psi^{(n1)} \sim \pm iv\gamma^2 e^{i\Omega\lambda}\left(\frac{4M}{\mathcal{R}} + \frac{7M^2}{\mathcal{R}^2}\right) + \frac{M^2(\lambda - \gamma\mathcal{R})(K^R + iK^I)}{\mathcal{R}^4} \tag{12e}$$

$$\psi^{(21)} \sim 0 \cdot \left(\frac{4M}{\mathcal{R}} + \frac{7M^2}{\mathcal{R}^2}\right) + \frac{M^2 z(K^R + iK^I)}{\mathcal{R}^4} \tag{12f}$$

$$\psi^{(22)} \sim -\frac{1}{2}v^2\gamma^2 e^{2i\Omega\lambda}\left(\frac{4M}{\mathcal{R}} + \frac{7M^2}{\mathcal{R}^2}\right) + \frac{M^2}{2\mathcal{R}^4}[(K^R)^2 - (K^I)^2 + 2iK^R K^I]. \tag{12g}$$

Worried about a possible sign discrepancy with the results in [1], we have also considered (12a–g) with all correction terms (those with $\mathcal{R}^4$ in the denominator) flipped by a sign.

References

1. Beetle, C., Bromley, B., Hernández, N., Price, R.H.: Periodic standing-wave approximation: post-Minkowski computations. Phys. Rev. **D76**, 084016 (2007)
2. Hernández, N., Price, R.H.: The periodic standing-wave approximation: computations in full general relativity. Phys. Rev. **D79**, 064008 (2009)
3. Lau, S.R., Price, R.H.: Multidomain spectral method for the helically reduced wave equation. J. Comput. Phys. **227**, 1126–1161 (2007). We regret an error in Eq. (42). The correct expressions are $\nu^{\pm} = \left[T_0'(\pm 1), T_1'(\pm 1), T_2'(\pm 1), T_3'(\pm 1), T_4'(\pm 1), \cdots \right] = \left[0, 1, \pm 4, 9, \pm 16, \cdots \right]$. The right-hand side of the second equation of (69) is also off by a sign
4. Lau, S.R., Price, R.H.: Sparse spectral-tau method for the three-dimensional helically reduced wave equation on two-center domains. J. Comput. Phys. **231**(2), 7695–7714 (2012)
5. Beroiz, M., Hagstrom, T., Lau, S.R., Price, R.H.: Multidomain, sparse, spectral-tau method for helically symmetric flow. Comput. Fluids **102**, 250–265 (2014)
6. Lau, S.R., Price, R.H.: Sparse modal tau-method for helical binary neutron stars. In: Proceedings of Spectral and High Order Methods for Partial Differential Equations ICOSAHOM 2014. Lecture Notes in Computational Science and Engineering, vol. 106, pp. 315–323 (2015)
7. Lau, S.R., Price, R.H.: Helically reduced wave equations and binary neutron stars. In: Proceedings of Spectral and High Order Methods for Partial Differential Equations ICOSAHOM 2016. Lecture Notes in Computational Science and Engineering, vol. 119, pp. 369–382 (2017)
8. Lau, S.R.: Stellar surface as low-rank modification in iterative methods for binary neutron stars. J. Comput. Phys. **348**, 460–481 (2017)
9. Pfeiffer, H.P., Kidder, L.E., Scheel, M.A., Teukolsky, S.A.: A multidomain spectral method for solving elliptic equations. Comput. Phys. Commun. **152**(3), 253–273 (2003)
10. Spectral Einstein Code (SpEC): http://www.black-holes.org/SpEC.html
11. Coutsias, E.A., Hagstrom, T., Hesthaven, J.S., Torres, D.: Integration preconditioners for differential operators in spectral τ-methods. In: Proceedings of the Third International Conference on Spectral and High Order Methods, pp. 21–38. Houston (1996)
12. Hill, E.L., Landshoff, R.: The dirac electron theory. Rev. Mod. Phys. **10**, 87–132 (1938)
13. Lau, S.R.: Direct, low-memory, spectral solution of harmonic problems on a block at near optimal complexity (in preparation)

Spectral Analysis of Isogeometric Discretizations of 2D Curl-Div Problems with General Geometry

Mariarosa Mazza, Carla Manni, and Hendrik Speleers

1 Introduction

In this paper we focus on isogeometric Galerkin discretizations of the weighted curl-div operator

$$\mathcal{L}_{\alpha,\beta} u := \alpha \nabla \times \nabla \times u - \beta \nabla \nabla \cdot u, \quad 0 < \alpha, \beta. \tag{1}$$

This parameter-dependent operator appears in several problems, including the Stokes equation and Maxwell equations [2]. Moreover, containing a weighting of the curl and div operators, it captures the essential features of the so-called Alfvén-like operator [14], which is of interest in magnetohydrodynamics [15]. We note that $\mathcal{L}_{\alpha,\beta}$ can be seen as a weighted Laplacian for vector fields (equivalently, Hodge Laplace for 1-forms). Indeed, when $\alpha = \beta = 1$, it is equal to the standard (negative) vector Laplace operator, i.e.,

$$\nabla \times \nabla \times u - \nabla \nabla \cdot u = -\nabla^2 u.$$

We assume that (1) is defined on a sufficiently smooth domain $\Omega \in \mathbb{R}^2$ that can be described through a geometry map $G : [0, 1]^2 \to \overline{\Omega}$, and we consider homogeneous

M. Mazza (✉)
Max Planck Institute for Plasma Physics, Garching, Germany
e-mail: mariarosa.mazza@ipp.mpg.de

C. Manni · H. Speleers
Department of Mathematics, University of Rome Tor Vergata, Rome, Italy
e-mail: manni@mat.uniroma2.it; speleers@mat.uniroma2.it

© The Author(s) 2020
S. J. Sherwin et al. (eds.), *Spectral and High Order Methods for Partial Differential Equations ICOSAHOM 2018*, Lecture Notes in Computational Science and Engineering 134, https://doi.org/10.1007/978-3-030-39647-3_19

251

Dirichlet (no-slip) boundary conditions, i.e., $u = 0$ on $\partial\Omega$. This leads us to the following variational formulation

$$(\mathcal{L}_{\alpha,\beta}u, v) = \alpha(\nabla \times u, \nabla \times v) + \beta(\nabla \cdot u, \nabla \cdot v), \quad u, v \in (H_0^1(\Omega))^2. \tag{2}$$

We refer the reader to [3, 15] for a discussion about well-posedness.

To find an approximate solution of the problem $\mathcal{L}_{\alpha,\beta}u = f$, with the stated boundary conditions, we consider the variational formulation (2) in a finite dimensional vector space $\mathbb{V}_h \subset (H_0^1(\Omega))^2$, i.e.,

$$(\mathcal{L}_{\alpha,\beta}u_h, v_h) = \alpha(\nabla \times u_h, \nabla \times v_h) + \beta(\nabla \cdot u_h, \nabla \cdot v_h), \quad u_h, v_h \in \mathbb{V}_h. \tag{3}$$

We focus on isogeometric analysis (IgA) as discretization technique, where the approximation space $\mathbb{V}_h$ is chosen to be composed of vector fields whose components are linear combinations of tensor-product B-splines mapped according to G.

The discretization (3) leads to solving linear systems, which turn out to be severely ill-conditioned and require ad hoc fast solvers for a proper treatment [4, 6, 15]. This requires a deep understanding of the spectral properties of the related matrices. They depend on many factors: the problem parameters α, β, the basic curl and div operators, the mesh-size, the degree of the B-spline approximation, and the map G used to describe the geometry of the computational domain.

In this paper we provide a spectral study of these matrices using the theory of (multilevel block) Toeplitz [13, 17, 19] and generalized locally Toeplitz [10–12] sequences. More precisely, we show that such matrices admit a spectral distribution which can be described in terms of a so-called *spectral symbol*. We determine this spectral symbol and we reveal its dependence on the characteristic parameters of the problem listed above. The spectral analysis presented in this paper extends the results of [15] to the case of non-trivial geometry and relies on the spectral theory developed for isogeometric discretizations of elliptic problems in [7, 8]. We also refer the reader to [16] for a spectral analysis of the curl-curl operator.

The remainder of the paper is organized as follows. In Sect. 2 we introduce notations and definitions relevant for our spectral analysis, and we recall the basics of B-splines. In Sect. 3 we detail the IgA discretization matrices and we perform a spectral analysis of them. We numerically illustrate those results in Sect. 4. Finally, we conclude the paper in Sect. 5.

2 Preliminaries

In this section we collect some preliminary tools on spectral analysis and IgA discretizations. In particular, we recall the formal definition of spectral distribution for a general matrix-sequence and the definition of (cardinal) B-splines.

2.1 Spectral Distribution

Throughout the paper, we follow the standard convention for operations with multi-indices (see e.g. [9, 18]). Given a multi-index $n := (n_1, \ldots, n_d) \in \mathbb{N}^d$, we say $n \to \infty$ if $n_i \to \infty$, $i = 1, \ldots, d$. Let $C_0(\mathbb{C})$ be the set of continuous functions $F : \mathbb{C} \to \mathbb{C}$ with compact support.

Definition 1 Let $f : D \to \mathbb{C}^{s \times s}$ be a measurable matrix-valued function, defined on a measurable set $D \subset \mathbb{R}^q$ with $q \geq 1$, $0 < \mu_q(D) < \infty$, where μ_q is the Lebesgue measure. Let $\{A_n\}_n$ be a matrix-sequence with $\dim(A_n) =: d_n$ and $d_n \to \infty$ as $n \to \infty$. Then, $\{A_n\}_n$ is distributed like the pair (f, D) in the sense of the eigenvalues, denoted by $\{A_n\}_n \sim_\lambda (f, D)$, if the following limit relation holds for all $F \in C_0(\mathbb{C})$:

$$\lim_{n \to \infty} \frac{1}{d_n} \sum_{j=1}^{d_n} F(\lambda_j(A_n)) = \frac{1}{\mu_q(D)} \int_D \frac{\sum_{i=1}^s F(\lambda_i(f(t)))}{s} dt, \qquad (4)$$

where $\lambda_j(A_n)$, $j = 1, \ldots, d_n$ are the eigenvalues of A_n and $\lambda_i(f)$, $i = 1, \ldots, s$ are the eigenvalues of f. We say that f is the (spectral) symbol of the matrix-sequence $\{A_n\}_n$.

If f is smooth enough and the matrix-size of A_n is sufficiently large, then the limit relation (4) has the following informal meaning: a first set of d_n/s eigenvalues of A_n is approximated by a sampling of $\lambda_1(f)$ on a uniform equispaced grid of the domain D, a second set of d_n/s eigenvalues of A_n is approximated by a sampling of $\lambda_2(f)$ on a uniform equispaced grid of the domain D, and so on, up to few outliers.

In general, understanding whether a matrix-sequence admits a symbol and how to compute it is not an easy task. On the other hand, any "reasonable" approximation of partial differential equations by local methods leads to matrix-sequences that are in the so-called generalized locally Toeplitz (GLT) algebra, and so admit a symbol [10–12]. The IgA discretization of our curl-div problem (3) fits in this frame.

2.2 B-Splines

For $p \geq 0$ and $n \geq 1$, consider the uniform knot sequence

$$\xi_1 = \cdots = \xi_{p+1} := 0 < \xi_{p+2} < \cdots < \xi_{p+n} < 1 =: \xi_{p+n+1} = \cdots = \xi_{2p+n+1},$$

where $\xi_{i+p+1} := \frac{i}{n}$, $i = 0, \ldots, n$. This knot sequence allows us to define $n + p$ B-splines of degree p. Let χ_I denote the characteristic function on the interval I.

Definition 2 The B-splines of degree p over a uniform mesh of $[0, 1]$, consisting of n intervals, are denoted by $N_i^p : [0, 1] \to \mathbb{R}$, $i = 1, \ldots, n + p$, and defined recursively as follows: for $1 \leq i \leq n + 2p$,

$$N_i^0(x) := \chi_{[\xi_i, \xi_{i+1})}(x);$$

for $1 \leq k \leq p$ and $1 \leq i \leq n + 2p - k$,

$$N_i^k(x) := \frac{x - \xi_i}{\xi_{i+k} - \xi_i} N_i^{k-1}(x) + \frac{\xi_{i+k+1} - x}{\xi_{i+k+1} - \xi_{i+1}} N_{i+1}^{k-1}(x),$$

where a fraction with zero denominator is assumed to be zero.

It is well known (see e.g. [1]) that the B-splines N_i^p, $i = 1, \ldots, n + p$, form a basis, and

$$N_i^p(0) = N_i^p(1) = 0, \quad i = 2, \ldots, n + p - 1. \tag{5}$$

The central B-splines N_i^p, $i = p + 1, \ldots, n$, are uniformly shifted and scaled versions of a single shape function, the so-called cardinal B-spline $\phi_p : \mathbb{R} \to \mathbb{R}$,

$$\phi_0(t) := \chi_{[0,1)}(t), \quad \phi_p(t) := \frac{t}{p}\phi_{p-1}(t) + \frac{p + 1 - t}{p}\phi_{p-1}(t - 1), \quad p \geq 1.$$

More precisely, we have

$$N_i^p(x) = \phi_p(nx - i + p + 1), \quad i = p + 1, \ldots, n.$$

The cardinal B-spline ϕ_p is a C^{p-1} function which is locally supported on the interval $[0, p + 1]$.

Finally, we recall the definition of tensor-product B-splines.

Definition 3 The tensor-product B-splines of bi-degree $p := (p_1, p_2)$ over a uniform mesh of $[0, 1]^2$, consisting of $n := (n_1, n_2)$ intervals in each direction, are denoted by $N_i^p : [0, 1]^2 \to \mathbb{R}$, $i = 1, \ldots, n + p$, and defined as

$$N_i^p := N_{i_1}^{p_1} \otimes N_{i_2}^{p_2},$$

where $1 := (1, 1)$ and $i := (i_1, i_2) \in \mathbb{N}^2$.

We define the tensor-product spline space $\mathbb{S}_n^p$ as

$$\mathbb{S}_n^p := \text{span}\left\{N_i^p : i = 2, \ldots, n + p - 1\right\}. \tag{6}$$

Note that all the elements of this space vanish at the boundary of $[0, 1]^2$; see (5). Hence, the space incorporates homogeneous Dirichlet boundary conditions.

3 Spectral Analysis of Isogeometric Discretizations in 2D

Suppose that the physical domain Ω can be described by a global geometry map, $\boldsymbol{G} := [G_1, G_2]^T$, $\boldsymbol{G} : \widehat{\Omega} \to \overline{\Omega}$, which is invertible in the parametric domain $\widehat{\Omega} := [0, 1]^2$ and satisfies $\boldsymbol{G}(\partial\widehat{\Omega}) = \partial\Omega$. Let

$$\mathbb{V}_h = \text{span}\left\{\phi_{i_1,i_2}^{p,1}, \phi_{j_1,j_2}^{p,2} : i_l, j_l = 2, \ldots, n+p-1; \; l = 1, 2\right\}, \tag{7}$$

where

$$\phi_{i_1,i_2}^{p,1} := \begin{bmatrix} \varphi_{i_1,i_2} \\ 0 \end{bmatrix}, \quad \phi_{j_1,j_2}^{p,2} := \begin{bmatrix} 0 \\ \varphi_{j_1,j_2} \end{bmatrix},$$

and for $k_l \in \{i_l, j_l\}, l = 1, 2$,

$$\varphi_{k_1,k_2}(x_1, x_2) := \hat{\varphi}_{k_1,k_2}(\boldsymbol{G}^{-1}(x_1, x_2)) = \hat{\varphi}_{k_1,k_2}(\hat{x}_1, \hat{x}_2), \quad (x_1, x_2) = \boldsymbol{G}(\hat{x}_1, \hat{x}_2).$$

Then, we set $\hat{\varphi}_{k_1,k_2} = N_{k_1}^p \otimes N_{k_2}^p$, i.e., the tensor-product B-splines in (6). For simplicity of notation, we have taken $n_1 = n_2 = n$ and $p_1 = p_2 = p$. Also note that

$$\nabla\varphi_{k_1,k_2} = (J_{\boldsymbol{G}})^{-T}\nabla(N_{k_1}^p \otimes N_{k_2}^p)$$

$$= \frac{1}{\det(J_{\boldsymbol{G}})}\begin{bmatrix} \frac{\partial G_2}{\partial\hat{x}_2}(N_{k_1}^p)' \otimes N_{k_2}^p - \frac{\partial G_2}{\partial\hat{x}_1}N_{k_1}^p \otimes (N_{k_2}^p)' \\ -\frac{\partial G_1}{\partial\hat{x}_2}(N_{k_1}^p)' \otimes N_{k_2}^p + \frac{\partial G_1}{\partial\hat{x}_1}N_{k_1}^p \otimes (N_{k_2}^p)' \end{bmatrix},$$

where

$$J_{\boldsymbol{G}} := \begin{bmatrix} \frac{\partial G_1}{\partial\hat{x}_1} & \frac{\partial G_1}{\partial\hat{x}_2} \\ \frac{\partial G_2}{\partial\hat{x}_1} & \frac{\partial G_2}{\partial\hat{x}_2} \end{bmatrix}.$$

In the following, we start by discussing the coefficient matrices arising from the IgA discretization of a generalized Poisson problem. Then, we construct the coefficient matrices related to the IgA discretization of our curl-div problem (3) using (7), and we perform a spectral analysis.

3.1 Matrices Related to a Generalized Poisson Problem

Let us focus on the following bivariate generalized Poisson operator:

$$\mathcal{L}_K u := -\nabla \cdot K\nabla u, \tag{8}$$

where $K : \Omega \to \mathbb{R}^{2\times 2}$, and consider homogeneous Dirichlet boundary conditions, i.e., $u = 0$ on $\partial\Omega$. From [8] we know that the Galerkin discretization of (8) using one component of the space (7) leads to the coefficient matrix $\mathcal{A}_{n,G}^{p,K}$ defined by

$$\left[\mathcal{A}_{n,G}^{p,K}\right]_{i,j} := \int_{\widehat{\Omega}} \left[\nabla(N_{j_1+1}^p \otimes N_{j_2+1}^p)^T K_G \nabla(N_{i_1+1}^p \otimes N_{i_2+1}^p)\right] |\det(J_G)|,$$

where

$$K_G := (J_G)^{-1} K(G)(J_G)^{-T}.$$

It has been proved in [8] that such matrices admit a spectral distribution according to Definition 1. To this end, let us define

$$H_p := \begin{bmatrix} \mathfrak{s}_p \otimes \mathfrak{m}_p & \mathfrak{a}_p \otimes \mathfrak{a}_p \\ \mathfrak{a}_p \otimes \mathfrak{a}_p & \mathfrak{m}_p \otimes \mathfrak{s}_p \end{bmatrix},$$

with

$$\mathfrak{m}_p(\theta) := \phi_{2p+1}(p+1) + 2\sum_{k=1}^{p} \phi_{2p+1}(p+1-k)\cos(k\theta),$$

$$\mathfrak{a}_p(\theta) := -2\sum_{k=1}^{p} \phi'_{2p+1}(p+1-k)\sin(k\theta),$$

$$\mathfrak{s}_p(\theta) := -\phi''_{2p+1}(p+1) - 2\sum_{k=1}^{p} \phi''_{2p+1}(p+1-k)\cos(k\theta).$$

Theorem 1 *Let G be a regular geometry map, i.e., $G \in C^1([0,1]^2)$ and $\det(J_G) \neq 0$ in $[0,1]^2$, and let K be a symmetric matrix. Then, the matrix-sequence $\{\mathcal{A}_{n,G}^{p,K}\}_n$ with $\boldsymbol{n} = (n,n)$ is distributed, in the sense of the eigenvalues, like the function*

$$f_G^{p,K}(\hat{\boldsymbol{x}}, \boldsymbol{\theta}) := [1\ 1]\left(|\det(J_G(\hat{\boldsymbol{x}}))| K_G(\hat{\boldsymbol{x}}) \circ H_p(\boldsymbol{\theta})\right)[1\ 1]^T, \tag{9}$$

where $\hat{\boldsymbol{x}} \in [0,1]^2$, $\boldsymbol{\theta} \in [-\pi,\pi]^2$, and $\circ$ is the Hadamard matrix product.

We refer the reader to [8, 9] for a detailed discussion about the symbol (9).

3.2 Matrices Related to Our Curl-Div Problem

We can reformulate (1) in 2D as

$$\mathcal{L}_{\alpha,\beta} u = \alpha \begin{bmatrix} \frac{\partial^2 u_2}{\partial x_1 x_2} - \frac{\partial^2 u_1}{\partial x_2^2} \\ \frac{\partial^2 u_1}{\partial x_1 x_2} - \frac{\partial^2 u_2}{\partial x_1^2} \end{bmatrix} - \beta \begin{bmatrix} \frac{\partial^2 u_2}{\partial x_1 x_2} + \frac{\partial^2 u_1}{\partial x_1^2} \\ \frac{\partial^2 u_1}{\partial x_1 x_2} + \frac{\partial^2 u_2}{\partial x_2^2} \end{bmatrix}, \tag{10}$$

where $u(x_1, x_2) := [u_1(x_1, x_2), u_2(x_1, x_2)]^T$. When discretizing the weak form (3) using the space (7) we arrive at the 2×2 block matrix

$$\mathcal{A}_{n,G}^{p,\alpha,\beta} := \alpha \begin{bmatrix} \mathcal{A}_{n,11}^{p,\text{curl}} & \mathcal{A}_{n,12}^{p,\text{curl}} \\ \mathcal{A}_{n,21}^{p,\text{curl}} & \mathcal{A}_{n,22}^{p,\text{curl}} \end{bmatrix} + \beta \begin{bmatrix} \mathcal{A}_{n,11}^{p,\text{div}} & \mathcal{A}_{n,12}^{p,\text{div}} \\ \mathcal{A}_{n,21}^{p,\text{div}} & \mathcal{A}_{n,22}^{p,\text{div}} \end{bmatrix}.$$

The blocks related to the curl-curl operator $(\nabla \times \cdot, \nabla \times \cdot)$ are given by

$$\left[\mathcal{A}_{n,11}^{p,\text{curl}} \right]_{i,j} = \int_{\widehat{\Omega}} \left[-\frac{\partial G_1}{\partial \hat{x}_2} (N_{j_1+1}^p)' \otimes N_{j_2+1}^p + \frac{\partial G_1}{\partial \hat{x}_1} N_{j_1+1}^p \otimes (N_{j_2+1}^p)' \right]$$
$$\left[-\frac{\partial G_1}{\partial \hat{x}_2} (N_{i_1+1}^p)' \otimes N_{i_2+1}^p + \frac{\partial G_1}{\partial \hat{x}_1} N_{i_1+1}^p \otimes (N_{i_2+1}^p)' \right] \frac{1}{|\det(J_G)|},$$

$$\left[\mathcal{A}_{n,12}^{p,\text{curl}} \right]_{i,j} = -\int_{\widehat{\Omega}} \left[\frac{\partial G_2}{\partial \hat{x}_2} (N_{j_1+1}^p)' \otimes N_{j_2+1}^p - \frac{\partial G_2}{\partial \hat{x}_1} N_{j_1+1}^p \otimes (N_{j_2+1}^p)' \right]$$
$$\left[-\frac{\partial G_1}{\partial \hat{x}_2} (N_{i_1+1}^p)' \otimes N_{i_2+1}^p + \frac{\partial G_1}{\partial \hat{x}_1} N_{i_1+1}^p \otimes (N_{i_2+1}^p)' \right] \frac{1}{|\det(J_G)|},$$

$$\left[\mathcal{A}_{n,22}^{p,\text{curl}} \right]_{i,j} = \int_{\widehat{\Omega}} \left[\frac{\partial G_2}{\partial \hat{x}_2} (N_{j_1+1}^p)' \otimes N_{j_2+1}^p - \frac{\partial G_2}{\partial \hat{x}_1} N_{j_1+1}^p \otimes (N_{j_2+1}^p)' \right]$$
$$\left[\frac{\partial G_2}{\partial \hat{x}_2} (N_{i_1+1}^p)' \otimes N_{i_2+1}^p - \frac{\partial G_2}{\partial \hat{x}_1} N_{i_1+1}^p \otimes (N_{i_2+1}^p)' \right] \frac{1}{|\det(J_G)|},$$

and $\mathcal{A}_{n,21}^{p,\text{curl}} = \mathcal{A}_{n,12}^{p,\text{curl}}$. Note that all those blocks are symmetric matrices. Similarly, the blocks related to the div-div operator $(\nabla \cdot, \nabla \cdot)$ are given by (see also (10))

$$\mathcal{A}_{n,11}^{p,\text{div}} = \mathcal{A}_{n,22}^{p,\text{curl}}, \quad \mathcal{A}_{n,12}^{p,\text{div}} = \mathcal{A}_{n,21}^{p,\text{div}} = -\mathcal{A}_{n,12}^{p,\text{curl}}, \quad \mathcal{A}_{n,22}^{p,\text{div}} = \mathcal{A}_{n,11}^{p,\text{curl}}.$$

In the next subsection we compute the symbol of the matrix-sequence $\{\mathcal{A}_{n,G}^{p,\alpha,\beta}\}_n$.

3.3 Spectral Symbol of Curl-Div Matrices $\mathcal{A}_{n,G}^{p,\alpha,\beta}$

We are now ready for the main contribution of the paper: we show that the matrix-sequence $\{\mathcal{A}_{n,G}^{p,\alpha,\beta}\}_n$ admits a spectral distribution according to Definition 1. This extends the symbol computation in [15] to the case of non-trivial geometry.

Theorem 2 *Let G be a regular geometry map, i.e., $G \in C^1([0,1]^2)$ and $\det(J_G) \neq 0$ in $[0,1]^2$. Then, the matrix-sequence $\{\mathcal{A}_{n,G}^{p,\alpha,\beta}\}_n$ with $\mathbf{n} = (n,n)$ is distributed, in the sense of the eigenvalues, like the 2×2 matrix-valued function*

$$f_G^{p,\alpha,\beta}(\hat{\mathbf{x}},\boldsymbol{\theta}) := \alpha f_G^{p,\mathrm{curl}}(\hat{\mathbf{x}},\boldsymbol{\theta}) + \beta f_G^{p,\mathrm{div}}(\hat{\mathbf{x}},\boldsymbol{\theta}), \tag{11}$$

where $\hat{\mathbf{x}} \in [0,1]^2$, $\boldsymbol{\theta} \in [-\pi,\pi]^2$, and

$$f_G^{p,\mathrm{curl}}(\hat{\mathbf{x}},\boldsymbol{\theta}) := \frac{1}{|\det(J_G(\hat{\mathbf{x}}))|} J_G(\hat{\mathbf{x}}) \, P \, H_p(\boldsymbol{\theta}) \, P^T \, (J_G(\hat{\mathbf{x}}))^T, \quad P := \begin{bmatrix} 0 & 1 \\ -1 & 0 \end{bmatrix},$$

$$f_G^{p,\mathrm{div}}(\hat{\mathbf{x}},\boldsymbol{\theta}) := |\det(J_G(\hat{\mathbf{x}}))| \, (J_G(\hat{\mathbf{x}}))^{-T} \, H_p(\boldsymbol{\theta}) \, (J_G(\hat{\mathbf{x}}))^{-1}.$$

Proof From (10) it follows that the block $\mathcal{A}_{n,11}^{p,\mathrm{curl}}$ corresponds to the isogeometric discretization of $-\frac{\partial^2 u_1}{\partial x_2^2}$. By means of a direct computation we can verify that Theorem 1, with

$$K = \begin{bmatrix} 0 & 0 \\ 0 & 1 \end{bmatrix},$$

ensures that the matrix-sequence $\{\mathcal{A}_{n,11}^{p,\mathrm{curl}}\}_n$ is distributed in the sense of the eigenvalues like the entry $(1,1)$ of the matrix $f_G^{p,\mathrm{curl}}$. The same argument (using a suitable matrix K) can also be applied to the remaining blocks. Then, it can be checked that all the considered blocks satisfy the hypotheses of [10, Theorem 5], which implies that $\mathcal{A}_{n,G}^{p,\alpha,\beta}$ is similar, via a proper permutation matrix, to a matrix $\mathcal{T}_{n,G}^{p,\alpha,\beta}$ such that the matrix-sequence $\{\mathcal{T}_{n,G}^{p,\alpha,\beta}\}_n$ has its symbol given by (11). $\qquad\square$

In the context of IgA, the geometry map G is expressed in terms of the same B-spline basis as used for the discretization space. However, as can be seen from the proof, the spectral result in the above theorem holds for any (smooth enough) geometry map.

Finally, we remark that the p-dependence of the symbol in (11) is completely captured by the matrix $H_p(\boldsymbol{\theta})$. As described in Sect. 3.1 this matrix also appears in the symbol expression of a generalized Poisson problem; its properties have been discussed in [5, 8].

4 Numerical Example

In this section we numerically illustrate the spectral results obtained in Sect. 3.3, using the same test problem as in [15, Sect. 5]. More precisely, we consider (3) defined on a quarter of an annulus,

$$\Omega = \{(x_1, x_2) \in \mathbb{R}^2 : r^2 < x_1^2 + x_2^2 < R^2, \ x_1 > 0, \ x_2 > 0\}, \quad r = 1, \quad R = 4,$$

with

$$G(\hat{x}_1, \hat{x}_2) = \begin{cases} x_1 = [r + \hat{x}_1(R - r)] \cos\left(\frac{\pi}{2}\hat{x}_2\right) \\ x_2 = [r + \hat{x}_1(R - r)] \sin\left(\frac{\pi}{2}\hat{x}_2\right) \end{cases}, \quad (\hat{x}_1, \hat{x}_2) \in [0, 1]^2.$$

Let us fix $n := (n, n) \in \mathbb{N}^2$, $p := (p, p) \in \mathbb{N}^2$ and $m \in \mathbb{N}^2$ such that $m^2 = n + p - 2$. We start by defining two equispaced grids on $[0, 1]^2$ and $[0, \pi]^2$:

$$x_j := \frac{j}{m-1}, \quad \theta_k := \frac{k\pi}{m-1}, \quad j, k = 0, \ldots, m-1.$$

Then, we denote by Λ_i the set of all evaluations of $\lambda_i(f_G^{p,\alpha,\beta})$ on $\Gamma := \{(x_j, \theta_k), j, k = 0, \ldots, m-1\}$ for a fixed $i \in \{1, 2\}$. Note that it suffices to consider only $[0, \pi]^2$ because the symbol (11) is symmetric on $[-\pi, \pi]^2$, and hence also its eigenvalue functions.

In Fig. 1 we numerically check relation (11) by comparing the eigenvalues of $\mathcal{A}_{n,G}^{p,\alpha,\beta}$ with the values collected in $\Lambda = \{\Lambda_1, \Lambda_2\}$, ordered in ascending way, for $\alpha = 1$ and $\beta = 0.1$. We observe that, in a complete agreement with the theory, the considered sampling of $\lambda_i(f_G^{p,\alpha,\beta})$, $i = 1, 2$, describes quite accurately the behavior of the eigenvalues of $\mathcal{A}_{n,G}^{p,\alpha,\beta}$, also for relatively small matrix-sizes, up to few outliers.

5 Conclusions

We have analyzed the spectral properties of matrix-sequences arising from isogeometric Galerkin methods for weighted curl-div operators on general planar domains, considering a non-trivial geometry map. More precisely, we have shown that an (asymptotic) spectral distribution exists and it is compactly described by a 2×2 spectral symbol. In other words, the eigenvalues of the matrices we are dealing with can be approximated accurately by a uniform sampling of the two eigenvalue functions of the 2×2 symbol matrix. The symbol depends on the characteristic parameters of the problem and on the geometry of the physical domain. Its formal

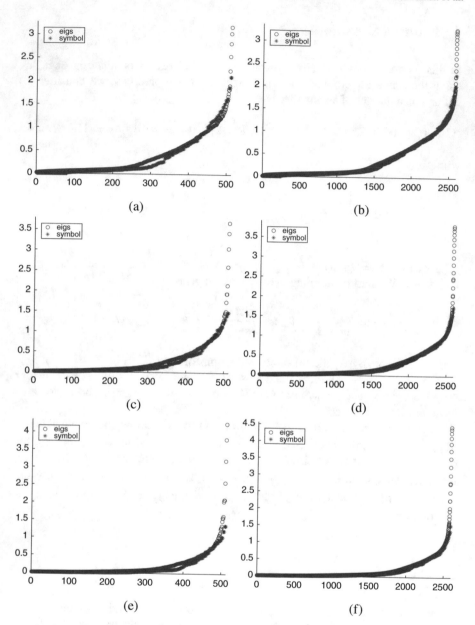

Fig. 1 Comparison of the eigenvalues of $\mathcal{A}_{n,G}^{p,\alpha,\beta}$ (open circle) with $\Lambda = \{\Lambda_1, \Lambda_2\}$ collecting uniform samples of $\lambda_i(f_G^{p,\alpha,\beta})$, $i = 1, 2$ (asterisk), ordered in ascending way, varying both n and p, and fixing $\alpha = 1$ and $\beta = 0.1$. (**a**) $p = 3, n = 15$. (**b**) $p = 3, n = 35$. (**c**) $p = 4, n = 14$. (**d**) $p = 4, n = 34$. (**e**) $p = 5, n = 13$. (**f**) $p = 5, n = 33$

structure nicely mimics the structure of the differential problem. The numerical results show a very good matching between the true eigenvalues and the estimates provided by the symbol, already for relatively small matrix-sizes.

The convergence of iterative solvers for linear systems strongly depends on the spectral behavior of the corresponding coefficient matrices. Since the symbol gives a precise description of the spectrum of the curl-div matrix $\mathcal{A}_{n,G}^{p,\alpha,\beta}$, it could be helpful in the design of good preconditioners that lead to better performance than current solution strategies, like the one in [15, Sect. 5].

Acknowledgements This work was partially supported by the INdAM research group GNCS, by the MIUR-DAAD Joint Mobility 2017 Programme through the project "ATOMA", and by the MIUR Excellence Department Project awarded to the Department of Mathematics, University of Rome Tor Vergata (CUP E83C18000100006).

References

1. de Boor, C.: A Practical Guide to Splines, Revised Edition. Springer, New York (2001)
2. Ciarlet, P.: Augmented formulations for solving Maxwell equations. Comput. Methods Appl. Mech. Eng. **194**, 559–586 (2005)
3. Costabel, M.: A coercive bilinear form for Maxwell's equations. J. Math. Anal. Appl. **157**, 527–541 (1991)
4. Donatelli, M., Garoni, C., Manni, C., Serra-Capizzano, S., Speleers, H.: Robust and optimal multi-iterative techniques for IgA Galerkin linear systems. Comput. Methods Appl. Mech. Eng. **284**, 230–264 (2015)
5. Donatelli, M., Garoni, C., Manni, C., Serra-Capizzano, S., Speleers, H.: Spectral analysis and spectral symbol of matrices in isogeometric collocation methods. Math. Comput. **85**, 1639–1680 (2016)
6. Donatelli, M., Garoni, C., Manni, C., Serra-Capizzano, S., Speleers, H.: Symbol-based multigrid methods for Galerkin B-spline isogeometric analysis. SIAM J. Numer. Anal. **55**, 31–62 (2017)
7. Garoni, C., Manni, C., Pelosi, F., Serra-Capizzano, S., Speleers, H.: On the spectrum of stiffness matrices arising from isogeometric analysis. Numer. Math. **127**, 751–799 (2014)
8. Garoni, C., Manni, C., Serra-Capizzano, S., Sesana, D., Speleers, H.: Spectral analysis and spectral symbol of matrices in isogeometric Galerkin methods. Math. Comput. **86**, 1343–1373 (2017)
9. Garoni, C., Manni, C., Serra-Capizzano, S., Sesana, D., Speleers, H.: Lusin theorem, GLT sequences and matrix computations: an application to the spectral analysis of PDE discretization matrices. J. Math. Anal. Appl. **446**, 365–382 (2017)
10. Garoni, C., Mazza, M., Serra-Capizzano, S.: Block generalized locally Toeplitz sequences: from the theory to the applications. Axioms **7**, 49 (2018)
11. Garoni, C., Serra-Capizzano, S.: Generalized Locally Toeplitz Sequences: Theory and Applications, vol. I. Springer Monographs. Springer, Berlin (2017)
12. Garoni, C., Serra-Capizzano, S.: Generalized Locally Toeplitz Sequences: Theory and Applications, vol. II. Springer Monographs. Springer, Berlin (2018)
13. Grenander, U., Szegö, G.: Toeplitz Forms and Their Applications, 2nd edn. Chelsea, New York (1984)
14. Jardin, S.C., Ferraro, N., Luo, X., Chen, J., Breslau, J., Jansen, K.E., Shephard, M.S.: The M3D-C^1 approach to simulating 3D 2-fluid magnetohydrodynamics in magnetic fusion experiments. J. Phys. Conf. Ser. **125**, 012044 (2008)

15. Mazza, M., Manni, C., Ratnani, A., Serra-Capizzano, S., Speleers, H.: Isogeometric analysis for 2D and 3D curl-div problems: spectral symbols and fast iterative solvers. Comput. Methods Appl. Mech. Eng. **344**, 970–997 (2019)
16. Mazza, M., Ratnani, A., Serra-Capizzano, S.: Spectral analysis and spectral symbol for the 2D curl-curl (stabilized) operator with applications to the related iterative solutions. Math. Comput. **88**, 1155–1188 (2019)
17. Tilli, P.: A note on the spectral distribution of Toeplitz matrices. Linear Multilinear Algebra **45**, 147–159 (1998)
18. Tyrtyshnikov, E.E.: A unifying approach to some old and new theorems on distribution and clustering. Linear Algebra Appl. **232**, 1–43 (1996)
19. Tyrtyshnikov, E.E., Zamarashkin, N.L.: Spectra of multilevel Toeplitz matrices: advanced theory via simple matrix relationships. Linear Algebra Appl. **270**, 15–27 (1998)

Performance of Preconditioners for Large-Scale Simulations Using Nek5000

N. Offermans, A. Peplinski, O. Marin, E. Merzari, and P. Schlatter

1 Introduction

The preconditioning of elliptic problems characterized by the propagation of information at infinite speed over the domain is a numerically challenging task. We study the case of the Poisson equation arising from the numerical resolution of the incompressible Navier–Stokes equations by operator splitting. We consider Nek5000, a code based on the spectral element method, as our framework. The current preconditioning strategy is based on an additive Schwarz method, which combines a domain decomposition method [5] and a so-called coarse grid problem [10]. The first step consists in solving directly local overlapping Poisson problems and is easily parallelizable. The second step corresponds to a Poisson-like problem over the whole domain and is hard to scale because of its relatively low number of degrees of freedom and the bottleneck induced by global communication.

A scalable solver for the coarse grid problem is critical to ensure strong scaling of the code. Existing strategies include a direct solution method similar to a Cholesky decomposition, called XX^T [14], and an algebraic multigrid (AMG) solver [6]. While the first choice works well for relatively small problems (typically $<100,000$ spectral elements on $<10,000$ cores), the second option is preferred for large scale simulations. The current AMG solver, which we will denote as the *in-house AMG*, is fast and scales well [11]. It has been shown that the use of AMG can speed up

N. Offermans (✉) · A. Peplinski · P. Schlatter
Linné FLOW Centre and Swedish e-Science Research Centre (SeRC), KTH Mechanics, Stockholm, Sweden
e-mail: nof@mech.kth.se; adam@mech.kth.se; pschlatt@mech.kth.se

O. Marin · E. Merzari
Argonne National Laboratory, Lemont, IL, USA
e-mail: oanam@mcs.anl.gov; emerzari@anl.gov

© The Author(s) 2020
S. J. Sherwin et al. (eds.), *Spectral and High Order Methods for Partial Differential Equations ICOSAHOM 2018*, Lecture Notes in Computational Science and Engineering 134, https://doi.org/10.1007/978-3-030-39647-3_20

large-scale simulations by up to 10%. In addition, the XX^T has been designed for optimal performance on a number of cores which is a power of 2, whereas the AMG is insensitive to this parameter.

However, the AMG solver requires a setup phase, performed once for each mesh, by an external and serial code. Besides inducing an unwanted overhead, it also limits the use of the *in-house AMG* solver in the framework of mesh refinement, which is the main motivation for this work. Therefore, we propose to replace the *in-house AMG* by *BoomerAMG*, a parallel AMG solver for arbitrary unstructured grids from the *hypre* library for linear algebra [1, 4, 8]. *BoomerAMG* offers a number of parallel algorithms for the coarsening, interpolation and smoothing steps of the AMG setup, to accommodate various types of problems, meshes and architectures. The *BoomerAMG* solver will be tested in terms of scalability and time to solution.

Scaling tests for the *BoomerAMG* solver have been performed up to 4096 cores by Baker et al. [1]. Matrices arising from the finite element and finite difference discretizations of 2D and 3D scalar diffusion problems were considered. The authors used HMIS coarsening and extended+i interpolation and showed that l_1-scaled Jacobi, l_1-scaled Gauss-Seidel and Chebyshev smoothers are good choices for such problems.

Weak scaling up to 125,000 cores has been presented in Ref. [2], where *BoomerAMG* was used as a preconditioner for a conjugate gradient solver. The test case considered is that of a 3D Laplace operator. The parameters for the AMG solver were again HMIS coarsening, extended+i interpolation and symmetric hybrid Gauss–Seidel for the smoother. Aggressive coarsening with multipass interpolation was used on the finest grid, while the problem on the coarsest level was solved by Gaussian elimination. The authors show the impact of additional parameters such as the use of 64 bits for the integers or the use of an hybrid parallel strategy with OpenMP and MPI.

In the present work, we use the *BoomerAMG* from *hypre* to precondition a GMRES solver for the pressure equation arising from the spectral element discretization of the Navier–Stokes equations. We study strong scaling up to 131,072 cores on two different supercomputers: Mira, based on the IBM Blue Gene/Q architecture, and Hazel Hen, a Cray XC40 system. The first test case considered is the flow around a NACA4412 airfoil, which we use to identify a set of best parameters for the *BoomerAMG* solver. A second test case is employed for a strong scaling study: the turbulent flow in wire-wrapped pin bundles [3, 13].

The paper is organized as follows. In Sect. 2, we introduce the discretization method and describe the preconditioning strategy and the *hypre* library. In Sect. 3, we study which set of parameters gives the fastest time to solution for the problem at hand. Using those parameters, we perform a strong scaling study for the flow in wire-wrapped pin bundles in Sect. 4. We finish with conclusions and outlook in Sect. 5

2   Problem Description

Considering an operator splitting strategy to solve the Navier–Stokes equations, the consistent pressure p is the solution to a Laplace problem of the form $\Delta p = r$. In Nek5000, this equation is discretized using the spectral element method [7] and it has been shown that it is well preconditioned by an overlapping additive Schwarz method. The preconditioner combines local problems $R_k^T A_k^{-1} R_k$ and a coarse grid problem $R_0^T A_0^{-1} R_0$ and is expressed as

$$M^{-1} = R_0^T A_0^{-1} R_0 + \sum_{k=1}^{K} R_k^T A_k^{-1} R_k \, ,$$

where R_0 and R_k are restriction operators, A_k are local stiffness matrices and K is the total number of spectral elements. The matrix A_0 corresponds to a Laplace operator defined on the element vertices only. Because of its low number of degrees of freedom and global extent, the scalability of the coarse grid problem is mostly limited by communication and latency. We note that the term "coarse grid" here refers to the fact that A_0 is defined on the vertices of the spectral elements only. The problem is therefore "coarse" in comparison to the solution fields, which are expanded on the Gauss–Lobatto–Legendre points inside each spectral element (typically order 10 quadrature points in each direction). When talking about the different levels arising from the coarsening phase of the AMG setup, we will use the term "coarse level" to avoid confusion.

As mentioned before, the solver of choice for large problems is currently an *in-house AMG*, developed specifically for Nek5000 [6], whose main drawback is a setup phase by an external and serial code. The default option for the setup step of the *in-house AMG* is a Matlab code, which uses Ostrowski coarsening with norm bound, a diagonal Chebyshev smoother, applied on the second branch of the V-cycle only, and an energy-minimizing interpolation, all described in Ref. [9]. Other properties of the AMG include no smoothing on the finest level, a number of smoothing steps predefined for each level during the setup phase and a coarsest level made of one variable only. The good scalability of the *in-house AMG* is due to the fact that it automatically chooses, at run time, the fastest communication strategy at each level of the coarsening process, between three options: a pairwise exchange, a crystal router method or an allreduce operation. Previous work has shown that, when far from the strong scaling limit, the total time spent in the pressure solver is typically 85–90% of the total computational time, including the time spent in the coarse grid solver, which amounts to about 5–10% of that [11].

As an intermediate step in a previous work [12], the coarsening and interpolation steps from this Matlab code have been transferred to *BoomerAMG*, while the smoother and the solver were left unchanged. This "hybrid" serial setup was shown to significantly reduce the setup time without affecting the rapidity and scalability of the *in-house AMG* solver.

In the present work, we completely replace the *in-house AMG* by *BoomerAMG*, which allows for the whole AMG problem (setup + solver) to be performed online and in parallel. The existing code requires only limited modifications. Local contributions to the coarse grid operator are built on each process and then handed over to *BoomerAMG*, which takes care of assembling the global operator and of communication. If the operator possesses a nullspace, the solution is normalized such that the mean of the solution entries is 0. Apart from that, the critical aspect of switching to *BoomerAMG* is the choice of parameters for the setup and for the solver that match the performance of the *in-house AMG*.

3   Optimal Parameter Selection

The choice of parameters for the *BoomerAMG* solver is done by testing a set of parameters on a medium-sized test case and looking for the optimal combination. We consider the turbulent flow around a NACA 4412 airfoil at $Re_c = 400,000$ [13] on a mesh made of 253,980 elements, with polynomial order 11, and we run the simulation for 30 timesteps. The best set of parameters is defined as the one which minimizes the time to solution for the pressure equation. This is achieved by balancing two competing aspects: the accuracy of the coarse grid solution and the total number of iterations of the GMRES solver used for the pressure equation. Since the AMG is used as a preconditioner, a high level of accuracy is not paramount. Yet, it should be sufficient to ensure efficient preconditioning. Based on results obtained with the *in-house AMG*, the initial error on the coarse grid problem should be reduced by approximately one order of magnitude. While the *in-house AMG* is designed to ensure that a given reduction in the error is attained at minimal cost, the *BoomerAMG* is designed to ensure the maximum reduction in the error occurs at a given cost. Therefore, the best choice of parameters for the *BoomerAMG* is case dependent; here we optimize this choice in the case of large 3D simulations, when the use of AMG is most relevant.

All tests are run on 4096 processors on the Blue Gene Mira at the Argonne National Laboratory. We test a total of 96 combinations of the following parameters:

- Coarsening type: classical Ruge–Stueben (C1), Falgout (C2), PMIS (C3), HMIS (C4), CGC (C5) and CGC-E (C6),
- Interpolation method: extended (I1) and extended+i (I2),
- Relaxation type: l_1-Gauss–Seidel forward solve on the down cycle + backward solve on the up cycle (R1), l_1-scaled hybrid symmetric Gauss–Seidel (R2), Chebyshev (R3) and l_1-scaled Jacobi (R4),
- AMG strength threshold: 0.25 and 0.5.

We assign a letter and a number to each option which will be used to identify the method when comparing the results.

Table 1 Five best timings for the *BoomerAMG* with corresponding parameters as compared to the *in-house AMG*

Run ID.	AMG solver	Coars.	Intp.	Relax.	Thresh.	No. pres. it.	Pres. time (s)	CGS time (s)
a	*in-house AMG*	–	–	–	–	3737	666.06	31.51
b	*BoomerAMG*	C3	I2	R2	0.25	3677	719.77	94.93
c	*BoomerAMG*	C4	I2	R1	0.5	3762	734.35	95.11
d	*BoomerAMG*	C3	I1	R1	0.25	3839	746.28	94.7
e	*BoomerAMG*	C3	I2	R4	0.25	3814	747.04	99.28
f	*BoomerAMG*	C3	I2	R1	0.5	3866	756.48	100.13

3 V-cycles for the *BoomerAMG*. Total number of iterations for the pressure solver and timings are reported for 30 timesteps

Table 2 Five best timings for the *BoomerAMG* with corresponding parameters as compared to the *in-house AMG*

Run ID.	AMG solver	Coars.	Intp.	Smo.	Thresh.	No. pres. it.	Pres. time (s)	CGS time (s)
a	*in-house AMG*	–	–	–	–	3737	666.06	31.51
d	*BoomerAMG*	C3	I1	R1	0.25	3752	675.52	38.09
c	*BoomerAMG*	C4	I2	R1	0.5	3786	683.65	41.0
b	*BoomerAMG*	C3	I2	R2	0.25	3877	698.44	40.31
f	*BoomerAMG*	C3	I2	R1	0.5	3990	721.03	44.35
e	*BoomerAMG*	C3	I2	R4	0.25	4476	803.74	46.24

1 V-cycle for the *BoomerAMG*. Total number of iterations for the pressure solver and timings are reported for 30 timesteps

In all cases, we set a relative tolerance of 0.1 on the solution of the coarse grid problem and a maximum of 3 V-cycles for the AMG. Moreover, the problem on the coarsest level is solved by Gaussian elimination. Total timings and number of pressure iterations for 30 timesteps are presented in Table 1 for the *in-house AMG* and the five fastest combinations of parameters for the *BoomerAMG*. We assign a letter to each run for comparison later. The time spent in the pressure solver is reported for process 0 and the time spent in the coarse grid solver (CGS) is the maximum value over all processors for a single run. Since an allocation on Mira is always made of cores that are physically contiguous and isolated from the rest of the network, the timings suffer little noise and uncertainty. We see that the number of pressure iterations is on par with the *in-house AMG*, but the time spent in the coarse grid solver is more that three times as much. As a result, the time spent in the pressure solver is between 8 and 13% higher.

To accelerate the *BoomerAMG* solver, we set the maximum number of V-cycles to 1, instead of 3, and perform another test using the optimal parameters. This should significantly accelerate the resolution of the coarse grid solver but reduce the accuracy of the solution, therefore increasing the number of pressure iterations. The corresponding results are presented in Table 2. A surprising result comes from run d, where the number of pressure iterations has actually decreased. Since this might be the sign of an unstable solver, we discard this choice of parameters. The second best set of parameters corresponds to HMIS coarsening, extended+i interpolation, l_1-Gauss–Seidel forward solve on the down cycle + backward solve on the up cycle for the smoother and an AMG strength threshold of 0.5, which we use for the rest of our simulations.

We also experimented with more aggressive non-Galerkin coarsening to change the communication pattern on the largest AMG levels and reduce communication time. However, this caused a drop of accuracy for the coarse grid solver and an increase of iterations for the pressure solver, which led to slower overall timings as a result.

We note that the time to setup the coarse grid problem, which includes building the matrix A_0 and performing the *BoomerAMG* setup, is negligible. In the present configuration, it amounts to less than a second, whereas a single timestep takes about 20 s. Since the setup is performed once at the beginning of the simulation, we do not discuss the matter further. Furthermore, it is orders of magnitude lower than the serial versions of the setup [12].

Further analysis of the results shows that the use of classical Ruge-Stueben, Falgout, CGC or CGC-E significantly slows down the solver. Moreover, another valid choice for the smoother could have been l_1-scaled hybrid symmetric Gauss–Seidel, whose speed is on par with our choice. Other relaxation methods, Chebyshev and l_1-scaled Jacobi, are also consistently slower. Finally, the choice of the interpolation method does not have a significant impact and both methods give very similar results.

4   Scaling Results

As is often the case with numerical simulations, we look for the fastest path to solution for a given problem. Therefore, a strong scaling study, where the total amount of work is fixed and the number of processes is increased, is a relevant measure of the efficiency of the *BoomerAMG*. This is opposed to a weak scaling analysis, where the amount of work per process is kept constant, which is not carried out here.

We consider the turbulent flow inside a reactor assembly made of 61 wire-wrapped pins, a configuration appearing in a nuclear reactor core [3]. The mesh consists of 1,650,240 elements, uses polynomial order 7 and has a complex, fully three-dimensional topology, making it a relevant test case for evaluating precondi-tioning strategies. The initial velocity field is turbulent and we run the simulation for 10 timesteps. Two series of tests were conducted on two supercomputers: Mira and Hazel Hen. The number of compute nodes considered is 512, 1024, 2048, 4096 and 8192 on the former machine and 256, 512, 1024, 2048, 4096 on the latter one. On both computers, the number of MPI processes per node is equal to the number of available compute cores, i.e. 16 on Mira and 24 on Hazel Hen. We use our previous defined optimal parameters for the setup of the *BoomerAMG* and we also include non-Galerkin coarsening. A drop-tolerance of 0.05 for sparsification is set as default on all levels, with the exception of the five finest levels, which have respective drop-tolerances of 0.0, 0.01, 0, 02, 0.03 and 0.04. This choice of parameters is motivated by the fact that, unlike with the wing case, the time for the coarse grid solver is reduced by about 25%, as tests on 6,144 cores on Hazel Hen have shown, without impacting the number of pressure iterations.

First, let us mention that the setup time for the *BoomerAMG* solver is once again negligible in comparison to the time for the entire simulation, requiring less than 3 s on any number of cores on any machine. It is also significantly lower than reading the data of the *in-house AMG*, a serial process, which takes about 80–90 s. Therefore, it does not represent a bottleneck and we do not investigate timings for the setup phase in details.

Next, we present the strong scaling results, based on a single run per core count. The reported value for the time spent in the pressure solver is the timing from core 0. The time spent in the coarse grid solver is measured on each processor and we consider the maximum value among all processes. The average time per timestep for the pressure solver on Mira is shown in Fig. 1, left plot. Unlike what was observed for the wing simulation, the choice of AMG solver does not impact the number of pressure iterations. The *in-house AMG* is slightly faster than the *BoomerAMG* on all core counts and it seems to scale marginally better. Since all other timings are the same, the reason is a faster coarse grid solver, as can be seen in Fig. 1, right plot. The *in-house AMG* achieves a better performance because it optimizes the communication process independently on each level of the AMG. On the coarsest levels, it is able to take advantage of the fast allreduce operation offered by the network of the Blue Gene architecture in hardware. On the finest levels, it picks up the fastest method between a crystal-router strategy or a pairwise exchange.

On 131,072 cores, the actual speed up for the coarse grid solver is 3.11 and the parallel efficiency is 0.194 for the *in-house AMG*, when the timings on 8,192

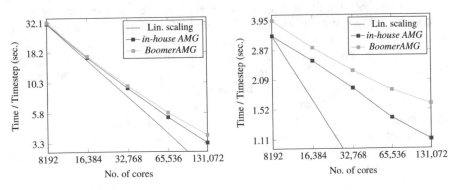

Fig. 1 Rod-bundle test case on Mira, with 1,650,240 elements, leading to 12 or 13 elements per core for the largest core count. AMG parameters: HMIS coarsening, extended+i interpolation, l_1-Gauss–Seidel forward solve on the down cycle + backward solve on the up cycle for the smoother and an AMG strength threshold of 0.5. Left: time spent in the pressure solver; value for process 0, averaged over the total number of timesteps. Right: time spent in the coarse grid solver; maximum value over all processes, normalized by the total number of timesteps

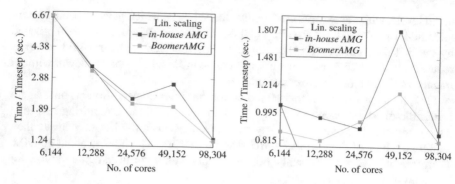

Fig. 2 Rod-bundle test case on Hazel Hen, with 1,650,240 elements, leading to 16 or 17 elements per core for the largest core count. AMG parameters: HMIS coarsening, extended+i interpolation, l_1-Gauss–Seidel forward solve on the down cycle + backward solve on the up cycle for the smoother and an AMG strength threshold of 0.5. Left: time spent in the pressure solver; value for process 0, averaged over the total number of timesteps. Right: time spent in the coarse grid solver; maximum value over all processes, normalized by the total number of timesteps

cores are used as references. These quantities are respectively 2.44 and 0.153 for the *BoomerAMG*. As mentioned before, the network on Mira is characterized by little noise and the uncertainty on the timings is low. Yet, these numbers are based on a single run and are only indications.

The same timings on Hazel Hen are shown in Fig. 2. Unfortunately, the node allocation on this machine can be scattered and the interconnect noise, which is shared with the rest of the computer, can be high and unpredictable. Therefore, a thorough analysis of the scaling results is not possible from a single run and the data from Fig. 2 are only indicative. Nevertheless, the runs for each AMG solver on the same number of cores are obtained using the same node allocation. This makes a comparison between the two solvers on a given core count somewhat relevant. In contrast to the results on Mira, the *BoomerAMG* is slightly faster than the *in-house AMG* on most core counts; the exception is on 6144 cores, where the timings are almost equal. Furthermore, it is quite clear that the coarse grid solver on Hazel Hen does not scale at all. Indeed, the time spent in the coarse grid solver is almost constant from the lowest amount of cores considered.

Based on the available data, we see that beyond 24,576 thousand cores, the time spent in the coarse grid solver accounts for roughly half of the time spent in the pressure solver. At that point, there is about 67 elements and 35,000 grid points per core, which is consistent with the strong scaling limit on a similar computer as identified in Ref. [11].

5 Conclusions

We used the *BoomerAMG* solver from the *hypre* library for linear algebra to solve a global coarse problem that is part of the preconditioner for the pressure equation arising when time-integrating the Navier–Stokes equations. The set of parameters for the *BoomerAMG* setup that leads to the lowest solver time for the pressure equation is HMIS coarsening, extended+i interpolation, l_1-Gauss–Seidel forward solve on the down cycle + backward solve on the up cycle for the smoother and an AMG strength threshold of 0.5. We also used non-Galerkin coarsening, with more aggressive drop-tolerance on the coarser levels, to speed up the solver. This new method replaces an existing AMG solver, which is fast and scales well but requires a setup phase done externally in serial. Strong scaling was assessed for both AMG solvers on a real large-scale test case on two supercomputers: an IBM Blue Gene/Q (Mira) and a Cray XC40 (Hazel Hen). On Mira, the *in-house AMG* leads to a faster pressure solver that the *BoomerAMG* on all core counts. The maximum difference is about 10% on 131,072 cores. This is because the *in-house AMG* is able to take advantage of the fast hardware allreduce operation on this machine at the coarsest levels of the AMG solver; in that sense the present result was expected. On Hazel Hen, however, the *BoomerAMG* is consistently faster than the *in-house AMG* and we observe the strong scaling limit to be reached at about 24,576 cores. Overall, the *BoomerAMG* is a valid alternative to the *in-house AMG*; both methods are close in terms of performance and the *BoomerAMG* has the advantage to be set up online and in parallel. In particular for modern architectures, the *BoomerAMG* is even faster than the *in-house AMG* with obvious advantages in the setup phase. All the codes developed in this work are available from the https://github.com/nicooff/nek5000/tree/amg_hypre_c Github repository.

Future work will extend the use of *BoomerAMG* to mesh refinement, where an online and parallel AMG setup phase is a requirement.

Acknowledgements We would like to thank Aleks Obabko for his advice and for sharing the wire-wrapped pin bundle case with us. Financial support by the H2020 EU Project "ExaFLOW: Enabling Exascale Fluid Dynamics Simulation" (grant reference 671571), and the Knut and Alice Wallenberg Foundation is gratefully acknowledged. This research used resources of the Argonne Leadership Computing Facility, which is a DOE Office of Science User Facility supported under Contract DE-AC02-06CH11357. Additional computer time was provided by ExaFLOW at HLRS Stuttgart, and by resources provided by the Swedish National Infrastructure for Computing (SNIC) at PDC Stockholm.

References

1. Baker, A.H., Falgout, R.D., Kolev, T.V., Yang, U.M.: Multigrid smoothers for ultraparallel computing. SIAM J. Sci. Comput. **33**(5), 2864–2887 (2011)
2. Baker, A.H., Falgout, R.D., Kolev, T.V., Yang, U.M.: Scaling hypre's multigrid solvers to 100,000 cores. In: Berry, M.W., Gallivan, K.A., Gallopoulos, E., Grama, A., Philippe, B., Saad,

Y., Saied, F. (eds.) High-Performance Scientific Computing: Algorithms and Applications, pp. 261–279. Springer, London (2012)

3. Brockmeyer, L.M., Sarikurt, F., Hassan, Y., Merzari, E.: CFD investigation of wire-wrapped fuel rod bundles and flow sensitivity to bundles size. In: Proceedings of the 16th International Topical Meeting on Nuclear Reactor Thermalhydraulics (NURETH-16) (2015)

4. Falgout, R.D., Jones, J., Yang, U.: The design and implementation of hypre, a library of parallel high performance preconditioners. In: Bruaset, A.M., Tveito, A. (eds.) Numerical Solution of Partial Differential Equations on Parallel Computers, pp. 267–294. Springer, Heidelberg (2006)

5. Fischer, P.F: An overlapping Schwarz method for spectral element solution of the incompressible Navier–Stokes equations. J. Comput. Phys. **133**(1), 84–101 (1997)

6. Fischer, P.F, Lottes, J.W., Pointer, D., Siegel, A.: Petascale algorithms for reactor hydrodynamics. J. Phys. Conf. Ser. **125**(1), 012076 (2008)

7. Fischer, P.F., Lottes, J.W., Kerkemeier, S.G.: Nek5000 (2008). http://nek5000.mcs.anl.gov

8. Henson, V.E., Yang, U.M.: BoomerAMG: a parallel algebraic multigrid solver and preconditioner. Appl. Numer. Math. **41**(1), 155–177 (2002)

9. Lottes, J.W.: Towards robust algebraic multigrid methods for nonsymmetric problems. Springer theses (2017)

10. Lottes, J.W., Fischer, P.F.: Hybrid multigrid/schwarz algorithms for the spectral element method. J. Sci. Comput. **24**(1), 45–78 (2005)

11. Offermans, N., et al.: On the strong scaling of the spectral element solver Nek5000 on petascale systems. In: Proceedings of the Exascale Applications and Software Conference. Stockholm (2016)

12. Offermans, N., Peplinski, A., Marin, O., Fischer, P.F., Schlatter, P.: Towards adaptive mesh refinement for the spectral element solver Nek5000. In: Proceedings of the DLES11 Conference. Pisa (2017)

13. Vinuesa, R., Negi, P.S., Atzori, M., Hanifi, A., Henningson, D.S., Schlatter, P.: Turbulent boundary layers around wing sections up to Rec=1,000,000. Int. J. Heat Fluid Flow **72**, 86–99 (2018)

14. Tufo, H.M., Fischer, P.F.: Fast parallel direct solvers for coarse grid problems. J. Parallel Distrib. Comput. **161**(2), 151–177 (2001)

Two Decades Old Entropy Stable Method for the Euler Equations Revisited

Björn Sjögreen and H. C. Yee

1 Introduction, Objectives and Preliminaries

The two decades old high order central differencing via entropy splitting and summation-by-parts (SBP) difference closure of Olsson and Oliger, Gerritsen and Olsson, and Yee et al. [2, 7, 25] is revisited. The entropy splitting is a form of skew-symmetric splitting in terms of the physical entropy of the nonlinear Euler flux derivatives. Central differencing applied to the entropy splitting form of the Euler flux derivatives together with SBP difference operators will, hereafter, be referred to as **entropy split schemes**.

The objective is to prove for the first time, in the recent definition of entropy stability based on the L_2-energy-like norm estimate, that entropy splitting for central schemes with SBP operators are entropy stable. The proof is to replace the spatial derivatives by summation-by-parts (SBP) difference operators in the entropy split form of the equations using the physical entropy of the Euler equations. The numerical boundary closure follows directly from the SBP operator. No additional numerical boundary procedure is required. In contrast, Tadmor-type entropy conserving schemes [18] using mathematical entropies do not naturally come with a numerical boundary closure. A generalized SBP operator has to be developed [8]. Standard high order spatial central differencing as well as high order central spatial DRP (dispersion relation preserving) spatial differencing is part of the entropy stable methodology. An entropy split scheme satisfies the L_2-energy

B. Sjögreen
Multid Analyses AB, Gothenburg, Sweden
e-mail: bjorn.sjogreen@multid.se

H. C. Yee (✉)
NASA Ames Research Center, Mountain View, CA, USA
e-mail: Helen.M.Yee@nas.nasa.gov

© The Author(s) 2020
S. J. Sherwin et al. (eds.), *Spectral and High Order Methods for Partial Differential Equations ICOSAHOM 2018*, Lecture Notes in Computational Science and Engineering 134, https://doi.org/10.1007/978-3-030-39647-3_21

norm estimate readily without an added numerical dissipation term for smooth flows. For flows containing discontinuities the Yee et al. nonlinear filter approach [10–12, 14, 15, 22–25] is employed at isolated computed locations. After each full time step of the entropy split method to suppress spurious oscillations while maintaining accuracy on the remaining flow field. Since the nonlinear filter step is executed as an Euler time discretization at isolated location after the completion of a full time step of the entropy stable central scheme, entropy conservation/stability is valid almost everywhere. The efficiency and performance of the entropy stable split schemes using the physical entropies are compared with Tadmor-type entropy conservative method [18] using mathematical entropies for long time integration of a 2D smooth flows and a 3D direct numerical simulation (DNS) of turbulence with shocklets. It is found that Tadmor-type entropy conservative methods required twice the CPU time than the entropy stable split schemes using the same order of the central scheme. Comparisons among the three skew-symmetric splittings (entropy splitting [19, 20, 25], Ducros et al. splitting [1] and the Kennedy and Grubber splitting [5]) on their nonlinear stability and accuracy performance without added numerical dissipations for smooth flows is included. See [16] for additional details and comparison.

Remarks It is noted that the Hughes et al. formulation [4] using the Harten's idea [3] but solving the flow equations in nonconservative form in terms of the entropy variables is completely different from the entropy split schemes. The entropy split scheme solve the entropy splitting form of the Euler flux derivatives consisting of a one parameter family of conservative and a non-conservative portions in terms of the entropy variables. If the parameter satisfies the energy estimate, entropy stability is immediate. The entropy split scheme has been generalized from a perfect gas to a thermally perfect gas and gas flows consisting of linear combination of perfect gases [21, 25]. In addition, these high order schemes have been formulated in time varying deforming curvilinear grids with free-stream preservation [17, 21].

2 Entropy Splitting of the Euler Flux Derivatives

We consider the 3D equations of inviscid compressible gas dynamics

$$\mathbf{q}_t + \mathbf{f}_x + \mathbf{g}_x + \mathbf{h}_x = 0$$

with conserved variables $\mathbf{q} = (\rho\ \rho u\ \rho v\ \rho w\ e)^T$ and fluxes in an arbitrary direction $\mathbf{k} = (k_1\ k_2\ k_3)$ with $|\mathbf{k}|^2 = 1$, and

$$\hat{\mathbf{f}} = k_1\mathbf{f} + k_2\mathbf{g} + k_3\mathbf{h} = (\rho\hat{u}\ \rho u\hat{u} + k_1 p\ \rho v\hat{u} + k_2 p\ \rho w\hat{u} + k_3 p\ \hat{u}(e+p))^T, \quad (1)$$

where $\hat{u} = k_1 u + k_2 v + k_3 w$. The total energy is related to the pressure p by the ideal gas law, $e = \frac{p}{\gamma-1} + \frac{1}{2}\rho|\mathbf{u}|^2$, where $\gamma > 1$ is a given constant, and $|\mathbf{u}|^2 = u^2 + v^2 + w^2$.

An entropy is a convex function, $E(\mathbf{q})$, of the conserved variables that allows an additional conservation law,

$$E_t + F_x + G_y + H_z = 0, \tag{2}$$

when the solution is smooth. The entropy fluxes in the x-, y-, and z-directions are denoted by F, G, and H, respectively. The entropy variables are defined by $\mathbf{v} = \nabla_\mathbf{q} E$ (the notation $E_\mathbf{q}$ for the gradient will sometimes be used). The convexity of E ensures that these are well-defined. The Entropy conservation law (2) follows if the relation $\mathbf{v}^T \frac{\partial \mathbf{f}}{\partial \mathbf{q}} = \nabla_\mathbf{q} F$ for the x-direction fluxes, and similarly for the y- and z-directions, holds. Moreover, the entropy variables symmetrize the equations; $\partial \mathbf{f}/\partial \mathbf{v}$ is a symmetric matrix.

Harten [3] considered the class of entropies

$$E = -\frac{\gamma + \alpha}{\gamma - 1} \rho (p \rho^{-\gamma})^{\frac{1}{\alpha + \gamma}}, \tag{3}$$

where α is a parameter. To ensure that E is convex, i.e., that the matrix $E_{\mathbf{q},\mathbf{q}}$ is positive definite, α is required to satisfy $\alpha > 0$ or $\alpha < -\gamma$. The full range for α was given in [25], while [3] only considered $\alpha > 0$, and [2] used only the special case $\alpha = 1 - 2\gamma$ from $\alpha < -\gamma$. The corresponding entropy flux in the direction $\mathbf{k} = (k_1\ k_2\ k_3)^T$ is

$$F = \hat{u} E.$$

The entropy variables $\mathbf{v} = E_\mathbf{q}$ are straightforwardly found to be

$$\mathbf{v} = \frac{\rho}{p} s^{\frac{1}{\alpha + \gamma}} \left(-\frac{\alpha}{\gamma - 1} \frac{p}{\rho} - \frac{1}{2}|\mathbf{u}|^2\ u\ v\ w\ -1 \right)^T, \tag{4}$$

where s denotes $p \rho^{-\gamma}$. The conserved variables are homogeneous functions of the entropy variables (4),

$$\mathbf{q}(\theta \mathbf{v}) = \theta^\beta \mathbf{q}(\mathbf{v}), \tag{5}$$

where $\beta = (\alpha + \gamma)/(1 - \gamma)$. From (5) it follows that

$$\mathbf{q}_\mathbf{v} \mathbf{v} = \beta \mathbf{q} \tag{6}$$

$$\hat{\mathbf{f}}_\mathbf{v} \mathbf{v} = \beta \hat{\mathbf{f}}. \tag{7}$$

See [3, 16] for the proof. The range of α, where $E_{\mathbf{q},\mathbf{q}}$ is positive definite, translates to β satisfying $\beta < -\frac{\gamma}{\gamma - 1}$ or $\beta > 0$.

Entropy splitting of the Euler flux derivative in the x-direction with the y- and z-directions suppressed [2, 25] is written as a weighted sum of a conservative part,

$\mathbf{f}_x$, and a non-conservative part, $\mathbf{f_v v}_x$, as

$$\mathbf{f}_x = \frac{\beta}{\beta + 1}\mathbf{f}_x + \frac{1}{\beta + 1}\mathbf{f_v v}_x.$$

Replacing $\mathbf{f}_x$ by this split flux derivative gives

$$\mathbf{q}_t + \frac{\beta}{\beta + 1}\mathbf{f}_x + \frac{1}{\beta + 1}\mathbf{f_v v}_x = \mathbf{0}. \tag{8}$$

The entropy splitting weights the non-conservative portion of the flux derivative by $\frac{1}{1+\beta}$. This means that the range $\beta > 0$ corresponds to a weight that is less than 1, whereas negative β leads, unphysically, to a weight that is greater than 1. The global entropy conservation can be rewritten as an L_2-like estimate. The entropy time derivative can be rewritten as

$$\frac{d}{dt}E(\mathbf{q}) = \frac{1}{\beta + 1}\frac{d}{dt}(\mathbf{v}^T(E_{\mathbf{q},\mathbf{q}})^{-1}\mathbf{v})$$

by using the homogeneity (5). Due to a page limit, see [16] for further discussion. Note that it is necessary to bound the eigenvalues of $E_{\mathbf{q},\mathbf{q}}^{-1}$ in order to make the L_2-like norm a valid estimate.

3 Semi-Discrete Entropy Split Discretization of the Euler Equations

Consider the 1D compressible gas dynamic equations discretized on a domain $a < x < b$ by a uniform grid $x_j = (j - 1)\Delta x + a$, $j = 1, \ldots, N$, and grid spacing $\Delta x = (b - a)/(N - 1)$. Define the semi-discrete entropy split approximation

$$\frac{d}{dt}\mathbf{q}_j + \frac{\beta}{\beta + 1}D\mathbf{f}_j + \frac{1}{\beta + 1}(\mathbf{f_v})_j D\mathbf{v}_j = 0, \quad j = 1, \ldots, N, \tag{9}$$

where D is a SBP difference operator. With *entropy split scheme*, we will always mean the entropy split form of Eqs. (8) discretized in space by a summation-by-parts finite difference operator. The flux Jacobian matrix with respect to the entropy variables, $\mathbf{f_v}$, is symmetric. The SBP scalar product is denoted by

$$(\mathbf{u}, \mathbf{v})_h = \Delta x \sum_{j=1}^{N}\omega_j \mathbf{u}_j^T \mathbf{v}_j,$$

where $\omega_j > 0$ are weights that are different from 1 only at a few points near the boundaries. The operator D satisfies the SBP property

$$(D\mathbf{u}, \mathbf{v})_h = -(\mathbf{u}, D\mathbf{v})_h - \mathbf{u}_1^T \mathbf{v}_1 + \mathbf{u}_N^T \mathbf{v}_N, \tag{10}$$

but is otherwise arbitrary. In the most common case D is a standard SBP centered difference operator, but other operators are possible.

A zero velocity, $u_1 = 0$, $u_N = 0$, boundary condition is enforced, corresponding to wall boundaries. Thanks to the SBP property of the difference approximation the derivation of entropy conservation for the continuous problem can be carried over to the discretization.

Theorem 1 *The approximation (9) together with the boundary conditions $u_1 = 0$ and $u_N = 0$ conserve the global entropy in the sense that $\frac{d}{dt} \sum_{j=1}^{N} \omega_j E_j = 0$.*

A method is entropy dissipative, or "entropy stable", if the computed solution satisfies (2) with inequality,

Proof Denote

$$\mathbf{r} = -\frac{\beta}{\beta + 1}(\mathbf{v}, D\mathbf{f})_h - \frac{1}{\beta + 1}(\mathbf{v}, (\mathbf{f}_\mathbf{v})D\mathbf{v})_h.$$

The scheme (9) can be written

$$\frac{d}{dt}\mathbf{q}_j = P\mathbf{r}_j, \tag{11}$$

where the projection P sets $u_1 = 0$ and $u_N = 0$. Because $P^2 = P$, applying P to both sides of (11) gives that

$$\frac{d}{dt}P\mathbf{q} = \frac{d}{dt}\mathbf{q},$$

i.e., that $P\mathbf{q} = \mathbf{q}$ if the initial data satisfy the boundary conditions. For the entropy

$$\frac{d}{dt}E = (\mathbf{v}, \mathbf{q}_t)_h = (\mathbf{v}, P\mathbf{r})_h = (\mathbf{v}, \mathbf{r})_h - (\mathbf{v}, (I - P)\mathbf{r})_h =$$

$$(\mathbf{v}, \mathbf{r})_h - (P\mathbf{v}, (I - P)\mathbf{r})_h = (\mathbf{v}, \mathbf{r})_h, \tag{12}$$

where we use that $P\mathbf{v} = \mathbf{v}$. This is due to the second component of $\mathbf{v}$ is zero when the x-velocity, u, is zero, and the orthogonality $(P\mathbf{v}, (I - P)\mathbf{r})_h = 0$. The entropy equation is now of the same form as for the continuous problem, but replacing with integration-by-parts by summation-by-parts gives

$$\frac{d}{dt}E(\mathbf{q}_j) = -F_N + F_1.$$

Entropy conservation follows by observing that $F = uE$, so that the boundary conditions imply that $F_1 = F_N = 0$.

If the boundary conditions are periodic, no SBP modification of the difference operator is needed. Entropy conservation is proved with periodic boundary conditions by direct application of the same technique as above. It can be shown that the result carries over directly to the semi-discrete approximation, since only time derivatives are used in the proof. Hence, the L^2-like estimate

$$\frac{d}{dt} \sum_{j=1}^{N} \omega_j \mathbf{v}_j (E_{\mathbf{q},\mathbf{q}})_j^{-1} \mathbf{v}_j = 0$$

is obtained for the approximation (9). It can be shown that Tadmor-type entropy conservative discretization using the Harten entropy and high order central spatial differencings are also entropy conservative methods. See Sjögreen and Yee [16] for the proof.

4 Numerical Experiments

More extensive numerical experiments are reported in the extended version of this paper [16]. Previous studies using SBP boundary closures for non-periodic boundary conditions can be found in [25]. Here selected summary results are presented.

Test Case 1: 2D Compressible Euler Simulation of Smooth Flow: Isentropic Vortex Convection
The compressible Euler equations in two space dimensions are solved with initial data

$$\rho(x, y) = (1 - \frac{(\gamma - 1)\beta^2}{8\gamma\pi^2} e^{1-r^2})^{\frac{1}{\gamma-1}} \tag{13}$$

$$u(x, y) = u_\infty - \frac{\beta(y - y_0)}{2\pi} e^{(1-r^2)/2} \tag{14}$$

$$v(x, y) = v_\infty + \frac{\beta(x - x_0)}{2\pi} e^{(1-r^2)/2} \tag{15}$$

$$p(x, y) = \rho(x, y)^\gamma, \tag{16}$$

where $r^2 = x^2 + y^2$, $\beta = 5$, $\gamma = 1.4$, $u_\infty = 1$, and $v_\infty = 0$. The exact solution is the initial data translated, $\mathbf{u}(x, t) = \mathbf{u}_0(x - u_\infty t, y - v_\infty t)$.

The computational domain is $0 \le x \le 18$, $0 \le y \le 18$ with periodic boundary conditions. The center of the vortex is chosen to be $(x_0, y_0) = (9, 9)$. The problem is solved in time with the classical fourth-order accurate explicit Runge–Kutta method to time $t = 72$, which corresponds to four revolutions of the vortex across the domain.

Comparisons of high order classical central split schemes with high order DRP schemes with grid refinements are reported in [13]. Due a space limitation only one grid with maximum and L_2 error norm compared with the exact solution is shown in Fig. 1. Here C08-DS represents eighth-order central differencing applied to the Ducros et al. splitting form of the Euler flux derivatives. The corresponding eighth-order entropy splitting, entropy conservative method and Kennedy Grubber splitting are indicated by "C08-ES", "C08-EC" and "C08-KGS". If the computed solutions by "C08-DS", "C08-ES", "C08-EC" and "C08-KGS" are nonlinearly filtered by a dissipative portion of WENO7 (seventh-order weighted essentially nonoscillatory spatial method) with an adaptive flow sensor, they are indicated by C08-DS+WENO7FI, C08-ES+WENO7FI, C08-EC+WENO7FI, and C08-KGS+WENO7FI [14, 15, 22–25]. For the smooth flow without any turbulent structure, $\beta = 1$ for the entropy split scheme. The β parameter studies are reported in [9, 16, 25]. In general, for compressible shock-free turbulence and turbulence with

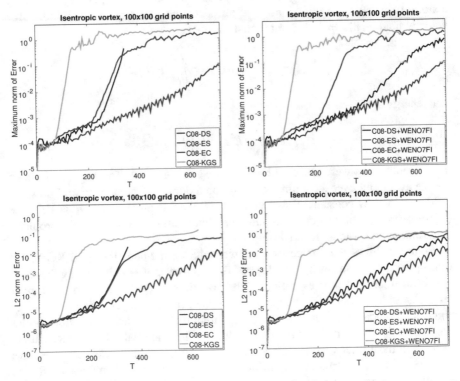

Fig. 1 Inviscid 2D compressible vortex convection with 100^2 grid points: comparison of maximum-norm of error vs. time for C08-DS, C08-ES, C08-EC, and C08-KGS (left, top), and C08-DS+WENO7FI, C08-ES+WENO7FI, C08-EC+WENO7FI, and C08-KGS+WENO7fFI (right top). Bottom left and bottom right are the corresponding L_2-norm of error vs. time

shocklets, β lies somewhere in the range $1.5 < \beta < 2.5$. In general, the optimal β is problem dependent. A general conclusion is that β should not be very large or below 1.

Other high resolution dissipative shock-capturing methods are also candidates for the nonlinear filter approach as well as other optimal WENO or ENO methods. However, with a good control of the numerical dissipation away from discontinuities, there is no need to use the more complicated and more CPU intensive shock-capturing methods. The non-split C08 without any added numerical dissipation diverges shortly after time evolution. Results by WENO5 or WENO7 are very diffusive with large maximum or L_2 errors. For this smooth long time integration flow, entropy splitting is the most accurate method.

Test Case 2: 3D Isotropic Turbulence with Eddy Shocklets

The second numerical test problem computes decaying compressible isotropic turbulence with eddy shocklets. For high enough turbulent Mach numbers weak shocks (shocklets) develop from the turbulent motion. Here the initial turbulent Mach number is 0.6. The Navier–Stokes equations are solved using $\gamma = 1.4$. The computational domain is a cube with side length 2π and periodic boundary conditions in all three directions. The initial datum is a random divergence free velocity field, $u_{i,0}, i = 1, 2, 3$, that satisfies

$$\frac{3}{2}u_{rms,0}^2 = \frac{1}{2}\langle u_{i,0}, u_{i,0} \rangle = \int_0^\infty E(k)\, dk$$

with energy spectrum

$$E(k) \sim k^4 e^{-2(k/k_0)^2}.$$

The computations were made with $u_{rms,0} = 1$ and $k_0 = 4$. The angular brackets denote averaging over the entire computational domain. The density and pressure fields are initially constant. The Taylor-scale Reynolds number, $Re_{\lambda,0}$, is 100. See [6] for definitions of the quantities and more details about the set up of the problem. The simulation is run to the final time 4.

Figure 2 shows the comparison of two splitting methods (DS and KGS), ES (entropy splitting and entropy stable) and EC (entropy conservative) using the same nonlinear filter. The time evolution of the domain averaged kinetic energy (upper left), enstrophy (upper right), temperature variance (lower left), and dilatation (lower right) are compared. All four forms of the nonlinear filter method provide similar resolution. All four schemes without the nonlinear filter are stable but not as accurate as the nonlinear filter versions. Over all, DS splitting is slightly less CPU intensive than ES. KGS skew-symmetric splitting is more CPU intensive than DS and ES. The EC method is around two times more expensive than DS. In addition, as the order of these methods increases, the gain in efficiency (CPU) by entropy split schemes increases.

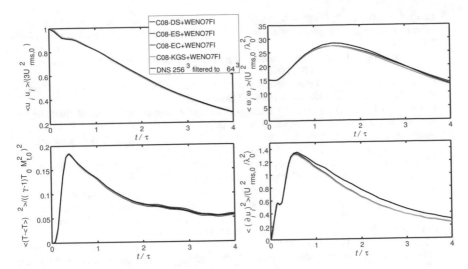

Fig. 2 3D Isotropic turbulence problem with 64^3 grid points. Comparison of two splitting method (DS and KGS), ES (entropy splitting and entropy stable) and EC (entropy conservative) using the same nonlinear filter. Evolution of kinetic energy (upper left), enstrophy (upper right), temperature variance (lower left), and dilatation (lower right) DNS computed on 256^3 grid points and filtered down to 64^3 resolution is considered as the reference solution

Although entropy split methods are not in conservation form but entropy conservative, Sect. 4 showed that they perform well on problems with shocklets. Over all, Extension of the entropy split scheme to other equations of state (non-perfect gas) and the MHD can be found in the original 2000 Yee et al. [25] paper. The entropies (3) can be used to construct entropy conserving schemes in conservative form. See [16] for the derivation.

References

1. Ducros, F., Laporte, F., Soulères, T., Guinot, V., Moinat, P., Caruelle, B.: High-order fluxes for conservative skew-symmetric-like schemes in structured meshes: application to compressible flows. J. Comput. Phys. **161**, 114–139 (2000)
2. Gerritsen, M, Olsson, P.: Designing an efficient solution strategy for fluid flows. I. A stable high order finite difference scheme and sharp shock resolution for the Euler equations. J. Comput. Phys. **129**, 245–262 (1996)
3. Harten, A: On the symmetric form of systems for conservation laws with entropy. J. Comput. Phys. **49**, 151 (1983)
4. Hughes, T., Franca, L., Mallet, M.: A new finite element formulation for computational fluid dynamics: K. Symmetric forms of the compressible Euler and Navier–Stokes equations and the second law of thermodynamics. Comput. Methods Appl. Mech. Eng. **54**, 223–234 (1986)
5. Kennedy, C.A., Gruber, A.: Reduced aliasing formulations of the convective terms Within the Navier–Stokes equations. J. Comput. Phys. **227**, 1676–1700 (2008)

6. Kotov, D.V., Yee, H.C., Wray, A.A., Sjögreen, B., Kritsuk, A.G.: Numerical dissipation control in high order shock-capturing schemes for LES of low speed flows. J. Comput. Phys. **307**, 189–202 (2016)
7. Olsson, P., Oliger, J.: Energy and maximum norm estimates for nonlinear conservation laws. RIACS Technical Report 94.01, 1994
8. Roanocha, H.: Generalized summation-by-parts operators and variable coefficients. J. Comput. Phys. **362**, 20–48 2018. aXiv:1705.10541v2 [math.NA]
9. Sandham, N.D., Li, Q., Yee, H.C.: Entropy splitting for high-order numerical simulation of compressible turbulence. J. Comput. Phys. **23**, 307–322 (2002)
10. Sandham, N.D., Li, Q., Yee, H.C.: Entropy splitting for high-order numerical simulation of compressible turbulence. J. Comput. Phys. **178**(2), 307–322 (2002)
11. Sjögreen, B., Yee, H.C.: Multiresolution wavelet based adaptive numerical dissipation control for high order methods. J. Sci. Comput. **20**, 211–255 (2004)
12. Sjögreen, B., Yee, H.C.: On skew-symmetric splitting and entropy conservation schemes for the Euler equations. In: Proceedings of the ENUMATH09. Uppsala University, Sweden (2009)
13. Sjögreen, B., Yee, H.C.: Accuracy consideration by DRP schemes for DNS and LES of compressible flow computations. Comput. Fluids **159**, 123–136 (2017)
14. Sjögreen, B., Yee, H.C.: Skew-symmetric splitting for multiscale gas dynamics and MHD turbulence flows. In: Extended Version of Proceedings of ASTRONUM-2016. Monterey (2018)
15. Sjögreen, B., Yee, H.C.: High order entropy conservative central schemes for wide ranges of compressible gas dynamics and MHD flows. J. Comput. Phys. **364**, 153–185 (2018)
16. Sjögreen, B., Yee, H.C.: Entropy stable method for Euler equations revisited: central differencing via entropy splitting and SBP. J. Sci. Comput. **81**(3), 1359–1385 (2019)
17. Sjögreen, B., Yee, H.C., Vinokur, M.: On high order finite-difference metric discretizations satisfying GCL on moving and deforming grids. J. Comput. Phys. **265** 211–220 (2014)
18. Tadmor, E.: Entropy stability theory for difference approximations of nonlinear conservation laws and related time-dependent problems. Acta Numer. **12**, 451–512 (2003)
19. Tauber, E., Sandham, N.D.: Comparison of three large-eddy simulations of shock-induced turbulent separation bubbles. Shock Waves **19**, 469–478 (2009)
20. Taylor, G., Green, A.: Mechanism of the production of small eddies from large ones. Proc. R. Soc. Lond. A **158**, 499–521 (1937)
21. Vinokur, M., Yee, H.C.: Extension of efficient low dissipation high-order schemes for 3D curvilinear moving grids. Front. Comput. Fluid Dyn. 129–164 (2002); Also, Proceedings of the Robert MacCormack 60th Birthday Conference (2000), Half Moon Bay, NASA/TM-2000-209598
22. Yee, H.C., Sjögreen, B.: Development of low dissipative high order filter schemes for multiscale Navier–Stokes and MHD systems. J. Comput. Phys. **225** 910–934 (2007)
23. Yee, H.C., Sjögreen, B.: High order filter methods for wide range of compressible flow speeds. In: Proceedings of the ICOSAHOM09. Trondheim (2009)
24. Yee, H. C., Sandham, N.D., Djomehri, M.J.: Low-dissipative high order shock-capturing methods using characteristic-based filters. J. Comput. Phys. **150**, 199–238 (1999)
25. Yee, H.C., Vinokur, M, Djomehri, M.J.: Entropy splitting and numerical dissipation. J. Comput. Phys. **162**, 33–81 (2000)

A Mimetic Spectral Element Method for Free Surface Flows

L. Nielsen and B. Gervang

1 Introduction

In the last decades, CFD simulations of free surface flows have become a key tool in engineering analysis in the design of marine structures. To be able to obtain valid estimates of environmental stress on ship-wave hydrodynamics, offshore wind turbines, wave energy converters, and offshore production systems the CFD tools need to be able to account for non-linear wave-wave and wave-body interaction. Traditionally free surface water simulation has been simulated using lower order methods, however recently spectral element methods have been used [2]. In contrast to earlier work, in the present article, we simulate 2D free surface waves using a mimetic spectral element method. This ensures that the invariants of the system mass, momentum, and energy are conserved throughout the simulation.

The governing equation for incompressible, Newtonian fluids is the Navier–Stokes equation. Free surface waves can be assumed to be governed by an inviscid and irrotational fluid flow. Assuming first the fluid to be inviscid we arrive at the Euler equations,

$$\rho \left[\frac{\partial \mathbf{u}}{\partial t} + \mathbf{u} \cdot \nabla \mathbf{u} \right] = -\nabla p + \rho \mathbf{g},$$

together with the continuity equation

$$\nabla \cdot \mathbf{u} = 0.$$

L. Nielsen · B. Gervang (✉)
Department of Engineering, Aarhus University, Aarhus, Denmark
e-mail: bge@ase.au.dk

© The Author(s) 2020

285

S. J. Sherwin et al. (eds.), *Spectral and High Order Methods for Partial Differential Equations ICOSAHOM 2018*, Lecture Notes in Computational Science and Engineering 134, https://doi.org/10.1007/978-3-030-39647-3_22

Using the vector identity $\frac{1}{2}\nabla(\mathbf{u} \cdot \mathbf{u}) = (\mathbf{u} \cdot \nabla)\mathbf{u} + \mathbf{u} \times (\nabla \times \mathbf{u})$ and using that the fluid is irrotational ($\nabla \times \mathbf{u} = 0$), we can rewrite the momentum and continuity equations,

$$\rho\frac{\partial \nabla \phi}{\partial t} = -\frac{\rho}{2}\nabla|\nabla\phi|^2 - \nabla p - \rho\mathbf{g}, \tag{1}$$

$$\nabla^2\phi = 0, \tag{2}$$

where ϕ is a vector potential defined as $\mathbf{u} = \nabla\phi$, $\phi = \phi(x, z, t)$. We can now rewrite the momentum equation as,

$$\nabla\left[\rho\frac{\partial \phi}{\partial t} + \frac{\rho}{2}|\nabla\phi|^2 + p + \rho gz\right] = 0,$$

which we can integrate in space to obtain the time dependent Bernoulli's equation.

$$\rho\frac{\partial \phi}{\partial t} + \frac{\rho}{2}|\nabla\phi|^2 + p + \rho gz = C(t),$$

where $C(t)$ is an arbitrary function of integration. We assign $C(t) = 0$ by recalling that ϕ and $\phi + \int C(t)dt$ yield exactly the same flow. Redefining ϕ and retaining the symbol $\phi := \phi + \int C(t)dt$ we obtain the time dependent Bernoulli's equation for the problem as,

$$\rho\frac{\partial \phi}{\partial t} + \frac{\rho}{2}|\nabla\phi|^2 + p + \rho gz = 0. \tag{3}$$

The governing equations for inviscid and irrotational flows for an incompressible fluid are stated through (2) and (3), where the unknowns are the velocity potential, ϕ, and the pressure, p. Equations (2) and (3) together with proper boundary conditions constitute a well-posed problem. The velocity potential, ϕ, can be solved from the Laplace equation and then substituted into the Bernoulli's equation to obtain the pressure field.

1.1 Boundary Conditions

The physical domain is shown in Fig. 1, where the notations are also illustrated. The fluid domain $\Omega \subset \mathbb{R}, d = 2$ is a bounded, connected domain with piecewise bathymetry $\Gamma^b \subset \mathbb{R}^{d-1}$. The time domain is taken as $T : t \geq 0$. The unknowns for the problem become the velocity potential and the free surface elevation $\eta(x, t) :$ $\Gamma^{FS} \times T \longrightarrow \mathbb{R}$. The pressure can hereafter be determined through (3).

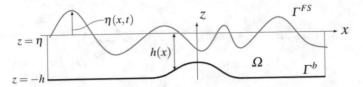

Fig. 1 Illustration of the physical domain with notation of the relevant quantities shown

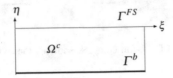

Fig. 2 Computational domain

The unsteady kinematic and dynamic free surface boundary conditions are given by Zakharov [8],

$$\partial_t \eta = -\partial_x \eta \partial_x \tilde{\phi} + \tilde{v}(1 + \partial_x \eta \partial_x \eta) \quad \in \quad \Gamma^{FS} \times T, \tag{4}$$

$$\partial_t \tilde{\phi} = -g\eta - \frac{1}{2}((\partial_x \tilde{\phi})^2 - \tilde{v}^2(1 + \partial_x \eta \partial_x \eta)) \quad \in \quad \Gamma^{FS} \times T, \tag{5}$$

where $\tilde{\ }$ signify functions defined only on the free surface. The vertical component of the velocity $\tilde{v} = \partial_z \phi|_{z=\eta}$ is calculated by solving the Laplace problem (2) together with the Zakharov boundary conditions (4) and (5) on the free surface. On the bottom we have the no penetration condition,

$$\partial_z \phi + \partial_x h \partial_x \phi = 0, \; for \; z = -h(x) \; on \; \Gamma^b. \tag{6}$$

On the inlet and outlet boundaries $(\Gamma \backslash \Gamma^{FS} \cup \Gamma^b)$ the gradient of the velocity potential is specified. The computational domain is shown in Fig. 2.

2 Discretization of Governing Equations

The developed method adopts elements from differential geometry. The unknowns of our system are described by use of *differential forms*. In a three-dimensional setting we are making use of four types of sub-manifolds: points, curves, surfaces, and volumes, both as inner and outer oriented objects, see an example in Fig. 3. The mimetic spectral element method uses an approach similar to the Galerkin method of the finite element method where the numerical residual is weighted by an arbitrary weight function. In contrast to the traditional finite element method the arbitrary weight functions are taken from the dual space of the function space used by the unknowns.

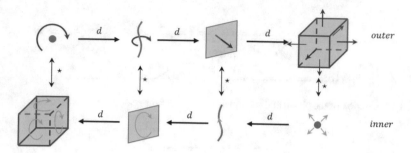

Fig. 3 Three-dimensional dual De Rahm complex showing the four types of sub-manifolds and their different orientations

2.1 Basis Functions

For the polynomial representation we use Lagrange polynomials $l_i(x)$ and edge polynomials $e_i(\xi)$, see [5]. The Lagrange polynomials are based on a Gauss-Lobatto-Legendre (GLL) point distribution for the nodal values. The Lagrange polynomials and edge polynomials satisfy the properties,

$$l_i(\xi_j) = \begin{cases} 1 & \text{if } i = j \\ 0 & \text{if } i \neq j, \end{cases} \qquad \int_{L_j} e_i(\xi) = \begin{cases} 1 & \text{if } i = j \\ 0 & \text{if } i \neq j, \end{cases}$$

and the edge polynomials are explicitly given in terms of the nodal Lagrange basis functions $l_i(x)$ as

$$e_i(\xi) = -\sum_{k=1}^{i-1} dl_k(\xi_i), \qquad (7)$$

where $dl_k(\xi)$ is the exterior derivative applied to the 0-form $l_k(\xi)$. This definition of the edge polynomial also implies, see [4] and [5],

$$dl_i = e_i - e_{i+1}. \qquad (8)$$

2.2 Mimetic Discretization in 2D

If we let the 0-form $\phi^{(0)} \in \Lambda^0(M)$ be expanded as

$$\phi_h^{(0)} = \sum_{i,j=0}^{N} \phi_{i,j} l_i(\xi) l_j(\eta), \qquad (9)$$

then we can write $\phi_h^{(0)}$ as a matrix-vector product

$$\phi_h^{(0)} = [\mathbf{L} \otimes \mathbf{L}]\, \phi = \mathbf{M}^{(0)} \cdot \phi, \tag{10}$$

where $\mathbf{L}_{i,j} = l_i(\xi_j)$ and ξ_j are the Gauss-Lobatto-Legendre points, GLL points.

If we let the 1-form $u^{(1)} \in \Lambda^1(M)$ be defined as

$$u^{(1)} = u^\xi \, d\xi + u^\eta \, d\eta, \tag{11}$$

we can expand u^ξ and u^η using edge polynomials as,

$$u_h^\xi = \sum_{i=1}^{N} \sum_{j=0}^{N} u_{i,j}^\xi e_i(\xi) l_j(\eta), \tag{12}$$

$$u_h^\eta = \sum_{i=0}^{N} \sum_{j=1}^{N} u_{i,j}^\eta l_i(\xi) e_j(\eta). \tag{13}$$

The discrete one-form $u^{(1)}$ can also be written as a matrix-vector product, where $\mathbf{u}$ is evaluated in the GLL points,

$$u_h^{(1)} = \begin{bmatrix} [\mathbf{L} \otimes \mathbf{E}] & \mathbf{0} \\ \mathbf{0} & [\mathbf{E} \otimes \mathbf{L}] \end{bmatrix} \cdot \begin{bmatrix} \mathbf{u}^\xi \\ \mathbf{u}^\eta \end{bmatrix} = \mathbf{M}^{(1)} \cdot \mathbf{u}, \tag{14}$$

where $\mathbf{E}_{i,j} = e_i(\xi_j)$.

The 2-form $P^{(2)} \in \Lambda^2(M)$ is expanded using only edge polynomials,

$$p_h^{(2)} = \sum_{i,j=1}^{N} p_{i,j} e_i(\xi) e_j(\eta) \Rightarrow [\mathbf{E} \otimes \mathbf{E}] \cdot \mathbf{p} = \mathbf{M}^{(2)} \cdot \mathbf{p}. \tag{15}$$

The Laplace equation can be reformulated using a mixed formulation, see [1], where the equilibrium equation and the constitutive relationship are separated into two equations.

$$\nabla \phi = u, \quad \nabla \cdot u = 0. \tag{16}$$

Writing (16) using differential geometry for a 3-D geometry we obtain,

$$d\phi^{(0)} = u^{(1)}, \tag{17}$$

$$dq^{(2)} = 0^{(3)}, \tag{18}$$

$$q^{(2)} = \star u^{(1)}, \tag{19}$$

where we have utilized the Hodge star operator. The Hodge star operator is a map, which maps p-forms onto $(n - p)$-forms, where n is the dimension of the domain, Ω. Given a p-form, $\lambda^{(p)}$, the hodge star maps as follows:

$$\star\lambda^{(p)}(\Omega^n) = \tilde{\lambda}^{(n-p)}(\Omega^n), \tag{20}$$

where $\tilde{\ }$ denotes the change of orientation of the new form. The Hodge star is also the coupling between the outer oriented domain and the inner oriented dual space, as seen in Fig. 3.

In 2-D, using differential geometry, equations (16) take the form,

$$d\phi^{(0)} = u^{(1)}, \quad \star u^{(1)} = \tilde{q}^{(1)}, \quad d\tilde{q}^{(1)} = \tilde{0}^{(2)}. \tag{21}$$

When the exterior derivative is applied to the balance equation of (21) we obtain, see [5]

$$d\tilde{q}_h^{(1)} = \sum_{i,j=1}^{N} (q_{i,j}^{\xi} - q_{i-1,j}^{\xi} + q_{i,j}^{\eta} - q_{i,j-1}^{\eta})e_i(\xi)e_j(\eta), \tag{22}$$

where we have utilized (8). The equilibrium equation, the first equation in (21), is equated to a zero valued 2-form. Expanding the last equation in (21) yields,

$$\sum_{i,j=1}^{N} f_{i,j}e_i(\xi)e_j(\eta) = \sum_{i,j=1}^{N} (q_{ij}^{\xi} - q_{i-1j}^{\xi} + q_{ij}^{\eta} - q_{ij-1}^{\eta})e_i(\xi)e_j(\eta), \tag{23}$$

where $f_{i,j} = 0$. The basis can then be cancelled and we can rewrite (23) as,

$$\mathbf{f} = \mathbb{E}^{(2,1)}\mathbf{q}, \tag{24}$$

where $\mathbb{E}^{(2,1)}$ is an incidence matrix, only consisting of 0, 1 and -1. This matrix relates the fluxes of $\mathbf{q}$ to the volume integral of the balance equation, see Fig. 4.

The first step in developing the discrete system is the formulation of the weak form, where we make use of duality pairing between an arbitrary k-form, $\alpha^{(k)}$, and

Fig. 4 Three-dimensional representation of surface fluxes making up the divergence of a volume integral

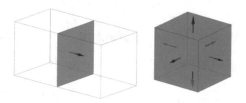

an arbitrary $(n-k)$-form, $\beta^{(n-k)}$. The duality pairing is defines as,

$$\left\langle \alpha^{(k)}, \tilde{\beta}^{(n-k)} \right\rangle_{\Omega^n} = \int_{\Omega^n} \alpha^{(k)} \wedge \tilde{\beta}^{(n-k)}. \tag{25}$$

The pairing with the $(n-k)$-form, $\beta^{(n-k)}$, takes the role of a weight function in traditional finite element analysis and lives in the dual space and carry the opposite orientation. The result of duality pairing can also be represented as a matrix-vector product,

$$\boldsymbol{\beta}^T \cdot \tilde{\mathbf{M}}^{(n-k),T} \cdot \mathbf{W} \cdot \mathbf{J} \cdot \mathbf{M}^{(k)} \cdot \boldsymbol{\alpha} = \boldsymbol{\beta}^T \cdot \mathbb{M}^{(k)} \cdot \boldsymbol{\alpha}, \tag{26}$$

where $\mathbf{W}$ contains the Gauss weights and $\mathbf{J}$ is the Jacobian matrix. $\mathbb{M}^{(k)}$ is a mass matrix of the corresponding discretized k- and $(n-k)$-form pairing, and $\tilde{(\ )}$ denotes a matrix of opposite orientation.

Using Stokes generalized theorem [3] and applying integration by parts to the balance equation (the last equation of (21)), we obtain.

$$\int_{\Omega} d\tilde{q}^{(1)} \wedge \alpha^{(0)} = \int_{\Omega} d\left(\tilde{q}^{(1)} \wedge \alpha^{(0)} \right) - \int_{\Omega} \tilde{q}^{(1)} \wedge d\alpha^{(0)} \tag{27}$$

$$= \int_{\partial\Omega} \left(\tilde{q}^{(1)} \wedge \alpha^{(0)} \right) - \int_{\Omega} \tilde{q}^{(1)} \wedge d\alpha^{(0)}. \tag{28}$$

Using duality pairing, an inner product projection for the term with the Hodge star operator, the expansions in (9)–(15), and appropriate boundary conditions we can set up the matrix system for the discrete Laplace operator as shown in (29).

$$\begin{bmatrix} \mathbf{0} & \mathbf{0} & \mathbb{E}^{(1,0),T}\tilde{\mathbb{M}}^{(1)} \\ \mathbf{0} & \mathbb{M}^{(1)} & \tilde{\mathbb{M}}^{(1)} \\ \mathbb{M}^{(1)}\mathbb{E}^{(1,0)} & \mathbb{M}^{(1)} & \mathbf{0} \end{bmatrix} \cdot \begin{bmatrix} \phi \\ \mathbf{u} \\ \tilde{\mathbf{q}} \end{bmatrix} = \begin{bmatrix} \mathbf{0} \\ \mathbf{0} \\ \mathbf{0} \end{bmatrix}. \tag{29}$$

Using the forward Euler scheme for the temporal term and pairing it with an arbitrary 0-form, $\tilde{\alpha}^{(0)}$, we can rewrite the Bernoulli's equation as,

$$\left\langle (\rho^{(2)} \wedge \frac{\phi_n^{(0)} - \phi_{n-1}^{(0)}}{\Delta t} + \frac{1}{2}\rho^{(2)} \wedge \left(i_{d\phi_{n-1}^{(0)}} d\phi_n^{(0)} \right) + p^{(2)} = -\rho^{(2)} \wedge h^{(0)} \, g)), \tilde{\alpha}^{(0)} \right\rangle_{\Omega}. \tag{30}$$

The density is considered a 2-form, which leaves (30) Hodge invariant. The interior product i is defined in [7]. The discrete version of (30) takes the form,

$$\left[\frac{\rho}{\Delta t}\mathbb{M}^{(2)} + \frac{\rho}{2}\mathbb{M}^{(2)}\mathbb{M}_i^{(1,1)} \, \mathbb{M}^{(2)} \right] \cdot \begin{bmatrix} \phi_n \\ \mathbf{P} \end{bmatrix} = -\rho \, g \, \mathbb{M}^{(2)}\mathbf{h} + \frac{\rho}{\Delta t}\mathbb{M}^{(2)}\phi_{n-1}. \tag{31}$$

$\mathbb{M}_i^{(1,1)}$ is derived from the interior product of the two 1-forms in the convective term, and contains information of ϕ from the previous time step and consequently has to be updated at each new time step.

The simulation is initialized by first solving the Laplace equation with the prescribed boundary conditions. The initial velocity potential ϕ on the free surface, is set to $\tilde{\phi}(x, t = 0) = x$, and the free surface height is set to $\eta(x, t = 0) = 0$. At the following time steps, the Zakharov free surface equations are solved to obtain new values of $\tilde{\phi}$ as well as the free surface elevation, η.

3 Numerical Results

The method is first applied to a non-temporal problem without a free surface. The geometry sketched in Fig. 5 contains a cylinder in the middle of a square. On the horizontal walls of the square and the cylinder wall the no penetration condition is applied. On the left vertical boundary a fully developed velocity profile is specified and on the right vertical boundary a constant velocity potential is defined. The velocity potential ϕ and streamlines are shown in the middle section of Fig. 5. In the right part of Fig. 5 the pressure field is shown. Figure 6 shows that we obtain spectral convergence for both unknowns.

Furthermore, the balance equation $\nabla \cdot \mathbf{u} = 0$ (conservation of mass) is satisfied both globally and point-wise independent of polynomial order as shown in Fig. 7.

Next we apply the method to a temporal and free surface problem where we have included a bump on the bottom boundary. The Zakharov free surface equations are applied on the top horizontal boundary. In Fig. 8 the pressure field and the free surface are plotted at $t = 1, 100, 200$.

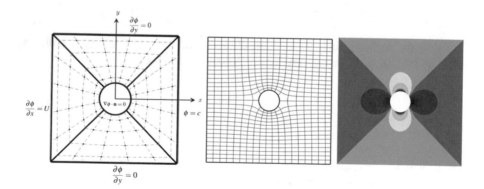

Fig. 5 Left: multi-element mesh of the cylinder problem with corresponding boundary conditions. Middle: solved velocity potential ϕ in black with corresponding streamlines in red. Right: Solved pressure field from the Bernoulli equation

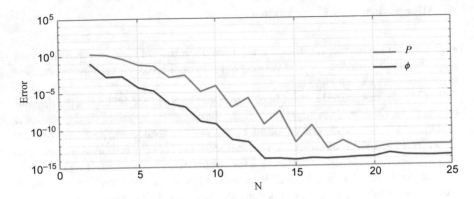

Fig. 6 The two unknowns of the system, the velocity potential, ϕ, and the pressure field, P, are shown to carry spectral convergence

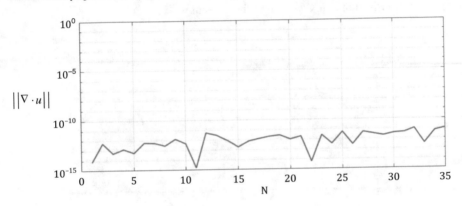

Fig. 7 The mass balance equation of (16) ($\nabla \cdot \mathbf{u} = 0$) is satisfied both globally and locally for any order of the expanding polynomial

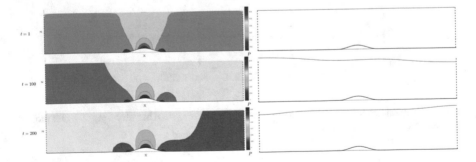

Fig. 8 Time progression of the pressure fields, P, at time steps $t = 1$, 100 and 200 are shown to the left. To the right the height of free surface wave η is shown (a scaling factor of 10 is used)

4   Discussion and Conclusion

Using an isoparametric, multi-element formulation the solution of the discretized Laplace equation shows spectral convergence. In addition, we observe that mass is conserved both globally and locally.

In (31), the discretized Bernoulli equation was kept Hodge-invariant, leaving the equation metric free. This suggests that the fundamental invariant of the equation is conserved. The Bernoulli equation conserves the total energy of the system. However, in Fig. 9 it is observed that a small amount of energy is gained and lost in a periodic manner. It is also observed that the mean energy is constant. It was possible to time integrate over very long time periods without noticing any degradation of data and we conclude that energy is conserved over long time periods even though fluctuations were observed for short time periods. In the future we plan on using a mimetic time integration scheme, which was used in [6], as well as the mimetic spatial discretization that was used in the present work.

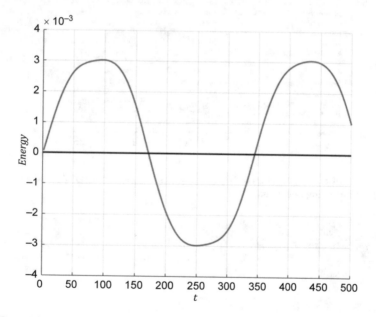

Fig. 9 Potential and kinetic energy is summed for the entire system at every time step and plotted against time

References

1. Boffi, D., Brezzi, F., Fortin, M.: Mixed Finite Element Methods and Applications. Springer, Berlin (2013). ISBN: 978-3642365188
2. Engsig-Karup, A.P., Eskilsson, C., Bigoni, D.: A stabilised nodal spectral element method for fully nonlinear water waves. J Comput. Phys. **318**, 1–21 (2016)
3. Frankel, T.: The Geometry of Physics, 3rd edn. Cambridge (2012). ISBN: 978-1107602601
4. Gerritsma, M.: Edge functions for spectral element methods. In: Spectral and High Order Methods for Partial Differential Equations, pp. 199–207 (2011)
5. Gerritsma, M.: An introduction to a compatible spectral element method. Mech. Adv. Mater. Struct. **19**(1–3), 48–67
6. Palha, A., Gerritsma, M.: Mimetic spectral element method for hamiltonian systems (2015). arXiv:1505.03422
7. Tu, L.W.: An Introduction to Manifolds, 2nd edn. Springer, Berlin (2011). ISBN: 978-1441973993
8. Zakharov, V.E.: Stability of periodic waves of finite amplitude on the surface of a deep fluid. J. Appl. Mech. Tech. Phys. **9**, 190–194 (1968)

Spectral/hp Methodology Study for iLES-SVV on an Ahmed Body

Filipe F. Buscariolo, Spencer J. Sherwin, Gustavo R. S. Assi, and Julio R. Meneghini

This work focuses on the correlation study between a computational and physical model of an Ahmed Body with slant angle of 25°, which generates a complex flow behaviour over the slant and back, with two vortices being generated from the side combined with separation on the slant. Physical results are from a wind tunnel test, performed by Strachan et al. [12] considering moving ground and Reynolds number of 1.7M, based on the length of the body.

CFD simulations were performed using the code Nektar++, which is an open source, spectral/hp element high-order solver, which methodology combine both mesh refinement (h), with higher polynomial order (p) for higher fidelity modelling. It employs an implicit type turbulence model using a Spectral Vanish Viscosity (iLES-SVV) model, which works as a filter for high frequencies. Same physical test conditions and tunnel test section were also considered, over a total time of 4 convective lengths, with same Reynolds number of 1.7 Million from reference experiments.

Considering the drag coefficient values for fully developed cases on the 5th and 6th polynomial order, the difference observed, compared with experimental results, was a maximum difference of 16%, however the simulation does not consider the upper support used in the experimental setup. Comparing the Spectral/hp element

F. F. Buscariolo (✉)
Imperial College London, NDF-USP, London, UK
e-mail: f.fabian-buscariolo16@imperial.ac.uk

S. J. Sherwin
Imperial College London, London, UK
e-mail: s.sherwin@imperial.ac.uk

G. R. S. Assi · J. R. Meneghini
NDF-USP, São Paulo, Brazil
e-mail: g.assi@usp.br; jmeneg@usp.br

© The Author(s) 2020
S. J. Sherwin et al. (eds.), *Spectral and High Order Methods for Partial Differential Equations ICOSAHOM 2018*, Lecture Notes in Computational Science and Engineering 134, https://doi.org/10.1007/978-3-030-39647-3_23

LES-SVV case from literature, the agreement with the experimental drag coefficient has been improved, reducing the gap from 45 to 16%. For the lift coefficient the maximum difference between the simulation results compared to experimental data is only 3%. There is also a good agreement between the LDA measurements on the end of the body with the results from the simulation. It is possible to observe a more intense vortex core on the simulation results, as compared to experimental data, which might well be explained by the upper support used to fix the Ahmed body in experimental test, which weakens the vortices.

The methodology shows promising results against the open literature once an appropriate validation study has been undertaken. Despite the relatively coarse resolution adopted the results are encouraging. Having identified an appropriate resolution, we will next consider other slant angles, to see how well these correlate with the experimental studies.

1 Introduction

Among all automotive bluff bodies in literature, the most studied one is the Ahmed Body. It was first proposed by Ahmed et al. [2], based on previous work from Morel [7], which was the first to study the behavior of slanted bluff bodies. The Ahmed body was designed to have shape similar to road vehicles and generate their main flow features, such as stagnation and separation points. The main dimensions of the Ahmed body are highlighted on Fig. 1.

Based on the results found by Ahmed et al. [2] on the variation of the slant inclination angle, Huminic and Huminic [4] states that three different flow configurations are found: from 0 to 12.5°, the airflow over the angled surface remains fully attached before separating from the model when it reaches the vertical surface of the back end. The flow from the angled section and the side walls produces a pair of counter rotating vortices, which continue downstream; from 12.5 to 30°, the flow over the angled section becomes highly complex. Two increased counter-rotating lateral vortices are shed from the sides of the angled section with increased size, which affects the flow over the whole back end, causing a three-dimensional wake. These vortices are also responsible for maintaining attached flow over angled surface up to an angle of 30°; from 30° and above, the flow is fully separated. There remains though a weak tendency of the flow to turn around the side edge of the model, a result of the relative separation positions of the flow over model top and that over the backlight side edges.

Due to some limitations on the wind tunnel and resources, Ahmed performed only force measurements on the bluff body during his experiments. In order to better understand the flow phenomena on an Ahmed Body, Lienhart and Becker [6] performed a study using Laser Doppler Anemometry (LDA), Hot-Wire Anemometry (HWA) and static pressure measurements in order to investigate the flow and turbulence structure around the Ahmed Body model for two slant angle conditions: 25 and 35°. The main scope was to supply a detailed data set acquired under

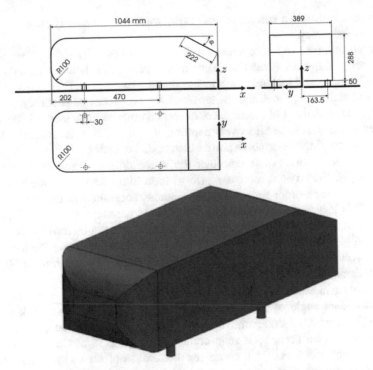

Fig. 1 Ahmed Body schematic drawing considering its main dimensions and 3D visualization

well-defined boundary conditions, similar to Ahmed first test, which considered a Reynolds number of 4.29 Million based on the length and static floor, to be used as reference data for numerical simulations.

Aiming to reproduce the real highway conditions of a vehicle, Strachan et al. [12], performed an Ahmed Body wind tunnel test with moving road conditions and both the aerodynamic forces and flow characteristics by time-averaged LDA were recorded. The flow conditions were slightly different from the ones used on Ahmed first test, by reducing the flow velocity to 25 m/s resulting in a Reynolds number of Re = 1.7 Million based on its length and the supports on the ground were replaced by a fixing system on the top of the tunnel, due to the rolling road simulation.

The Ahmed Body stands as one of the most used validation cases for CFD codes employed for automotive applications. Simulations employing a Reynolds Averaged Navier-Stokes (RANS) methodology are able to predict with good accuracy the drag coefficient, even for cases with complex flow topology, such as the slant angle of 25°, with correlation factor of around 95% compared to experimental results, however the flow physics does not agree, usually under-estimating the flow features. Attempts considering more refined methodologies such as Detached Eddy Simulations (DES) and Large Eddies Simulation (LES) provide better correlation with experiments when comparing the flow structures but aerodynamic quantities values lose accuracy.

A trend that rose to improve the confidence level of CFD simulations was the high-order or high-fidelity methods, such as the spectral/hp element method [5]. The spectral/hp elemental method combines, according to Xu et al. [13], the advantages of the spectral element method, in terms of the properties of accuracy and rapid convergence, with those of the classical h-version finite element method, that allows complex geometries to be effectively captured. It also provides an attractive higher-precision approximation to solve partial differential equations.

One of the software that employs the spectral/hp element methodology is Nektar++ [9]. Nektar++ is a cross-platform spectral/hp element framework which aims to make high-order finite element methods accessible to the broader community. This is achieved by providing a structured hierarchy of C++ components, encapsulating the complexities of these methods, which can be readily applied to a range of application areas, as stated by Cantwell et al. [3]. It allows the use of high complex solution such as implicit LES (iLES) using a Spectral Vanish Viscosity (SVV) technique to stabilize the solution.

The latest achievements in the high-fidelity turbulence models around an Ahmed Body with slant angle of 25° are summarized in the compilation work of Serre et al. [11], in which a comparative analysis of recent simulations, conducted in the framework of a French–German collaboration on LES of Complex Flows at Reynolds number of 768,000. It compares the results obtained with different eddy-resolving modelling approaches, with two LES on body-fitted curvilinear grids: LES with Smagorinsky model and wall function (LES-NWM) and Wall-resolving LES with dynamic Smagorinsky model (LES-NWR), a stabilized spectral method known as iLES-SVV, similar to the one used in this present work, which is the base of the Nektar++ code and a DES-SST approach on an unstructured grid with element number ranging from 18.5 to 40 Million. Results of the flow field shows good agreement with results measured by Lienhart and Becker [6] by a gap on the drag coefficient values of 17% for the best case and 45% for the one using iLES-SVV.

2 Objectives

The main objective of this work is to evaluate the aerodynamic behaviour in terms of the drag and lift coefficients, considering an Ahmed Body with slant angle of 25° using a spectral/hp elements method methodology as shown on Fig. 2. To achieve this, we first present a mesh study, evaluating two different size refinements referred as h-refinement for each of those, we employ three high-order surface mesh values to improve curvature representation. As the spectral/hp element method has also the possibility to improve the solution by increasing the polynomial order and consequently the number of degrees of freedom, we also evaluated three high polynomial orders for each mesh case, in a total of eighteen load cases.

Fig. 2 Representation of the Ahmed Body with slant angle of 25°

All load cases employ moving ground condition and Reynolds number of 1.7 Million, based on the length of the body. Due to the same conditions considered, results are compared with experiments performed in the study of Strachan et al. [12].

3 Spectral/hp iLES-SVV

In this work, Nektar++ is used to run an implicit LES simulation using spectral/hp method. In this method, the domain is first divided into non-overlapping elements, offering geometric flexibility and allows for local refinement. Simulations were performed using the incompressible Navier–Stokes solver employing a velocity correction scheme, combined with a Continuous Galerkin (CG) projection. More details are presented by Cantwell et al. [3].

The mathematics behind Nektar++ basically considers the numerical solution of partial differential equations (PDEs) of on a domain Ω, which may be geometrically complex, for some solution u. Practically, Ω takes the form of a d-dimensional finite element mesh consisting of elements K_i, embedded in a space of dimension dc, such that $d \le dc \le 3$, with $\Omega = u_i K_i$ is an empty set or an interface between elements of dimension dbar $< d$. The PDE problem is solved then in the weak sense, considering that u|K_i must be smooth with at least a 1st-order derivative. Therefore is required that u|K_i is in the Sobolev space $W_{1,2}(K_i)$ equivalent to H_1 (K_i), according to Adams [1]. For a continuous discretisation, we impose C^0 continuity along element interfaces.

We assume the solution can be represented as $u_\delta(x) = \sum_n \hat{u}_n \Phi_n(x)$, a weighted sum of N trial functions $\Phi_n(x)$ defined on Ω and the problem becomes that of finding the coefficients $\hat{u}_n$. The approximation u_δ does not directly give unique choices for the coefficients $\hat{u}_n$. To achieve this, a restriction is placed on the residual so that its L2 inner product, with respect to the test functions $\Psi_n(x)$, is zero. For a Galerkin projection it is chosen that the test functions are the same as the trial functions, that is $\Psi_n = \Phi_n$. As outlined previously, to construct the global basis Φ_n it is first considered the contributions from each element in the domain. Each K_i is mapped from a standard reference space K is between $[-1, 1]$ by a parametric mapping χe: K becomes K_i given by $x = \chi e(\xi)$, where K is one of the supported region shapes, and ξ are d-dimensional coordinates representing positions in a reference element, distinguishing them from x which are d-dimensional coordinates in the Cartesian coordinate space.

The next step is to construct a local polynomial basis on each reference element with which to represent solutions. For 3D regions, a tensorial basis may be used, where the polynomial space is constructed as the tensor-product of one-dimensional bases on segments, quadrilaterals or hexahedral regions.

Spectral/hp element discretisation generally lead to approximations that have low dissipation and low dispersion per degree of freedom when compared to lower-order methods. As stated by Xu et al. [13], in solving advection-diffusion equations and nonlinear partial differential equations such as advection-dominated flows, at marginal resolutions, oscillations appear that may render the computation unstable. Artificial viscosity has been used in may discretisation methods to suppress wiggles associated with high wavenumbers has been broadly and effectively used in simulations using the Fourier method. A related concept is the so-called SVV, which was originally proposed based on a second-order diffusion operator for spectral Fourier methods. SVV has been explicitly regarded as a turbulent model of implementing iLES under the assumption that the action of subgrid scales on the resolved scales is equivalent to strictly dissipative action stated by Sagaut [10], even though SVV is not explicitly designed as a subgrid-scale model. An example of a 1-D SVV kernel is:

$$Df = \begin{cases} 0, & p \leq \text{Pcut} \\ \exp\left(-\frac{(p-P^2)}{(p-Pcut)^2}\right., & p > \text{Pcut} \end{cases} \tag{1}$$

where P is the total number of modes employed and P_{cut} is the cutoff polynomial order. SVV with the kernel function Df can be regarded as a low-pass filter. We see that the SVV dissipation added to the high mode numbers with respect to the spectral element discretisation does indeed yield dissipation at the global high wave number scales of the solution.

For this work, we employed a novel CG-SVV scheme with DGKernel, proposed by Moura et al. [8] where he dissipation curves of CG of order p are match to those of DG with order $p - 2$, eliminating non-smooth dissipation characteristics arising from CG dissipation when considering high Reynolds number.

4 Simulation Methodology

We first define the coordinate system as X the streamwise direction, Y the vertical direction and Z the spamwise direction. The Ahmed Body length of 1.044 m is defined as 1 AL. The virtual wind tunnel dimensions are 2.74 × 1.66 m for the test section and total length of the domain of 4 AL, similar to Strachan et al. [12] study. The Ahmed Body model back in placed on X = 0, inlet position at X = −2 AL and outlet position at X = 2 AL. A schematic setup is shown on Fig. 3.

In terms of boundary conditions, velocity was normalized to 1 in order to match the Reynolds number previously stated and set as the inlet boundary condition. The outlet was set as pressure high-order outlet condition and the floor was also set with the same velocity of the free stream in order to reproduce the moving floor effect. The top and outer side wall and the Ahmed Body wall are set as no slip condition and a symmetry condition. Total simulated time is 7 convective lengths AL, which means that the flow is able to cross the whole domain.

This study evaluates two mesh configuration considering different h-refinements and referred as Original and Refined meshes and for each of those, three high-order surface mesh settings: 4th, 5th, and 6th order, generating six different meshes. All mesh files were generated by NekMesh, which is Nektar++ high-order mesh generator. In both Original and Refined meshes, cases two refinement zones were generated, where the first one, defined as the Ahmed Body refinement, ranges from 0.3 AL before the beginning of the geometry and 0.3 AL after the end of the body, in a total length of 1.6 AL. The second refinement, defined as the Wake Refinement region, intercepting the first refinement in 0.3 AL before the end of the body, to 1.3 AL after the end of the body, in order to fully capture the flow phenomena in the separation region, with same total length of 1.6 AL, as illustrated on Fig. 4.

The Original Mesh has total number of elements for half model around 95,000. For the Refined mesh, the boundary layer setup was the same and the dimensions

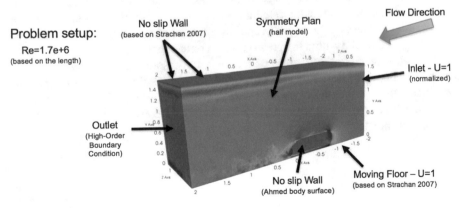

Fig. 3 Schematic representation of the boundary condition on the Ahmed Body simulation

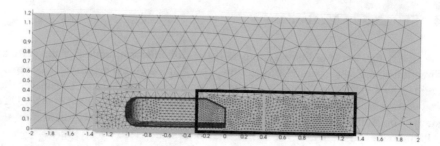

Fig. 4 Plane Z = 0 representation of mesh refinement regions. Ahmed Body Refinement region highlighted in yellow and Wake Refinement region highlighted in black

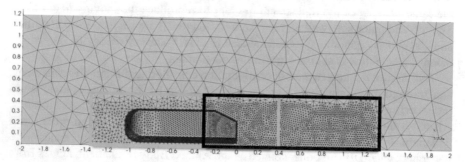

Fig. 5 Plane Z =0 mesh refinement comparison between two h-refinement cases. (**a**) Original mesh. (**b**) Refined mesh

were kept the same in terms of sizing. Refined mesh setup, giving a total of 310,00 elements. Details of both meshes are shown in Fig. 5.

Most of the commercial CFD code employ low order methods and the highest order polynomial interpolation for the solutions usually seem is 3rd. The mesh plays the major role for complex simulations such as LES, leading to elevate number of elements to reach a reliable result. To make use of the flexibility of the spectral/hp element methods, we proposed solutions considering polynomials with order higher than 3rd within the previous mesh refinement studies as the higher order polynomials increase the degrees of freedom and resolution of the mesh. For the Nektar++ implicit LES simulations using the Incompressible Navier–Stokes solver evaluated three different polynomial expansions, 4th, 5th and 6th orders, referred here as P4, P5 and P6. In summary, 18 load cases were evaluated using HPC with 432 CPUs for each case.

5 Results

5.1 Drag Coefficient Comparison Results

The drag coefficient for the 18 cases evaluated, considering 9 from the Original mesh with 95,000 elements considering fourth, fifth and sixth polynomial order and the Refined mesh case with 310,000 elements also considering fourth, fifth and sixth polynomial order expansion, with maximum RMS and compared with experimental results are shown on Fig. 6.

From Fig. 6 it is possible to observe that for the drag coefficient, P4 polynomial expansion considering both mesh cases presented mean drag results around 35% higher than the experimental results. For the P5 cases, considering again both Original and Refined mesh cases, the error was reduced to 5% however results change the trend from over-predicted to under-predicted when the mesh is refined further. The cases considering P6 polynomial expansion presented the same trend for both mesh cases, highlighting its consistency although the mean error when compared to experiments increases to 16%.

5.2 Lift Coefficient Comparison Results

Similar to the drag coefficient graph, in Fig. 7 the lift coefficient for the all evaluated cases is shown, considering Original and Refined meshes and fourth, fifth and sixth polynomial order expansion. Maximum RMS is also plotted for all cases and compared with experimental results from Strachan et al. [12].

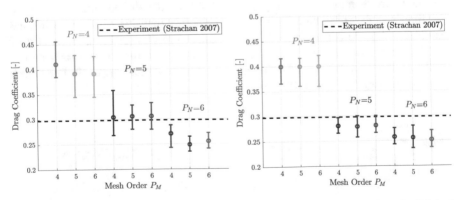

Fig. 6 Drag coefficient for the 18 evaluated test cases. On the left, average values for Original mesh, considering fourth, fifth and sixth polynomial expansions (P4, P5 and P6). On the right average, values for Refined mesh, considering fourth, fifth and sixth polynomial expansions (P4, P5 and P6)

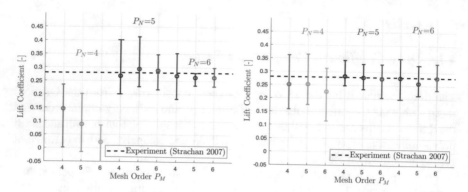

Fig. 7 Lift coefficient for the 18 evaluated test cases. On the left, average values for Original mesh, considering fourth, fifth and sixth polynomial expansions (P4, P5 and P6). On the right average, values for Refined mesh, considering fourth, fifth and sixth polynomial expansions (P4, P5 and P6)

Analyzing Fig. 7, we observe that the h-refinement from Original mesh to Refined mesh lead to results closer to experimental values when adopting P4 as the polynomial expansion basis. For both P5 and P6 polynomial expansions, lift coefficient results present good agreement with experimental data, with maximum mean error of 5%.

5.3 Flow Structure Comparison

In terms of the polynomial order expansions for the solution, combined with the 6th order surface mesh, the results present focus on the Refined mesh case, once they improved the correlation for the P4 polynomial expansion within experimental results and kept similar trend for P5 and consistent results for P6 in terms of drag and lift coefficient prediction. An initial comparison in terms of flow structures is present by the Q-Criterion of 350 coloured by pressure, comparing the Refined mesh case, considering P4, P5 and P6 polynomial expansions in Fig. 8.

Fig. 8 Iso-Surface of Q-Criterion = 350 colored by pressure on the Ahmed Body with slant angle of 25°, considering Refined mesh and 6th order surface mesh for fourth, fifth and sixth polynomial expansions (P4, P5 and P6)

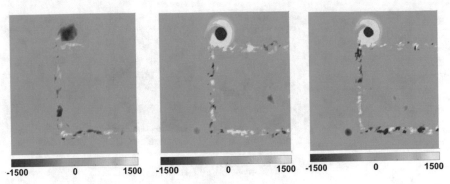

-1500   0   1500 -1500   0   1500 -1500   0   1500

Fig. 9 Contour of Lambda 2 on the plane x/L = 0 on the back of the Ahmed Body with slant angle of 25°, considering Refined mesh and 6th order surface mesh for fourth, fifth and sixth polynomial expansions (P4, P5 and P6)

From Fig. 8 it is possible to visualize that P4 is unable to define the vortex on the side of the slant, explaining also the difference in terms of both drag and lift coefficients, compared to experimental results. Results for P5 show the side vortex clearly defined and P6 is also able to capture the lower vortex, detailed on the lower image, which is not present in the studies considering the Ahmed Body, but they are important to understand the behaviour with the moving floor. Figure 9 shows a contour of Lambda 2 to illustrate the lower vortex detail on the plane x/L = 0, on the back of the Ahmed Body.

We next focus only on the Refined mesh with 6th order surface mesh for P5 and P6, once they were able to predict both lower and top vortices. Due to nature of the wind tunnel with moving ground used by Strachan et al. [12], the model had to be fixed on the top by a steam, which can be removed in the drag coefficient calculations, however it might change the flow topology over the slant, as stated by the authors themselves.

Comparing the plane x/L = 0.076 with the measurements of the flow velocity on x direction U normalized by the free stream velocity of Lienhart and Becker [6] with static floor without the steam on the upper portion with results of Strachan et al. [12], it is possible to notice intensity changes in the U normalized velocity and this is attributed by the last due to the upper support.

As the simulations do not included the upper support, but do include the moving ground, the expected results are the top portion to be similar to Lienhart and Becker [6] measurements and the lower part, correlated with measurements of Strachan et al. [12], which both P5 and P6, proved to have good agreement in terms of normalized U velocity, shown in Fig. 10.

Similar comparison is presented on Fig. 11 for the vortex intensity on the slant, on the plane x/L = 0, on the back of the Ahmed Body for vertical velocity V normalized by the free stream velocity. In this, the simulations close correlate to Lienhart and Becker [6] study, due to the absence of the steam support but for this case, the higher polynomial order expansion P6 is able to capture more scales than the P5 for the core of the main vortex, highlighting the gain of resolution of the high-order simulations.

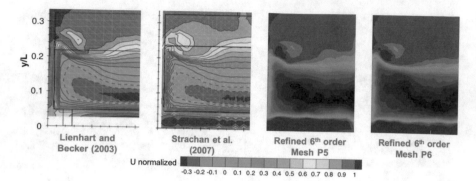

Fig. 10 Contour of U velocity normalized by free stream velocity on the plane x/L = 0.076 of the Ahmed Body with slant angle of 25°, comparing LDA measurements of Lienhart and Becker [6] with static floor without the steam (left), results of Strachan et al. [12] with moving floor and steam support (middle left), Refined mesh and 6th order surface P5 (middle right) and Refined mesh and 6th order surface P6 (right)

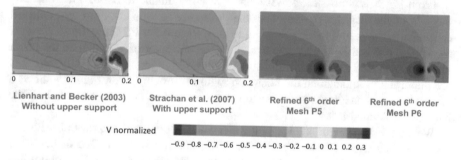

Fig. 11 Contour of V velocity normalized by free stream velocity on the plane x/L = 0 of the Ahmed Body with slant angle of 25°, comparing LDA measurements of Lienhart and Becker [6] with static floor without the steam (left), results of Strachan et al. [12] with moving floor and steam support (middle left), Refined mesh and 6th order surface P5 (middle right) and Refined mesh and 6th order surface P6 (right)

6 Conclusions

Within the advances in CFD codes, confidence level and computational power, aerodynamic simulations are applied in almost every automotive company. The reason is very simple: reduced development cost and time, which is an enormous advantage in a competitive market.

High-fidelity simulations are becoming a reality for complex industrial cases in order to improve resolution and results in a reliable response time, such as presented for the Ahmed Body on this work.

On the meshing definition study, the surface mesh order seems not to influence the results in terms of aerodynamic quantities, presenting similar trend for same polynomial order, as the Ahmed Body geometry has curved surfaces only on the front portion.

Still on the mesh definition, as the h-refinement increases from Original to Refined mesh, the drag coefficient values for P4 and P6 remains unchanged and P5 values switched from positive to negative. We conclude that consistency is shown for P4 and P6 cases but P6 presented the most reliable results, with a maximum deviation of 16%. For the lift coefficient, results for P4 improved as the h-refinement increased and kept similar values for both P5 and P6, where the best agreement was found for the case considering Refined mesh with 6th order surface mesh and P6 as the polynomial expansion.

Flow structure results focus only the Refined 6th order surface mesh, where the main expected features were captured by P5 and P6 cases. It was confirmed by those two simulation cases that the lower portion has similar behaviour of the moving ground test conducted by Strachan et al. [12], however the top portion close correlate to Lienhart and Becker [6] experiments, as the simulation cases allow the body to be fixed without the upper support used in the experiment. This fact might also explain the difference from the simulation results with the literature experiments, as the simulation allows idealized configurations.

For all simulation cases, half of the body is being simulated and a symmetry plane is set on the middle portion. From Fig. 12, which shows the normalized U velocity on a line of coordinate y/L=0.15 on the plane x/L = 0, we observe that

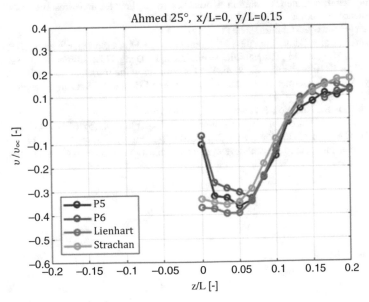

Fig. 12 Normalized U velocity distribution over a line at coordinate y/L = 0.15 on the plane x/L = 0 at the back of the Ahmed Body with slant angle of 25°, comparing LDA measurements of P5 (red), P6 (orange), Lienhart and Becker [6] (dark green) and Strachan et al. [12] (light green)

simulation has good agreement with experimental results, with a small distortion as it gets closer to the symmetry plane.

Acknowledgements We should acknowledge the HPC facilities at Imperial and also under the UK Turbulence Consortium.

References

1. Adams, R.A.: Sobolev Spaces. Academic Press, New York (1975)
2. Ahmed, S.R., Ramm, G., Faltin, G.: Some salient features of the time-averaged ground vehicle wake. Technical report, SAE Technical Paper (1984)
3. Cantwell, C.D., Moxey, D., Comerford, A., Bolis, A., Rocco, G., Mengaldo, G., De Grazia, D., Yakovlev, S., Lombard, J-E., Ekelschot, D., et al.: Nektar++: an open-source spectral/hp element framework. Comput. Phys. Commun. **192**, 205–219 (2015)
4. Huminic, A., Huminic, G.: Computational study of flow in the underbody diffuser for a simplified car model. Technical report, SAE Technical Paper (2010)
5. Karniadakis, G.E., Sherwin, S.J.: Spectral/hp Element Methods for Computational Fluid Dynamics. Oxford Science Publications, Oxford (2013)
6. Lienhart, H., Becker, S.: Flow and turbulence structure in the wake of a simplified car model. Technical report, SAE Technical Paper (2003)
7. Morel, T.: Aerodynamic drag of bluff body shapes characteristic of hatch-back cars. Technical report, SAE Technical Paper (1978)
8. Moura, R.C., Mengaldo, G., Peiro, J., Sherwin, S.J.: On the eddy-resolving capability of high-order discontinuous Galerkin approaches to implicit les/under-resolved DNS of euler turbulence. J. Comput. Phys. **330**, 615–623 (2017)
9. Nektar++. www.nektar.info
10. Sagaut, P.: Large Eddy Simulation for Incompressible Flows. Springer, Berlin (2001)
11. Serre, E., Minguez, M., Pasquetti, R., Guilmineau, E., Deng, G.B., Kornhaas, M., Schaefer, M., Froehlich, J., Hinterberger, C., Rodi, W.: On simulating the turbulent flow around the ahmed body: A French–German collaborative evaluation of LES and DES. Comput. Fluids **78**, 10–23 (2013)
12. Strachan, R.K., Knowles, K., Lawson, N.J.: The vortex structure behind an Ahmed reference model in the presence of a moving ground plane. Exp. Fluids **42**(5), 659–669 (2007)
13. Xu, H., Cantwell, C.D., Monteserin, C., Eskilsson, C., Engsig-Karup, A.P., Sherwin, S.J.: Spectral/hp element methods: recent developments, applications, and perspectives. J. Hydrodyn. **30**, 1–22 (2018)

A High-Order Discontinuous Galerkin Solver for Multiphase Flows

Juan Manzanero, Carlos Redondo, Gonzalo Rubio, Esteban Ferrer,
Eusebio Valero, Susana Gómez-Álvarez, and Ángel Rivero-Jiménez

1 Introduction

Multiphase flow is not a canonical problem, therefore different models can be found in the literature. Volume Of Fluid (VOF) model [9] is amongst the simplest. It defines a single set of momentum equations shared by all phases, whilst the volume fraction (fraction of a particular infinitesimal control volume which is occupied by each phase) is tracked throughout the domain following an advection equation. Phase-field methods [11] conserve the simplicity of VOF whilst increasing the physical meaning of the evolution equation of the fluids present in the simulation. The volume fraction is substituted by a phase-field parameter, which identifies each phase. In this work, the Cahn–Hilliard equation [4] is chosen to model the evolution of the phase-field parameter.

The introduced model is discretised in space using a high-order discontinuous Galerkin method. These methods have been gaining popularity for the discretisation of conservation laws, such as the Navier–Stokes equations [5–7, 13, 16, 22, 26]. Specifically, we use a Discontinuous Galerkin Spectral Element Method (DGSEM) [2] that allows the generation of provably stable schemes [8]. These schemes provide enhanced robustness when compared to classical high-order methods [17–20]. As far as the temporal discretisation is concerned, we use an efficient implicit-

J. Manzanero (✉) · C. Redondo · G. Rubio · E. Ferrer · E. Valero
ETSIAE-UPM – School of Aeronautics, Universidad Politécnica de Madrid, Madrid, Spain

Center for Computational Simulation, Universidad Politécnica de Madrid, Boadilla del Monte, Madrid, Spain
e-mail: juan.manzanero@upm.es

S. Gómez-Álvarez · Á. Rivero-Jiménez
Repsol Technology Lab, Móstoles, Madrid, Spain

© The Author(s) 2020
S. J. Sherwin et al. (eds.), *Spectral and High Order Methods for Partial Differential Equations ICOSAHOM 2018*, Lecture Notes in Computational Science and Engineering 134, https://doi.org/10.1007/978-3-030-39647-3_24

313

explicit approach that permits maintaining the time step restriction of a typical one phase Navier–Stokes solver. It should be noticed that similar approaches to model multiphase flows have been proposed in the past, see for example [29], where an algorithm to model N immiscible incompressible fluids with high-order methods is described. However, according to the authors knowledge, this is the first implementation using the DGSEM.

The rest of the paper is organised as follows: in Sect. 2 the governing equations of the model are described. In Sect. 3 the numerical techniques to discretise the described model are introduced. Finally, in Sect. 4 the results of two validation test cases are shown.

2 Governing Equations

In this work we model multiphase flows with a phase field approach. The flow field is modelled by means of the incompressible Navier–Stokes equations. The evolution of each of the fluids is modelled with the Cahn–Hilliard equation, which defines a phase field variable, $\phi \in [-1, 1]$, that identifies spatial coordinates occupied by fluid 1, $\phi = -1$, fluid 2, $\phi = 1$, or an interface $\phi \in (-1, 1)$. The value of the thermodynamic properties of the fluids at each spatial coordinate can be computed as:

$$\rho(\phi) = \rho_1 \left(\frac{1-\phi}{2}\right) + \rho_2 \left(\frac{1+\phi}{2}\right), \quad \eta(\phi) = \eta_1 \left(\frac{1-\phi}{2}\right) + \eta_2 \left(\frac{1+\phi}{2}\right), \quad (1)$$

where ρ_i is the density of fluid i whilst η_i is the dynamic viscosity of fluid i. The complete system is built considering first the momentum equation,

$$\frac{\partial (\rho \mathbf{v})}{\partial t} + \nabla \cdot (\rho \mathbf{v}\mathbf{v}) = -\nabla p + \frac{1}{Re} \nabla \cdot \left(\eta \left(\nabla \mathbf{v} + \nabla \mathbf{v}^T\right)\right) + \frac{3}{\sqrt{2}\varepsilon ReCa} \mu \nabla \phi + \frac{1}{Fr^2} \rho \mathbf{e}_g, \quad (2)$$

with velocity $\mathbf{v}$, static pressure p, Reynolds number $Re = \frac{\rho_1 u_0 L}{\eta_1}$ (where u_o is a reference velocity whilst L is a reference length), Capillary number $Ca = \frac{\eta_1 u_0}{\sigma}$ (where σ represents the surface tension), Froude number $Fr = \frac{u_0}{\sqrt{gL}}$, (where g is the gravity acceleration) and $\mathbf{e}_g$ is the gravity direction. Second, an artificial compressibility method [25] is used to couple the divergence-free condition,

$$\frac{\partial p}{\partial t} + \frac{\rho_0}{\rho_1} \frac{1}{M_0^2} \nabla \cdot \mathbf{v} = 0, \quad (3)$$

where $\rho_0 = \max (\rho_1, \rho_2)$ is a reference density, and M_0 is the artificial compressibility Mach number. Third, the Cahn–Hilliard equation for the phase field,

$$\frac{\partial \phi}{\partial t} + \nabla \cdot (\phi \mathbf{v}) = M \nabla^2 \mu, \quad \mu = -\phi + \phi^3 - \varepsilon^2 \nabla^2 \phi, \tag{4}$$

with M the mobility, and ε the interface width, the two free parameters of the model. In (2) and (4), μ represents the chemical potential. Moreover, this equation is designed to minimize the free-energy functional [4], $\mathcal{F}$,

$$\mathcal{F}(\phi, \nabla \phi) = \int_\Omega \left(\frac{1}{4} (1 - \phi)^2 (1 + \phi)^2 + \frac{1}{2} \varepsilon^2 |\nabla \phi|^2 \right) dx. \tag{5}$$

Note that the set of Eqs. (2)–(4) is written in non-dimensional form, where the thermodynamic variables of fluid 1 are taken as reference values, e.g.,

$$\rho(\phi) = \left(\frac{1 - \phi}{2} \right) + \frac{\rho_2}{\rho_1} \left(\frac{1 + \phi}{2} \right), \quad \eta(\phi) = \left(\frac{1 - \phi}{2} \right) + \frac{\eta_2}{\eta_1} \left(\frac{1 + \phi}{2} \right). \tag{6}$$

The set (2)–(4) can be written as an advection-diffusion system:

$$\frac{\partial u}{\partial t} + \nabla \cdot \mathbf{F}(u) = \nabla \cdot \mathbf{F}_v(u, \mathbf{g}) + S(u, \mathbf{g}), \tag{7}$$

where $u = (\phi, \rho \mathbf{v}, p)$ is the state vector, $\mathbf{g} = (\mathbf{g}_\phi, \mathbf{g}_v, \mathbf{g}_\mu) = (\nabla \phi, \nabla \mathbf{v}, \nabla \mu)$ is the gradients vector, $\mathbf{F}(u)$ and $\mathbf{F}_v(u, \mathbf{g})$ are the inviscid and viscous fluxes respectively, and $S(u, \mathbf{g})$ is a source term,

$$\mathbf{F}(u) = \begin{bmatrix} \phi \mathbf{v} \\ \rho \mathbf{v} \mathbf{v} + p \mathcal{I}_3 \\ \frac{\rho_0}{\rho_1} \frac{1}{M_0^2} \mathbf{v} \end{bmatrix}, \quad \mathbf{F}_v(u, \mathbf{g}) = \begin{bmatrix} \mathbf{g}_\mu \\ \eta \left(\mathbf{g}_v + \mathbf{g}_v^T \right) \\ 0 \end{bmatrix},$$

$$S(u, \mathbf{g}) = \begin{bmatrix} 0 \\ \frac{3}{\sqrt{2} \varepsilon \, ReCa} \mu \mathbf{g}_\phi + \frac{1}{Fr^2} \rho \mathbf{e}_g \\ 0 \end{bmatrix}. \tag{8}$$

3 Numerical Methods

The numerical implementation of (2)–(4) is performed using a high-order discontinuous Galerkin scheme for the spatial discretisation (DGSEM variant) and an implicit-explicit Euler scheme for the time discretisation.

3.1 Spatial Discretisation Using a Nodal Discontinuous Galerkin Scheme (DGSEM)

Discontinuous Galerkin (DG) schemes (see [15]) are constructed by tessellating the domain in non-overlapping elements, where the solution is approximated using polynomials of an arbitrary order, N. In this particular implementation, we use a nodal variant of the DG method, and we restrict ourselves to hexahedral elements.

In each element we approximate the solution using polynomials written in a set of local spatial coordinates $\boldsymbol{\xi} = (\xi, \eta, \zeta) \in [-1, 1]^3$, which are related to the physical space by a transfinite mapping,

$$\mathbf{x} = (x, y, z) = \mathbf{X}(\boldsymbol{\xi}) = \mathbf{X}\left(\xi, \eta, \zeta\right). \tag{9}$$

Using the local coordinates, we write the solution using tensor product Lagrange polynomials,

$$u(\mathbf{x})\big|_E \approx U(\boldsymbol{\xi}) = \sum_{i,j,k=0}^{N} U^{ijk}(t) l_i(\xi) l_j(\eta) l_k(\zeta), \tag{10}$$

where the time-dependent coefficients $U^{ijk}(t)$ are the nodal values of the solution U, and $l_j(\xi)$ are the Lagrange polynomials based on a set of Gauss points $\{\xi_j\}_{j=0}^{N}$. To handle curvilinear geometries, we use a mapping $\mathbf{X}$ that transforms local and physical spaces. With this mapping, we can construct covariant $\mathbf{a}_i$ and contravariant $\mathbf{a}^i$ basis, and their associated Jacobian J, and metrics matrix $\mathcal{M}$:

$$\mathbf{a}_i = \frac{\partial \mathbf{X}(\boldsymbol{\xi})}{\partial \xi_i}, \quad \mathbf{a}^i = \nabla \xi_i = \frac{1}{J} \mathbf{a}_j \times \mathbf{a}_k, \quad J = \mathbf{a}_i \cdot \left(\mathbf{a}_j \wedge \mathbf{a}_k\right), \quad \mathcal{M} = [J\mathbf{a}^\xi, J\mathbf{a}^\eta, J\mathbf{a}^\zeta]. \tag{11}$$

Following [14], we transform the system of Eqs. (7) to local coordinates,

$$\frac{\partial}{\partial t} \left\{ \begin{array}{c} J\phi \\ J\rho\mathbf{v} \\ Jp \end{array} \right\} + \nabla_\xi \cdot \left\{ \begin{array}{c} \mathcal{M}^T \mathbf{v}\phi \\ \mathcal{M}^T \rho\mathbf{v}\mathbf{v} + \mathcal{M}^T p\mathcal{I}_3 \\ \mathcal{M}^T \frac{1}{M_0^2}\mathbf{v} \end{array} \right\} = \nabla_\xi \cdot \left\{ \begin{array}{c} \mathcal{M}\mathcal{M}^T \mathbf{g}_\mu \\ \frac{1}{Re}\mathcal{M}^T \left(\eta \left(\mathbf{g}_\mathbf{v} + \mathbf{g}_\mathbf{v}^T\right)\right) \\ 0 \end{array} \right\}$$

$$+ J \left\{ \begin{array}{c} 0 \\ \frac{1}{Fr^2}\rho\mathbf{e}_g + \frac{3}{\sqrt{2}ReCa\varepsilon}\mu\mathbf{g}_\phi \\ 0 \end{array} \right\}, \tag{12}$$

with gradients,

$$J\mathbf{g}_\mathbf{v} = \mathcal{M}\nabla_\xi \mathbf{v}, \quad J\mathbf{g}_\phi = \mathcal{M}\nabla_\xi \phi, \quad J\mathbf{g}_\mu = \mathcal{M}\nabla_\xi \mu, \tag{13}$$

and the chemical potential definition,

$$J\mu = -J\phi + J\phi^3 - \varepsilon^2 \nabla_\xi \cdot \left(M^T \mathbf{g}_\phi \right).$$ (14)

We obtain the DG scheme replacing the continuous solution by their polynomial counterpart (10), then multiplying (12), written in compact form (7), by a polynomial test function (with same order N as the solution) ϑ, and we integrate the result in one element $E = [-1, 1]^3$,

$$\int_E J\vartheta \frac{\partial U}{\partial t} + \int_E \vartheta \nabla_\xi \cdot \mathbf{F}(U) = \int_E \vartheta \nabla_\xi \cdot \mathbf{F}_v(U, \mathbf{G}) + \int_E J\vartheta S(U, \mathbf{G}).$$ (15)

Next, we integrate by parts the terms containing divergences, which yields surface integrals. Since the solution is discontinuous at the inter-element faces, we replace the surface flux by a *numerical flux*, $\mathbf{F}^\star$,

$$\int_E J\vartheta \frac{\partial U}{\partial t} + \int_{\partial E} \vartheta \mathbf{F}^\star \cdot \hat{n} dS - \int_E \nabla_\xi \vartheta \cdot \mathbf{F} = \int_{\partial E} \vartheta \mathbf{F}_v^\star \cdot \hat{n} dS$$

$$- \int_E \nabla_\xi \vartheta \cdot \mathbf{F}_v + \int_E J\vartheta S(U, \mathbf{G}),$$ (16)

where ∂E represents the six surfaces of the element E. For the inviscid numerical flux $\mathbf{F}^\star$, we use the exact Riemann solver derived in [1], whilst for the viscous numerical flux we use the Symmetric Interior Penalty (SIP) method [27], with the penalty parameter value derived in [24] and recently discussed for the DGSEM in [21]. In (16), $\hat{n}$ is the surface outward normal vector in local coordinates. To obtain the evolution equations for each nodal degree of freedom U^{ijk}, we let $\vartheta = l_i(\xi)l_j(\eta)l_k(\zeta)$, and compute the integrals using the Gauss quadrature points (and weights $\{w_i\}$) associated to the interpolation points (which provide an accuracy of $2N + 1$),

$$J^{ijk} \frac{dU^{ijk}}{dt} + \frac{F_x^\star}{w_i}(\xi, \eta_j, \zeta_k)l_i(\xi)\Big|_{\xi=-1}^{\xi=1} + \frac{F_y^\star}{w_j}(\xi_i, \eta, \zeta_k)l_j(\eta)\Big|_{\eta=-1}^{\eta=1}$$

$$+ \frac{F_z^\star}{w_k}(\xi_i, \eta_j, \zeta)l_k(\zeta)\Big|_{\zeta=-1}^{\zeta=1}$$

$$- \sum_{m=0}^{N} \left(\frac{w_m}{w_i} D_{mi} F_x^{mjk} + \frac{w_m}{w_j} D_{mj} F_y^{imk} + \frac{w_m}{w_k} D_{mk} F_z^{ijm} \right) =$$

$$\frac{F_{v,x}^{\star ijk}}{w_i}(\delta_{iN} - \delta_{i0}) + \frac{F_{v,y}^{\star ijk}}{w_j}(\delta_{jN} - \delta_{j0}) + \frac{F_{v,z}^{\star ijk}}{w_k}(\delta_{kN} - \delta_{k0})$$

$$- \sum_{m=0}^{N} \left(\frac{w_m}{w_i} D_{mi} F_{v,x}^{mjk} + \frac{w_m}{w_j} D_{mj} F_{v,y}^{imk} + \frac{w_m}{w_k} D_{mk} F_{v,z}^{ijm} \right) + J^{ijk} S^{ijk},$$ (17)

where $F^{ijk} = F(U^{ijk})$ and $F_v^{ijk} = F_v(U^{ijk}, \mathbf{G}^{ijk})$, being $\mathbf{G}^{ijk}$ the nodal values of the gradient $\mathbf{G}$. The symbol δ_{ik} represents the Kronecker delta. The derivation matrix D_{ij} is defined as $D_{ij} = l_j'(\xi_i)$. To compute the gradient $\mathbf{G}$, we perform the weak formulation of (13),

$$\int_E J\boldsymbol{\tau} \cdot \mathbf{G} = \int_{\partial E} U^\star M^T \cdot \boldsymbol{\tau} dS - \int_E U \nabla_\xi \cdot \left(M^T \cdot \boldsymbol{\tau}\right), \tag{18}$$

where $\boldsymbol{\tau}$ is an arbitrary vector test function (from the order N polynomials space). Since we use the SIP method, we use solution averages to couple inter-element fluxes, $U^\star = \{\!\{U\}\!\}$. All the integrals involved in (18) are computed discretely similar to those in (16), i.e.,

$$J^{ijk} \tau_{ijk}^d G_d^{ijk} = \left. \frac{U^\star(\xi, \eta_j, \zeta_k) Ja_d^\xi(\xi, \eta_j, \zeta_k)}{w_i} l_i(\xi) \right|_{\xi=-1}^{\xi=1}$$

$$+ \left. \frac{U^\star(\xi_i, \eta, \zeta_k) Ja_d^\eta(\xi_i, \eta, \zeta_k)}{w_j} l_j(\eta) \right|_{\eta=-1}^{\eta=1} + \left. \frac{U^\star(\xi_i, \eta_j, \zeta) Ja_d^\zeta(\xi_i, \eta_j, \zeta)}{w_k} l_k(\zeta) \right|_{\zeta=-1}^{\zeta=1}$$

$$- \sum_{m=0}^N \left(\frac{w_m}{w_i} Ja_d^{\xi,ijk} D_{mi} U^{mjk} + \frac{w_m}{w_j} Ja_d^{\eta,ijk} D_{mj} U^{imk} + \frac{w_m}{w_k} Ja_d^{\zeta,ijk} D_{mk} U^{ijm} \right). \tag{19}$$

The gradient nodal values G_d^{ijk} are introduced in the viscous fluxes $F_v(U^{ijk}, \mathbf{G}^{ijk})$ of (17) hence completing the discretisation of (16). Note that one needs to compute $\mathbf{g}_\phi$ before computing μ and its gradient $\mathbf{g}_\mu$.

3.2 Time Integration Using IMplicit–EXplicit (IMEX) and Runge–Kutta Schemes

The time integration of (17) is performed with a combination of forward and backwards Euler and explicit Runge–Kutta schemes. On the one hand, the Navier–Stokes equations are integrated by means of a third order explicit Runge–Kutta (RK3) scheme [28]. On the other hand, the Cahn–Hilliard equation is integrated with a combination of explicit RK3 for the phase field advection, forward Euler for the chemical free-energy, and backwards Euler for the interfacial energy,

$$\frac{\phi^{n+1} - \phi^n}{\Delta t} + \nabla \cdot \left(\mathbf{v}\phi\right)^{RK3} = \nabla^2 \left(-\phi^n + \left(\phi^n\right)^3 - \varepsilon^2 \nabla^2 \phi^{n+1}\right). \tag{20}$$

The reason behind this choice, is that the numerical stiffness of the bi-Laplacian ($\nabla^4\phi$) operator prevents from using an explicit method, as restricts the time-step Δt

to unpractical values. We only treat implicitly the interfacial energy since it yields a constant Jacobian matrix, represented by J^{∇_2}. In particular, the linear system to solve is,

$$\left[J^{\nabla_2} + \frac{I}{\Delta t} \right] \phi^{n+1} = \frac{\phi^n}{\Delta t} - \nabla \cdot (\mathbf{v}\phi)^{RK3} + \nabla^2 \left(-\phi^n + (\phi^n)^3 \right). \tag{21}$$

The Jacobian matrix is computed numerically (see [3]) and a LU factorisation is performed only at the first time step. In each following iteration, the RHS of (21) is computed and the linear system is solved by means of forward and backward substitutions. Both the LU factorisation and the forward and backward substitutions are performed with the library MKL-PARDISO [23].

4 Validation

The proposed methodology is tested with two test cases. First, the validity of the discontinuous Galerkin discretisation of the Cahn–Hilliard equation is tested with a benchmark spinodal decomposition problem [12]. Second, the validity of the coupled Cahn–Hilliard/Navier–Stokes system is tested with a two dimensional rising bubble test [10].

4.1 Spinodal Decomposition

This test problem considers an initial mixture of two fluids. These fluids are immiscible, therefore they tend to separate to minimise their free energy (5). As stated before, the geometry, initial condition and fluid parameters are taken from [12]. In particular, the initial condition for this benchmark problem is:

$$\phi(x, y) = -0.05 \Big[\cos{(0.105x)} \cos{(0.11y)} + \big[\cos{(0.13x)} \cos{(0.087y)} \big]^2$$
$$+ \cos{(0.025x - 0.15y)} \cos{(0.07x - 0.02y)} \Big]. \tag{22}$$

The physical domain is a "T" shape with a total height of 120 units, a total width of 100 units, and horizontal and vertical section widths of 20 units (Fig. 1). No-flux boundary conditions are applied at the boundaries. Following [12] mobility is set to $M = 10$, whilst the interface width is set to $\varepsilon = 3.16$. The physical domain is discretised with an unstructured mesh of 326 elements and a polynomial order of $N = 4$. For the time discretization, we use a time step $\Delta t = 10^{-3}$.

Figure 1 shows qualitatively how the different phases separate, whilst Fig. 2 shows quantitatively the evolution of the total free energy with time. In Fig. 2

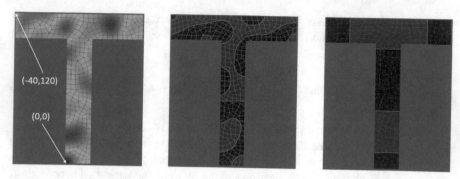

Fig. 1 "T" domain for the spinodal decomposition. Initial condition (left figure) and evolution with time (the right figure is the steady-state solution)

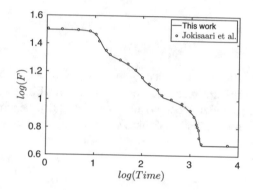

Fig. 2 Evolution of total free energy (5) with time

the results of this work are compared with those obtained in [12], validating the proposed method.

4.2 Rising Bubble

This test case considers a bubble of light fluid submerged in a heavy fluid, both subjected to a gravitational field. Following [10] the initial configuration, see Fig. 3, consists of a bubble of radius $r = 0.25$ centred at $[0.5, 0.5]$ in a $[1 \times 2]$ domain. A no-slip boundary condition is used at the top and the bottom of the domain whilst a free slip condition is enforced at the vertical walls. Following [10], the Reynolds number is set to $Re = 35$ whilst σ and ε are set to 24.5 and 0.03125 respectively (this gives a Eötvös number $Eo = 10$) whilst both density and viscosity ratios are set to $\rho_1/\rho_2 = \mu_1/\mu_2 = 10$. The gravitational acceleration is $g = 0.98$. The problem is discretised with 16×32 elements with a polynomial order of $N = 4$, and a time step $\Delta t = 4 \cdot 10^{-6}$.

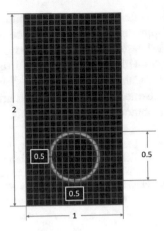

Fig. 3 Initial condition of the rising bubble test problem

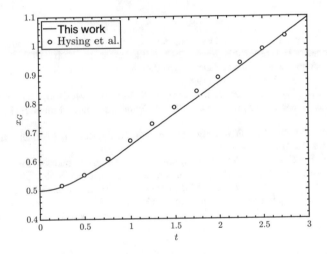

Fig. 4 Evolution of the center of mass of the bubble with time

This test case is quantitively compared with the results of [10] in Fig. 4 with satisfactory results. It should be mentioned that the benchmark results of [10] are obtained with a sharp-interface model which may explain the small disagreement in the evolution of the center of mass shown in Fig. 4.

5 Conclusions

A method to model incompressible two phases flows is introduced. The model solves the incompressible Navier–Stokes equations coupled with the Cahn–Hilliard equation to track the evolution of the different fluids. The model is discretised in space using a discontinuous Galerkin spectral element method (DGSEM) whilst an efficient implicit-explicit approach is used to advance in time. The validity of the model is shown with two test cases. A spinodal decomposition benchmark problem is solved to validate the Cahn–Hilliard solver whilst a rising-bubble test problem is solved to validate the coupled Cahn–Hilliard–Navier–Stokes system. Both test cases are solved showing good agreement with the literature, and proving the accuracy and robustness of the proposed method.

Acknowledgements This work has been partially supported by REPSOL under the research grant P180021090. This work has been partially supported by Ministerio de Economía y Competitividad under the research grant TRA2015-67679-C2-2-R and under the research grant EUIN2017-88294 (Gonzalo Rubio). The authors acknowledge the computer resources and technical assistance provided by the Centro de Supercomputación y Visualización de Madrid (CeSViMa).

References

1. Bassi, F., Massa, F., Botti, L., Colombo, A.: Artificial compressibility Godunov fluxes for variable density incompressible flows. Comput. Fluids **169**, 186–200 (2018)
2. Black, K.: A conservative spectral element method for the approximation of compressible fluid flow. Kybernetika **35**(1), 133–146 (1999)
3. Browne, O.M., Rubio, G., Ferrer, E., Valero, E.: Sensitivity analysis to unsteady perturbations of complex flows: a discrete approach. Int. J. Numer. Methods Fluids **76**(12), 1088–1110 (2014)
4. Cahn, J.W., Hilliard, J.E.: Free energy of a nonuniform system. I. Interfacial free energy. J. Chem. Phys. **28**(2), 258–267 (1958)
5. Cockburn, B., Shu, C.W.: The local discontinuous Galerkin method for time-dependent convection-diffusion systems. SIAM J. Numer. Anal. **35**(6), 2440–2463 (1998)
6. Ferrer, E.: An interior penalty stabilised incompressible discontinuous Galerkin–Fourier solver for implicit large eddy simulations. J. Comput. Phys. **348**, 754–775 (2017)
7. Fraysse, F., Redondo, C., Rubio, G., Valero, E.: Upwind methods for the Baer–Nunziato equations and higher-order reconstruction using artificial viscosity. J. Comput. Phys. **326**, 805–827 (2016)
8. Gassner, G.J., Winters, A.R., Kopriva, D.A.: Split form nodal discontinuous Galerkin schemes with summation-by-parts property for the compressible Euler equations. J. Comput. Phys. **327**, 39–66 (2016)
9. Hirt, C.W., Nichols, B.D.: Volume of fluid (VOF) method for the dynamics of free boundaries. J. Comput. Phys. **39**(1), 201–225 (1981)
10. Hysing, S., Turek, S., Kuzmin, D., Parolini, N., Burman, E., Ganesan, S., Tobiska, L.: Quantitative benchmark computations of two-dimensional bubble dynamics. Int. J. Numer. Methods Fluids **60**(11), 1259–1288
11. Jacqmin, D.: Calculation of two-phase Navier–Stokes flows using phase-field modeling. J. Comput. Phys. **155**(1), 96–127 (1999)
12. Jokisaari, A., Voorhees, P., Guyer, J., Warren, J., Heinonen, O.: Benchmark problems for numerical implementations of phase field models. Comput. Mater. Sci. **126**, 139–151 (2017)
13. Kompenhans, M., Rubio, G., Ferrer, E., Valero, E.: Adaptation strategies for high order discontinuous Galerkin methods based on Tau-estimation. J. Comput. Phys. **306**, 216–236 (2016)
14. Kopriva, D.A.: Metric identities and the discontinuous spectral element method on curvilinear meshes. J. Sci. Comput. **26**(3), 301 (2006)

15. Kopriva, D.A.: Implementing Spectral Methods for Partial Differential Equations: Algorithms for Scientists and Engineers. Springer, Berlin (2009)
16. Manzanero, J., Ferrer, E., Rubio, G., Valero, E.: Design of a Smagorinsky Spectral Vanishing Viscosity turbulence model for discontinuous Galerkin methods. Comput. Fluids, 104440 (2020)
17. Manzanero, J., Rubio, G., Ferrer, E., Valero, E., Kopriva, D.A.: Insights on aliasing driven instabilities for advection equations with application to Gauss–Lobatto discontinuous Galerkin methods. J. Sci. Comput. **75**(3), 1262–1281 (2018)
18. Manzanero, J., Rubio, G., Kopriva, D.A., Ferrer, E., Valero, E.: Entropy-stable discontinuous Galerkin approximation with summation-by-parts property for the incompressible Navier-Stokes/Cahn-Hilliard system (2019). arXiv:1910.11252.
19. Manzanero, J., Rubio, G., Kopriva, D.A., Ferrer, E., Valero, E.: A free–energy stable nodal discontinuous Galerkin approximation with summation–by–parts property for the Cahn–Hilliard equation. J. Comput. Phys. **403**, 109072 (2020)
20. Manzanero, J., Rubio, G., Kopriva, D.A., Ferrer, E., Valero, E.: An entropy–stable discontinuous Galerkin approximation for the incompressible Navier–Stokes equations with variable density and artificial compressibility. J. Comput. Phys. **408**, 109241 (2020)
21. Manzanero, J., Rueda-Ramírez, A.M., Rubio, G., Ferrer, E.: The Bassi Rebay 1 scheme is a special case of the symmetric interior penalty formulation for discontinuous Galerkin discretisations with Gauss–Lobatto points. J. Comput. Phys. **363**, 1–10 (2018)
22. Rueda-Ramírez, A.M., Manzanero, J., Ferrer, E., Rubio, G., Valero, E.: A p-multigrid strategy with anisotropic p-adaptation based on truncation errors for high-order discontinuous Galerkin methods. J. Comput. Phys. **378**, 209–233 (2019)
23. Schenk, O., Gärtner, K.: Solving unsymmetric sparse systems of linear equations with PARDISO. Futur. Gener. Comput. Syst. **20**(3), 475–487 (2004)
24. Shahbazi, K.: Short note: an explicit expression for the penalty parameter of the interior penalty method. J. Comput. Phys. **205**(2), 401–407 (2005)
25. Shen, J.: Pseudo-compressibility methods for the unsteady incompressible Navier–Stokes equations. In: Proceedings of the 1994 Beijing Symposium on Nonlinear Evolution Equations and Infinite Dynamical Systems, pp. 68–78 (1997)
26. Wang, Z.J., Fidkowski, K., Abgrall, R., Bassi, F., Caraeni, D., Cary, A., Deconinck, H., Hartmann, R., Hillewaert, K., Huynh, H.T., et al.: High-order CFD methods: current status and perspective. Int. J. Numer. Methods Fluids **72**(8), 811–845 (2013)
27. Wheeler, M.F.: An elliptic collocation-finite element method with interior penalties. SIAM J. Numer. Anal. **15**(1), 152–161 (1978)
28. Williamson, J.: Low-storage Runge–Kutta schemes. J. Comput. Phys. **35**(1), 48–56 (1980)
29. Yang, Z., Dong, S.: Multiphase flows of N immiscible incompressible fluids: an outflow/open boundary condition and algorithm. J. Comput. Phys. **366**, 33–70 (2018)

High-Order Propagation of Jet Noise on a Tetrahedral Mesh Using Large Eddy Simulation Sources

M. A. Moratilla-Vega, V. Saini, H. Xia, and G. J. Page

1 Introduction

Due to the rapid expansion of the commercial aviation industry, authorities have been tightening the legislation for aircraft noise. For instance, the European Commission has set a 65% reduction goal of overall aircraft noise from the year 2000 to 2050 [1]. The noise generated by the jet exhaust is one of the main contributors to the overall aircraft noise, especially during take-off [2]. Moreover, in new generation ultra-high by-pass ratio turbofan engines the increased interaction between the engine jet and the high-lift devices can potentially affect the noise field [3]. Thus, our overall aim is to develop and investigate an accurate and efficient method for the prediction of far-field jet noise in installed jet configurations.

Rapid growth in computing power during the last decades has enabled the use of scale resolving numerical simulations for jet noise research at a reduced cost than most experimental campaigns. Conventionally, 2nd-order numerical schemes combined with surface integral techniques, particularly the Ffowcs Williams-Hawkings (FW-H) method [4] have been widely adopted for predicting the far-field noise, due to its simplicity and low cost. However, defining the envelope surface used in the FW-H method is not always trivial in complex configurations [5], for example, installed jets on aircraft wings. Also, the results may be overly sensitive to the size, shape and location of these surfaces. Now, directly resolving the Navier-Stokes (NS) equations for sufficiently accurate far-field jet noise results is prohibitively expensive [6]. LES using finite volume 2nd-order accurate schemes has proven to be reliable and robust for solving jets' near field, but large numerical dispersion and dissipation error makes them less suitable for the propagation of the

M. A. Moratilla-Vega · V. Saini (✉) · H. Xia · G. J. Page
Rolls-Royce University Technology Centre, Loughborough University, Loughborough, UK
e-mail: m.a.moratilla-vega@lboro.ac.uk; v.saini@lboro.ac.uk; g.j.page@lboro.ac.uk

© The Author(s) 2020
S. J. Sherwin et al. (eds.), *Spectral and High Order Methods for Partial Differential Equations ICOSAHOM 2018*, Lecture Notes in Computational Science and Engineering 134, https://doi.org/10.1007/978-3-030-39647-3_25

sound waves to the far field. High-order methods provide more accurate propagation due to their reduced numerical error but are insufficiently robust for simulating complex jet flows. Therefore, we have used a coupled approach in which a finite volume LES solver is used to obtain the acoustic sources, which are then transferred to a high-order acoustic solver that propagates noise to the far-field.

The spectral/*hp* DG method [7] is capable of providing high-order accuracy and handling mixed mesh elements types such as tetrahedra and hexahedra, thus providing a potential solution to geometrically complex acoustic problems. The solver based on this approach is *AcousticSolver* of the Nektar++ framework [8, 9]. The LES code HYDRA and acoustic code *AcousticSolver* have been coupled and validated using hexahedral elements [10, 11]. A similar coupling strategy has been used previously for jet noise [12] and combustion noise on tetrahedral grids [9].

In this paper, our focus is on two aspects: (1) estimates of mesh design for the high-order solver using a canonical two-dimensional (2D) case and (2) comparison of three-dimensional (3D) turbulent isolated jet-noise results on a tetrahedral grid and a comparable hexahedral grid using the coupling approach. From the perspective of our near future work, the tetrahedral grid results provide motivation and parameters for the set-up of the coupled methodology for jet-flap interactions.

2 Numerical Methods and Solvers

In this section, the details of the high-order spectral/*hp* DG solver employed to solve the APE equations are provided followed by a brief description of the LES code that solves the filtered compressible NS equations. Finally, the coupling of the two is briefly mentioned.

2.1 APE Solver

Equations for Propagation The acoustic perturbation equations (APE) solved here are the ones proposed by Ewert and Schröder [6] in the APE-4 form. These equations describe the transport of acoustic fluctuations in a linearized form, where the source terms can be non-linear, and can be written as:

$$\partial_t p' + \bar{c}^2 \nabla \cdot \left(\bar{\rho} \mathbf{u}' + \bar{\mathbf{u}} \frac{p'}{\bar{c}^2} \right) = \bar{c}^2 q_c, \tag{1}$$

$$\partial_t \mathbf{u}' + \nabla \left(\bar{\mathbf{u}} \cdot \mathbf{u}' \right) + \nabla \left(\frac{p'}{\bar{\rho}} \right) = \mathbf{q}_m, \tag{2}$$

where p', $\mathbf{u}$ are the acoustic pressure and acoustic velocity vector respectively and c is the speed of sound. The time-averaged quantities are denoted by the over-bar and acoustic fluctuations are primed. The left-hand side of (1) and (2) represents the advection of waves in the mean flow. The right-hand side describes different sources that may be present in a generic aeroacoustic problem.

Finally, the source terms, q_c and $\mathbf{q}_m$ are defined as:

$$q_c = -\nabla \cdot \left(\rho' \mathbf{u'} \right)' + \frac{\overline{\rho}}{c_p} \frac{\overline{D}s'}{Dt}, \tag{3}$$

$$\mathbf{q}_m = -\left(\boldsymbol{\omega} \times \mathbf{u} \right)' + T'\nabla \overline{s} - s'\nabla \overline{T} - \left(\frac{\nabla \left(\mathbf{u'} \right)^2}{2} \right)' + \left(\frac{\nabla \cdot \underline{\tau}}{\rho} \right)'. \tag{4}$$

These terms are classified into four categories:

1. the non-linear terms: $-\nabla \cdot \left(\rho' \mathbf{u'} \right)'$ and $-(\nabla \left(\mathbf{u'} \right)^2 /2)'$,
2. the heat/entropy terms: $(\overline{\rho}/c_p) \cdot (\overline{D}s'/Dt)$ and $T'\nabla \overline{s} - s'\nabla \overline{T}$,
3. the viscous term: $(\nabla \cdot \underline{\tau}/\rho)'$ and
4. the vortical term, known as the Lamb vector, $\mathbf{L'} = -(\boldsymbol{\omega} \times \mathbf{u})'$.

In this paper, only the Lamb vector $\mathbf{L'}$ is considered as a source term because it is the dominant contributor for isothermal applications with strong vortical motions (shear layers and wakes), as demonstrated in [12, 13].

Numerical Solver The solver used for the above APE equations is called *AcousticSolver*, which is part of the open-source Nektar++ framework [8]. The solver employs a high-order, spectral/hp element method with a DG formulation [7]. In short (for details see [9]), the present DG method works as follows:

1. The computational domain is divided into non-overlapping elements.
2. The governing equations are discretised in each element by a weighted sum of basis functions where the coefficients of the expansion are the unknowns. In case of tetrahedral elements, the basis functions are modified hierarchical Jacobi basis [8].
3. The discretised equation is then multiplied by a test function (same as the basis function) followed by integration over each element in order to obtain the variational form of the governing equations.
4. The flux terms in the variational equation are responsible for communicating the information across the elements. The interface fluxes are calculated using the immediate left- and right-side values with a Riemann solver.

The scheme used here to solve the Riemann problem is a local Lax-Friederichs scheme as defined in [9]. The temporal discretisation is performed using a 4th-order Runge-Kutta scheme. A numerical sponge layer [14] is set up using source terms to dampen out the outgoing acoustic waves smoothly, thus minimising reflections from the boundaries of the domain.

2.2 LES Solver

The LES is performed using the in-house code of Rolls-Royce plc., HYDRA [15] that solves Favre-filtered unsteady compressible Navier-Stokes equations [10]. It is a density-based, spatially 2nd-order accurate finite volume cell-vertex code used for propulsion and turbomachinery applications. More details on the set-up of the spatial scheme used can be found in [10]. For the temporal discretisation, a 2nd-order, four-stage Runge-Kutta explicit algorithm is employed. The size of the time step is chosen to keep the Courant number less than unity. The code is capable of solving arbitrary mesh topologies which is beneficial for complex geometries. The sub-grid scale model is chosen as σ-model [16] with model constant $C_\sigma = 1.35$ [17].

2.3 Coupling of Solvers

The 3D data from LES mesh is transferred and interpolated onto the APE mesh in real time. The interpolation is necessary because two solvers have different meshes designed specifically to capture flow and acoustics. The transfer-interpolation process takes place in parallel. This is achieved using an MPI based coupling strategy with the open-source library CWIPI [18]. More details on the coupling mechanism are provided in Lackhove et al. [9] and Moratilla-Vega et al. [10]. Note that larger time steps can be used for *AcousticSolver* since it is not restricted to resolve the small flow structures.

3 Test Cases

Two cases are presented here. First, a canonical noise propagation case due to a well-defined vortex-pair source run on *AcousticSolver* alone. A study of numerical error by changing the mesh and polynomial expansion order (P) is performed. The second case uses the mesh parameters from the first to propagate noise generated by an isolated jet in an LES simulation. This case provides validation of the coupling for a 3D turbulent jet noise case on a fully tetrahedral mesh. The results are then compared to the ones obtained with the FW-H technique.

3.1 Spinning Vortex Pair

The case is an acoustic wave propagation problem in two dimensions where the source is mathematically well-defined, as in the original work on APE [6]. The case is run with standalone *AcousticSolver*. The source is in the form of two-point vortices at a distance of r_0 from the origin, rotating with a circulation Γ. An analytical solution of the induced acoustic field was found by Müller and Obermeier [19] as:

$$\tilde{p}' = \frac{\rho_\infty \Gamma^4}{64\pi^3 r_0^4 c_\infty^2} H_2^{(2)}(kr),$$ (5)

where, $H_2^{(2)}$ is the Hankel function of 2nd-order and second kind, the rotation period is defined as $T = 8\pi^2 r_0^2 / \Gamma$; the angular velocity as $\omega = \Gamma/4\pi r_0^2$ and the Mach number as $M_r = \Gamma/4\pi r_0 c_\infty$. The real part of Eq. (5) gives the pressure fluctuations. Ewert and Schröder [6] found the source-term based on the Lamb vector that represents the acoustic field for this case as:

$$\boldsymbol{q}_m = -\frac{\Gamma^2 \boldsymbol{e}_r(t)}{8\pi^2 \sigma^2 r_0} \sum_{i=1}^{2} (-1)^i \exp\left(-\frac{|\boldsymbol{r} + (-1)^i \boldsymbol{r}_0(5)|^2}{2\sigma^2}\right), \quad \sigma \approx r_0,$$ (6)

where, $\boldsymbol{r} = (x, y)^T$, $\boldsymbol{r}_0 = r_0 \boldsymbol{e}_r$, $\boldsymbol{e}_r = (\cos\theta, \sin\theta)^T$ and $\theta = \omega t$.

The computational domain considered is circular and extends to $250 r_0$. The source parameters are set as in [6] i.e. $\Gamma/(c_\infty r_0) = 1.6$ and $M_r = 0.1273$. Simulations are run until the pressure fluctuations reach $r = 200 r_0$ in order to minimise the boundary effects. All the elemental meshes consist of triangles and a modified hierarchical Jacobi polynomial basis [8].

First, a reference simulation with polynomial order 4 (P4) is run on a fine uniform mesh. The resulting acoustic pressure field is plotted along a diagonal line in Fig. 1a. The result from this simulation matches its analytical counterpart well. Further comparisons are made with respect to this well-resolved P4 numerical

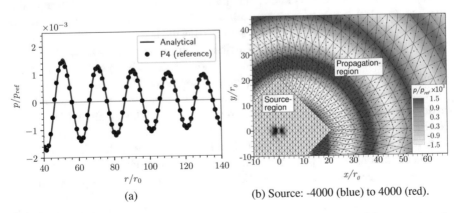

(a) (b) Source: -4000 (blue) to 4000 (red).

Fig. 1 (a) Pressure field along $x = y$ line, (b) solution points, and contours of the source term in the source region and the pressure field in the propagation region

Table 1 Details of the test cases run

Simulation	Poly. order	$N_r \times N_\theta$	ppw at $\frac{r}{r_0} = 150$	CPU cost
P4 (reference)	4	260×152	90	6.53
P1 1× coarse	1	64×76	5.20	1.00
P2 2× coarse	2	32×76	5.15	0.82
P4 4× coarse	4	15×76	5.00	0.68

N_r is the elements in the radial direction

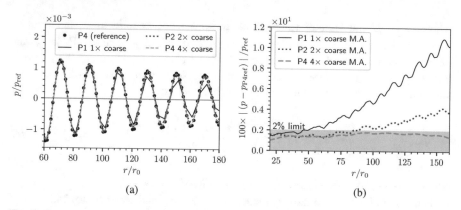

Fig. 2 Acoustic pressure and relative error comparison of the test cases. (**a**) Pressure along $x=y$ line. (**b**) Relative error (as moving average)

result, henceforth called as "P4 reference". For the test cases, elemental meshes are coarsened radially and the polynomial order is elevated in the propagation region, such that, the solution points-per-wavelength (referred as "ppw") distribution is similar in the radial direction. The radial growth rate is kept $\sim$1.023 with geometric distribution in all the cases. In the source region, the mesh is kept the same with P1 expansion for all cases. This allows having a smooth transition of solution points distribution when crossing from one region to the other. A sample P2 mesh and contours are shown in Fig. 1b. Table 1 summarises the test runs.

Figure 2a compares the pressure fields in different test cases with the P4 reference. As expected, the P1 simulation shows a considerable reduction in the amplitude. P2 and P4 preserve this quantity more accurately. For simplification, we unify the dissipation and dispersion error by calculating the overall relative error as a moving average (M.A.) over bins of $\sim$30r_0. This is plotted in Fig. 2b. For P1 and P2 simulations, a 2% error limit is reached around 50r_0 (ppw $\sim$ 9) and 85r_0 respectively (ppw $\sim$ 6.4). P4 simulations remain below this limit under the present conditions (note ppw $\sim$ 5 at 150r_0). The values of ppw for different P agree with those suggested in [20] and provide an estimate for mesh design in different polynomial order setting.

Note that for the given ppw, mesh expansion rate and Riemann solver, we did not observe reflections of the acoustic signals on the inter-element boundaries. A caveat of the present study is that we calculate total numerical error here for brevity, however dissipation and dispersion error could be studied separately as done in [20] on a one-dimensional advection study.

3.2 3D Turbulent Isolated Jet Noise

As a step forward towards noise prediction of installed jets, an isolated jet is simulated using a tetrahedral mesh for *AcousticSolver* to verify the capabilities of the present methodology for complex 3D cases. Note that the coupling is already validated on a cylinder in cross flow and a cylinder-airfoil interaction case in [11].

Jet Flow The LES performed is described here briefly since the same is detailed in [10]. An isothermal turbulent jet issuing from a circular cross-section nozzle at Mach 0.9 and Reynolds number $Re = 10,000$ (based on jet bulk velocity U_j and jet diameter D_j) is considered. Following Shur et al. [21], the present LES domain is cylindrical in shape and extends as $x/D_j = [-5, 100]$ and $r/D_j = [0, 50]$. The mesh has $190 \times 75 \times 49$ nodes in the axial, radial and azimuthal directions respectively. It is refined in the shear layer development area and coarsened towards the outer boundaries. Figure 3a shows a central cross-section of the LES mesh. The inlet boundary condition is a total pressure profile.

Jet Acoustics The acoustics domain is cubical to facilitate control on mesh growth. It extends as $[-5, 40]D_j$ in streamwise direction and $[-25, 25]D_j$ in transverse directions. Noise propagation on two different grids is compared: fully hexahedral ("hexa") and fully tetrahedral ("tetra"). The former mesh consists of $107 \times 69 \times 69$ elements in the streamwise and transverse directions respectively [10]. The tetra grid is generated to give a similar distribution as the hexa mesh in the vicinity of the

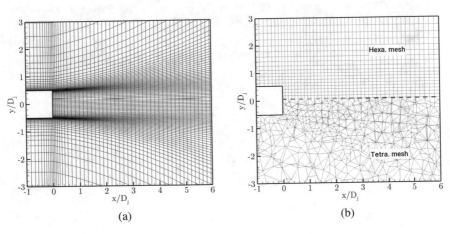

(a) (b)

Fig. 3 Cross-section view through the centre of the jet nozzle. (**a**) LES mesh elements. (**b**) *AcousticSolver* meshes

jet, providing 300,000 elements in total. Figure 3b shows the two meshes where it is seen that the nominal element size in the tetra mesh is slightly larger away from the jet nozzle. Results on the hexa grid (P4) are available from [10] and calculations are performed on the tetra grid in this study. The expansion type utilised is a P4 modified Jacobi basis [8]. A numerical sponge layer [14] of thickness $3D_j$ is applied at the outer boundaries to avoid reflections of the outgoing waves. A factor of 3 in time step size is used as compared to the compressible LES. In line with Sect. 3.1 and [20], a value of ppw ~ 5 is chosen for accurately resolving frequencies up to a Strouhal number $St = 0.9$.

It is already demonstrated in [10] that the LES flow quantities are in acceptable agreement with the high-order LES study of Shur et al. [21]. The noise propagation is calculated using the FW-H method [4] in addition to the present coupled approach. The nominal cut-off St for the integral surface defined is ~ 0.3 based on the 22 ppw criterion [6]. Figure 4 shows a visual comparison between the acoustic pressure field computed by LES alone and coupled LES-APE (on two meshes). Figure 4a, b qualitatively show that the coupled LES-APE has retained more acoustic content (especially at higher frequencies) due to lower numerical error. This difference is more pronounced in the direction perpendicular to the jet centre-line. Qualitatively comparable results are obtained on the tetra mesh as depicted in Fig. 4c.

Figure 5 shows a quantitative comparison in terms of power-spectral-density (PSD) at two observer locations at a distance of $120D_j$. The PSD for FW-H is calculated over the surface indicated by the dashed line in Fig. 4a (details in [10]). Comparison is done with the LES of Shur et al. [21] and the experiment of Tanna [22] ($Re = 10^6$, Mach $= 0.9$). As previously observed in Fig. 4, the difference between FW-H and the present coupled approach is significant at higher frequencies. For the 30° location, the tetra mesh results match the hexa results well. At 90°

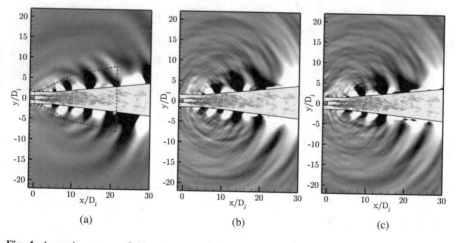

Fig. 4 Acoustic pressure field at the same time instant (in grayscale $[-30, 30]$ Pa). (**a**) LES. (**b**) LES-APE (hexa). (**c**) LES-APE (tetra)

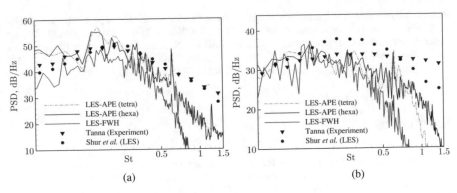

Fig. 5 PSD at $120D_j$ at two observer locations with respect to the jet centre-line. (**a**) $30°$. (**b**) $90°$

location, there is an improvement of cut-off St from 0.3 to 0.8. A small discrepancy is seen at $90°$ for $St > 0.8$ (close to the cut-off $St = 0.9$). This may be improved by using a finer mesh in the far-field. Overall, the APE results are an improvement over the present FW-H prediction in the high frequency domain. Moreover, the results from the tetra mesh are comparable to ones from the hexa mesh. This implies that the present methodology using tetra grids can be extended to more complex cases (such as installed jets).

4   Conclusions

A spectral/hp code *AcousticSolver* (under Nektar++ framework) has been employed for acoustic waves propagation. The favourable properties of this solver are high-order accuracy and capability to handle unstructured mesh elements. A study on a canonical test case with an analytical solution provided estimates for designing the mesh for the jet application. For polynomial order expansion P4, 5 solution points-per-wavelength is found to provide a low overall error. This value is close to the one reported in a related study [20]. These estimates are used to design a tetrahedral mesh for prediction of noise from an isolated jet ($Re = 10^4$, Mach $= 0.9$). The noise sources are calculated from a 2nd-order accurate finite volume LES solver and interpolated onto *AcousticSolver* mesh on-the-fly for noise propagation. The noise results thus obtained offer an improvement over the traditional FW-H method due to high-order accuracy. The power-spectral-density (PSD) results of the noise signal at two different locations relative to the jet nozzle show that the PSDs obtained on the tetrahedral mesh agree with the ones obtained on a slightly finer hexahedral mesh. Further improvements may be achieved by refining the former mesh in the radial direction. These results are encouraging for noise-prediction of more complex industrially relevant geometries such as installed jets.

Acknowledgements The authors would like to thank EPSRC for the UK supercomputing facility ARCHER via the UK Turbulence Consortium (EP/L000261/1). This work was partially funded under the embedded CSE programme of the ARCHER UK National Supercomputing Service. The first author would like to acknowledge K. Lackhove for the useful discussions on the use of CWIPI and Nektar++ and the support given by HPC at Loughborough University.

References

1. Darecki, M., et al.: Flightpath 2050 Europe's vision for aviation. European Commission (2011)
2. NASA Glenn: Making Future Commercial Aircraft Quieter (FS-1999-07-003-GRC) (2004)
3. Jordan, P., et al.: Jet-flap interaction tones. J. Fluid Mech. **853**, 333–358 (2018)
4. Mendez, S., Shoeybi, M., Lele, S.K., Moin, P.: On the use of the Ffowcs Williams-Hawkings equation to predict far-field jet noise from large-eddy simulations. Int. J. Aeroacoust. **12**(1), 1–20 (2013)
5. Tyacke, J.C., Wang, Z.-N., Tucker, P.G.: LES-RANS of installed ultra-high bypass-ratio coaxial jet aeroacoustics with a finite span wing-flap geometry and flight stream – Part 1: round nozzle. 23rd AIAA/CEAS Aeroacoustics Conference (2017)
6. Ewert, R., Schröder, W.: Acoustic perturbation equations based on flow decomposition via source filtering. J. Comput. Phys. **188**(2), 365–398 (2003)
7. Karniadakis, G., Sherwin, S.: Spectral/hp Element Methods for Computational Fluid Dynamics. Oxford University Press, Oxford (2013)
8. Cantwell, C.D., et al.: Nektar++: an open-source spectral/hp element framework. Comput. Phys. Commun. **192**, 205–219 (2015)
9. Lackhove, K., Sadiki, A., Janicka, J.: Efficient three dimensional time-domain combustion noise simulation of a premixed flame using acoustic perturbation equations and incompressible LES. In: ASME Turbo Expo 2017: Turbomachinery Technical Conference and Exposition. American Society of Mechanical Engineers, New York (2017)
10. Moratilla-Vega, M., Lackhove, K., Janicka, J., Xia, H., Page, G.J.: An efficient LES-Acoustic coupling method for sound generation and high order propagation from jets. In: Tenth International Conference on Computational Fluid Dynamics (ICCFD10) (2018)
11. Moratilla-Vega, M., Xia, H., Page, G.J.: A coupled LES-APE approach for jet noise prediction. In: 46th INTER-NOISE Conference (2017)
12. Gröschel, E., Schröder, W., Renze, P., Meinke, M., Comte, P.: Noise prediction for a turbulent jet using different hybrid methods. Comput. Fluids **37**(4), 414–426 (2008)
13. Ewert, R., Schröder, W.: On the simulation of trailing edge noise with a hybrid LES/APE method. J. Sound Vib. **270**(3), 509–524 (2004)
14. Moratilla-Vega, M.A.: A coupled LES/high-order acoustic method for jet noise analysis. Ph.D. Thesis, Loughborough University (2019)
15. Moinier, P.: Algorithm developments for an unstructured viscous flow solver. Ph.D. Thesis, Oxford University (1999)
16. Nicoud, F., Toda, H.B., Cabrit, O., Bose, S., Lee, J.: Using singular values to build a subgrid-scale model for large eddy simulations. Phys. Fluids **23**(8), 085–106 (2011). https://doi.org/10.1063/1.3623274
17. Mahak, M., Moratilla-Vega, M., Page, G.J., Xia, H.: Assessment of WALE and Sigma sub-grid scale models for jet noise prediction. In: 22nd AIAA/CEAS Aeroacoustics Conference (2016)
18. Refloch, A., et al.: CEDRE Software. Technical report, AerospaceLab (2011)
19. Müller, E., Obermeier, F.: The spinning vortices as a source of sound. In: AGARD CP-22 (1967)
20. Moura, R.C., Sherwin, S.J., Peiró, J.: Linear dispersion–diffusion analysis and its application to under-resolved turbulence simulations using discontinuous galerkin spectral/hp methods. J. Comput. Phys. **298**, 695–710 (2015)

21. Shur, M.L., Spalart, P.R., Strelets, M.Kh., Travin, A.K.: Towards the prediction of noise from jet engines. Int. J. Heat Fluid Flow **24**(4), 551–561 (2003)
22. Tanna, H.K.: An experimental study of jet noise Part I: turbulent mixing noise. J. Sound Vib. **50**, 405–428 (1977)

Dynamical Degree Adaptivity
for DG-LES Models

M. Tugnoli, A. Abbà, and L. Bonaventura

1 Introduction

Discontinuous Galerkin spatial discretizations of compressible flows allow to perform local degree adaptation (shortly, p-adaptation) in a very straightforward way and almost without computational overhead, as shown e.g. in [6]. Dynamical adaptation was also applied successfully to inviscid geophysical flows in [11, 12]. All the previous works relied however on a refinement criterion which essentially estimates the L^2 norm approximation error. In [10], we have argued that such a criterion may not be optimal for LES and we have proposed a different, physically based criterion that was shown to be more effective in a number of numerical experiments. The goal of this work, which summarizes some of the results presented in [9], is to extend the above approach to dynamical adaptation and to test the new criterion also in a dynamically adaptive framework.

2 The DG-LES Approach and Its Numerical Implementation

The DG-LES model for compressible flows employed in this work, based on a Local Discontinuous Galerkin (LDG) discretization of the viscous terms [3], is fully described in [1], to which we refer for all the details on the model

M. Tugnoli · A. Abbà
Dipartimento di Ingegneria Aerospaziale, Politecnico di Milano, Milano, Italy
e-mail: matteo.tugnoli@polimi.it; antonella.abba@polimi.it

L. Bonaventura (✉)
MOX – Modelling and Scientific Computing, Dipartimento di Matematica, Politecnico di Milano, Milano, Italy
e-mail: luca.bonaventura@polimi.it

S. J. Sherwin et al. (eds.), *Spectral and High Order Methods for Partial Differential Equations ICOSAHOM 2018*, Lecture Notes in Computational Science and Engineering 134, https://doi.org/10.1007/978-3-030-39647-3_26

equations and numerical discretization approach. Here, only a short description of the discretization elements necessary to introduce dynamical adaptivity will be reported. On the computational domain $\Omega \subset \mathbb{R}^3$ a tessellation $\mathcal{T}_h$ is defined, composed of non overlapping simplicial elements. A discontinuous finite element space $\mathcal{V}_h$ is defined as

$$\mathcal{V}_h = \left\{ v_h \in L^2(\Omega) : v_h|_K \in \mathbb{P}^{q_K}(K), \forall K \in \mathcal{T}_h \right\}, \tag{1}$$

where $\mathbb{P}^{q_K}(K)$ denotes the space of polynomial functions of total degree q_K. The degree can vary arbitrarily from element to element, and the definition of a suitable way to assign such polynomial degree will be discussed in the following. The numerical approximation of the generic variable a can be expressed as

$$a_h|_K = \sum_{l=0}^{n_\phi(K)} a^{(l)} \phi_l^K, \tag{2}$$

where ϕ_l^K are the basis functions on element K, $a^{(l)}$ are the modal coefficients of the basis functions and $n_\phi(K) + 1$ is the number of basis functions required to span the polynomial space $\mathbb{P}^{q_K}(K)$ of degree q_K, defined in $\mathbb{R}^3$ as:

$$n_\phi(K) = \frac{1}{6}(q_K + 1)(q_K + 2)(q_K + 3) - 1 \tag{3}$$

It is worth noting that the expression in (2) can be rewritten, thanks to the hierarchical nature of the basis, as

$$a_h|_K = \sum_{p=0}^{q_K} \sum_{l \in d_p} a^{(l)} \phi_l^K, \tag{4}$$

where $d_0 = \{0\}$ and $d_p = \left\{ l \in 1 \ldots n_\phi(K) \mid \phi_l \in \mathbb{P}^p(K) \backslash \mathbb{P}^{p-1}(K) \right\}$ is the set of indices of the basis functions of degree p. Obtaining a more or less accurate approximation can be done through increasing or decreasing the limit q_K of the sum over p. It is also worth noticing that the basis normalization implies that the first coefficient of the polynomial expansion $a^{(0)}$ coincides with the mean value of $a_h|_K$ over K.

In the present DG-LES approach, as discussed extensively in [1], the LES filtering operators are built directly into the DG discretization, in a spirit similar to the VMS approach [4]. Considering $\Pi_{\mathcal{V}} : L^2(\Omega) \to \mathcal{V}$ the L^2 projector over the subspace $\mathcal{V} \subset L^2(\Omega)$, defined by

$$\int_\Omega \Pi_{\mathcal{V}} u \, v \, d\mathbf{x} = \int_\Omega u \, v \, d\mathbf{x}, \qquad \forall u, v \in \mathcal{V}.$$

it is possible to define the LES filtering ¯· as the projection over the finite dimensional solution subspace $\mathcal{V}_h$ in the following way:

$$\overline{a} = \Pi_{\mathcal{V}_h} a. \tag{5}$$

The application of the main LES filtering is purely formal, since it coincides with the discretization of the equations. In this way, simply discretizing the equations leads to solving them for the filtered quantities.

Another parameter to be defined is the filter characteristic dimension, $\overline{\Delta}$, employed in the definition of all the eddy-viscosity based subgrid model. The definition of the filter size is constant over each element, since the projection is performed elementwise. While more refined definitions can be employed, see e.g. [2], the simple definition

$$\overline{\Delta}(K) = \sqrt[3]{\frac{Vol(K)}{n_\phi(K) + 1}} \tag{6}$$

was employed with success. For the time discretization, the five stages, fourth order Strong Stability Preserving Runge-Kutta method proposed in [8] is employed. The numerical implementation of the previously sketched approach is built in the solver dg-comp using the finite elements toolkit FEMilaro [7].

A first attempt to introduce static p-adaptivity in a DG-LES framework has been presented in [10]. In order to overcome the limitations of classical error estimations in LES, a novel indicator based on the classical structure function

$$D_{ij} = \left\langle \left[u_i(\boldsymbol{x} + \boldsymbol{r}, t) - u_i(\boldsymbol{x}, t)\right]\left[u_j(\boldsymbol{x} + \boldsymbol{r}, t) - u_j(\boldsymbol{x}, t)\right]\right\rangle \tag{7}$$

was proposed. Large values of the structure function calculated inside the element denote a poorly correlated velocity field and the need of higher resolution, while a low structure function value denotes a highly correlated velocity field, which is an indication of a well resolved turbulent region or laminar conditions and of the possibility to employ a lower resolution. However, most of the subgrid models (and in particular the Smagorinsky model) perform adequately in a regime of homogeneous isotropic turbulence, if the filter cut-off length is inside the inertial range. Therefore, in such conditions excessive refinement is not necessary and one can let the subgrid scale model simulate the turbulent dissipation. For this reason, the contribution due to homogeneous isotropic turbulence is removed from the structure function (7). This contribution, as discussed in detail in [10], can be written as

$$D_{ij}^{iso}(\boldsymbol{r}, t) = D_{NN}(r, t)\delta_{ij} + \left(D_{LL}(r, t) - D_{NN}(r, t)\right)\frac{r_i r_j}{r^2} \tag{8}$$

where $r = \|\mathbf{r}\|$ and D_{LL}, D_{NN} are the longitudinal and transverse structure functions, respectively. Once $\mathbf{r}$ is known, only D_{LL} and D_{NN} need to be determined. The procedure to compute the error indicator can then be described as follows:

1. choose a pair of points defining $\mathbf{x}$ and $\mathbf{r}$ in K
2. compute the structure function $D_{ij}(K)$ based on $\mathbf{x}$, $\mathbf{r}$ and the simulated velocity field
3. compute D_{NN} and D_{LL} by a least square fit of (8) to the structure function values within the element
4. define the degree adaptation indicator as:

$$Ind_{SF}(K) = \sqrt{Q(K)} = \sqrt{\sum_{ij}\left[D_{ij}(K) - D_{ij}(K)^{iso}\right]^2}. \tag{9}$$

The static adaptivity procedure presented in [10] is able to produce accurate results with a significant reduction in computational cost. For the simulation of transient phenomena, however, a dynamic adaptivity approach must be applied. The goal of this work, which summarizes results presented in [9], is to extend the above approach to dynamical adaptation, which was successfully employed in the inviscid case in [11, 12].

In those papers, in which special time discretizations approaches were employed that allow the use of very long time steps, the adaptation process was performed at each time step. In the dynamically adaptive simulations presented here, instead, which are carried out with a relatively small time step, the structure function indicator $Ind_{SF}(K)$ is computed every $n_i(K)$ time steps and the average of $s_i(K)$ subsequent values of this quantity is computed. Then, every $n_i(K) \times s_i(K)$ time steps, based on the resulting indicator value in each element, either the polynomial degree is left unchanged or it is updated along with the solution representation. Since the solution is expressed in terms of a hierarchical basis (4), when lowering the polynomial degree, the contribution bound to the removed modes is simply discarded, while when raising the polynomial degree the contribution of the newly added mode is left to zero, to be populated when the integrals over the element and faces couple the old modes with the newly introduced ones.

Notice that, in the present implementation, no dynamic load balancing has been implemented for parallel runs. This means that, during the parallel execution, the dynamic change of number of degrees of freedom could potentially lead to unbalances between the load of different processors. At the moment the balancing is generally executed using a static polynomial distribution. While avoiding excessive unbalancing, this is definitely not the optimal approach and more effective load balancing techniques will have to be investigated in the future.

3 Dynamical Adaptivity Experiments

The proposed dynamic adaptation criterion has been tested in the simulation of a isolated vortex superimposed on a uniform horizontal flow [5]. This simple test has been chosen for the preliminary study reported here, in anticipation the more complex tests already discussed in [9], in which the same isolated vortex impinges on an obstacle. The DG-LES approach described in [1] was applied, as in [10], with a standard Smagorinsky model for the subgrid stresses. A coarser and a finer mesh have been employed, both based on fully unstructured tetrahedra of constant characteristic length equal to $l_h = 1$ and $l_h = 0.5$, respectively. The indicator (9) is computed every $n_i(K) = 2$ time steps and $s_i(K) = 10$ subsequent values are averaged, in order to adapt the resolution every 20 time steps. The sensitivity analysis of the results with respect to these parameters has not yet been carried out and will be the focus of future study. As in [10], two threshold values ϵ_1, ϵ_2 are used to determine p-refinement and p-derefinement. More specifically. the cells with indicator values smaller than ϵ_1 are assigned polynomial degree 2, those with indicator values larger than ϵ_2 are assigned polynomial degree 4, while the others are polynomial degree 3. The threshold values employed are given by $\epsilon_1 = 1 \times 10^{-4}$, $\epsilon_2 = 1 \times 10^{-2}$. Following [10], these values were chosen so as to achieve on average a total number of degrees of freedom slightly smaller than that required by a uniform degree simulation with $p = 3$. The dynamic adaptation procedure is able to effectively increase the polynomial degree around the vortex and follow it as it is advected downstream, leaving all the elements with no vortex activity at the lowest resolution. A map of the polynomial degrees in the domain during the advection of the vortex is shown in Fig. 1.

The profiles of velocity magnitude recorded during time, along the path of the vortex, at different distances from the vortex starting point, employing the coarsest mesh, are presented in Fig. 2. The simulations obtained at different uniform polynomial orders are compared with the adaptive results. It can be observed that, even at the highest uniform resolution of degree 4 the velocity profile is distorted during the advection, due to the very limited grid resolution. However, the vortex does not diffuse and dissipate excessively, as opposed to the low resolution uniform

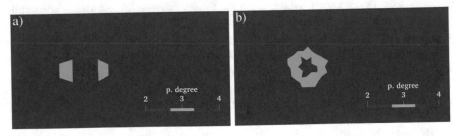

Fig. 1 Polynomial degree values following the advected vortices on the (**a**) coarse and (**b**) fine mesh; green color corresponds to polynomial degree 3, red color corresponds to polynomial degree 4

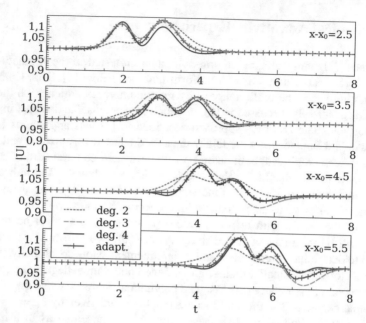

Fig. 2 Profiles of velocity magnitude recorded during time in the vortex path centreline at different distances from vortex starting point, comparison of uniform degree simulations and dynamic adaptive one on the coarse mesh

degree 2 simulation in which the vortex is quickly dissipated. The behaviour of the adaptive simulation is generally mid way between the uniform degree 4 and the uniform degree 3 results.

The comparison with the uniform high degree simulations can be more easily observed in Fig. 3, which show the difference of the velocity magnitude profiles with respect to the uniform degree 4 results, still for the coarse mesh case. In the locations nearer to the starting position of the vortex the adaptive simulation appears close to the degree 4 solution when the first part of the vortex is passing, while a slight difference appears in the second part of the vortex, which is however always within the error of the uniform degree 3 simulation. In the locations farther from the initial starting point of the vortex, which sense the vortex passage after a longer advection time, the adapted simulation is always very close to the uniform degree 4 solution. It has to be noted that the average number of degrees of freedom of the adaptive simulation is 41,488, which remain almost constant throughout the simulation. This is 10.8% more than the 37,430 degrees of freedom needed for the uniform degree 2 solution, 44.6% less than the 74,860 degrees of freedom of the uniform degree 3 resolution, which is always outperformed by the adaptive one, and 68.3% less than the uniform degree 4 simulation.

To correctly assess the effects of adaptivity in the case of the refined mesh, we study the difference of the various results with respect to the uniform degree 4 one, presented in Fig. 4. The differences are generally very small, even for the

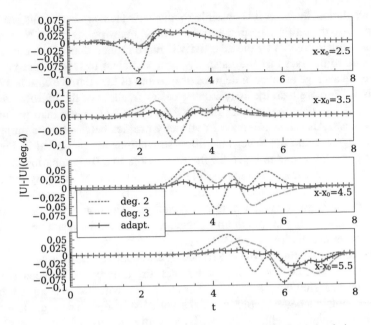

Fig. 3 Difference of velocity magnitude with respect to the most refined simulation at uniform degree 4, recorded during time in the vortex path centreline at different distances from vortex starting point, on coarse mesh

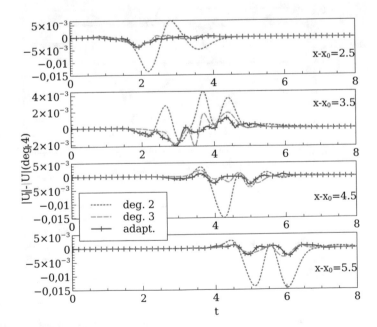

Fig. 4 Difference of velocity magnitude with respect to the most refined simulation at uniform degree 4, recorded during time in the vortex path centreline at different distances from vortex starting point, on fine mesh

lowest resolution, however it is possible to note how the adaptive results are always comparable to the uniform degree 3 results, and in many points better. Nonetheless, the improvement created by the adaptivity is more limited than in the coarse case, mainly due to the fact that the mesh by itself sufficient to resolve the vortex. In this case the average number of degrees of freedom of the adaptive case is 170,470, which is 5.7% more than the 161,320 degrees of freedom of the uniform degree 2 case, 47.2% less than the uniform degree 3 case and 70.0% less than the uniform degree 4 case. Also the difference in vorticity profiles between the simulation at uniform degree 4 and the lower resolution simulations are presented in Fig. 5 for the coarse resolution and in Fig. 6 for the finer resolution. By comparing the results at the two different resolution is possible to note also for the vorticity that, at the finer resolution, the large scale phenomenon is correctly represented by almost all polynomial degrees, with a minimal vorticity dissipation, while at the coarser resolution only the higher polynomial degree, as well as the adaptive simulation, avoid an excessive dissipation of vorticity.

At the coarser resolution, the difference of the adaptive simulations with respect to the uniform degree 4 ones is smaller than the differences between the other uniform degree simulations (Fig. 5), showing that with the adaptation is also possible to obtain a better resolution of the vorticity profiles. The same is true also at the finer resolution (Fig. 5). In the dynamically adaptive simulations spurious acoustic waves seem to be produced by the dynamical adaptation process, see

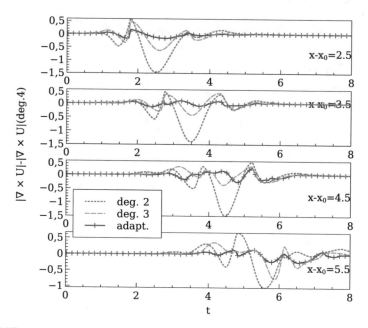

Fig. 5 Difference of vorticity magnitude with respect to the most refined simulation at uniform degree 4, recorded during time in the vortex path centreline at different distances from vortex starting point, on coarse mesh

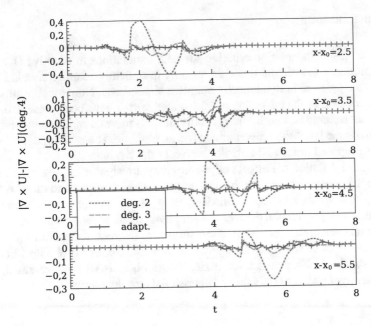

Fig. 6 Difference of vorticity with respect to the most refined simulation at uniform degree 4, recorded during time in the vortex path centreline at different distances from vortex starting point, on fine mesh

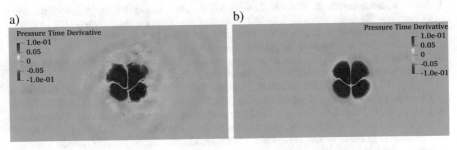

Fig. 7 Pressure time derivative in the adaptive simulation of vortex advection on (**a**) coarse mesh, (**b**) finer mesh, at time $T = 4$; in both plots, the represented quantity takes values in the interval $[-0.1, 0.1]$

Fig. 7. These spurious disturbances were not observed in the dynamically adaptive tests presented in [11, 12], which employed an implicit time discretization, thus strongly damping these high frequency solution components. However, as it can be seen inspecting the time series of the pressure values (not reported here due to the limited space available), these disturbances decrease rapidly in amplitude on the finer mesh and do not seem to propagate through the domain but rather follow the advected vortex. This spurious feature warrants further investigation of the dynamical adaptation approach if a correct approximation of acoustic waves is desired.

4 Conclusions

The novel degree adaptation criterion for LES simulations in adaptive DG frame-works proposed in [10] and tested so far only in statically adaptive simulations has been also employed in dynamically adaptive simulations. Numerical results in the benchmark case of the advection of an isolated vortex have been presented. These results are meant to be a preliminary for the study of more complex configurations in which the same isolated vortex impinges on an obstacle. The presented results show that the proposed criterion is also effective in the dynamical case. With a coarse basic mesh resolution the effects of p-adaptivity are significant, leading to results close to the ones obtained with the maximum resolution allowed to the polynomial base, while when the mesh resolution is already suitable to represent the vortex even with the lowest polynomial degrees the adaptivity leads anyway to accurate results, but with an even higher reduction of the number of degrees of freedom with respect to the non-adaptive solutions. In a subsequent work, the results obtained in [9] for the case of the isolated vortex impinging on an obstacle will be presented, along with other application to fully three-dimensional turbulent flows.

References

1. Abbà, A., Bonaventura, L., Nini, M., Restelli, M.: Dynamic models for Large Eddy Simulation of compressible flows with a high order DG method. Comput. Fluids. **122**, 209–222 (2015)
2. Abbà, A., Campaniello, D., Nini, M.: Filter size definition in anisotropic subgrid models for large eddy simulation on irregular grids. J. Turb. **18**, 589–610 (2017)
3. Cockburn, B., Shu, C.-W.: The local discontinuous Galerkin method for time-dependent convection-diffusion systems. SIAM J. Numer. Anal. **35**, 2440–2463 (1998)
4. Hughes, T.J.R., Feijóo, G.R., Mazzei, L., Quincy, J.-B.: The variational multiscale method—a paradigm for computational mechanics. Comput. Methods Appl. Mech. Eng. **166**, 3–24 (1998)
5. Lodato, G., Domingo, P., Vervisch, L.: Three-dimensional boundary conditions for direct and large-eddy simulation of compressible viscous flows. J. Comput. Phys. **227**, 5105–5143 (2008)
6. Remacle, J.-F., Flaherty, J. E., Shephard, M. S.: An adaptive discontinuous Galerkin technique with an orthogonal basis applied to compressible flow problems. SIAM Rev. **45**, 53–72 (2003)
7. Restelli, M.: FEMilaro, a finite element toolbox. Available at https://bitbucket.org/mrestelli/femilaro/wiki/Home
8. Spiteri, R.J., Ruuth, S.J.: A new class of optimal high-order strong-stability-preserving time discretization methods. SIAM J. Numer. Anal. **40**, 469–491 (2002)
9. Tugnoli, M.: Polynomial Adaptivity for Large Eddy Simulation of Compressible Turbulent Flows. Ph.D. Thesis, Politecnico di Milano, Department of Aerospace Science and Technology, Doctoral Programme in Aerospace Engineering (2017)
10. Tugnoli, M., Abbà, A., Bonaventura, L., Restelli, M.: A locally p-adaptive approach for LES of compressible flows in a DG framework. J. Comput. Phys. **349**, 33–58 (2017)

11. Tumolo, G., Bonaventura, L.: A semi-implicit, semi-Lagrangian, DG framework for adaptive numerical weather prediction. Q. J. R. Meteorol. Soc. **141**, 2582–2601 (2015)
12. Tumolo, G., Bonaventura, L., Restelli, M.: A semi-implicit, semi-Lagrangian, p−adaptive DG method for the shallow water equations. J. Comput. Phys. **232**, 46–67 (2013)

A Novel Eighth-Order Diffusive Scheme for Unstructured Polyhedral Grids Using the Weighted Least-Squares Method

Duarte M. S. Albuquerque, Artur G. R. Vasconcelos, and Jose C. F. Pereira

1 Introduction

The numerical solution of transport phenomena in complex geometrical domains is a subject of continuous development regarding three characteristics: accuracy, robustness and efficiency. The geometrical complexity can be handled with different grid topologies and the understanding of their issues is relevant for industrial applications. High-order computation is a demanding issue, motivated by a potential reduction of computational cost for complex computational fluid dynamics (CFD) problems.

High-order accurate methods for unstructured grids have historically been focused on hyperbolic equations, see e.g. Lê et al. [1]. Barth and Frederickson [2] developed a high-order Finite Volume Methods (FVM) for the resolution of the Euler equations, using a quadratic polynomial. The coupling of Euler system with viscous terms, which requires diffusive schemes was achieved by Ollivier-Gooch et al. [3].

In the last years, the development of high-order methods was applied for the resolution of parabolic and elliptic problems in unstructured grids, see e.g. Boularas et al. [4]. The range of possible applications varies from Poisson problems, see Batty [5], heat transfer problems, see e.g. Chantasiriwan [6], diffusion equations with variable coefficients, see Zhai [7], or discontinuous coefficients, see e.g. Clain et al. [8].

Several polynomial reconstruction techniques applied to FVM can be highlighted: the fourth-order methods of Ollivier-Gooch et al. [9], Cueto-Felgueroso

D. M. S. Albuquerque (✉) · A. G. R. Vasconcelos · J. C. F. Pereira
LAETA, IDMEC, IST, Universidade de Lisboa, Lisboa, Portugal
e-mail: DuarteAlbuquerque@tecnico.ulisboa.pt; Artur.Vasconcelos@tecnico.ulisboa.pt; jcfpereira@tecnico.ulisboa.pt

© The Author(s) 2020
S. J. Sherwin et al. (eds.), *Spectral and High Order Methods for Partial Differential Equations ICOSAHOM 2018*, Lecture Notes in Computational Science and Engineering 134, https://doi.org/10.1007/978-3-030-39647-3_27

349

et al. [10], and Nogueira et al. [11], also sixth-order results have been reported by Clain et al. [12]. The objective of this work is to extend the weighted least-squares (WLS) method to very high-order schemes and polyhedral unstructured grids.

In terms of other applications with the weighted least-squares technique. Magalhaes et al. [13] and Albuquerque et al. [14] have developed, respectively, relative and absolute error estimators for second-order finite volume schemes with unstructured grids. Martins et al. [15, 16] has created a third-order interpolation method with divergence free constraint for immersed boundary applications, respectively, for Cartesian and unstructured polyhedral grids.

The following manuscript is divided in four sections: in Sect. 2 the implemented method for two dimensions is briefly described, in Sect. 3 the verification of the implemented schemes, with Cartesian and perturbed grids, is carried out. Section 4 shows the results for a case with irregular polyhedral and triangular grids and proposes a novel method to treat the Neumann boundary conditions, Sect. 5 concludes the manuscript with a summary of the principal achievements of this work.

2   Elliptical Operator for Unstructured Grids with the Least Squares Technique

In this work, the Poisson equation will be solved, which is defined by:

$$\nabla \cdot \nabla \phi = \varphi_\phi, \tag{1}$$

where ϕ is the transported variable and φ_ϕ is the source term that is required when using manufactured analytical solutions and it is equal to its own Laplacian. After applying the classic Finite Volume method in a Poisson equation the following equation is obtained:

$$\sum_{f \in \mathcal{F}(P)} \sum_{g \in \mathcal{G}(f)} \nabla \phi_g w_{G_g} \cdot \mathbf{S_f} = \int_{CV} \varphi_\phi dV, \tag{2}$$

where $\mathcal{F}(P)$ is the set of faces of cell P, $\mathcal{G}(f)$ is the set of Gauss points of the face f, $\mathbf{S_f}$ is the face normal vector and w_G is the weight of Gauss-Legendre Quadrature. The important part of this method is how the calculation of the face gradient $\nabla \phi_g$ is carried out at each Gauss point. This will be explained in the next subsection.

2.1   Polynomial Reconstructions

To obtain the gradients values at the integration points, a reconstruction of the unknown primitive variable is performed at the face centroid, using a polynomial expansion.

Table 1 Number of terms of the Taylor expansion required for a pth order polynomial at two dimensional (2D) cases

p^{th} Order polynomial	1	3	5	7
Number of terms	3	10	21	36

The number of terms of the polynomial has to take into account the required order of the scheme and it has the following form:

$$\phi_f^R(x, y) = C_1 + C_2(x - x_f) + C_3(y - y_f) +$$
$$+C_4(x - x_f)^2 + C_5(x - x_f)(y - y_f) + C_6(y - y_f)^2 + \cdots . \tag{3}$$

Expression (3) can be written in a more compact form, a vectorial one, as:

$$\phi_f^R(\mathbf{x}) = \mathbf{d}_f(\mathbf{x})\,\mathbf{c}_f. \tag{4}$$

where the subscript f refers that the reconstruction is made at the face f and $\mathbf{d}_f(\mathbf{x}) = \left[1, (x - x_f), (y - y_f), (x - x_f)^2, (x - x_f)(y - y_f), (y - y_f)^2, \cdots \right]$, $\mathbf{x}_f = (x_f, y_f)$ is the face centroid coordinates vector, $\mathbf{x} = (x, y)$ is the coordinates vector of a point used for the reconstruction and $\mathbf{c}_f = [C_1, C_2, C_3, C_4, C_5, C_6, \cdots]^T$ are the reconstruction constants.

Table 1 lists the number of terms of the expansion for each polynomial used in this work.

The order of accuracy of the numerical scheme is $p + 1$, consequently the linear reconstruction will be second order accurate, the cubic reconstruction will be fourth order accurate, the fifth polynomial will have sixth order accurate and finally the seventh polynomial will be eighth order accurate. The numerical schemes will be called of $FLS(p + 1)$ according to the global order of the implemented method. For each order a minimum number of Gauss points are required to maintain the respective Quadrature order.

2.2 General Approach

The Weighted Least Squares (WLS) method is a technique used to solve overdetermined problems, where there are more independent equations than unknowns.

Equation (4) results in a system of linear equations, which the form as:

$$\mathbf{D}_f\,\mathbf{c}_f = \boldsymbol{\phi}_s, \tag{5}$$

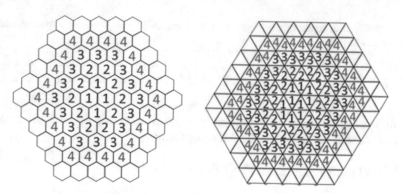

Fig. 1 Examples of different vertex neighbours order from the red face

where $\mathbf{D}_f$ is a combination of $\mathbf{d}_f(\mathbf{x})$ for every point of the reconstruction resulting in a matrix with $n_s \times n_{coefs}$ entries. The $\mathbf{c}_f$ is a column vector with n_{coefs} entries, $\boldsymbol{\phi}_s$ is a column vector with n_s entries, n_{coefs} is the number of constants of the pth polynomials and n_s the size of the computational stencil, which is the set of the computational values and points used in the reconstruction and is made of cell neighbours of the face. Since $n_s > n_{coefs}$, the problem is overdetermined and so the WLS technique is used in order to minimize the weighted residual of the problem.

To solve this problem, specific stencils must be used for each scheme order. This is done by using vertex neighbours according to the experience of the Authors in a previous work [14]. Each successive order scheme requires an higher stencil to respect the $n_s > n_{coefs}$ condition. Figure 1 shows examples of these stencils for a regular polyhedral and triangular grid. Basically each successive vertex neighbours (from 1 to 4) is used for a scheme with an even order accuracy. For example the second order scheme only needs a first order of vertex neighbours from the face marked in red.

Other details used in the global matrix construction A_{ij} are described in the work of Vasconcelos et al. [17]. Each line of the global matrix A_{ij} corresponds to the diffusive discretization of the cell i and has to consider the diffusive flux integral for each face of the cell. This flux integral is computed from the polynomial reconstruction centered in the respective face and which was described previously. Finally the high-order diffusive fluxes can be written in the following matrix form:

$$\mathbf{A}_{ij}\phi_j = \sum_{f \in \mathcal{F}(i)} \left(\sum_{g \in \mathcal{G}(f)} w_{Gg} \mathbf{t}_{\mathbf{f}j}(\mathbf{x}_g) \phi_j \right) \cdot \mathbf{S}_f, \tag{6}$$

where $\mathcal{F}(i)$ is the set of faces from cell i, $\mathcal{G}(f)$ is the set of Gauss-Legendre points of face f, w_{Gg} is the weight, $\mathbf{x}_g$ are the coordinates of each Gauss-Legendre point

g and $\mathbf{t}_{\mathbf{f}j}$ is the contribution from cell j to the face f diffusive flux from the reconstructed polynomial. The set of cells j is defined by the used stencil in the polynomial reconstructed at each face f, globally each line i will have contributions from all cells that result from the junction of sets from all faces of cell i.

3 Order Convergence Verification with Cartesian and Perturbed Non-uniform Grids

A numerical test is performed in non-uniform grids with an certain imposed displacement. This perturbation is done by moving randomly the grid lines in a range between zero and a % (γ) of the grid size from the Cartesian grid counterpart. This perturbation can be done in either a positive or negative direction. The cells of the grids are always squares and a reference grid without any perturbation is used, i.e. a Cartesian one. For this case, the following analytical solution was used and solved in a 1×1 square domain:

$$\phi(x, y) = \exp\left(-\frac{(x - 0.5)^2 + (y - 0.5)^2}{0.0175}\right), \tag{7}$$

Table 2 lists the error ratios, r, between a grid with an imposed perturbation and a regular one. Showing the ratio for both the mean and maximum error of the finest grid at study. Particularly for the FLS6 and FLS8 schemes, the error could be one order of magnitude greater than the obtained with the Cartesian grid. It is also shown that an imposed perturbation up to 20% has a low numerical error penalization.

Figure 2 shows the convergence curves obtained but only with the FLS4 and FLS8 schemes. It is possible to observe that the theoretical convergence orders is achieved for every perturbed grid.

Figure 3 shows the error distribution for the FLS8 scheme with a Cartesian and perturbed grid with $\gamma = 30\%$. It is shown that the error distribution is severally changed by the imposed perturbation at the grid.

Table 2 Ratio of mean and maximum error norms for all schemes between grids with an imposed perturbation and a Cartesian one with 25,600 cells

$\gamma\%$	FLS2		FLS4		FLS6		FLS8	
	r_1	r_∞	r_1	r_∞	r_1	r_∞	r_1	r_∞
10	1.01	1.54	1.52	4.44	1.86	10.35	2.20	8.19
20	1.09	1.55	2.33	4.72	2.81	15.15	3.62	11.05
30	1.22	1.78	3.24	7.53	3.89	22.30	17.28	30.61

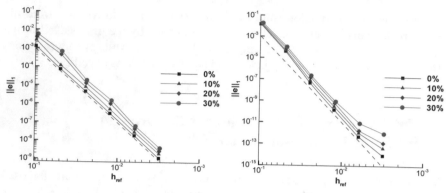

Fig. 2 Convergence curves for the imposed perturbed grids with FLS4 and FLS8

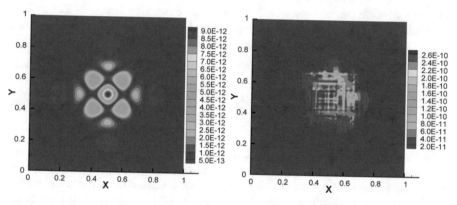

Fig. 3 Error distribution for FLS8 scheme and two grid with 25,600 cells: one without any perturbation (left) and one with an imposed perturbation (right)

4   Results for Several Grid Types and with Neumann Boundary Conditions

To verify the applicability of the proposed schemes to other grid types and Neumann boundary conditions. The numerical verification was performed with an analytical solution in a square domain, $[0, 1]$. A Neumann boundary condition were imposed at the vertical faces and a Dirichlet boundary condition at the remaining ones.

Two different approaches were used when considering Neumann boundary conditions. The first approach is the classic one, which consists in simply derivation of the respective line from the least-squares matrix that represent the boundary face. It will be defined as the general case (GC) approach.

The second approach which is new to the Author's knowledge and it consists on the multiplication of each line of the $\mathbf{D}_f$ matrix referent to Neumann boundary face, b, by the respective face area, $\mathbf{S}_b$. That line will be written by $\nabla \mathbf{d}_f(\mathbf{x}_b)\mathbf{S}_b$, instead of $\nabla \mathbf{d}_f(\mathbf{x}_b)\,\mathbf{n}_b$ and the entry for the vector $\boldsymbol{\phi}_f$ is given by $\nabla \boldsymbol{\phi}_b S_b$, instead of $\nabla \boldsymbol{\phi}_b \mathbf{n}_b$. Consequently, the problem will have the following aspect:

$$
\begin{bmatrix}
1 & (x_1 - x_f) & (y_1 - y_f) & (x_1 - x_f)^2 & \cdots \\
1 & (x_2 - x_f) & (y_2 - y_f) & (x_2 - x_f)^2 & \cdots \\
\vdots & \vdots & \vdots & \vdots & \ddots \\
\hline
0 & S_{b_x} & S_{b_y} & 2(x_b - x_f)\,S_{b_x} & \cdots \\
\vdots & \vdots & \vdots & \vdots & \ddots
\end{bmatrix}
\begin{bmatrix}
C_1 \\
C_2 \\
\vdots \\
C_{n_{coefs}}
\end{bmatrix}
=
\begin{bmatrix}
\phi_1 \\
\phi_2 \\
\vdots \\
\hline
\nabla \boldsymbol{\phi}_b S_b \\
\vdots
\end{bmatrix},
\tag{8}
$$

where the line in the matrix separates the contribution of the stencil cells and Dirichlet faces from the contribution of the Neumann faces of the current considered stencil.

The goal of this operation is to ensure that the vector $\boldsymbol{\phi}_s$ and each line of the least-squares matrix $\mathbf{D}_f$ have the same unit dimensions, something that does not happen with the classic approach. This approach will be designed as dimensional correction (DC$_N$) for Neumann boundary condition.

Numerical tests were performed for two grid types: irregular polyhedral and triangular grids. The analytical solution is given by:

$$
\phi(x, y) = \sin(3\pi x)\sin(3\pi y),
\tag{9}
$$

where in the Neumann boundaries the face flux will be $\nabla \phi_b \cdot \mathbf{S}_b \neq 0$.

Figure 4 shows the convergence curves to both approaches applied for all schemes with the irregular polyhedral (left) and triangular grids (right). The solid line represents the DC$_N$ approach and the dotted one represents the classic GC approach. The results point out that the theoretical convergence order is always

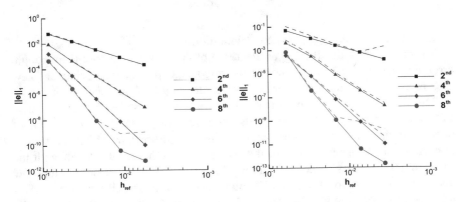

Fig. 4 Convergence curves of the mean error for mixed boundary conditions with irregular polyhedral and triangular grids for all schemes. The dotted lines are the convergence curves for the GC approach and the solid ones represent the convergence curves with the DC$_N$ approach

Table 3 Comparison between the two approaches for a problem with an imposed Neumann BC for all schemes applied to the irregular polyhedral grids

Polyhedral		FLS2		FLS4		FLS6		FLS8	
n_{cells}	h_{ref}	r_1	r_∞	r_1	r_∞	r_1	r_∞	r_1	r_∞
145	8.31E−02	1.12	1.65	0.97	1.40	0.92	1.48	1.11	0.96
545	4.34E−02	1.11	1.84	1.06	1.04	1.07	1.02	1.24	1.26
2113	2.22E−02	1.04	1.63	1.15	1.39	1.08	1.45	1.13	2.81
8321	1.12E−02	0.99	1.04	1.08	1.16	1.04	1.08	21.57	74.12
33,024	5.65E−03	0.98	1.10	1.04	1.12	1.02	1.04	178.69	261.82

Table 4 Comparison between the two approaches for a problem with an imposed not-null Neumann BC for all schemes applied to the triangular grids

Triangular		FLS2		FLS4		FLS6		FLS8	
n_{cells}	h_{ref}	r_1	r_∞	r_1	r_∞	r_1	r_∞	r_1	r_∞
211	5.98E−02	2.24	2.85	1.80	1.51	1.44	1.25	0.88	0.91
899	2.91E−02	1.70	4.08	1.48	1.74	1.05	0.89	1.35	0.53
3638	1.45E−02	1.36	7.47	1.49	1.94	1.73	0.94	1.74	1.98
14,632	7.25E−03	1.10	7.91	1.42	1.68	2.30	0.95	190.13	297.64
58,698	3.62E−03	12.73	357.14	1.65	1.73	3.70	1.44	790.73	915.16

achieved for both grids and indicates that the DC_N approach improves the schemes performance, being more evident for the FLS2 and for FLS8 schemes, specially to the last one. The behaviour of the finest grids are more stable with the DC_N.

Table 3 lists the comparison of the two approaches used for the Neumann BC for the irregular polyhedral grid, the comparison is made through the ratio between both approaches and using the mean and maximum error norm, r is computed by:

$$r_i = \frac{\|e\|_i^{GC}}{\|e\|_i^{DC_N}}, \tag{10}$$

where i is the error norm used for the calculation.

The results show that the biggest decrease of the error occurs for the maximum error. For the FLS8 scheme the error can be reduced up to 21 times, since the new method avoids the truncation error issue presented in the GC and showed in Fig. 4.

Table 4 lists the comparison between the two approaches used for the Neumann BC with the triangular grids. The results obtained allow to conclude that the major decrease of the numerical error occurs for the maximum error, which can be reduced almost one order of magnitude for the second-order scheme and to half with the fourth-order scheme. For the sixth-order scheme the maximum error with this new approach is slightly worse, almost 10%, than the general approach, however in terms of mean error the gain is evident since the mean error is reduced to half with the DC_N approach. For the eighth-order scheme, it is possible to reduce the error in about three orders of magnitude since it avoids the truncation error issue.

5 Conclusions

Verifications tests have been performed for a new high-order scheme based on the weighted least-squares technique and the Finite Volume method. The convergence curves have showed an excellent behaviour indicating that the theoretical order is achieved for all cases at study. Also the new reconstruction method is not very sensitive to the imposed perturbations in the grid or either the topology of the cells.

Additionally, the results allowed the novel proposed approach to treat the Neumann boundary conditions, improving the quality of the solution. These results are the expected ones, since in the WLS problem the dimensions of the matrices are identical to each other, when using this proposed approach.

References

1. Lê, T.H., Le Gouez, J.M., Garnier, E.: High accuracy flow simulations: advances and challenges for future needs in aeronautics. Comput. Fluids **43**, 90–97 (2011)
2. Barth, T., Frederickson, P.: Higher order solution of the Euler equations on unstructured grids using quadratic reconstruction. In: 28th Aerospace Sciences Meeting from American Institute of Aeronautics and Astronautics (1990)
3. Sejekan, C.B., Ollivier-Gooch, C.F.: Improving finite volume diffusive fluxes through better reconstruction. Comput. Fluids **139**, 216–232 (2016)
4. Boularas, A., Clain, S.L., Baudoin, F.: A sixth-order finite volume method for diffusion problem with curved boundaries. Appl. Math. Model. **42**, 401–422 (2017)
5. Batty, C.: A cell-centred finite volume method for the Poisson problem on non-graded quadtrees with second order accurate gradients. J. Comput. Phys. **331**, 49–72 (2017)
6. Chantasiriwan, S.: Methods of fundamental solutions for time-dependent heat conduction problems. Int. J. Numer. Methods Eng. **66**, 147–165 (2006)
7. Zhai, S., Weng, Z., Feng, X.: An adaptive local grid refinement method for 2D diffusion equation with variable coefficients based on block-centered finite differences. Appl. Math. Comput. **268**, 284–294 (2015)
8. Costa, R., Clain, S., Stéphane, Machado, G.J.: Improving finite-volume diffusive fluxes through better reconstruction. Comput. Fluids **139**, 216–232 (2016)
9. Jalali, A., Ollivier-Gooch, C.: Higher-order unstructured finite volume RANS solution of turbulent compressible flows. Comput. Fluids **147**, 32–47 (2017)
10. Nogueira, X., Colominas, I., Cueto-Felgueroso, L., Khelladi, S.: On the simulation of wave propagation with a higher-order finite volume scheme based on Reproducing Kernel Methods. Comput. Methods Appl. Mech. Eng. **199**, 1471–1490 (2010)
11. Chassaing, J., Khelladi, S., Nogueira, X.: Accuracy assessment of a high-order moving least squares finite volume method for compressible flows. Comput. Fluids **71**, 41–53 (2013)
12. Clain, S.L., Machado, G.J., Pereira, R.M.S.: A new very high-order finite volume method for the 2D convection diffusion problem on unstructured meshes. IV Conferência Nacional em Mecânica dos Fluidos, Termodinâmica e Energia (2012)
13. Magalhães, J.P.P., Albuquerque, D.M.S., Pereira, J.M.C., Pereira, J.C.F.: Adaptive mesh finite-volume calculation of 2D lid-cavity corner vortices. J. Comput. Phys. **243**, 365–381 (2013)
14. Albuquerque, D.M.S., Pereira, J.M.C., Pereira, J.C.F.: Residual least-squares error estimate for unstructured h-adaptive meshes. Numer. Heat Transfer, Part B Fundam. **67**, 187–210 (2015)

15. Martins, D.M.C., Albuquerque, D.M.S., Pereira, J.C.F.: Continuity constrained least-squares interpolation for SFO suppression in immersed boundary methods. J. Comput. Phys. **336**, 608–626 (2017)
16. Martins, D.M.C., Albuquerque, D.M.S., Pereira, J.C.F.: On the use of polyhedral unstructured grids with moving immersed boundary method. Comput. Fluids **174**, 78–88 (2018)
17. Vasconcelos, A.G.R., Albuquerque, D.M.S., Pereira, J.C.F.: A very high-order finite volume method based on weighted least squares for elliptic on polyhedral unstructured grids. Comput. Fluids **181**, 383–402 (2019)

An Explicit Mapped Tent Pitching Scheme for Maxwell Equations

Jay Gopalakrishnan, Matthias Hochsteger, Joachim Schöberl, and Christoph Wintersteiger

1 Introduction

Electromagnetic waves propagate at the speed of light. Thus, the field at a certain point in space and time depends only on field values within a dependency cone. A tent pitching method introduces a special "causal" spacetime mesh that respects this finite speed of propagation. It is not limited to Maxwell equations, but can be applied to general hyperbolic equations. A tent pitching method requires a numerical scheme to discretize the equation on that mesh. Discontinuous Galerkin (DG) methods are of particular interest since they offer a systematic avenue to build high order methods. For a given initial condition at the bottom of a tent, the discrete equations may be solved within each individual tent, up to the tent top. The computed solution at the tent top provides initial conditions for the tents that follow later in time. This method is highly parallel, since many tents can be solved independently. Methods using such tent-pitched meshes may be traced back to [5, 7]. More recent works [1, 6, 8] develop Spacetime DG (SDG) methods within tents by formulating local variational problems, for which linear systems are set up and solved. Although these systems are local, the matrix size can grow rapidly with the polynomial order, especially in four-dimensional spacetime tents. In this context

J. Gopalakrishnan (✉)
Fariborz Maseeh Department of Mathematics & Statistics, Portland State University, Portland, OR, USA
e-mail: gjay@pdx.edu

M. Hochsteger · J. Schöberl · C. Wintersteiger
Institute for Analysis and Scientific Computing, Technische Universität Wien, Wien, Austria
e-mail: matthias.hochsteger@tuwien.ac.at; joachim.schoeberl@tuwien.ac.at; christoph.wintersteiger@tuwien.ac.at

© The Author(s) 2020
S. J. Sherwin et al. (eds.), *Spectral and High Order Methods for Partial Differential Equations ICOSAHOM 2018*, Lecture Notes in Computational Science and Engineering 134, https://doi.org/10.1007/978-3-030-39647-3_28

it is natural to ask if one can develop explicit schemes (which usually perform well under low memory bandwidth) that take advantage of tents.

A key ingredient to answer this question was presented in [2], where Mapped Tent Pitching (MTP) schemes were introduced. The MTP discretization, which proceeds by mapping tents to a spacetime cylinder, allows one to evolve the solution either implicitly or explicitly within tents. The memory requirements of the explicit MTP scheme are limited to what is needed for storing the spatial mesh, the solution coefficients at one time step, and the topology of the tents.

In this work, we show that notwithstanding the above-mentioned advantages of the explicit MTP scheme, one may lose higher order convergence if a naive time stepping strategy (involving a standard explicit Runge-Kutta scheme) is used. We then develop a new Taylor time-stepping for the local problems within tents. Despite its simplicity, our numerical experiments show that it delivers optimal order of convergence.

2 Mesh Generation by Tent Pitching

We start with a conforming spatial mesh consisting of elements $\mathscr{T} = \{T\}$ and vertices $\mathscr{V} = \{V\}$. We progress in time by defining a sequence of advancing fronts τ_i. A front τ_i is given as a standard nodal finite element function on this mesh. It is defined by storing the current time for every vertex of the mesh. We move from τ_i to the next front τ_{i+1} by moving one vertex forward in time, while keeping all other vertices fixed. The spacetime domain between τ_i and τ_{i+1} we call a tent. In Fig. 1, the red domain is the tent between τ_i and τ_{i+1}.

Its projection to the spatial domain is exactly the vertex patch ω_V around V of the original mesh. The data to be stored for one tent are the bottom and top-times of the central vertex, plus the times for all neighboring vertices.

Note that although the algorithm is described sequentially, it is highly parallel. Vertices with graph-distance of at least two can be moved forward independently. For example, in Fig. 1, all blue tents can be built and processed in parallel.

The distance for advancing a vertex is limited by the speed of light, a constraint often referred to in the literature as the *causality condition*. Under this condition, the Maxwell problem inside the tent is solvable using the initial conditions at the tent bottom. Thus, the top boundary is an outgoing boundary and no boundary conditions are needed there.

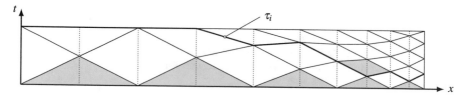

Fig. 1 Tent pitched spacetime mesh for a one-dimensional spatial mesh

Note that the spatial mesh in Fig. 1 is refined towards the right boundary, which leads to smaller tent heights at the right boundary. Hence, smaller time steps in locally refined regions is a very natural feature of tent pitching methods.

3   The MTP Discretization

Now, we consider the discretization method for one tent domain $K = \{(x, t) : x \in \omega_V, \varphi_b(x) \le t \le \varphi_t(x))\}$, where ω_V is the union of elements containing the vertex V, and φ_b and φ_t are the bottom and top fronts, respectively, restricted to ω_V. Our aim is to numerically solve the Maxwell system on K, namely

$$\partial_t \varepsilon E = \nabla \times H , \qquad \partial_t \mu H = -\nabla \times E , \qquad (1)$$

where boundary values for both fields are given at the tent bottom and $\nabla = \nabla_x$ denotes the spatial gradient.

The approach of MTP schemes is to map the tent domain to a spacetime cylinder $\omega_V \times (0, 1)$ and solve the transformed equation there. The transformation from the cylinder to the tent is denoted by $\Phi : \omega_V \times (0, 1) \to K$ and is defined by $\Phi(x, \hat{t}) = (x, \varphi(x, \hat{t}))$ where

$$\varphi(x, \hat{t}) = (1 - \hat{t})\varphi_b(x) + \hat{t}\varphi_t(x) .$$

It is similar to the Duffy transformation mapping a square to a triangle (see Fig. 2). With the notation

$$\text{skew } E = \begin{pmatrix} 0 & E_z & -E_y \\ -E_z & 0 & E_x \\ E_y & -E_x & 0 \end{pmatrix} ,$$

we can rephrase the curl operator as $\nabla \times E = \text{div skew } E$, where the divergence of the matrix function is taken row-wise. To simplify notation further, we define $u : K \to \mathbb{R}^6$ by $u = (E, H)$, and set $g : K \to \mathbb{R}^6$ and $f : K \to \mathbb{R}^{6\times3}$ by

$$g(u) = \begin{bmatrix} \varepsilon E \\ \mu H \end{bmatrix} , \qquad f(u) = \begin{bmatrix} -\text{ skew } H \\ \text{skew } E \end{bmatrix} . \qquad (2)$$

Then (1) may be rewritten as the conservation law $\partial_t g(u) + \text{div}_x f(u) = 0$. Furthermore, we define $F(u) \in \mathbb{R}^{6\times4}$ as

$$F(u) = \begin{bmatrix} f(u) & g(u) \end{bmatrix} = \begin{bmatrix} -\text{ skew } H & \varepsilon E \\ \text{skew } E & \mu H \end{bmatrix} ,$$

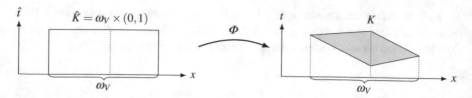

Fig. 2 Tent mapped from a tensor product domain

which allows us to write Maxwell's system (1) as the spacetime conservation law

$$\text{div}_{x,t}\, F(u) = 0 \, . \tag{3}$$

For each row of F, the spacetime divergence $\text{div}_{x,t}$ sums the spatial divergence of the first three components with the time-derivative of the last component.

Now, we apply the Piola transformation to pull back F from the tent K to the cylinder using the mapping Φ. The derivative of Φ and its transposed inverse are

$$\Phi' = \begin{bmatrix} I & 0 \\ \nabla\varphi^T & \delta \end{bmatrix} \quad \text{and} \quad (\Phi')^{-T} = \begin{bmatrix} I & -\delta^{-1}\nabla\varphi \\ 0 & \delta^{-1} \end{bmatrix} .$$

The Piola transform of F is $\widehat{F}(\hat{u}) = \mathcal{P}\{F\} = (\det \Phi')(F \circ \Phi)(\Phi')^{-T}$ with $\hat{u} = u \circ \Phi$. Since the Piola transform provides an algebraic transformation of the divergence, Eq. (3) is simply transformed to $\text{div}_{x,\hat{t}}\, \widehat{F}(\hat{u}) = 0$ on the spacetime cylinder. Then, inserting the Jacobian of Φ leads us to the transformed equation

$$\partial_{\hat{t}}(g(\hat{u}) - f(\hat{u})\nabla\varphi) + \text{div}_x(\delta f(\hat{u})) = 0 \, , \tag{4}$$

where $\delta(x) = \varphi_t(x) - \varphi_b(x)$ is the local height of the tent. Note that $\nabla\varphi$ is an affine-linear function in quasi-time $\hat{t}$. Equation (4) describes the evolution of $\hat{u}$ along quasi-time from $\hat{t} = 0$ to $\hat{t} = 1$. Details of the calculations are given in [2].

The next step is the space discretization of (4) by a standard discontinuous Galerkin method. Let $V_h \subset [L_2]^6$ be the DG finite element space of degree p on $\mathcal{T}$. On each tent we search for $\hat{u} : [0, 1] \to V_h$ such that

$$\int_{\omega_V} \partial_{\hat{t}}\big[g(\hat{u}) - f(\hat{u})\nabla\varphi\big] v_h - \sum_{T \subset \omega_V} \int_T \delta f(\hat{u})\nabla v_h + \sum_{F \subset \omega_V} \int_F \delta f_n(\hat{u}^+, \hat{u}^-)[\![v]\!] = 0$$

holds for all $v_h \in V_h$ and all $\hat{t} \in [0, 1]$. Only the restriction of V_h on the patch ω_V is used in this equation. The numerical flux $f_n(\hat{u}^+, \hat{u}^-)$ depends on the positive trace $\lim_{s\to 0+} \hat{u}(x + sn)$ and negative trace $\lim_{s\to 0+} \hat{u}(x - sn)$, where n is a unit normal vector of arbitrary orientation to the face. The jump is defined as usual by

$[\![\hat{u}]\!] := \hat{u}^+ - \hat{u}^-$ and the mean value by $\{\hat{u}\} := \frac{1}{2}(\hat{u}^+ + \hat{u}^-)$. One example is the upwind flux [3, p. 434]

$$f_n(\hat{u}^+, \hat{u}^-) = \begin{bmatrix} \{\hat{H}\} \times n + [\![\hat{E}_t]\!] \\ -\{\hat{E}\} \times n + [\![\hat{H}_t]\!] \end{bmatrix},$$

with the tangential components $\hat{E}_t = -(\hat{E} \times n) \times n$ and $\hat{H}_t = -(\hat{H} \times n) \times n$ of $\hat{E} = E \circ \Phi$ and $\hat{H} = H \circ \Phi$. Note that the local tent height δ enters the boundary integrals as a multiplicative factor. At the outer boundary of the vertex patch we have $\delta = 0$, so the facet integrals on the outer boundary disappear. For the above semidiscrete system, initial values for the tent problem are given finite element functions at the tent bottom. The finite element solution on the tent top provides the initial conditions for the next level tent. Therefore, no projection of initial values is needed when propagating from one tent to the next.

After the semi-discretization, as usual, we are left to solve a system of $N = \dim V_h(\omega_V)$ ordinary differential equations for $U : [0, 1] \to \mathbb{R}^N$,

$$\frac{d}{d\hat{t}} [MU](\hat{t}) - AU(\hat{t}) = 0, \qquad \hat{t} \in (0, 1), \tag{5}$$

given $U(0)$. The non-standard feature of (5) is that M is an affine-linear function of the quasi-time $\hat{t}$ (since our mapping enters the mass matrix M through $\nabla\varphi$). The matrix A is independent of $\hat{t}$. A straightforward approach is to substitute $Y = MU$ and solve

$$\frac{d}{d\hat{t}} Y - AM^{-1}Y = 0,$$

instead of (5). Although first order convergence was observed with this strategy, further numerical studies showed reduced order of convergence if the stage-order of the Runge Kutta (RK) method is not high enough—see Fig. 3 (right). While the implicit MTP schemes discussed in [2] do not show this problem, the issue remains critical for explicit schemes. Thus, we propose to use a new type of explicit time-stepping for time discretization, discussed next.

4 Structure-Aware Taylor Time-Stepping

Returning to the ordinary differential equation (5) and continuing to make the substitution $Y = MU$, we now reconsider the previous equation as the following differential-algebraic system:

$$\frac{d}{d\hat{t}} Y = AU, \qquad Y = MU. \tag{6}$$

We begin by subdividing the interval $(0, 1)$ into $m \in \mathbb{N}$ smaller intervals of size $\frac{1}{m}$, defined by $(\hat{t}_i, \hat{t}_{i+1}) = (\frac{i}{m}, \frac{i+1}{m})$, for $i \in \mathbb{N}$ and $0 \le i \le m - 1$. Recall that A is independent of quasi-time $\hat{t}$, and M is an affine function of $\hat{t}$, i.e.,

$$M(\hat{t}) = M_i + (\hat{t} - \hat{t}_i)M', \qquad \hat{t} \in (\hat{t}_i, \hat{t}_{i+1})$$

where $M_i = M(\hat{t}_i)$ and the derivative M' is a constant matrix. We want to design a time-stepping scheme that is aware of this structure.

Consider the approximations to Y, U on $(\hat{t}_i, \hat{t}_{i+1})$ in the form of Taylor polynomials Y_i, U_i of degree q, defined by

$$Y_i(\hat{t}) = \sum_{n=0}^{q} \frac{(\hat{t} - \hat{t}_i)^n}{n!} Y_{i,n} \qquad U_i(\hat{t}) = \sum_{n=0}^{q-1} \frac{(\hat{t} - \hat{t}_i)^n}{n!} U_{i,n} , \qquad \hat{t} \in (\hat{t}_i, \hat{t}_{i+1}) ,$$

$$\tag{7}$$

where $Y_{i,n} = Y_i^{(n)}(\hat{t}_i)$ and $U_{i,n} = U_i^{(n)}(\hat{t}_i)$. To find these derivatives, we differentiate both equations of (6) n times to get

$$Y^{(n+1)}(\hat{t}) = AU^{(n)}(\hat{t}) , \qquad\qquad n \ge 0 ,$$

$$Y^{(n)}(\hat{t}) = M(\hat{t})U^{(n)}(\hat{t}) + nM'U^{(n-1)}(\hat{t}) , \qquad\qquad n \ge 1 .$$

For the second equation we used Leibnitz' formula $(fg)^{(n)} = \sum_{i=0}^{n} \binom{n}{i} f^{(i)} g^{(n-i)}$, and the fact that M is affine-linear. Evaluating these equations for the Taylor polynomials Y_i, U_i at $\hat{t} = \hat{t}_i$, we obtain a recursive formula for $Y_{i,n}$ and $U_{i,n}$ in terms of $U_{i,n-1}$, namely

$$Y_{i,n} = AU_{i,n-1} , \qquad\qquad 1 \le n \le q ,$$
$$M_i U_{i,n} = Y_{i,n} - nM'U_{i,n-1} , \qquad\qquad 1 \le n \le q - 1 , \tag{8}$$

for all $0 \le i \le m - 1$. Given $Y_{0,0} = Y(\hat{t}_0)$, $M_0 U_{0,0} = Y_{0,0}$, applying (8) with $i = 0$ gives the approximate functions $Y_0(\hat{t}), U_0(\hat{t})$ in the first subinterval $(\hat{t}_0, \hat{t}_1)$. The recursive formulas are initiated for later subintervals at $n = 0$ by

$$Y_{i,0} = Y_{i-1}(\hat{t}_i), \qquad M_i U_{i,0} = Y_{i,0} , \qquad 1 \le i \le m - 1 . \tag{9}$$

After the final subinterval, we get $Y_{m-1}(t_m)$, our approximation to $Y(1)$. We shall refer to the new time-stepping scheme generated by (8) as the q-stage SAT (structure-aware Taylor) time-stepping.

Note that $Y_{m-1}(t_m)$ is our approximation to $Y = MU$ at the top of the tent. This value is then passed to the next tent in time. The time dependence of M arises from the time dependence of $\nabla\varphi$. This gradient is continuous along spacetime lines of constant spatial coordinates. Therefore, when passing from one element of a tent to the same element within the next tent in time, Y is continuous (since the solution U is continuous). Of course, on flat fronts $\nabla\varphi = \nabla\tau = 0$, so there M is just a diagonal matrix containing the material parameters.

To briefly remark on the expected convergence rate of a q-stage SAT time-stepping, recall that due to the mapping of the MTP method we solve for $\hat{u} = u \circ \Phi$, which satisfies $\partial_{\hat{t}}^n \hat{u} = \delta^n(\partial_t^n u) \circ \Phi$. The causality condition implies that $\delta \to 0$ if the mesh size $h \to 0$. Thus we may expect the nth temporal derivative of $\hat{u}$, and correspondingly $U^{(n)}$, to go to zero at the rate $\mathcal{O}(h^n)$. By using a q-stage SAT time-stepping, we approximate the first $q - 1$ terms of the exact Taylor expansion of U. Thus we expect the convergence rate to be $O(h^q)$, the size of the remainder term involving $U^{(q)}$. The next section provides numerical evidence for this.

Before concluding this section, we should note that in (8) and (9), we tacitly assumed that M_i is invertible. Let us show that this is indeed the case whenever the causality condition (see Sect. 2) $|\nabla\varphi| < \sqrt{\varepsilon\mu}$ is fulfilled. At any quasi-time $\hat{t}$, given a $\hat{w} = (\hat{w}_E, \hat{w}_H) \in V_h$ whose coefficient vector in the basis expansion is $W \in \mathbb{R}^N$, consider the equation $M(\hat{t})U = W$ for the coefficient vector U of $\hat{u} \in V_h$. This equation, in variational form, is

$$\int_{\omega V} [g(\hat{u}) - f(\hat{u})\nabla\varphi] \cdot \hat{v} = \int_{\omega V} (\hat{w}_E, \hat{w}_H) \cdot \hat{v}, \qquad \text{for all } \hat{v} \in V_h. \tag{10}$$

Let $a(\hat{u}, \hat{v})$ denote the left hand side of (10). To prove solvability of (10), it suffices to prove that $a(\cdot, \cdot)$ is a coercive bilinear form on $[L_2]^6$ for any $\hat{t}$. By inserting $g(\hat{u}) = [\varepsilon\hat{E}, \mu\hat{H}]^T$ and $f(\hat{u}) = [-\operatorname{skew}\hat{H}, \operatorname{skew}\hat{E}]^T$ into $a(\hat{u}, \hat{u})$,

$$a(\hat{u}, \hat{u}) = \int_{\omega V} (\varepsilon\hat{E} - \hat{H} \times \nabla\varphi) \cdot \hat{E} + (\mu\hat{H} + \hat{E} \times \nabla\varphi) \cdot \hat{H}$$

$$= \int_{\omega V} \varepsilon\hat{E} \cdot \hat{E} + \mu\hat{H} \cdot \hat{H} + 2(\hat{E} \times \nabla\varphi) \cdot \hat{H}$$

$$\geq \int_{\omega V} \varepsilon\hat{E} \cdot \hat{E} + \mu\hat{H} \cdot \hat{H} - 2\frac{|\nabla\varphi|}{\sqrt{\varepsilon\mu}}\sqrt{\varepsilon}|\hat{E}|\sqrt{\mu}|\hat{H}| \, ,$$

where we used the Cauchy-Schwarz inequality and inserted $\sqrt{\varepsilon}$ and $\sqrt{\mu}$ to achieve the desired scaling. By applying Young's inequality and $|\nabla\varphi| < \sqrt{\varepsilon\mu}$,

$$a(\hat{u}, \hat{u}) \geq \int_{\omega V} \varepsilon\hat{E} \cdot \hat{E} + \mu\hat{H} \cdot \hat{H} - \frac{|\nabla\varphi|}{\sqrt{\varepsilon\mu}}(\varepsilon\hat{E} \cdot \hat{E} + \mu\hat{H} \cdot \hat{H})$$

$$= \int_{\omega V} \left(1 - \frac{|\nabla\varphi|}{\sqrt{\varepsilon\mu}}\right)(\varepsilon\hat{E} \cdot \hat{E} + \mu\hat{H} \cdot \hat{H}) \geq C \min(\varepsilon, \mu)\|\hat{u}\|_{L_2}^2 \, ,$$

form some constant $C > 0$. Thus M_i is invertible and the SAT time-stepping is well defined on all tents respecting the causality condition.

One may exploit the specific details of the Maxwell problem to avoid the assembly and the inversion of matrices M_i (as we have done in our implementation). In fact, instead of (10), we can explicitly solve the corresponding exact undiscretized equation obtained by replacing V_h by $[L_2]^6$ in (10). The solution $\hat{u} = (\hat{E}, \hat{H})$ in closed form reads

$$\hat{E} = \frac{1}{\varepsilon\mu - |\nabla\varphi|^2} \left(I - \frac{1}{\varepsilon\mu}\nabla\varphi\nabla\varphi^T\right)(\mu\hat{w}_E + \hat{w}_H \times \nabla\varphi),$$

$$\hat{H} = \frac{1}{\varepsilon\mu - |\nabla\varphi|^2} \left(I - \frac{1}{\varepsilon\mu}\nabla\varphi\nabla\varphi^T\right)(\varepsilon\hat{w}_H - \hat{w}_E \times \nabla\varphi).$$

We then perform a projection of these into V_h to obtain the coefficients $U(\hat{t}_i)$. For uncurved elements, this just involves the inversion of a diagonal mass matrix. For the small number of curved elements, we use a highly optimized algorithm which uses an approximation instead of the exact inverse mass matrix.

5 Numerical Results

The MTP discretization in combination with the SAT time-stepping on tents is implemented within the Netgen/NGSolve finite element library. In this section numerical results concerning accuracy as well as performance are reported.

5.1 Convergence Studies in Two Space Dimensions

We consider the model problem in two space dimensions

$$\partial_t \varepsilon E_z = \partial_x H_y - \partial_y H_x, \qquad \partial_t \mu H_x = -\partial_y E_z, \qquad \partial_t \mu H_y = \partial_x E_z,$$

on the spacetime cube $[0, \pi]^2 \times [0, \sqrt{2}\pi]$. Parameters are set $\varepsilon = \mu = 1$ such that speed of light is $c = 1$. Initial and boundary values are set such that the exact solution is given by

$$E_z = \sin(x)\sin(y)\cos(\sqrt{2}t),$$
$$H_x = -\frac{1}{\sqrt{2}}\sin(x)\cos(y)\sin(\sqrt{2}t),$$
$$H_y = \frac{1}{\sqrt{2}}\cos(x)\sin(y)\sin(\sqrt{2}t).$$

Based on a spatial mesh with mesh size h, we generate a tent pitched mesh such that the maximal slope $|\nabla\varphi|$ is bounded by $(2c)^{-1}$ and apply a discontinuous

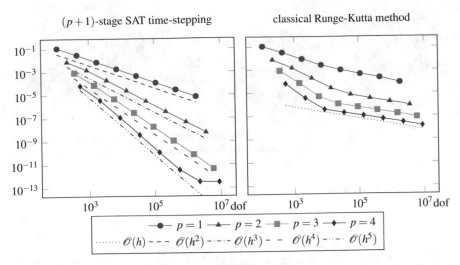

Fig. 3 Spatial L_2 error of all field components over degrees of freedom (dof) for the $(p+1)$-stage SAT time-stepping (left) and the classical Runge-Kutta (right)

Galerkin method in space using polynomials of order p, with $1 \leq p \leq 4$. On each cylinder we perform a $(p+1)$-stage SAT time-stepping with $m = 2p$ intervals. The spatial L_2 error of all field components at the final time is reported in the left plot of Fig. 3. We observe that the error goes to zero at the optimal rate of $\mathcal{O}(h^{p+1})$ until we are close to machine precision.

In contrast, the right plot in Fig. 3 illustrates the previously mentioned loss of convergence rates when the classical Runge-Kutta method is used. The convergence rates stagnate at first order no matter what p is used. A similar behavior was also observed for other explicit Runge-Kutta methods.

5.2 Large Scale Problem in Three Space Dimensions

As a second example we present a simulation on a domain similar to the resonator shown in [4]. The geometry is given as body of revolution of smooth B-spline curves. The mesh consisting of 489,593 curved tetrahedral elements is shown in Fig. 4. Due to higher curvature the mesh is refined along the inner roundings, where the ratio of the largest to the smallest element is approximately 5:1. We used a Gaussian peak (located at the axis of revolution and the position of the fifth inner rounding) for the electric field as initial data. The explicit MTP scheme with SAT time-stepping then computed the solution at $t = 260$ using time slabs of height 1, with each slab composed of $N_{\text{tents}} = 149,072$ tents. On each tent we used a $(p+1)$-stage SAT time-stepping with $m = 2p$ intervals, where p denotes the spatial polynomial order. With the spatial degrees of freedom $N_{\text{dof},i}$ of the ith tent and the

Fig. 4 Tetrahedral mesh with 489 k curved elements, ratio of the largest to the smallest element of approximately 5:1 and the H_y component of solution at $t = 260$ calculated with spatial polynomial order $p = 3$

Table 1 Number of degrees of freedom and simulation times for spatial polynomial orders $p = 2, 3$

	$p = 2$	$p = 3$
Number of spatial dof	2.938×10^7	5.875×10^7
Number of spacetime dof per slab	1.908×10^9	7.632×10^9
Simulation time per slab	4.6 s	49.2 s
Total simulation time	20 min	3 h 33 min

This data was generated using a shared memory server with 4 E7-8867 CPUs with 16 cores each

number of stages $q = p + 1$, we obtain the total spacetime degrees of freedom per time slab

$$\sum_{i=1}^{N_{\text{tents}}} N_{\text{dof},i}\, m\, q = \left(\sum_{i=1}^{N_{\text{tents}}} N_{\text{dof},i} \right) 2p(p+1) .$$

The corresponding numbers of degrees of freedom and the simulation times are shown in Table 1. In [4] a similar problem is solved using a discontinuous Galerkin method with quadratic elements, combined with a polynomial Krylov subspace method in time. Using 96 cores it took them 7:10 h to reach the final time. Our simulation with polynomial order $p = 3$, which has a comparable number of unknowns, took 3:33 h on 64 cores. This significant speed up is an illustration of the capability of the new method. The H_y component of the obtained solution at $t = 260$, using third order polynomials in space, is shown in Fig. 4.

Acknowledgement This work was supported in part by the National Science Foundation grant DMS-1912779.

References

1. Abedi, R., Petracovici, B., Haber, R.B.: A spacetime discontinuous Galerkin method for elastodynamics with element-wise momentum balance. Comput. Methods Appl. Mech. Eng. **195**, 3247–3273 (2006)

2. Gopalakrishnan, J., Schöberl, J., Wintersteiger, C.: Mapped tent pitching schemes for hyperbolic systems. SIAM J. Sci. Comput. **39**, B1043–B1063 (2017)
3. Hesthaven, J.S., Warburton, T.: Nodal Discontinuous Galerkin Methods. Texts in Applied Mathematics, vol. 54. Springer, New York (2008). Algorithms, analysis, and applications
4. Hochbruck, M., Pažur, T., Schulz, A., Thawinan, E., Wieners, C.: Efficient time integration for discontinuous Galerkin approximations of linear wave equations. ZAMM Z. Angew. Math. Mech. **95**(3), 237–259 (2015)
5. Lowrie, R.B., Roe, P.L., van Leer, B.: A space-time discontinuous Galerkin method for the time-accurate numerical solution of hyperbolic conservation laws. In: Proceedings of the 12th AIAA Computational Fluid Dynamics Conference, pp. 95–1658 (1995)
6. Monk, P., Richter, G.R.: A discontinuous Galerkin method for linear symmetric hyperbolic systems in inhomogeneous media. J. Sci. Comput. **22/23**, 443–477 (2005)
7. Richter, G.R.: An explicit finite element method for the wave equation. Appl. Numer. Math. **16**(1–2), 65–80 (1994)
8. Yin, L., Acharia, A., Sobh, N., Haber, R.B., Tortorelli, D.A.: A spacetime discontinuous Galerkin method for elastodynamics analysis. In: Cockburn, B., Karniadakis, G., Shu, C.W. (eds.) Discontinuous Galerkin Methods: Theory, Computation and Applications, pp. 459–464 (2000)

Viscous Diffusion Effects in the Eigenanalysis of (Hybridisable) DG Methods

Rodrigo C. Moura, Pablo Fernandez, Gianmarco Mengaldo,
and Spencer J. Sherwin

1 Introduction

When numerically solving partial differential equations, numerical errors are likely to impact not only solution accuracy, but also the stability/robustness of the computation. This is particularly the case in eddy-resolving approaches to turbulent flows, such as large-eddy simulation (LES) and direct numerical simulation (DNS). Also, in the so-called implicit LES / under-resolved DNS strategies [1], where numerical error (specifically dissipation) provides small-scale regularisation in lieu of a turbulence model, understanding the nature of numerical errors is crucial. These typically appear in the form of dispersion and diffusion errors, where the former distorts the solution, while the latter is responsible for its damping. A useful framework for the assessment of such numerical errors is the eigensolution analysis technique [2, 3].

R. C. Moura (✉)
Instituto Tecnológico de Aeronáutica, São José dos Campos, Brazil
e-mail: moura@ita.br

P. Fernandez
Massachusetts Institute of Technology, Cambridge, MA, USA
e-mail: pablof@mit.edu

G. Mengaldo
California Institute of Technology, Pasadena, CA, USA
e-mail: mengaldo@caltech.edu

S. J. Sherwin
Imperial College, London, UK
e-mail: s.sherwin@imperial.ac.uk

© The Author(s) 2020
S. J. Sherwin et al. (eds.), *Spectral and High Order Methods for Partial Differential Equations ICOSAHOM 2018*, Lecture Notes in Computational Science and Engineering 134, https://doi.org/10.1007/978-3-030-39647-3_29

371

We present the first eigenanalysis of hybridisable discontinuous Galerkin (HDG) methods. This is also one of the first studies to consider viscous diffusion effects in the eigenanalysis of discontinuous SEM (spectral element methods), as it addresses the advection-diffusion equation in one dimension. Focus is given to the temporal analysis approach [2, 5], which is suited for problems with periodic boundary conditions. The spatial analysis [3, 4], suited for inflow-outflow problems, will be considered in subsequent studies. Here, we offer preliminary results on (i) the effects of the Peclét number (a cell-based Reynolds number), and (ii) the interplay between upwind (numerical) dissipation and viscous (physical) diffusion. We highlight how these results improve upon our understanding and practice of implicit LES / under-resolved DNS approaches.

We note that, although a non-modal eigenanalysis strategy better suited for turbulence computations has been recently proposed [6], the present work will focus on more fundamental aspects and follow therefore the classical eigenanalysis. Finally, the results presented here are representative of a broader class of discontinuous SEM, given the well established connections within this class—see e.g. [7].

This paper is organized as follows. Section 2 introduces the HDG discretisation as applied to the linear advection-diffusion equation in one dimension. Section 3 details the temporal eigenanalysis framework and presents our preliminary results. Finally, in Sect. 4, our conclusions are summarised and future research topics are outlined.

2 HDG Discretisation

In one dimension, the linear advection-diffusion equation is given by

$$\frac{\partial u}{\partial t} + a \frac{\partial u}{\partial x} = \mu \frac{\partial^2 u}{\partial x^2} \,, \tag{1}$$

where the advection velocity a and the viscosity μ are positive constants. This equation can be written in conservation form through the flux function $f(u, g) = au - \mu g$, as the system

$$\frac{\partial u}{\partial t} + \frac{\partial f}{\partial x} = 0 \,, \tag{2}$$

$$g - \frac{\partial u}{\partial x} = 0 \,, \tag{3}$$

where g is the auxiliary gradient variable. The discretisation procedure is similar to that of traditional DG methods.

After the (1D) physical domain is partitioned into non-overlapping elemental regions Ω of size h, the numerical solution and its gradient are locally approximated by polynomial expansions in the form

$$u|_\Omega = \sum_{j=0}^{P} \hat{u}_j(t)\,\phi_j(\xi)\,, \quad g|_\Omega = \sum_{j=0}^{P} \hat{g}_j(t)\,\phi_j(\xi)\,, \tag{4}$$

where ϕ_j are polynomial basis functions of degree up to P, defined in the standard domain $\Omega_{st} = [-1, 1]$. A linear mapping relation is assumed between the physical coordinate x of element Ω and the coordinate $\xi \in \Omega_{st}$.

Multiplying Eqs. (2)–(3) by ϕ_i, integrating over element Ω and applying integration by parts leads respectively to

$$\frac{h}{2} \int_{\Omega_{st}} \frac{\partial u}{\partial t} \phi_i \, d\xi + \left(\tilde{f}\phi_i\right)_{\ominus}^{\oplus} = \int_{\Omega_{st}} f \frac{\partial \phi_i}{\partial \xi} \, d\xi \,, \tag{5}$$

$$\frac{h}{2} \int_{\Omega_{st}} g\phi_i \, d\xi + \int_{\Omega_{st}} u \frac{\partial \phi_i}{\partial \xi} \, d\xi = \left(\tilde{u}\phi_i\right)_{\ominus}^{\oplus} \,, \tag{6}$$

where $\ominus$ and $\oplus$ denote the left and right boundaries of element Ω, in that order. As typical, expansions in (4) are to be inserted into (5)–(6), which are then required to hold for $i = 0, \ldots, P$. Note that the integrals above have been moved to Ω_{st} and interface quantities $\tilde{u}$ and $\tilde{f}$ have been introduced. The state average $\tilde{u}$ is peculiar to HDG in that it represents a uniquely defined interface variable whose value stems indirectly from the enforced continuity of the numerical flux $\tilde{f}$. This continuity ensures local conservation for HDG methods, regardless of the chosen flux formula.

For the advection-diffusion problem at hand, the interface fluxes on either side of a given element (cf. Fig. 1, left diagram) can be taken in the form

$$\tilde{f}_\oplus = f(\tilde{u}_\oplus, g_\oplus) - \tau(\tilde{u}_\oplus - u_\oplus)\,, \tag{7}$$

$$\tilde{f}_\ominus = f(\tilde{u}_\ominus, g_\ominus) - \tau(u_\ominus - \tilde{u}_\ominus)\,, \tag{8}$$

Fig. 1 Notation adopted for the element viewpoint (left) and the interface viewpoint (right)

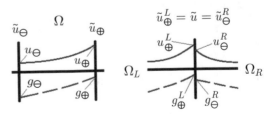

in which

$$u_\oplus = \sum_{j=0}^{P} \hat{u}_j \, \phi_j(+1), \quad g_\oplus = \sum_{j=0}^{P} \hat{g}_j \, \phi_j(+1), \tag{9}$$

$$u_\ominus = \sum_{j=0}^{P} \hat{u}_j \, \phi_j(-1), \quad g_\ominus = \sum_{j=0}^{P} \hat{g}_j \, \phi_j(-1). \tag{10}$$

Also, $\tau = \beta|a| + \sigma$ is a stabilisation constant combining an upwinding parameter β and a penalty term σ that accounts for the partially diffusive character of the model equation considered. This work however assumes $\sigma = 0$ as it focuses on advection-dominated cases, which are typically stable without the penalty term σ, even within the context of turbulence simulations [8].

Flux formulas (7)–(8) are inspired in Ref. [9]. In the case of pure advection (with $\sigma = 0$), the interface solution variable becomes the simple average $\tilde{u} = u_\oplus^L + u_\ominus^R$ of the adjacent states from the left (L) and right (R) elements sharing the considered interface. Under this case, it is also easy to show that the fluxes in (7)–(8) recover those used in traditional DG methods, whereby HDG exactly reproduces DG. This does not hold, however, when diffusion is taken into account, in which case $\tilde{u}$ is only implicitly defined from the flux continuity condition enforced at interfaces, $\tilde{f}_\oplus^L = \tilde{f}_\ominus^R$, namely

$$a\tilde{u} - \mu g_\oplus^L - \tau \left(\tilde{u} - u_\oplus^L \right) = a\tilde{u} - \mu g_\ominus^R - \tau \left(u_\ominus^R - \tilde{u} \right), \tag{11}$$

where $g_\oplus^L$ and $g_\ominus^R$ depend on values of $\tilde{u}$ at two other interfaces via (6). The diagram on the right-hand-side of Fig. 1 should help clarify the notation adopted.

Using vectors $\hat{u} = \{\hat{u}_0, \ldots, \hat{u}_P\}^T$ and $\hat{g} = \{\hat{g}_0, \ldots, \hat{g}_P\}^T$, the flux continuity condition (11) becomes

$$\tilde{u} = \frac{1}{2} \left(\hat{\phi}_\oplus^T \hat{u}^L + \hat{\phi}_\ominus^T \hat{u}^R \right) + \frac{\mu}{2\tau} \left(\hat{\phi}_\ominus^T \hat{g}^R - \hat{\phi}_\oplus^T \hat{g}^L \right), \tag{12}$$

where $\hat{\phi}_\oplus = \{\hat{\phi}_0(+1), \ldots, \hat{\phi}_P(+1)\}^T$ and $\hat{\phi}_\ominus = \{\hat{\phi}_0(-1), \ldots, \hat{\phi}_P(-1)\}^T$. Likewise, (6) can be written as

$$\frac{h}{2} M\hat{g} + D\hat{u} = \hat{\phi}_\oplus \tilde{u}_\oplus - \hat{\phi}_\ominus \tilde{u}_\ominus, \tag{13}$$

in which matrices M and D have been introduced, namely

$$M_{i,j} = \int_{\Omega_{st}} \phi_i \phi_j \, d\xi, \quad D_{i,j} = \int_{\Omega_{st}} \frac{\partial \phi_i}{\partial \xi} \phi_j \, d\xi. \tag{14}$$

Finally, (5) becomes

$$\frac{h}{2} M \frac{d\hat{u}}{dt} + \hat{\phi}_\oplus \tilde{f}_\oplus - \hat{\phi}_\ominus \tilde{f}_\ominus = a D \hat{u} - \mu D \hat{g},$$ (15)

with

$$\tilde{f}_\oplus = a \tilde{u}_\oplus - \mu \hat{\phi}_\oplus^T \hat{g} - \tau(\tilde{u}_\oplus - \hat{\phi}_\oplus^T \hat{u}),$$ (16)

$$\tilde{f}_\ominus = a \tilde{u}_\ominus - \mu \hat{\phi}_\ominus^T \hat{g} - \tau(\hat{\phi}_\ominus^T \hat{u} - \tilde{u}_\ominus).$$ (17)

Note that (12) is a scalar equation written from the point of view of a given interface, whereas (13) and (15) are vector equations written from the viewpoint of an arbitrary element Ω of size h.

It is now convenient to eliminate $\hat{g}$ and work with variables $\hat{u}$ and $\tilde{u}$ alone. This can be done by solving (13) for $\hat{g}$ and substituting the resulting expression in both (12) and (15). The former substitution leads, after some algebra, to

$$\left(\beta + \frac{m_\oplus^\oplus + m_\ominus^\ominus}{\text{Pe}}\right) \tilde{u} - \frac{m_\oplus^\ominus}{\text{Pe}} \tilde{u}_\ominus^L - \frac{m_\ominus^\oplus}{\text{Pe}} \tilde{u}_\oplus^R = \hat{\phi}_\oplus^T B_\oplus^L \hat{u}^L + \hat{\phi}_\ominus^T B_\ominus^R \hat{u}^R,$$ (18)

where $\text{Pe} = |a| h / \mu$ denotes the Péclet number, for which a uniform mesh spacing is assumed. Moreover, four scalar constants 'm' have been introduced, defined as

$$m_\oplus^\oplus = \hat{\phi}_\oplus^T M^{-1} \hat{\phi}_\oplus, \quad m_\ominus^\ominus = \hat{\phi}_\ominus^T M^{-1} \hat{\phi}_\ominus, \quad m_\oplus^\ominus = \hat{\phi}_\oplus^T M^{-1} \hat{\phi}_\ominus, \quad m_\ominus^\oplus = \hat{\phi}_\ominus^T M^{-1} \hat{\phi}_\oplus.$$ (19)

In addition, the following matrices appear in (18)

$$B_\oplus^L = \frac{\beta}{2} I + \frac{M^{-1} D}{\text{Pe}}, \quad B_\ominus^R = \frac{\beta}{2} I - \frac{M^{-1} D}{\text{Pe}}.$$ (20)

Note that (18) relates the solution vectors $\hat{u}$ of two adjacent elements (Ω_L and Ω_R) with the three interface states $\tilde{u}$ associated to the boundaries of these elements.

The second step consists in using $\hat{g}$ from (13) into (15), not forgetting to take the fluxes (16)–(17) into account. After some more algebra, one arrives at

$$\frac{h}{2a} M \frac{d\hat{u}}{dt} + A \hat{u} = A_\oplus \hat{\phi}_\oplus \tilde{u}_\oplus + A_\ominus \hat{\phi}_\ominus \tilde{u}_\ominus,$$ (21)

whose matrices now introduced are given by

$$A = \beta \left(\Phi_\oplus^\oplus + \Phi_\ominus^\ominus\right) + \left(2 \text{Pe}^{-1} N - I\right) D,$$ (22)

$$A_\oplus = (\beta - 1) I + 2 \text{Pe}^{-1} N, \quad A_\ominus = (\beta + 1) I - 2 \text{Pe}^{-1} N,$$ (23)

where

$$\Phi_\oplus^\oplus = \hat{\phi}_\oplus \hat{\phi}_\oplus^T, \quad \Phi_\ominus^\ominus = \hat{\phi}_\ominus \hat{\phi}_\ominus^T, \quad N = \left(\Phi_\oplus^\oplus - \Phi_\ominus^\ominus - D\right) M^{-1}. \tag{24}$$

Note that (21) links the solution vector $\hat{u}$ and its time derivative to the two interface variables $\tilde{u}$ at the boundaries of the considered element.

In the actual context of simulations, (21) would be first solved (analytically) for $\hat{u}$ after an implicit time-stepping scheme is chosen. This is possible since it entails expressing $d\hat{u}/dt$ in terms of $\hat{u}$ at the current as well as previous time levels. The next step would be to insert the resulting expression for $\hat{u}$ into (18), from which a scalar equation whose only unknowns are $\tilde{u}$ at various interfaces is obtained. This equation is finally used for the assembly of a global system given suitable boundary conditions, which can be solved via direct or iterative techniques. Since the system's solution grants $\tilde{u}$ for all interfaces, $\hat{u}$ can be obtained locally for each element from the time-discrete version of (21). The reader is referred to [9] for the details of this procedure. In this work, however, as we are interested in the eigenanalysis of HDG, a different strategy is adopted, as outlined next.

3 Temporal Eigenanalysis

In the eigenanalysis of spectral element methods [2, 5], it is typical to assume wave-like solutions in the form $\hat{u} \propto \exp[i(\kappa x - \omega t)]$, whereby $\hat{u}^L = \hat{u}\exp(-i\kappa h)$ and $\hat{u}^R = \hat{u}\exp(+i\kappa h)$. Here, $\hat{u}$ is the solution vector of a "central" element, whereas $\hat{u}^L$ and $\hat{u}^R$ refer to solution vectors of neighbouring elements from the left (L) and from the right (R), respectively. For the HDG formulation, an additional assumption can be made regarding a wave-like behaviour for $\tilde{u}$. We assume that $\tilde{u}_\ominus^L = \tilde{u}\exp(-i\kappa'h)$ and $\tilde{u}_\oplus^R = \tilde{u}\exp(+i\kappa'h)$, where now $\tilde{u}$ is the interface variable shared by two adjacent elements, whereas $\tilde{u}_\ominus^L$ and $\tilde{u}_\oplus^R$ refer to interface variables at the nearest interfaces from the left/right (L/R). This second assumption is only natural given the connection between $\hat{u}$ and $\tilde{u}$. Actually, we now show that $\kappa' = \kappa$, which is not surprising.

We start from (21) assuming wave-like behaviour for $\hat{u}$, obtaining

$$\left(-i\frac{\omega h}{2a}M + A\right)\hat{u} = A_\oplus\hat{\phi}_\oplus\tilde{u}_\oplus + A_\ominus\hat{\phi}_\ominus\tilde{u}_\ominus, \tag{25}$$

which uniquely defines $\hat{u}$ from $\tilde{u}_\oplus$ and $\tilde{u}_\ominus$. If the above is written for another element, say, the adjacent element from the right (a translation $x \mapsto x + h$), one has

$$\left(-i\frac{\omega h}{2a}M + A\right)\hat{u}\exp(i\kappa h) = A_\oplus\hat{\phi}_\oplus\tilde{u}_\oplus\exp(i\kappa'h) + A_\ominus\hat{\phi}_\ominus\tilde{u}_\ominus\exp(i\kappa'h), \tag{26}$$

which then implies

$$\left(-i\frac{\omega h}{2a}M + A\right)\hat{u}\frac{\exp(i\kappa h)}{\exp(i\kappa' h)} = A_\oplus\hat{\phi}_\oplus\tilde{u}_\oplus + A_\ominus\hat{\phi}_\ominus\tilde{u}_\ominus = \left(-i\frac{\omega h}{2a}M + A\right)\hat{u},$$

(27)

where (25) has been used on the right-hand side. Comparing the left- and right-most expressions above leads to $\exp(i\kappa' h) = \exp(i\kappa h)$, which means $\kappa' h = \kappa h + 2n\pi$, for n integer. This phase ambiguity can be sorted out by the evaluation of the x-derivative of (25) at $x + h$, given by

$$i\kappa\left(-i\frac{\omega h}{2a}M + A\right)\hat{u}\exp(i\kappa h) = i\kappa'\left(A_\oplus\hat{\phi}_\oplus\tilde{u}_\oplus + A_\ominus\hat{\phi}_\ominus\tilde{u}_\ominus\right)\exp(i\kappa' h),$$

(28)

which yields $\kappa' = \kappa$. This last step about the phase is, however, not really necessary to the eigenanalysis because only the complex exponential factors appear throughout the relevant equations, hence knowing that $\exp(i\kappa' h) = \exp(i\kappa h)$ is sufficient.

In the remainder of the study, orthonormal Legendre basis functions are assumed, whereby $M = I$. We note that numerical dispersion and diffusion eigencurves, which are the focus of the study, do not change depending on the basis functions adopted, provided that exact integrations are used in the spatial discretisation.

In the temporal analysis, an eigenvalue problem is set where, given a real-valued wavenumber κ, multiple $(P + 1)$ eigenvalues of the relevant eigenmatrix are associated to admissible complex-valued numerical frequencies $\omega = \omega(\kappa)$. The procedure to obtain this eigenvalue problem is described below.

We begin from (18), assuming $\tilde{u}^L_\ominus = \tilde{u}\exp(-i\kappa h)$ and $\tilde{u}^R_\oplus = \tilde{u}\exp(i\kappa h)$, to find

$$\tilde{u} = \left(\hat{\phi}^T_\oplus B^L_\oplus\hat{u}^L + \hat{\phi}^T_\ominus B^R_\ominus\hat{u}^R\right)b^{-1},$$

(29)

with scalar $b = b(\kappa h; \text{Pe}, \beta)$ defined as

$$b = \beta + \left[m^\oplus_\oplus + m^\ominus_\ominus - m^\oplus_\ominus\exp(i\kappa h) - m^\ominus_\oplus\exp(-i\kappa h)\right]\text{Pe}^{-1}.$$

(30)

Then, (29) is used into (21), relating the solution vector $\hat{u}$ at a given element to the state vectors of its left ($\hat{u}^L$) and right ($\hat{u}^R$) neighbours. From the wave-like behaviour of $\hat{u}$ and the relations $\hat{u}^L = \hat{u}\exp(-i\kappa h)$ and $\hat{u}^R = \hat{u}\exp(+i\kappa h)$, one can arrive at

$$-i\varpi h\hat{u} = Z\hat{u},$$

(31)

where $\varpi = \omega/a$ and matrix $Z = Z(\kappa h; \text{Pe}, \beta)$ is given by

$$Z = 2b^{-1}[A_\oplus\Phi^\oplus_\ominus B_\ominus\exp(i\kappa h) + A_\ominus\Phi^\ominus_\oplus B_\oplus\exp(-i\kappa h)+$$
$$+A_\oplus\Phi^\oplus_\oplus B_\oplus + A_\ominus\Phi^\ominus_\ominus B_\ominus - Ab],$$

(32)

in which $\Phi_\oplus^\oplus$ and $\Phi_\ominus^\ominus$ are given by (24), whereas

$$\Phi_\ominus^\oplus = \hat{\phi}_\oplus \hat{\phi}_\ominus^T , \quad \Phi_\oplus^\ominus = \hat{\phi}_\ominus \hat{\phi}_\oplus^T .$$ (33)

In (31), we have the desired eigenvalue problem of size $P + 1$, which thus supports this same number of eigenvalues λ_j. These are related to the (normalised) numerical frequencies ϖ_j via

$$\varpi_j h = i\lambda_j \{Z(\kappa h)\} .$$ (34)

Typically, one of the eigenvalues represents the so-called primary eigenmode, while the remaining ones can be regarded as secondary as they simply replicate the behaviour of the primary mode on shifted wavenumber ranges. This formally allows us to focus on the analysis of the primary eigenmode and on its dispersion and diffusion eigencurves. The reader is referred to [2, 5] for the concepts relevant to the separation of primary and secondary modes adopted in this work.

Once the primary mode is identified, the scheme's numerical diffusion behaviour can be assessed in wavenumber space through the imaginary part of $\varpi_* h$, where the asterisk subscript denotes the primary mode from (34). Note that numerical diffusion is especially relevant to turbulence computations as it impacts not only accuracy, but also stability. Note that eigencurves are entirely defined by the polynomial order P, the upwinding parameter β and, in case viscosity is present, the normalised Péclet number $\mathrm{Pe}^\star = |a| \hbar/\mu$, with $\hbar = h/(P+1)$. Standard upwinding is here assumed.

Figure 2 depicts a comparison between HDG's primary dissipation curves for pure advection and for advection-diffusion at $\mathrm{Pe}^\star = 100$ for $P = 1, 4$ and 7. As explained further below, this is about the lowest value of $\mathrm{Pe}^\star$ one achieves (domain-wise) in a turbulent flow computation. However, at this $\mathrm{Pe}^\star$, viscous effects are still somewhat weak in regular (linear-scale) plots of $\varpi_i \hbar$ vs. $\kappa \hbar$, where ϖ_i is the absolute value of ϖ's imaginary part. This is especially true for $P \leq 4$. Hence, Fig. 2 also shows these plots in log-log scale, highlighting what happens at well-resolved wavenumbers.

The log-log plots in Fig. 2 are revealing. They make clear that HDG's numerical diffusion follows the correct diffusive behaviour up to a certain wavenumber, here-inafter named κ_c, beyond which upwind dissipation overcomes viscous diffusion. The exact diffusive behaviour, as derived from our model problem, is given by

$$\varpi_i \hbar = (\kappa \hbar)^2 / \mathrm{Pe}^\star \quad \text{or} \quad \log_{10}(\varpi_i \hbar) = 2 \log_{10}(\kappa \hbar) - \log_{10}(\mathrm{Pe}^\star) ,$$ (35)

showing that, as $\mathrm{Pe}^\star$ increases, the reference line of exact diffusive behaviour shifts downwards, reducing the value of $\kappa_c \hbar$. Also, for a given number of DOFs, i.e. fixed $\hbar$, increasing the discretisation order increases κ_c. This type of analysis reveals how upwind dissipation and viscous diffusion complement each other, allowing also for the estimation of the wavenumber κ_c after which upwinding dominates. The

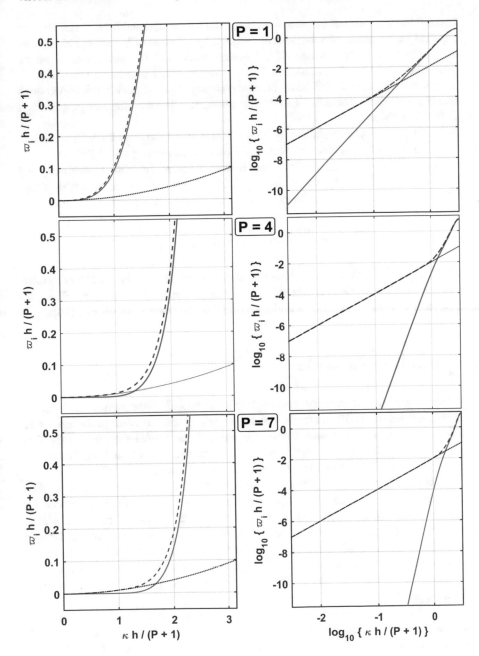

Fig. 2 Normalised numerical diffusion in bilinear (left) and log-log plots (right) for $P = 1$, $P = 4$ and $P = 7$ (top to bottom), with/without viscosity (dashed/full curve), the former considering $Pe^\star = 100$. The exact diffusive behaviour is shown as a dotted parabola/line (left/right plots)

latter, though important for small-scale regularisation and stability, is not entirely physical in the sense of subgrid-scale modelling. Hence, k_c values could be used as quality criteria for implicit LES / under-resolved DNS approaches based on discontinuous SEM. For transitional flows, where small numerical dissipation is particularly important, this kind of analysis might prove very useful. Although specific estimates would be needed for different schemes, the analysis strategy should be similar.

Finally, it is now explained why $Pe^\star = 100$ is about the lowest Péclet value one may find in a turbulent flow simulation. As candidates for very small $Pe^\star$, one could think of the near-wall region of turbulent boundary layers, given the low velocity and small mesh spacing in typical wall-resolved LES. For the viscous sublayer, where $u^+ < 5$, the streamwise Peclét number can be evaluated using wall quantities:

$$Pe^\star = \frac{u(y)\,\hbar}{\nu} = \frac{u^+\,\hbar^+}{\nu}\,u_\tau\,\delta_\nu = u^+\,\hbar^+\,, \tag{36}$$

where by definition $\nu = u_\tau\,\delta_\nu$, being u_τ the friction velocity and δ_ν the associated viscous lengthscale. Our argument is then concluded since $50 < \Delta x^+ = \hbar^+ < 150$ in typical wall-resolved LES or under-resolved DNS approaches, cf. e.g. [10].

4   Concluding Remarks

We presented a preliminary study of the numerical dispersion and diffusion characteristics of HDG methods for linear advection-diffusion problems using the temporal eigenanalysis technique. To the authors' knowledge, this is the first eigenanalysis of HDG methods, and also one of the first of such analyses of a discontinuous SEM to consider viscous diffusion effects, cf. also [11].

It was shown that, for the range of Péclet numbers encountered in under-resolved turbulence simulations, upwind (numerical) dissipation dominates viscous (physical) diffusion in the smallest resolved scales. Only in the large scales, the effect of viscous diffusion becomes significant. The wavenumber beyond which upwind dissipation overcomes viscous diffusion, and its dependence on the polynomial order, can be estimated through eigenanalysis, and this can be used as quality criterion for LES and DNS in general, and for implicit LES/under-resolved DNS in particular.

Future work includes further analysing the interplay between viscous and upwind diffusion, investigating other numerical fluxes (e.g. over-upwinding $\beta \gg 1$, nearly central fluxes $\beta \approx 0$, non-zero viscous stabilization $\sigma \neq 0$), and testing eigenanalysis against actual turbulence simulations. Finally, the dispersion-diffusion characteristics of HDG methods for spatially developing simulations could be investigated using spatial eigenanalysis techniques.

Acknowledgements Pablo Fernandez would like to acknowledge financial support from the MIT Zakhartchenko Fellowship. Spencer J. Sherwin acknowledges support from EPSRC Platform grant EP/R029423/1.

References

1. Moura, R.C., Mengaldo, G., Peiró, J., Sherwin, S.J.: On the eddy-resolving capability of high-order discontinuous Galerkin approaches to implicit LES / under-resolved DNS of Euler turbulence. J. Comput. Phys. **330**, 615–623 (2017)
2. Moura, R.C., Sherwin, S.J., Peiró, J.: Linear dispersion-diffusion analysis and its application to under-resolved turbulence simulations using discontinuous Galerkin spectral/hp methods. J. Comput. Phys. **298**, 695–710 (2015)
3. Mengaldo, G., Moura, R.C., Giralda, B., Peiró, J., Sherwin, S.J.: Spatial eigensolution analysis of discontinuous Galerkin schemes with practical insights for under-resolved computations and implicit LES. Comput. Fluids **169**, 349–364 (2017)
4. Moura, R.C., Aman, M., Peiró, J., Sherwin, S.J.: Spatial eigenanalysis of spectral/hp continuous Galerkin schemes and their stabilisation via DG-mimicking spectral vanishing viscosity for high Reynolds number flows. J. Comput. Phys. **406**, 109112 (2020)
5. Moura, R.C., Sherwin, S.J., Peiró, J.: Eigensolution analysis of spectral/hp continuous Galerkin approximations to advection-diffusion problems: insights into spectral vanishing viscosity. J. Comput. Phys. **307**, 401–422 (2016)
6. Fernandez, P., Moura, R.C., Mengaldo, G., Peraire, J.: Non-modal analysis of spectral element methods: Towards accurate and robust large-eddy simulations. Comput. Methods Appl. Mech. Eng. **346**, 43–62 (2019)
7. Mengaldo, G., De Grazia, D., Vincent, P.E., Sherwin, S.J.: On the connections between discontinuous Galerkin and flux reconstruction schemes: extension to curvilinear meshes. J. Sci. Comput. **67**(3), 1272–1292 (2016)
8. Fernandez, P., Nguyen, N.C., Peraire, J.: The hybridized discontinuous Galerkin method for implicit large-eddy simulation of transitional turbulent flows. J. Comput. Phys. **336**, 308–329 (2017)
9. Peraire, J., Nguyen, N.C., Cockburn, B.: A hybridizable discontinuous Galerkin method for the compressible Euler and Navier-Stokes equations. In: Proceedings of the 48th AIAA Aerospace Sciences Meeting (AIAA Paper 2010-363), Orlando (2010)
10. Georgiadis, N.J., Rizzetta, D.P., Fureby, C.: Large-eddy simulation: current capabilities, recommended practices, and future research. AIAA J. **48**(8), 1772–1784 (2010)
11. Manzanero, J., Ferrer, E., Rubio, G., Valero, E.: Design of a Smagorinsky Spectral Vanishing Viscosity turbulence model for discontinuous Galerkin methods. Comput. Fluids **200**, 104440 (2020)

Spectral Galerkin Method for Solving Helmholtz and Laplace Dirichlet Problems on Multiple Open Arcs

Carlos Jerez-Hanckes and José Pinto

1 Introduction

We seek solutions of Helmholtz and Laplace equations in a two-dimensional plane after removing a finite collection of open finite curves—also called arcs. This setting can be found in areas such as structural and mechanical engineering [2], or biomedical imaging [11] to name a few. Such problems pose the following challenges: (1) *unbounded domains*, which call for boundary integral methods with carefully chosen radiation conditions; (2) *singular behaviors* of solutions near arc endpoints; and (3) *large number of degrees of freedom* when the wavenumber or number of arcs increase.

Our approach is to recast the problem as a system of boundary integral equations defined on the arcs, so as to obtain an integral representation of the volume solution. Well-posedness for a single arc was proven in [9], with an extension to the multiple arcs case given in [5]. We will consider numerical approximations of the resulting surface densities based on Galerkin-Bubnov discretizations of the corresponding system of boundary integral equations.

In the present note, we start by briefly introducing a spectral scheme to account for general arcs as well as for a wide wavenumber range. We show that significant reduction in both memory consumption and computational work can be achieved by an *ad hoc* matrix compression algorithm. Moreover, we establish detailed interde-

C. Jerez-Hanckes (✉)
Faculty of Engineering and Sciences, Universidad Adolfo Ibañez, Santiago, Chile
e-mail: carlos.jerez@uai.cl

J. Pinto
School of Engineering, Pontificia Universidad Católica de Chile, Santiago, Chile
e-mail: jspinto@uc.cl

© The Author(s) 2020
S. J. Sherwin et al. (eds.), *Spectral and High Order Methods for Partial Differential Equations ICOSAHOM 2018*, Lecture Notes in Computational Science and Engineering 134, https://doi.org/10.1007/978-3-030-39647-3_30

pendencies between compression parameters and accuracy. Numerical experiments validate our claims and point out further improvements.

2 Continuous Model Problem

Let the canonical domain $(-1, 1) \times \{0\}$ be denoted by $\widehat{\Gamma}$. We say that $g : \widehat{\Gamma} \to \mathbb{C}$ is ρ-analytic if the function $t \mapsto g(t, 0)$ can be extended to an analytic function on the Bernstein ellipse of parameter $\rho > 1$ (cf. [10, Chapter 8]). We say that $\Lambda \subset \mathbb{R}^2$ is a regular Jordan arc of class C^m, for $m \in \mathbb{N}$, if it is the image of a bijective parametrization, denoted by $\mathbf{r} = (r_1, r_2)$, such that its components are $C^m(\widehat{\Gamma})$-functions, $\mathbf{r} : \widehat{\overline{\Gamma}} \to \overline{\Lambda}$ and $\|\mathbf{r}'(t)\|_2 > 0$, $\forall\, t \in \widehat{\overline{\Gamma}}$, where $\|\cdot\|_2$ is the Euclidean norm. Similarly, we define ρ-analytic arcs as those whose components are ρ-analytic. Throughout, we will assume that for any Λ regular Jordan arc, there exists an extension of Λ to $\tilde{\Lambda}$, which is a closed and keep the same regularity.

Consider a finite number $M \in \mathbb{N}$ of at least C^1-arcs, written $\{\Gamma_i\}_{i=1}^M$, such that their closures are mutually disjoint. Moreover, we assume that there are disjoint domains Ω_i whose boundaries are given by extensions $\partial\Omega_i = \tilde{\Gamma}_i$, for $i = 1, \ldots, M$. Let us define

$$\Gamma := \bigcup_{i=1}^M \Gamma_i \quad \text{and} \quad \Omega := \mathbb{R}^2 \setminus \overline{\Gamma}.$$

We say that Γ is of class C^m, $m \in \mathbb{N}$, if each arc Γ_i is of class C^m and analogously for the ρ-analytic case. For $i \in \{1, \ldots, M\}$, let $\mathbf{r}_i : \widehat{\Gamma} \to \Gamma_i$ and $g_i : \overline{\Gamma}_i \to \mathbb{C}$. We claim that $\mathbf{g} = (g_1, \ldots, g_M)$ is of class $C^m(\Gamma)$ if $g_i \circ \mathbf{r}_i \in C^m(\widehat{\Gamma})$, for $i \in \{1, \ldots, M\}$. A similar definition holds for the analytic case.

Let $G \subseteq \mathbb{R}^d$, $d = 1, 2$, be an open domain. For $s \in \mathbb{R}$, we denote by $H^s(G)$ the standard Sobolev spaces, by $H^s_{loc}(G)$ their locally integrable counterparts [8, Section 2.3], and by $\tilde{H}^{-s}(G)$ the corresponding dual spaces. The corresponding duality product (when the dual space of $L^2(G)$ is identified with itself) is denoted $\langle \cdot, \cdot \rangle_G$. Finally, $\tilde{H}^s_{(0)}(G)$ refers to mean-zero spaces [5, Section 2.3]. We will also make use of the following Hilbert space in $\mathbb{R}^2$:

$$W(G) := \left\{ U \in \mathcal{D}^*(G) : \frac{U(\mathbf{x})}{\sqrt{1 + \|\mathbf{x}\|_2^2}\, \log(2 + \|\mathbf{x}\|_2^2)} \in L^2(G), \nabla U \in L^2(G) \right\},$$

where $\mathcal{D}^*(G)$ is the dual space of $C^\infty(G) = \cap_{n>1} C^n(G)$. For $s \in \mathbb{R}$ and for the finite union of disjoint open arcs Γ, we define Cartesian product spaces as

$$\mathbb{H}^s(\Gamma) := H^s(\Gamma_1) \times H^s(\Gamma_2) \times \cdots \times H^s(\Gamma_M).$$

Spaces $\widetilde{\mathbb{H}}^s(\Gamma)$ and $\widetilde{\mathbb{H}}^s_{(0)}(\Gamma)$ are defined similarly. Also, $\mathbb{H}^s(\widehat{\Gamma})$ is to be understood as the Cartesian product $\prod_{i=1}^M H^s(\widehat{\Gamma})$. Finally, given an open bounded neighborhood G_i such that $\Gamma_i \subset \partial G_i$, Dirichlet traces are defined as extensions to $H^s(G_i)$, for $s \geq 1/2$, of the following operator (applied to smooth functions):

$$\gamma_i^{\pm} u(\mathbf{y}) := \lim_{\epsilon \downarrow 0} u(\mathbf{y} \pm \epsilon \mathbf{n}_i(\mathbf{y})),$$

where $\mathbf{n}_i(\mathbf{y})$ is the unitary vector with direction $(r'_{i,2}(t), -r'_{i,1}(t))$ and t such that $\mathbf{r}(t) = \mathbf{y}$. For a function u defined in an open neighborhood of Γ_i such that $\gamma_i^+ u = \gamma_i^- u$, we denote $\gamma_i u := \gamma_i^{\pm} u$.

Problem 1 (Volume Problem) Let $\mathbf{g} \in \mathbb{H}^{\frac{1}{2}}(\Gamma)$ and $\kappa \geq 0$. We seek $U \in H^1_{loc}(\Omega)$ such that

$$-\Delta U - \kappa^2 U = 0 \qquad\qquad \text{in } \Omega, \tag{1}$$

$$\gamma_i^{\pm} U = g_i \qquad\qquad \text{for } i = 1, \dots, M, \tag{2}$$

$$\text{Condition at infinity}(\kappa). \tag{3}$$

The behavior at infinity (3) depends on κ in the following way: if $\kappa > 0$, we employ the classical Sommerfeld condition [8, Section 3.9]. If $\kappa = 0$, we seek for solutions $U \in W(\Omega)$. This last condition was discussed in detail in [5, Remarks 3.9, 4.2 and 4.5] with uniqueness proofs for $\kappa \geq 0$ provided in [5, Propositions 3.8 and 3.10].

For $\kappa \geq 0$, we can express U solution of Problem 1 as

$$U(\mathbf{x}) = \sum_{i=1}^M (\mathsf{SL}_i[\kappa]\lambda_i)(\mathbf{x}), \quad \forall \mathbf{x} \in \Omega, \tag{4}$$

where

$$(\mathsf{SL}_i[\kappa]\lambda_i)(\mathbf{x}) := \int_{\Gamma_i} G_\kappa(\mathbf{x}, \mathbf{y})\lambda_i(\mathbf{y}) d\Gamma_i(\mathbf{y}), \quad \forall \mathbf{x} \in \Omega,$$

denotes the single layer potential generated at a curve Γ_i with G_κ the corresponding fundamental solution, defined as in [8, Section 3.1]. It is direct from (4) that U solves (1)–(2) in Ω (see [8, Theorem 3.1.1]). Also, it displays the desired behavior at infinity as long as each λ_i lies in the right functional space [5, Section 4]. In order to find the surface densities λ_i, we take Dirichlet traces $\gamma_i^{\pm}$ of the SL_j and impose boundary conditions (2). This naturally defines of weakly singular boundary integral operators:

$$\mathcal{L}_{ij}[\kappa] := \frac{1}{2}\left(\gamma_i^+ \mathsf{SL}_j[\kappa] + \gamma_i^- \mathsf{SL}_j[\kappa]\right) = \gamma_i \mathsf{SL}_j[\kappa],$$

and an equivalent boundary integral equation problem to Problem 1.

Problem 2 (Boundary Integral Problem) Let $g \in \mathbb{H}^{\frac{1}{2}}(\Gamma)$. For $\kappa > 0$, we seek $\lambda = (\lambda_1, \ldots, \lambda_M) \in \widetilde{\mathbb{H}}^{-\frac{1}{2}}(\Gamma)$ such that

$$\mathcal{L}[\kappa]\lambda = g,$$

where $\mathcal{L}[\kappa] : \widetilde{\mathbb{H}}^{-\frac{1}{2}}(\Gamma) \to \mathbb{H}^{\frac{1}{2}}(\Gamma)$ is a matrix operator with entries $\mathcal{L}[\kappa]_{ij} = \mathcal{L}_{ij}[\kappa]$, for $i, j \in \{1, \ldots M\}$. If $\kappa = 0$, we seek $\lambda \in \widetilde{\mathbb{H}}^{-\frac{1}{2}}_{\langle 0 \rangle}(\Gamma)$, given g in the dual space of the aforementioned space.

Theorem 1 (Theorem 4.13 in [5]) *For $\kappa > 0$, Problem 2 has a unique solution $\lambda \in \widetilde{\mathbb{H}}^{-\frac{1}{2}}(\Gamma)$, whereas for $\kappa = 0$ a unique solution exists in the subspace $\widetilde{\mathbb{H}}^{-\frac{1}{2}}_{\langle 0 \rangle}(\Gamma)$. Also, the following continuity estimate holds*

$$\|\lambda\|_{\widetilde{\mathbb{H}}^{-\frac{1}{2}}(\Gamma)} \leq C(\Gamma, \kappa) \|g\|_{\mathbb{H}^{\frac{1}{2}}(\Gamma)}.$$

3 Spectral Discretization

We present a family of finite dimensional subspaces in $\widetilde{\mathbb{H}}^{-\frac{1}{2}}(\Gamma)$ that can be used to approximate the solution of Problem 2 (*cf.* [4, 6]). Let $\mathbb{T}_N(\widehat{\Gamma})$ denote the space spanned by first kind Chebyshev polynomials, denoted by $\{T_n\}_{n=0}^N$, of degree lower or equal than N on $\widehat{\Gamma}$, orthogonal with the $L^2(-1, 1)$ inner product, under the weight w^{-1} with $w(t) := \sqrt{1 - t^2}$. Now, let us construct elements $p_n^i = T_n \circ \mathbf{r}_i^{-1}$ over each arc Γ_i spanning the space $\mathbb{T}_N(\Gamma_i)$. For practical reasons, we define the normalized space:

$$\overline{\mathbb{T}}_N(\Gamma_i) := \left\{ \bar{p}^i \in C(\Gamma_i) : \bar{p}_n^i := \frac{p_n^i}{\left\| \mathbf{r}_i' \circ \mathbf{r}_i^{-1} \right\|_2}, \quad p_n^i \in \mathbb{T}_N(\Gamma_i) \right\}.$$

We account for edge singularities by multiplying the basis $\{\bar{p}_n^i\}_{n=0}^N$ by a suitable weight:

$$\mathbb{Q}_N(\Gamma_i) := \left\{ q_n^i := w_i^{-1} \bar{p}_n^i : \bar{p}_n^i \in \overline{\mathbb{T}}_N(\Gamma_i) \right\},$$

wherein $w_i := w \circ \mathbf{r}_i^{-1}$. The corresponding basis for $\mathbb{Q}_N(\Gamma_i)$ will be denoted $\{q_n^i\}_{n=0}^N$. By Chebyshev orthogonality, we can easily define the mean-zero subspace $\mathbb{Q}_{N,\langle 0 \rangle}(\Gamma_i) := \mathbb{Q}_N(\Gamma_i) \setminus \mathbb{Q}_0(\Gamma_i)$, spanned by $\{q_n^i\}_{n=1}^N$. With these definitions, we set the discretization space for a Galerkin-Bubnov solution of Problem 2 as

$$\mathbb{H}_N[\kappa] := \begin{cases} \prod_{i=1}^{M} \mathbb{Q}_{N,\langle 0 \rangle}(\Gamma_i) & \text{for } \kappa = 0, \\ \prod_{i=1}^{M} \mathbb{Q}_N(\Gamma_i) & \text{for } \kappa > 0. \end{cases}$$

Problem 3 (Linear System) For $\kappa > 0$, let $N \in \mathbb{N}$ and $\mathbf{g} \in \mathbb{H}^{\frac{1}{2}}(\Gamma)$ be the same as in Problem 2. Then, we seek coefficients $\mathfrak{u} = (\mathfrak{u}_1, \ldots, \mathfrak{u}_M) \in \mathbb{C}^{M(N+1)}$, such that

$$\mathbf{L}[\kappa]\mathfrak{u} = \mathfrak{g}.$$

Therein, we have defined the Galerkin matrix $\mathbf{L}[\kappa] \in \mathbb{C}^{M(N+1) \times M(N+1)}$ composed of matrix blocks $\mathsf{L}_{ij}[\kappa] \in \mathbb{C}^{(N+1) \times (N+1)}$ whose entries are

$$(\mathsf{L}_{ij}[\kappa])_{lm} = \left\langle \mathcal{L}_{ij}[\kappa] q_m^j, q_l^i \right\rangle_{\Gamma_i} = \left\langle \widehat{\mathcal{L}}_{ij}[\kappa] w^{-1} T_m, w^{-1} T_l \right\rangle_{\widehat{\Gamma}}.$$

There, $\widehat{\mathcal{L}}_{ij}[\kappa]$ is the weakly-singular operator whose kernel is parametrized by $\mathbf{r}_i$, $\mathbf{r}_j$ and right-hand $\mathfrak{g} = (\mathfrak{g}_1, \ldots, \mathfrak{g}_M) \in \mathbb{C}^{M(N+1)}$ with components

$$(\mathfrak{g}_i)_l = \left\langle g_i, q_l^i \right\rangle_{\Gamma_i} = \left\langle \widehat{g}_i, w^{-1} T_l \right\rangle_{\widehat{\Gamma}},$$

where $\widehat{g}_i = g_i \circ \mathbf{r}_i$. The approximation $\lambda_N \in \mathbb{H}_N[\kappa]$ is constructed as

$$(\lambda_N)_i = \sum_{m=0}^{N} (\mathfrak{u}_i)_m q_m^i \quad \text{in } \Gamma_i, \quad \text{for all } i \in \{1, \ldots, M\}.$$

For $k = 0$ we need $\mathbf{g}$ as in Problem 2; we also have $\mathfrak{u} \in \mathbb{C}^{MN}$, and $\mathbf{L}[0] \in \mathbb{C}^{MN \times MN}$ since the approximation space is $\mathbb{H}_N[0]$. By conformity and density of these spaces in $\widetilde{\mathbb{H}}^{-\frac{1}{2}}(\Gamma)$, one derives the following result:

Theorem 2 (Theorem 4.23 [4]) *Let $\kappa \geq 0$, $m \in \mathbb{N}$ with $m > 2$, $\Gamma \in C^m$, $\mathbf{g} \in C^m(\Gamma)$, and λ be the only solution of Problem 2. Then, there exists $N_0 \in \mathbb{N}$ such that for every $N > N_0 \in \mathbb{N}$ there is a unique $\lambda_N \in \mathbb{H}_N[\kappa]$ solution of Problem 3. Moreover, the following error convergence rates hold*

$$\|\lambda - \lambda_N\|_{\widetilde{\mathbb{H}}^{-\frac{1}{2}}(\Gamma)} \leq C(\Gamma, \kappa) N^{-m+1}.$$

Moreover, if Γ and $\mathbf{g}$ are ρ-analytic with $\rho > 1$, we have the following super-algebraic convergence rates

$$\|\lambda - \lambda_N\|_{\widetilde{\mathbb{H}}^{-\frac{1}{2}}(\Gamma)} \leq C(\Gamma, \kappa) \rho^{-N+2} \sqrt{N},$$

where $C(\Gamma, \kappa)$ is a positive constant, which does not depend on N.

Remark 1 Observe that the constants $C(\Gamma, \kappa)$ and N_0 depend on the geometry and frequency. To the best of our knowledge previous convergence results for 2D arcs are somehow limited. For intervals, the result was established in [6] whereas for more general arc results are only obtained for the Laplace case [1]. Super-algebraic convergence rates can be achieved by the method detailed in [3], though their scheme is limited to intervals and to the case of elliptic problems ($N_0 = 0$). More complex cases are still an open problem.

4 Numerical Implementation and Compression Algorithm

Before fleshing out our proposed compression technique, we explain how $\mathbf{L}[\kappa]$ and $\mathfrak{g}$ of Problem 3 are computed. For the right-hand side, one must compute integrals of the form:

$$\int_{-1}^{1} \widehat{g}(t) w^{-1}(t) T_l(t) dt, \quad \forall l \in \mathbb{N}_0,$$

which corresponds to Fourier-Chebyshev coefficients of $\widehat{g}(t)$ and can be approximated using the Fast Fourier Transform [10]. Computations for matrix terms $\mathbf{L}_{ij}[\kappa]$ are split into two groups: (a) *cross-interactions*, where test and trial functions supports lie along curves Γ_i, Γ_j with $i \neq j$; and (b) *self-interactions*, where both trial and test functions are defined on the same curve. As for cross-interactions the integral kernel is smooth, we use the same computational procedure for the right-hand side.

For self-interactions, the kernel function has a singularity that can be characterized as

$$G_k(\mathbf{r}(t), \mathbf{r}(s)) = (2\pi)^{-1} \log |t - s| J_0(k \| \mathbf{r}(t) - \mathbf{r}(s) \|_2) + G_r(t, s), \quad t \neq s,$$

for $t, s \in \widehat{\Gamma}$, where J_0 is the zeroth-order first kind Bessel function, and G_r is a regular function. Thus, integration for the regular part is done as in the cross-interaction case, while integrals with the first term as kernel are obtained by convolution as integrals for $\log |t - s|$ are known (see [6, Remark 4.2]).

Yet, as κ increases, larger values of N will be required, and thus, the need to compress the resulting matrix terms. As stated in [10, Chapters 7 and 8], the regularity of a function controls the decay of its Fourier-Chebyshev coefficients. Hence, as the entries of the matrix $\mathbf{L}[\kappa]$ are precisely such coefficients, for a smooth kernel one observes fast decaying terms. This implies that we can select small blocks to approximate the matrix and obtain a sparse approximation by discarding the remaining entries, based on a predetermined tolerance $\epsilon > 0$. Specifically, the kernel function is smooth when we compute cross-interactions. Let the routine Quadrature(l,m) compute the term (l, m) of this interaction matrix using a 2D

Gauss-Chebyshev quadrature. Given a tolerance $\epsilon > 0$, we minimize the number of computations needed by performing the following binary search:

Matrix Compression Algorithm

```
INPUT: Tolerance (Tol), Max level of search (Lmax)
OUTPUT: Number of columns to use (Ncols)
INITIALIZE: Ncols = N, level = 0, a = 0, b = N
While{level < Lmax}
    m = (a+b)/2
    Tleft = m-1
    Tcenter = m
    Tright =  m+1
    Veft = abs(Quadrature(0,Teft))
    Vcenter = abs(Quadrature(0,Tcenter))
    Vright = abs(Quadrature(0,Tright))
    If{Vright & Vcenter < 0.5*Tol} or {Vleft & Vcenter < 0.5*Tol}
        b = m
    Else
        a = m
    EndIF
    level++
EndWhile
Ncols = b
```

The algorithm returns the minimum number of columns required, N_{cols}, by searching in the first row the minimum index such that the matrix entries' absolute value is lower than ϵ. The binary search is restricted to a depth $L_{max} \in \mathbb{N}$. The same procedure is used to estimate the number of rows, N_{rows}, by executing a binary search in the first column. Once N_{cols} and N_{rows} are selected, we define $N_\epsilon := \max\{N_{rows}, N_{cols}\}$ and compute the block of size $N_\epsilon \times N_\epsilon$ as in the full matrix implementation.

The matrix compression percentage will strongly depend on the regularity of the arcs involved. For ρ-analytic arcs, using [10, Theorem 8.1] we can prove the lower bound:

$$N_\epsilon \geq \frac{-\log \epsilon}{2 \Upsilon \log \rho},$$

where Υ is an upper bound for the absolute value of the kernel in the corresponding Bernstein ellipse. However, since compression is done by a binary search, the bound for the compression rate depends on L_{max} as

$$N_\epsilon \geq \frac{N}{2^{L_{max}}}.$$

Compression of self-interaction blocks does not follow the same ideas. In fact, these blocks can be characterized as two perturbations over the canonical case, $\Gamma = \widehat{\Gamma}$ for $\kappa = 0$, leading to a diagonal matrix. Namely, these are

1. A low frequency perturbation caused by the mapping $r_i : \widehat{\Gamma} \mapsto \Gamma$, similar to the cross-interaction case.
2. A frequency perturbation that creates banded matrices.

In order to reduce memory consumption—though not computational time—we discard the entries of the self-interaction matrices lower than the given tolerance.

As expected, matrix compression induces an extra error as it perturbs the original linear system solved by λ_N in Problem 3. We denote by $\mathbf{L}_\epsilon[k]$ the matrix generated by the compression algorithm with tolerance ϵ, and define the matrix difference $\Delta \mathbf{L}_\epsilon[k] := \mathbf{L}_\epsilon[k] - \mathbf{L}[k]$. We seek to control the solution $\mathfrak{u}^\epsilon = \mathfrak{u} + \Delta \mathfrak{u}$ of

$$(\mathbf{L}[k] + \Delta \mathbf{L}_\epsilon[k])\mathfrak{u}^\epsilon = \mathfrak{g},$$

where $\mathfrak{u}$ and $\mathfrak{g}$ are the same as in Problem 3. In order to bound this error, we will assume that, for every pair of indices (i, j) in the matrix $\mathbf{L}[k]$, we have,

$$|(\Delta \mathbf{L}_\epsilon[k])_{ij}| < \epsilon. \tag{5}$$

Theorem 3 *Let $N \in \mathbb{N}$ be such there is only one λ_N solution of Problem 3. Then, there is a constant $C(\Gamma, \kappa) > 0$, not depending on N, such that*

$$\frac{\|\Delta \mathfrak{u}\|_2}{\|\mathfrak{u}\|_2} \leq \left| \frac{N\epsilon}{C(\kappa, \Gamma) - N\epsilon} \right|.$$

Proof By [7, Section 1.13.2] we have that

$$\frac{\|\Delta \mathfrak{u}\|_2}{\|\mathfrak{u}\|_2} \leq \frac{\|\Delta \mathbf{L}_\epsilon[k]\|_2}{\|(\mathbf{L}[k])^{-1}\|_2 - \|\Delta \mathbf{L}_\epsilon[k]\|_2},$$

and thus, we need to estimate $\|\Delta \mathbf{L}_\epsilon[k]\|_2$ and $\|(\mathbf{L}[k])^{-1}\|_2$. The bound for the first term is direct from (5) and matrix norm definitions. By the classical bound of a matrix inverse and the continuity of the associated boundary integral operator, it holds that

$$\left\| \mathbf{L}[k]^{-1} \mathfrak{g} \right\|_2 \geq \left\| \mathbf{L}[k]\mathfrak{g} \right\|_2^{-1} \geq C(\kappa, \Gamma),$$

from where the result follows directly. $\qquad\qquad\qquad\qquad\qquad\qquad\qquad\qquad\qquad \square$

We can also estimate the error introduced by the compression algorithm in terms of the energy norm. In order to do so, define $(\lambda_N^\epsilon)_i := \sum_{m=0}^{N}(\mathbf{u}_i^\epsilon)_m q_m^i$ in Γ_i. By the same arguments in the above proof, we obtain

$$\left\| \lambda_N - \lambda_N^\epsilon \right\|_{\widetilde{\mathbb{H}}^{-\frac{1}{2}}(\Gamma)} \leq C_1(\kappa, \Gamma) \left\| \mathbf{g} \right\|_{\mathbb{H}(\Gamma)^{\frac{1}{2}}} \frac{\epsilon N^{3/2}}{C_2(\kappa, \Gamma) - \epsilon N},$$

where $\mathbf{g}$ is the same that in Problem 2 and $C_1(\kappa, \Gamma)$, $C_2(\kappa, \Gamma)$ are two different constants.

Remark 2 Our compression algorithm produces a faster and less memory demanding implementation of the spectral Galerkin method at the cost of accuracy loss, similar to fast multipole or hierarchical matrices methods. Moreover, once we have compressed the matrix, we can implement a fast matrixvector product.

5 Numerical Results

To illustrate the above claims, Fig. 1 presents convergence results for different wavenumbers, $\kappa = 0, 25, 50, 100$ for a configuration of $M = 28$ arcs. As the chosen geometry and excitation are given by analytic functions, Theorem 2 predicts exponential rate of convergence as observed numerically.

Table 1 provides matrix compression results for $\kappa = 100$ and for the same geometry of Fig. 1. It presents the percentage of non-zero entries (%NNZ) and relative errors as bounded in Theorem 3 as functions of the maximum level of binary

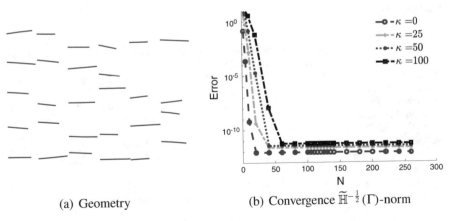

(a) Geometry   (b) Convergence $\widetilde{\mathbb{H}}^{-\frac{1}{2}}(\Gamma)$-norm

Fig. 1 (a) Smooth geometry with $M = 28$ open arcs parametrized as $\mathbf{r}_i(t) = (a_i t, c_i \sin(b_i t) + d_i)$, with $a_i \in [0.14, 0.25]$, $b_i \in [0, 0.2]$, $c_i \in [1, 2]$, $d_i \in [0, 20]$, $t \in [-1, 1]$. (b) Convergence results for different wavenumbers and a planewave excitation along $(1, 1)$. Errors computed against an overkill solution using $N = 660$ per arc

Table 1 Compression performance for $\kappa = 100$

Order	$L_{max} = 2$		$L_{max} = 3$		$L_{max} = 4$	
	% NNZ	Rel. error	% NNZ	Rel. error	% NNZ	Rel. error
	$\epsilon = 1e-6$					
5	65.24	5.05e−01	65.24	5.05e−01	65.24	5.05e−01
10	81.62	5.32e−01	81.62	5.32e−01	81.62	5.32e−01
20	89.41	2.33e−01	88.62	2.33e−01	88.31	2.33e−01
40	77.63	9.10e−04	70.63	9.10e−04	67.11	9.10e−04
60	45.25	2.02e−07	36.68	2.76e−07	33.36	3.31e−07
80	27.20	1.97e−07	21.97	3.17e−07	19.50	3.35e−07
	$\epsilon = 1e-10$					
5	65.29	5.05e−01	65.29	5.05e−01	65.29	5.05e−01
10	81.68	5.32e−01	81.68	5.32e−01	81.68	5.32e−01
20	89.86	2.33e−01	89.59	2.33e−01	89.44	2.33e−01
40	83.46	9.10e−04	78.70	9.10e−04	76.28	9.10e−04
60	51.94	2.14e−09	44.87	3.19e−09	40.70	3.89e−09
80	33.86	2.31e−09	26.89	1.73e−08	23.78	1.73e−10

search (L_{max}), tolerances (ϵ), and polynomial order per arc (Order). For low orders (Order < 60), relative errors are quite large, and therefore, most of the matrix terms are kept. This is due to an insufficient number of matrix entries to solve the problem with good accuracy (see Fig. 1), rendering compression pointless. On the other hand, once convergence is achieved, the compression error drastically decreases along with the percentage of matrix terms stored.

References

1. Atkinson, K.E., Sloan, I.H.: The numerical solution of first-kind logarithmic-kernel integral equations on smooth open arcs. Math. Comput. **56**(193), 119–139 (1991)
2. Costabel, M., Dauge, M.: Crack singularities for general elliptic systems. Math. Nachr. **235**(1), 29–49 (2002)
3. Hewett, D.P., Langdon, S., Chandler-Wilde, S.N.: A frequency-independent boundary element method for scattering by two-dimensional screens and apertures. IMA J. Numer. Anal. **35**(4), 1698–1728 (2014)
4. Jerez-Hanckes, C., Pinto, J.: High-order Galerkin method for Helmholtz and Laplace problems on multiple open arcs. Technical Report 2018-49, Seminar for Applied Mathematics, ETH Zürich (2018)
5. Jerez-Hanckes, C., Pinto, J.: Well-posedness of Helmholtz and Laplace problems in unbounded domains with multiple screens. Technical Report 2018-45, Seminar for Applied Mathematics, ETH Zürich (2018)
6. Jerez-Hanckes, C., Nicaise, S., Urzúa-Torres, C.: Fast spectral Galerkin method for logarithmic singular equations on a segment. J. Comput. Math. **36**(1), 128–158 (2018)

7. Saad, Y.: Iterative Methods for Sparse Linear Systems. Computer Science Series. PWS Publishing Company, Boston (1996)
8. Sauter, S., Schwab, C.: Boundary Element Methods. Springer Series in Computational Mathematics. Springer, Berlin (2010)
9. Stephan, E.P.: A boundary integral equation method for three-dimensional crack problems in elasticity. Math. Methods Appl. Sci. **8**(4), 609–623 (1986)
10. Trefethen, L.: Approximation Theory and Approximation Practice. Other Titles in Applied Mathematics. SIAM, Philadelphia (2013)
11. Verrall, G., Slavotinek, J., Barnes, P., Fon, G., Spriggins, A.: Clinical risk factors for hamstring muscle strain injury: a prospective study with correlation of injury by magnetic resonance imaging. Br. J. Sports Med. **35**(6), 435–439 (2001)

Explicit Polynomial Trefftz-DG Method for Space-Time Elasto-Acoustics

H. Barucq, H. Calandra, J. Diaz, and E. Shishenina

1 Trefftz-DG Formulation for the Elasto-Acoustic Equation

Trefftz methods are particular finite element methods where the basis and test functions are locally solutions to the partial differential equation that governs the problem to be solved. Compared to the existing literature for solving frequency problems, space-time Trefftz methods are still not widely used. One reason could be that they require using space-time meshes [6, 12]. To our knowledge, few references on Trefftz approximations of time-dependent wave equations are available and they mainly address theoretical properties in the case of Acoustics and Electromagnetism [4, 8, 10, 11]. They provide convergence and stability studies and some numerical results are displayed by using plane wave bases in 1D + time dimension. Numerical in 2D + time dimensions are proposed in [4] for electromagnetism. There are also some studies devoted to the second-order formulation of the acoustic wave equation approximated in Trefftz spaces by the mean of Lagrange multipliers [1, 13]. In [3], we have proposed a Trefftz-DG formulation for elasto-acoustic. The method required the inversion of a huge sparse matrix. The goal of this paper is to show how to derive a semi-explicit scheme, requiring only the inversion of a block-diagonal matrix on each element of the mesh.

H. Barucq · J. Diaz · E. Shishenina (✉)
Magique-3D, Inria, E2S UPPA, CNRS, Université de Pau et des Pays de l'Adour, Pau, France
e-mail: helene.barucq@inria.fr; julien.diaz@inria.fr; elvira.shishenina@inria.fr

H. Calandra
Total SA, CSTJF Total, Pau, France
e-mail: henri.calandra@total.com

S. J. Sherwin et al. (eds.), *Spectral and High Order Methods for Partial Differential Equations ICOSAHOM 2018*, Lecture Notes in Computational Science and Engineering 134, https://doi.org/10.1007/978-3-030-39647-3_31

In this section, following [10] and the framework therein, we propose a formulation of the elasto-acoustic coupling reading as a first-order system. Here and further the sub-scripts F and S corresponds to the acoustic (fluid) and elastodynamic (solid) domains.

1.1 Elasto-Acoustic Equations

We introduce a space-time domain $Q \equiv (\Omega_F \cup \Omega_S) \times I$, where $\Omega_F \subset \mathbb{R}^d$ is a bounded Lipschitz domain of dimension d filled with fluid, $\Omega_S \subset \mathbb{R}^d$ is a bounded Lipschitz elastodynamic domain of dimension d filled with solid, and $I \equiv [0, T]$ is the time interval. All medium parameters $c_F \equiv c_F(\mathbf{x})$ and $\rho_F \equiv \rho_F(\mathbf{x})$, standing for the acoustic wave propagation velocity and fluid density respectively, as well as the inverted stiffness tensor $\underline{\underline{\mathbf{C}}}^{-1}(\mathbf{x}) \equiv \underline{\underline{\mathbf{A}}}(\mathbf{x})$ and the solid density $\rho_S \equiv \rho_S(\mathbf{x})$, are assumed to be piecewise constant and positive. We denote by $\Gamma_{FS} = \Omega_F \cap \Omega_S$ the fluid-solid interface. The elasto-acoustic system of equations is based on the coupling of the first-order acoustic equation, written in terms of velocity $\mathbf{v}_F \equiv \mathbf{v}_F(\mathbf{x}, t)$ and pressure $p \equiv p(\mathbf{x}, t)$ fields:

$$
\begin{cases}
\dfrac{1}{c_F^2 \rho_F} \dfrac{\partial p}{\partial t} + \mathrm{div} \mathbf{v}_F = f & \text{in } Q_F, \\[2mm]
\rho_F \dfrac{\partial \mathbf{v}_F}{\partial t} + \nabla p = 0 & \text{in } Q_F, \\[2mm]
\mathbf{v}_F(\cdot, 0) = \mathbf{v}_{F0}, \ p(\cdot, 0) = p_0 & \text{in } \Omega_F, \\[2mm]
\mathbf{v}_F \cdot \mathbf{n}_{\Omega_F} = g_F & \text{in } \partial\Omega_F \backslash \Gamma_{FS} \times I,
\end{cases}
\tag{1}
$$

where $\mathbf{n}_{\Omega_F}$ is the normal vector to $\partial\Omega_F$, the source term $f \equiv f(\mathbf{x}, t)$, the boundary condition g_F, the velocity $\mathbf{v}_{F0}$ and the pressure p_0 are the initial data, with the first-order elastodynamic system, written in terms of velocity $\mathbf{v}_S \equiv \mathbf{v}_S(\mathbf{x}, t)$ and stress tensor (symmetrical and positive) $\underline{\underline{\sigma}} \equiv \underline{\underline{\sigma}}(\mathbf{x}, t)$ fields:

$$
\begin{cases}
\underline{\underline{\mathbf{A}}} \dfrac{\partial \underline{\underline{\sigma}}}{\partial t} - \underline{\underline{\varepsilon}}(\mathbf{v}_S) = 0 & \text{in } Q_S, \\[2mm]
\rho_S \dfrac{\partial \mathbf{v}_S}{\partial t} - \mathbf{div}\, \underline{\underline{\sigma}} = 0 & \text{in } Q_S, \\[2mm]
\mathbf{v}_S(\cdot, 0) = \mathbf{v}_{S0}, \ \underline{\underline{\sigma}}(\cdot, 0) = \underline{\underline{\sigma}}_0 & \text{in } \Omega_S, \\[2mm]
\underline{\underline{\sigma}} \mathbf{n}_{\Omega_S} = \mathbf{g}_S & \text{in } \partial\Omega_S \backslash \Gamma_{FS} \times I,
\end{cases}
\tag{2}
$$

where $\mathbf{n}_{\Omega_S}$ is the normal vector to $\partial\Omega_S$, the boundary condition g_F, the velocity $\mathbf{v}_{S0}$ and the stress tensor $\boldsymbol{\sigma}_0$ are the initial data. The transmission conditions between the two systems (2) and (1) represent the continuity of velocity and stress normal components Γ_{FS}:

$$\begin{cases} \mathbf{v}_F \cdot \mathbf{n}_{\Gamma_{FS}} = \mathbf{v}_S \cdot \mathbf{n}_{\Gamma_{FS}} & \text{at } \Gamma_{FS}, \\ -p\mathbf{n}_{\Gamma_{FS}} = \underline{\boldsymbol{\sigma}}\mathbf{n}_{\Gamma_{FS}} & \text{at } \Gamma_{FS}. \end{cases} \tag{3}$$

The velocities aligned with the interface and the tangential stress remain unconstrained.

1.2 Space-Time Trefftz-DG Formulation

We introduce a non-overlapping space-time mesh $\mathcal{T}_h$ on Q composed of space-time Lipschitz elements $K_F \subset \Omega_F \times I$ and $K_S \subset \Omega_S \times I$. We denote by $\mathcal{T}_{Fh}$ (resp. $\mathcal{T}_{Sh}$) the restriction of $\mathcal{T}_h$ to the fluid (resp. solid) domain. Let $\mathbf{n}_{K_F} \equiv (\mathbf{n}^x_{K_F}, n^t_{K_F})$ be the outward-pointing unit normal vector on ∂K_F, and $\mathbf{n}_{K_S} \equiv (\mathbf{n}^x_{K_S}, n^t_{K_S})$ be the outward-pointing unit normal vector on ∂K_S. We assume that all medium parameters are constant in K_F and K_S respectively. The mesh skeleton $\mathcal{F}_h \equiv \bigcup_{K_{F,S}\in\mathcal{T}_h} \partial K_{F,S}$ can be decomposed into families of the internal $\mathcal{F}^Q_h$ faces, the fluid-solid $\mathcal{F}^{FS}_h$ faces, the boundary $\mathcal{F}^D_h$ faces, the initial and final time $\mathcal{F}^0_h$ and $\mathcal{F}^T_h$ element faces respectively, as it shown in Fig. 1. We introduce the space $V_h(\mathcal{T}_h)$ as a subspace of $L^2(Q)$ defined by $V_h(\mathcal{T}_h) = \{\phi \in L^2(Q), \phi_{|K_{F,S}} \in \mathbb{P}^p(K_{F,S})\}$. The unknowns $(\mathbf{v}_{Fh}, p_h, \mathbf{v}_{Sh}, \underline{\boldsymbol{\sigma}}_h)$ are supposed to be in $\mathbf{V}_h(\mathcal{T}_h) \equiv V_h(\mathcal{T}_{Fh})^d \times V_h(\mathcal{T}_{Fh}) \times V_h(\mathcal{T}_{Sh})^d \times V_h(\mathcal{T}_{Sh})^{d^2}$. We consider the test functions $\boldsymbol{\omega}_F, q, \boldsymbol{\omega}_S, \underline{\boldsymbol{\xi}}$ in

Fig. 1 Example of 1D + time mesh $\mathcal{T}_h$ covering Q. The internal element faces $\mathcal{F}^Q_h$ are represented by dotted line, the element faces of fluid-solid interface $\mathcal{F}^{FS}_h$—by dash-dotted line, the boundary element faces $\mathcal{F}^D_h$—by thick line, the initial $\mathcal{F}^0_h$ and the final $\mathcal{F}^T_h$ time element faces—by double and dashed line respectively

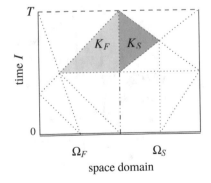

$\mathbf{T}(\mathcal{T}_h)$ for $\mathbf{v}_F$, p, $\mathbf{v}_S$ and $\underline{\underline{\sigma}}$ respectively, where the Trefftz space $\mathbf{T}(\mathcal{T}_h)$ is defined on the mesh $\mathcal{T}_h$ as follows:

$$\mathbf{T}(\mathcal{T}_h) \equiv \left\{ (\boldsymbol{\omega}_F,\, q,\, \boldsymbol{\omega}_S,\, \underline{\xi}) \in \mathbf{V}_h(\mathcal{T}_h) \text{ s. t. } \frac{1}{c_F^2 \rho_F} \frac{\partial q}{\partial t} + \operatorname{div} \boldsymbol{\omega}_F = 0,\ \rho_F \frac{\partial \boldsymbol{\omega}_F}{\partial t} + \nabla q = 0 \right.$$

$$\left. \forall K_F \in \mathcal{T}_{Fh}, \text{ and } \underline{\underline{\mathbf{A}}} \frac{\partial \underline{\xi}}{\partial t} - \underline{\underline{\varepsilon}}(\boldsymbol{\omega}_S) = 0,\ \rho_S \frac{\partial \boldsymbol{\omega}_S}{\partial t} - \operatorname{\mathbf{div}} \underline{\xi} = 0 \ \forall K_S \in \mathcal{T}_{Sh} \right\}.$$

This space is of Trefftz type since it is a subspace of the regular space $\mathbf{V}_h(\mathcal{T}_h)$ composed of local solutions of the volumic governing equations (1) and (2) set in each element K_F and K_S respectively.

As in the standard DG methods, the next step in order to obtain the variational formulation consists in multiplying the equations of (1) by the test functions q and $\boldsymbol{\omega}_F$ in $\mathbf{T}(\mathcal{T}_h)$, and the equations of (2) by the test functions $\underline{\xi}$ and $\boldsymbol{\omega}_S$ in $\mathbf{T}(\mathcal{T}_h)$ respectively, and, as is standard in space-time DG methods, we integrate by parts the obtained equations not only in space but also in time:

$$\sum_{K_F} \int_{\partial K_F} \left[\frac{1}{c_F^2 \rho_F} \check{p}_h q\, n_{K_F}^t + q \hat{\mathbf{v}}_{Fh} \cdot \mathbf{n}_{K_F}^x + \rho_F \check{\mathbf{v}}_{Fh} \cdot \boldsymbol{\omega}_F\, n_{K_F}^t + \hat{p}_h \boldsymbol{\omega}_F \cdot \mathbf{n}_{K_F}^x \right] ds +$$

$$\sum_{K_S} \int_{\partial K_S} \left[\underline{\underline{\mathbf{A}}} \check{\underline{\sigma}}_h : \underline{\xi}\, n_{K_S}^t - \underline{\xi} \hat{\mathbf{v}}_{Sh} \cdot \mathbf{n}_{K_S}^x + \rho_S \check{\mathbf{v}}_{Sh} \cdot \boldsymbol{\omega}_S\, n_{K_S}^t - \hat{\underline{\sigma}}_h : (\boldsymbol{\omega}_S \otimes \mathbf{n}_{K_S}^x) \right] ds =$$

$$\sum_{K_F} \int_{K_F} f q\, dv. \tag{4}$$

Thanks to the choice of test functions the left hand side of the above space-time formulation contains only surface integrals. The numerical fluxes in time $\hat{\mathbf{v}}_{Fh}$, $\hat{p}_h$, $\hat{\mathbf{v}}_{Sh}$, $\hat{\underline{\sigma}}_h$ and in space $\check{\mathbf{v}}_{Fh}$, $\check{p}_h$, $\check{\mathbf{v}}_{Sh}$, $\check{\underline{\sigma}}_h$ are defined in the standard DG notations [2, 3, 7] as follows:

$$\begin{pmatrix} \hat{\mathbf{v}}_{Fh} \cdot \mathbf{n}_{K_F}^x \\ \hat{p}_h \\ \hat{\mathbf{v}}_{Sh} \\ \hat{\underline{\sigma}}_h \mathbf{n}_{K_S}^x \end{pmatrix} \equiv \begin{pmatrix} \mathbf{v}_{Sh} \cdot \mathbf{n}_{K_F}^x + \delta_1 (\underline{\sigma}_h \mathbf{n}_{K_F}^x + p_h \mathbf{n}_{K_F}^x) \cdot \mathbf{n}_{K_F}^x \\ p_h + \alpha_1 (\mathbf{v}_{Fh} \cdot \mathbf{n}_{K_F}^x - \mathbf{v}_{Sh} \cdot \mathbf{n}_{K_F}^x) \\ \mathbf{v}_{Sh} - \delta_1 (\underline{\sigma}_h \mathbf{n}_{K_S}^x + p_h \mathbf{n}_{K_S}^x) \\ -p_h \mathbf{n}_{K_S}^x + \alpha_1 (\mathbf{v}_{Fh} \cdot \mathbf{n}_{K_S}^x - \mathbf{v}_{Sh} \cdot \mathbf{n}_{K_S}^x) \mathbf{n}_{K_S}^x \end{pmatrix} \text{ on } \mathcal{F}_h^{FS},$$

$$\begin{pmatrix} \hat{\mathbf{v}}_{Fh} \cdot \mathbf{n}_{K_F}^x \\ \hat{p}_h \end{pmatrix} \equiv \begin{pmatrix} g_F \\ p_h + \alpha_1 (\mathbf{v}_{Fh} \cdot \mathbf{n}_{K_F}^x - g_F) \end{pmatrix},$$

$$\begin{pmatrix} \hat{\mathbf{v}}_{Sh} \\ \hat{\underline{\sigma}}_h \mathbf{n}_{K_S}^x \end{pmatrix} \equiv \begin{pmatrix} \mathbf{v}_{Sh} - \delta_1 (\underline{\sigma}_h \mathbf{n}_{K_S}^x - g_S) \\ g_S \end{pmatrix} \qquad \text{on } \mathcal{F}_h^D.$$

$$\begin{pmatrix} \hat{\mathbf{v}}_{Fh} \\ \hat{p}_h \end{pmatrix} \equiv \begin{pmatrix} \{\mathbf{v}_{Fh}\} + \beta_1 [\![p_h]\!]_x \\ \{p_h\} + \alpha_1 [\![\mathbf{v}_{Fh}]\!]_x \end{pmatrix}, \qquad \begin{pmatrix} \hat{\mathbf{v}}_{Sh} \\ \hat{\underline{\sigma}}_h \end{pmatrix} \equiv \begin{pmatrix} \{\mathbf{v}_{Sh}\} - \delta_1 [\![\underline{\sigma}_h]\!]_x \\ \{\underline{\xi}\} - \gamma_1 [\![\mathbf{v}_{Fh}]\!]_x \end{pmatrix} \quad \text{on } \mathcal{F}_h^Q,$$

$$\begin{pmatrix} \check{\mathbf{v}}_{Fh} \\ \check{p}_h \end{pmatrix} \equiv \begin{pmatrix} \{\mathbf{v}_{Fh}\} + \alpha_2 [\![\mathbf{v}_{Fh}]\!]_t \\ \{p_h\} + \beta_2 [\![p_h]\!]_t \end{pmatrix}, \qquad \begin{pmatrix} \check{\mathbf{v}}_{Sh} \\ \check{\underline{\sigma}}_h \end{pmatrix} \equiv \begin{pmatrix} \{\mathbf{v}_{Sh}\} + \gamma_2 [\![\mathbf{v}_{Fh}]\!]_t \\ \{\underline{\sigma}_h\} + \delta_2 [\![\underline{\sigma}_h]\!]_t \end{pmatrix} \quad \text{on } \mathcal{F}_h^Q,$$

$$\begin{pmatrix} \check{\mathbf{v}}_{Fh} \\ \check{p}_h \end{pmatrix} \equiv \begin{pmatrix} \mathbf{v}_{Fh} \\ p_h \end{pmatrix}, \qquad \begin{pmatrix} \check{\mathbf{v}}_{Sh} \\ \check{\underline{\sigma}}_h \end{pmatrix} \equiv \begin{pmatrix} \mathbf{v}_{Sh} \\ \underline{\sigma}_h \end{pmatrix} \quad \text{on } \mathcal{F}_h^T,$$

$$\begin{pmatrix} \check{\mathbf{v}}_{Fh} \\ \check{p}_h \end{pmatrix} \equiv \begin{pmatrix} (\frac{1}{2} - \alpha_2)\mathbf{v}_{Fh} + (\frac{1}{2} + \alpha_2)\mathbf{v}_{F0} \\ (\frac{1}{2} - \beta_2)p_h + (\frac{1}{2} + \beta_2)p_0 \end{pmatrix},$$
$$\begin{pmatrix} \check{\mathbf{v}}_{Sh} \\ \check{\underline{\sigma}}_h \end{pmatrix} \equiv \begin{pmatrix} (\frac{1}{2} - \gamma_2)\mathbf{v}_{Sh} + (\frac{1}{2} + \gamma_2)\mathbf{v}_{S0} \\ (\frac{1}{2} - \delta_2)\underline{\sigma}_h + (\frac{1}{2} + \delta_2)\underline{\sigma}_0 \end{pmatrix} \quad \text{on } \mathcal{F}_h^0,$$

Here, α_1, α_2, β_1, β_2, δ_1, δ_2, γ_1, and γ_2 are positive penalty parameters. As in standard DG methods, a suitable choice of these penalty parameters allows one to prove stability of the overall method. It is shown in [2, 3] that they contribute to the accuracy and convergence of the numerical method. We refer to [2, 3] for more details on the definition of the numerical fluxes.

Summing the contribution (4) of all elements K_F, $K_S \in \mathcal{T}_h$, and introducing the bilinear $\mathcal{A}_{TDG}(\cdot\,;\,\cdot)$ and the linear $\ell_{TDG}(\cdot)$ forms for the left-hand side and the right-hand side expressions respectively, we obtain the Trefftz-DG formulation for the elasto-acoustic problem:

Seek $(\mathbf{v}_{Fh}, p_h, \mathbf{v}_{Sh}, \underline{\sigma}_h) \in \mathbf{T}(\mathcal{T}_h)$ such that, for all $(\omega_F, q, \omega_S, \underline{\xi}) \in \mathbf{T}(\mathcal{T}_h)$ it holds true:

$$\mathcal{A}_{TDG}\big((\mathbf{v}_{Fh}, p_h, \mathbf{v}_{Sh}, \underline{\sigma}_h); (\omega_F, q, \omega_S, \underline{\xi})\big) = \ell_{TDG}\big(\omega_F, q, \omega_S, \underline{\xi}\big). \tag{5}$$

The analysis of well-posedness of (5) is based on the coercivity and continuity estimates of the bilinear and linear forms in mesh-dependent norms [2, 3]. The proof is similar to the one given in [10] where the acoustic wave equation is addressed. In Sect. 2 we provide the algorithm of the Trefftz-DG formulation (5), and we discuss different analytical and numerical approaches for its optimization.

2 Implementation of the Algorithm

The numerical implementation of the Trefftz-DG formulation is different from the standard DG ones which address the space and time integration separately. Standard DG space integrations have the interesting feature of leading to a block-diagonal

mass matrix and allow then the use of explicit time integration. The computational
costs thus depend on a CFL condition which sets the value of the time step as a
function of the space step. On the other hand, a naive implementation of Trefftz-
DG methods require performing a space-time integration which leads to invert a
sparse matrix whose size tends to be huge. It is thus not obvious that a crude
implementation of the Trefftz-DG algorithm does not generate additional cost as
compared to standard DG ones.

In this section we provide some important steps of implementation of Trefftz-DG
formulation (5) and discuss optimization techniques. The complete algorithm with
more numerical details can be found in [2, 3].

2.1 Change-Over Between the Time Slabs

To simplify the presentation, we assume here that we use the same order of
approximation on each cells, so that we have N_{dof}^f degrees of freedom on fluid
cells and N_{dof}^s degrees of freedom on solid cells. Once we have defined the discrete
approximation space, we can solve the problem inside each element K_F and K_S,
communicating the corresponding values at the boundaries ∂K_F and ∂K_S by the
incoming and outgoing fluxes. Thus, the variational problem is represented by a
algebraic linear system, with a sparse matrix M, of size equals to the total number
of elements $N_{el}^{f,s}$ multiplied by the number of degrees of freedom per element $N_{dof}^{f,s}$,
that is $N_{el}^f \times N_{dof}^f + N_{el}^s \times N_{dof}^s$. When compared to the computational cost of
standard DG implementation, the corresponding Trefftz-DG cost is thus increased
and it is mainly due to the need of inverting the large-sized matrix. The most obvious
way to reduce the size of the matrix, which is classically used in most work on
space-time Trefftz method, is to consider time slabs. We restrict ourselves to the
case of cartesian meshes, but this methodology can also be applied to unstructured
meshed. An alternative is to use tent-pitched meshes that respect the causality,
this will be the topic of a future work. In order to optimize the execution of the
algorithm, we propose to divide the space-time domain Q into N_t elementary time
slabs $Q_1, Q_2, \ldots, Q_{N_t}$ and to solve the problem slab by slab, considering the final
results, computed in the current time slab at time t, as initial values for the next
slab at time $t + \Delta t$ (see Fig. 2). Thus the size of matrix inside each time slab is N_t

Fig. 2 Example of 1D + time
mesh $\mathcal{T}_h$ on Q decomposed
into N_t time slabs

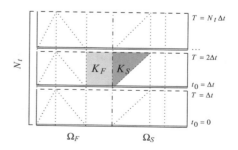

times smaller, compared to the initial one. Moreover, if the medium parameters are fixed in time, and the space discretization is preserved from slab to slab, the matrix can be computed and inverted once, and then re-projected onto the next time slabs, reducing thus the global numerical cost.

2.2 Polynomial Basis

One of the important advantages of Trefftz type methods is the flexibility in the choice of basis functions provided they satisfy the Trefftz property locally in each element. To perform the numerical simulations, we have extended the algorithm proposed by Maciag in [9] for computing wave polynomials, solutions of the second order transient wave equation, to the first order acoustic and elastodynamic systems of dimension one and higher. It consists in computing a polynomial basis, defined in the reference element, using Taylor expansions of generating exponential functions which are local solutions of the initial system of equations. An example of space-time wave polynomial basis for the first-order acoustic wave equation reads as follows (approximation degree p=3, dimension of the physical space $d = 1$):

$$\hat{\phi}_1^v = 0 \qquad \hat{\phi}_2^v = 1 \qquad \hat{\phi}_3^v = x \qquad \hat{\phi}_4^v = c_F t$$
$$\hat{\phi}_1^p = -c_F \qquad \hat{\phi}_2^p = 0 \qquad \hat{\phi}_3^p = -c_F^2 t \qquad \hat{\phi}_4^p = -c_F x$$

$$\hat{\phi}_5^v = -\frac{x^2}{2} - \frac{c_F^2 t^2}{2} \quad \hat{\phi}_6^v = -c_F x t \qquad \hat{\phi}_7^v = -\frac{x^3}{6} - \frac{x c_F^2 t^2}{2} \qquad \hat{\phi}_8^v = -\frac{c_F^3 t^3}{6} - \frac{x^2 c_F t}{2}$$
$$\hat{\phi}_5^p = c_F^2 x t \qquad \hat{\phi}_6^p = c_F(\frac{x^2}{2} + \frac{c_F^2 t^2}{2}) \quad \hat{\phi}_7^p = c_F(\frac{c_F^2 t^3}{6} + \frac{x^2 c_F t}{2}) \quad \hat{\phi}_8^p = c_F(\frac{x^3}{6} + \frac{x c_F^2 t^2}{2})$$

This basis contains the couples of polynomial functions $(\hat{\phi}^v, \hat{\phi}^p)$, corresponding to the velocity and pressure respectively, which are locally defined and satisfy the Trefftz property inside each element of the mesh, and of degrees less or equal to p ($p = 0, 1, 2, 3$) to provide an approximation of order p. By their construction, the Trefftz basis functions are not attached to the coordinates of the degrees of freedom inside the element, contrary to the Lagrange polynomials. Even if we compute only surface integrals, we can evaluate the final approximation solution in any point of the element refinement. We refer to [2] for more numerical details as well as for the acoustic and elastodynamic basis examples of higher dimensions.

2.3 Inversion of the Matrix M Inside a Time Slab

The inversion of the matrix inside the time slab can be explicitly reduced to the inversion of its block-diagonal component, which corresponds to the integration at the bottom and top of the time slab (initial and final time faces $\mathcal{F}_h^0$ and $\mathcal{F}_h^T$), thanks to the Taylor expansion formulas. More precisely, let us recall the expression for the bilinear form $\mathcal{A}_{TDG}(\cdot\,;\,\cdot)$ from Sect. 1.2:

$$\mathcal{A}_{TDG}(\cdot\,;\,\cdot) \equiv \underbrace{\int_{\mathcal{F}_h^T} + \int_{\mathcal{F}_h^0}}_{\mathcal{A}_{TDG}^{\Omega}} + \underbrace{\int_{\mathcal{F}_h^Q} + \int_{\mathcal{F}_h^D} + \int_{\mathcal{F}_h^{FS}}}_{\mathcal{A}_{TDG}^I}.$$

It consists of $\mathcal{A}_{TDG}^{\Omega}(\cdot\,;\,\cdot)$, that corresponds to the integration at the initial and final time element faces of the time slab, and $\mathcal{A}_{TDG}^I(\cdot\,;\,\cdot)$, that corresponds to the integration at the internal, boundary and fluid-solid element faces. Thus, the matrix M can be represented by the sum of two matrices $\Delta_{\Omega} M_{\Omega}$ and $\Delta_I M_I$ corresponding to $\mathcal{A}_{TDG}^{\Omega}$ and $\mathcal{A}_{TDG}^I$ respectively, as follows:

$$M = \Delta_{\Omega} M_{\Omega} + \Delta_I M_I.$$

Here, $\Delta_{\Omega} \propto (\Delta x)^d$ represents the area of the local faces in $\mathcal{F}_h^0$ and $\mathcal{F}_h^T$, and $\Delta_I \propto (\Delta x)^{d-1} \Delta t$ represents the area of the local faces in $\mathcal{F}_h^Q$, $\mathcal{F}_h^D$ and $\mathcal{F}_h^{FS}$ respectively. We refer to [2] for more details.

This decomposition is of particular interest since M_{Ω} is block-diagonal, each block corresponding to one element. Indeed, we have:

$$\Delta_{\Omega} M_{\Omega} + \Delta_I M_I = \left(\Delta_{\Omega} M_{\Omega}\right)\left(I + \frac{\Delta_I}{\Delta_{\Omega}} M_{\Omega}^{-1} M_I\right) = \left(\Delta_{\Omega} M_{\Omega}\right)\left(I + \kappa P\right),$$

Here I is the identity matrix, $\kappa \equiv \frac{\Delta_I}{\Delta_{\Omega}} \propto \frac{\Delta t}{\Delta x}$, and $P \equiv M_{\Omega}^{-1} M_I$.

If $||\kappa P||$ is sufficiently small, we can apply the Maclaurin formula in order to obtain the polynomial expansion for M^{-1} as follows:

$$M^{-1} \equiv \left(I + \kappa P\right)^{-1}\left(\Delta_{\Omega} M_{\Omega}\right)^{-1} = \left(\sum_{n=0}^{\infty} (-1)^n \kappa^n P^n\right)\left(\Delta_{\Omega} M_{\Omega}\right)^{-1}.$$

This representation reduces the inversion of the sparse matrix M to the inversion of its block-diagonal component M_{Ω}^{-1} and the multiplication of the inverted block-diagonal M_{Ω} by the sparse M_I. It provides an explicit way for solving the initial linear system approximately. Even though it requires a CFL—type condition related to value of $||\kappa P||$, justifying the approximate solution of the system, it significantly accelerates the algorithm execution.

In Table 1 we compare the numerical accuracy (L^2-norm in time and space of numerical error as a function of cell size Δx) of the TDG method in a 2D homogeneous acoustic case for both the exact and approximate matrix inversions as a function of the mesh size and of the number n of terms in the Taylor expansion.

Table 1 Accuracy (L^2-error in space and in time) of the solution when using the approximate inversion with $n = 3, 4, 5$ and the exact inversion

n	$\Delta x = 10^{-2}$	$\Delta x = 2 \cdot 10^{-2}$	$\Delta x = 5 \cdot 10^{-2}$	$\Delta x = 10^{-1}$
The approximate inversion ($\kappa = 10^{-2}$). Accelerating factor ≈ 18 times				
3	1.4166e−05	4.3741e−05	2.8780e−04	2.5772e−03
4	3.1623e−07	1.2656e−06	5.3868e−05	1.2674e−03
5	2.8903e−07	9.1744e−07	4.1029e−05	1.3010e−03
The exact inversion ($\kappa = 10^{-2}$)				
.	2.2540e−07	8.9583e−07	5.5811e−05	1.3004e−03

The accelerating factor is the ratio of the computational costs of the two methods for reaching the same accuracy

3   Numerical Tests

For the numerical implementation of the Trefftz-DG method we have considered a 2D medium composed of two homogeneous rectangular layers: the acoustic one and the elastodynamic one. We have set a source term at the fluid-solid interface, and two receivers in the acoustic layer and in the elastodynamic one. The numerical signals at both receivers have been validated with the analytical solutions computed with *Gar6more* code [5]. In Fig. 3 we show the convergence of the numerical velocity as a function of cell size for different degrees of approximation ($p = 0, 1, 2, 3$) computed at receivers in (a) 2D acoustic layer and (b) 2D elastodynamic layer. In each case, the convergence rate is higher than the corresponding approximation degree. We refer to [2], where we provide more examples.

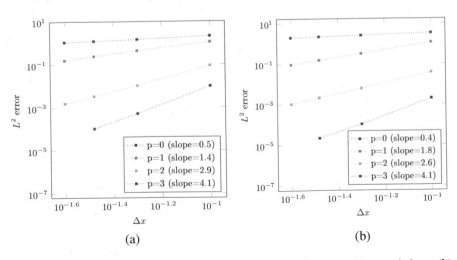

Fig. 3 Convergence of numerical velocity in function of cell size Δx. (**a**) 2D acoustic layer. (**b**) 2D elastodynamic layer

4   Conclusion

The Trefftz-DG methodology for solving the first order elasto-acoustic system has demonstrated the important advantages, such as the use of degrees of freedom evaluated at the element faces only, the flexibility in the choice of the basis functions and the unconditional stability. However, in its initial form, it still shows some limitations due to the space-time integration that leads to the representation of the discrete system by a huge sparse matrix whose straightforward inversion is very expensive, even when using time slabs. We find ourselves in a situation of using an implicit scheme for solving the forward problem that risks to overload the iterative process of the corresponding inverse problem in order to reconstruct very large propagation domains. Fortunately, thanks to the decomposition of the matrix by separating the time variables from the space ones, we could benefit from the block-diagonal structure of the standard DG formulation ending up with an explicit scheme, that is more convenient from the numerical point of view. The performed numerical tests clearly illustrate the interest of the split version of discrete problem.

Acknowledgements This project is supported by the Inria—Total SA strategic action "Depth Imaging Partnership" (http://dip.inria.fr), and has received funding from the European Union's Horizon 2020 research and innovation program under the Marie Sklodowska-Curie grant agreement Number 777778 (Rise action Mathrocks).

References

1. Banjai, L., Georgoulis, E.H., Lijoka, O.: A Trefftz polynomial space-time discontinuous Galerkin method for the second order wave equation. SIAM J. Numer. Anal. **55**(1), 63–86 (2017)
2. Barucq, H., Calandra, H., Diaz, J., Shishenina, E.: Space–Time Trefftz - Discontinuous Galerkin Approximation for Elasto-Acoustics. RR-9104, Inria Bordeaux Sud-Ouest, UPPA (LMA-Pau), Total SA, <hal-01614126> (2017)
3. Barucq, H., Calandra, H., Diaz, J., Shishenina, E.: Space–time Trefftz-DG approximation for elasto-acoustics. Appl. Anal. 1–14 (2018)
4. Egger, H., Kretzschmar, F., Schnepp, S.M., Weiland, T.: A space-time discontinuous Galerkin Trefftz method for time dependent Maxwell's equations. SIAM J. Sci. Comput. **37**(5), B689–B711 (2015)
5. Gar6more2D. Magique-3D (2013). https://gforge.inria.fr/projects/gar6more2d/
6. Herrera, I.: Trefftz method: a general theory. Numer. Methods Partial Differ. Equ. Int. J. **16**(6), 561–580 (2000)
7. Hesthaven, J.S., Warburton, T.: Nodal Discontinuous Galerkin Methods: Algorithms, Analysis, and Applications. Springer Science & Business Media, Berlin (2007)
8. Kretzschmar, F., Moiola, A., Perugia, I., Schnepp, S.M.: A priori error analysis of space–time Trefftz discontinuous Galerkin methods for wave problems. IMA J. Numer. Anal. **36**(4), 1599–1635 (2015)
9. Maciag, A.: The usage of wave polynomials in solving direct and inverse problems for two-dimensional wave equation. Int. J. Numer. Methods Biomed. Eng. **27**(7), 1107–1125 (2011)
10. Moiola, A., Perugia, I.: A space–time Trefftz discontinuous Galerkin method for the acoustic wave equation in first-order formulation. Numer. Math. **138**(2), 389–435 (2018)

11. Petersen, S., Farhat, C., Tezaur, R.: A space-time discontinuous Galerkin method for the solution of the wave equation in the time domain. Int. J. Numer. Methods Eng. **78**(3), 275–295 (2009)
12. Trefftz, E.: Ein Gegenstück zum Ritzschen Verfahren. Proceedings of the 2nd International Congress for Applied Mechanics, Zurich, pp. 131–137 (1926)
13. Wang, D., Tezaur, R., Farhat, C.: A hybrid discontinuous in space and time Galerkin method for wave propagation problems. Int. J. Numer. Methods Eng. **99**(4), 263–289 (2014)

An *hp*-Adaptive Iterative Linearization Discontinuous-Galerkin FEM for Quasilinear Elliptic Boundary Value Problems

Paul Houston and Thomas P. Wihler

1 Introduction

In this article, we consider the *a posteriori* error analysis, in a natural mesh-dependent energy norm, for a class of interior-penalty *hp*-version discontinuous Galerkin finite element methods (DGFEMs) for the numerical solution of the following quasilinear elliptic boundary value problem:

$$-\nabla \cdot (\mu(x, |\nabla u|)\nabla u) = f \quad \text{in } \Omega, \qquad u = 0 \quad \text{on } \Gamma. \tag{1}$$

Here, $\Omega \subset \mathbb{R}^2$ is a bounded polygon with a Lipschitz continuous boundary Γ, and $f \in L^2(\Omega)$, where for an open set $D \subseteq \Omega$, we signify by $L^2(D)$ the space of all square integrable functions on D. Additionally, we assume that the nonlinearity μ satisfies the following assumptions: (A1) $\mu \in C^0(\overline{\Omega} \times [0, \infty))$; (A2) there exist positive constants m_μ, M_μ such that $m_\mu(t-s) \leq \mu(x, t)t - \mu(x, s)s \leq M_\mu(t-s)$, $t \geq s \geq 0$, $x \in \overline{\Omega}$. We remark that, if μ satisfies (A2), there exist constants $\beta \geq \alpha > 0$, such that for all vectors $v, w \in \mathbb{R}^2$, and all $x \in \overline{\Omega}$,

$$|\mu(x, |v|)v - \mu(x, |w|)w| \leq \beta|v - w|, \tag{2}$$

$$\alpha|v - w|^2 \leq \left(\mu(x, |v|)v - \mu(x, |w|)w\right) \cdot (v - w);$$

P. Houston (✉)
School of Mathematical Sciences, University of Nottingham, Nottingham, UK
e-mail: Paul.Houston@nottingham.ac.uk

T. P. Wihler
Mathematics Institute, University of Bern, Bern, Switzerland
e-mail: wihler@math.unibe.ch

© The Author(s) 2020
S. J. Sherwin et al. (eds.), *Spectral and High Order Methods for Partial Differential Equations ICOSAHOM 2018*, Lecture Notes in Computational Science and Engineering 134, https://doi.org/10.1007/978-3-030-39647-3_32

see [14, Lemma 2.1]. For ease of notation, in the sequel, we will simply write $\mu(s)$ instead of $\mu(x, s)$, thereby suppressing the explicit dependence of μ on $x \in \Omega$.

The weak formulation of (1) is to find $u \in H_0^1(\Omega)$ such that

$$A(u; u, v) = (f, v)_{L^2(\Omega)} \qquad \forall v \in H_0^1(\Omega), \tag{3}$$

where, given $w \in H_0^1(\Omega)$, we define the bilinear form $A(w; u, v) = \int_\Omega \mu(|\nabla w|) \nabla u \cdot \nabla v \, dx$, $u, v \in H_0^1(\Omega)$, as well as the $L^2(\Omega)$-inner product $(v, w)_{L^2(\Omega)} = \int_\Omega vw \, dx$, $v, w \in L^2(\Omega)$. Here, $H_0^1(\Omega)$ is the standard Sobolev space of first order, with zero trace along Γ, equipped with the norm $\|v\|_{H_0^1(\Omega)} = \|\nabla v\|_{L^2(\Omega)}$, $v \in H_0^1(\Omega)$. Under the assumptions (A1)–(A2) above, it is elementary to show that the form A is strongly monotone and Lipschitz continuous in the sense that

$$A(u; u, u - v) - A(v; v, u - v) \geq \alpha \|u - v\|_{H_0^1(\Omega)}^2 \qquad \forall u, v \in H_0^1(\Omega), \tag{4}$$

and

$$|A(u; u, v) - A(w; w, v)| \leq \beta \|u - w\|_{H_0^1(\Omega)} \|v\|_{H_0^1(\Omega)} \qquad \forall u, v, w \in H_0^1(\Omega),$$

respectively. From these properties, classical monotone operator theory implies existence and uniqueness of a solution of (3); see, e.g., [17, Theorem 3.3.23].

The exploitation of automatic adaptive hp-refinement algorithms has the potential to compute numerical solutions to partial differential equations (PDEs) in a highly efficient manner, often leading to exponential rates of convergence as the underlying finite element space is enriched; see, e.g., [11, 16]. The key tool required to design such strategies is the derivation of *a posteriori* estimates for the Galerkin discretization errors; in recent years such bounds have been extended to the context of linearization and/or linear solver errors, cf. [1, 2, 4, 5, 7, 9]. In the present article we consider the derivation of an hp-version *a posteriori* error bound for the DGFEM approximation of the second-order quasilinear elliptic PDE problem stated in (1). To this end, we employ the interior penalty DGFEM proposed in [10], cf. also [12], together with a discrete Kačanov iterative linearization scheme, cf. [6]. Based on the analysis undertaken in [12], together with the use of a suitable reconstruction operator, cf. [13, 15], we derive a fully computable bound for the error, measured in terms of a suitable DGFEM energy norm, which separately accounts for the three main sources of error: discretization, linearization, and linear solver errors. On the basis of this *a posteriori* bound, we design and implement an hp-adaptive refinement algorithm which automatically controls each of these error contributions as the underlying finite element space is enriched. Numerical experiments highlighting the practical performance of the proposed adaptive strategy are presented.

2 Iterative Discontinuous Galerkin Methods

2.1 Discrete hp-Discontinuous Galerkin Spaces

Let $\mathscr{T}_h$ be a partition of Ω into disjoint open and shape-regular elements κ such that $\overline{\Omega} = \bigcup_{\kappa \in \mathscr{T}_h} \overline{\kappa}$. We assume that each $\kappa \in \mathscr{T}_h$ is an affine image of a given master element $\widehat{\kappa}$, which is either the open triangle $\{(x, y) : -1 < x < 1, -1 < y < -x\}$) or the open square $(-1, 1)^2$ in $\mathbb{R}^2$. By h_κ we denote the element diameter of $\kappa \in \mathscr{T}_h$, and $\boldsymbol{n}_\kappa$ signifies the unit outward normal vector to κ. We allow $\mathscr{T}_h$ to be *1-irregular*, i.e., each edge of any one element $\kappa \in \mathscr{T}_h$ contains at most one hanging node (which, for simplicity, we assume to be the midpoint of the corresponding edge). In this context, we suppose that $\mathscr{T}_h$ is *regularly reducible* (cf. [18, Section 7.1] and [12]), i.e., there exists a shape-regular conforming (regular) mesh $\widetilde{\mathscr{T}}_h$ (consisting of triangles and parallelograms) such that the closure of each element in $\mathscr{T}_h$ is a union of closures of elements of $\widetilde{\mathscr{T}}_h$, and that there exists a constant $C > 0$, independent of the element sizes, such that for any two elements $\kappa \in \mathscr{T}_h$ and $\widetilde{\kappa} \in \widetilde{\mathscr{T}}_h$ with $\widetilde{\kappa} \subseteq \kappa$ we have $h_\kappa / h_{\widetilde{\kappa}} \leq C$. Note that these assumptions imply that $\mathscr{T}_h$ is of *bounded local variation*, i.e., there exists a constant $\rho_1 \geq 1$, independent of the element sizes, such that $\rho_1^{-1} \leq h_{\kappa_\sharp} / h_{\kappa_\flat} \leq \rho_1$, for any pair of elements $\kappa_\sharp, \kappa_\flat \in \mathscr{T}_h$ which share a common edge $e = \partial \kappa_\sharp \cap \partial \kappa_\flat$. Moreover, let us consider the set $\mathscr{E}$ of all one-dimensional open edges of all elements $\kappa \in \mathscr{T}_h$. Further, we denote by $\mathscr{E}_{\mathscr{I}}$ the set of all edges $e \in \mathscr{E}$ that are contained in the open domain Ω (interior edges). Additionally, we introduce $\mathscr{E}_{\mathscr{B}}$ to be the set of boundary edges consisting of all $e \in \mathscr{E}$ that are contained in Γ.

For any integer $p \in \mathbb{N}_0$, we denote by $\mathbb{P}_p(\kappa)$ the set of polynomials of total degree p on κ. Similarly, when κ is a quadrilateral, we also consider $\mathbb{Q}_p(\kappa)$, the set of all tensor-product polynomials on κ of degree p in each coordinate direction. To each $\kappa \in \mathscr{T}_h$ we assign a polynomial degree p_κ (local approximation order). We collect the local polynomial degrees in a vector $\boldsymbol{p} = \{p_\kappa : \kappa \in \mathscr{T}_h\}$, and then introduce the *hp*-DGFEM space

$$\mathbb{V}_{\text{DG}}(\mathscr{T}_h, \boldsymbol{p}) = \{v \in L^2(\Omega) : v|_\kappa \in \mathbb{S}_{p_\kappa}(\kappa) \quad \forall \kappa \in \mathscr{T}_h\},$$

with $\mathbb{S}$ being either $\mathbb{P}$ or $\mathbb{Q}$. We shall suppose that the polynomial degree vector $\boldsymbol{p}$, with $p_\kappa \geq 1$ for each $\kappa \in \mathscr{T}$, has *bounded local variation*, i.e., there exists a constant $\rho_2 \geq 1$, independent of the local element sizes and $\boldsymbol{p}$, such that, for any pair of neighbouring elements $\kappa_\sharp, \kappa_\flat \in \mathscr{T}_h$, we have $\rho_2^{-1} \leq p_{\kappa_\sharp} / p_{\kappa_\flat} \leq \rho_2$.

We also define the L^2-projection $\Pi_{\mathscr{T}_h, \boldsymbol{p}} : L^2(\Omega) \to \mathbb{V}_{\text{DG}}(\mathscr{T}_h, \boldsymbol{p})$ by

$$(\Pi_{\mathscr{T}_h, \boldsymbol{p}} v - v, w)_{L^2(\Omega)} = 0 \quad \forall w \in \mathbb{V}_{\text{DG}}(\mathscr{T}_h, \boldsymbol{p}).$$

Evidently, since functions in $\mathbb{V}_{\text{DG}}(\mathscr{T}_h, \boldsymbol{p})$ do not need to be continuous, we have that $\Pi_{\kappa, p_\kappa} = \Pi_{\mathscr{T}_h, \boldsymbol{p}}|_\kappa$, where, for $\kappa \in \mathscr{T}_h$, we let Π_{κ, p_κ} be the L^2-projection onto $\mathbb{S}_{p_\kappa}(\kappa)$.

2.2 Nonlinear hp-DGFEM Formulation

Let $\kappa_\sharp$ and $\kappa_\flat$ be two adjacent elements of $\mathcal{T}_h$, and x an arbitrary point on the interior edge $e \in \mathcal{E}_{\mathcal{I}}$ given by $e = (\partial \kappa_\sharp \cap \partial \kappa_\flat)^\circ$. Furthermore, let v and q be scalar- and vector-valued functions, respectively, that are sufficiently smooth inside each element $\kappa_\sharp$, $\kappa_\flat$. Then, the averages of v and q at $x \in e$ are given by $\langle\!\langle v \rangle\!\rangle = 1/2(v|_{\kappa_\sharp} + v|_{\kappa_\flat})$, $\langle\!\langle q \rangle\!\rangle = 1/2(q|_{\kappa_\sharp} + q|_{\kappa_\flat})$, respectively. Similarly, the jumps of v and q at $x \in e$ are given by $[\![v]\!] = v|_{\kappa_\sharp} \, n_{\kappa_\sharp} + v|_{\kappa_\flat} \, n_{\kappa_\flat}$, $[\![q]\!] = q|_{\kappa_\sharp} \cdot n_{\kappa_\sharp} + q|_{\kappa_\flat} \cdot n_{\kappa_\flat}$, respectively. On a boundary edge $e \in \mathcal{E}_{\mathcal{B}}$, we set $\langle\!\langle v \rangle\!\rangle = v$, $\langle\!\langle q \rangle\!\rangle = q$ and $[\![v]\!] = vn$, with n denoting the unit outward normal vector on the boundary Γ.

Furthermore, we introduce the edge functions $\mathfrak{h}, \mathfrak{p} \in L^\infty(\mathcal{E})$, which, for an edge $e \in \mathcal{E}$, are given by $\mathfrak{h}|_e := h_e$ and $\mathfrak{p}|_e := \langle\!\langle p \rangle\!\rangle|_e$, with h_e denoting the length of e. In addition, we define the discontinuity penalisation function $\sigma \in L^\infty(\mathcal{E})$ given by $\sigma = \gamma \mathfrak{p}^2 \mathfrak{h}^{-1}$, where $\gamma \geq 1$ is a (sufficiently large) constant. Then, we equip the DGFEM space $\mathbb{V}_{DG}(\mathcal{T}_h, p)$ with the DGFEM norm $\|v\|_{DG}^2 := \|\nabla_{\mathcal{T}_h} v\|_{L^2(\Omega)}^2 + \int_{\mathcal{E}} \sigma |[\![v]\!]|^2 \, ds$, $v \in \mathbb{V}_{DG}(\mathcal{T}_h, p)$, where $\nabla_{\mathcal{T}_h}$ is the element-wise gradient operator.

With this notation, following [10], we introduce the interior penalty DGFEM discretization of (3) by: find $u_{DG} \in \mathbb{V}_{DG}(\mathcal{T}_h, p)$ such that

$$A_{DG}(u_{DG}; u_{DG}, v) = (f, v)_{L^2(\Omega)} \qquad \forall v \in \mathbb{V}_{DG}(\mathcal{T}_h, p), \tag{5}$$

where, for given $w \in \mathbb{V}_{DG}(\mathcal{T}_h, p)$, we define the DGFEM bilinear form

$$A_{DG}(w; u, v) = \int_\Omega \mu(|\nabla_{\mathcal{T}_h} w|) \nabla_{\mathcal{T}_h} u \cdot \nabla_{\mathcal{T}_h} v \, dx$$

$$- \int_{\mathcal{E}} \langle\!\langle \mu(|\nabla_{\mathcal{T}_h} w|) \nabla_{\mathcal{T}_h} u \rangle\!\rangle \cdot [\![v]\!] \, ds + \theta \int_{\mathcal{E}_{\mathcal{B}}} \langle\!\langle \mu(\mathfrak{h}^{-1}|[\![w]\!]|) \nabla_{\mathcal{T}_h} v \rangle\!\rangle \cdot [\![u]\!] \, ds$$

$$+ \int_{\mathcal{E}_{\mathcal{B}}} \sigma \, [\![u]\!] \cdot [\![v]\!] \, ds, \qquad u, v \in \mathbb{V}_{DG}(\mathcal{T}_h, p),$$

where $\theta \in [-1, 1]$ is a method parameter. Referring to [10, Theorem 2.5], provided that $\gamma \geq 1$ is chosen sufficiently large (independent of the local element sizes and of the polynomial degree distribution), the existence and uniqueness of the DGFEM solution $u_{DG} \in \mathbb{V}_{DG}(\mathcal{T}_h, p)$ satisfying (5) is guaranteed.

Assumption 1 *In the sequel, we suppose that there exists a computable* a posteriori *error estimate of the form* $\|u - u_{DG}\|_{DG} \leq \eta(u_{DG}, f)$, *where* $u \in H_0^1(\Omega)$ *is the solution of* (1), *and* u_{DG} *is its hp-DGFEM approximation defined in* (5).

Remark 1 In the article [12, Theorem 3.2] it has been proved that such a bound does indeed exist. More precisely, we have that

$$\|u - u_{DG}\|_{DG} \leq C \left(\sum_{\kappa \in \mathcal{T}_h} \eta_\kappa^2 + \mathcal{O}(f, u_{DG}) \right)^{1/2} =: \eta(u_{DG}, f), \tag{6}$$

where, the local error indicators η_κ, $\kappa \in \mathcal{T}_h$, are defined by

$$\eta_\kappa^2 := h_\kappa^2 p_\kappa^{-2} \| \Pi_{\mathcal{T}_h, \mathbf{p}-1}(f + \nabla \cdot (\mu(|\nabla u_{\text{DG}}) \nabla u_{\text{DG}})) \|_{L^2(\kappa)}^2$$
$$+ h_\kappa p_\kappa^{-1} \| \Pi_{\mathcal{E}, \overline{\mathbf{p}}-1}(\llbracket \mu(|\nabla u_{\text{DG}}|) \nabla u_{\text{DG}} \rrbracket) \|_{0, \partial \kappa \backslash \Gamma}^2 + \gamma^2 h_\kappa^{-1} p_\kappa^3 \| \llbracket u_{\text{DG}} \rrbracket \|_{L^2(\partial \kappa)}^2, \tag{7}$$

and $\mathcal{O}(f, u_{\text{DG}}) := \sum_{\kappa \in \mathcal{T}_h} \mathcal{O}_\kappa^{(1)} + \sum_{e \in \mathcal{E}_{\mathscr{I}}} \mathcal{O}_e^{(2)}$ is a data oscillation term. For $\kappa \in \mathcal{T}_h$ and $e \in \mathcal{E}_{\mathscr{I}}$, we have $\mathcal{O}_\kappa^{(1)} := h_\kappa^2 p_\kappa^{-2} \| (\mathsf{I} - \Pi_{\mathcal{T}_h, \mathbf{p}-1})|_\kappa (f + \nabla \cdot (\mu(|\nabla u_{\text{DG}}|) \nabla u_{\text{DG}})) \|_{0,\kappa}^2$, and $\mathcal{O}_e^{(2)} := h_e \overline{p}_e^{-1} \| (\mathsf{I} - \Pi_{\mathcal{E}, \overline{\mathbf{p}}-1})|_e (\llbracket \mu(|\nabla_{\mathcal{T}_h} u_{\text{DG}}|) \nabla_{\mathcal{T}_h} u_{\text{DG}} \rrbracket) \|_{0,e}^2$, where I denotes a generic identity operator. Here, we write $\mathbf{p} - 1 := \{p_\kappa - 1\}_{\kappa \in \mathcal{T}_h}$. Additionally, we denote by $\Pi_{\mathcal{E}, \overline{\mathbf{p}}-1}|_e$ the L^2-projector onto $\mathbb{P}_{\overline{p}_e - 1}(e)$, where we let $\overline{p}_e = \max\{p_{\kappa_\sharp}, p_{\kappa_\flat}\}$, with $\kappa_\sharp, \kappa_\flat \in \mathcal{T}_h$, $e = \partial \kappa_\sharp \cap \partial \kappa_\flat$. Moreover, $C > 0$ in (6) is a constant that is independent of the local element sizes, the polynomial degree vector $\mathbf{p}$, and the parameters γ and θ.

2.3 Iterative DGFEM

In order to provide a practical solution scheme for the nonlinear hp-DGFEM system (5) we propose a linearization approach based on a discrete *Kačanov fixed point iteration*, see, e.g., [6]. To this end, we begin by selecting an initial guess $u_{\text{DG}}^0 \in \mathbb{V}_{\text{DG}}(\mathcal{T}_h, \mathbf{p})$. Then, for $n \geq 1$, given $u_{\text{DG}}^{n-1} \in \mathbb{V}_{\text{DG}}(\mathcal{T}_h, \mathbf{p})$, we solve the *linear hp-DGFEM formulation*, defined by

$$A_{\text{DG}}(u_{\text{DG}}^{n-1}; u_{\text{DG}}^n, v) = (f, v)_{L^2(\Omega)} \qquad \forall v \in \mathbb{V}_{\text{DG}}(\mathcal{T}_h, \mathbf{p}), \tag{8}$$

for $u_{\text{DG}}^n \in \mathbb{V}_{\text{DG}}(\mathcal{T}_h, \mathbf{p})$. We emphasize that, in actual computations, the linear system (8) may be solved by an iterative algorithm, thereby generating an approximate numerical solution $\widehat{u}_{\text{DG}}^n \in \mathbb{V}_{\text{DG}}(\mathcal{T}_h, \mathbf{p})$, with $\widehat{u}_{\text{DG}}^n \approx u_{\text{DG}}^n$. This means that, in practice, instead of computing the sequence $\{u_{\text{DG}}^n\}_{n \geq 0}$ obtained from the iteration (8), an inexact sequence $\{\widehat{u}_{\text{DG}}^n\}_{n \geq 0}$ is generated such that

$$A_{\text{DG}}(\widehat{u}_{\text{DG}}^{n-1}; \widehat{u}_{\text{DG}}^n, v) \approx (f, v)_{L^2(\Omega)} \qquad \forall v \in \mathbb{V}_{\text{DG}}(\mathcal{T}_h, \mathbf{p}). \tag{9}$$

From a mathematical view point, this (inexact) iterative linearization DGFEM approach gives rise to three different sources of error:

1. *Discretization error*, which is expressed by the residual

$$\rho_{\text{DG}}^n := A_{\text{DG}}(\widehat{u}_{\text{DG}}^n; \widehat{u}_{\text{DG}}^n, \cdot) - (f, \cdot)_{L^2(\Omega)}. \tag{10}$$

2. *Linearization error*, which is given in terms of the residual $\psi_{\text{DG}}^n \in \mathbb{V}_{\text{DG}}(\mathscr{T}_h, \boldsymbol{p})$:

$$(\psi_{\text{DG}}^n, v)_{\text{L}^2(\Omega)} := A_{\text{DG}}(\widehat{u}_{\text{DG}}^n; \widehat{u}_{\text{DG}}^n, v) - A_{\text{DG}}(\widehat{u}_{\text{DG}}^{n-1}; \widehat{u}_{\text{DG}}^n, v) \qquad \forall v \in \mathbb{V}_{\text{DG}}(\mathscr{T}_h, \boldsymbol{p}). \tag{11}$$

We observe that, if (1) is linear, then we immediately obtain $\psi_{\text{DG}}^n = 0$.

3. *Linear solver error*, which is described by a residual $\lambda_{\text{DG}}^n \in \mathbb{V}_{\text{DG}}(\mathscr{T}_h, \boldsymbol{p})$:

$$(\lambda_{\text{DG}}^n, v)_{\text{L}^2(\Omega)} := A_{\text{DG}}(\widehat{u}_{\text{DG}}^{n-1}; \widehat{u}_{\text{DG}}^n, v) - (f, v)_{\text{L}^2(\Omega)} \qquad \forall v \in \mathbb{V}_{\text{DG}}(\mathscr{T}_h, \boldsymbol{p}). \tag{12}$$

Note that, if (8) is solved exactly, then we have $\widehat{u}_{\text{DG}}^{n-1} = u_{\text{DG}}^{n-1}$ and $\widehat{u}_{\text{DG}}^n = u_{\text{DG}}^n$, and it follows that $\lambda_{\text{DG}}^n = 0$.

Remark 2 Since $\mathbb{V}_{\text{DG}}(\mathscr{T}_h, \boldsymbol{p})$ may not need to be continuous along element interfaces, the linearization and linear solver residuals ψ_{DG}^n and λ_{DG}^n, respectively, can be computed elementwise, i.e., in parallel, and, hence, at a low computational cost.

The aim of the analysis in the following section is to investigate the above residuals, and then to provide a computable *a posteriori* error estimate for the error $\|u - \widehat{u}_{\text{DG}}^n\|_{\text{DG}}$ between the solution u of (1) and $\widehat{u}_{\text{DG}}^n \in \mathbb{V}_{\text{DG}}(\mathscr{T}_h, \boldsymbol{p})$.

2.4 A Posteriori *Error Estimation*

In order to bound the residual ρ_{DG}^n in (10), we apply an elliptic reconstruction technique along the lines of the works [13, 15], see also [7]. Specifically, we define an auxiliary function $\widetilde{u}^n \in \text{H}_0^1(\Omega)$ to be the unique solution of the weak formulation

$$A(\widetilde{u}^n; \widetilde{u}^n, v) = (f + \psi_{\text{DG}}^n + \lambda_{\text{DG}}^n, v)_{\text{L}^2(\Omega)} \qquad \forall v \in \text{H}_0^1(\Omega),$$

where ψ_{DG}^n and λ_{DG}^n are the linearization and linear solver residuals from (11) and (12), respectively. Upon adding (11) and (12), we notice that

$$A_{\text{DG}}(\widehat{u}_{\text{DG}}^n; \widehat{u}_{\text{DG}}^n, v) = (f + \psi_{\text{DG}}^n + \lambda_{\text{DG}}^n, v)_{\text{L}^2(\Omega)} \qquad \forall v \in \mathbb{V}_{\text{DG}}(\mathscr{T}_h, \boldsymbol{p}).$$

In particular, we observe that $\widehat{u}_{\text{DG}}^n$ is the DGFEM approximation of $\widetilde{u}^n$ based on employing the (nonlinear) DGFEM scheme defined in (5). In particular, we may exploit the *a posteriori* error estimate in Assumption 1 to infer the computable bound

$$\|\widetilde{u}^n - \widehat{u}_{\text{DG}}^n\|_{\text{DG}} \leq \eta(\widehat{u}_{\text{DG}}^n, f + \psi_{\text{DG}}^n + \lambda_{\text{DG}}^n). \tag{13}$$

We now turn to bounding the elliptic reconstruction error $u - \widetilde{u}^n \in \mathrm{H}_0^1(\Omega)$; to this end, we first observe that $\|u - \widetilde{u}^n\|_{\mathrm{DG}} = \|u - \widetilde{u}^n\|_{\mathrm{H}_0^1(\Omega)}$. Then, employing the strong monotonicity property (4), and recalling the weak formulation (3), we obtain

$$\alpha \|u - \widetilde{u}^n\|_{\mathrm{DG}}^2 \leq A(u; u; u - \widetilde{u}^n) - A(\widetilde{u}^n; \widetilde{u}^n, u - \widetilde{u}^n)$$
$$= -(\psi_{\mathrm{DG}}^n, u - \widetilde{u}^n)_{\mathrm{L}^2(\Omega)} - (\lambda_{\mathrm{DG}}^n, u - \widetilde{u}^n)_{\mathrm{L}^2(\Omega)}.$$

Employing the Cauchy-Schwarz inequality, together with the Poincaré-Friedrichs inequality, $\|v\|_{\mathrm{L}^2(\Omega)} \leq C_{\mathrm{PF}} \|\nabla v\|_{\mathrm{L}^2(\Omega)}$ for all $v \in \mathrm{H}_0^1(\Omega)$, where $C_{\mathrm{PF}} > 0$ is a constant, we deduce that

$$\|u - \widetilde{u}^n\|_{\mathrm{DG}} \leq \Psi_{\mathrm{DG}}^n + \Lambda_{\mathrm{DG}}^n, \tag{14}$$

where the linearization and linear solver residuals are given, respectively, by

$$\Psi_{\mathrm{DG}}^n := C_{\mathrm{PF}}/\alpha \left(\sum_{\kappa \in \mathscr{T}_h} \|\psi_{\mathrm{DG}}^n\|_{\mathrm{L}^2(\kappa)}^2 \right)^{1/2}, \quad \Lambda_{\mathrm{DG}}^n := C_{\mathrm{PF}}/\alpha \left(\sum_{\kappa \in \mathscr{T}_h} \|\lambda_{\mathrm{DG}}^n\|_{\mathrm{L}^2(\kappa)}^2 \right)^{1/2}.$$

Summarizing the above analysis leads to the following result.

Theorem 1 *Suppose that Assumption 1 is satisfied. Then, given a sequence of (possibly inexact) DGFEM approximations $\{\widehat{u}_{\mathrm{DG}}^n\}_{n \geq 0} \subset \mathbb{V}_{\mathrm{DG}}(\mathscr{T}_h, p)$, cf. (9), for $n \geq 1$, the following a posteriori error bound holds:*

$$\|u - \widehat{u}_{\mathrm{DG}}^n\|_{\mathrm{DG}} \leq \eta(\widehat{u}_{\mathrm{DG}}^n, f + \psi_{\mathrm{DG}}^n + \lambda_{\mathrm{DG}}^n) + \Psi_{\mathrm{DG}}^n + \Lambda_{\mathrm{DG}}^n.$$

Here, u is the analytical solution of (1), ψ_{DG}^n and λ_{DG}^n are the residuals defined in (11) and (12), respectively, and $\alpha > 0$ is the constant occurring in (2) and (4).

Proof The result follows immediately upon application of the triangle inequality, i.e., $\|u - \widehat{u}_{\mathrm{DG}}^n\|_{\mathrm{DG}} \leq \|u - \widetilde{u}^n\|_{\mathrm{DG}} + \|\widetilde{u}^n - \widehat{u}_{\mathrm{DG}}^n\|_{\mathrm{DG}}$, and inserting the bounds (13) and (14).

Remark 3 We note that the above analysis naturally applies to other finite element schemes, provided that Assumption 1 is satisfied.

2.5 Adaptive Iterative hp-DGFEM Procedure

In this section we introduce an automatic hp-refinement algorithm which ensures that each of the three components of the error, namely discretization, linearization, and linear solver, are controlled in a suitable fashion. To this end, we propose the following strategy, cf. [9].

Algorithm 1 *Given a (coarse) starting mesh* $\mathcal{T}_h$, *with an associated (low-order) polynomial degree distribution* p, *and an initial guess* $\hat{u}_{DG}^0 \in \mathbb{V}_{DG}(\mathcal{T}_h, p)$. *Set* $n \leftarrow 1$.

1: *Compute the DGFEM solution* $\hat{u}_{DG}^n$ *from (9) based on employing an iterative linear solver. Furthermore, evaluate the corresponding error indicators* $\eta(\hat{u}_{DG}^n, f + \psi_{DG}^n + \lambda_{DG}^n)$, Ψ_{DG}^n, *and* Λ_{DG}^n.
2: **if**

$$\Psi_{DG}^n + \Lambda_{DG}^n \leq \Upsilon \, \eta(\hat{u}_{DG}^n, f + \psi_{DG}^n + \lambda_{DG}^n) \tag{15}$$

holds, for some given parameter $\Upsilon > 0$, **then** *hp-adaptively refine the space* $\mathbb{V}_{DG}(\mathcal{T}_h, p)$; *go back to step (1:) with the new mesh* $\mathcal{T}_h$ *(and based on the previously computed solution* $\hat{u}_{DG}^n$ *interpolated on the refined mesh).*
3: **else,** *i.e., if (15) is not fulfilled, then set* $n \leftarrow n + 1$, *and perform another linearization step by going back to (1:).*
4: **end if**

In Step 2 of Algorithm 1, if (15) is fulfilled then the space $\mathbb{V}_{DG}(\mathcal{T}_h, p)$ is adaptively hp-refined based on first marking elements for refinement according to the size of the local element indicators η_κ, cf. (7). To this end, we exploit the maximal strategy whereby elements are marked for refinement which satisfy the condition $\eta_\kappa > 1/3 \max_{\kappa \in \mathcal{T}_h} \eta_\kappa$. Secondly, once an element $\kappa \in \mathcal{T}_h$ has been marked for refinement, we undertake either local mesh subdivision or local polynomial enrichment based on employing the hp-refinement criterion developed within the article [8]. Finally, when (15) is not fulfilled, rather than determining which source of error, i.e., the (computable) quantities Ψ_{DG}^n or Λ_{DG}^n from (11) and (12), respectively, is dominant, we choose to always undertake a further linearization step, and hence a further linear solver step is also computed, since this ensures that the most up to date approximation $\hat{u}_{DG}^n$ is employed at all times.

3   Application to Quasilinear Elliptic PDEs

In this section we present numerical experiments to highlight the performance of the proposed iterative hp-refinement procedure outlined in Algorithm 1. To this end, we set the interior penalty parameter constant γ to 10 and the steering parameter Υ to $1/4$. The solution of the resulting set of linear equations is computed using an ILU(0) preconditioned GMRES algorithm.

For the first numerical experiment, we let $\Omega = (0, 1)^2$ and define the nonlinear coefficient as $\mu(|\nabla u|) = 2 + (1 + |\nabla u|)^{-1}$. The right-hand forcing function f is selected so that the analytical solution to (1) is given by $u(x, y) = x(1 - x)y(1 - y)(1 - 2y)e^{-20(2x-1)^2}$. In Fig. 1 we present a comparison of the actual error measured in terms of the energy norm versus the square root of the number of degrees of freedom in $\mathbb{V}_{DG}(\mathcal{T}_h, p)$. From Fig. 1a we clearly observe exponential convergence of the proposed hp-refinement strategy as $\mathbb{V}_{DG}(\mathcal{T}_h, p)$ in enriched.

Furthermore, in Fig. 1b we plot the individual residual error indicators; for this smooth problem, we notice that the discretization indicator (denoted as η^n in the figure) is always dominant, while the linearization and linear solver residuals (denoted as Ψ^n and λ^n, respectively) are roughly of the same magnitude.

Secondly, we let Ω denote the L-shaped domain $(-1, 1)^2 \setminus [0, 1) \times (-1, 0] \subset \mathbb{R}^2$ and select $\mu(|\nabla u|) = 1 + \exp(-|\nabla u|^2)$. By writing (r, φ) to denote the system of polar coordinates, we choose the forcing function f and an inhomogeneous boundary condition such that the analytical solution to (1) is $u = r^{2/3} \sin(2/3\varphi)$, cf. [3]. In Fig. 2 we now present a comparison of the actual error measured in terms of the energy norm versus the third root of the number of degrees of freedom in $\mathbb{V}_{\mathrm{DG}}(\mathscr{T}_h, \boldsymbol{p})$; as before we again attain exponential convergence of the proposed

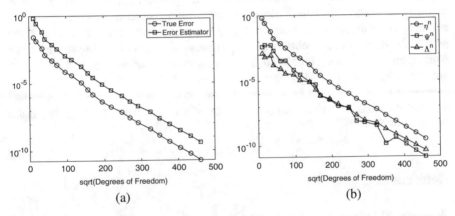

Fig. 1 Example 1. (**a**) Comparison of the DGFEM norm of the error and the *a posteriori* bound, with respect to the square root of the number of degrees of freedom; (**b**) individual error estimators

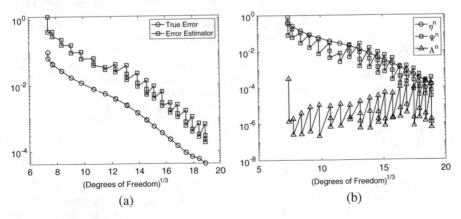

Fig. 2 Example 2. (**a**) Comparison of the DGFEM norm of the error and the *a posteriori* bound, with respect to the third root of the number of degrees of freedom; (**b**) individual error estimators

hp-refinement strategy as $\mathbb{V}_{DG}(\mathcal{T}_h, \boldsymbol{p})$ is adaptively refined, though convergence of the *a posteriori* error estimator is no longer monotonic. Indeed, from Fig. 2b, we observe that once an hp-mesh refinement has been undertaken, then several linearization/solver steps may be required to ensure that the numerical solution has been computed to a sufficient accuracy before future refinements may be undertaken.

4 Conclusions

In this article we have derived a computable hp-version *a posteriori* error bound for the DGFEM approximation of a second-order quasilinear elliptic PDE problem, whereby a discrete Kačanov iterative linearization scheme is employed. The resulting computable upper bound directly takes into account discretization error, as well as the errors stemming from linearization and the underlying linear solver. Numerical experiments highlighting the performance of this bound within an automatic hp-refinement algorithm are presented.

Acknowledgements TW acknowledges the support of the Swiss National Science Foundation (SNF), Grant No. 200021_162990.

References

1. Amrein, M., Wihler, T.P.: Fully adaptive Newton-Galerkin methods for semilinear elliptic partial differential equations. SIAM J. Sci. Comput. **37**(4), A1637–A1657 (2015)
2. Amrein, M., Melenk, J.M., Wihler, T.P.: An hp-adaptive Newton-Galerkin finite element procedure for semilinear boundary value problems. Math. Methods Appl. Sci. **40**(6), 1973–1985 (2017)
3. Congreve, S., Houston, P., Wihler, T.P.: Two-grid hp-version discontinuous Galerkin finite element methods for second-order quasilinear elliptic PDEs. J. Sci. Comput. **55**(2), 471–497 (2013)
4. El Alaoui, L., Ern, A., Vohralík, M.: Guaranteed and robust a posteriori error estimates and balancing discretization and linearization errors for monotone nonlinear problems. Comput. Methods Appl. Mech. Eng. **200**(37–40), 2782–2795 (2011)
5. Ern, A., Vohralík, M.: Adaptive inexact Newton methods with a posteriori stopping criteria for nonlinear diffusion PDEs. SIAM J. Sci. Comput. **35**(4), A1761–A1791 (2013)
6. Garau, E.M., Morin, P., Zuppa, C.: Convergence of an adaptive Kačanov FEM for quasi-linear problems. Appl. Numer. Math. **61**(4), 512–529 (2011)
7. Heid, P., Wihler, T.P.: Adaptive iterative linearization Galerkin methods for nonlinear problems (2018). Technical Report 1808.04990. arxiv.org
8. Houston, P., Süli, E.: A note on the design of hp-adaptive finite element methods for elliptic partial differential equations. Comput. Methods Appl. Mech. Eng. **194**(2–5), 229–243 (2005)
9. Houston, P., Wihler, T.P.: An hp-adaptive Newton-discontinuous-Galerkin finite element approach for semilinear elliptic boundary value problems. Math. Comp. **87**(314), 2641–2674 (2018)

10. Houston, P., Robson, J., Süli, E.: Discontinuous Galerkin finite element approximation of quasilinear elliptic boundary value problems I: the scalar case. IMA J. Numer. Anal. **25**, 726–749 (2005)
11. Houston, P., Schötzau, D., Wihler, T.P.: Energy norm a posteriori error estimation of *hp*-adaptive discontinuous Galerkin methods for elliptic problems. Math. Models Methods Appl. Sci. **17**(1), 33–62 (2007)
12. Houston, P., Süli, E., Wihler, T.P.: A posteriori error analysis of *hp*-version discontinuous Galerkin finite-element methods for second-order quasi-linear elliptic PDEs. IMA J. Numer. Anal. **28**(2), 245–273 (2008)
13. Lakkis, O., Makridakis, C.: Elliptic reconstruction and a posteriori error estimates for fully discrete linear parabolic problems. Math. Comp. **75**(256), 1627–1658 (2006)
14. Liu, W.B., Barrett, J.W.: Quasi-norm error bounds for the finite element approximation of some degenerate quasilinear elliptic equations and variational inequalities. RAIRO Modél. Math. Anal. Numér. **28**(6), 725–744 (1994)
15. Makridakis, C., Nochetto, R.H.: Elliptic reconstruction and a posteriori error estimates for parabolic problems. SIAM J. Numer. Anal. **41**(4), 1585–1594 (2003)
16. Melenk, J.M., Wohlmuth, B.I.: On residual-based a posteriori error estimation in *hp*-FEM. Adv. Comput. Math. **15**(1–4), 311–331 (2001). A posteriori error estimation and adaptive computational methods
17. Nečas, J.: Introduction to the Theory of Nonlinear Elliptic Equations. John Wiley and Sons, Hoboken (1986)
18. Ortner, C., Süli, E.: Discontinuous Galerkin finite element approximation of nonlinear second-order elliptic and hyperbolic systems. SIAM J. Numer. Anal. **45**(4), 1370–1397 (2007)

Erosion Wear Evaluation Using Nektar++

Manuel F. Mejía, Douglas Serson, Rodrigo C. Moura, Bruno S. Carmo, Jorge Escobar-Vargas, and Andrés González-Mancera

1 Introduction

Wear is a common phenomenon on many machines and devices, it is characterised by the removal or loss of material. Erosion wear is a particular wear process which occurs when solid particles or droplets, carried by a fluid (liquid or gas), impact on a solid surface [1]. Turbomachinery such as pumps, turbines and pipe accessories (i.e. tees, elbows, nozzles, valves), are examples of elements affected by the erosion wear, decreasing the performance and the lifetime. In many industrial sectors e.g. energy and mining, and oil & gas; massive amounts of resources are used for maintenance and replacement of affected parts [2–4]. Despite this phenomenon have been broadly investigated [5–14] there are still unsolved challenges in establishing the influence of small eddies during the erosion process leading to modest accuracy levels in the simulation results.

M. F. Mejía (✉)
Universidad de Los Andes, Bogotá, D.C., Colombia

Universidad Central, Bogotá, D.C., Colombia
e-mail: mf.mejia@uniandes.edu.co

D. Serson · B. S. Carmo
Universidade de São Paulo (USP), São Paulo, Brazil

R. C. Moura
Instituto Tecnológico de Aeronáutica, São Paulo, Brazil

J. Escobar-Vargas
Pontificia Universidad Javeriana, Bogotá, D.C., Colombia

A. González-Mancera
Universidad de Los Andes, Bogotá, D.C., Colombia

© The Author(s) 2020
S. J. Sherwin et al. (eds.), *Spectral and High Order Methods for Partial Differential Equations ICOSAHOM 2018*, Lecture Notes in Computational Science and Engineering 134, https://doi.org/10.1007/978-3-030-39647-3_33

419

Due to the microscopic nature of erosion, the smallest scales in the flow play a fundamental role in the complete process. One of the aspects which has not been carefully studied in erosion wear modelling is the effect that the smaller eddies and secondary flows have on the particles interactions with the surface. In general, these secondary flows could not be represented using linear Reynolds Average Navier Stokes (RANS) simulations, this is mainly because the Reynolds stress imbalance is neglected and the secondary flow does not develop. As was mention by Gross and Fasel [15], predictions of the secondary flow require non-linear Reynolds stress, full Reynolds-stress models, Large Eddy Simulations (LES) or Direct Numerical Simulation (DNS). Due to their relatively low computational cost, RANS models often used to predict on erosion using CFD in industrial simulations. The inclusion of smaller eddies and secondary flows in the simulation could be a major breakthrough in the modelling of erosion process. In order to capture in an accurate way the physics related with the small eddies and secondary flows, a numerical technique capable to represent those processes, is needed. As emphasised by Jacobs [16], the use of spectral methods could allow increased accuracy in the simulation due to the potential to simulate a wider range of scales. With this in mind, the purpose of this work is to assess the impact of higher resolution methods on the prediction of erosion wear rate and distribution.

1.1 Spectral Methods

Several numerical techniques are used to solve Navier Stokes (NS) equations. Some of them are finite differences, finite volumes and finite elements. Nevertheless, when high accuracy is required the use of a lot of elements is needed in the modelling, which significantly increase the computational cost [17, 18]. Hence novel methods are subject of research to offer a better rate accuracy and computational cost.

Among novel numerical methods considered nowadays are spectral methods, which have shown to be a powerful tool with high level of accuracy for solving large problems in computational fluid dynamics (CFD), according to the available literature, especially in the studies developed by Boyd [19], Canuto et al. [20–22], Trefethen [23, 24] and Sherwin [25, 26]. Nektar++ is an open-source software framework designed to support the development of high-performance scalable solvers for partial differential equations using the spectral/hp element method[27]. High Order CFD methods have been receiving considerable attention in the past two decades. Traditional CFD software could be replaced by high order code in many applications in few years [28].

1.2 Particles Tracking

To the best of the authors' knowledge, there is no work that uses high order methods to evaluate erosion wear rate. This research aims to assess the impact of higher resolution methods on the prediction of erosion wear rate and distribution. It comprises the solution of fluid flow using incompressible NS solver with implicit LES modelling, the implementation of a Lagrangian particle tracking model and the later data processing through traditional erosion rate models but using the available high order information. This could allow the evaluation of traditional rate models with more spatial resolution and accuracy.

The Lagrangian particle tracking model is based on one-way coupling approach, that is the most simple case when just the iteration between the fluid and each particle is taking into account in just one way. That means that the particles are moved by the fluid but the fluid flow is not perturbed by the particles. Moreover, the effects of the collision between particles are also neglected. The one-way coupling model is valid for volume concentrations of particles lower than 10^{-6} [29, 30].

The problem of predicting particle motion in a fluid flow can be predicted by solving an evolution equation in time:

$$\frac{d\mathbf{v}_p}{dt} = F(\mathbf{u}, \rho, \rho_p, C_d, \ldots) \quad ; \quad \frac{d\mathbf{x}_p}{dt} = \mathbf{v}_p \tag{1}$$

where $\mathbf{v}_p$ and $\mathbf{x}_p$ are the particle velocity and position and F is a function of the velocity of the fluid $\mathbf{u}$, the density of the fluid ρ and particle density ρ_p, among others.

To start, it is necessary to obtain the velocity on a certain point from the eulerian velocity field. This process consists of finding the element containing the particle and interpolating the velocity with the element information. In a higher-order velocity element field, the use of linear interpolation is inaccurate and could vanish the advantage won with the use of high order methods. On the other hand, using high order interpolation could be computationally expensive. Therefore, special attention to this procedure is required [16, 31, 32].

1.3 Erosion Wear Evaluation

Once the information about the collisions is complete, the erosion wear model is used to predict the pattern of material removed. The general erosion equation, based on the work of Finnie [33–38] can be presented as

$$W = kF_s V_p^n f(\theta) \tag{2}$$

W is the erosion rate or material removed by collision, k is a wall material dependent constant, F_s is the particle geometric factor, V_p and n are the collision velocity and the velocity exponent, and $f(\theta)$ is a function of collision angle. Several authors define these values for different materials configurations and test cases. Three of the most used models, which include experimental results are the jet impingement test [39–45], elbow erosion [46–49], and the works of the Wong et al. [4, 46, 50, 51].

2 Implementation

This section describes the implementation of erosion wear in Nektar++. To achieve this objective is important to have in mind the partition of the problems into two parts. The first one is the particle tracking as a filter within the Nektar++ incompressible Navier Stokes solver. A filter in Nektar++ is a module for calculating a variety of useful quantities from the field variables as the solution evolves in time [27].

The second one is implemented as a FieldConvert module to evaluate the erosion of each collision and generate the fields on the boundaries walls. FieldConvert is a utility embedded in Nektar++ with the primary aim of allowing the user to work with the Nektar++ output files, some of the modules within FieldConvert allow the user to postprocess the output data [27].

2.1 Particles Tracking

The first step was the implementation of a ODE time solver. Several options are available, but having into account the discrete time flow fields calculated with the Navier Stokes incompressible solver, and to avoid the use of temporal interpolation, the selected option was the Adams-Bashforth (AB) and Adams-Moulton (AM) schemes.

The implementation was tested with a benchmark case presented in [31]. In this model, the particle velocity is the fluid velocity at certain point and the evolution equation is reduced to one equation; Eq. 1 is reduced to:

$$\frac{d\,\mathbf{x}_p}{d\,t} = \mathbf{u} \tag{3}$$

To solve this system a Time-Marching Method was implemented, meaning that the future values are evaluated using the present and past values of the variables. Explicit AB and Implicit AM methods were implemented using first to fourth integration order. The error values obtaining using AB and AM with different order presents features from this kind of methods.

The next step was the implementation of the solid particles. In this case, the momentum equation is evaluated on each particle, resulting:

$$\frac{d\,\mathbf{v}_p}{d\,t} = F_d\,(\mathbf{u} - \mathbf{v}_p) + \mathbf{g}\frac{\rho_p - \rho}{\rho_p} \quad ; \quad \frac{d\,\mathbf{x}_p}{d\,t} = \mathbf{v}_p \tag{4}$$

$$F_d = \frac{3}{4}\frac{C_d\,Re_p\,\rho_p}{v d_p^2} \tag{5}$$

$$Re_p = \frac{(\mathbf{u}-\mathbf{v}_p)d_p}{\nu} \tag{6}$$

$$C_d = \begin{cases} 24/Re_p, & Re_p < 0.5 \\ 24/Re_p\left(1+0.15Re_p^{0.687}\right), & 0.5 < Re_p < 1000 \\ 0.44, & Re_p > 1000 \end{cases} \tag{7}$$

where F_d is the drag force, Re_p is the Reynolds Number based on the diameter of the particle, $\mathbf{g}$ is the gravity acceleration, and C_d is the drag force evaluated on each particle.

Figure 1 shows a diagram of the evolution equation. Current position, velocities and forces are evaluated to get the future positions (BP, OP) until the next position is located outside of the domain (NP). When this happens, the evolution algorithm stop and a function is used to evaluate the collision point (CP) and the position after of collision (NP') using the high order information about the walls.

2.2 Erosion Wear Evaluation

Erosion rate per collision (Eq. 2) has to be integrated over each element of the eroded surface. For each particle collision, more material is removed from the surface, the elemental erosion rate has to take into account this cumulative effect over the surface.

As mentioned before, the set of parameters used in this work, has been based on experimental data. One of the most used parameter set is the one proposed by Erosion group of the University of Tulsa [38, 48, 52]. The erosion rate takes the form of Eq. 2, $F_s = 1$ for sharp (angular), 0.53 for semi-rounded, or 0.2 for fully rounded sand particles. V_p is the impact velocity and $n = 1.73$. The angle function has the form:

$$f(\theta) = \begin{cases} a\theta^2 + b\theta, & \text{for } \theta \leq \phi \\ x\cos^2(\theta)\sin(w\theta) + y\sin^2(\theta) + z & \text{for } \theta > \phi \end{cases} \tag{8}$$

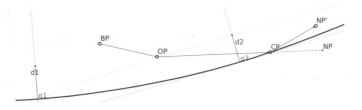

Fig. 1 Evolution of particle tracking

All the parameters and empirical constants depend on the material being eroded. For velocity in ft/s, the steel-sand parameters are: $a = 38.4$, $b = 22.7$ $\phi = 1$, $x = 0.3147$, $y = 0.03609$, $w = 0.2532$ and $z = 0$ [53].

3 Test Case

To test the new feature in Nektar++, a Backward Facing Step (BFS) model was developed based on the experimental setup of [30, 32, 54] showed in Fig. 2. In the model developed in this work, the simulations were done with the addition of gravitational effects on the $-y$ direction. In original experiments the air at the inlet is a well development turbulent flow ($\bar{u} = 10.5$ m/s), this is used a inlet condition and, to complete the model, a zero pressure condition at the output. The additional boundaries were set as walls. A zero velocity field was set as initial condition. The particles used have a 70 μm diameter and 8808 kg/m^3 density.

Figure 3 (top) shows a snapshot of the velocity field when the statistically stationary regime is reached ($t = 8$ s), next the particles are released and were convected by the flow. Particular trajectories are shown in grey lines in Fig. 3 (bottom). In the same figure, results of the particle collision with the walls, computed with Eq. 2, are also shown.

From the results presented Fig. 3, the typical BFS velocity profiles can be recognised. It is important to note the details behind the step, the main flow originates the secondary eddies and defines the limit of the recirculation zone ($x/H = 7$ from the step) where backflow occurs. Additionally, interesting details appear in between each main velocity flow ripple and the walls along the x-direction.

It is noteworthy that particle tracking is evaluated using a steady velocity field, therefore the existence of several irregularities is expected, for instance, particle trajectories inside recirculation zone. Erosion rate depends on the number of collisions at specific points. It is a localised phenomenon that does not occur continuously in the domain. Its distribution shows a strong dependence on the flow dynamics.

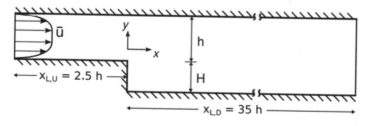

Fig. 2 Geometry of the Backward Facing Step setup. The initial velocity was set to get a $Re = 18,600$

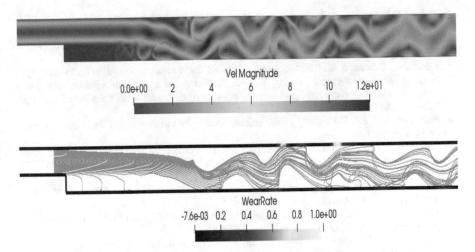

Fig. 3 BFS case results. Top: Velocity field at a statistically stationary condition. Bottom: Distribution of the particles inside the flow (gray lines). The colours in the walls indicate the location of the normalised erosion rate

4 Conclusion and Future Work

This work presented a method developed to asses the erosion wear rate using a high-order (spectral) element based technique on a modified test case implemented in Nektar++. The methodology proposed in this study have a potential to increase the accuracy when solving this kind of problems. Future research activities are going to be focused on the determination of accuracy improvements and optimisation of the proposed methodology. Several more cases have to be tested to produce solid conclusions about the implemented methodology, as well as a detailed comparison with experimental test cases.

Despite the methodology implemented had several important simplifications, as the use of one-way coupling and the few forces taken into account, allowed quicker implementations and results. This work would be an interesting starting to implement this kind of simulations using Nektar++. However, to run more realistic cases, additional research efforts are required for the implementation of two-way and four-way coupling and the effects of other forces over each particle.

References

1. Kulu, P., Kleis, I.: Solid Particle Erosion: Occurrence, Prediction and Control. Springer, London (2008)
2. Neopane, H., Cervantes, M.: Sediment erosion in hydraulic turbines. Global J. Res. Eng. Mech. Mech. Eng. **11**, 17–26 (2011)

3. Walker, C.I., Robbie, P.: Comparison of some laboratory wear tests and field wear in slurry pumps. Wear **302**(1–2), 1026–1034 (2013)
4. Wong, C.Y., Solnordal, C., Graham, L., Short, G., Wu, J.: Slurry erosion of surface imperfections in pipeline systems. Wear **336**, 72–85 (2015)
5. Karimi, A., Schmid, R.K.: Ripple formation in solid-liquid erosion. Wear **156**(1), 33–47 (1992)
6. Zeng, C., Xiao, Y., Wang, Z., Zhang, J., Luo, Y.: Numerical analysis of a Pelton bucket free surface sheet flow and dynamic performance affected by operating head. Proc. IMechE **231**(3), 182–196 (2017)
7. Padhy, M.K., Saini, R.P.: Study of silt erosion mechanism in Pelton turbine buckets. Energy **39**(1), 286–293 (2012)
8. Bajracharya, T.R., Acharya, B., Joshi, C.B., Saini, R.P., Dahlhaug, O.G.: Sand erosion of Pelton turbine nozzles and buckets: a case study of Chilime Hydropower Plant. Wear **264**(3–4), 177–184 (2008)
9. Kumar, P., Saini, R.P.: Study of cavitation in hydro turbines—a review. Renew. Sust. Energ. Rev. **14**(1), 374–383 (2010)
10. Xiao, Y.-X., Han, F.-Q., Zhou, J.-L., Kubota, T.: Numerical prediction of dynamic performance of Pelton turbine. J. Hydrodyn. B **19**(3), 356–364 (2007)
11. Zhu, H., Pan, Q., Zhang, W., Feng, G., Li, X.: CFD simulations of flow erosion and flow-induced deformation of needle valve: effects of operation, structure and fluid parameters. Nucl. Eng. Des. **273**, 396–411 (2014)
12. Chongji, Z., Yexiang, X., Wei, Z., Yangyang, Y., Lei, C., Zhengwei, W.: Pelton turbine needle erosion prediction based on 3D three-phase flow simulation. IOP Conf. Ser. Earth Environ. Sci. **22**(5), 052019 (2014)
13. Finnie, I., Kabil, Y.H.: On the formation of surface ripples during erosion. Wear **8**(1), 60–69 (1965)
14. Dutta, S.: Bulle-effect and its implications for morphodynamics of river diversions. Ph.D. Thesis. University of Illinois at Urbana-Champaign (2017)
15. Gross, A., Fasel, H.: Hybrid RANS/LES simulations of turbulent channel and diffuser flows. In: 48th AIAA Aerospace Sciences Meeting Including the New Horizons Forum and Aerospace Exposition, Aerospace Sciences Meetings. American Institute of Aeronautics and Astronautics (2010)
16. Jacobs, G.: Inertial particle behavior in an unsteady separated flow. In: 46th AIAA Aerospace Sciences Meeting and Exhibit, Aerospace Sciences Meetings. American Institute of Aeronautics and Astronautics (2008)
17. Sengupta, K., Mashayek, F., Jacobs, B.: Direct numerical simulation of turbulent flows using spectral methods. In: 46th AIAA Aerospace Sciences Meeting and Exhibit, pp. 7–10 (2008)
18. Olige, J., Kreiss, H.O.: Comparison of accurate methods for integration of hyperbolic equations. Tellus **24**, 199–215 (1972)
19. Boyd, J.P.: Chebyshev and Fourier Spectral Methods, 2nd edn. Dover Publications, Mineola (2000)
20. Canuto, C., Hussaini, M.Y., Quarteroni, A., Zang, T.A.: Spectral Methods: Fundamentals in Single Domains. Scientific Computation. Springer, Berlin (2007)
21. Canuto, C, Hussaini, M.Y., Quarteroni, A., Zang, T.A.: Spectral Methods in Fluid Dynamics. Springer, New York (1987)
22. Zang, T.A., Canuto, C., Hussaini, M.Y., Quarteroni, A.: Spectral Methods: Complex Geometries and Applications to Fluid Dynamics, 1st edn. Springer, Berlin (2007)
23. Trefethen, L.N.: Finite difference and spectral methods for ordinary and partial differential equations, 1st edn. Cornell University, Ithaca (1996)
24. Trefethen, L.N.: Spectral Methods in Matlab, 1st edn. Cornell University, Ithaca (2000)
25. Karniadakis, G.E.M., Sherwin, S.: Spectral/hp Element Methods for Computational Fluid Dynamics. Oxford University Press, New York (2005)
26. Karniadakis, G.E., Sherwin, S.J.: Spectral/hp Element Methods for CFD. Oxford University Press, New York (1999)

27. Cantwell, C.D., Moxey, D., Comerford, A., Bolis, A., Rocco, G., Mengaldo, G., De Grazia, D., Yakovlev, S., Lombard, J.-E., Ekelschot, D., Jordi, B., Xu, H., Mohamied, Y., Eskilsson, C., Nelson, B., Vos, P., Biotto, C., Kirby, R.M., Sherwin, S.J.: Nektar++: an open-source spectral/element framework. Comput. Phys. Commun. **192**, 205–219 (2015)

28. Wang, Z.J., Fidkowski, K., Abgrall, R., Bassi, F., Caraeni, D., Cary, A., Deconinck, H., Hartmann, R., Hillewaert, K., Huynh, H.T., Kroll, N., May, G., Persson, P.-O., van Leer, B., Visbal, M.: High-order CFD methods: current status and perspective. Int. J. Numer. Methods Fluids **72**(8), 811–845 (2013)

29. Elghobashi, S: On predicting particle-laden turbulent flows. Appl. Sci. Res. **52**(4), 309–329 (1994)

30. Greifzu, F., Kratzsch, C., Forgber, T., Lindner, F., Schwarze, R.: Assessment of particle-tracking models for dispersed particle-laden flows implemented in OpenFOAM and ANSYS FLUENT. Eng. Appl. Comput. Fluid Mech. **10**(1), 30–43 (2016)

31. Coppola, G., Sherwin, S.J., Peiro, J.: Nonlinear particle tracking for high-order elements. J. Comput. Phys. **172**, 356–386 (2001)

32. Jacobs, G.B., Kopriva, D.A., Mashayek, F.: Towards efficient tracking of inertial particles with high-order multidomain methods. J. Comput. Appl. Math. **206**(1), 392–408 (2007)

33. Finnie, I.: Erosion of surfaces by solid particles. Wear **3**(2), 87–103 (1960)

34. Finnie, I.: Some observations on the erosion of ductile metals. Wear **19**(1), 81–90 (1972)

35. Padhy, M.K., Saini, R.P.: A review on silt erosion in hydro turbines. Renew. Sust. Energ. Rev. **12**(7), 1974–1987 (2008)

36. Finnie, I.: Some reflections on the past and future of erosion. Wear **186–187**, 1–10 (1995)

37. Hutchings, I., Shipway, P.: Tribology: Friction and Wear of Engineering Materials. Butterworth-Heinemann, London (2017)

38. Parsi, M., Najmi, K., Najafifard, F., Hassani, S., McLaury, B.S., Shirazi, S.A.: A comprehensive review of solid particle erosion modeling for oil and gas wells and pipelines applications. J. Nat. Gas Sci. Eng. **21**, 850–873 (2014)

39. Sugiyama, K., Harada, K., Hattori, S.: Influence of impact angle of solid particles on erosion by slurry jet. Wear **265**(5), 713–720 (2008)

40. Gnanavelu, A., Kapur, N., Neville, A., Flores, J.F., Ghorbani, N.: A numerical investigation of a geometry independent integrated method to predict erosion rates in slurry erosion. Wear **271**(5), 712–719 (2011)

41. Efird, K.D., Wright, E.J., Boros, J.A., Hailey, T.G.: Correlation of steel corrosion in pipe flow with jet impingement and rotating cylinder tests. Corrosion **49**(12), 992–1003 (1993)

42. López, A., Nicholls, W., Stickland, M.T., Dempster, W.M.: CFD study of jet impingement test erosion using Ansys Fluent® and OpenFOAM®. Comput. Phys. Commun. **197**, 88–95 (2015)

43. Gnanavelu, A., Kapur, N., Neville, A., Flores, J.F.: An integrated methodology for predicting material wear rates due to erosion. Wear **267**(11), 1935–1944 (2009)

44. Sugiyama, K., Nakahama, S., Hattori, S., Nakano, K.: Slurry wear and cavitation erosion of thermal-sprayed cermets. Wear **258**(5–6), 768–775 (2005)

45. Oka, Y.I., Hayashi, H.: Evaluation of erosion resistance for metal–ceramic composites and cermets using a water-jet testing apparatus. Wear **271**(9–10), 1397–1403 (2011)

46. Solnordal, C.B., Wong, C.Y., Boulanger, J.: An experimental and numerical analysis of erosion caused by sand pneumatically conveyed through a standard pipe elbow. Wear **336–337**, 43–57 (2015)

47. Asgharpour, A., Zahedi, P., Khanouki, H.A., Shirazi, S.A., McLaury, B.S.: Experimental and numerical study on solid particle erosion in elbows mounted in series. In: ASME 2017 Fluids Engineering Division Summer Meeting, p. V01BT06A015. American Society of Mechanical Engineers, New York (2017)

48. Shirazi, S.A., Shadley, J.R., McLaury, B.S., Rybicki, E.F.: A procedure to predict solid particle erosion in elbows and tees. J. Press. Vessel. Technol. **117**(1), 45–52 (1995)

49. Vieira, R.E., Mansouri, A., McLaury, B.S., Shirazi, S.A.: Experimental and computational study of erosion in elbows due to sand particles in air flow. Powder Technol. **288**, 339–353 (2016)

50. Wong, C.Y., Solnordal, C., Swallow, A., Wu, J.: Experimental and computational modelling of solid particle erosion in a pipe annular cavity. Wear **303**(1), 109–129 (2013)
51. Boulanger, J., Wong, C., Solnordal, C., Zamberi, MSA., Shaffee, A., Johar, Z., Jadid, M.: A simplified computational approach to multi-phase erosion. International Conference on CFD in the Minerals and Process Industries, CSIRO, vol. 11 (2015)
52. Chen, X., McLaury, B.S., Shirazi, S.A.: Application and experimental validation of a computational fluid dynamics (CFD)-based erosion prediction model in elbows and plugged tees. Comput. Fluids **33**(10), 1251–1272 (2004)
53. Ahlert, K.R.: Effects of particle impingement angle and surface wetting on solid particle erosion of AISI 1018 steel (1994)
54. Fessler, J.R., Eaton, J.K.: Turbulence modification by particles in a backward-facing step flow. J. Fluid Mech. **394**, 97–117 (1999)

An Inexact Petrov-Galerkin Approximation for Gas Transport in Pipeline Networks

Herbert Egger, Thomas Kugler, and Vsevolod Shashkov

1 Introduction

The flow of gas in a horizontal pipeline of constant cross section is described by [2]

$$A\partial_t \rho + \partial_x m = 0 \tag{1}$$

$$\partial_t m + \partial_x \left(\frac{m^2}{A\rho} + Ap \right) = -\frac{\lambda}{2D} \frac{|m|}{A\rho} m. \tag{2}$$

Here A and D are the cross section and diameter of the pipe, and λ is a dimensionless friction parameter. The functions ρ, p, and m describe the density, pressure, and mass flow rate of the gas. Under isothermal flow conditions, one has

$$p = c^2 \rho \tag{3}$$

with constant c denoting the speed of sound. In practically relevant scaling regimes, the nonlinear term on the left hand side of (2) is usually neglected, which can be justified by an asymptotic analysis [2, 7]. Using this simplification and Eq. (3) to

H. Egger (✉) · T. Kugler
TU Darmstadt, Department of Mathematics, Darmstadt, Germany
e-mail: egger@mathematik.tu-darmstadt.de; kugler@mathematik.tu-darmstadt.de

V. Shashkov
TU Darmstadt, GSC Computational Engineering, Darmstadt, Germany
e-mail: shashkov@gsc.tu-darmstadt.de

© The Author(s) 2020

S. J. Sherwin et al. (eds.), *Spectral and High Order Methods for Partial Differential Equations ICOSAHOM 2018*, Lecture Notes in Computational Science and Engineering 134, https://doi.org/10.1007/978-3-030-39647-3_34

eliminate the density, one arrives at evolution problems of the general form

$$a\partial_t p + \partial_x m = 0 \tag{4}$$

$$b\partial_t m + \partial_x p = -dm \tag{5}$$

where a and b are positive constants and $d = d(p, m)$ denotes a state dependent friction coefficient. For our analysis, we will consider $d = d(x)$ as a function depending only on space which can be justified, e.g., by linearization around a steady state. Corresponding models for the gas flow on pipe networks are obtained by coupling the flow equations for single pipes via algebraic conditions [9, 10]; see below.

The discretization of (4)–(5) and its extension to pipeline networks has been discussed intensively in the literature. In [9], a Galerkin approximation for (1)–(2) with cubic Hermite polynomials is investigated numerically. The discretization of transient gas flow models is also studied [2, 5, 8]. An entropy stable finite volume method is proposed in [10], and an energy stable mixed finite element approximation is investigated in [3]. Apart from [9], all methods discussed above are of lowest order and no rigorous convergence analysis is given.

In this paper, we study the discretization of (4)–(5) by a Petrov-Galerkin approach of potentially high order. The resulting scheme is shown to be stable which allows us to prove order optimal convergence rates. By using an appropriate functional analytic setting, the convergence results can be generalized almost verbatim to pipeline networks. A hybridization strategy will be discussed that facilitates the implementation and that allows to incorporate non-standard coupling conditions. The proposed method formally also allow to treat nonlinear models of gas transport and, in principle, high order convergence can be obtained in practically relevant regimes.

2 Notation and Preliminaries

Let $x_L < x_R$ and denote by $L^p(x_L, x_R)$ and $W^{k,p}(x_L, x_R)$, $k \geq 0$ the standard Lebesgue and Sobolev spaces. The scalar product and norm of $L^2(x_L, x_R)$ are written as (v, w) and $\|v\| = \|v\|_{L^2}$. Other norms will be designated by subscripts. We write $H^k(x_L, x_R) = W^{k,2}(x_L, x_R)$ for the Hilbert spaces and define

$$H_0^1 = \{v \in H^1(x_L, x_R) : v(x_L) = v(x_R) = 0\} \quad \text{and} \quad H(\text{div}) = H^1(x_L, x_R)$$

for convenience. The reason for introducing the space $H(\text{div})$ will become clear when considering networks, where the spaces H^1 and $H(\text{div})$ have different continuity properties across junctions. By $L^p(0, T; X)$ and $W^{k,p}(0, T; X)$ we denote the Bochner spaces of functions $f : [0, T] \to X$ with values in X. The value of $f(t)$ may then itself be a function. In the following, we consider the linear system

$$a\partial_t p(x, t) + \partial_x m(x, t) = f(x, t), \tag{6}$$

$$b\partial_t m(x, t) + \partial_x p(x, t) + d(x)m(x, t) = g(x, t), \tag{7}$$

for $x_L < x < x_R$ and $t > 0$ with homogeneous boundary conditions

$$p(x_L, t) = p(x_R, t) = 0. \tag{8}$$

Inhomogeneous and more general boundary conditions can be considered as well and our analysis applies with minor modifications. We will assume that

(A1) a, b are positive constants, and
(A2) $d \in L^\infty(x_L, x_R)$ with $0 < \underline{d} \le d(x) \le \overline{d}$ and constants $\underline{d}, \overline{d}$.

For given $f, g \in L^2(0, T; L^2(x_L, x_R))$ and initial values $p(0) \in H_0^1$, $m(0) \in H(\text{div})$, existence of a unique solution follows from semigroup theory. Any smooth solution of problem (6)–(8) also satisfies $p(t) \in H_0^1$, $m(t) \in H(\text{div})$, and

$$(a\partial_t p(t), \tilde{q}) + (\partial_x m(t), \tilde{q}) = (f(t), \tilde{q}) \tag{9}$$

$$(b\partial_t m(t), \tilde{v}) + (\partial_x p(t), \tilde{v}) + (dm(t), \tilde{v}) = (g(t), \tilde{v}) \tag{10}$$

for all $\tilde{v}, \tilde{q} \in L^2(x_L, x_R)$ and all $0 < t < T$. This variational characterization will be the starting point for our discretization approach introduced in the next section.

3   Petrov-Galerkin Approximation

Let $x_L = x_0 < x_1 < \ldots < x_N = x_R$ be a partition of the interval $[x_L, x_R]$ into elements $T_n = [x_{n-1}, x_n]$. We call $T_h := \{T_n : 1 \le n \le N\}$ the mesh and denote by $h_n = |x_n - x_{n-1}|$ and $h = \max_n h_n$ the local and global mesh size, respectively. Let

$$P_k(T_h) := \{v \in L^2(x_L, x_R) : v|_T \in P_k(T) \ \forall T \in T_h\} \tag{11}$$

be the space of piecewise polynomials on the mesh T_h. We fix $k \ge 1$ and search for approximations for the solutions $p(t), m(t)$ of problem (6)–(8) in the spaces

$$Q_h = P_k(T_h) \cap H_0^1 \quad \text{and} \quad V_h = P_k(T_h) \cap H(\text{div}) \tag{12}$$

of continuous piecewise polynomials with appropriate boundary conditions. As finite dimensional test spaces for the variational problem (9)–(10), we choose

$$\tilde{Q}_h = P_{k-1}(T_h) \quad \text{and} \quad \tilde{V}_h = P_{k-1}(T_h) \tag{13}$$

consisting of discontinuous piecewise polynomials of lower order $k - 1$. We denote by $I_h^k : H^1(x_L, x_R) \rightarrow P_k(T_h) \cap H^1(x_L, x_R)$ the H^1-projection operator, defined by

$$(I_h^k v)(x_k) = v(x_k) \qquad \text{for all } 0 \le k \le N, \tag{14}$$

$$\text{and} \qquad (\partial_x I_h^k v, \widetilde{v}_h) = (\partial_x v, \widetilde{v}_h) \qquad \text{for all } \widetilde{v}_h \in P_{k-1}(T_h), \tag{15}$$

and let $\pi_h^{k-1} : L^2(x_L, x_R) \rightarrow P_{k-1}(T_h)$ be the L^2-orthogonal projection, satisfying

$$(\pi_h^{k-1} v, \widetilde{v}_h) = (v, \widetilde{v}_h) \qquad \text{for all } \widetilde{v}_h \in P_{k-1}(T_h). \tag{16}$$

Note that both projection operators I_h^k and π_h^{k-1} can be defined locally on every element. Moreover, they are mutually related to each other by the *commuting diagram* property

$$\partial_x I_h^k v = \pi_h^{k-1} \partial_x v \qquad \text{for all } v \in H^1(x_L, x_R). \tag{17}$$

For the approximation of problem (6)–(8), we then use the following approximation.

Problem 1 (Inexact Petrov-Galerkin Method) Find functions $p_h \in H_0^1(0, T; Q_h)$, $m_h \in H^1(0, T; V_h)$ with $p_h(0) = I_h^k p(0)$ and $m_h(0) = I_h^k m(0)$, and such that

$$(a \partial_t p_h(t), \widetilde{q}_h) + (\partial_x m_h(t), \widetilde{q}_h) = (f(t), \widetilde{q}_h) \tag{18}$$

$$(b \partial_t m_h(t), \widetilde{v}_h) + (\partial_x p_h(t), \widetilde{v}_h) + (d \pi_h^{k-1} m_h(t), \widetilde{v}_h) = (g(t), \widetilde{v}_h) \tag{19}$$

for all $\widetilde{q}_h \in \widetilde{Q}_h = P_{k-1}(T_h)$ and $\widetilde{v}_h \in \widetilde{V}_h = P_{k-1}(T_h)$, and for all $0 \le t \le T$.

The well-posedness of this problem follows from the results of the next section.

4 Discrete Stability Estimates

We now derive some discrete stability estimates that yield well-posedness of the semidiscrete method and that allow us to establish error estimates of optimal order.

Lemma 1 *Let p_h, m_h denote a solution of Problem 1. Then*

$$a \|\pi_h^{k-1} p_h(t)\|^2 + b \|\pi_h^{k-1} m_h(t)\|^2$$

$$\le C(T) \left(a \|\pi_h^{k-1} p_h(0)\|^2 + b \|\pi_h^{k-1} m_h(0)\|^2 + \int_0^t \frac{1}{a} \|\pi_h^{k-1} f(s)\|^2 + \frac{1}{b} \|\pi_h^{k-1} g(s)\|^2 ds \right)$$

with constant $C(T) \le CT$ and C independent of T and the solution.

Proof Let us first note that $(\pi_h^{k-1} q_h, \pi_h^{k-1} q) = (q_h, \pi_h^{k-1} q)$ for all $q \in H^1(x_L, x_R)$. By testing (18)–(19) with $q_h = \pi_h^{k-1} p_h(t)$ and $v_h = \pi_h^{k-1} m_h(t)$, we then get

$$\frac{d}{dt}\left(\frac{a}{2}\|\pi_h^{k-1} p_h(t)\|^2 + \frac{b}{2}\|\pi_h^{k-1} m_h(t)\|^2\right)$$

$$= (a\partial_t p_h(t), \pi_h^{k-1} p_h(t)) + (b\partial_t m_h(t), \pi_h^{k-1} m_h(t))$$

$$= -(\partial_x m_h(t), \pi_h^{k-1} p_h(t)) - (\partial_x p_h(t), \pi_h^{k-1} m_h(t)) - (d\pi_h^{k-1} m_h(t), \pi_h^{k-1} m_h(t))$$

$$+ (\pi_h^{k-1} f(t), \pi_h^{k-1} p_h(t)) + (\pi_h^{k-1} g(t), \pi_h^{k-1} m_h(t)).$$

By identity (17), integration-by-parts, and the boundary conditions (8), one can verify that $(\partial_x m_h(t), \pi_h^{k-1} p_h(t)) + (\partial_x p_h(t), \pi_h^{k-1} m_h(t)) = 0$. Via Cauchy-Schwarz and Young inequalities, and using positivity of d, we then obtain the estimate

$$\frac{d}{dt}\left(\frac{a}{2}\|\pi_h^{k-1} p_h(t)\|^2 + \frac{b}{2}\|\pi_h^{k-1} m_h(t)\|^2\right)$$

$$= -(d\pi_h^{k-1} m_h(t), \pi_h^{k-1} m_h(t)) + (\pi_h^{k-1} f(t), \pi_h^{k-1} p_h(t)) + (\pi_h^{k-1} g(t), \pi_h^{k-1} m_h(t))$$

$$\leq \frac{\alpha}{2}(a\|\pi_h^{k-1} p_h(t)\|^2 + b\|\pi_h^{k-1} m_h(t)\|^2) + \frac{1}{2\alpha}(\frac{1}{a}\|\pi_h^{k-1} f(t)\|^2 + \frac{1}{b}\|\pi_h^{k-1} g(t)\|^2).$$

The Gronwall lemma and the choice $\alpha = 1/T$ finally yields the assertion.   □

Note that the above estimate does not yet give full control over the solution. A repeated application, however, allows us to prove the following stability estimate.

Lemma 2 *Let p_h, m_h denote a solution of Problem 1. Then*

$$\|p_h(t)\|^2 + \|m_h(t)\|^2$$

$$\leq C'(T)\Big(\|\pi_h^{k-1} p_h(0)\|^2 + \|\pi_h^{k-1} m_h(0)\|^2 + h\|\pi_h^{k-1}\partial_t p_h(0)\|^2 + h\|\pi_h^{k-1}\partial_t m_h(0)\|^2$$

$$+ \int_0^t \|\pi_h^{k-1} f(s)\|^2 + \|\pi_h^{k-1} g(s)\|^2 ds + h\|\pi_h^{k-1}\partial_t f(s)\|^2 + h\|\pi_h^{k-1}\partial_t g(s)\|^2\Big)$$

for all $0 \leq t \leq T$ with $C'(T) = C'T$ and C' independent of T and of the solution.

Proof As a direct consequence of the Poincaré inequality, one has

$$\|p_h\| \leq \|\pi_h^{k-1} p_h\| + h\|\partial_x p_h\| \quad \text{and} \quad \|m_h\| \leq \|\pi_h^{k-1} m_h\| + h\|\partial_x m_h\|.$$

The first terms in these estimates are already covered by Lemma 1. From the two Eqs. (18)–(19) with $\widetilde{q}_h = \partial_x m_h(t)$ and $\widetilde{v}_h = \partial_x p_h(t)$, we further deduce that

$$\|\partial_x m_h(t)\|^2 \le (\|\pi_h^{k-1} f(t)\| + a\|\pi_h^{k-1}\partial_t p_h(t)\|)\|\partial_x m_h(t)\| \qquad \text{and}$$

$$\|\partial_x p_h(t)\|^2 \le (\|\pi_h^{k-1} g(t)\| + b\|\pi_h^{k-1}\partial_t m_h(t)\| + \overline{d}\|\pi_h^{k-1} m_h(t)\|)\|\partial_x p_h(t)\|.$$

Bounds for $\|\pi_h^{k-1}\partial_t p_h(t)\|$ and $\|\pi_h^{k-1}\partial_t m_h(t)\|$ can be obtained by formally differentiating (18)–(19) with respect to time and applying Lemma 1 for the resulting system. A combination of the above estimates then yields the assertion of the lemma.

$\square$

Remark 1 Problem 1 formally amounts to a finite dimensional system of differential algebraic equations. From the stability estimates of Lemma 2 and [6, Theorem 4.12], one can deduce that this system is solvable for any choice of admissible initial values. The semidiscretization is thus well-defined. Further note that the stability constants in Lemma 1 and 2 are independent of the polynomial degree k.

5 Error Estimates

As usual, we decompose the error according to $\|p - p_h\| \le \|p - I_h^k p\| + \|I_h^k p - p_h\|$ and $\|m - m_h\| \le \|m - I_h^k m\| + \|I_h^k m - m_h\|$ into approximation and discrete error components. The first part can be handled by the following estimates [11]. To simplify notation, we assume that the mesh is quasi-uniform in the following.

Lemma 3 Let $w \in H^{s+1}(T_h)$, $0 \le s \le k$. Then

$$\|w - I_h^k w\| \le C \left(\frac{h}{k}\right)^{s+1} |w|_{s+1;h}. \tag{20}$$

For any $w \in L^2(x_L, x_R) \cap H^s(T_h)$, $0 \le s \le k$, one has

$$\|w - \pi_h^{k-1} w\| \le C \left(\frac{h}{k}\right)^{s} |w|_{s;h}. \tag{21}$$

Here $H^s(T_h) = \{w \in L^2(x_L, x_R) : w|_T \in H^s(T)\}$ is the space of piecewise smooth functions and $|w|_{s;h} := (\sum_T \|\partial_x^s w\|_{L^2(T)}^2)^{1/2}$ is the corresponding seminorm. Moreover, the constant C in the estimates is independent of h and k.

Using Eqs. (9)–(10) and (18)–(19) characterizing the continuous and the discrete solutions, one can see that the discrete error components $\widehat{p}_h(t) := I_h^k p(t) - p_h(t)$ and $\widehat{m}_h(t) := I_h^k m(t) - m_h(t)$ satisfy Eqs. (18)–(19) with initial values $\widehat{p}_h(0) = 0$

and $\widehat{m}_h(0) = 0$, and right hand sides given by

$$\widehat{f}(t) := a(I_h^k \partial_t p(t) - \partial_t p(t)) \quad \text{and}$$

$$\widehat{g}(t) := b(I_h^k \partial_t m(t) - \partial_t m(t)) + d(\pi_h^{k-1} I_h^k m(t) - m(t)).$$

By the a-priori estimates of Lemma 2, one then obtains the following result.

Lemma 4 *Let $d \in P_0(T_h)$ be piecewise constant. Then for all $0 \le t \le T$ one has*

$$\|I_h^k p(t) - p_h(t)\|^2 + \|I_h^k m(t) - m_h(t)\|^2$$

$$\le C''(T)\Big(h\|I_h^k \partial_t p(0) - \partial_t p(0)\|^2 + h\|I_h^k \partial_t m(0) - \partial_t m(0)\|^2$$

$$+ \int_0^t \|I_h^k m(s) - m(s)\|^2 + \|I_h^k \partial_t p(s) - \partial_t p(s)\|^2 + \|I_h^k \partial_t m(s) - \partial_t m(s)\|^2$$

$$+ h\|I_h^k \partial_{tt} p(s) - \partial_{tt} p(s)\|^2 + h\|I_h^k \partial_{tt} m(s) - \partial_{tt} m(s)\|^2 ds\Big),$$

with a constant $C''(T) = C''T$ and C'' independent of h, k, T, and of the solution.

Proof We apply Lemma 2 for $\widehat{p}_h(t) = I_h^k p(t) - p_h(t)$ and $\widehat{m}_h(t) = I_h^k m(t) - m_h(t)$ and then estimate the terms on the right hand side of the result step by step. By definition of the initial values, we have $\widehat{p}_h(0) = \widehat{m}_h(0) = 0$. Moreover,

$$\pi_h^{k-1} \partial_t p_h(0) = \pi_h^{k-1} f(0) - \partial_x m_h(0) = \pi_h^{k-1} f(0) - \partial_x I_h^k m(0)$$

$$= \pi_h^{k-1} f(0) - \pi_h^{k-1} \partial_x m(0) = \pi_h^{k-1} \partial_t p(0),$$

where we used the definition of the initial value $m_h(0)$ in the second and (17) in the third step. Thus $\|\pi_h^{k-1} \partial_t \widehat{p}_h(0)\| \le \|I_h^k \partial_t p(0) - \partial_t p(0)\|$, and in a similar manner, one can show $\|\pi_h^{k-1} \partial_t \widehat{m}_h(0)\| \le \|I_h^k \partial_t m(0) - \partial_t m(0)\|$. This explains the first two terms in the estimate in the lemma. The terms under the integral are derived by estimating $\|\pi_h^{k-1} \widehat{f}(t)\|$, $\|\pi_h^{k-1} \widehat{g}(t)\|$ and the derivatives $\|\pi_h^{k-1} \partial_t \widehat{f}(t)\|$, $\|\pi_h^{k-1} \partial_t \widehat{g}(t)\|$ via the triangle inequality, and noting that

$$\pi_h^{k-1}(d\pi_h^{k-1} I_h^k m(t) - dm(t)) = d\pi_h^{k-1}(I_h^k m(t) - m(t)),$$

where we used that d is piecewise constant. $\qquad\square$

Remark 2 A similar result can be proven for piecewise smooth $d \in W^{1,\infty}(T_h)$ and additional terms of the form $\|d - \pi_h^0 d\|\|\pi_h^{k-1} p(t) - p(t)\|$ arise. For $d \in W^{1,\infty}(T_h)$, the product of the two terms again has optimal approximation order.

By combination of the above estimates, we finally obtain the following result.

Theorem 1 *Let (A1)–(A2) hold and $d \in W^{1,\infty}(T_h)$. Furthermore, let (p, m) be a sufficiently smooth solution of (6)–(8). Then for all $0 \leq t \leq T$, one has*

$$\|p(t) - p_h(t)\| + \|m(t) - m_h(t)\| \leq C(u, p, T) \frac{h^{k+1}}{k^k}.$$

For sufficiently smooth solutions, the proposed method thus converges at optimal order in h and at almost optimal order in the polynomial degree k.

6 Extension to Networks

We now illustrate that our method and the convergence results of the previous section can be generalized easily to pipe networks. Let $(\mathcal{V}, \mathcal{E})$ denote a directed graph with vertices $v \in \mathcal{V}$ and edges $e \in \mathcal{E}$; see Fig. 1 for illustration. For any edge $e = (v_1, v_2)$, we define $n^e(v_1) = -1$ and $n^e(v_2) = 1$. The matrix N with entries $N_{ij} = n_{ej}(v_i)$ then is the incidence matrix of the graph. For any vertex $v \in \mathcal{V}$, we define $\mathcal{E}(v) = \{e : e = (v, \cdot) \text{ or } e = (\cdot, v)\}$, and we set $\mathcal{V}_0 = \{v \in \mathcal{V} : |\mathcal{E}(v)| > 1\}$ and $\mathcal{V}_\partial = \{v \in \mathcal{V} : |\mathcal{E}(v)| = 1\}$ which gives a decomposition $\mathcal{V} = \mathcal{V}_0 \cup \mathcal{V}_\partial$ into interior and boundary vertices.

To every edge e, we associate a positive length ℓ^e, and we identify e with $[0, \ell^e]$ in the sequel. This allows us to define spaces $L^p(\mathcal{E}) = \{v : v|_e \in L^p(e)\}$ and $H^1(\mathcal{E}) = \{v \in L^p(\mathcal{E}) : v|_e \in H^1(e)\}$ of, respectively, integrable and piecewise smooth functions on the graph. The flow of gas in a pipe network is then described as follows: On every edge e representing a pipe, we require that

$$a^e \partial_t p^e + \partial_x m^e = f^e \tag{22}$$

$$b^e \partial_t m^e + \partial_x p^e + d^e m^e = g^e, \tag{23}$$

where $f^e = f|_e$ denotes the restriction of a function $f \in L^p(\mathcal{E})$ to one edge. The equations for the individual pipes are coupled by algebraic conditions

$$\sum_{e \in \mathcal{E}(v)} m^e(v) n^e(v) = 0 \qquad v \in \mathcal{V}_0 \tag{24}$$

$$p^e(v) = p^{e'}(v) \qquad v \in \mathcal{V}_0, \, e, e' \in \mathcal{E}(v) \tag{25}$$

at the pipe junctions, and at the boundary vertices, we assume that

$$p^e(v) = 0 \qquad v \in \mathcal{V}_\partial. \tag{26}$$

Fig. 1 Directed graph $(\mathcal{V}, \mathcal{E})$ modeling the pipe network topology used for numerical tests

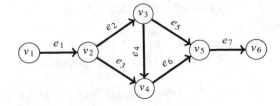

Inhomogeneous and other types of boundary conditions can again be incorporated with minor modifications. For the analysis of the problem, we now utilize the spaces

$$H_0^1 := \{p \in H^1(\mathcal{E}) : (25) \text{ and } (26) \text{ are valid}\} \tag{27}$$

$$H(\text{div}) := \{m \in H^1(\mathcal{E}) : (24) \text{ is valid}\} \tag{28}$$

which are the natural generalization of those used for the analysis on a single pipe. Any solution (p, m) of (22)–(26) then again satisfies $p(t) \in H_0^1$, $m(t) \in H(\text{div})$, and

$$(a\partial_t p(t), \tilde{q}) + (\partial_x m(t), \tilde{q}) = (f(t), \tilde{q}) \tag{29}$$

$$(b\partial_t m(t), \tilde{v}) + (\partial_x p(t), \tilde{v}) + (dm(t), \tilde{v}) = (g(t), \tilde{q}) \tag{30}$$

for all $\tilde{q} \in L^2(\mathcal{E})$, $\tilde{v} \in L^2(\mathcal{E})$, and all $0 < t < T$. Here $(v, w) = \sum_e (v^e, w^e)_e$ with $(v^e, w^e)_e = \int_e v^e w^e dx$ denotes the scalar product on $L^2(\mathcal{E})$.

Remark 3 Let us note that (29)–(30) has exactly the same form as the variational problem (18)–(19) on a single pipe. The inexact Petrov-Galerkin method and all results derived in the previous sections therefore translate almost verbatim to the network setting; let us refer to [4] for details and similar results for a different method, and to Sect. 9 for numerical illustration.

7 Remarks on the Efficient Implementation

In the discretization of (29)–(30), also compare with (18)–(19), the continuity and boundary conditions (24)–(26) are directly incorporated in the definition of the spaces $Q_h \subset H_0^1$ and $V_h \subset H(\text{div})$. For the implementation, it may be more convenient to use larger spaces $Q_h, V_h \subset H^1(\mathcal{E})$, and to enforce some of the boundary and coupling conditions (24)–(26) explicitly by additional equations. Using the wording of [1], this approach of relaxing continuity conditions might be called *hybridization*. Since the resulting method is algebraically equivalent to the original scheme based on function spaces with incorporated coupling and boundary conditions, all results of the previous sections apply verbatim also to the method obtained after hybridization.

8 Nonlinear Problems

The formal extension of the Petrov-Galerkin method to nonlinear problems is straight-forward. The discrete variational formulation for (1)–(2), for instance, reads

$$(A \partial_t \rho_h(t), \widetilde{q}_h) + (\partial_x m_h(t), \widetilde{q}_h) = 0$$

$$(\partial_t m_h(t), \widetilde{v}_h) + (\partial_x \left(\frac{m_h(t)^2}{A\rho_h(t)} + Ap_h(t) \right), \widetilde{v}_h) = -(\frac{\lambda}{2D} \frac{|m_h(t)|}{A\rho_h(t)} \pi_h^{k-1} m_h(t), \widetilde{v}_h).$$

Numerical quadrature can be used in practice to facilitate the handling of the nonlinear terms. We do not give a complete convergence analysis here, but instead, we will demonstrate by numerical tests that for smooth solutions, the convergence results of Theorem 1 remain valid, at least in the practically relevant case of nonlinear friction.

9 Numerical Results

We now illustrate the theoretical results of Sect. 5 by numerical tests. For our computations, we consider the pipe network depicted in Fig. 1. As a first test case, we consider the linear problem (22)–(25) with inhomogeneous boundary conditions

$$p|_v(t) = p_v(t) \qquad v \in \mathcal{V}_\partial \tag{31}$$

and we set $p_{v_1}(t) = 1$ and $p_{v_6}(t) = 1 + \frac{1}{2} \sin(\pi t)$ in the following. All pipes are chosen of unit length $\ell = 1$ and the model parameters are set to $a \equiv b \equiv d \equiv 1$. The simulation is started from a stationary state for the boundary values at initial time. The results of the computations are summarized in the left column of Table 1. As predicted by our theoretical results, we observe second order convergence.

We now repeat our numerical tests for the same network but with a semilinear gas flow model resulting from (1)–(3) by dropping the nonlinear term $\partial_x(\frac{m^2}{A\rho})$ in

Table 1 Errors $e_h = (a\|p_h(T) - p_{h/2}(t)\|^2 + b\|m_h(T) - m_{h/2}(T)\|^2)^{1/2}$ at time $T = 10$ obtained with the Petrov-Galerkin approximation for the network problem with different gas flow models: linear model (left), semilinear model (middle), and quasilinear model (right)

h	Linear	eoc	Semilinear	eoc	Quasilinear	eoc
0.10000	0.01936	–	0.02359	–	0.02534	–
0.05000	0.00482	2.00	0.00660	1.83	0.00693	1.87
0.02500	0.00120	2.00	0.00168	1.97	0.00200	1.79
0.01250	0.00030	2.00	0.00042	1.99	0.00076	1.40
0.00625	0.00008	2.00	0.00011	2.00	0.00036	1.09

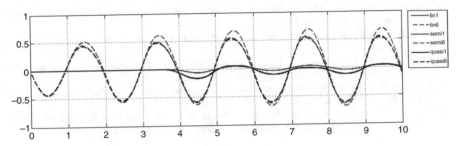

Fig. 2 Flow rates at boundary vertices v_1 and v_6 for linear, semilinear, and quasilinear flow models

Eq. (2). The model parameters are chosen as $A = 1$, $c = 1$, and $\lambda/(2D) = 7/2$; the latter was selected such that average of the resulting mass flow was similar to that of the linear model considered above. The computational results are depicted in the middle column of Table 1. Also for this nonlinear friction model, we observe second order convergence. These results can be explained theoretically in a similar way as those for the linear case by using a perturbation argument. In the right column of Table 1, we display the corresponding results for the quasilinear flow model (1)–(3) with the same parameters as used in the semilinear case. Note that a decrease in the convergence rates to first order is observed here. This is no surprise, since our analysis heavily relied on the anti-symmetry of the spatial derivative terms in (18)–(19), which is no longer valid for the quasilinear model (1)–(2).

In Fig. 2, we display the flow rates $m|_v$ at the boundary vertices v_1 and v_6 for the three different gas flow models discussed above as function of time. The results are in reasonable agreement. In summary, the semilinear model seems to yield the best compromise between modelling errors and convergence order.

Acknowledgements Support by the German Science Foundation (DFG) via grants TRR 154 project C4, TRR 146 project C3, GSC 233, and Eg-331/1-1 is gratefully acknowledged.

References

1. Boffi, D., Brezzi, F., Fortin, M.: Mixed Finite Element Methods and Applications. Springer Series in Computational Mathematics, vol. 44. Springer, Heidelberg (2013)
2. Brouwer, J., Gasser, I., Herty, M.: Gas pipeline models revisited: model hierarchies, non-isothermal models and simulations of networks. Multiscale Model. Simul. **9**, 601–623 (2011)
3. Egger, H.: A robust conservative mixed finite element method for compressible flow on pipe networks. SIAM J. Sci. Comput. **40**, A108–A129 (2018)
4. Egger, H., Kugler, T.: Damped wave systems on networks: exponential stability and uniform approximations. Numer. Math. **138**, 839–867 (2018)
5. Kiuchi, T.: An implicit method for transient gas flows in pipe networks. Int. J. Heat Fluid Flow **15**, 378–383 (1994)
6. Kunkel, P., Mehrmann, V.: Differential-algebraic equations: analysis and numerical solution. EMS Textbooks in Mathematics. European Mathematical Society (EMS), Zürich (2006)

7. Marcati, P., Milani, A.: The one-dimensional Darcy's law as the limit of a compressible Euler flow. J. Diff. Equ. **84**, 129–147 (1990)
8. Osiadacz, A.: Simulation of transient gas flows in networks. Int. J. Numer. Meth. Fluids **4**, 13–24 (1984)
9. Rachford Jr., H.H., Dupont, T.: A fast, highly accurate means of modeling transient flow in gas pipeline systems by variational methods. Soc. Pet. Eng. J. **14**, 165–178 (1974)
10. Reigstad, G.A.: Numerical network models and entropy principles for isothermal junction flow. Netw. Heterog. Media **9**, 65–95 (2014)
11. Schwab, C.: p- and hp-finite element methods. Numerical Mathematics and Scientific Computation. The Clarendon Press/Oxford University Press, Oxford (1998)

New Preconditioners for Semi-linear PDE-Constrained Optimal Control in Annular Geometries

Lasse Hjuler Christiansen and John Bagterp Jørgensen

1 Introduction

Large-scale optimization problems that are constrained by partial differential equations (PDEs) play a key role in various fields of science and engineering [2, 10]. As a challenge, the size and complexity of the PDE-constraints presents severe computational difficulties that often prevent the use of general-purpose black-box optimizers. As a consequence, cost efficient, specialized solvers become essential [1, 3, 7, 8]. As a contribution in this direction, this paper demonstrates how to extend seminal ideas of Shen [15–17] to construct fast and memory-efficient optimizers for the class of semi-linear PDE-constrained optimization problems with non-linear reaction kinetics

$$\min_{y,\, u \in U_{ad}} \quad \frac{1}{2} \int_{\Omega} (y(x) - y_d(x))^2 dx + \frac{\rho}{2} \int_{\Omega} u(x)^2 dx, \tag{1a}$$

$$\text{s.t.} \quad -\Delta y + G(y) = u \quad \text{in} \quad \Omega. \tag{1b}$$

The paper focuses on the specific cases of either homogeneous (1) Dirichlet or (2) Neumann boundary conditions, where $\Omega \subset \mathbb{R}^2$ is an annular domain of the type

$$\Omega := \{(x, y) \in \mathbb{R}^2 \mid a \leq x^2 + y^2 \leq b\}, \ 0 < a < b. \tag{2}$$

For a given non-linear reaction term, $G(\cdot)$, and Tikhonov regularization parameter, $\rho > 0$, the control problem (1) aims to determine the optimal state and control

L. H. Christiansen (✉) · J. B. Jørgensen
Department of Applied Mathematics and Computer Science & Center for Energy Resources
Engineering, Technical University of Denmark (DTU), Kgs. Lyngby, Denmark
e-mail: lhch@dtu.dk; jbjo@dtu.dk

© The Author(s) 2020
S. J. Sherwin et al. (eds.), *Spectral and High Order Methods for Partial Differential Equations ICOSAHOM 2018*, Lecture Notes in Computational Science and Engineering 134, https://doi.org/10.1007/978-3-030-39647-3_35

441

variables, (y^*, u^*), that minimize the objective (1a). Here the optimal solution must belong to the set of feasible pairs, (y, u), that satisfy the PDE-constraints (1b) and the additional admissibility condition, $u \in U_{\text{ad}}$. To be concrete, this paper focuses on the case of bi-lateral point-wise control constraints

$$U_{\text{ad}} := \{u \in L^2(\Omega) : u_a \le u(x) \le u_b \text{ a.e. in } \Omega_d\}. \tag{3}$$

Point-wise bounds of the type (3) appear in a number of practical applications, where the control must satisfy, e.g., operational limitations that are not naturally captured by the underlying PDE (1b). In the limiting case, where $u_a := -\infty$ and $u_b := \infty$, the admissible set becomes $U_{\text{ad}} = L^2(\Omega_d)$. This corresponds to the case where the PDE (1b) constitutes the only constraint.

1.1 Main Contributions and Outline

This paper contributes to a recent series of efforts by the authors that seek to construct fast, iterative solvers for a range of PDE-constrained optimization problems by exploiting the properties of customized spectral bases [4–6]. This series of work aims to introduce a high-order alternative to the widely-used constellation of low-order finite-element methods and Schur-complement preconditioners that currently predominates the literature on PDE control [12–14]. Previous efforts have mainly considered distributed control of elliptic and parabolic non-linear diffusion-reaction systems. The main focus has been on problems in rectangular domains, where PDEs constitute the only constraints. As a natural extension, this paper investigates how to modify the existing methods to account for (1) bound constraints of the type (3) and (2) different geometries. For the sake of brevity, the paper restricts attention to annular domains (2). However, with slight modifications, the approach generalizes to cylindrical geometries of the type

$$\Omega_C := \{(x, y, z) \in \mathbb{R}^3 \mid a \le x^2 + y^2 \le b, \ z \in (0, h)\}, \ 0 < a < b. \tag{4}$$

As the main contribution, this work proposes a collection of Poisson-like precon-ditioners that are customized for efficient solution of the control problems (1) by a semi-smooth Newton (SSN) strategy [9]. Similar to a traditional Newton method, the SSN scheme solves (1) iteratively by finding a locally optimal solution to the non-linear Karuhn-Kush-Tucker (KKT) optimality conditions by solving a sequence of linearized, variable-coefficient subproblems. Direct solution of the subproblems is often time consuming and requires considerable memory-allocation. To this end, the new preconditioners are designed to promote efficient solution of the SSN subproblems by appropriate Krylov subspace (KSP) methods. Following seminal ideas of Shen [16], the preconditioners rely on fast direct solvers for constant-coefficient problems that exploit (1) the structure of boundary-adapted spectral bases and (2) the separable nature of annular domains. As the main feature, inversion of the preconditioners decouples to form to a sequence of *independent* 2×2 systems. This implies that the preconditioners can be applied matrix-free and scale linearly with the problem size. In addition, the independence of the 2×2 systems makes the preconditioners amenable to parallelization. To establish proof-of-concept, a numerical case study solves (1), where $G(\cdot)$ is given by a cubic non-linearity. The results demonstrate computational efficiency and show that the preconditioners respond well to different problem sizes, boundary conditions, point-wise bound constraints and various choices of the regularization parameter, $\rho > 0$.

To establish the necessary background, Sect. 2 outlines how to solve the optimal control problem (1) using the SSN scheme. Further, to motivate the contributions of this paper, the section discusses some of the computational challenges that arise from discretization of the associated linearized subproblems. These challenges naturally leads to the construction of the new Poisson-like preconditioners in Sect. 3. Section 4 presents numerical results, while Sect. 5 draws overall conclusions and addresses future work.

2 Motivation: A Semi-smooth Newton Method

This paper solves the control problem (1) by a semi-smooth Newton strategy [9]. The SSN scheme seeks to generate a locally optimal solution, $(\overline{y}, \overline{u})$, by solving the first-order necessary optimality system

$$-\Delta \overline{y} + G(\overline{y}) - H(p) = 0 \quad \text{in} \quad \Omega, \tag{5a}$$

$$-\Delta p + G_y(\overline{y})p + \varphi_y(\overline{y}) = 0 \quad \text{in} \quad \Omega. \tag{5b}$$

Here the boundary conditions of the original problem (1) are preserved, G_y denotes the Fréchet derivative of G with respect to the state variable, y, and the optimal control satisfies $\overline{u} = H(p) = \max(u_a, \min(\rho^{-1}p(x), u_b))$. In the special case $U_{ad} := L^2(\Omega)$, it can be shown that $\overline{u} = H(p) = \rho^{-1}p$ [18]. In the concrete case of annular domains (2), the system (5) can be recast to polar coordinates. To this end, define the functions

$$Y(t, \theta) := y(r(t)\cos(\theta), r(t)\sin(\theta)), \quad P(t, \theta) := p(r(t)\cos(\theta), r(t)\sin(\theta)), \tag{6}$$

where $r(t) := \frac{b-a}{2}(t+c)$, $t \in [-1, 1]$, $c = \frac{b+a}{b-a}$. The optimality system then reads

$$-\Delta_t Y + \kappa G(Y) - \kappa H(P) = 0 \quad \text{in} \quad \mathcal{Q}_R \tag{7a}$$

$$-\Delta_t P + \kappa G_Y(Y)P + \kappa \varphi_Y(Y) = 0 \quad \text{in} \quad \mathcal{Q}_R, \tag{7b}$$

where $\Delta_t Y := \left(((t+c)Y_t)_t + \frac{1}{(t+c)} Y_{\theta\theta} \right)$, $\kappa = \frac{(t+c)(b-a)^2}{4}$ and $\mathcal{D}_R := [-1, 1] \times [0, 2\pi)$. To solve the KKT conditions (7), the SSN scheme considers the system as an operator equation $F(y, p) = 0$ and solves it by generating a recursive sequence of iterates, $x_i := (Y_i, P_i)$, $1 \leq i \leq k$, where the next iterate, $x_{k+1} := (Y, P)$, is found by solution of the linearized optimality conditions:

$$-\Delta_t Y + C_0(x_k)Y - C_1(x_k)P = f(x_k) \quad \text{in} \quad \Omega, \tag{8a}$$

$$-\Delta_t P + C_0(x_k)P + C_2(x_k)Y = g(x_k) \quad \text{in} \quad \Omega. \tag{8b}$$

Here $C_0(x_k) := \kappa G_Y(Y_k)$, $C_1(x_k) := \kappa H_P(P_k)$, $C_2(x_k) := \kappa(G_{YY}(Y_k)P_k + \varphi_{YY}(Y_k))$ and

$$f(x_k) := \kappa(G_Y(Y_k)Y_k - G(Y_k) - (H_P(P_k)P_k - H(P_k))), \tag{9a}$$

$$g(x_k) := \kappa(G_{YY}(Y_k)P_kY_k + \varphi_{YY}(Y_k)Y_k - \varphi_Y(Y_k)), \tag{9b}$$

where H_p denotes the generalized Newton derivative of H with respect to the adjoint variable, P, i.e.,

$$H_P(P) = \frac{1}{\rho} \begin{cases} 1 & \text{if } u_a \leq \frac{1}{\rho}P \leq u_b, \\ 0 & \text{otherwise.} \end{cases} \tag{10}$$

2.1 Numerical Challenges: Discretization of the SSN Subproblems

As a numerical challenge, the SSN scheme relies on successive solution of coupled PDEs in the form (8). Upon discretization, this leads to repeated solution of large saddle-point problems. To illustrate the associated difficulties, consider a spectral-Galerkin discretization of the linear subproblems (8). To this end, define the boundary-adapted approximation spaces

$$V_N := \{v \in \mathbb{P}_N : av(\pm 1) + bv'(\pm 1) = 0\}, \quad \mathbb{F}_M := \text{span}\{e^{ik(\cdot)}, M/2 \leq k \leq M/2-1\}. \tag{11}$$

Let $K := N \cdot M$ and define $S_K := V_N \times \mathbb{F}_M$. The discrete Galerkin approximation of (8) then seeks to find $Y, P \in S_K$ such that

$$\langle (t+c)Y_t, v_t \rangle + \langle (t+c)^{-1}Y_\theta, v_\theta \rangle + \langle C_0 Y - C_1 P, v \rangle = \langle f, v \rangle \quad \forall v \in S_K, \tag{12a}$$

$$\langle (t+c)P_t, v_t \rangle + \langle (t+c)^{-1}P_\theta, v_\theta \rangle + \langle C_0 P + C_2 Y, v \rangle = \langle g, v \rangle \quad \forall v \in S_K. \tag{12b}$$

where $\langle v, w \rangle := \int_0^{2\pi} \int_{-1}^1 v\overline{w}\, dt d\theta$. To represent the approximate solutions, $Y_{N,M}$ and $P_{N,M}$, consider the truncated series expansions

$$Y_{N,M}(t, \theta) := \sum_{k=-M/2}^{M/2-1} \sum_{m=0}^{N-2} \widehat{y}_{l(k)m}\psi_m(t)e^{ik\theta}, \quad P_{N,M}(t, \theta) := \sum_{k=-M/2}^{M/2-1} \sum_{m=0}^{N-2} \widehat{p}_{l(k)m}\psi_m(t)e^{ik\theta},$$

(13)

where $l(k) := k + \frac{M}{2}$. Now, define the $(N-1) \times (N-1)$ matrices associated with the basis $\{\psi_k\}_{k=0}^{N-2}$:

$$a_{ij} = \langle (c+t)\psi_j', \psi_i' \rangle, \qquad A = (a_{ij})_{i,j=0..N-2}, \tag{14}$$

$$b_{ij} = \langle (c+t)^{-1}\psi_j, \psi_i \rangle, \quad B = (b_{ij})_{i,j=0..N-2}. \tag{15}$$

Note that appropriate choices of the basis functions $\{\psi_k\}_{k=0}^{N-2} \in V_n$ will be constructed in Sect. 3. Further, let Γ and $\varXi$ denote the $M \times M$ diagonal matrices defined by

$$\gamma_{mn} = \langle e^{in(\cdot)}, e^{im(\cdot)} \rangle = 2\pi\delta_{mn}, \; \xi_{mn} = mn\langle e^{in(\cdot)}, e^{im(\cdot)} \rangle = 2\pi nm\delta_{mn}, \tag{16}$$

where δ_{mn} denotes the Kronecker delta. Finally, consider the $(MN \times 1)$ vectors

$$\widehat{y} := (\widehat{y}_0, \ldots, \widehat{y}_{M-1}), \widehat{y}_k = \{\widehat{y}_{jk}\}_{j=0}^{N-2}, \tag{17}$$

$$\widehat{p} := (\widehat{p}_0, \ldots, \widehat{p}_{M-1}), \widehat{p}_k = \{\widehat{p}_{jk}\}_{j=0}^{N-2}, \tag{18}$$

$$\widehat{G} := (\widehat{g}_0, \ldots, \widehat{g}_{M-1}), \widehat{g}_k = \{\langle g, \psi_j e^{ik(\cdot)} \rangle\}_{j=0}^{N-2}, \tag{19}$$

$$\widehat{F} := (\widehat{f}_0, \ldots, \widehat{f}_{M-1}), \widehat{f}_k = \{\langle f, \psi_j e^{ik(\cdot)} \rangle\}_{j=0}^{N-2}. \tag{20}$$

The discretized linear subproblem (8) can then be written in matrix form

$$\underbrace{\begin{bmatrix} M_{C_2} & \mathbb{B} + M_{C_0} \\ \mathbb{B} + M_{C_0} & -M_{C_1} \end{bmatrix}}_{\mathscr{A}} \underbrace{\begin{bmatrix} \widehat{y} \\ \widehat{p} \end{bmatrix}}_{x} = \underbrace{\begin{bmatrix} \widehat{G} \\ \widehat{F} \end{bmatrix}}_{b}, \tag{21}$$

where $\mathbb{B} = \Gamma \otimes A + \varXi \otimes B$. Here the matrices $M_{C_\ell}, \ell = 1, 2, 3$ are defined by the elements

$$(mc_\ell)_{ij} = \langle C_i \psi_k e^{im(\cdot)}, \psi_l e^{in(\cdot)} \rangle, \tag{22}$$

where i, j satisfy that

$$i = n(N - 1) + (l + 1), \quad j = m(N - 1) + (k + 1), \tag{23a}$$

$$0 \le k, l \le N - 2, \quad\quad\quad 0 \le n, m \le M - 1. \tag{23b}$$

3 New Poisson-Like Preconditioners

As a significant challenge to the numerical solution of (7), the SSN scheme relies on repeated solution of saddle-point problems (21) of dimension $2(N - 1)M \times 2(N - 1)M$. Consequently, direct solution strategies often become computational intractable. As a cost efficient alternative, the following introduces new preconditioners that seek to accelerate the inner SSN subproblems (8) by using appropriate Krylov subspace methods to solve the associated preconditioned linear systems

$$P_k^{-1} \mathscr{A}_k x_k = P_k^{-1} b_k. \tag{24}$$

Concretely, this paper proposes approximative *constraint* preconditioners of the type

$$P_k = \begin{bmatrix} \widehat{M}_{C_2} & \widehat{\mathbb{B}} + \widehat{M}_{C_0} \\ \widehat{\mathbb{B}} + \widehat{M}_{C_0} & -\widehat{M}_{C_1} \end{bmatrix}. \tag{25}$$

Following ideas of traditional Poisson preconditioners, the new preconditioners are constructed by approximating each block of the SSN subproblem (21) by the matrices, $\widehat{\mathbb{B}}$ and $\widehat{M}_{c_\ell}$, $\ell = 0, 1, 2$, that come from a spectral Galerkin discretization of the corresponding *constant-coefficient* problem that determines $Y, P \in S_K$ such that

$$\langle C_A Y_t, v_t \rangle + \langle C_B Y_\theta, v_\theta \rangle + \langle \overline{C}_0 Y - \overline{C}_1 P, v \rangle = \langle f, v \rangle \quad \forall v \in S_K, \tag{26a}$$

$$\langle C_A P_t, v_t \rangle + \langle C_B P_\theta, v_\theta \rangle + \langle \overline{C}_0 P + \overline{C}_2 Y, v \rangle = \langle g, v \rangle \quad \forall v \in S_K, \tag{26b}$$

where $C_A = c$, $C_B = \dfrac{c}{c^2 - 1}$ and $\overline{C}_i = \frac{1}{2} \left(\max\limits_{\Omega} C_i(x_k) + \min\limits_{\Omega} C_i(x_k) \right)$, $i = 0, 1, 2$.

To be efficient, the new preconditioners crucially rely on carefully chosen basis functions $\{\psi_k\}_{k=0}^{N-2}$ for the discrete approximation space, V_N (11). To this end, this paper uses Fourier-like (FL) bases that were originally introduced by Shen and Wang in the context of traditional initial-boundary-value problems [17]. As a key property to construction of the preconditioners, the FL bases lead to diagonal mass- and stiffness matrices, i.e.,

$$\mathbb{M}_{ij} = ((\psi_j, \psi_i))_{ij} = \lambda_j \delta_{j,i}, \quad \mathbb{S}_{ij} = ((\partial_t \psi_j, \partial_t \psi_i))_{ij} = \delta_{j,i}. \tag{27}$$

The FL bases can be constructed as part of an offline preprocessing stage in two steps:

1. Let $\{L_k(\cdot)\}_{k=0}^{N}$ be the Legendre polynomials. Then there exists a unique set of coefficients $\{a_k, b_k\}_{k=0}^{N-2}$ such that

$$\phi_k := c_k \left(L_k + a_k L_{k+1} + b_k L_{k+2}\right) \in V_{k+2}, \quad c_k := (\sqrt{-b_k(4k+6)})^{-1}.$$

Furthermore, the mass matrix, $M_A = (\langle \phi_j, \phi_i \rangle)_{ij}$, is penta-diagonal and symmetric positive definite, whereas the stiffness matrix, $S_A = (\langle \partial_x \phi_j, \partial_x \phi_i \rangle)_{ij}$, becomes diagonal [15]. In the concrete cases of Dirichlet and Neumann boundary conditions, the coefficients, $\{a_k, b_k\}_{k=0}^{N-2}$ are given by respectively

$$a_k = 0, \ b_k = -1 \text{ and } a_k = 0, \ b_0 = 1/2, \ b_k = -k(k+1)/((k+2)(k+3)). \tag{28}$$

2. The second step computes the diagonalization $\Lambda = Q^T M_A Q$, where $Q = (q_{ij})$ denotes the matrix of eigenvectors and $\{\lambda_i\}_{i=1}^{N-2}$ are the associated eigenvalues. Using the matrix Q, the FL basis can be constructed by the linear combinations:

$$\psi_k(x) = \sum_{j=0}^{N-2} q_{jk} \phi_j(x), \ 0 \le k \le N - 2. \tag{29}$$

3.1 Efficient Inversion of the Preconditioners

As the main feature of the preconditioners, P_k, the following describes an efficient inversion procedure that exploits the orthogonal structures of the FL bases (27). To this end, consider the following preconditioning problem that is solved during each iteration of the KSP method:

$$\underbrace{\begin{bmatrix} \widehat{M}_{C_2} & \widehat{\mathbb{B}} + \widehat{M}_{C_0} \\ \widehat{\mathbb{B}} + \widehat{M}_{C_0} & -\widehat{M}_{C_1} \end{bmatrix}}_{P_k} \underbrace{\begin{bmatrix} \widehat{y}^k \\ \widehat{p}^k \end{bmatrix}}_{z_k} = \underbrace{\begin{bmatrix} \widehat{G}^k \\ \widehat{F}^k \end{bmatrix}}_{\mathscr{A}_k x_k}. \tag{30}$$

Note that (30) corresponds to the discrete first-order necessary optimality conditions associated with the *constant-coefficient* optimal control problem (26). Hence, by definition (22), it follows that

$$\widehat{\mathbb{B}} = C_A \Gamma \otimes \mathbb{S} + C_B \Xi \otimes \mathbb{M}, \quad \widehat{M}_{C_\ell} = \overline{C}_\ell \Gamma \otimes \mathbb{M}, \tag{31}$$

where $\mathbb{S}_{ij} = ((\partial_t \psi_j, \partial_t \psi_i))_{ij}$ and $\mathbb{M}_{ij} = ((\psi_j, \psi_i))_{ij}$. Further, by the orthogonal properties of the Fourier bases (16), the matrices, Γ and Ξ, are diagonal. Therefore, using the notation,

$$\widehat{y}_l^k = \{\widehat{y}_{lm}^k\}_{m=0}^{N-2}, \quad \widehat{p}_l^k = \{\widehat{p}_{lm}^k\}_{m=0}^{N-2}, \quad \widehat{G}_l^k = \{\widehat{G}_{lm}^k\}_{m=0}^{N-2}, \quad \widehat{F}_l^k = \{\widehat{F}_{lm}^k\}_{m=0}^{N-2},$$

it follows that the preconditioning problem (30) can be written as M independent linear systems

$$\begin{bmatrix} 2\pi \overline{C}_2 \mathbb{M} & \Sigma_l \\ \Sigma_l & -2\pi \overline{C}_1 \mathbb{M} \end{bmatrix} \begin{bmatrix} \widehat{y}_l^k \\ \widehat{p}_l^k \end{bmatrix} = \begin{bmatrix} \widehat{G}_l^k \\ \widehat{F}_l^k \end{bmatrix}, \quad 0 \le l \le M - 1, \tag{32}$$

where $\Sigma_l := C_A \mathbb{S} + (C_B k(l)^2 + 2\pi \overline{C}_0) \mathbb{M}$. In addition, the properties of the FL basis, $\{\psi_k\}_{k=0}^{N-2}$, implies that $\mathbb{S}$ and $\mathbb{M}$ become diagonal (29). Hence, the system (32) reduces to $M(N-1)$ independent 2×2 linear systems in the form

$$\begin{bmatrix} 2\pi \overline{C}_2 \lambda_m & \sigma_{nm} \\ \sigma_{nm} & -2\pi \overline{C}_1 \lambda_m \end{bmatrix} \begin{bmatrix} \widehat{y}_{lm}^k \\ \widehat{p}_{lm}^k \end{bmatrix} = \begin{bmatrix} \widehat{G}_{nm}^k \\ \widehat{F}_{nm}^k \end{bmatrix}, \quad 0 \le l \le M - 1, \ 0 \le m \le N - 2, \tag{33}$$

where $\sigma_{lm} := C_A + (C_B k(l)^2 + 2\pi \overline{C}_0) \lambda_m$. By (33), it follows that the original preconditioning problem (30) decouples into $(N-1)M$ *independent* 2×2 subsystems. As a consequence, the Poisson-like preconditioners (25) scale linearly with the problem size and can be applied matrix-free.

4   Numerical Results

To investigate the potential of the Poisson-like preconditioners, the following case study solves the control problem (1), where the reaction term is given by the cubic non-linearity $G(y) := y^3$. The corresponding problem serves as a recurring example

in the control literature [18]. In this case study, the goal is to track the desired state
of the type

$$z_d(r, \theta) = \begin{cases} Z, & (r, \theta) \in [\alpha, \beta] \times [0, \pi/2] \cup [\pi, \pi/3] \\ 0, & \text{otherwise} \end{cases}, \qquad (34)$$

where $a \leq \alpha < \beta \leq b$. The following example uses the parameters, $Z = 4$, $a = 30$, $\alpha = 40$ and $\beta = b = 60$. The main purpose of the study is to
investigate efficiency and robustness of the preconditioners (25). To this end, the
study solves (1) for different choices of (1) problem size, (2) boundary conditions,
(3) regularization parameter, and (4) point-wise bound constraints of the type (3).[1]
As a benchmark reference, the results are compared to MATLABs state-of-the-art
direct solver. All computations are carried out in [11] on a 2.9 GHz Intel processor.
The SSN scheme is said to have converged when the 2-norm difference between
successive iterates is below $\eta = 10^{-4}$. The KSP iterations are performed using the
MATLAB function GMRES with a tolerance of $\epsilon = 10^{-9}$. The direct solver relies
on MATLABs backslash command. Table 1 lists the results, where KSP iter
denotes the average number of KSP iterations required for each SSN step. Note
also that DOF denotes the number of degrees of freedom for *each individual* SSN
subproblem. Hence, the total degrees of freedom, DOF_T, is therefore given by
#SSN steps $\times$ DOF. The results reflect some overall tendencies that generalize
to other choices of the parameters, Z, a, α, β and b. Firstly, the preconditioners
provide significant reductions in CPU-time compared to the direct strategy. In
particular, the results show that the non-linear control problem with up to $DOF_T = 875,000$ unknowns can be solved in less than a minute using modest hardware.
Secondly, the preconditioners prove robust with respect to the problem size and
the choice of boundary conditions. Thirdly, as a drawback, the number of SSN
steps and KSP iterations increase as the point-wise bounds become more strict. The
authors suspect that these increases in SSN steps and KSP iterations are caused by
the combination of a decrease in regularity of the solution and an increase in non-
linearity of the KKT system (Fig. 1).

---

[1] By the choices of parameters, the study strives to provide a representative example of the general
tendencies of performance and robustness that can be expected from the preconditioners. To allow
for more diverse and elaborate experiments, the MATLAB source code of this study has been made
publicly available from https://github.com/LHCH-DK/PDE_Control_Annular.git.

Table 1 For comparison, [·] denotes the CPU time (s) required by MATLABs direct solver

N,M	DOF	$u_a = -\infty, u_b = \infty, \rho = 10^{-5}$			$u_a = -35, u_b = 35, \rho = 10^{-4}$			$u_a = -10, u_b = 10, \rho = 10^{-3}$		
		Time (s)	SSN steps	KSP iter.	Time (s)	SSN steps.	KSP iter.	Time (s)	SSN steps	KSP iter.
Optimal tracking problem: Dirichlet boundary conditions										
50	5000	0.54 [59.95]	4	9	1.42 [76.13]	6	25	1.71 [82.27]	7	28
100	20,000	1.23 [–]	4	9	4.26 [–]	6	26	5.53 [–]	7	30
250	125,000	7.21 [–]	4	9	36.35 [–]	6	28	51.9 [–]	7	33
Optimal tracking problem: Neumann boundary conditions										
50	5000	0.65 [53.26]	4	10	1.57 [83.91]	6	24	1.77 [86.51]	7	28
100	20,000	1.25 [–]	4	10	4.25 [–]	6	26	5.42 [–]	7	29
250	125,000	7.43 [–]	4	8	34.23 [–]	6	27	51.81 [–]	7	32

A horizontal lines indicates that the computations were manually terminated after 300 s, without reaching convergence

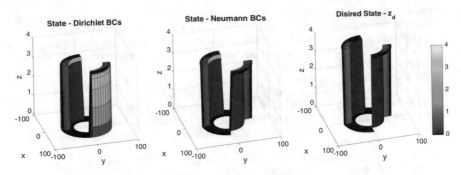

Fig. 1 The computed states for (1) Dirichlet boundary conditions, (2) Neumann boundary conditions and (3) the desired state for $u_a = -35, u_b = 35, \rho = 10^{-4}$. Note that both solutions manage to approximate the desired state well, despite of the bound constraints

5 Conclusions and Outlook

This paper has proposed new Poisson-like preconditioners for semi-linear PDE-constrained optimization problems with non-linear reaction kinetics and point-wise bound constraints. The preconditioners specifically target problems in annular domains. Inspired by [16], the new preconditioners exploit the orthogonal properties of customized, boundary-adapted spectral bases. This leads to matrix-free preconditioners that scale linearly with the problem size. Numerical results have demonstrated that the preconditioners lead to fast solution of large-scale optimization problems with significant computational benefits compared to MATLABs state-of-the-art direct methods. Furthermore, the preconditioners have proven to be robust with respect to the problem size for both homogeneous Dirichlet and Neumann boundary conditions. As a challenge, numerical experiments indicated that the non-linearity of the problem increases as the point-wise bound constraints become more strict. In turn, this leads to an increase in the number of SSN steps and KSP iterations that are required to reach convergence. A future study seeks to improve this situation by providing the SSN scheme with an educated starting guess that uses a coarse-grid solution to a similar control problem with less restrictive constraints.

References

1. Biegler, L.: Efficient solution of dynamic optimization and NMPC problems. Prog. Syst. C **26**, 219–243 (2000)
2. Biegler, L.T.: Real-Time PDE-Constrained Optimization, vol. 3. Society for Industrial and Applied Mathematics, Philadelphia (2007)
3. Borzi, A.: Multigrid methods for parabolic distributed optimal control problems. J. Comput. Appl. Math. **157**(2), 365–382 (2003)

4. Christiansen, L.H., Jørgensen, J.B.: A fast and memory-efficient spectral Galerkin scheme for distributed elliptic optimal control problems (2017). Preprint. arXiv:1712.08225
5. Christiansen, L.H., Jørgensen, J.B.: A fast PDE-constrained optimization solver for nonlinear diffusion-reaction processes. In: Proceedings of 2018 IEEE Conference on Decision and Control, pp. 2635–2640. IEEE (2018)
6. Christiansen, L.H., Jørgensen, J.B.: A new Lagrange-Newton-Krylov solver for PDE-constrained nonlinear model predictive control. IFAC-PapersOnLine **51**(20), 325–330 (2018)
7. Diehl, M., Bock, H., Schloder, J.: Newton-type methods for the approximate solution of nonlinear programming problems in real-time. Appl. Optim. **82**, 177–200 (2003)
8. Herzog, R., Kunisch, K.: Algorithms for PDE-constrained optimization. Gamm Mitteilungen **33**(2), 163–176 (2010)
9. Ito, K., Kunisch, K.: Lagrange Multiplier Approach to Variational Problems and Applications. Society for Industrial and Applied Mathematics, Philadelphia (2008)
10. Leugering, G., Benner, P., Engell, S., Griewank, A., Harbrecht, H., Hinze, M., Rannacher, R., Ulbrich, S.: Trends in PDE Constrained Optimization. International Series of Numerical Mathematics. Springer International Publishing, Basel (2014)
11. MATLAB.: version 8.6.0 (R2015b). The MathWorks Inc., Natick, Massachusetts (2015)
12. Pearson, J.W., Stoll, M.: Fast iterative solution of reaction-diffusion control problems arising from chemical processes. SIAM J. Sci. Comput. **35**(5), B987–B1009 (2013)
13. Rees, T., Wathen, A.J.: Preconditioning iterative methods for the optimal control of the Stokes equations. SIAM J. Sci. Comput. **33**(5), 2903–2926 (2011)
14. Rees, T., Dollar, H.S., Wathen, A.J.: Optimal solvers for PDE-constrained optimization. SIAM J. Sci. Comput. **32**(1), 271–298 (2010)
15. Shen, J.: Efficient spectral-Galerkin method. I. Direct solvers of second- and fourth-order equations using Legendre polynomials. SIAM J. Sci. Comput. **15**(6), 1489–1505 (1994)
16. Shen, J.: Efficient spectral-Galerkin methods III: polar and cylindrical geometries. SIAM J. Sci. Comput. **18**(6), 1583–1604 (1997)
17. Shen, J., Wang, L.L.: Fourierization of the Legendre-Galerkin method and a new space-time spectral method. Appl. Numer. Math. **57**(5–7), 710–720 (2007)
18. Tröltzsch, F.: Optimal Control of Partial Differential Equations: Theory, Methods, and Applications. Graduate Studies in Mathematics. American Mathematical Society, Providence (2010)

DIRK Schemes with High Weak Stage Order

David I. Ketcheson, Benjamin Seibold, David Shirokoff, and Dong Zhou

1 Introduction

Runge-Kutta (RK) methods achieve high-order accuracy in time by means of combining approximations to the solution at multiple stages. An s-stage RK scheme can be represented via the Butcher tableau

$$\frac{\mathbf{c} \; A}{\mathbf{b}^T} = \begin{array}{c|ccc} c_1 & a_{11} & \cdots & a_{1s} \\ \vdots & \vdots & & \vdots \\ c_s & a_{s1} & \cdots & a_{ss} \\ \hline & b_1 & \cdots & b_s \end{array} .$$

D. I. Ketcheson
Applied Mathematics and Computational Science, King Abdullah University of Science and Technology, Thuwal, Saudi Arabia
e-mail: david.ketcheson@kaust.edu.sa

B. Seibold (✉)
Department of Mathematics, Temple University, Philadelphia, PA, USA
e-mail: seibold@temple.edu

D. Shirokoff
Department of Mathematical Sciences, New Jersey Institute of Technology, Newark, NJ, USA
e-mail: david.g.shirokoff@njit.edu

D. Zhou
Department of Mathematics, California State University Los Angeles, Los Angeles, CA, USA
e-mail: dzhou11@calstatela.edu

© The Author(s) 2020
S. J. Sherwin et al. (eds.), *Spectral and High Order Methods for Partial Differential Equations ICOSAHOM 2018*, Lecture Notes in Computational Science and Engineering 134, https://doi.org/10.1007/978-3-030-39647-3_36

Throughout the whole paper we assume that $\mathbf{c} = A\mathbf{e}$, where $\mathbf{e}$ is the vector of all ones. The scheme's stability function [12] $R(\zeta) = 1 + \zeta \mathbf{b}^T (I - \zeta A)^{-1} \mathbf{e}$ measures the growth u^{n+1}/u^n per step Δt, when applying the scheme to the linear model equation $u'(t) = \lambda u$, with $\zeta = \lambda \Delta t$.

A particular interest lies in the accuracy of the RK scheme for stiff problems, i.e., problems in which a larger time step is chosen than the fastest time scale of the problem's dynamics. A standard stiff model problem [8] is the scalar linear ordinary differential equation (ODE)

$$u' = \lambda(u - \phi(t)) + \phi'(t) \,, \tag{1}$$

with i.c. $u(0) = \phi(0)$ and $\mathrm{Re}\,\lambda \leq 0$. The true solution $y(t) = \phi(t)$ evolves on an $O(1)$ time scale. Hence, λ-values with large negative real part result in stiffness. Considering a family of test problems (parametrized by λ), one can now establish the scheme's convergence via two different limits: (a) the non-stiff limit $\Delta t \to 0$ and $\zeta \to 0$; and (b) the stiff limit $\Delta t \to 0$ and $\zeta \to -\infty$. A characteristic property of most RK schemes is that, while the non-stiff limit recovers the scheme's order (as given by the order conditions [2, 5]), the error decays at a reduced order in the stiff limit. This phenomenon is called "order reduction" (OR) [1, 3, 7, 10, 11] and it manifests in various ways for more complex problems, including numerical boundary layers [6]. The OR phenomenon can be seen by studying the RK scheme applied to (1). The approximation error at time t_{n+1} reads [12, Chapter IV.15]

$$\epsilon^{n+1} = R(\zeta)\,\epsilon^n + \zeta \mathbf{b}^T (I - \zeta A)^{-1} \delta_s^{n+1} + \delta^{n+1} \,, \tag{2}$$

where $R(\zeta)$ is the growth factor, and

$$\delta_s^{n+1} = \sum_{j \geq 2} \frac{\Delta t^j}{(j-1)!}\, \tau^{(j)} \phi^{(j)}(t_n) \,, \quad \delta^{n+1} = \sum_{j \geq 1} \frac{\Delta t^j}{(j-1)!} \left(\mathbf{b}^T \mathbf{c}^{j-1} - \frac{1}{j} \right) \phi^{(j)}(t_n)$$

are the truncation errors incurred at the intermediate stages and at the end of the step, respectively. Here, $\phi^{(j)}$ denotes the j-th derivative of the solution, and the vectors

$$\tau^{(j)} = A\mathbf{c}^{j-1} - \tfrac{1}{j}\mathbf{c}^j \,, \quad j = 1, 2, \ldots$$

we call the *stage order residuals* or *stage order vectors*. The condition $\tau^{(\eta)} = 0$ for $0 \leq \eta \leq j$ appears often in the literature and is also referred to as the simplifying assumption $C(\eta)$ [12]. In (2), the step error δ^{n+1} is of the formal order (in Δt) of the scheme (due to the order conditions). Moreover, the growth factor carries over (more or less, see [4]) the accuracy from one to the next step. Hence, the critical expression for OR is the term involving the stage error δ_s^{n+1}. Specifically, the asymptotic behavior of the expression

$$g^{(j)} = \zeta \mathbf{b}^T (I - \zeta A)^{-1} \tau^{(j)} \tag{3}$$

matters. In the non-stiff limit ($\zeta \ll 1$), a Neumann expansion yields $\zeta(I - \zeta A)^{-1} = \zeta I + \zeta^2 A + \zeta^3 A^2 + \ldots$, leading to expressions $\mathbf{b}^T A^\ell \boldsymbol{\tau}^{(j)}$ with $\ell > 0$. And in fact the order conditions guarantee that $\mathbf{b}^T A^\ell \boldsymbol{\tau}^{(j)} = 0$ for $0 \le \ell + j \le p - 1$ to ensure the formal order of the scheme.

Conversely, in the stiff limit we can treat ζ^{-1} as the small parameter and expand $\zeta(I - \zeta A)^{-1} = -A^{-1}(I - \zeta^{-1} A^{-1})^{-1} = -A^{-1} - \zeta^{-1} A^{-2} - \zeta^{-2} A^{-3} - \ldots$, leading to expressions $\mathbf{b}^T A^\ell \boldsymbol{\tau}^{(j)}$ with $\ell < 0$. The order conditions do *not* imply that these quantities vanish, and in general one may observe a reduced rate of convergence.

A key question is therefore whether additional conditions can be imposed on the RK scheme that recover the scheme's order in the stiff regime. A well-known answer to the question is:

Definition 1 Let $\hat{p}$ denote the order of the quadrature rule of an RK scheme. Let $\hat{q}$ denote the largest integer such that $\boldsymbol{\tau}^{(j)} = 0$ for $1 \le j \le \hat{q}$. The *stage order* of a RK scheme is $q = \min(\hat{p}, \hat{q})$.

Having stage order q implies that the error decays at an order of (at least) q in the stiff regime (see also [12]). This work focuses particularly on diagonally-implicit Runge-Kutta (DIRK) schemes, for which A is lower diagonal. A known drawback of DIRK schemes is that they cannot have high stage order:

Theorem 1 *The stage order of an irreducible DIRK scheme is at most 2. The stage order of a DIRK scheme with non-singular A is at most 1.*

Proof Since $\mathbf{c} = A\mathbf{e}$, we have $\tau_1^{(2)} = a_{11}c_1 - \frac{1}{2}(c_1)^2 = \frac{1}{2}(a_{11})^2$. Thus if A is non-singular, one has $\boldsymbol{\tau}^{(2)} \neq 0$, so $q \le 1$. Consider now the case that $a_{11} = c_1 = 0$, and suppose that the method has stage order 3. The conditions $\tau_2^{(2)} = \tau_2^{(3)} = 0$ then imply $a_{21} = a_{22} = c_2 = 0$, which would render the scheme reducible. Hence, $q \le 2$. $\qquad\square$

Hence, while DIRK schemes possess an implementation-friendly structure (each stage is a backward-Euler-type solve), their potential to avoid OR by means of high stage order is limited. We therefore move to a weaker condition that can avoid OR in some situations for higher order in the context of DIRK schemes.

2 Weak Stage Order

To avoid order reduction, the expressions $g^{(j)}$ in (3) need to vanish in the stiff limit. In line with [9], we define the following criteria:

Definition 2 (Weak Stage Order) A RK scheme has weak stage order (WSO) $\tilde{q}$ if there is an A-invariant subspace that is orthogonal to $\mathbf{b}$ and that contains the stage order vectors $\boldsymbol{\tau}^{(j)}$ for $1 \le j \le \tilde{q}$.

Theorem 2 (WSO Is the Most General Condition that Ensures $g^{(j)} = 0$ for All $\zeta > 0$) *Let coefficients A, b be given. Then $g^{(j)} = 0$ for all $\zeta > 0$ and $1 \leq j \leq \tilde{q}$ if and only if the corresponding RK scheme has weak stage order $\tilde{q}$.*

Proof Let $C(G)$ denote the column space of

$$G := \left[\tau^{(1)}, A\tau^{(1)}, A^2\tau^{(1)}, \ldots, A^{s-1}\tau^{(1)}, \tau^{(2)}, A\tau^{(2)}, \ldots, A^{s-1}\tau^{(\tilde{q})} \right].$$

From the Cayley-Hamilton theorem it follows that WSO $\tilde{q}$ is equivalent to

$$\mathbf{b}^T A^\ell \tau^{(j)} = 0, \qquad 0 \leq \ell \leq s - 1, \ 1 \leq j \leq \tilde{q}. \tag{4}$$

$\boxed{\Longrightarrow}$ Because $C(G)$ is A-invariant, $C(G)$ is invariant under multiplication by $(1 - \zeta A)^{-1}$, i.e. if $\mathbf{v} \in C(G)$ then for any $\zeta > 0$, the product $(1 - \zeta A)^{-1}\mathbf{v} \in C(G)$. Since $\mathbf{b}$ is orthogonal to $C(G)$, we have $g^{(j)} = 0$ for all $1 \leq j \leq \tilde{q}$.

$\boxed{\Longleftarrow}$ If $g^{(j)} = 0$, then $\zeta^{-1} g^{(j)} = \mathbf{b}^T (1 - \zeta A)^{-1} \tau^{(j)} = 0$ for all $\zeta > 0$. Differentiating both sides of this equation ℓ-times, with respect to ζ, and taking the limit as $\zeta \to 0^+$, yields the conditions in Eq. (4). $\qquad\square$

Definition 3 (Weak Stage Order Eigenvector Criterion) A RK scheme satisfies the WSO eigenvector criterion of order $\tilde{q}_e$ if for each $1 \leq j \leq \tilde{q}_e$, there exists μ_j such that $A\tau^{(j)} = \mu_j \tau^{(j)}$, and moreover, $\mathbf{b}^T \tau^{(j)} = 0$.

The WSO eigenvector criterion of order $\tilde{q}_e$ implies WSO (of at least) $\tilde{q}_e$. For a given scheme, let p denote the classical order, q the stage order, and $\tilde{q}$ the weak stage order. Then we have $\tilde{q} \geq q$ and $p \geq q$. Note however that a method with WSO $\tilde{q} \geq 1$ need not even be consistent; order conditions must be imposed separately.

The WSO eigenvector criterion may serve to avoid OR because it implies that

$$g^{(j)} = \zeta \mathbf{b}^T (1 - \zeta \mu_j)^{-1} \tau^{(j)} = \frac{\zeta}{1 - \zeta \mu_j} \mathbf{b}^T \tau^{(j)},$$

i.e., it allows one to "push" the stage order residuals past the matrix $(1 - \zeta A)^{-1}$, and then use $\mathbf{b}^T \tau^{(j)} = 0$. Note that the condition $\mathbf{b}^T \tau^{(j)} = 0$ that is required in Definition 3 is actually automatically satisfied (due to the order conditions) if $p > \tilde{q}_e$ (or $p \geq \tilde{q}_e$ for stiffly accurate schemes).

It must be stressed that the concept of WSO (both criteria) is based on the linear test equation (1), hence it is not clear to what extent WSO will remedy OR for nonlinear problems or problems with time-dependent coefficients. In Sect. 4 we numerically investigate some nonlinear test problems.

Finally, we present a limitation theorem on the WSO eigenvector criterion.

Theorem 3 *DIRK schemes with invertible A have $\tilde{q}_e \leq 3$.*

Proof Because the $\tau^{(j)}$ only depend on A, the eigenvector relation in Definition 3 depends only on A, not on $\mathbf{b}$. With A lower triangular, the first k components of $\boldsymbol{\tau}^{(j)}$ depend only on the upper k rows of A; and the same is true for the eigenvector relation as well. Hence, for a scheme to have an A that allows for the WSO eigenvector criterion of order $\tilde{q}_e$, all upper sub-matrices of A must admit the same, too. We can therefore study A row by row. The first component of $\boldsymbol{\tau}^{(j)}$ equals $(1 - \frac{1}{j})a_{11}^j$, which is nonzero for $j > 1$. Hence, the first row of the equation $A\boldsymbol{\tau}^{(j)} = \mu_j \boldsymbol{\tau}^{(j)}$ is equivalent to $\mu_j = a_{11}$. With that, we can move to the second row of the equation, which reads

$$(1-\tfrac{1}{j})a_{11}^j a_{21} + (a_{22}-a_{11})\left(a_{11}^{j-1}a_{21} + (a_{21}+a_{22})^{j-1}a_{22} - \tfrac{1}{j}(a_{21}+a_{22})^j\right) = 0 .$$
(5)

To determine the set of solutions (a_{11}, a_{21}, a_{22}) of (5), we first observe that (5) is homogeneous, i.e., if (a_{11}, a_{21}, a_{22}) solves (5), then $(\mu a_{11}, \mu a_{21}, \mu a_{22})$ solves (5) as well for any $\mu \in \mathbb{R}$. It therefore suffices to consider the solutions of (5) in the 2D-plane $(\frac{a_{11}}{a_{21}}, \frac{a_{22}}{a_{21}})$. Figure 1 shows the resulting solution curves for $j \in \{2, 3, 4\}$.

One class of solutions lies on the straight line of slope 1 passing through $(1, 0)$. Those schemes are *equal-time* methods, i.e., RK schemes that have $\mathbf{c} = \nu\mathbf{e}$, where $\nu \in \mathbb{R}$ is a constant. In fact, equal-time schemes satisfy the eigenvector relation for all j. However, they are not particularly useful RK methods, because—among other limitations—they are restricted to second order. This follows because the order 1 and 2 conditions require $\mathbf{b}^T\mathbf{e} = 1$ and $\mathbf{b}^T\mathbf{c} = \frac{1}{2}$. Thus $\nu = \frac{1}{2}$, and $\mathbf{b}^T\mathbf{c}^2 = \nu^2 = \frac{1}{4}$, which contradicts the order 3 condition $\mathbf{b}^T\mathbf{c}^2 = \frac{1}{3}$. Note that the equal-time scenario also covers the points at infinity in Fig. 1, i.e., the schemes with $a_{21} = 0$.

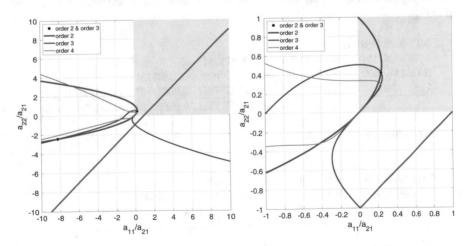

Fig. 1 Curves of WSO orders 2, 3, and 4 as functions of the re-scaled parameters $\frac{a_{11}}{a_{21}}$ and $\frac{a_{22}}{a_{21}}$. Left panel: scale 10; right panel: scale 1. All orders are satisfied along the line of slope 1 going through $(1,0)$, corresponding to equal-time DIRK schemes. Moreover, there are two further points (other than the origin), where orders 2 and 3 are satisfied. Neither of these two points satisfies order 4

Non-equal-time schemes that satisfy (5) for $j = 2$ and $j = 3$ are the following two points in the $(\frac{a_{11}}{a_{21}}, \frac{a_{22}}{a_{21}})$ plane: $P_1 = (-4 + 3\sqrt{2}, \sqrt{2} - 1) = (0.2426, 0.4142)$ and $P_2 = (-(\sqrt{2}+1)(\sqrt{2}+2), -(\sqrt{2}+1)) = (-8.2426, -2.4142)$. None of these two points satisfies (5) for $j = 4$ (green curve in Fig. 1). Therefore $\tilde{q}_e \leq 3$.   □

Among the two sets of solutions found in the proof, P_1 implies that a_{11}, a_{21}, and a_{22} all have the same sign, which is a desirable property. In contrast, P_2 implies that $a_{21} < 0$. Both WSO 3 schemes presented below correspond to the P_1 solution.

3 DIRK Schemes with High Weak Stage Order

Imposing the classical order conditions [2, 5], together with the WSO eigenvector relation (Definition 3), we determine RK schemes by searching the parameter space of DIRK schemes (with all diagonal entries non-zero). A stiffly accurate structure ($\mathbf{b}^T$ equals the last row of A) is imposed, as is A-stability (verified by evaluating the stability function $R(\zeta)$ along the imaginary axis). Together this implies that the resulting scheme is L-stable; i.e., it ensures that unresolved stiff modes decay [5]. The number of stages is chosen so that the constraints admit solutions. The optimization itself is carried out using MATLAB's optimization toolbox, using multiple local optimization algorithms included in the function fmincon. An effort was made to minimize the L_2 norm of the local truncation error coefficients. However, in multiple cases the solver exhibited bad convergence properties; so while the schemes below yield reasonable truncation errors, it should not be expected that they are optimal. We find an order 3 scheme with WSO 2 (see also [9]),

0.01900072890	0.01900072890			
0.78870323114	0.40434605601	0.38435717512		
0.41643499339	0.06487908412	−0.16389640295	0.51545231222	
1	0.02343549374	−0.41207877888	0.96661161281	0.42203167233
	0.02343549374	−0.41207877888	0.96661161281	0.42203167233

an order 3 scheme with WSO 3,

0.13756543551	0.13756543551			
0.80179011576	0.56695122794	0.23483888782		
2.33179673002	−1.08354072813	2.96618223864	0.44915521951	
1	0.59761291500	−0.43420997584	−0.05305815322	0.88965521406
	0.59761291500	−0.43420997584	−0.05305815322	0.88965521406

and an order 4 scheme with WSO 3,

0.079672377876931	0.079672377876931	0	0	0	0	0
0.464364648310935	0.328355391763968	0.136009256546967	0	0	0	0
1.348559241946724	−0.650772774016417	1.742859063495349	0.256472952467792	0	0	0
1.312664210308764	−0.714580550967259	1.793745752775934	−0.078254785672497	0.311753794172585	0	0
0.989469293495897	−1.120092779092918	1.983452339867353	3.117393885836001	−3.761930177913743	0.770646024799205	0
1	0.214823667785537	0.536367363903245	0.154488125726409	−0.217748592703941	0.072226422925896	0.239843012362853
1	0.214823667785537	0.536367363903245	0.154488125726409	−0.217748592703941	0.072226422925896	0.239843012362853

4 Numerical Results

In this section we verify the order of accuracy of the schemes above and demonstrate that WSO remedies order reduction for linear problems. We confirm that WSO p is required for ODEs, and WSO $p - 1$ is required for PDE IBVPs. In addition, we study the effect of WSO for two nonlinear problems.

4.1 Linear ODE Test Problem

We consider the linear ODE test problem (1) with the true solution $\phi(t) = \sin(t + \frac{\pi}{4})$, the stiffness parameter $\lambda = -10^4$, and the initial condition $u(0) = \sin(\frac{\pi}{4})$. The problem is solved using three 3rd order DIRK schemes (with WSO 1, 2, and 3) and two 4th order DIRK schemes (with WSO 1 and 3)[1] up to the final time $T = 10$. The convergence results are shown in Fig. 2. In the stiff regime where $|\zeta| = |\lambda|\Delta t \gg 1$, first order convergence is observed for the WSO 1 schemes as expected, the WSO 2 scheme improves the convergence rate to 2, and the WSO 3 schemes exhibit 3rd order convergence. In addition to yielding better convergence orders in the stiff regime, the schemes with higher WSO also turn out to yield substantially smaller error constants in the non-stiff regime ($\Delta t \ll 1/|\lambda|$). For comparison, we also display a DIRK scheme with explicit first stage (EDIRK), that is, $a_{11} = 0$, of stage order 2 (see Theorem 1). The left panel of Fig. 2 shows that the WSO 2 scheme exhibits the same convergence behavior as the stage order 2 EDIRK scheme and performs equally well in terms of accuracy.

4.2 Linear PDE Test Problem: Schrödinger Equation

As a linear PDE test problem, we study the dispersive Schrödinger equation. The method of manufactured solutions is used, i.e., the forcing, the boundary conditions (b.c.) and initial conditions (i.c.) are selected to generate a desired true solution. The

---

[1] We do not construct an order 4 scheme with WSO 2, as we see no role for such a method.

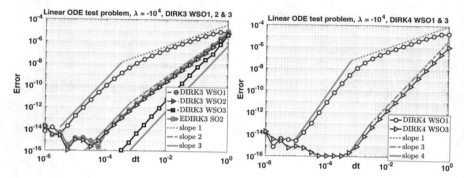

Fig. 2 Error convergence for linear ODE test problem (1). Left: 3rd order DIRK schemes with WSO 1 (blue circles), WSO 2 (red triangles), WSO 3 (black squares), and a 3rd order EDIRK scheme with stage order 2 (light red dots). Right: 4th order DIRK schemes with WSO 1 (blue circles) and WSO 3 (red triangles)

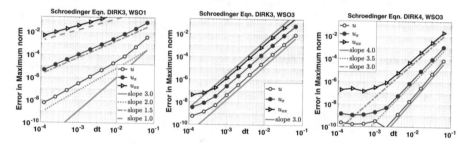

Fig. 3 Error convergence for the Schrödinger equation using 3rd order DIRK schemes with WSO 1 (left) and WSO 3 (middle), and a 4th order DIRK with WSO 3 (right)

spatial approximation is carried out using 4th order centered differences on a fixed spatial grid of 10,000 cells. This renders spatial approximation errors negligible and thus isolates the temporal errors due to DIRK schemes. The errors are measured in the maximum norm in space.

We consider

$$u_t = \frac{i\omega}{k^2} u_{xx} \text{ for } (x, t) \in (0, 1) \times (0, 1.2], \quad u = g \text{ on } \{0, 1\} \times (0, 1.2], \quad (6)$$

with the true solution $u(x, t) = e^{i(kx - \omega t)}$, $\omega = 2\pi$ and $k = 5$. Figure 3 shows the convergence orders of u, u_x and u_{xx} for 3rd order DIRK schemes with WSO 1 (left), WSO 3 (middle) and a 4th order DIRK scheme with WSO 3 (right). For IBVPs, spatial boundary layers are produced by RK methods, thus limiting the convergence order in u to $\tilde{q} + 1$, with an additional half an order loss per derivative when $\tilde{q} < p$ [9]. As a result, the 4th order WSO 3 scheme recovers 4th order convergence in u and improves the convergence in u_x and u_{xx}. When $\tilde{q} = p$, the full convergence order in u, u_x and u_{xx} is achieved, as seen in the middle panel in Fig. 3.

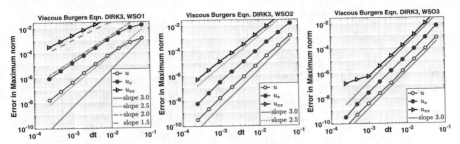

Fig. 4 Error convergence for the viscous Burgers' equation using 3rd order DIRK schemes with WSO 1 (left), WSO 2 (middle) and WSO 3 (right)

4.3 Nonlinear PDE Test Problem: Burgers' Equation

This example demonstrates that WSO avoids order reduction for certain nonlinear IBVPs as well. We consider the viscous Burgers' equation with pure Neumann b.c.

$$u_t + uu_x = \nu u_{xx} + f \quad \text{for } (x, t) \in (0, 1) \times (0, 1], \quad u_x = h \text{ on } \{0, 1\} \times (0, 1] . \tag{7}$$

Here $\nu = 0.1$ and $u(x, t) = \cos(2 + 10t) \sin(0.2 + 20x)$. The nonlinear implicit equations arising at each time step are solved using a standard Newton iteration. The choice of Neumann b.c. distinguishes this example from the one given in [9]. With Neumann b.c., the convergence order in u is limited to $\tilde{q} + 1.5$ (half an order better than with Dirichlet b.c.). Figure 4 shows that order reduction arises with the stage order 1 scheme, and that the WSO 2 scheme recovers 3rd order convergence for u and u_x, and the 3rd order WSO 3 scheme yields 3rd order convergence for u, u_x and u_{xx}.

4.4 Stiff Nonlinear ODE: Van der Pol Oscillator

This example illustrates that DIRK schemes with high WSO may not remove order reduction for all types of nonlinear problems. Consider the Van der Pol oscillator

$$x' = y \quad \text{and} \quad y' = \mu(1 - x^2)y - x , \tag{8}$$

with i.c. $(x(0), y(0)) = (2, 0)$, stiffness parameter $\mu = 500$, and final time $T = 10$. The nonlinear system at each time step is solved via MATLAB's built-in nonlinear system solver. The "exact" solution is computed using explicit RK4 with a time step $\Delta t = 10^{-6}$. In this case, the presented DIRK schemes with high WSO do not improve the convergence rates in the stiff regime and they perform worse than the WSO 1 scheme in terms of accuracy (see Fig. 5). On the other hand, an EDIRK with

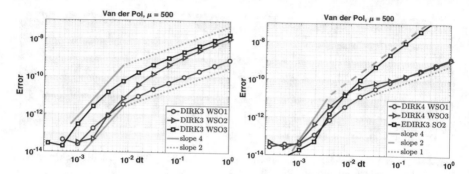

Fig. 5 Error convergence for Van der Pol's equation. Left: 3rd order DIRK schemes with WSO 1 (blue circles), WSO 2 (red triangles) and WSO 3 (black squares). Right: 4th order DIRK schemes with WSO 1 (blue circles) and WSO 3 (red triangles), and a 3rd order EDIRK scheme with stage order 2 (black squares)

stage order 2 improves the rate of convergence in the stiff regime (see right panel in Fig. 5). However, it does so, interestingly, by yielding larger errors for large time steps.

5 Conclusions and Outlook

This study demonstrates that it is possible to overcome order reduction (OR) for certain classes of problems in the context of DIRK schemes, even though these are limited to low stage order. A specific *weak stage order* (WSO) "eigenvector" criterion has been presented, analyzed, and applied to determine DIRK schemes with WSO up to 3. The numerical results confirm that the schemes avoid OR for linear problems and for some nonlinear problems in which the mechanism for order reduction is linear (i.e., boundary conditions). The key limitation found herein is that the eigenvector criterion cannot go beyond WSO 3 for DIRK schemes. Hence, a key question of future research is how high WSO is admitted by the general criterion in Definition 2. Another important future research task is to devise further DIRK schemes that are truly optimized in terms of truncation error coefficients or other criteria.

Acknowledgements This work was supported by the National Science Foundation via grants DMS-1719640 (BS&DZ) and DMS-1719693 (DS); and the Simons Foundation (#359610) (DS).

References

1. Burrage, K., Petzold, L.: On order reduction for Runge-Kutta methods applied to differential/algebraic systems and to stiff systems of ODEs. SIAM J. Numer. Anal. **27**(2), 447–456 (1990)
2. Butcher, J.C.: Numerical Methods for Ordinary Differential Equations, 2nd edn. Wiley, New York (2008)
3. Carpenter, M.H., Gottlieb, D., Abarbanel, S., Don, W.-S.: The theoretical accuracy of Runge-Kutta time discretizations for the initial boundary value problem: a study of the boundary error. SIAM J. Sci. Comput. **16**(6), 1241–1252 (1995)
4. Ditkowski, A., Gottlieb, S.: Error inhibiting block one-step schemes for ordinary differential equations. J. Sci. Comput. **73**(2–3), 691–711 (2017)
5. Hairer, E., Nørsett, S.P., Wanner, G.: Solving Ordinary Differential Equations I (2nd Revised. Ed.): Nonstiff Problems. Springer, New York (1993)
6. Minion, M.L.: Semi-implicit spectral deferred correction methods for ordinary differential equations. Commun. Math Sci. **1**(3), 471–500 (2003)
7. Ostermann, A., Roche, M.: Runge-Kutta methods for partial differential equations and fractional orders of convergence. Math. Comput. **59**(200), 403–420 (1992)
8. Prothero, A., Robinson, A.: On the stability and accuracy of one-step methods for solving stiff systems of ordinary differential equations. Math. Comput. **28**(125), 145–162 (1974)
9. Rosales, R.R., Seibold, B., Shirokoff, D., Zhou, D.: Order reduction in high-order Runge-Kutta methods for initial boundary value problems (2017). Preprint. arXiv:1712.00897
10. Sanz-Serna, J.M., Verwer, J.G., Hundsdorfer, W.H.: Convergence and order reduction of Runge-Kutta schemes applied to evolutionary problems in partial differential equations. Numer. Math. **50**(4), 405–418 (1986)
11. Verwer, J.G.: Convergence and order reduction of diagonally implicit Runge-Kutta schemes in the method of lines. In: Numerical Analysis: Proceedings of the Dundee Conference on Numerical Analysis, 1985, pp. 220–237 (1986)
12. Wanner, G., Hairer, E.: Solving Ordinary Differential Equations II: Stiff and Differential-Algebraic Problems, vol. 1. Springer, Berlin (1991)

Scheme for Evolutionary Navier-Stokes-Fourier System with Temperature Dependent Material Properties Based on Spectral/hp Elements

Jan Pech

1 Introduction

This work presents a numerical algorithm for the system of the Navier-Stokes equations coupled with the balance of internal energy

$$\rho \left(\frac{\partial \mathbf{v}}{\partial t} + \mathbf{v} \cdot \nabla \mathbf{v} \right) = -\nabla p + \frac{1}{\mathrm{Re}} \nabla \cdot \left[2\mu \mathbb{D} + \lambda \left(\nabla \cdot \mathbf{v} \right) \mathbb{I} \right] + \mathbf{f_v} \tag{1a}$$

$$\frac{\partial \rho}{\partial t} + \nabla \cdot (\rho \mathbf{v}) = m \tag{1b}$$

$$\frac{\partial T}{\partial t} + \mathbf{v} \cdot \nabla T = \frac{1}{\mathrm{Re}\,\mathrm{Pr}} \nabla \cdot (\kappa \nabla T) + f_T , \tag{1c}$$

where $\mathbf{v} = [u, v, w]^{\mathrm{T}}$ is the velocity vector (by setting $w = \mathrm{const.} = 0$ we restrict to 2D problem), p is a variable related to the thermodynamic pressure,[1] T denotes the temperature, $\mathbb{D} = \frac{1}{2} \left[\nabla \mathbf{v} + (\nabla \mathbf{v})^{\mathrm{T}} \right]$ is the symmetric part of the rate of strain

---

[1] We call *thermodynamic pressure* the variable acting in the equation of state, e.g. $p = \rho R T$ for ideal gas. Quantities with physical units (superscript star) are normalized by its farfield values (subscript infinity), e.g. $\mathbf{v} = \frac{\mathbf{v}^*}{|\mathbf{v}_\infty^*|}$, $T = \frac{T^*}{T_\infty^*}$, etc. The dimensionless pressure in (1a) is $p = \frac{p^*}{\rho_\infty^* |\mathbf{v}_\infty^*|^2}$.

J. Pech (✉)
Institute of Thermomechanics of the Czech Academy of Sciences, Praha 8, Czech Republic
e-mail: jpech@it.cas.cz

© The Author(s) 2020
S. J. Sherwin et al. (eds.), *Spectral and High Order Methods for Partial Differential Equations ICOSAHOM 2018*, Lecture Notes in Computational Science and Engineering 134, https://doi.org/10.1007/978-3-030-39647-3_37

tensor, constant Re is the Reynolds number and constant Pr is the Prandtl number (for sake of simplicity we set Re = Pr = 1 for the testing on exact solution).

The fluid is expected to be (calorically) perfect,[2] Newtonian,[3] whose heat flux obeys the Fourier law.[4] In system (1), we consider those fluids, which become nonhomogeneous in variable temperature fields due to temperature dependence of its material parameters, namely the density $\rho = \rho(T)$, dynamic viscosity $\mu = \mu(T)$ and thermal conductivity $\kappa = \kappa(T)$.

Instead of (1a), we solve

$$\rho \left(\frac{\partial \mathbf{v}}{\partial t} + \mathbf{v} \cdot \nabla \mathbf{v} \right) = -\nabla \tilde{p} + \frac{1}{\text{Re}} \nabla \cdot \left[2\mu \mathbb{D} - \frac{2}{3}\mu \left(\nabla \cdot \mathbf{v} \right) \mathbb{I} \right] + \mathbf{f_v}, \qquad (2)$$

where $\tilde{p} = p - \mu_b \nabla \cdot \mathbf{v}$ is *mean* or *mechanical* pressure, while $\mu_b = \lambda + \frac{2}{3}\mu$ is the *bulk viscosity*. Equation (2) has the same structure as (1a) while setting $\lambda = -\frac{2}{3}\mu$ (or equivalently $\mu_b = 0$, c.f. Stokes hypothesis), but physical interpretation of pressure changes.

Without loss of generality, solving (2) instead of (1a), we avoid specification of the second viscosity coefficient λ.

The forcing terms $\mathbf{f_v}$, f_T, may represent action of volumetric forces, e.g. gravity or viscous heating, but m is set zero in most of realistic situations. In case of testing of our algorithm on a given solution $[\mathbf{v}_e, p_e, T_e]^T$, we construct the forcing terms such, that Eqs. (2), (1b) and (1c) are satisfied.

Our computational scheme is developed for simulations based on the spectral/hp element approximation in spatial coordinates. We use the polynomial approximations of degree 15 in our tests, what eliminates the numerical error in spatial coordinates and we are getting an overview of error production, which belongs directly to the algorithm/discretisation in time. The high order spatial approximations also naturally include approximations of higher-order derivatives, what is utilized in the scheme.

The previous results from literature are, up to the authors knowledge, restrictions of (1) setting at least one of the material parameters constant, the velocity field to be divergence-free or modelling a stationary flow, see Table 1.

---

[2] Internal energy e of the *calorically perfect* fluids obeys $e = c_V T$, where specific heat at constant volume is independent of temperature ($c_V = $ const.).

[3] We use the term Newtonian fluid in a general sense for fluids, whose stress tensor is linearly dependent on the strain rate tensor. However, the viscous part of the stress tensor is not traceless as often expected if fluid is called Newtonian.

[4] The Fourier law relates the heat flux $\mathbf{q}$ to the thermal conductivity κ and the temperature gradient ∇T as $\mathbf{q} = -\kappa \nabla T$.

Table 1 Chosen results concerning equation systems with variable material parameters

	Eq. type	$\nabla \cdot \mathbf{v}$	μ	κ	ρ
[1]	nonst.	0	$\mu(\rho)$	const.	var.
[4, 5]	nonst.	0	const.	const.	var.
[6]	nonst.	0	var.	const.	const.
[9]	stat.	0	$\mu(T)$	$\kappa(T)$	const.
[10]	nonst.	0	$\mu(T)$	$\kappa(T)$	const.
[11, 12]	stat.	0	$\mu(T)$	$\kappa(T)$	const.
[13]	nonst.	$\neq 0$	const.	const.	var.
[14]	nonst.	0	$\mu(T)$	$\kappa(T)$	const.
[16]	nonst.	0	$\mu(T)$	$\kappa(T)$	const.

Stationary and non-stationary models are denoted *stat.* and *nonst.*, unspecified variability of a property is denoted *var.*

2 Algorithm

Our approach is inspired by the velocity-correction scheme with the high order pressure boundary condition (HOPBC) proposed for the incompressible Navier-Stokes equations in [7]. The constant property case, [7], is widely used for its efficiency and was already extended to problems with variable viscosity in [6]. Its modification was used also to the incompressible Navier-Stokes-Fourier system with temperature dependent viscosity and thermal conductivity in [10]. Efficiency of the approach comes from the implicit-explicit (IMEX) formulation, which allows decoupling of the system.

The main contribution of the present work, which is a continuation of [10], is in extension to the problems with temperature dependent density. However, the velocity divergence cannot be further neglected in the momentum balance, what is the substantial difference from the previously discussed models and algorithms.

2.1 Decoupled System

We use the IMEX scheme in which the *Backward difference formula* (BDF) of order Q approximates the temporal derivative and a consistent extrapolation is applied to chosen terms ($\mathcal{N}$)

$$\frac{\partial u}{\partial t} = \mathcal{L}(u) + \mathcal{N}(u) \xrightarrow{IMEX} \frac{\gamma u_{n+1} - \sum_{q=0}^{Q-1} \alpha_q u_{n-q}}{\Delta t} = \mathcal{L}_{n+1} + \sum_{q=0}^{Q-1} \beta_q \mathcal{N}_{n-q} .$$

$$(3)$$

In (3), u is the searched solution, $\mathcal{L}$ denotes the terms solved implicitly, which we expect to be constant in time. Subscript $n + 1$ (or operator in square brackets with subscript) denotes evaluation at time $t_{n+1} = t_0 + (n+1)\Delta t$, where Δt is the discrete time step. Coefficients $\{\alpha_q\}_{q=0}^{Q-1}$, $\{\beta_q\}_{q=0}^{Q-1}$ and γ for particular Q can be found, e.g., in [10]. Henceforward, we use '$*$' in the superscript to denote extrapolation, $\mathcal{N}^* \equiv [\mathcal{N}]^* := \sum_{q=0}^{Q-1} \beta_q \mathcal{N}_{n-q}$.

The extrapolated terms are evaluated using data from previous time steps, $\{\mathcal{N}_{n-q}\}_{q=0}^{Q-1}$ and $\{u_{n-q}\}_{q=0}^{Q-1}$, what allows separate/decoupled solution of the (generalized) Navier-Stokes equations (2)–(1b) and the non-linear energy equation (1c).

Solution during one time step may be summarized to the scheme

1. Update μ, κ, ρ, $\nabla \cdot \mathbf{v}$, and HOPBC using already known values $\{\mathbf{v}_{n-q}\}_{q=0}^{Q-1}$, $\{T_{n-q}\}_{q=0}^{Q-1}$.
2. Solve the system of momentum and mass balance

   (a) Solve the pressure-Poisson equation for $\tilde{p}_{n+1}$
   (b) Solve velocity-correction for $\mathbf{v}_{n+1}$

3. Solve the non-linear advection-diffusion problem for T_{n+1}.

2.2   Balance of Momentum and Mass

The scheme decouples solution of the Navier-Stokes system (2)–(1b) to the pressure-Poisson equation and an elliptic equation for velocity. The equation for pressure is derived as a projection to the irrotational space by application of the divergence operator to (2)

$$
\nabla^2 p = \frac{\gamma}{\Delta t}\left(\left[\frac{\partial \rho}{\partial t}\right]^{**} - m_{n+1}\right) - \frac{1}{\mathrm{Re}}\left[\nabla\mu \cdot (\nabla \times \nabla \times \mathbf{v})\right]^*
$$
$$
+ \nabla \cdot \left\{ \frac{1}{\Delta t}\rho^*\hat{\mathbf{v}} + \frac{1}{\mathrm{Re}}\left[\nabla\mu \cdot \left(\nabla\mathbf{v} + (\nabla\mathbf{v})^{\mathrm{T}}\right)\right]^* \right.
$$
$$
\left. + \frac{1}{\mathrm{Re}}\left(-\frac{2}{3}[\nabla\mu]^*[\nabla \cdot \mathbf{v}]_{n+1} + \frac{4}{3}\mu^*\nabla[\nabla \cdot \mathbf{v}]_{n+1}\right) + \mathbf{f}_{n+1} \right\}
$$

$$(4)$$

where we applied (1b), identities $\nabla \times \nabla \times \mathbf{v} = \nabla\nabla \cdot \mathbf{v} - \nabla^2\mathbf{v}$ and $\nabla \cdot \nabla\times \equiv 0$, $\partial\mathbf{v}/\partial t$ was substituted by BDF and $\hat{\mathbf{v}} = \sum_{q=0}^{Q-1} \alpha_q \mathbf{v}_{n-q} - \Delta t\,[\mathbf{v} \cdot \nabla\mathbf{v}]^*$. The temporal derivative of the density, which is extrapolated in (4), is approximated by Q-th order BDF

$$
\left[\frac{\partial \rho}{\partial t}\right]_n \approx \frac{\gamma\rho_n - \sum_{q=0}^{Q-1}\alpha_q\rho_{n-1-q}}{\Delta t}.
$$

$$(5)$$

We denote the extrapolation of the derivative approximation by superscript '**'. Note, that we have to specify the initial value $\left[\frac{\partial \rho}{\partial t}\right]_0$ or both ρ_0, ρ_{-1} to initialise the scheme of the lowest order $Q = 1$.

Our model assumes, that the density is entirely determined by the temperature distribution. Then, the divergence of velocity, whose forward estimate, $[\nabla \cdot \mathbf{v}]_{n+1}$, is required in (4), follows from (1b)

$$[\nabla \cdot \mathbf{v}]_{n+1} \approx \frac{1}{\rho^*} \left\{ m_{n+1} - [\mathbf{v} \cdot \nabla \rho]^* - \left[\frac{\partial \rho}{\partial t}\right]^{**} \right\}. \tag{6}$$

The forward estimate of velocity divergence is the crucial step in the proposed scheme.

HOPBC is the natural boundary condition for (4). It is derived as projection of the momentum equation (2) to the direction of normal $\mathbf{n}$ to the domain boundary $\partial \Omega$

$$\frac{\partial \tilde{p}}{\partial \mathbf{n}} = \mathbf{n} \cdot \left\{ -\rho^* \left[\frac{\partial \mathbf{v}}{\partial t}\right]^{**} + \left[-\rho \mathbf{v} \cdot \nabla \mathbf{v} + \frac{1}{Re} \left(-\mu \nabla \times \nabla \times \mathbf{v} + \nabla \mu \cdot \left[\nabla \mathbf{v} + (\nabla \mathbf{v})^{\mathrm{T}}\right] \right) \right]^* \right.$$
$$\left. + \frac{1}{Re} \left(-\frac{2}{3} [\nabla \mu]^* [\nabla \cdot \mathbf{v}]_{n+1} + \frac{4}{3} \mu^* \nabla [\nabla \cdot \mathbf{v}]_{n+1} \right) + \mathbf{f}_{n+1} \right\} \tag{7}$$

The forward estimate of velocity divergence follows from (6) again. Similarly to (4), we approximate the acceleration term $\frac{\partial \mathbf{v}}{\partial t}$ by the BDF of Q-th order, whose initialisation requires value $\left[\frac{\partial \mathbf{v}}{\partial t}\right]_0$ or both the values $\mathbf{v}_0$ and $\mathbf{v}_{-1}$. The problem of initialisation of $\left[\frac{\partial \mathbf{v}}{\partial t}\right]^{**}$ and $\left[\frac{\partial \rho}{\partial t}\right]^{**}$ is circumvented in many realistic simulations, which begin from a constant fields.

The solution of (4) gives estimate/prediction of $\tilde{p}_{n+1}$ and we can solve (2) as an elliptic problem for $\mathbf{v}_{n+1}$. However, in the case of temperature dependent viscosity and density, the algebraic system derived for operators with variable coefficients has time dependent matrices, whose direct solution is inefficient. To preserve efficiency of the scheme, we split such operators to the time independent part, which is solved implicitly using a direct method and a variable part, which is extrapolated together with the non-linear terms. We introduce material properties in form

$$\mu = \mu(T) = \bar{\mu} + \mu_i, \quad \kappa = \kappa(T) = \bar{\kappa} + \kappa_i, \quad \frac{1}{\rho(T)} = \overline{\left(\frac{1}{\rho}\right)} + \left(\frac{1}{\rho}\right)_i, \tag{8}$$

where $\bar{\mu}$ and $\bar{\kappa}$ are time-independent, while $\mu_i = \mu_i(\mathbf{x}, t)$ and $\kappa_i = \kappa_i(\mathbf{x}, t)$. The variable density $\rho = \rho(T)$ acts in our scheme as an inverse value, c.f. (10), so the splitting is done accordingly.

To demonstrate the splitting, we consider the second order operator with variable viscosity and density

$$\frac{1}{\rho(T)} \nabla \cdot \left[\mu(T) (\nabla \mathbf{v})^{\mathrm{T}} \right] = \overline{\left(\frac{1}{\rho}\right)} \nabla \cdot \left[\bar{\mu} (\nabla \mathbf{v})^{\mathrm{T}} \right] + \left(\frac{1}{\rho}\right)_i \nabla \cdot \left[\bar{\mu} (\nabla \mathbf{v})^{\mathrm{T}} \right]$$

$$+ \frac{1}{\rho} \nabla \cdot \left[\mu_i (\nabla \mathbf{v})^{\mathrm{T}} \right].$$
(9)

Only the term with time independent operator $\overline{\left(\frac{1}{\rho}\right)} \nabla \cdot \left[\bar{\mu} (\nabla \mathbf{v})^{\mathrm{T}} \right]$ is solved implicitly, while we apply extrapolation to terms containing variable parameters $\left(\frac{1}{\rho}\right)_i$ and μ_i. This approach is valid for $\bar{\mu} = \bar{\mu}(\mathbf{x})$, $\bar{\kappa} = \bar{\kappa}(\mathbf{x})$, resp. $\overline{\left(\frac{1}{\rho}\right)} = \overline{\left(\frac{1}{\rho}\right)}(\mathbf{x})$, but if $\bar{\mu}$ is constant in space, the constant operator simplifies to $\left(\frac{\mu}{\rho}\right) \nabla^2 \mathbf{v}$ (resp. $\nabla^2 \mathbf{v}$ if the properties are normalized to $\bar{\mu} = \bar{\kappa} = \bar{\rho} = 1$, what is the case of (1), the balance equations in form independent of physical units).

The final form of the equation for velocity becomes

$$\nabla^2 \mathbf{v}_{n+1} - \frac{\gamma}{\Delta t} \frac{\bar{\rho}}{\bar{\mu}} \mathrm{Re} \mathbf{v}_{n+1} =$$

$$\frac{\bar{\rho}}{\bar{\mu}} \left\{ -\mathrm{Re} \frac{\gamma}{\Delta t} \hat{\mathbf{v}} + \frac{1}{\rho^*} \left[\mathrm{Re}(\nabla \tilde{p}_{n+1} - \mathbf{f}_{n+1}) - \left[\nabla \mu \cdot [\nabla \mathbf{v} + (\nabla \mathbf{v})^{\mathrm{T}}] \right]^* \right. \right.$$

$$\left. \left. + \frac{2}{3} [\nabla \mu]^* [\nabla \cdot \mathbf{v}]_{n+1} - \frac{1}{3} \mu^* \nabla [\nabla \cdot \mathbf{v}]_{n+1} \right] \right\}$$

$$- \left(\nabla [\nabla \cdot \mathbf{v}]_{n+1} - [\nabla \times \nabla \times \mathbf{v}]^* \right) \left[\frac{\bar{\rho}}{\bar{\mu}} \left(\frac{1}{\rho^*}\right)_i \mu^* + \frac{1}{\bar{\mu}} \mu_i^* \right]$$
(10)

2.3 Balance of Energy

The energy equation with temperature dependent thermal conductivity is strongly non-linear. We split the diffusion operator to the time independent and the variable part, following the technique shown for the velocity-correction (10). We set $\bar{\kappa} = 1$ for simplicity and the discretized energy equation (1c) gets form

$$\nabla^2 T_{n+1} - \frac{\gamma \mathrm{Re} \mathrm{Pr}}{\Delta t} T_{n+1} = \mathrm{Re} \mathrm{Pr} \left(-\frac{\hat{T}}{\Delta t} - f_{T\,n+1} \right) - [\nabla \cdot \kappa_i \nabla T]^* ,$$
(11)

where $\hat{T} = \sum_{q=0}^{Q-1} \alpha_q T_{n-q} - \Delta t \left[\mathbf{v} \cdot \nabla T \right]^*$. Operator $\left(\nabla^2 - \frac{\gamma \mathrm{Re}\,\mathrm{Pr}}{\Delta t} \right)$ is time independent and allow inversion using a direct method, what results in good performance in computations on long time intervals.

3 Temporal Convergence on Manufactured Solution and Application

Our convergence tests are based on the method of manufactured solutions, alternative to estimates of numerical analysis on simplified system. This approach lacks generality, because we always restrict to particular data and some representative of the solution space, but we get rough convergence estimate for unrestricted equation system (1), while proving also the correctness of method implementation.

As an exact solution, we take a smooth functions $\mathbf{v}_e : \Omega \times (0 : T) \rightarrow \mathbb{R}^n$, $\tilde{p}_e : \Omega \times (0 : T) \rightarrow \mathbb{R}$, $T_e : \Omega \times (0 : T) \rightarrow \mathbb{R}$

$$
\begin{pmatrix} \mathbf{v}_e \\ \tilde{p}_e \\ T_e \end{pmatrix} = \begin{pmatrix} u_e \\ v_e \\ \tilde{p}_e \\ T_e \end{pmatrix} = \begin{pmatrix} 2\cos(\pi x)\cos(\pi y)\sin(t) \\ \sin(\pi x)\sin(\pi y)\sin(t) \\ 2\sin(\pi x)\sin(\pi y)\cos(t) \\ \sin(x)\sin(y)\cos(t) \end{pmatrix}
\tag{12}
$$

and derive the forcing terms $\mathbf{f_v}$, m and f_T such, that Eqs. (2), (1b), (1c) are fulfilled (in all cases we set $\mathrm{Re} = \mathrm{Pr} = 1$). Divergence of velocity in (12) is $\nabla \cdot \mathbf{v}_e = -\pi \sin(\pi x)\cos(\pi y)\sin(t)$, variable in both the spatial and temporal coordinates and with amplitude comparable with the solution itself. We choose a computational domain $\Omega = [0:2] \times [0.5:2.5]$ consisting of two elements $\Omega = [0:1] \times [0.5:2.5] \cup [1:2] \times [0.5:2.5]$. Extent of Ω and form of the exact solution, is inspired by [3], where the velocity-correction scheme of [7] was tested on a similar, manufactured solution.

The incompressible Navier-Stokes equations define the pressure up to a constant value and only the boundary condition for velocity is needed. In this sense, we set the Dirichlet boundary condition for velocity on whole $\partial \Omega$. However, the pressure-Poisson equation (4) requires setting a boundary condition as a consequence of the decoupling. We set HOPBC (7) at $\partial \Omega$ and solve the fully Neumann problem, which defines the solution up to a constant value, which we set by fixing the solution to zero in one of the grid points. The boundary condition for pressure is an artificial element of the computational scheme and its existence is related to the splitting error. The boundary condition for energy equation (11) is of Dirichlet type for whole $\partial \Omega$.

We present the first and second order schemes in time in the convergence tests. The technique is applicable to higher-order schemes as well. A multi step schemes use data from multiple time steps, what complicates its initialisation. We apply the first order BDF method for initialisation of the second order scheme. The first order scheme needs data of only one backward time step, but the time step must be appropriately shortened.

As mentioned already, the acceleration in HOPBC (7) and the term $\frac{\partial \rho}{\partial t}$ in (4) require an initial value or one other backward value for proper initialisation also in case of the first order scheme, what is in contradiction to standard initial conditions for system (1), which require only the initial values. However, setting the correct values for calculation of the first time step is crucial for the final accuracy of the solution.

Finally, we trace appropriate norms of difference between the exact and computed solutions on a set of computations with time steps $\Delta t = \Delta \tilde{t}/2^n$, $\Delta \tilde{t} = 0.2, n = 0, \ldots, 9$ for $t \in [0 : 1]$.

We use the power laws for approximation of dependence of material parameters on temperature

$$\mu(T) = (\alpha_m T + 1.0)^{\beta_m}, \quad \kappa(T) = (\alpha_k T + 1.0)^{\beta_k}, \quad \rho(T) = (\alpha_r T + 1.0)^{\beta_r}.$$

$$(13)$$

The temporal convergence of the above scheme for $\alpha_m = \alpha_k = \alpha_r = 0.1$, $\beta_m = \beta_k = \beta_r = 2$ is shown in Fig. 1.

A detail view of error production, Fig. 2, shows, that the dominant error production arises at the grid point, which was used to set the unknown constant for the Neumann problem.

The scheme was successfully applied in a 2D simulation of flow around the heated cylinder and the results were compared with experimental data [15], where the dependence of the vortex shedding frequency (Strouhal number St) on the wall temperature of the cylinder, T_W, was observed. Figure 3 shows the substantial difference in results between the model neglecting the thermal expansion, [10], and the present one. Fig. 4 shows value range and structure of velocity divergence in a chosen realistic simulation.

Fig. 1 The temporal convergence for the Navier-Stokes-Fourier system with temperature dependent material properties. The number of steps in BDF $Q = 1$ or $Q = 2$, is denoted by subscripts $I1$ and $I2$. The label "Errors" refers to $\|a - a_e\|_{L_\infty}$, where a is the computed function and a_e the exact value from (12), at $t = 1$

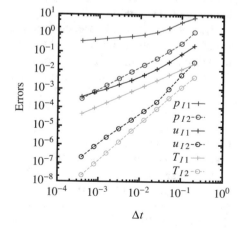

Fig. 2 Test on manufactured solution: difference $v\_err = v_e - v$ at the final time $t = 1$ for computation with $Q = 2$, $\Delta t = 0.2/2^7$, c.f. Fig. 1. Polynomial approximation of degree 15

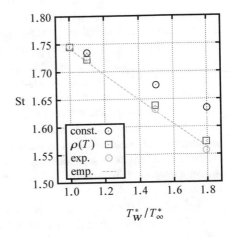

Fig. 3 Frequency of the vortex shedding (Strouhal no. "St") as dependent on the normalized wall temperature T_W^* in the flow around heated cylinder (Re $\approx$ 121.2). Comparison of the data from [10] ($\rho = $ const.) "const.", the present scheme with $Q = 1$ "$\rho(T)$", experimental data of [15] "exp." and empirical formula "emp."[8]

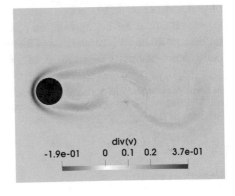

Fig. 4 Computed field of divergence, div(v) $= \nabla \cdot \mathbf{v}$, caused by the thermal expansion in the flow around heated cylinder (Re $= 121.2$, $T_W/T_\infty = 1.494$), c.f. Fig. 3

4 Conclusion

The numerical scheme proposed for the Navier-Stokes-Fourier system with variable parameters allows to solve the highly complex mathematical model, which has an impact to understanding the processes connected with the heat exchange, transport and energy storage in fluids.

The computational scheme for a fluid flows influenced by temperature as modelled by system (2), (1b), (1c) was developed and tested. The scheme was primarily constructed for spatial discretisations based on spectral/hp finite elements and presented results were obtained after implementation to the Nektar++ framework [2], modified version 3.3.

We did not impose restrictions to the type of functional dependency of the material parameters on temperature. Graph of error convergence in L_∞ norm, Fig. 1, results from testing on a manufactured solution and shows a good convergence properties of the scheme, what is promising for applications.

Considered model neglects compressibility in the sense of direct dependence of density on pressure, but the velocity field is not divergence free as a consequence of the thermal expansion. A forward estimate of velocity divergence is needed in the proposed scheme and its successful approximation is one of the main contributions presented in this work. For these reasons, the scheme is unique among numerical schemes based on the finite element approximations in space.

Proposed scheme is an extension of the efficient semi-implicit solver for Incompressible Navier-Stokes system [7] and it is suitable for a fast and highly accurate simulations of problems on long time intervals. The present results inspire implementation of high order BDF schemes and extension of the solver to 3 spatial coordinates.

Derivation of the scheme includes a number of sub-steps, whose detail description is beyond the scope of this article and will be published separately, together with extension of the scheme for energy equation with variable density and further testing of performance as dependent on various physical parameters in the equations.

Also the results from application of the scheme to computations of a physically realistic problem and comparison of its results with experimental data exhibit a good coincidence and will be presented with detail description in a separate article.

Acknowledgements The author wishes to acknowledge the support of the research programme no. 3 "Efficient energy conversion and storage" of the Strategy AV21 initiative of the Czech Academy of Sciences.

References

1. Axelsson, O., He, X., Neytcheva, M.: Numerical solution of the time-dependent Navier-Stokes equation for variable density-variable viscosity. Part I. Math. Model. Anal. **20**(2), 232–260 (2015)

2. Cantwell, C., Moxey, D., Comerford, A., Bolis, A., Rocco, G., Mengaldo, G., Grazia, D.D., Yakovlev, S., Lombard, J.-E., Ekelschot, D., Jordi, B., Xu, H., Mohamied, Y., Eskilsson, C., Nelson, B., Vos, P., Biotto, C., Kirby, R., Sherwin, S.: Nektar++: an open-source spectral/hp element framework. Comput. Phys. Commun. **192**, 205–219 (2015)
3. Dong, S., Shen, J.: A pressure correction scheme for generalized form of energy-stable open boundary conditions for incompressible flows. J. Comput. Phys. **291**, 254–278 (2015)
4. Guermond, J.-L., Salgado, A.: A splitting method for incompressible flows with variable density based on a pressure Poisson equation. J. Comput. Phys. **228**(8), 2834–2846 (2009)
5. Guermond, J.-L., Salgado, A.: Error analysis of a fractional time-stepping technique for incompressible flows with variable density. SIAM J. Numer. Anal. **49**(3/4), 917–944 (2011)
6. Karamanos, G.-S., Sherwin, S.: A high order splitting scheme for the Navier-Stokes equations with variable viscosity. Appl. Numer. Math. **33**(1), 455–462 (2000)
7. Karniadakis, G.E., Orszag, S.A., Israeli, M.: High-order splitting methods for the incompressible Navier-Stokes equations. J. Comput. Phys. **97**, 414–443 (1991)
8. Maršík, F., Trávníček, Z., Yen, R.-H., Wang, A.-B.: Sr-Re-Pr relationship for a heated/cooled cylinder in laminar cross flow. In: Proceedings of CHT-08 ICHMT International Symposium on Advances in Computational Heat Transfer (2008)
9. Oyarzúa, R., Zúniga, P.: Analysis of a conforming finite element method for the Boussinesq problem with temperature-dependent parameters. J. Comput. Appl. Math. **323**, 71–94 (2017)
10. Pech, J.: On computations of temperature dependent incompressible flows by high order methods. EPJ Web Conf. **114**, 02089 (2016)
11. Pérez, C.E., Thomas, J.-M., Blancher, S., Creff, R.: The steady Navier-Stokes/energy system with temperature-dependent viscosity-Part 1: analysis of the continuous problem. Int. J. Numer. Methods Fluids **56**(1), 63–89 (2008)
12. Pérez, C.E., Thomas, J.-M., Blancher, S., Creff, R.: The steady Navier-Stokes/energy system with temperature-dependent viscosity-Part 2: the discrete problem and numerical experiments. Int. J. Numer. Methods Fluids **56**(1), 91–114 (2008)
13. Shunn, L., Ham, F., Moin, P.: Verification of variable-density flow solvers using manufactured solutions. J. Comput. Phys. **231**(9), 3801–3827 (2012)
14. Tabata, M., Tagami, D.: Error estimates of finite element methods for nonstationary thermal convection problems with temperature-dependent coefficients. Numer. Math. **100**(2), 351–372 (2005)
15. Wang, A.-B., Trávníček, Z., Chia, K.-C.: On the relationship of effective Reynolds number and Strouhal number for the laminar vortex shedding of a heated circular cylinder. Phys. Fluids **12**(6), 1401–1410 (2000)
16. Zhang, T., Feng, X., Yuan, J.: Implicit-explicit schemes of finite element method for the non-stationary thermal convection problems with temperature-dependent coefficients. Int. Commun. Heat Mass Transf. **76**, 325–336 (2016)

Implicit Large Eddy Simulations for NACA0012 Airfoils Using Compressible and Incompressible Discontinuous Galerkin Solvers

Esteban Ferrer, Juan Manzanero, Andres M. Rueda-Ramirez, Gonzalo Rubio, and Eusebio Valero

1 Introduction

High order Discontinuous Galerkin (DG) methods provide accurate solutions by enabling arbitrarily high polynomial approximations inside each grid element. For high order polynomials, the numerical errors are not distributed along all wave-numbers but localised at high wave-numbers [1–5]. This characteristic of high order methods results in very accurate simulations with low dissipative and dispersive errors. Although this characteristic seems a-priori beneficial for well resolved simulations, when computing under-resolved Large Eddy Simulations (LES), it can prove difficult to obtain stable simulations. In implicit (or under-resolved) Large Eddy Simulations (iLES), the smallest numerical eddies are larger than would have been in a finer mesh, leading to numerical under-resolution (i.e. coarse grid or low polynomial order) and aliasing [6]. Various methods have been proposed to stabilise under-resolved computations with aliasing. Among others, split forms or skew symmetric variants [7, 8]), localised interior penalty fluxes [9], over-integration [10–12] or filtering [13] may be incorporated into the solver to stabilize the computations and remove or alleviate the aliasing.

Contrarily to low order methods, high order methods do not have enough inherent numerical dissipation in under-resolved simulations, to dissipate large flow structures (when compared to Kolmogorov scales). Therefore, computation of iLES flows using high order DG solvers require localised dissipative mechanisms to dissipate flow structures close to cut-off size. In what follows, we compare two

E. Ferrer (✉) · J. Manzanero · A. M. Rueda-Ramirez · G. Rubio · E. Valero
ETSIAE-UPM (School of Aeronautics - Universidad Politécnica de Madrid), Madrid, Spain

CCS-UPM (Centre for Computational Simulation - Universidad Politécnica de Madrid), Madrid, Spain
e-mail: esteban.ferrer@upm.es

S. J. Sherwin et al. (eds.), *Spectral and High Order Methods for Partial Differential Equations ICOSAHOM 2018*, Lecture Notes in Computational Science and Engineering 134, https://doi.org/10.1007/978-3-030-39647-3_38

dissipative stabilising mechanisms that enable the simulation of turbulent under-resolved flows. On the one hand, we use a compressible formulation with an energy conserving split-form and dissipation through Roe fluxes [14]. On the other hand, the incompressible solver uses the viscous discretisation through interior penalty formulation to enhance stability [9]. We challenge both formulations with a NACA0012 airfoil at various angles of attack in turbulent regimes, to explore both accuracy and stability. We compare simulated results to experimental data and simulations using low order methods (Xfoil and Ansys-Fluent).

2 Methodologies

We first introduce the two different mechanisms used to stabilise both compressible and incompressible high order DG formulations. The explanation included here is brief and aims only at introducing the fundamental concepts and motivating ideas. Further details can be found in the following references by the authors [9, 14].

The 3D Navier-Stokes equations can be written as:

$$\mathbf{u}_t + \nabla \cdot \boldsymbol{F}_e = \nabla \cdot \boldsymbol{F}_v, \tag{1}$$

where $\mathbf{u}$ is the vector of conservative variables $\boldsymbol{u} = (\rho, \rho v_1, \rho v_2, \rho v_3, \rho e)^T$ in compressible solvers. For incompressible solvers $\boldsymbol{u} = (v_1, v_2, v_3)^T$ and Eq. (1) is complemented with $\nabla \cdot \boldsymbol{u}$. Details on the definition of inviscid and viscous solvers can be found in [9, 14]. To derive discontinuous Galerkin schemes, we consider Eq. (1) for one mesh element el, multiply by a locally smooth test function ϕ_j, for $0 \leq j \leq P$, where P is the polynomial degree, and integrate on el:

$$\int_{el} \mathbf{u}_t \phi_j + \int_{el} \nabla \cdot \boldsymbol{F}_e \phi_j = \int_{el} \nabla \cdot \boldsymbol{F}_v \phi_j. \tag{2}$$

We can now integrate by parts the inviscid fluxes, $\boldsymbol{F}_e$, integral to obtain a local weak form of the equations (one per mesh element):

$$\int_{el} \mathbf{u}_t \phi_j + \int_{\partial el} \boldsymbol{F}_e \cdot \mathbf{n} \phi_j - \int_{el} \boldsymbol{F}_e \cdot \nabla \phi_j = \int_{el} \nabla \cdot \boldsymbol{F}_v \phi_j, \tag{3}$$

where $\mathbf{n}$ is the normal vector at element boundaries ∂el. We replace discontinuous fluxes at inter-element faces by a numerical inviscid flux, $\boldsymbol{F}_e^*$, to obtain a weak form for the equations for each element,

$$\int_{el} \mathbf{u}_t \cdot \phi_j + \int_{\partial el} \boldsymbol{F}_e^* \cdot \mathbf{n} \phi_j - \int_{el} \boldsymbol{F}_e \cdot \nabla \phi_j = \int_{el} \nabla \cdot \boldsymbol{F}_v \phi_j, \tag{4}$$

where, we have omitted the fluxes at external boundaries, for simplicity. This set of equations for each element is coupled through the inviscid fluxes F_e^* and governs flow behaviour. Note that one can proceed similarly and integrate by parts the viscous terms (see [9, 15]), but here for simplicity we retain the volume integral.

$$\int_{el} \mathbf{u}_t \cdot \phi_j + \int_{\partial el} \underbrace{F_e^* \cdot \mathbf{n}}_{\text{Riemann solver}} \phi_j - \int_{el} F_e \cdot \nabla \phi_j = \int_{el} (\underbrace{\nabla \cdot F_v}_{\text{Viscous term}}) \cdot \phi_j \qquad (5)$$

The non-linear inviscid and viscous terms that can be discretised to control dissipation in the numerical scheme have been underlined.

Riemann solvers are the classic option to include numerical dissipation in DG schemes [16, 17], since they naturally arise when discretising the non-linear terms. Comparison of different fluxes for homogeneous turbulence can be found in [14, 18]. A different option is to modify the viscous terms to enhance its dissipative properties. The latter has been proposed in [9] using an increased penalty parameter (compared to the minimum required to ensure coercivity of the scheme) when discretising the viscous terms using a interior penalty formulation.

2.1 Compressible DGSEM Solver

The compressible solver uses conservative variables to solve the Navier-Stokes equations. We use a particular nodal variant of DG methods: the Discontinuous Galerkin Spectral Element Method (DGSEM), see for example [19]. In addition, the compressible formulation is modified to be energy preserving [20]. The required split-form necessitate Gauss–Lobatto points to cancel out boundary terms using the summation-by-parts simultaneous-approximation-term property (SBP-SAT). The interested reader is referred to [5, 20–22]. These energy conserving schemes are designed to remain stable and energy conserving and consequently do not necessitate additional localised numerical dissipation. Nonetheless, in this work we introduce dissipation through Roe fluxes, to enhance robustness at high Reynolds numbers. Additionally, viscous terms are discretised using the Bassi-Rebay 1 (BR1) scheme, which is equivalent to the interior penalty formulation when using Gauss-Lobatto points and hexahedral elements [23]. Let us note that this formulation for the viscous fluxes is neutrally stable [24] and adds the minimum dissipation required to achieve a stable scheme, whilst others may introduce some extra dissipation. Other techniques are available to discretise second order derivatives and can be found in the classic review by Arnold et al. [15].

2.2 *Incompressible DG-Fourier Solver*

Flow solutions of the incompressible Navier-Stokes equations, are obtained from the 3D unsteady high order h/p Discontinuous Galerkin-Fourier solver [9, 25–28]. The solver uses a second order stiffly stable approach to discretise the NS equations in time whilst spatial discretisation is provided by the discontinuous Galerkin-Symmetric Interior Penalty formulation with modal basis functions in the x-y plane. Here, x represents the streamwise flow direction and y is the normal direction. Spatial discretisation in the z-direction (here defining the spanwise airfoil length) is provided by a purely spectral method that uses Fourier series and allows computation of spanwise periodic three-dimensional flows. Since high order methods (e.g. discontinuous Galerkin and Fourier) are unable to provide enough numerical dissipation to enable under-resolved high Reynolds computations (e.g. as necessary in Large Eddy Simulations), we have adapted the original laminar version of the solver to increase (controllably) the dissipation and enhance the stability in under-resolved simulations [9]. This dissipative formulation has minimal impact on well resolved flow regions and its implicit treatment does not restrict the use of relatively large time steps, thus providing an efficient stabilization mechanism for Large Eddy Simulations. The solver has been widely validated for a variety of flows, including bluff body flows, airfoil and blade aerodynamics and vertical axis turbines under static and rotating conditions [9, 25–30].

3 Numerical Results

This section considers a NACA0012 airfoil at $Re = 1 \times 10^4$, $Re = 1 \times 10^5$ and $Re = 1 \times 10^6$ (based on the airfoil chord c) for a range of Angles of Attack (AoA): $0° \leq AoA \leq 10°$. In what follows we compare incompressible and compressible simulations using polynomial orders $P = 3$ and $P = 4$. The averaged values have been computed after the development of three dimensional flow. The compressible solver uses a hexahedral mesh with 18,000 elements, which for P = 3 and 4 result in 1.1 and 2.2 million degrees of freedom. The incompressible solver, uses a mixed tri-quad 2D mesh and is expanded using Fourier in the homogeneous third direction (here 16 Fourier modes). Depending on the angle of attack, the resulting meshes include 0.6 to 1 million degrees of freedom. Meshes for the two solvers and for $AoA = 0°$ are depicted in Fig. 1. Finally, all the simulations are computed with both DG solvers and consider a periodic spanwise lengths of $L_z/c = 0.1$. Note that we have not observed significant differences in the results when increasing the spanwise length. Statistics are accumulated during at least 40 convective time scales (based on the airfoil chord) and starting after the turbulent flow has developed (typically an initial transient of 10 convective time scales).

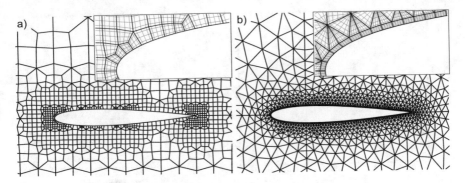

Fig. 1 Meshes for NACA0012 airfoil: (**a**) Hexahedral mesh for compressible solver and (**b**) mixed tri-quad mesh for incompressible solver. Inset figures show high order polynomial mesh for order $P = 4$

Re=1.000.000
AoA = 0 deg

Re=1.000.000
AoA = 5 deg

Re=1.000.000
AoA = 10 deg

contours of velocity: [0.85; 1.2]

Fig. 2 NACA0012 airfoil at $Re = 1 \times 10^6$, from left to right: AoA: 0°, AoA: 5° and AoA: 10°. Simulations are obtained using the incompressible DG solver

3.1 $Re = 1 \times 10^6$ and Various Angles of Attack

We start by illustrating the highest Reynolds number case, which is the most challenging in terms of stability and robustness. To illustrate the range of the flow behaviour at various AoAs, we show in Fig. 2, velocity contours for AoA: 0°,5° and 10°, computed using the incompressible DG solver. It can be seen that at $Re = 1 \times 10^6$ the flow remains attached for all angles, and that only mild separation is seen near the trailing edge. We will see in the next section that at lower Reynolds numbers this is not necessarily the case.

Figure 3 compares the aerodynamic coefficients with experimental data for various angles of attack and the two solvers. Figure 3a shows the lift coefficient against the AoA and Fig. 3b depicts the Lift-Drag Polar for $Re = 1 \times 10^6$. We observe very good agreement with experimental data for both solvers.

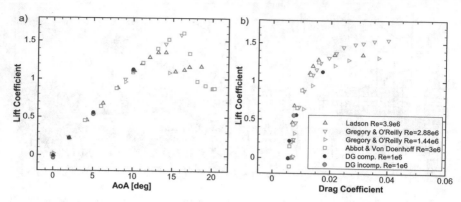

Fig. 3 NACA0012 airfoil at $Re = 1 \times 10^6$: (**a**) Lift coefficient vs angle of attack and (**b**) Lift-Drag Polar. Compressible (comp.) and incompressible (incomp.) DG simulations are compared to experimental data sets of Ladson [31], Gregory and O'Reilly [32], Abbot and Von Doenhoff [33]

3.2   *AoA* = 5° *and Various Reynolds Numbers*

Having shown the overall good performance in terms of aerodynamic quantities at the most challenging Reynolds numbers, we now focus our attention on the angle $AoA = 5°$ and compare the usability of the solvers to study the NACA0012 boundary layer evolution.

First, we compare the aerodynamic coefficients for $AoA = 5°$, and Reynolds numbers $Re = 1 \times 10^5$ and $Re = 1 \times 10^6$, using the incompressible and compressible solvers, both with polynomial order P = 3 and P = 4, in Table 1. We observe good agreement for the highest polynomial order. Small discrepancies are attributed to post-processing of statistics and lack of near wall resolution when using P = 3, which influences mainly the drag coefficient and particularly viscous drag. For completeness, we depict the flow evolution within the boundary layer using both solvers in Fig. 4. It can be seen that detachment near the trailing edge is similar for both solvers. Regarding transition to turbulence (represented by fluctuations in velocity contour), both solvers capture transition on the suction side. The compressible solver shows a transition location near the maximum thickness

Table 1 NACA0012 airfoil at $AoA = 5°$ for $Re = 1 \times 10^5$ and $Re = 1 \times 10^6$

	$Re = 1 \times 10^5$		$Re = 1 \times 10^6$	
	Cl	Cd	Cl	Cd
DG comp. P = 3	0.588	0.028	0.567	0.005
DG comp. P = 4	0.575	0.025	0.558	0.008
DG incomp. P = 3	0.484	0.028	0.538	0.017
DG incomp. P = 4	0.545	0.018	0.551	0.007

Comparison of Lift and Drag using the DG compressible and DG incompressible solvers and two polynomial orders P = 3 and P = 4

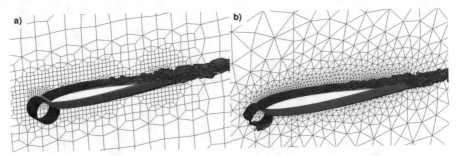

Fig. 4 NACA0012 airfoil at $Re = 1 \times 10^5$ and $AoA = 5°$ for P=4: (**a**) Compressible DG solver. (**b**) Incompressible DG solver

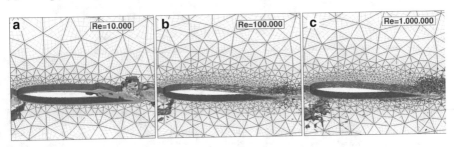

Fig. 5 NACA0012 airfoil at AoA: $5°$ for (**a**) $Re = 1 \times 10^4$, (**b**) $Re = 1 \times 10^5$ and (**c**) $Re = 1 \times 10^6$. Velocity magnitude isocontours and unstructured mesh details are included

($x/c \approx 0.4$), whilst the incompressible solver shows transition closer to the leading edge ($x/c \approx 0.2$). We have observed significant variations of the transition location for the compressible solver when varying the polynomial order, that we have not seen in the incompressible solver. Further studies are necessary to completely assess the influence of discretisation in the transition location for the two solvers.

Second, we explore the pressure coefficient distribution along the airfoil profile when varying the Reynolds number. We only depict results for the incompressible DG solver since these are very similar to the results provided by the compressible solver. Note that this is not surprising, since the lift coefficients at $Re = 1 \times 10^5$ and $Re = 1 \times 10^6$ are very similar for P=4 at $AoA = 5°$, see Table 1. Figure 5 shows velocity contours for $Re = 1 \times 10^4$, $Re = 1 \times 10^5$ and $Re = 1 \times 10^6$ at $AoA = 5°$. It can be seen that for the lowest Reynolds, the boundary layer remains laminar until it detaches after the maximum thickness, showing a highly unsteady wake. When the Reynolds number increases, the boundary layer shows transition to turbulence before the maximum thickness, as appreciated by the fluctuations and small scales appearing in Fig. 5.

To quantify these results, we depict in Fig. 6, the pressure distribution (Cp) for the three Reynolds numbers. In the top row, we show instantaneous Cp against averaged for incompressible DG solver. In the bottom row, we compare mean Cp distributions against Xfoil [34] (with critical N-factor $N_{cr} = 1$) and Fluent SST (fully turbulent

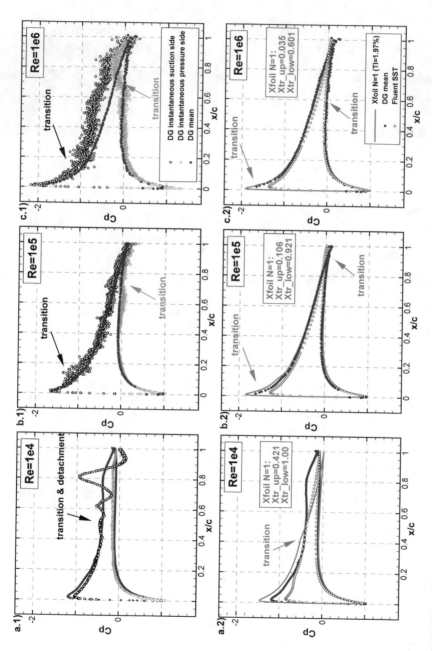

Fig. 6 NACA0012 airfoil at *AoA*: 5° for (**a**) $Re = 1 \times 10^4$, (**b**) $Re = 1 \times 10^5$ and (**c**) $Re = 1 \times 10^6$. Top row shows instantaneous and mean Cp for DG-Fourier solver, and bottom row shows comparison of mean Cp values to other solvers: Xfoil and Fluent SST (fully turbulent simulation)

simulation) [35]. At $Re = 1 \times 10^4$, the top figure shows that the boundary layer detaches before transition occurs and after the maximum thickness, as shown by the velocity contours in Fig. 5. Since the flow detaches leading to a highly unsteady wake, there is little hope that the averaged Cp captures the actual behaviour of the boundary layer. This is why, in the bottom figure, the mean values obtained using the incompressible DG solver do not agree with the mean Xfoil and Fluent values that assume steady turbulent flow. At $Re = 1 \times 10^5$ and At $Re = 1 \times 10^6$, the instantaneous Cp values (top row) show scattering in the data associated to transition. This occurs close to the leading edge on the suction side, whilst it is delayed towards the trailing edge on the pressure side. The bottom row shows that the DG results compare very well to Xfoil when using a critical $N = 1$ (to set the transition point close to the leading edge), whilst Fluent SST (fully turbulent) shows lower Cp values associated to simulating the complete boundary layer as turbulent (no laminar region). This results suggest that DG solvers using iLES approaches (compressible and incompressible) can capture transitional behaviour in boundary layers even when relatively coarse meshes are selected.

4 Conclusions

In this contribution, we have presented results for turbulent flows over a NACA0012 airfoil. High order discontinuous Galerkin formulations require localised dissipation to remain stable for under-resolved turbulent flow conditions, often referred to as implicit Large Eddy Simulations. Here we have presented compressible and an incompressible DG formulations (with different stabilising mechanisms) that are able to cope with high Reynolds number flows. Both DG formulations provide aerodynamic coefficients and boundary layer information that compare favorably to experimental data and well established low order solvers. We conclude that the compressible and incompressible formulations included in this work can be very useful in aeronautical applications.

Acknowledgements This project has received funding from the European Union Horizon 2020 Research and Innovation Program under the Marie Skłodowska-Curie grant agreement No 675008 for the SSeMID project. The authors acknowledge the computer resources and technical assistance provided by the *Centro de Supercomputación y Visualización de Madrid* (CeSViMa).

References

1. Hesthaven, J.S., Warburton, T.: Nodal Discontinuous Galerkin Methods - Algorithms, Analysis, and Applications. Springer, Berlin (2008)
2. Moura, R.C., Sherwin, S.J., Peiro, J.: Eigensolution analysis of spectral/hp continuous Galerkin approximations to advection-diffusion problems: insights into spectral vanishing viscosity. J. Comput. Phys. **307**, 401–422 (2016)

3. Gassner, G.J., Beck, A.D.: On the accuracy of high-order discretizations for underresolved turbulence simulations. Theor. Comput. Fluid Dyn. **27**(3–4), 221–237 (2013)
4. Sherwin, S.: Dispersion analysis of the continuous and discontinuous Galerkin formulations. In: International Symposium on Discontinuous Galerkin Methods, pp. 425–431. Springer, Berlin (1999)
5. Manzanero, J., Rubio, G., Ferrer, E., Valero, E.: Dispersion-dissipation analysis for advection problems with nonconstant coefficients: applications to discontinuous Galerkin formulations. SIAM J. Sci. Comput. **40**(2), A747–A768 (2018)
6. Canuto, C., Hussaini, M.Y., Quarteroni, A., Zang, T.A.: Spectral Methods in Fluid Dynamics. Springer, Berlin (1988)
7. Zang, T.A.: On the rotation and skew-symmetric forms for incompressible flow simulations. Appl. Numer. Math. **7**(1), 27–40 (1991)
8. Blaisdell, G.A., Spyropoulos, E.T., Qin, J.H.: The effect of the formulation of nonlinear terms on aliasing errors in spectral methods. Appl. Numer. Math. **21**(3), 207–219 (1996)
9. Ferrer, E.: An interior penalty stabilised incompressible discontinuous Galerkin-Fourier solver for implicit large eddy simulations. J. Comput. Phys. **348**, 754–775 (2017)
10. Kirby, R.M., Sherwin, S.J.: Aliasing errors due to quadratic non-linearities on triangular spectral h/p element discretisations. J. Eng. Math. **56**(3), 273–288 (2006)
11. Kirby, R.M., Karniadakis, G.E.: De-aliasing on non-uniform grids: algorithms and applications. J. Comput. Phys. **191**, 249–264 (2003)
12. Mengaldo, G., De Grazia, D., Moxey, D., Vincent, P.E., Sherwin, S.J.: Dealiasing techniques for high-order spectral element methods on regular and irregular grids. J. Comput. Phys. **299**, 56–81 (2015)
13. Blackburn, H.M., Schmidt, S.: Spectral element filtering techniques for large eddy simulation with dynamic estimation. J. Comput. Phys. **186**(2), 610–629 (2003)
14. Rubio, G., Valero, E., Manzanero, J., Ferrer, E.: On the role of numerical dissipation in stabilising under-resolved turbulent simulations using discontinuous Galerkin methods (2018). arXiv:1805.10519
15. Arnold, D.N., Brezzi, F., Cockburn, B., Marini, L.D.: Unified analysis of discontinuous Galerkin methods for elliptic problems. SIAM J. Numer. Anal. **39**(5), 1749–1779 (2001)
16. Wang, Z.J., Fidkowski, K., Abgrall, R., Bassi, F., Caraeni, D., Cary, A., Deconinck, H., Hartmann, R., Hillewaert, K., Huynh, H.T., Kroll, N., May, G., Persson, P.O., van Leer, B., Visbal, M.: High-order CFD methods: current status and perspective. Int. J. Numer. Methods Fluids **72**(8), 811–845 (2013)
17. Beck, A.D., Bolemann, T., Flad, D., Frank, H., Gassner, G.J., Hindenlang, F., Munz, C.D.: High-order discontinuous Galerkin spectral element methods for transitional and turbulent flow simulations. Int. J. Numer. Methods Fluids **76**(8), 522–548 (2014)
18. Flad, D., Gassner, G.: On the use of kinetic energy preserving DG-schemes for large Eddy simulation. J. Comput. Phys. **350**, 782–795 (2017)
19. Kopriva, D.A.: Implementing Spectral Methods for Partial Differential Equations. Springer Netherlands, Dordrecht (2009)
20. Gassner, G.J., Winters, A.R., Kopriva, D.A.: Split form nodal discontinuous Galerkin schemes with Summation-By-Parts property for the compressible Euler equations. J. Comput. Phys. **327**, 39–66 (2016)
21. Gassner, G.J., Beck, A.D.: On the accuracy of high-order discretizations for underresolved turbulence simulations. Theor. Comput. Fluid Dyn. **27**, 221 (2013)
22. Kopriva, D.A., Gassner, G.J.: An energy stable discontinuous Galerkin spectral element discretization for variable coefficient advection problems. SIAM J. Sci. Comput. **36**(4), A2076–A2099 (2014)
23. Manzanero, J., Rueda-Ramirez, A.M., Rubio, G., Ferrer, E.: The Bassi Rebay 1 scheme is a special case of the symmetric interior penalty formulation for discontinuous Galerkin discretisations with Gauss–Lobatto points. J. Comput. Phys. **363**, 1–10 (2018)
24. Gassner, G.J., Winters, A.R., Hindenlang, F.J., Kopriva, D.A.: The BR1 scheme is stable for the compressible Navier–Stokes equations. J. Sci. Comput. **77**, 154 (2018)

25. Ferrer, E., Willden, R.H.J.: A high order discontinuous Galerkin - Fourier incompressible 3D Navier-Stokes solver with rotating sliding meshes. J. Comput. Phys. **231**(21), 7037–7056 (2012)
26. Ferrer, E.: A high order discontinuous Galerkin - Fourier incompressible 3D Navier-Stokes solver with rotating sliding meshes for simulating cross-flow turbines. PhD thesis, University of Oxford (2012)
27. Ferrer, E., Willden, R.H.J.: A high order discontinuous Galerkin finite element solver for the incompressible Navier-Stokes equations. Comput. Fluids **46**(1), 224–230 (2011)
28. Ferrer, E., Moxey, D., Willden, R.H.J., Sherwin, S.: Stability of projection methods for incompressible flows using high order pressure-velocity pairs of same degree: continuous and discontinuous Galerkin formulations. Commun. Comput. Phys. **16**(3), 817–840 (2014)
29. Le Clainche, S., Ferrer, E.: A reduced order model to predict transient flows around straight bladed vertical axis wind turbines. Energies **11**(3), 566 (2018)
30. Ferrer, E., de Vicente, J., Valero, E.: Low cost 3D global instability analysis and flow sensitivity based on dynamic mode decomposition and high-order numerical tools. Int. J. Numer. Methods Fluids **76**(3), 169–184 (2014)
31. Ladson, C.L., United States National Aeronautics and Space Administration, Scientific and Technical Information Division: Effects of independent variation of Mach and Reynolds numbers on the low-speed aerodynamic characteristics of the NACA 0012 airfoil section. NASA Technical Memorandum. National Aeronautics and Space Administration, Scientific and Technical Information Division (1988)
32. Gregory, N., O'Reilly, C.L., H.M.S.O.: Low-speed aerodynamic characteristics of NACA 0012 aerofoil section, including the effects of upper-surface roughness simulating hoar frost. Aeronautical Res. Council. Reports and Memoranda. National Physical Laboratory (1970)
33. Abbott, I.H.A., Von Doenhoff, A.E.: Theory of Wing Sections, Including a Summary of Airfoil Data. Dover Books on Engineering. Dover Publications, New York (1959)
34. Drela, M.: Xfoil: an analysis and design system for low Reynolds number airfoils. In: Mueller, T.J. (ed.) Low Reynolds Number Aerodynamics, pp. 1–12. Springer, Berlin (1989)
35. Fluent-ANSYS Academic Research Mechanical, Release 18.1, ANSYS Inc., Canonsburg (2017)

SAV Method Applied to Fractional Allen-Cahn Equation

Xiaolan Zhou, Mejdi Azaiez, and Chuanju Xu

1 Introduction

The Allen-Cahn equation was originally introduced to describe the motion of anti-phase boundaries in crystalline solids [1]. There have been a large body of work on numerical analysis of Allen-Cahn equations (cf. [2–5] and the references therein). We aim in this paper to use the SAV scheme, recently introduced and analyzed by a number of researchers; see, e.g., [5] and the references therein, to approximate the solution of the fractional version of the Allen-Cahn model. It consists in finding $\phi : \Omega \times (0, T] \to \mathbb{R}$ solution of

$$
\begin{cases}
\frac{\partial \phi}{\partial t} + \gamma\left((-\Delta)^s \phi + f(\phi)\right) = 0, & \forall (x, t) \in \Omega \times (0, T], \\
\nabla \phi \cdot n\big|_{\partial\Omega} = 0, & \forall t \in (0, T] \\
\phi(t = 0) = \phi_0(x), & \forall x \in \Omega.
\end{cases}
\tag{1.1}
$$

In the above, γ is a positive kinetic coefficient, $s \in (0, 1)$, $\Omega \subset \mathbb{R}^d$ is a bounded domain, n is the outward normal, $f(\phi) = F'(\phi)$ with a given function $F(\phi) = \frac{1}{4\varepsilon^2}(\phi^2 - 1)^2$ being the Ginzburg-Landau double-well potential. The phase field ϕ

X. Zhou · C. Xu (✉)
School of Mathematical Sciences, Xiamen University, Xiamen, China
e-mail: xlzhou@stu.xmu.edu.cn; cjxu@xmu.edu.cn

M. Azaiez
School of Mathematical Sciences, Xiamen University, Xiamen, China

Bordeaux INP, I2M (UMR CNRS 5295), Université de Bordeaux, Pessac, France
e-mail: azaiez@enscbp.fr

© The Author(s) 2020
S. J. Sherwin et al. (eds.), *Spectral and High Order Methods for Partial Differential Equations ICOSAHOM 2018*, Lecture Notes in Computational Science and Engineering 134, https://doi.org/10.1007/978-3-030-39647-3_39

is such that

$$\phi = \begin{cases} 1, & \text{phase 1,} \\ -1, & \text{phase 2,} \end{cases}$$

and ε represents the thickness of the smooth transition layer connecting the two phases, which is small compared to the characteristic length of the system scale. The homogeneous Neumann boundary condition implies that no mass loss occurs across the boundary walls.

Among the different definitions of fractional Laplacians (see [6, 7] for a quantitative assessment of new numerical methods as well as available state-of-the-art methods for discretizing the fractional Laplacians problems), we choose in this paper to focus on the fractional spectral definition. It is defined by

$$(-\Delta)^s u := \sum_{i \in \mathbb{N}} a_i \lambda_i^s e_i,$$

where λ_i, e_i are the eigenvalues and eigenfunctions of the Laplace operator $-\Delta$ in Ω with homogeneous Neumann boundary condition, i.e., they satisfy

$$\begin{cases} -\Delta e_i = \lambda_i e_i, & x \in \Omega, \\ \nabla e_i \cdot n\big|_{\partial\Omega} = 0. \end{cases}$$

While, a_i represents the projection of u on the direction e_i, $a_i = (u, e_i)_{L_\Omega^2}$. The spectral fractional Laplacian is nonlocal on the interior for noninteger $s \in (0, 1)$. We see that to compute the inner product $a_i = (u, e_i)_{L_\Omega^2}$, it suffices for u to be defined on the interior of Ω. No information about u on the exterior $R^d \setminus \Omega$ is required. Thus, from a conceptual viewpoint, in boundary value problems the spectral fractional Laplacian can admit the same type of boundary conditions as the standard, local Laplacian $-\Delta$. In this paper, we let $\Omega =]-1, 1[^2$. Set $u(x, t) = \sum_{n=1}^{\infty} \sum_{m=1}^{\infty} a_{m,n}(t) e_{m,n}(x)$, where $e_{m,n}$ are the orthogonal eigenfunctions of the Laplace operator with homogeneous Neumann boundary conditions and $\lambda_{m,n}$ are the corresponding eigenvalues. Then we define the spectral fractional Laplacian as,

$$(-\Delta)^s u(x, t) := \sum_{n=1}^{\infty} \sum_{m=1}^{\infty} \lambda_{m,n}^s a_{m,n}(t) e_{m,n}(x), \quad 0 < s < 1, \ \forall u \in H^s(\Omega).$$

$$(1.2)$$

Here

$$H^s(\Omega) := \left\{ u = \sum_{n=1}^{\infty} \sum_{m=1}^{\infty} a_{m,n} e_{m,n} \in L^2(\Omega) : |u|_s := \left(\sum_{n=1}^{\infty} \sum_{m=1}^{\infty} \lambda_{m,n}^s a_{m,n}^2 \right)^{1/2} < \infty \right\}.$$

The rest of this paper in organized as follows. In Sect. 2, we present briefly the spectral method by giving some notations and reminders. The fractional Laplace operator and its possible applications is discussed in Sect. 3. To demonstrate the applicability of the approximative fractional Laplacian for real applications, we consider a fractional Allen-Cahn equation (FACE). Based on the scalar auxiliary variable (SAV) approach, we construct an unconditionally second-order energy stable BDF scheme (SAV/BDF2) for FACE. We present numerical results for a test case as well as a benchmark example in Sect. 4.

2 Spatial Discretizations

We limit here the description of the spectral approximation to the introduction of some notations and reminders (see [8, 9]). For complex domain, we can use spectral element method [10]. Let $\Sigma = \{(\xi_i, \rho_i); 0 \leq i \leq N\}$ denote the sets of Gauss-Lobatto-Legendre quadrature nodes and weights associated to polynomials of degree N. These quantities are such that on $\Lambda :=]-1, +1[$

$$\forall \phi \in \mathbb{P}_{2N-1}(\Lambda), \qquad \int_{-1}^{+1} \phi(\xi) \, d\xi = \sum_{j=0}^{N} \phi(\xi_j) \, \rho_j, \qquad (2.1)$$

where $\mathbb{P}_N(\Lambda)$ denotes the space of polynomials of degree $\leq N$. We recall that the nodes ξ_i $(0 \leq i \leq N)$ are solution to $(1 - x^2) L_N'(x) = 0$, where L_N denotes the Legendre polynomial of degree N.

The canonical polynomial interpolation basis $h_i(x) \in \mathbb{P}_N(\Lambda)$ built on Σ is given by the relationships:

$$h_i(x) = -\frac{1}{N(N+1)} \frac{1}{L_N(\xi_i)} \frac{(1 - x^2) L_N'(x)}{(x - \xi_i)}, \qquad -1 \leq x \leq +1, \quad 0 \leq i \leq N, \qquad (2.2)$$

with the elementary cardinality property

$$h_i(\xi_j) = \delta_{ij}, \qquad 0 \leq i, j \leq N, \qquad (2.3)$$

where δ_{ij} is Kronecker's delta symbol.

In the sequel the phase field ϕ will be approximated in space variable by suitable polynomial functions ϕ_N as follows

$$\phi_N(x, t) = \sum_{i=0}^{N} \sum_{j=0}^{N} \alpha_{i,j}(t) h_i(x) h_j(y). \tag{2.4}$$

The L^2-inner products involved in the calculation will be achieved using Gauss-Lobatto-Legendre quadrature, which reads: for all continuous functions φ and ψ in $\bar{\Omega}$,

$$(\varphi, \psi) \approx (\varphi, \psi)_N := \sum_{i=0}^{N} \sum_{j=0}^{N} \varphi(\xi_i) \psi(\xi_j) \rho_i \rho_j. \tag{2.5}$$

3 Scalar Auxiliary Variable (SAV) Approach for FACE

SAV approach was introduced in [4, 5] to solve gradient flows. The main purpose of this section is to construct efficient unconditionally stable scheme based on this approach for (1.1).

Throughout the paper, we assume there exists a constant C_0 such that $\int_\Omega F(\phi) dx + C_0 > 0$. We first introduce a *scalar* auxiliary variable

$$r(t) := \sqrt{\int_\Omega F(\phi) \, dx + C_0}.$$

Then, we rewrite the phase-field equation (1.1) under an equivalent form as: find $\phi : (0, T] \times \Omega \to \mathbb{R}$ and $r : (0, T] \to \mathbb{R}$, such that

$$\begin{cases} \frac{\partial \phi}{\partial t} = \gamma \mu, \qquad \nabla \phi \cdot n\big|_{\partial\Omega} = 0, \\[2mm] \mu = -(-\Delta)^s \phi - \frac{r(t)}{\sqrt{\int_\Omega F(\phi) dx + C_1}} f(\phi), \\[2mm] \frac{dr}{dt} = \frac{1}{2\sqrt{\int_\Omega F(\phi) dx + C_1}} \int_\Omega f(\phi) \frac{\partial \phi}{\partial t} \, dx. \end{cases} \tag{3.1}$$

Theorem 3.1 *If $\phi \in L^2((0, T], H^s(\Omega))$, $0 < s < 1$, is the solution of equations (3.1), then we have the following energy dissipation law*

$$\frac{d}{dt}\left(r^2 + \frac{1}{2}|\phi|_s^2\right) = -\gamma \|\mu\|_0^2. \tag{3.2}$$

Proof Taking the inner product of the first two equations with μ, $\frac{\partial\phi}{\partial t}$ respectively, and multiplying the third equation with $2r(t)$, then adding them together, We obtain

$$-\gamma\|\mu\|_0^2 = \frac{d}{dt}(r^2) + ((-\Delta)^s\phi, \phi_t). \tag{3.3}$$

Let $\phi(x, t) = \sum_{m,n=1}^{\infty} a_{m,n}(t)e_{m,n}(x)$ and taking advantage of the orthogonality of $\{e_{m,n}\}$, we verify

$$((-\Delta)^s\phi, \phi_t) = \sum_{m,n=1}^{\infty} \lambda_{m,n}^s a_{m,n}(t)a'_{m,n}(t) = \frac{1}{2}\frac{d}{dt}\left(\sum_{m,n=1}^{\infty} \lambda_{m,n}^s a_{m,n}^2(t)\right) = \frac{1}{2}\frac{d}{dt}|\phi|_s^2. \tag{3.4}$$

Then combining (3.3) and (3.4) proves (3.2). □

The energy law (3.2) means that the SAV approach (3.1) makes the modified energy

$$H(\phi) = r^2 + \frac{1}{2}|\phi|_s^2$$

decay in time.

Now we construct a second-order semi-implicit scheme for the system (3.1). Given initial conditions $\phi^0 = \phi_0$, and let $r^0 = \sqrt{\int_\Omega F(\phi^0)dx + C_0}$, find $\phi^{n+1} \in H^s(\Omega)$ and $r^{n+1} \in \mathbb{R}$, $n = 1, \ldots$, such that

$$\frac{3\phi^{n+1} - 4\phi^n + \phi^{n-1}}{2\Delta t} = \gamma\mu^{n+1} \tag{3.5}$$

$$\mu^{n+1} = -(-\Delta)^s\phi^{n+1} - \frac{r^{n+1}}{\sqrt{\int_\Omega F\left(\bar{\phi}^{n+1}\right)dx + C_0}}f\left(\bar{\phi}^{n+1}\right) \tag{3.6}$$

$$\frac{3r^{n+1} - 4r^n + r^{n-1}}{2\Delta t} = \frac{1}{2\sqrt{\int_\Omega F\left(\bar{\phi}^{n+1}\right)dx + C_0}}\int_\Omega f\left(\bar{\phi}^{n+1}\right)\frac{3\phi^{n+1} - 4\phi^n + \phi^{n-1}}{2\Delta t}dx \tag{3.7}$$

In the above, $\bar{\phi}^{n+1}$ can be any explicit approximation of $\phi(t^{n+1})$ with an error of $\mathcal{O}(\Delta t^2)$. For instance, we may choose the following one

$$\bar{\phi}^{n+1} = 2\phi^n - \phi^{n-1}.$$

Theorem 3.2 *The scheme* (3.5)–(3.7) *is unconditionally stable in the sense that*

$$\frac{1}{\Delta t}\left(\tilde{H}\left[(\phi^{n+1}, r^{n+1}), (\phi^n, r^n)\right] - \tilde{H}\left[(\phi^n, r^n), (\phi^{n-1}, r^{n-1})\right]\right) \le -\gamma\|\mu^{n+1}\|_0^2,$$

$$(3.8)$$

with the modified energy

$$\tilde{H}\left[(\phi^{n+1}, r^{n+1}), (\phi^n, r^n)\right] = \frac{1}{4}\left(|\phi^{n+1}|_s^2 + |2\phi^{n+1} - \phi^n|_s^2\right) + \frac{1}{2}\left((r^{n+1})^2 + (2r^{n+1} - r^n)^2\right).$$

$$(3.9)$$

Proof The result can be directly deduced from taking the inner product of the first two equations (3.5) and (3.6) with μ^{n+1} and $\frac{3\phi^{n+1} - 4\phi^n + \phi^{n-1}}{2\Delta t}$ respectively, and multiplying the third equation (3.7) with $2r^{n+1}$, then using the following identity:

$$2(a^{k+1}, 3a^{k+1} - 4a^k + a^{k-1}) = \|a^{k+1}\|^2 + \|2a^{k+1} - a^k\|^2 + \|a^{k+1} - 2a^k + a^{k-1}\|^2$$
$$- \|a^k\|^2 - \|2a^k - a^{k-1}\|^2.$$

□

3.1 Implementation

Besides its unconditional stability, a most remarkable feature of the above scheme is that it can be solved very efficiently. Indeed, by inserting (3.6) and (3.7) into (3.5), and let $\bar{\mathcal{F}}^{n+1} := \int_\Omega F(\bar{\phi}^{n+1})d\mathbf{x} + C_0$, we obtain

$$\left(\frac{3}{2\gamma\Delta t}I + (-\Delta)^s\right)\phi^{n+1} + \frac{(f(\bar{\phi}^{n+1}), \phi^{n+1})}{2\bar{\mathcal{F}}^{n+1}}f(\bar{\phi}^{n+1}) = g^n, \qquad (3.10)$$

where

$$g^n := \frac{4\phi^n - \phi^{n-1}}{2\gamma\Delta t} - \left(\frac{4r^n - r^{n-1}}{3\sqrt{\bar{\mathcal{F}}^{n+1}}} - \frac{(f(\bar{\phi}^{n+1}), \frac{4\phi^n + \phi^{n-1}}{3})}{2\bar{\mathcal{F}}^{n+1}}\right)f(\bar{\phi}^{n+1}). \qquad (3.11)$$

We shall first determine $(f(\bar{\phi}^{n+1}), \phi^{n+1})$ from (3.10). To this end we multiply (3.10) by $\left(\frac{3}{2\gamma\Delta t}I + (-\Delta)^s\right)^{-1}$ and take the inner product by $f(\bar{\phi}^{n+1})$ to get

$$(f(\bar{\phi}^{n+1}), \phi^{n+1}) + \frac{(f(\bar{\phi}^{n+1}), \phi^{n+1})}{2\bar{\mathcal{F}}^{n+1}}(f(\bar{\phi}^{n+1}), \beta^{n+1}) = (f(\bar{\phi}^{n+1}), \alpha^{n+1}),$$

$$(3.12)$$

with

$$\alpha^{n+1} = \left(\frac{3}{2\gamma \Delta t} I + (-\Delta)^s\right)^{-1} g^n, \qquad \beta^{n+1} = \left(\frac{3}{2\gamma \Delta t} I + (-\Delta)^s\right)^{-1} f(\bar{\phi}^{n+1}).$$

(3.13)

Then, we have

$$(f(\bar{\phi}^{n+1}), \phi^{n+1}) = \frac{(f(\bar{\phi}^{n+1}), \alpha^{n+1})}{1 + \frac{1}{2\bar{\mathcal{F}}^{n+1}}(f(\bar{\phi}^{n+1}), \beta^{n+1})}.$$

(3.14)

Thus we obtain an expression to compute ϕ^{n+1} by bringing back (3.14) into (3.10):

$$\phi^{n+1} = \alpha^{n+1} - \frac{(f(\bar{\phi}^{n+1}), \alpha^{n+1})}{2\bar{\mathcal{F}}^{n+1} + (f(\bar{\phi}^{n+1}), \beta^{n+1})} \beta^{n+1}.$$

(3.15)

Finally we compute r^{n+1} through

$$r^{n+1} = \frac{4r^n - r^{n-1}}{3} + \frac{(f(\bar{\phi}^{n+1}), \phi^{n+1}) - (f(\bar{\phi}^{n+1}), \frac{4\phi^n - \phi^{n-1}}{3})}{2\sqrt{\bar{\mathcal{F}}^{n+1}}}.$$

We now summarize the algorithm of the Scalar Auxiliary Variable approach/Semi-Implicit Second-Order Scheme (3.5)–(3.7) as follows:

---

1. Set $\bar{\phi}^{n+1} = 2\phi^n - \phi^{n-1}$, $\quad \bar{\mathcal{F}}^{n+1} = \int_\Omega F(\bar{\phi}^{n+1}) dx + C_0$,

   $\tilde{c}_0 = (f(\bar{\phi}^{n+1}), 4\phi^n - \phi^{n-1})/3$, $\quad \tilde{c}_1 = \frac{4r^n - r^{n-1}}{3\sqrt{\bar{\mathcal{F}}^{n+1}}} - \frac{\tilde{c}_0}{2\bar{\mathcal{F}}^{n+1}}$,

   $g^n = \frac{4\phi^n - \phi^{n-1}}{2\gamma \Delta t} - \tilde{c}_1 f(\bar{\phi}^{n+1})$;

2. Solve $\dfrac{3}{2\gamma \Delta t} \beta^{n+1} + (-\Delta)^s \beta^{n+1} = f(\bar{\phi}^{n+1})$;

3. Solve $\dfrac{3}{2\gamma \Delta t} \alpha^{n+1} + (-\Delta)^s \alpha^{n+1} = g^n$;

4. Compute $\tilde{c}_2 = (f(\bar{\phi}^{n+1}), \beta^{n+1})$, $\quad \tilde{c}_3 = (f(\bar{\phi}^{n+1}), \alpha^{n+1})$, $\quad \tilde{c}_4 = \tilde{c}_3/(2\bar{\mathcal{F}}^{n+1} + \tilde{c}_2)$;

5. Compute $\phi^{n+1} = \alpha^{n+1} - \tilde{c}_4 \beta^{n+1}$, $\quad r^{n+1} = \frac{4r^n - r^{n-1}}{3} +$

   $\dfrac{(f(\bar{\phi}^{n+1}), \phi^{n+1}) - \tilde{c}_0}{2\sqrt{\bar{\mathcal{F}}^{n+1}}}$.

---

4 Numerical Results and Discussion

In this section, we first present a numerical example to illustrate the efficiency of the SAV scheme in terms of stability and accuracy. We then use the proposed scheme to simulate a benchmark problem.

4.1 Test of the Convergence Order

In order to validate the proposed SAV/BDF2 scheme for the fractional phase-field equation, we consider a fabricated forcing term so that the exact solution to (1.1) is $\phi(x, t) = \sin(t) \cos(\pi x) \cos(\pi y)$. In this test we set $\gamma = 1$, $\Omega =] - 1, 1[^2$, and the nonlinear term is given by $f(\phi) = \phi(\phi^2 - 1)$.

In the calculation we use polynomial degree 32×32 for the spatial discretization, which is large enough so that the spatial discretization error is negligible compared to the temporal error. Figure 1 shows the L^2-errors at $T = 1.0$ in log-log scale as a function of the time step size for several fractional orders. It is observed from this figure that the convergence rate of the time stepping scheme is exactly second order as expected for all tested values of s. It is worthy to mention that no numerical instability was observed for all time step sizes used in the calculation. This implies that the proposed scheme is unconditionally stable.

Fig. 1 L^2-errors at $T = 1.0$ in log-log scale with respect to the time step size Δt for different fractional order s

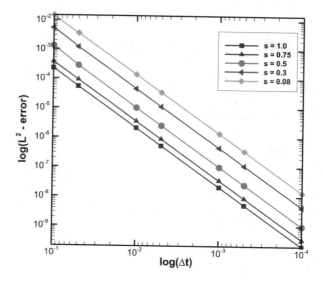

4.2 Benchmark Test

In this subsection, we apply the SAV/BDF2 scheme to the fractional version of a classical benchmark problem (cf. [11]) that we describe below. Our main purpose in this test is to demonstrate the applicability of the constructed method for the FACE. We are particularly interested in numerically investigating the impact of the fractional order on the evolution of the phase interface.

At the initial state, there is a circular phase interface of the radius $R_0 = 100$ in the rectangular domain $] - 128, 128[^2$. In other words, the initial condition is given by

$$\phi(x, 0) = \begin{cases} 1, & |x|^2 < 100^2, \\ -1, & |x|^2 \geq 100^2. \end{cases}$$

Such a circular interface is unstable and the driving force will make it shrink and eventually disappear. It has been shown that in the limit that the radius of the circle is much larger than the interfacial thickness, the velocity and the radius of the moving interface are given (see [1]) by

$$V = \frac{dR}{dt} = -\frac{1}{R}, \quad R(t) = \sqrt{R_0^2 - 2t}.$$

In the implementation we map the computational domain $] - 128, 128[^2$ to $] - 1, 1[^2$. Therefore actually we are led to solve the fractional Allen-Cahn equation (1.1) with the coefficients $\gamma = 1/128^2$ and $\varepsilon = 0.0078$. In the simulation, the space resolution is set to $N = 512$, and the time step size is $\Delta t = 0.1$. The computed radius $R(t)$ for $s = 1$ using the SAV/BDF2 scheme is plotted in Fig. 2. We observe that $R(t)$ keeps monotonously decreasing and very close to the sharp interface limit value. This confirms the accuracy of the proposed method, at least in the case $s = 1$.

Next we apply the proposed scheme to investigate the impact of the fractional order on the radius behavior. In Fig. 3 we present the numerical radius evolution for a number of the fractional orders. Specifically, Fig. 4 shows the circle shrinking for fractional orders $s = 1.0, 0.9, 0.8$. It is clearly indicated that the radius decay rate slow down when the fractional order decreases. However, for the time being the physical meaning and mathematical explanation of this phenomena remain unknown. We plan to address this issue in future work.

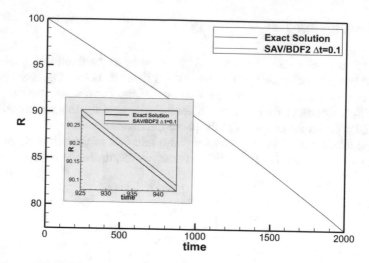

Fig. 2 The evolution of radius $R(t)$: comparison of the exact solution and numerical result in the case $s = 1$

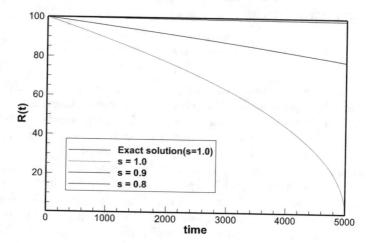

Fig. 3 Evolution of the radius for different fractional order s: impact of the order on the radius decay rate

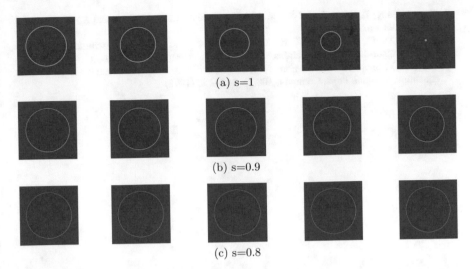

Fig. 4 Temporal evolution of a circular domain from left to right at times $t = 1000, 2000, 3000, 4000, 5000$, for fractional order $s = 1$ **(a)**,0.9 **(b)**, 0.8 **(c)**, for the top, middle and bottom rows, respectively

Acknowledgements The work was supported by NSFC/ANR joint program 51661135011 - PHASEFIELD. The second author has received financial support from the French State in the frame of the "Investments for the future" Programme Idex Bordeaux, reference ANR-10-IDEX-03-02. The third author was also supported by NSFC grant 11971408, NNW2018-ZT4A06 project.

References

1. Allen, S.M., Cahn, J.W.: A microscopic theory for antiphase boundary motion and its application to antiphase domain coarsening. Acta Metall. **27**(6), 1085–1095 (1979)
2. Shen, J., Yang, X.: Numerical approximations of Allen-Cahn and Cahn-Hilliard equations. Discrete Contin. Dynam. Syst. **28**(4), 1669–1691 (2010)
3. Feng, X., Prohl, A.: Numerical analysis of the Allen-Cahn equation and approximation for mean curvature flows. Numer. Math. **94**(1), 1669–1691 (2003)
4. Shen, J., Xu, J., Yang, J.: A new class of efficient and robust energy stable schemes for gradient flows (2017). arXiv:1710.01331 [math.NA]
5. Shen, J., Xu, J., Yang, J.: The scalar auxiliary variable (SAV) approach for gradient flows. J. Comput. Phys. **353**, 407–416 (2018)
6. Lischke, A., Pang, G., Gulian, M., Song, F., Glusa, C., Zheng, X., Mao, Z., Cai, W., Meerschaert, M.M., Ainsworth, M., Karniadakis, G.E.: What is the fractional Laplacian?. ArXiv:1801.09767V1 [MATH. NA] (29 Jan 2018)
7. Song, F., Xu, C., Karniadakis, G.E.: A fractional phase-field model for two-phase flows with tunable sharpness: algorithms and simulations. Comput. Methods Appl. Mech. Eng. **305**, 376–404 (2016)
8. Deville, M., Fischer, P.F., Mund, E.: High-order methods for incompressible fluid flow. Appl. Mech. Rev. **56**(3), B43 (2002)

9. Shen, J., Tang, T., Wang, L.-L.: Spectral Methods: Algorithms, Analysis and Applications, 1st edn., pp. 300–305. Springer, Berlin (2011)
10. Song, F., Xu, C., Karniadakis, G.E.: Computing fractional Laplacians on complex-geometry domains: algorithms and simulations. SIAM J. Sci. Comput. **39**(4), A1320–A1344 (2017)
11. Chen, L.Q., Shen, J.: Applications of semi-implicit Fourier-spectral method to phase field equations. Comput. Phys. Commun. **108**(2), 147–158 (1998)

A First Meshless Approach to Simulation of the Elastic Behaviour of the Diaphragm

Nicola Cacciani, Elisabeth Larsson, Alberto Lauro, Marco Meggiolaro, Alessio Scatto, Igor Tominec, and Pierre-Frédéric Villard

1   Introduction

When intensive care patients are subjected to mechanical ventilation, this is part of the life support. At the same time the ventilator causes damage to the muscles that govern the normal breathing. Normally, the muscles contract when we inhale, and air is pulled into the lungs. During controlled mechanical ventilation, the ventilator instead pushes the air into the lungs that then exert a pressure on the muscles. The function of the muscle tissue can deteriorate quite rapidly, leading to Ventilator

N. Cacciani
Division of Basic and Clinical Muscle Biology, Karolinska Institutet, Stockholm, Sweden
e-mail: nicola.cacciani@ki.se

E. Larsson (✉) · I. Tominec
Scientific Computing, Department of Information Technology, Uppsala University, Uppsala, Sweden
e-mail: elisabeth.larsson@it.uu.se; igor.tominec@it.uu.se

A. Lauro
UOC di Radiologia, University-Hospital of Padova, Padova, Italy

M. Meggiolaro
UOC di Anestesia e Rianimazione, ULSS 3 Sereneness, Saints Giovanni and Paolo Hospital, Venice, Italy
e-mail: meggiomarco@libero.it

A. Scatto
UOC di Anestesia e Rianimazione, University-Hospital of Padova, Padova, Italy

P.-F. Villard
Computer Science, Université de Lorraine, CNRS, Inria, LORIA, Nancy, France
e-mail: pierrefrederic.villard@loria.fr

© The Author(s) 2020
S. J. Sherwin et al. (eds.), *Spectral and High Order Methods for Partial Differential Equations ICOSAHOM 2018*, Lecture Notes in Computational Science and Engineering 134, https://doi.org/10.1007/978-3-030-39647-3_40

501

Induced Diaphragmatic Dysfunction (VIDD) [2]. Because of this, the rehabilitation process, including the weaning from the ventilator, is more difficult and takes longer.

The Individual Virtual Ventilator (INVIVE) project [5] aims to study the mechanics of respiration through numerical simulation in order to learn more about the onset of VIDD, and the factors that influence its progress in a patient. This work is the first publication from the project and is a pilot study for the numerical techniques that we plan to use.

The diaphragm is the main respiratory muscle. It has not been studied as much in the literature as other muscles, and not with detailed models. However, there are a few studies that uses continuum mechanical descriptions of the muscle tissue and simulate its behaviour using FEM [6, 10]. The main drawbacks of the FEM solvers are that they are time-consuming, and that meshing of complex geometries can be difficult. We instead propose to use a meshfree RBF-FD method [4] for the numerical simulation. Some of the potential advantages are that meshing can be replaced with scattered node generation, which in some respects is easier, and allows for a lot of flexibility; that it is easy to construct high-order accurate approximations that can reduce the computational cost; and that the method is easy to implement and modify, providing flexibility when performing experiments. The objectives of the paper are

- to show the feasibility of using the RBF-FD method for this type of problems,
- to work with real medical data such that the results will be relevant,
- to investigate how the high aspect ratio of the geometry affects the simulation and if this can be mitigated by using high aspect ratio node sets.

The paper is organized as follows: In Sect. 2 we describe the linear elasticity equations in three dimensions. Section 3 briefly introduces the RBF-FD method. The process from medical images to input data for the simulation is described in Sect. 4, which is followed by Sect. 5 on Numerical experiments.

2 The Elasticity Equations

The constitutive relations that describe the real behaviour of muscle tissue are non-linear. The displacement of the diaphragm is large, and should therefore also be modeled by non-linear elasticity equations. For our final simulation tool, we aim to solve the fully non-linear equations. However, for the initial development of meshless numerical methods for the diaphragm simulations, we use a linear elasticity test case.

2.1 The Linearized Equations of Motion

For the linear test problem, the following simplifying assumptions are made: The relationship between stress and strain is *linear*, the material is *isotropic* and *homogeneous*, and displacements are *small*.

We define the displacement $u(X) = (u_1(X), u_2(X), u_3(X))^T \in \mathbb{R}^3$ of the tissue from the initial configuration $X = (x, y, z)^T \in \mathbb{R}^3$ to a later configuration $X^* \in \mathbb{R}^3$ as

$$u(X) = X^* - X. \tag{1}$$

The strain-displacement relationship for small displacements, $\|\nabla u\| \ll 1$, has the form

$$\varepsilon = \frac{1}{2}\left[\nabla u + (\nabla u)^T\right], \tag{2}$$

where the strain $\varepsilon \in \mathbb{R}^{3\times3}$ is a tensor. For a linear material, the constitutive relation between the strain and the stress $\sigma \in \mathbb{R}^{3\times3}$ is characterized by the Lamé parameters λ, and μ, leading to

$$\sigma = 2\mu\varepsilon + \lambda tr(\varepsilon)I. \tag{3}$$

In tissue mechanics, the acceleration is typically small compared with the forces, and can be neglected. The equations of motion (Newton's second law) can then be written as

$$\nabla \cdot \sigma + f = 0, \tag{4}$$

where $f \in \mathbb{R}^3$ represents body forces. We assume that (4) holds for all points $X \in \Omega$, where Ω is the domain of interest, which for our problem is the diaphragm. To close the problem formulation, we also need boundary conditions. The first type is displacement boundary conditions

$$u = g, \quad X \in \Omega_D. \tag{5}$$

These are applied where the geometry is attached, for example where the diaphragm is attached to the ribs and the spine. Traction boundary conditions are given in terms of the stress as

$$\sigma \cdot n = h, \quad X \in \Omega_T. \tag{6}$$

These represent forces applied to the surface of the domain of interest, such as the pressure against the diaphragm from below generated by the abdominal compliance.

2.2 The Lamé-Navier PDE Formulation

The Lamé-Navier equations gives the steady-state motion equation in terms of the displacement field [13]. This means we are solving a system of three PDEs with three unknowns. We rewrite (4) and (6) in terms of u using relations (2), (3), and the identity $tr\left(\nabla u + (\nabla u)^T\right) = 2(\nabla \cdot u)$ to get

$$(\lambda + \mu)\nabla(\nabla \cdot u) + \mu\nabla^2 u + f = 0, \quad u \in \Omega \tag{7}$$

$$u = g, \quad u \in \partial\Omega_D \tag{8}$$

$$\left[\lambda(\nabla \cdot u)I + \mu(\nabla u + (\nabla u)^T)\right] \cdot n = h, \quad u \in \partial\Omega_T \tag{9}$$

When we later discretize the system, it is more convenient to work with the operators and the displacement in component form. The two operators in the PDE (7) applied to u expand to

$$\nabla(\nabla \cdot u) = \begin{pmatrix} \nabla_{xx} & \nabla_{xy} & \nabla_{xz} \\ \nabla_{xy} & \nabla_{yy} & \nabla_{yz} \\ \nabla_{xz} & \nabla_{yz} & \nabla_{zz} \end{pmatrix} \begin{pmatrix} u_1 \\ u_2 \\ u_3 \end{pmatrix}, \quad \nabla^2 u = \begin{pmatrix} \mathcal{L} & 0 & 0 \\ 0 & \mathcal{L} & 0 \\ 0 & 0 & \mathcal{L} \end{pmatrix} \begin{pmatrix} u_1 \\ u_2 \\ u_3 \end{pmatrix},$$

where $\mathcal{L} = \nabla_{xx} + \nabla_{yy} + \nabla_{zz}$. Rewriting the two terms in the traction condition (9) in the same way yields

$$(\nabla \cdot u)I \cdot n = \begin{pmatrix} n_1\nabla_x & n_1\nabla_y & n_1\nabla_z \\ n_2\nabla_x & n_2\nabla_y & n_2\nabla_z \\ n_3\nabla_x & n_3\nabla_y & n_3\nabla_z \end{pmatrix} \begin{pmatrix} u_1 \\ u_2 \\ u_3 \end{pmatrix},$$

$$(\nabla u + (\nabla u)^T) \cdot n = \begin{pmatrix} \mathcal{T}_{1x} & n_2\nabla_x & n_3\nabla_x \\ n_1\nabla_y & \mathcal{T}_{2y} & n_3\nabla_y \\ n_1\nabla_z & n_2\nabla_z & \mathcal{T}_{3z} \end{pmatrix} \begin{pmatrix} u_1 \\ u_2 \\ u_3 \end{pmatrix},$$

where $\mathcal{T}_{iq} = n_i\nabla_q + n_1\nabla_x + n_2\nabla_y + n_3\nabla_z$.

3 The RBF-FD Numerical Method

In the RBF-FD method [4], scattered node stencil approximations are used for representing the differential operators in the PDE and the boundary conditions. Let $X_1, \ldots, X_N$ be a global set of node points, and let $u_{ij} \approx u_i(X_j)$. We collect the unknown displacement values in the vectors $U_i = (u_{i1}, \ldots, u_{iN})^T$. When we want to approximate the result of a differential operator $\mathcal{D}$ applied to u_i, we first find a local neighbourhood $X_1^{(j)}, \ldots, X_n^{(j)}$ with local unknowns $u_{ik}^{(j)}$ to the point X_j, where we want to evaluate the result. The stencil approximation then takes the form

$$\mathcal{D}u_i(X_j) \approx \sum_{k=1}^{n} w_k u_{ik}^{(j)}. \tag{10}$$

The weights are computed for each point in the global node set by solving a linear system of size $n \times n$, where the stencil size $n \ll N$. In this work, we consider stencil approximations where RBFs augmented by a polynomial basis are used. The small linear systems then take the form

$$\begin{pmatrix} A & P \\ P^T & 0 \end{pmatrix} \begin{pmatrix} w \\ \gamma \end{pmatrix} = \begin{pmatrix} b \\ c \end{pmatrix}, \tag{11}$$

where $A(i, k) = \phi(\|X_k^{(j)} - X_i^{(j)}\|)$, where $\phi(r)$ is an RBF, for $i, k = 1, \ldots, N$, and where $P(i, k) = p_k(X_i^{(j)})$, for $i = 1, \ldots, N$ and $k = 1, \ldots, m$. The polynomials p_k are chosen as the lowest degree monomial basis with dimension m, and m is usually chosen such that a full basis for a certain maximum degree K is obtained. The right hand side vectors are defined by $b(i) = \mathcal{D}\phi(X_j - X_i^{(j)})$ for $i = 1, \ldots, N$, and $c(i) = \mathcal{D}p_i(X_j)$ for $i = 1, \ldots, m$. The vector γ can be seen as a Lagrange multiplier in this problem and is discarded. The stencil approximation is exact for polynomials up to degree K as can be seen from the last block row in the system, and it is also exact for the RBFs centered at the stencil nodes.

A global differentiation matrix D is assembled by inserting the weights corresponding to X_j in the jth row of the matrix, and in the columns corresponding to the global indices of the nodes $X_k^{(j)}$ in the local neighbourhood (X_j is normally one of the points in the neighbourhood). Then we can compute

$$(\mathcal{D}u_i(X_1), \ldots, \mathcal{D}u_i(X_N))^T \approx DU_i. \tag{12}$$

When solving the PDE problem (7)–(9), u is replaced with the discrete field variables, and the differential operators are replaced with the corresponding differentiation matrices. The PDE operator is applied for interior node points, and the boundary operators at boundary node points.

In recent work on RBF-FD methods it has been found that a combination of polyharmonic spline RBFs $\phi(r) = |r|^{2k+1}$, $k > 0$ with polynomials up to degree K has excellent approximation properties [1, 3]. The (asymptotic) convergence rate is guided by the polynomial degree K, and oscillations near boundaries, which are common both with pure RBF and pure polynomial approximations, are suppressed as soon as K is large enough. In this work, we use the cubic polyharmonic spline

$$\phi(r) = |r|^3. \tag{13}$$

4 The Medical Image Input Data

The medical research questions are the motivation for the INVIVE project, and it is important that the numerical simulations can emulate what is seen in the medical image data. To start with, we use medical images to extract the real diaphragm geometry. We also use image data to find the displacement of the diaphragm at different times during the respiratory cycle. Later in the project medical image data will also be used for validation of the numerical simulations.

4.1 Medical Image Acquisition

The type of medical image data that is available to us is thoracic 3-D CT images acquired using a TOSHIBA Aquilion ONE CT scan machine. The images were captured at Azienda Ospedaliera di Padova from adult patients that were subjected to the CT scan for medical reasons (the CT scans were not performed only for research). The images were made and are used in anonymous form. The computed 3-D images are associated with two specific times in the breathing cycle or, equivalently, with two different states of lung inflation. The images have a pixel size of $0.927 \times 0.927 \, \text{mm}^2$ and a slice thickness of $0.3 \, \text{mm}$. They have a resolution of $512 \times 512 \times 1500$ that includes the thoracic and abdominal regions. Examples of image views are shown in Fig. 1.

4.2 Converting Image Data to Mesh-Based Geometry Data

Automated segmentation methods are currently not able to identify the diaphragm that is barely visible in the images. Therefore, the diaphragm was manually segmented on a Wacom tablet using a method similar to the description in [14]. The segmentation time is roughly 6 h for one 3-D image. The manual segmentation

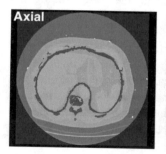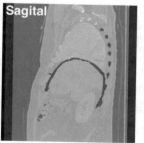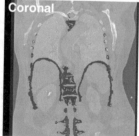

Fig. 1 Manual segmentation of the diaphragm. Red: diaphragm, yellow: lungs, blue: bones

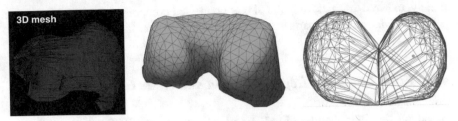

Fig. 2 Left: The initial 3D-mesh and the decimated mesh with 1000 vertices. Right: The sagittal cut and centers of gravity (green)

method consists in following the organs that are known to surround the diaphragm such as the bottom of the lungs, the top of the liver, and the inside of the ribs. Figure 1 shows the result of the segmentation.

The labelized voxel data is then converted into a mesh with the marching cube algorithm. It contains around $1.5 \cdot 10^6$ vertices due to the CT scan resolution. The initial mesh is then decimated using Vorpaline [9], a fast and automatic method, where the only input is the number of final points comprising the mesh. The initial mesh and a decimated mesh are shown in Fig. 2.

Both when implementing the boundary conditions and for node generation, it is necessary to be able to identify vertices belonging to different parts of the surface of the geometry. Two relevant sections are the upper thoracic surface and the lower abdominal surface. These correspond to two different pressure regions.

To separate the surface components, we employ the following algorithm: First the whole diaphragm is separated into a left and right part. If we orient the diaphragm such that the parameter $t \in [t_{\min}, t_{\max}]$ describes a position from left to right, and we let $V(\cdot)$ denote the volume of a convex region, we let $C(\cdot)$ denote the convex hull of a node set, and we let $\Omega(t_1, t_2)$ be the part of the diaphragm that falls within that range of t. Then we can find the sagittal cut t_{sep} as the position that maximizes the sum of the left and right volume

$$t_{\text{sep}} = \underset{t_{\min} \le t \le t_{\max}}{\arg\max} \; V\left(C\left(\{X_j | X_j \in \Omega(t_{\min}, t)\}\right)\right) + V\left(C\left(\{X_j | X_j \in \Omega(t, t_{\max})\}\right)\right).$$

The result is illustrated in the right panel of Fig. 2, where also the two centres of gravity c_L and c_R, for the left and right part respectively, are indicated.

For each surface vertex $X_j \in \Omega_i$, for $i = L, R$, of the diaphragm, a vertex location tag is given by the dot product between the diaphragm vertex normal n_j and the normalized vector $v_j = (X_j - c_i)/\|X_j - c_i\|$ in the direction from the center of gravity to the vertex.

$$\text{tag}(X_j) = \begin{cases} \text{thorax}, & \text{if } n_j \cdot v_j \ge 0 \\ \text{abdomen}, & \text{otherwise} \end{cases} \tag{14}$$

Finally, to avoid artifacts, only the bigger connected component of tagged locations is kept and disconnected parts the are changed to the other location.

4.3 Final Geometry Representation and Node Generation

Based on the OGr method [11, 12] and a least-squares RBF-partition of unity method [7], the mesh-based geometry is smoothed and parametrized. The details of this process are described in a forthcoming paper [8].

Scattered nodes sets of different resolutions are generated from the smoothed geometry. A level set function inside the volume is used for anisotropic node placement such that the resolution in the direction normal to the surface is higher then along the surface.

4.4 A Test Problem with Real Displacements

We are still working with the analyses of the images shown in the previous section. Therefore, we use an older data set with a bit lower resolutions for the test case and the numerical experiments. We only have the end of inhalation state segmented at this point. To define a realistic displacement function, we have identified nine different landmarks on the diaphragm. There are four insertion points of the diaphragm that we take as immobile. These are the left and right transverse processes of the two lowest thoracic vertebra T11 and T12. The five moving landmarks and their displacements are given in Table 1. We augment this information by also requiring the extremal points of the lower edge of the diaphragm to be immobile, and the thickness change from contracted to relaxed state at the two domes to be 66%. We then interpolate the displacements at the augmented landmarks by the $|r|^7$ polyharmonic spline. The initial and displaced states are shown in Fig. 3. As the first test problem, we solve for the interior displacement given that the boundary displacement changes from the relaxed to the contracted state.

Table 1 Displacements of five landmark points

	Right dome	Left dome	Right costophrenic recess	Left costophrenic recess	Xiphoid process
u_1	−1.08	1.08	−1.08	−2.16	0
u_2	4.32	4.32	0	−2.16	−1.08
u_3	−2.50	−7.50	−2.50	−10.00	0

Fig. 3 Initial node locations (higher, red) and displaced node locations (lower, blue), using the constructed displacement function for a node set with $N = 8404$ nodes

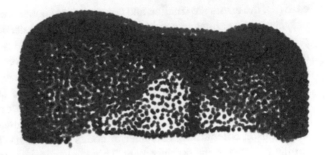

5 Numerical Experiments

A main concern when solving the linear elasticity problem for the diaphragm is the high aspect ratio of the geometry. The overall size of the diaphragm is around $30 \times 20 \times 15$ cm, while the thickness is just a few mm. In the experiments we want to test how important the resolution in the normal direction is for the results. Our hypothesis is that it needs to be large enough to allow for a stencil with a similar number of nodes in each dimension. That is, we need at least $\sqrt[3]{n}$ nodes in the normal direction. We compare two cases, (i) using uniform node sets with similar distances in the normal and tangential directions, and (ii) using node sets that are refined in the normal direction according to the stencil size. Convergence is tested against a reference solution computed at a higher resolution.

The left part of Fig. 4 shows the convergence of the displacements. The errors are larger for case (i), and no convergence trend is observed for the largest stencil size. The number of points in the normal direction increases gradually as N increases. For

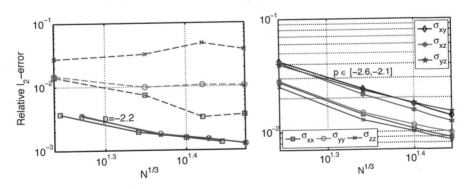

Fig. 4 Left: Convergence of the displacement against the reference solution for uniform nodes (dashed) and nodes refined in the normal direction (solid) for $n = 50$, $K = 3$ (square), $n = 78$, $K = 4$ (circle), and $n = 120$, $K = 5$ (x), where n is the stencil size and K is the order of the polynomial basis augmenting the polyharmonic spline functions. Right: Convergence of the stresses against the reference solution for $n = 50$. The slopes p, with $-p$ indicating the order of convergence are also shown

case (ii), the errors are smaller, and convergence is observed in all cases. When using polyharmonic splines in combination with polynomials, we expect the convergence rate to be of order h^{K+1}, where h is a measure of the node spacing and K is the maximum degree of the polynomial terms [3]. However, for case (ii), we get the same rate of convergence for all K. One reason can be that the normal refinement is constant when the tangential refinement is increased. There may also be issues concerning the smoothness of the node distribution and/or the solution.

In the right part of Fig. 4, we display the convergence of the functions in the stress tensor, computed for the interior nodes for case (ii). The convergence rates are similar to those of the displacement. This is also unexpected, as we would normally expect a derivative of order ℓ to converge as $h^{K+1-\ell}$ [3].

In Fig. 5, we show the components of the stress tensor, computed for the interior nodes for case (ii). We can see that the magnitude of the stresses is large at the domes where we enforce compression of the muscle.

6 Conclusions

We have developed a pipeline for converting CT image data into input data for numerical simulation. The main bottleneck is the manual segmentation of the diaphragm. One thing that will be investigated in future work is if a mapping from a reference geometry can be used to simplify this step.

When the thin dimension is resolved with enough node points, the RBF-FD approximations converge as the number of nodes increase. Also the stresses can be computed with similar accuracy. This shows that it is possible to use this type of discretization, but further work is needed on how to generate smooth non-uniform node sets, and also on the implementation of more advanced test problems.

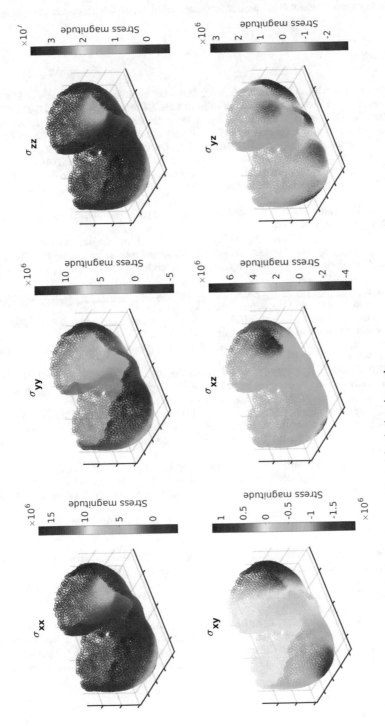

Fig. 5 The six components of the stress tensor evaluated for the interior nodes

Acknowledgements The INVIVE project is funded by the Swedish Research Council, grant no. 2016-04849.

References

1. Bayona, V., Flyer, N., Fornberg, B., Barnett, G.A.: On the role of polynomials in RBF-FD approximations: II. Numerical solution of elliptic PDEs. J. Comput. Phys. **332**, 257–273 (2017)
2. Cacciani, N., Ogilvie, H., Larsson, L.: Age related differences in diaphragm muscle fiber response to mid/long term controlled mechanical ventilation. Exp. Gerontol. **59**, 28–33 (2014)
3. Flyer, N., Fornberg, B., Bayona, V., Barnett, G.A.: On the role of polynomials in RBF-FD approximations: I. Interpolation and accuracy. J. Comput. Phys. **321**, 21–38 (2016)
4. Fornberg, B., Flyer, N.: A primer on radial basis functions with applications to the geosciences. In: CBMS-NSF Regional Conference Series in Applied Mathematics, vol. 87. SIAM, Philadelphia (2015)
5. INVIVE: The Individal Virtual Ventilator Project. http://www.it.uu.se/research/scientific_computing/project/rbf/biomech
6. Ladjal, H., Azencot, J., Beuve, M., Giraud, P., Moreau, J.M., Shariat, B.: Biomechanical modeling of the respiratory system: human diaphragm and thorax. In: Doyle, B., Miller, K., Wittek, A., Nielsen, P.M. (eds.) Computational Biomechanics for Medicine, pp. 101–115. Springer, Cham (2015)
7. Larsson, E., Shcherbakov, V., Heryudono, A.: A least squares radial basis function partition of unity method for solving PDEs. SIAM J. Sci. Comp. **39**(6), A2538–A2563 (2017)
8. Larsson, E., Tominec, I., Villard, P.F., Cacciani, N.: A Isogeometric Radial Basis Function Partition of Unity Method for PDEs in Thin Volumes (2020)
9. Lévy, B., Liu, Y.: L_p centroidal Voronoi tessellation and its applications. ACM Trans. Graph. **29**(4), 119:1–119:11 (2010)
10. Pato, M.P., Santos, N.J., Areias, P., Pires, E.B., de Carvalho, M., Pinto, S., Lopes, D.S.: Finite element studies of the mechanical behaviour of the diaphragm in normal and pathological cases. Comput. Methods Biomech. Biomed. Engin. **14**(6), 505–513 (2011)
11. Piret, C.: The orthogonal gradients method: a radial basis functions method for solving partial differential equations on arbitrary surfaces. J. Comput. Phys. **231**(14), 4662–4675 (2012)
12. Piret, C., Dunn, J.: Fast RBF OGr for solving PDEs on arbitrary surfaces. AIP Conf. Proc. **1776**(1), 070005 (2016)
13. Slaughter, W.S.: The Linearized Theory of Elasticity. Birkhäuser, Basel (2002)
14. Villard, P.F., Boshier, P., Bello, F., Gould, D.: Virtual reality simulation of liver biopsy with a respiratory component. In: Takahashi, H. (ed.) Liver Biopsy, pp. 315–334. IntechOpen, Rijeka (2011)

An Explicit Hybridizable Discontinuous Galerkin Method for the 3D Time-Domain Maxwell Equations

Georges Nehmetallah, Stéphane Lanteri, Stéphane Descombes, and Alexandra Christophe

1 Motivations and Objectives

The DGTD method is nowadays a very popular numerical method in the computational electromagnetics community. A lot of works are mostly concerned with time explicit DGTD methods relying on the use of a single global time step computed so as to ensure stability of the simulation. It is however well known that when combined with an explicit time integration method and in the presence of an unstructured locally refine mesh, a high order DGTD method suffers from a severe time step size restriction. An alternative approach that has been considered in [5, 7, 16] is to use a hybrid explicit-implicit (or locally implicit) time integration strategy. Such a strategy relies on a component splitting deduced from a partitioning of the mesh cells in two sets respectively gathering coarse and fine elements. The computational efficiency of this locally implicit DGTD method depends on the size of the set of fine elements that directly influences the size of the sparse part of the matrix system to be solved at each time. Therefore, an approach for reducing the size of the subsystem of globally coupled (i.e. implicit) unknowns is worth considering if one wants to solve very large-scale problems.

A particularly appealing solution in this context is given by the concept of hybridizable discontinuous Galerkin (HDG) method. The HDG method has been first introduced by Cockbrun et al. in [4] for a model elliptic problem and has been subsequently developed for a variety of PDE systems in continuum mechanics [13]. The essential ingredients of a HDG method are a local Galerkin projection of the underlying system of PDEs at the element level onto spaces of polynomials

G. Nehmetallah (✉) · S. Lanteri · S. Descombes · A. Christophe
Université Côte d'Azur, INRIA, CNRS, LJAD, Sophia-Antipolis Cedex, France
e-mail: georges.nehmetallah@inria.fr; stephane.lanteri@inria.fr;
stephane.descombes@univ-cotedazur.fr; alexandra.christophe@inria.fr

© The Author(s) 2020
S. J. Sherwin et al. (eds.), *Spectral and High Order Methods for Partial Differential Equations ICOSAHOM 2018*, Lecture Notes in Computational Science and Engineering 134, https://doi.org/10.1007/978-3-030-39647-3_41

to parameterize the numerical solution in terms of the numerical trace; a judicious choice of the numerical flux to provide stability and consistency; and a global jump condition that enforces the continuity of the numerical flux to arrive at a global weak formulation in terms of the numerical trace. The HDG methods are fully implicit, high-order accurate and most importantly, they reduce the globally coupled unknowns to the approximate trace of the solution on element boundaries, thereby leading to a significant reduction in the degrees of freedom. HDG methods for the system of time-harmonic Maxwell equations have been proposed in [9, 10, 14]. We have only developed the implicit HDG method for the time-domain Maxwell equations [3]. In view of devising a hybrid explicit-implicit HDG method, a preliminary step is therefore to elaborate on the principles of a fully explicit HDG formulation. It happens that fully explicit HDG methods have been studied recently for the acoustic wave equation by Kronbichler et al. [8] and Stanglmeier et al. [15]. In [15] the authors present a fully explicit, high order accurate in both space and time HDG method. In this paper we outline the formulation of this explicit HDGTD, present numerical results including a preliminary assessment of its superconvergence properties. We adopt a low storage Runge-Kutta scheme [2] for the time integration of the semi-discrete HDG equations. This work is a first step towards the construction of a hybrid explicit-implicit HDG method for time-domain electromagnetics.

2 Problem Statement and Notations

We consider the system of 3D time-domain Maxwell equations on a bounded polyhedral domain $\Omega \subset \mathbb{R}^3$

$$\begin{cases} \varepsilon \partial_t \mathbf{E} - \mathbf{curl}\mathbf{H} = -\mathbf{J}, \text{ in } \Omega \times [0, T], \\ \mu \partial_t \mathbf{H} + \mathbf{curl}\mathbf{E} = 0, \text{ in } \Omega \times [0, T], \end{cases} \quad (1)$$

where the symbol ∂_t denotes a time derivative, $\mathbf{J}$ the current density, T a final time, $\mathbf{E}(\mathbf{x}, t)$ and $\mathbf{H}(\mathbf{x}, t)$ are the electric and magnetic fields. The dielectric permittivity ε and the magnetic permeability μ are varying in space, time-invariant and both positive functions. The boundary of Ω is defined as $\partial \Omega = \Gamma_m \cup \Gamma_a$ with $\Gamma_m \cap \Gamma_a = \emptyset$. The boundary conditions are chosen as

$$\begin{cases} \mathbf{n} \times \mathbf{E} = 0, \text{ on } \Gamma_m \times [0, T], \\ \mathbf{n} \times \mathbf{E} + \mathbf{n} \times (\mathbf{n} \times \mathbf{H}) = \mathbf{n} \times \mathbf{E}^{\text{inc}} + \mathbf{n} \times (\mathbf{n} \times \mathbf{H}^{\text{inc}}) \\ \qquad\qquad = \mathbf{g}^{\text{inc}}, \text{ on } \Gamma_a \times [0, T]. \end{cases} \quad (2)$$

Here $\mathbf{n}$ denotes the unit outward normal to $\partial\Omega$ and $(\mathbf{E}^{\text{inc}}, \mathbf{H}^{\text{inc}})$ a given incident field. The first boundary condition is often referred as a metallic boundary condition and is applied on a perfectly conducting surface. The second relation is an absorbing boundary condition and takes here the form of the Silver-Müller condition. It is applied on a surface corresponding to an artificial truncation of a theoretically unbounded propagation domain. Finally, the system is supplemented with initial conditions: $\mathbf{E}_0(\mathbf{x}) = \mathbf{E}(\mathbf{x}, 0)$ and $\mathbf{H}_0(\mathbf{x}) = \mathbf{H}(\mathbf{x}, 0)$. For sake of simplicity, we omit the volume source term $\mathbf{J}$ in what follows.

We introduce now the notations and approximation spaces. We first consider a partition $\mathscr{T}_h$ of $\Omega \subset \mathbb{R}^3$ into a set of tetrahedron. Each non-empty intersection of two elements K^+ and K^- is called an interface. We denote by $\mathscr{F}_h^I$ the union of all interior interfaces of $\mathscr{T}_h$, by $\mathscr{F}_h^B$ the union of all boundary interfaces of $\mathscr{T}_h$, and $\mathscr{F}_h = \mathscr{F}_h^I \cup \mathscr{F}_h^B$. Note that $\partial \mathscr{T}_h$ represents all the interfaces ∂K for all $K \in \mathscr{T}_h$. As a result, an interior interface shared by two elements appears twice in $\partial \mathscr{T}_h$, unlike in $\mathscr{F}_h$ where this interface is evaluated once. For an interface $F \in \mathscr{F}_h^I, F = \overline{K}^+ \cap \overline{K}^-$, let $\mathbf{v}^\pm$ be the traces of $\mathbf{v}$ on F from the interior of $K^\pm$. On this interior face, we define mean values as $\{\mathbf{v}\}_F = (\mathbf{v}^+ + \mathbf{v}^-)/2$ and jumps as $[\![\mathbf{v}]\!]_F = \mathbf{n}^+ \times \mathbf{v}^+ + \mathbf{n}^- \times \mathbf{v}^-$ where the unit outward normal vector to K is denoted by $\mathbf{n}^\pm$. For the boundary faces these expressions are modified as $\{\mathbf{v}\}_F = \mathbf{v}^+$ and $[\![\mathbf{v}]\!]_F = \mathbf{n}^+ \times \mathbf{v}^+$ since we assume $\mathbf{v}$ is single-valued on the boundaries. In the following, we introduce the discontinuous finite element spaces and some basic operations on these spaces for later use. Let $\mathbb{P}_{p_K}(K)$ denotes the space of polynomial functions of degree at most p_K on the element $K \in \mathscr{T}_h$. The discontinuous finite element space is introduced as

$$\mathbf{V}_h = \left\{ \mathbf{v} \in \left[L^2(\Omega) \right]^3 \text{ such that } \mathbf{v}|_K \in \left[\mathbb{P}_{p_K}(K) \right]^3, \quad \forall K \in \mathscr{T}_h \right\}, \tag{3}$$

where $L^2(\Omega)$ is the space of square integrable functions on the domain Ω. The functions in $\mathbf{V}_h$ are continuous inside each element and discontinuous across the interfaces between elements. In addition, we introduce a traced finite element space

$$\mathbf{M}_h = \left\{ \boldsymbol{\eta} \in \left[L^2(\mathscr{F}_h) \right]^3 \text{ such that } \boldsymbol{\eta}|_F \in \left[\mathbb{P}_{p_F}(F) \right]^3 \right.$$

$$\left. \text{and } (\boldsymbol{\eta} \cdot \mathbf{n})|_F = 0, \quad \forall F \in \mathscr{F}_h \right\}. \tag{4}$$

For two vectorial functions $\mathbf{u}$ and $\mathbf{v}$ in $\left[L^2(D) \right]^3$, we denote $(\mathbf{u}, \mathbf{v})_D = \int_D \mathbf{u} \cdot \mathbf{v} \, d\mathbf{x}$ provided D is a domain in $\mathbb{R}^3$, and we denote $< \mathbf{u}, \mathbf{v} >_F = \int_F \mathbf{u} \cdot \mathbf{v} \, ds$ if F is a two-dimensional face. Accordingly, for the mesh $\mathscr{T}_h$ we have

$$(\cdot, \cdot)_{\mathscr{T}_h} = \sum_{K \in \mathscr{T}_h} (\cdot, \cdot)_K, \qquad \langle \cdot, \cdot \rangle_{\partial \mathscr{T}_h} = \sum_{K \in \mathscr{T}_h} \langle \cdot, \cdot \rangle_{\partial K},$$

$$\langle \cdot, \cdot \rangle_{\mathscr{F}_h} = \sum_{F \in \mathscr{F}_h} \langle \cdot, \cdot \rangle_F, \qquad \langle \cdot, \cdot \rangle_{\Gamma_a} = \sum_{F \in \mathscr{F}_h \cap \Gamma_a} \langle \cdot, \cdot \rangle_F.$$

We set $\mathbf{v}^t = -\mathbf{n} \times (\mathbf{n} \times \mathbf{v})$, $\mathbf{v}^n = \mathbf{n}(\mathbf{n} \cdot \mathbf{v})$ where $\mathbf{v}^t$ and $\mathbf{v}^n$ are the tangential and normal components of $\mathbf{v}$ such as $\mathbf{v} = \mathbf{v}^t + \mathbf{v}^n$.

3 Principles and Formulation of the HDG Method

Following the classical DG approach, approximate solutions $(\mathbf{E}_h, \mathbf{H}_h)$, for all $t \in [0, T]$, are seeked in the space $\mathbf{V}_h \times \mathbf{V}_h$ satisfying for all K in $\mathscr{T}_h$

$$\begin{cases} \left(\varepsilon \partial_t \mathbf{E}_h, \mathbf{v}\right)_K - \left(\mathbf{curl}\mathbf{H}_h, \mathbf{v}\right)_K = 0, \ \forall \mathbf{v} \in \mathbf{V}_h, \\ \left(\mu \partial_t \mathbf{H}_h, \mathbf{v}\right)_K + \left(\mathbf{curl}\mathbf{E}_h, \mathbf{v}\right)_K = 0, \ \forall \mathbf{v} \in \mathbf{V}_h. \end{cases} \tag{5}$$

Applying Green's formula, on both equations of (5) introduces boundary terms which are replaced by numerical traces $\hat{\mathbf{E}}_h$ and $\hat{\mathbf{H}}_h$ in order to ensure the connection between element-wise solutions and global consistency of the discretization. This leads to the global formulation for all $t \in [0, T]$

$$\begin{cases} \left(\varepsilon \partial_t \mathbf{E}_h, \mathbf{v}\right)_K - \left(\mathbf{H}_h, \mathbf{curl}\mathbf{v}\right)_K + \left\langle \hat{\mathbf{H}}_h, \mathbf{n} \times \mathbf{v} \right\rangle_{\partial K} = 0, \ \forall \mathbf{v} \in \mathbf{V}_h, \\ \left(\mu \partial_t \mathbf{H}_h, \mathbf{v}\right)_K + \left(\mathbf{E}_h, \mathbf{curl}\mathbf{v}\right)_K - \left\langle \hat{\mathbf{E}}_h, \mathbf{n} \times \mathbf{v} \right\rangle_{\partial K} = 0, \ \forall \mathbf{v} \in \mathbf{V}_h. \end{cases} \tag{6}$$

It is straightforward to verify that $\mathbf{n} \times \mathbf{v} = \mathbf{n} \times \mathbf{v}^t$ and $< \mathbf{H}, \mathbf{n} \times \mathbf{v} > = - < \mathbf{n} \times \mathbf{H}, \mathbf{v} >$. Therefore, using numerical traces defined in terms of the tangential components $\hat{\mathbf{H}}_h^t$ and $\hat{\mathbf{E}}_h^t$, we can rewrite (6) as

$$\begin{cases} \left(\varepsilon \partial_t \mathbf{E}_h, \mathbf{v}\right)_K - \left(\mathbf{H}_h, \mathbf{curl}\mathbf{v}\right)_K + \left\langle \hat{\mathbf{H}}_h^t, \mathbf{n} \times \mathbf{v} \right\rangle_{\partial K} = 0, \ \forall \mathbf{v} \in \mathbf{V}_h, \\ \left(\mu \partial_t \mathbf{H}_h, \mathbf{v}\right)_K + \left(\mathbf{E}_h, \mathbf{curl}\mathbf{v}\right)_K - \left\langle \hat{\mathbf{E}}_h^t, \mathbf{n} \times \mathbf{v} \right\rangle_{\partial K} = 0, \ \forall \mathbf{v} \in \mathbf{V}_h. \end{cases} \tag{7}$$

The hybrid variable Λ_h introduced in the setting of a HDG method [4] is here defined for all the interfaces of $\mathscr{F}_h$ as

$$\Lambda_h := \hat{\mathbf{H}}_h^t, \quad \forall F \in \mathscr{F}_h. \tag{8}$$

We want to determine the fields $\hat{\mathbf{H}}_h^t$ and $\hat{\mathbf{E}}_h^t$ in each element K of $\mathscr{T}_h$ by solving system (7) and assuming that Λ_h is known on all the faces of an element K. We consider a numerical trace $\hat{\mathbf{E}}_h^t$ for all K given by

$$\hat{\mathbf{E}}_h^t = \mathbf{E}_h^t + \tau_K \mathbf{n} \times (\Lambda_h - \mathbf{H}_h^t) \text{ on } \partial K, \tag{9}$$

where τ_K is a local stabilization parameter which is assumed to be strictly positive. We recall that $\mathbf{n} \times \mathbf{H}_h^t = \mathbf{n} \times \mathbf{H}_h$. The definitions of the hybrid variable (8) and numerical trace (9) are exactly those adopted in the context of the formulation of HDG methods for the 3D time-harmonic Maxwell equations [10–12, 14].

Following the HDG approach, when the hybrid variable Λ_h is known for all the faces of the element K, the electromagnetic field can be determined by solving the local system (7) using (8) and (9).

From now on we will note by g^{inc} the L^2 projection of g^{inc} on $\mathbf{M}_h$. Summing the contributions of (7) over all the elements and enforcing the continuity of the tangential component of $\hat{\mathbf{E}}_h$, we can formulate a problem which is to find $(\mathbf{E}_h, \mathbf{H}_h, \Lambda_h) \in \mathbf{V}_h \times \mathbf{V}_h \times \mathbf{M}_h$ such that for all $t \in [0, T]$

$$
\begin{aligned}
(\varepsilon \partial_t \mathbf{E}_h, \mathbf{v})_{\mathscr{T}_h} - (\mathbf{H}_h, \mathbf{curl}\,\mathbf{v})_{\mathscr{T}_h} + \langle \Lambda_h, \mathbf{n} \times \mathbf{v} \rangle_{\partial \mathscr{T}_h} &= 0, \ \forall \mathbf{v} \in \mathbf{V}_h, \\
(\mu \partial_t \mathbf{H}_h, \mathbf{v})_{\mathscr{T}_h} + (\mathbf{E}_h, \mathbf{curl}\,\mathbf{v})_{\mathscr{T}_h} - \langle \hat{\mathbf{E}}_h^t, \mathbf{n} \times \mathbf{v} \rangle_{\partial \mathscr{T}_h} &= 0, \ \forall \mathbf{v} \in \mathbf{V}_h, \\
\left\langle [\![\hat{\mathbf{E}}_h]\!], \eta \right\rangle_{\mathscr{F}_h} - \langle \Lambda_h, \eta \rangle_{\Gamma_a} - \left\langle \mathbf{g}^{inc}, \eta \right\rangle_{\Gamma_a} &= 0, \ \forall \eta \in \mathbf{M}_h,
\end{aligned}
\tag{10}
$$

where the last equation is called the conservativity condition with which we ask the tangential component of $\hat{\mathbf{E}}_h$ to be weakly continuous across any interface between two neighboring elements.

We now reformulate the system with numerical fluxes. We can deduce from the third equation of (10) that

$$
\Lambda_h = \begin{cases}
\dfrac{1}{\tau_{K+} + \tau_{K-}} \left(2 \{ \tau_K \mathbf{H}_h^t \}_F + [\![\mathbf{E}_h^t]\!]_F \right), & \text{if } F \in \mathscr{F}_h^I, \\[2ex]
\dfrac{1}{\tau_K} \mathbf{n} \times \mathbf{E}_h^t + \mathbf{H}_h^t, & \text{if } F \in \mathscr{F}_h \cap \Gamma_m, \\[2ex]
\dfrac{1}{\tau_K + 1} \left(\tau_K \mathbf{H}_h^t + \mathbf{n} \times \mathbf{E}_h^t - \mathbf{g}^{inc} \right). & \text{if } F \in \mathscr{F}_h \cap \Gamma_a.
\end{cases}
\tag{11}
$$

By replacing (11) in (9) we obtain $\hat{\mathbf{E}}_h^t = \hat{\mathbf{E}}_h^{t,+} = \hat{\mathbf{E}}_h^{t,-}$ with

$$
\hat{\mathbf{E}}_h^t = \begin{cases}
\dfrac{\tau_K + \tau_{K-}}{\tau_{K+} + \tau_{K-}} \left(2 \left\{ \dfrac{1}{\tau_K} \mathbf{E}_h^t \right\}_F - [\![\mathbf{H}_h^t]\!]_F \right), & \text{if } F \in \mathscr{F}_h^I, \\[2ex]
0, & \text{if } F \in \mathscr{F}_h \cap \Gamma_m, \\[2ex]
\dfrac{1}{\tau_K + 1} \left(\mathbf{E}_h^t - \tau_K \mathbf{n} \times \mathbf{H}_h^t - \tau_K \mathbf{n} \times \mathbf{g}^{inc} \right). & \text{if } F \in \mathscr{F}_h \cap \Gamma_a.
\end{cases}
\tag{12}
$$

Thus, the numerical traces (8) and (9) have been reformulated from the conservativity condition. This means that the conservativity condition is now included in the new formulation of the numerical fluxes and can be neglected in the global system of equations. Hence, the local system (6) takes the form of a classical DG formulation, $\forall \mathbf{v} \in \mathbf{V}_h$

$$
\begin{cases}
\left(\varepsilon \partial_t \mathbf{E}_h, \mathbf{v}\right)_K - \left(\mathbf{H}_h, \mathbf{curlv}\right)_K + \left\langle \hat{\mathbf{H}}_h^t, \mathbf{n} \times \mathbf{v} \right\rangle_{\partial K} = 0, \\
\left(\mu \partial_t \mathbf{H}_h, \mathbf{v}\right)_K + \left(\mathbf{E}_h, \mathbf{curlv}\right)_K - \left\langle \hat{\mathbf{E}}_h^t, \mathbf{n} \times \mathbf{v} \right\rangle_{\partial K} = 0.
\end{cases} \tag{13}
$$

where the numerical fluxes are defined by (11) and (12).

Remark 3 Let $Y_K = \sqrt{\varepsilon_K}/\sqrt{\mu_K}$ be the local admittance associated to cell K and $Z_K = 1/Y_K$ the corresponding local impedance. If we set $\tau_K = Z_K$ in (11) and $1/\tau_K = Y_K$ in (12), the obtained numerical traces coincide with those adopted in the classical upwind flux DGTD method [6].

4 Numerical Results

In order to validate and study the numerical convergence of the proposed HDG method, we consider the propagation of an eigenmode in a closed cavity (Ω is the unit square) with perfectly metallic walls. The frequency of the wave is $f = \sqrt{3}/\sqrt{2}c_0$ where c_0 is the speed of light in vacuum. The electric permittivity and the magnetic permeability are set to the constant vacuum values. The exact time-domaine solution is given in [6].

We start our study by assuming that the penalization parameter τ is equal to 1. In order to insure the stability of the method, numerical CFL conditions are determined for each value of the interpolation order p_K. In our particular case we have ε_K and μ_k are constant $= 1 \, \forall K \in \mathcal{T}_h$, so we have verified that, as we said in Remark 3, for $\tau = 1$, the values of CFL number correspond to the classical upwind flux-based DG method. In Table 1 we summarize the maximum Δt obtained numerically to insure the stability of the scheme

Given these values of Δt max, the L^2-norm of the error is calculated for a uniform tetrahedral mesh with 3072 elements which is constructed from a finite difference grid with $n_x = n_y = n_z = 9$ points, each cell of this grid yielding 6 tetrahedrons. The wave is propagated in the cavity during a physical time t_{max} corresponding to 8 periods (as shown in Fig. 1). Figure 2 depicts a comparison of

Table 1 Numerically obtained values of Δt max

Interpolation order	$\mathbb{P}_1$	$\mathbb{P}_2$	$\mathbb{P}_3$	$\mathbb{P}_4$
Δt max (s.)	0.32×10^{-9}	0.19×10^{-9}	0.13×10^{-9}	0.94×10^{-10}

Fig. 1 Time evolution of the exact and the numerical solution of E_x at point $A(0.25, 0.25, 0.25)$ with a $\mathbb{P}_3$ interpolation

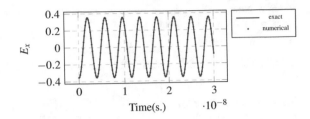

Fig. 2 Time evolution of the L^2-norm of the error for P4

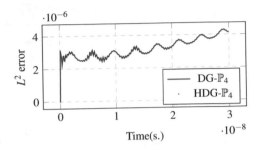

Fig. 3 Numerical convergence order of the time explicit HDG method for $\tau = 1$

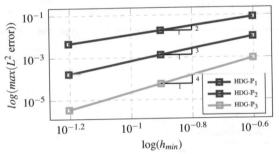

the time evolution of the L^2-norm of the error between the solution obtained with an HDG method and a classical upwind flux-based DG method for $p_K = 4$. An optimal convergence with order $p_K + 1$ is obtained as shown in Fig. 3.

Now, we keep the same case than previously and we assess the behavior of the HDG method for various values of the penalization parameter τ. We observe that the time evolution of the electromagnetic energy for any order of interpolation, for different values of the parameter $\tau \neq 1$ and when the Δt used is fixed to the values defined in Table 1, the energy increases in time. In fact, It is necessary to decrease the Δt max for each value of τ to assure the stability (see Table 2 and Fig. 4). For this example, the optimal cost will be for the parameter $\tau = 1$ (having the same cost as an upwind flux for a DG method) otherwise we will spend more time to finish our simulation. On Fig. 5, we show the time evolution of the L^2-error for several values of τ with respect to the maximal time step for the considered parameters. In addition, Table 3 sums up numerical results in term of maximum L^2 errors and convergence rates. It appears that the order of convergence is not affected when the stabilization parameter is varied from 1 (with their associated CFL conditions).

Table 2 Numerically obtained values of the CFL number as a function of the stabilization parameter τ for a $\mathbb{P}1$ interpolation

τ	0.1	1.0	2.0	5.0	10.0
Δt max (s.)	0.31×10^{-10}	3.2×10^{-10}	1.7×10^{-10}	0.66×10^{-10}	0.32×10^{-10}

Fig. 4 Variation of the Δt max as a function of τ

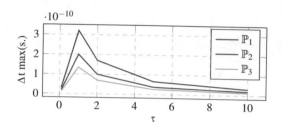

Fig. 5 Time evolution of the L^2-error as a function of τ with a $\mathbb{P}_3$ interpolation

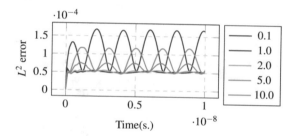

Table 3 Maximum L2-errors and convergence orders

1/h	$\tau = 1.0$					
	$\mathbb{P}_1, \Delta t = 0.16 \times 10^{-09}$		$\mathbb{P}_2, \Delta t = 0.99 \times 10^{-10}$		$\mathbb{P}_3, \Delta t = 0.66 \times 10^{-10}$	
1/4	8.29e−02	–	9.87e−03	–	9.34e−04	–
1/8	1.90e−02	2.13	1.34e−03	2.88	5.68e−05	4.04
1/16	4.74e−03	2.00	1.72e−04	2.97	3.46e−06	4.04
1/h	$\tau = 0.1$					
	$\mathbb{P}_1, \Delta t = 0.16 \times 10^{-10}$		$\mathbb{P}_2, \Delta t = 0.96 \times 10^{-11}$		$\mathbb{P}_3, \Delta t = 0.66 \times 10^{-11}$	
1/4	2.14e−01	–	1.78e−02	–	2.19e−03	–
1/8	5.46e−02	1.97	2.85e−03	2.65	1.68e−04	3.70
1/16	1.18e−02	2.21	4.06e−04	2.81	1.14e−05	3.88
1/h	$\tau = 10.0$					
	$\mathbb{P}_1, \Delta t = 0.16 \times 10^{-10}$		$\mathbb{P}_2, \Delta t = 0.96 \times 10^{-11}$		$\mathbb{P}_3, \Delta t = 0.68 \times 10^{-11}$	
1/4	1.74e−01	–	1.53e−02	–	1.68e−03	–
1/8	4.24e−02	2.04	2.23e−03	2.76	1.17e−04	3.84
1/16	9.4e−03	2.16	3.10e−04	2.87	7.81e−06	3.91

5 Local Postprocessing

We define here, following the ideas of the local postprocessing developed in [1], new approximations for electric and magnetic field and expect that both $\mathbf{E}_h^{n*}$ and $\mathbf{H}_h^{n*}$ converge with order $k + 1$ in the $H^{curl}(\mathscr{T}_h)$-norm, whereas $\mathbf{E}_h^n$ and $\mathbf{H}_h^n$ converge with order k in the $H^{curl}(\mathscr{T}_h)$-norm. To postprocess E_h^{n*} we first compute an approximation $(\mathbf{p}_{1,h}^n, \mathbf{p}_{2,h}^n) \in \mathbf{V}(K) \times \mathbf{V}(K)$ to the curl of $\mathbf{E}$, $\mathbf{p}_1(t^n) = \nabla \times \mathbf{E}(t^n)$ and the curl of $\mathbf{H}$, $\mathbf{p}_2(t^n) = \nabla \times \mathbf{H}(t^n)$ by locally solving the below system

$$(\mathbf{p}_{1,h}^n, \mathbf{v})_K = (\mathbf{E}_h^n, \nabla \times \mathbf{v})_K - \langle \hat{\mathbf{E}}_h^{t,n}, \mathbf{n} \times \mathbf{v} \rangle_{\partial K} \quad \forall \mathbf{v} \in \mathbf{V}(K)$$

and,

$$(\mathbf{p}_{2,h}^n, \mathbf{v})_K = (\mathbf{H}_h^n, \nabla \times \mathbf{v})_K - \langle \hat{\mathbf{H}}_h^{t,n}, \mathbf{n} \times \mathbf{v} \rangle_{\partial K} \quad \forall \mathbf{v} \in \mathbf{V}(K)$$

We then find $(\mathbf{E}_h^{n*}, \mathbf{H}_h^{n*}) \in [\mathscr{P}_{k+1}(K)]^3 \times [\mathscr{P}_{k+1}(K)]^3$ such that

$$\begin{cases} (\nabla \times \mathbf{E}_h^{n*}, \nabla \times \mathbf{W})_K = (\mathbf{p}_{h,1}^n, \nabla \times \mathbf{W})_K, & \forall \mathbf{W} \in [\mathscr{P}_{k+1}(K)]^3, \\ (\mathbf{E}_h^{n*}, \nabla Y)_K = (\mathbf{E}_h^n, \nabla Y)_K & \forall Y \in \mathscr{P}_{k+2}(K) \end{cases}$$

and,

$$\begin{cases} (\nabla \times \mathbf{H}_h^{n*}, \nabla \times \mathbf{W})_K = (\mathbf{p}_{h,2}^n, \nabla \times \mathbf{W})_K, & \forall \mathbf{W} \in [\mathscr{P}_{k+1}(K)]^3, \\ (\mathbf{H}_h^{n*}, \nabla Y)_K = (\mathbf{H}_h^n, \nabla Y)_K & \forall Y \in \mathscr{P}_{k+2}(K) \end{cases}$$

It is important to point out that we can compute $\mathbf{E}_h^{n*}$ and $\mathbf{H}_h^{n*}$ at any time step without advancing in time. Hence, the local postprocessing can be performed whenever we need higher accuracy at particular time steps. Numerical results given in Table 4 shows that a second order convergence rate is obtained for the post-processed solution.

6 Conclusion

In this paper we have presented an explicit HDG method to solve the system of Maxwell equations in 3D. The next step is to couple explicit and implicit HDG methods to treat the case of a locally refined mesh.

Table 4 Errors and orders of convergence before and after postprocessing

P_k	$1/h$	$\tau = 1.0$			
		$\|E - E_h\|_{H_{curl}}$		$\|E - E_h^*\|_{H_{curl}}$	
		Error	Order	Error	Order
P_1	1/4	9.30e−01	–	6.83e−01	–
	1/6	5.84e−01	1.14	3.10e−01	1.95
	1/8	4.34e−01	1.03	1.67e−01	2.15
P_2	1/4	1.67e−01	–	4.28e−02	–
	1/6	7.46e−02	1.98	1.19e−02	3.16
	1/8	4.29e−02	1.92	4.90e−03	3.06
P_3	1/4	2.30e−02	–	5.00e−03	–
	1/6	7.10e−03	2.90	1.10e−03	3.79
	1/8	3.00e−03	2.99	3.58e−04	3.84

References

1. Abgrall, R., Shu, C.-W.: Handbook of Numerical Methods for Hyperbolic Problems, vol. 17, pp. 190–194. Elsevier/North-Holland, Amsterdam (2016)
2. Carpenter, M.H., Kennedy, C.A.: Fourth-Order 2N-Storage Runge-Kutta Schemes. NASA, Washington (1994)
3. Christophe, A., Descombes, S., Lanteri, S.: An implicit hybridized discontinuous Galerkin method for the 3D time-domain Maxwell equations. Appl. Math. Comput. **319**, 395–408 (2018)
4. Cockburn, B., Gopalakrishnan, J., Lazarov, R.: Unified hybridization of discontinuous Galerkin, mixed, and continuous Galerkin methods for second order elliptic problems. SIAM J. Numer. Anal. **47**, 1319–1365 (2009)
5. Descombes, S., Lanteri, S. Moya, L.: Locally implicit discontinuous Galerkin time domain method for electromagnetic wave propagation in dispersive media applied to numerical dosimetry in biological tissues. SIAM J. Sci. Comput. **38**, A2611–A2633 (2016)
6. Hesthaven, J.S., Warburton, T.: Nodal high-order methods on unstructured grids. I. Time-domain solution of Maxwell's equations. Int. J. Numer. Methods Eng. **181**, 186–221 (2002)
7. Hochbruck, M. Sturm, A.: Error analysis of a second-order locally implicit method for linear Maxwell's equations. SIAM J. Numer. Anal. **54**, 3167–3191 (2016)
8. Kronbichler, M., Schoeder, S., Müller, C., Wall, W.A.: Comparison of implicit and explicit hybridizable discontinuous Galerkin methods for the acoustic wave equation. Int. J. Numer. Methods Eng. **270**, 330–342 (2014)
9. Li, L., Lanteri, S. Perrussel, R.: Numerical investigation of a high order hybridizable discontinuous Galerkin method for 2D time-harmonic Maxwell's equations. COMPEL **2**, 1112–1138 (2013)
10. Li, L., Lanteri, S., Perrussel, R.: A hybridizable discontinuous Galerkin method combined to a Schwarz algorithm for the solution of 3D time-harmonic Maxwell's equations. J. Comput. Phys. **256**, 563–581 (2014)
11. Moya, L.: Temporal convergence of a locally implicit discontinuous Galerkin method for Maxwell's equations. ESAIM Math. Model. Numer. Anal. (M2AN) **46**, 1225–1246 (2012)
12. Moya, L., Descombes, S. Lanteri, S.: Locally implicit time integration strategies in a discontinuous Galerkin method for Maxwell's equations. J. Sci. Comp. **56**, 190–218 (2013)
13. Nguyen, N.C., Peraire, J.: Hybridizable discontinuous Galerkin methods for partial differential equations in continuum mechanics. J. Comput. Phys. **231**, 5955–5988 (2012)
14. Nguyen, N.C., Peraire, J., Cockburn, B.: Hybridizable discontinuous Galerkin methods for the time-harmonic Maxwell's equations. J. Comput. Phys. **231**, 7151–7175 (2011)

15. Stanglmeier, M., Nguyen, N.C., Peraire, J., Cockburn, B.: An explicit hybridizable discontinuous Galerkin method for the acoustic wave equation. Comput. Methods Appl. Mech. Eng. **300**, 748–769 (2016)
16. Verwer, J.G.: Component splitting for semi-discrete Maxwell equations. BIT Numer. Math. **51**, 427–445 (2011)

Entropy Conserving and Kinetic Energy Preserving Numerical Methods for the Euler Equations Using Summation-by-Parts Operators

Hendrik Ranocha

1 Introduction

Considering the solution of hyperbolic conservation laws, high order methods can be very efficient, providing accurate numerical solutions with relatively low computational effort [21]. In order to make use of this accuracy, stability has to be established. Mimicking estimates obtained on the continuous level via integration-by-parts, summation-by-parts (SBP) operators [22, 37] can be used. In short, SBP operators are discrete derivative operators equipped with a compatible quadrature providing a discrete analogue of the L^2 norm. The compatibility of discrete integration and differentiation mimics integration-by-parts on a discrete level. Combined with the weak enforcement of boundary conditions via simultaneous approximation terms (SATs) [1], highly efficient and stable semidiscretisations can be obtained at least for linear problems, see e.g. [6, 14, 39] and references cited therein.

In recent years, there has been an enduring and increasing interest in the basic ideas of SBP operators and their application in various frameworks including finite volume (FV) [25, 26], discontinuous Galerkin (DG) [2, 4, 10, 11, 13, 20, 27, 28, 30], and the recent flux reconstruction/correction procedure via reconstruction framework [15, 16, 42] as described in [31, 32]. While there is only a limited amount of well-posedness theory for nonlinear conservation laws, mimicking properties such as entropy stability semidiscretely has received much interest. Building on the seminal work of Tadmor [40, 41], entropy stability of second order schemes using symmetric numerical fluxes has been investigated, resulting in well-defined properties that numerical fluxes have to satisfy in order to result in entropy conservative

H. Ranocha (✉)
TU Braunschweig, Institute Computational Mathematics, Braunschweig, Germany
e-mail: h.ranocha@tu-bs.de

© The Author(s) 2020
S. J. Sherwin et al. (eds.), *Spectral and High Order Methods for Partial Differential Equations ICOSAHOM 2018*, Lecture Notes in Computational Science and Engineering 134, https://doi.org/10.1007/978-3-030-39647-3_42

525

schemes. Decomposing general semidiscretisations into a non-dissipative central part and an additional dissipative part, suitable artificial dissipation or filtering can be added afterwards, cf. [7, 9, 38]. Second order methods based on symmetric numerical fluxes can be extended to high order in a conservative way, cf. [4, 7, 28] and [8, 23, 34–36].

Another property of numerical methods for the Euler equations that has received much interest in the literature concerns the kinetic energy. A structural property of numerical fluxes described by Jameson [18] has been used to construct so-called kinetic energy preserving (KEP) numerical fluxes inter alia by Chandrashekar [3]. However, schemes using these fluxes do not preserve the kinetic energy as expected in numerical experiments by Gassner et al. [12]. They had to change the discretisation of the pressure to reduce undesired changes of the kinetic energy. However, this resulted in a loss of entropy conservation. Motivated by these results, some analytical insights into this behaviour have been developed in [29, Section 7.4] and will be presented here.

This chapter is structured as follows. At first, some basic results about SBP operators and corresponding semidiscretisations of hyperbolic conservation laws are reviewed in Sect. 2. Afterwards, the Euler equations are considered in Sect. 3. After demonstrating that the property that has been used to characterise numerical fluxes as KEP is not well-defined, the new concept of KEP numerical methods is introduced. Moreover, a numerical flux that is both entropy conservative and kinetic energy preserving in the new sense is developed. Thereafter, results of a numerical experiment comparing entropy conservative numerical fluxes are described in Sect. 4. Finally, a brief summary is given in Sect. 5.

2 Discretisations Using Summation-by-Parts Operators

Consider the Euler equations in two space dimensions

$$
\partial_t \underbrace{\begin{pmatrix} \rho \\ \rho v_x \\ \rho v_y \\ \rho e \end{pmatrix}}_{=u} + \partial_x \underbrace{\begin{pmatrix} \rho v_x \\ \rho v_x^2 + p \\ \rho v_x v_y \\ (\rho e + p) v_x \end{pmatrix}}_{=f^x(u)} + \partial_y \underbrace{\begin{pmatrix} \rho v_y \\ \rho v_x v_y \\ \rho v_y^2 + p \\ (\rho e + p) v_y \end{pmatrix}}_{=f^y(u)} = 0, \tag{1}
$$

where ρ is the density, v the velocity, e the specific total energy, and p the pressure. For a perfect gas, $p = (\gamma - 1)\left(\rho e - \frac{1}{2}\rho v^2\right)$. The usual entropy is $U = -\frac{\rho s}{\gamma - 1}$, where $s = \log p - \gamma \log \rho$ is the specific (physical) entropy.

With the entropy fluxes F^j fulfilling $\partial_u U \cdot \partial_u f^j = \partial_u F^j$, smooth solutions of the Euler equations in d space dimensions satisfy $\partial_t U(u) + \sum_{j=1}^{d} \partial_j F^j(u) = 0$ and the entropy inequality

$$\partial_t U(u) + \sum_{j=1}^{d} \partial_j F^j(u) \le 0 \tag{2}$$

is used as additional admissibility criterion for weak solutions, cf. [5].

In order to discretise (1), the domain Ω is divided into several non-overlapping sub-domains $\Omega_l \subseteq \Omega$ and SBP operators will be used on each element. SBP operators consist of discrete derivative operators D_j, approximating the partial derivative in direction j, and a symmetric and positive definite mass/norm matrix M, approximating the $L^2(\Omega_l)$ scalar product via $u^T M v = \langle u, v, \rangle_M \approx \langle u, v, \rangle_{L^2(\Omega_l)} = \int_{\Omega_l} u\, v$. Moreover, an interpolation operator R approximates the restriction of functions on Ω_l to the boundary $\partial \Omega_l$ and a symmetric and positive definite boundary mass matrix B approximate the $L^2(\partial \Omega_l)$ scalar product. Representing the multiplication by the j-th component of the outer unit normal v at $\partial \Omega_l$ by the diagonal matrix n_j, the SBP property

$$M D_j + D_j^T M = R^T B n_j R \tag{3}$$

has to be satisfied in order to mimic integration-by-parts discretely via

$$\underbrace{u^T M D_j v + u^T D_j^T M v}_{\wr\wr} = \underbrace{u^T R^T B N_j R v}_{\wr\wr}, \tag{4}$$
$$\underbrace{\int_{\Omega_l} u\,(\partial_j v) + \int_{\Omega_l} (\partial_j u)\, v} = \underbrace{\int_{\partial \Omega_l} u\, v\, n_j}.$$

Semidiscretisation of (1) will be constructed as follows. Each sub-domain $\Omega_l \subseteq \Omega$ is mapped onto a reference element and all computations are performed there. On each element, the resulting semidiscretisation is of the form

$$\partial_t u + \text{VOL} + \text{SURF} = 0, \tag{5}$$

where the volume terms VOL discretise the flux divergence in the interior of Ω_l and the surface terms SURF couple elements or impose boundary conditions. Here, u is the vector of the nodal values of the numerical solution at specified nodes ξ_i in Ω_l and a collocation approach is used. Thus, nonlinear operations are performed pointwise and the discrete fluxes f^j are given by their nodal values $f_i^j = f^j(u_i) = f^j(u(\xi_i))$. As in (nodal) discontinuous Galerkin methods, the surface terms will be built using numerical fluxes $f^{\text{num},j}$ in the j-th coordinate direction as

$$\text{SURF} = \sum_{j=1}^{d} M^{-1} R^T B n_j \left(f^{\text{num},j} - R f^j \right). \tag{6}$$

Finally, the volume terms are constructed using symmetric (two-point) numerical fluxes $f^{\text{vol},j}$ (volume fluxes) that are consistent with f^j as

$$\text{VOL}_i = \sum_{j=1}^{d} \sum_{k} 2(D_j)_{i,k} f^{\text{vol},j}(u_i, u_k), \tag{7}$$

where VOL_i is the volume term at ξ_i [7]. If $f^{\text{vol},j}$ are smooth fluxes, the discretisation (7) is of the same order of accuracy as the derivative matrices D_j [4, 28]. Moreover, if the mass matrix M is diagonal, this approximation can be written in a conservative form [7]. Finally, if the boundary operators $R^T B n_j R$ are also diagonal and $f^{\text{vol},j}$ are entropy conservative in the sense of Tadmor [40, 41], the semidiscretisation (5) is entropy conservative/stable across elements if the numerical surface fluxes $f^{\text{num},j}$ are entropy conservative/stable. Moreover, some results on the kinetic energy can be transferred as well [12]. In the following, the focus will lie on the fluxes $f^{\text{vol},j}$.

3   Euler Equations and Kinetic Energy

The kinetic energy $E_{\text{kin}} = \frac{1}{2}\rho v^2$ fulfils (for sufficiently smooth solutions)

$$\partial_t E_{\text{kin}} + \text{div}\left(\frac{1}{2}\rho v^2 v\right) + v \cdot \text{grad } p = 0. \tag{8}$$

Jameson [18] investigated the kinetic energy in a one-dimensional semidiscrete setting using finite volume methods. To simplify the notation, this setup will be used in the following; its extension to multiple dimensions is straightforward. Jameson proposed to mimic (8) semidiscretely by using numerical momentum fluxes of the form $f_{\rho v}^{\text{num}} = f_{\rho v}^{\text{num}}(u_-, u_+) = \{\!\{v\}\!\} f_\rho^{\text{num}} + p^{\text{num}}$, where $\{\!\{v\}\!\}$ is the arithmetic mean of v_- and v_+, f_ρ^{num} is the numerical density flux, and p^{num} is a consistent numerical approximation of the pressure. Later, this has been used as a kind of "definition" of kinetic energy preserving (KEP) numerical fluxes, e.g. in [3, 12]. However, this is not a well-defined concept, cf. [28, 29]. Indeed, every numerical momentum flux can be written as

$$f_{\rho v}^{\text{num}} = \{\!\{v\}\!\} f_\rho^{\text{num}} + \underbrace{\left(f_{\rho v}^{\text{num}} - \{\!\{v\}\!\} f_\rho^{\text{num}} \right)}_{=:\ p^{\text{num}}?}. \tag{9}$$

Since the numerical fluxes are consistent, $p^{\text{num}} := f_{\rho v}^{\text{num}} - \{\!\{v\}\!\} f_{\rho}^{\text{num}}$ is a consistent approximation of the pressure. The insufficiency of the condition $f_{\rho v}^{\text{num}} = \{\!\{v\}\!\} f_{\rho}^{\text{num}} + p^{\text{num}}$ is in accordance with observations of Gassner et al. [12]. They investigated a Taylor-Green vortex problem and compared several numerical fluxes for the Euler equations. There, numerical fluxes of the form $f_{\rho v}^{\text{num}} = \{\!\{v\}\!\} f_{\rho}^{\text{num}} + p^{\text{num}}$ with $p^{\text{num}} \neq \{\!\{p\}\!\}$ resulted in a clear loss of kinetic energy compared to other KEP fluxes using the arithmetic average $p^{\text{num}} = \{\!\{p\}\!\}$ as approximation of the pressure. They observed that "the discretisation of the pressure plays a crucial role for the kinetic energy" and that the choice of the arithmetic average $p^{\text{num}} = \{\!\{p\}\!\}$ "seems to be important for the kinetic energy equation" [12, Section 4.2]. However, they had no (theoretical) explanations for this observation.

3.1 New Approach to Kinetic Energy Preservation

By a heuristic argument, the balance law (8) may not be suitable in the incompressible limit: Indeed, for smooth solutions, (8) can be rewritten as

$$\partial_t E_{\text{kin}} + \text{div}\left(\frac{1}{2}\rho v^2 v + pv\right) - p \, \text{div} \, v = 0, \qquad (10)$$

which becomes a conservation law for smooth solutions of the incompressible Euler equations due to $\text{div}(v) = 0$ or an energy inequality similar to the entropy inequality (2). Since the kinetic energy is plays a crucial role in the incompressible limit [24], the second form (10) might be considered the "better" one. Thus, a semidiscretisation mimicking this equation might be desirable near the incompressible limit.

Definition 1 A numerical flux $f^{\text{num}} = (f_{\rho}^{\text{num}}, f_{\rho v}^{\text{num}}, f_{\rho e}^{\text{num}})$ for the Euler equations is called *kinetic energy preserving* (KEP), if the momentum flux can be written as $f_{\rho v}^{\text{num}} = \{\!\{v\}\!\} f_{\rho}^{\text{num}} + \{\!\{p\}\!\}$.

Definition 1 results in a well-defined concept of KEP numerical fluxes.

Theorem 1 (Corollary 7.5 of [29]) *If a kinetic energy preserving numerical flux is used in a semidiscrete FV method, the resulting semidiscrete kinetic energy equation mimics both the conservative and the non-conservative terms of Eq. (10).*

Proof (Sketch) Using the chain rule in a one dimensional finite volume setting, the time derivative of the kinetic energy in cell i becomes

$$\partial_t \left(\frac{1}{2}\rho v^2\right)_i = -\frac{1}{\Delta x_i}\left(\left(\frac{1}{2}\rho v^2 v + pv\right)^{\text{num}}(u_i, u_{i+1}) - \left(\frac{1}{2}\rho v^2 v + pv\right)^{\text{num}}(u_{i-1}, u_i)\right)$$

$$+ p_i \frac{\{\!\{v\}\!\}_{i,i+1} - \{\!\{v\}\!\}_{i-1,i}}{\Delta x_i},$$

where $\left(\frac{1}{2}\rho v^2 v + pv\right)^{\text{num}}(u_i, u_j) = v_i v_j f_{\rho}^{\text{num}}(u_i, u_j) + \frac{p_i v_j + p_j v_i}{2}$. □

Using the momentum flux $f_{\rho v}^{num} = \{\{v\}\} f_\rho^{num} + \{\{p\}\}$ in the volume terms (7) in one dimension, the arithmetic average of the pressure yields the volume term Dp, i.e. a straightforward discretisation of $\partial_x p$. Analogous results hold in multiple space dimensions, cf. Sect. 2.

The kinetic energy preserving DG methods presented in [11, 27] use volume terms corresponding to the numerical fluxes $f_\rho^{num} = \{\{\rho v\}\}$, $f_{\rho v}^{num} = \{\{\rho v\}\}\{\{v\}\} + \{\{p\}\}$, which are kinetic energy preserving in the sense of Definition 1.

3.2 Entropy Conservative and KEP Numerical Fluxes

Since entropy stability has received much interest and the entropy conservative numerical fluxes of [3, 17] are not KEP in the sense of Definition 1, it is interesting whether both concepts can be fulfilled simultaneously. The logarithmic mean value $\{\{\rho\}\}_{\log} = [\![\rho]\!]/[\![\log \rho]\!]$ has been proposed by Roe [33] in the context of entropy conservative numerical fluxes and is described in [17]. Many useful entropy conservative numerical density fluxes are of the form $f_\rho^{num} = \{\{\rho\}\}_{\log}\{\{v\}\}$, e.g. the one presented in [3]. This form seems to be preferable, since positivity preservation of the density can be achieved using local Lax-Friedrichs/Rusanov dissipation operators [28, Section 6.2]. Using this ansatz for f_ρ^{num} and Definition 1, the following entropy conservative and kinetic energy preserving numerical flux ($f^{num,y}$ analogously) has been constructed in [29, Section 7.4]

$$f_\rho^{num,x} = \{\{\rho\}\}_{\log}\{\{v_x\}\}, \quad f_{\rho v_x}^{num,x} = \{\{v_x\}\} f_\rho^{num,x} + \{\{p\}\}, \quad f_{\rho v_y}^{num,x} = \{\{v_y\}\} f_\rho^{num,x},$$

(11)

$$f_{\rho e}^{num,x} = \left(\{\{\rho\}\}_{\log} \left(\{\{v_x\}\}^2 + \{\{v_y\}\}^2 - \frac{\{\{v_x^2 + v_y^2\}\}}{2} \right) + \frac{1}{\gamma - 1} \frac{\{\{\rho\}\}_{\log}}{\{\{\rho/p\}\}_{\log}} + \{\{p\}\} \right) \{\{v_x\}\}$$
$$- \frac{[\![p]\!][\![v]\!]}{4}.$$

4 Numerical Results

Since the kinetic energy is an important quantity for the incompressible Euler equations, a Taylor-Green vortex given by

$$\rho(t, x, y) = 1, \qquad\qquad v_x(t, x, y) = \sin(x)\cos(y),$$

$$v_y(t, x, y) = -\cos(x)\sin(y), \quad p(t, x, y) = \frac{100}{\gamma} + \frac{\cos(2x) + \cos(2y)}{4},$$

(12)

for $(x, y) \in [0, 2\pi]^2$ with periodic boundary conditions is considered, which is a stationary solution of the incompressible Euler equations. Using tensor product Lobatto bases for polynomials of degree $p = 5$ on $N = 16$ elements per coordinate direction, the numerical solutions have been computed in the time interval $t \in [0, 30]$ with the fourth order, ten-stage, strong stability preserving Runge-Kutta method of [19]. The time step Δt has been chosen as $\Delta t = $ cfl min $\{\Delta x / (2p + 1)\lambda\}$, where λ is the greatest absolute value of the eigenvalues of f' and the minimum is taken over all cells and nodes. As in [12], the given numerical fluxes have been used for both the volume terms (7) and as surface fluxes in (6), without additional dissipation.

The evolution of the entropy U and the kinetic energy E_{kin} using a CFL number cfl = 0.9 for the entropy conservative fluxes of Ismail and Roe [17], Chandrashekar [3], and the new flux (11) are visualised in Fig. 1. As can be seen there, the entropy remains approximately constant and the kinetic energy oscillates uniformly until $t \approx 20$. Afterwards, the kinetic energy drops for the fluxes of [3, 17] and there is a relative change of the entropy of order 10^{-5}. Contrary, there is no visible change for the new flux (11).

The entropy loss for the fluxes of Ismail and Roe [17] and Chandrashekar [3] is caused by the time integration scheme, as can be seen in Fig. 2, where the time step is reduced by an order of magnitude (cfl = 0.09). However, the behaviour of the kinetic energy is nearly unchanged.

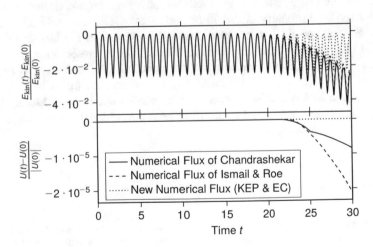

Fig. 1 Total entropy and kinetic energy of numerical solutions using different entropy conservative numerical fluxes with cfl = 0.9

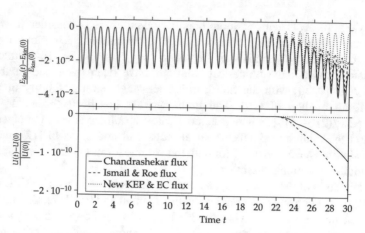

Fig. 2 Total entropy and kinetic energy of numerical solutions using different entropy conservative numerical fluxes with cfl = 0.09

5   Summary and Discussion

Using summation-by-parts operators, high order numerical schemes with specific properties can be constructed using symmetric (two-point) numerical fluxes. While several "kinetic energy preserving" methods have been proposed, they have been characterised by a property of the numerical fluxes that is not well-defined. Such numerical fluxes resulted in schemes that did not preserve the kinetic energy as expected [12]. Here, a new approach to kinetic energy preservation inspired by the incompressible Euler equations and developed in [29, Section 7.4] has been described. This results in a well-defined property numerical fluxes have to satisfy in order mimic the balance law for the kinetic energy more reliably. Moreover, new entropy conservative numerical fluxes have been developed that are kinetic energy preserving in the new sense.

References

1. Carpenter, M.H., Gottlieb, D., Abarbanel, S.: Time-stable boundary conditions for finite-difference schemes solving hyperbolic systems: methodology and application to high-order compact schemes. J. Comput. Phys. **111**(2), 220–236 (1994). https://doi.org/10.1006/jcph.1994.1057
2. Chan, J.: On discretely entropy conservative and entropy stable discontinuous Galerkin methods. J. Comput. Phys. **362**, 346–374 (2018). https://doi.org/10.1016/j.jcp.2018.02.033
3. Chandrashekar, P.: Kinetic energy preserving and entropy stable finite volume schemes for compressible Euler and Navier-Stokes equations. Commun. Comput. Phys. **14**(5), 1252–1286 (2013). https://doi.org/10.4208/cicp.170712.010313a

4. Chen, T., Shu, C.W.: Entropy stable high order discontinuous Galerkin methods with suitable quadrature rules for hyperbolic conservation laws. J. Comput. Phys. **345**, 427–461 (2017). https://doi.org/10.1016/j.jcp.2017.05.025
5. Dafermos, C.M.: Hyperbolic Conservation Laws in Continuum Physics. Springer, Heidelberg (2010). https://doi.org/10.1007/978-3-642-04048-1
6. Fernández, D.C.D.R., Hicken, J.E., Zingg, D.W.: Review of summation-by-parts operators with simultaneous approximation terms for the numerical solution of partial differential equations. Comput. Fluids **95**, 171–196 (2014). https://doi.org/10.1016/j.compfluid.2014.02.016
7. Fisher, T.C., Carpenter, M.H.: High-order entropy stable finite difference schemes for nonlinear conservation laws: finite domains. J. Comput. Phys. **252**, 518–557 (2013). https://doi.org/10.1016/j.jcp.2013.06.014
8. Fisher, T.C., Carpenter, M.H., Nordström, J., Yamaleev, N.K., Swanson, C.: Discretely conservative finite-difference formulations for nonlinear conservation laws in split form: theory and boundary conditions. J. Comput. Phys. **234**, 353–375 (2013). https://doi.org/10.1016/j.jcp.2012.09.026
9. Fjordholm, U.S., Mishra, S., Tadmor, E.: Arbitrarily high-order accurate entropy stable essentially nonoscillatory schemes for systems of conservation laws. SIAM J. Numer. Anal. **50**(2), 544–573 (2012). https://doi.org/10.1137/110836961
10. Gassner, G.J.: A skew-symmetric discontinuous Galerkin spectral element discretization and its relation to SBP-SAT finite difference methods. SIAM J. Sci. Comput. **35**(3), A1233–A1253 (2013). https://doi.org/10.1137/120890144
11. Gassner, G.J.: A kinetic energy preserving nodal discontinuous Galerkin spectral element method. Int. J. Numer. Methods Fluids **76**(1), 28–50 (2014). https://doi.org/10.1002/fld.3923
12. Gassner, G.J., Winters, A.R., Kopriva, D.A.: Split form nodal discontinuous Galerkin schemes with summation-by-parts property for the compressible Euler equations. J. Comput. Phys. **327**, 39–66 (2016). https://doi.org/10.1016/j.jcp.2016.09.013
13. Gassner, G.J., Winters, A.R., Kopriva, D.A.: A well balanced and entropy conservative discontinuous Galerkin spectral element method for the shallow water equations. Appl. Math. Comput. **272**, 291–308 (2016). https://doi.org/10.1016/j.amc.2015.07.014
14. Gustafsson, B., Kreiss, H.O., Oliger, J.: Time-Dependent Problems and Difference Methods. Wiley, Hoboken (2013)
15. Huynh, H.T.: A flux reconstruction approach to high-order schemes including discontinuous Galerkin methods. In: 18th AIAA Computational Fluid Dynamics Conference. American Institute of Aeronautics and Astronautics (2007). https://doi.org/10.2514/6.2007-4079
16. Huynh, H.T., Wang, Z.J., Vincent, P.E.: High-order methods for computational fluid dynamics: a brief review of compact differential formulations on unstructured grids. Comput. Fluids **98**, 209–220 (2014). https://doi.org/10.1016/j.compfluid.2013.12.007
17. Ismail, F., Roe, P.L.: Affordable, entropy-consistent Euler flux functions II: entropy production at shocks. J. Comput. Phys. **228**(15), 5410–5436 (2009). https://doi.org/10.1016/j.jcp.2009.04.021
18. Jameson, A.: Formulation of kinetic energy preserving conservative schemes for gas dynamics and direct numerical simulation of one-dimensional viscous compressible flow in a shock tube using entropy and kinetic energy preserving schemes. J. Sci. Comput. **34**(2), 188–208 (2008). https://doi.org/10.1007/s10915-007-9172-6
19. Ketcheson, D.I.: Highly efficient strong stability-preserving Runge-Kutta methods with low-storage implementations. SIAM J. Sci. Comput. **30**(4), 2113–2136 (2008). https://doi.org/10.1137/07070485X
20. Kopriva, D.A., Gassner, G.J.: An energy stable discontinuous Galerkin spectral element discretization for variable coefficient advection problems. SIAM J. Sci. Comput. **36**(4), A2076–A2099 (2014). https://doi.org/10.1137/130928650
21. Kreiss, H.O., Oliger, J.: Comparison of accurate methods for the integration of hyperbolic equations. Tellus **24**(3), 199–215 (1972). https://doi.org/10.1111/j.2153-3490.1972.tb01547.x

22. Kreiss, H.O., Scherer, G.: Finite element and finite difference methods for hyperbolic partial differential equations. In: de Boor, C. (ed.) Mathematical Aspects of Finite Elements in Partial Differential Equations, pp. 195–212. Academic, New York (1974)

23. LeFloch, P.G., Mercier, J.M., Rohde, C.: Fully discrete, entropy conservative schemes of arbitrary order. SIAM J. Numer. Anal. **40**(5), 1968–1992 (2002). https://doi.org/10.1137/S003614290240069X

24. Lions, P.L.: Mathematical topics in fluid mechanics. Incompressible Models, vol. 1. Oxford University, Oxford (1996)

25. Nordström, J., Björck, M.: Finite volume approximations and strict stability for hyperbolic problems. Appl. Numer. Math. **38**(3), 237–255 (2001). https://doi.org/10.1016/S0168-9274(01)00027-7

26. Nordström, J., Forsberg, K., Adamsson, C., Eliasson, P.: Finite volume methods, unstructured meshes and strict stability for hyperbolic problems. Appl. Numer. Math. **45**(4), 453–473 (2003). https://doi.org/10.1016/S0168-9274(02)00239-8

27. Ortleb, S.: A kinetic energy preserving DG scheme based on Gauss-Legendre points. J. Sci. Comput. **71**(3), 1135–1168 (2017). https://doi.org/10.1007/s10915-016-0334-2

28. Ranocha, H.: Comparison of some entropy conservative numerical fluxes for the Euler equations. J. Sci. Comput. (2017). https://doi.org/10.1007/s10915-017-0618-1

29. Ranocha, H.: Generalised summation-by-parts operators and entropy stability of numerical methods for hyperbolic balance laws. Ph.D. Thesis, TU Braunschweig (2018)

30. Ranocha, H.: Generalised summation-by-parts operators and variable coefficients. J. Comput. Phys. **362**, 20–48 (2018). https://doi.org/10.1016/j.jcp.2018.02.021

31. Ranocha, H., Öffner, P., Sonar, T.: Summation-by-parts operators for correction procedure via reconstruction. J. Comput. Phys. **311**, 299–328 (2016). https://doi.org/10.1016/j.jcp.2016.02.009

32. Ranocha, H., Öffner, P., Sonar, T.: Summation-by-parts and correction procedure via reconstruction. In: Bittencourt, M.L., Dumont, N.A., Hesthaven, J.S. (eds.) Spectral and High Order Methods for Partial Differential Equations ICOSAHOM 2016. Lecture Notes in Computational Science and Engineering, vol. 119, pp. 627–637. Springer, Cham (2017). https://doi.org/10.1007/978-3-319-65870-4_45

33. Roe, P.L.: Affordable, entropy-consistent Euler flux functions. In: Talk presented at the Eleventh International Conference on Hyperbolic Problems: Theory, Numerics, Applications (2006). http://www2.cscamm.umd.edu/people/faculty/tadmor/references/files/Roe_Affordable_entropy_Hyp2006.pdf

34. Sjögreen, B., Yee, H.C.: On skew-symmetric splitting and entropy conservation schemes for the Euler equations. In: Kreiss, G., Lötstedt, P., Målqvist, A., Neytcheva, M. (eds.) Numerical Mathematics and Advanced Applications 2009: Proceedings of ENUMATH 2009, the 8th European Conference on Numerical Mathematics and Advanced Applications, Uppsala, July 2009, pp. 817–827. Springer, Heidelberg (2010). https://doi.org/10.1007/978-3-642-11795-4_88

35. Sjögreen, B., Yee, H.: High order entropy conservative central schemes for wide ranges of compressible gas dynamics and MHD flows. J. Comput. Phys. **364**, 153–185 (2018). https://doi.org/10.1016/j.jcp.2018.02.003

36. Sjögreen, B., Yee, H.C., Kotov, D.: Skew-symmetric splitting and stability of high order central schemes. In: Journal of Physics: Conference Series, vol. 837, p. 012019. IOP Publishing, Philadelphia (2017). https://doi.org/10.1088/1742-6596/837/1/012019

37. Strand, B.: Summation by parts for finite difference approximations for d/dx. J. Comput. Phys. **110**(1), 47–67 (1994). https://doi.org/10.1006/jcph.1994.1005

38. Svärd, M., Mishra, S.: Shock capturing artificial dissipation for high-order finite difference schemes. J. Sci. Comput. **39**(3), 454–484 (2009). https://doi.org/10.1007/s10915-009-9285-1

39. Svärd, M., Nordström, J.: Review of summation-by-parts schemes for initial-boundary-value problems. J. Comput. Phys. **268**, 17–38 (2014). https://doi.org/10.1016/j.jcp.2014.02.031

40. Tadmor, E.: The numerical viscosity of entropy stable schemes for systems of conservation laws. I. Math. Comput. **49**(179), 91–103 (1987). https://doi.org/10.1090/S0025-5718-1987-0890255-3
41. Tadmor, E.: Entropy stability theory for difference approximations of nonlinear conservation laws and related time-dependent problems. Acta Numer. **12**, 451–512 (2003). https://doi.org/10.1017/S0962492902000156
42. Wang, Z.J., Gao, H.: A unifying lifting collocation penalty formulation including the discontinuous Galerkin, spectral volume/difference methods for conservation laws on mixed grids. J. Comput. Phys. **228**(21), 8161–8186 (2009). https://doi.org/10.1016/j.jcp.2009.07.036

Multiwavelet Troubled-Cell Indication: A Comparison of Utilizing Theory Versus Outlier Detection

Mathea J. Vuik

1 Introduction

Solutions to nonlinear hyperbolic PDEs develop discontinuities in time. The generation of spurious oscillations in such regions can be prevented by applying a limiter in the troubled zones. In [16, 18], two different multiwavelet troubled-cell indicators were introduced, one based on a parameter, the other using outlier detection. We present this comparison in order to begin to understand in which regime these tools are effective. In this paper, we investigate the effectiveness of a different detection scheme, based on the theoretical detection of troubled cells using multiwavelet approaches. It uses the cancelation property [6] and the theory about thresholding [8]. This technique was originally used for a multiwavelet-based adaptive strategy in combination with the DG method. However, we are specifically interested in its application for troubled-cell indication. In the troubled cells, the moment limiter is applied [11]. We demonstrate the performance of this new indicator and show that it works very well when very fine meshes are used (the asymptotic regime). For coarser meshes, it seems that the existing multiwavelet troubled-cell indicators perform better.

The outline of this paper is as follows: in Sect. 2, some background information about the multiwavelet theory is given. The existing multiwavelet troubled-cell indicators, as well as the cancelation property and the derived thresholding technique are described in Sect. 3. Numerical results are shown in Sect. 4, and some concluding remarks are given in Sect. 5.

M. J. Vuik (✉)
VORtech, Delft, The Netherlands
e-mail: thea.vuik@vortech.nl

© The Author(s) 2020
S. J. Sherwin et al. (eds.), *Spectral and High Order Methods for Partial Differential Equations ICOSAHOM 2018*, Lecture Notes in Computational Science and Engineering 134, https://doi.org/10.1007/978-3-030-39647-3_43

2 Multiwavelets and DG

In this section, we consider the multiwavelet theory that is used to design the different troubled-cell indicators. For the sake of brevity, we neglect discussion of the DG scheme [4, 5], that is used in the computations.

The relation between the DG scheme and multiwavelets was shown in [16]. Any global one-dimensional DG approximation of degree k can be written as

$$u_h(x) = 2^{-\frac{n}{2}} \sum_{j=0}^{2^n-1} \sum_{\ell=0}^{k} u_j^{(\ell)} \phi_{\ell j}^n(x),$$

where $\phi_{\ell j}^n$ are the scaling functions related to the orthonormal Legendre polynomials. The corresponding multiwavelet decomposition is

$$u_h(x) = \sum_{\ell=0}^{k} s_{\ell 0}^0 \phi_\ell(x) + \sum_{m=0}^{n-1} \sum_{j=0}^{2^m-1} \sum_{\ell=0}^{k} d_{\ell j}^m \psi_{\ell j}^m(x),$$

where $s_{\ell 0}^0$ are the scaling-function coefficients belonging to u_h, and $d_{\ell j}^m$ are the corresponding multiwavelet coefficients, [2, 16]. The multiwavelets ψ_ℓ have been developed by Alpert [1].

3 Utilizing Multiwavelet Coefficients for Troubled-Cell Indication

In this section, we show different troubled-cell indicators that utilize multiwavelet coefficients. Note that, as the detectors are solely based on the underlying approximation space, the ideas do not need to be modified in order to be applied to other types of model problems than those included in this paper. First, the existing indicators that use either a parameter or the boxplot method are presented. Next, the cancelation property and thresholding technique are used to design a different indication technique.

3.1 Boxplots for Outlier Detection

In [16, 17], we have shown that the coefficients d_{kj}^{n-1} are very useful for troubled-cell indication. With this knowledge, we have designed two different troubled-cell indicators. The first indicator is the so-called *parameter-based* multiwavelet troubled-cell indicator [16]. Here, we detect an element as troubled when

$$|d_{kj}^{n-1}| > C \cdot \max\{|d_{kj}^{n-1}|, j = 0, \ldots, 2^n - 1\}, \ C \in [0, 1]. \tag{1}$$

The value of C is a useful tool to prescribe the strictness of the limiter.

Another option is to use *outlier detection* on the multiwavelet coefficients d_{kj}^{n-1} to detect the troubled cells [18]. Here, Tukey's boxplot method [14] is applied locally to prevent the need for a problem-dependent parameter. The different steps are presented in Algorithm 1.

Algorithm 1 Outlier-detection algorithm using local vectors

Send in a suitable troubled-cell indication vector **D**.
Split this vector into local vectors, **d**.
for all local vectors **do**
    Sort **d** to obtain $\mathbf{d}^s$.
    Compute the quartiles Q_1 and Q_3.
    Detect d_j^s in the smallest 25% of $\mathbf{d}^s$ if $d_j^s < Q_1 - 3(Q_3 - Q_1)$, and d_j^s in the biggest 25% of
    $\mathbf{d}^s$ if $d_j^s > Q_3 + 3(Q_3 - Q_1)$.
end for
Ignore the detected outliers in the left half of the local region when they are not detected with respect to the left-neighboring vector, and similarly test the detected coefficients in the right half of the local region.

Outliers are the coefficients in the vector that are straying far out beyond the others. In order to pick out certain coefficients as outliers, the outer fences are constructed, which were originally defined by Tukey [14]. The outer fences of a vector are $[Q_1 - 3(Q_3 - Q_1), Q_3 + 3(Q_3 - Q_1)]$ (coefficients outside are called *extreme outliers*). The coverage for this whisker length is 99.9998%, such that only 0.0002% of the data in a normally distributed vector is detected as an extreme outlier (asymptotically) [9].

In our computations, we always use local vectors of length 16.

3.2 Cancelation Property

In this section, the *cancelation property* is stated and proved for the one-dimensional case [6]. Here, we assume that the multiwavelets have $M + 1$ vanishing moments. In our case, we have $M = \ell + k$ [1, 15]. If the solution satisfies the continuity requirement $u|_{I_j^m} \in C^{M+1}(I_j^m)$ (where I_j^m is the j-th element in level m), then

$$d_{\ell j}^m \leq \frac{1}{(M + 1)!} \cdot ||u^{(M+1)}||_{L^\infty(I_j^m)} \cdot 2^{(-m+1)(M+3/2)}, \tag{2}$$

$m = 0, \ldots, n, \ j = 0, \ldots, 2^m - 1, \ell = 0, \ldots, k.$

The proof uses a Taylor expansion of u about element center x_j^m: there exists a ξ between x and x_j^m such that

$$u(x) = u(x_j^m) + u'(x_j^m)(x - x_j^m) + \ldots + \frac{u^{(M)}(x_j^m)}{M!}(x - x_j^m)^M + \frac{u^{(M+1)}(\xi)}{(M+1)!}(x - x_j^m)^{M+1}.$$

Using that the first $M + 1$ moments of the multiwavelets vanish, we find

$$d_{\ell j}^m = \langle u, \psi_{\ell j}^m \rangle_{I_j^m} = \left\langle \frac{u^{(M+1)}(\xi)}{(M+1)!}(x - x_j^m)^{M+1}, \psi_{\ell j}^m \right\rangle_{I_j^m}$$

$$\leq \frac{1}{(M+1)!} \|u^{(M+1)}\|_{L^\infty(I_j^m)} \langle (x - x_j^m)^{M+1}, \psi_{\ell j}^m \rangle_{I_j^m}. \tag{3}$$

Next, we use Cauchy-Schwarz's inequality to find

$$\langle (x - x_j^m)^{M+1}, \psi_{\ell j}^m \rangle_{I_j^m} \leq \|(x - x_j^m)^{M+1}\|_{L^2(I_j^m)} \cdot \|\psi_{\ell j}^m\|_{L^2(I_j^m)} = \|(x - x_j^m)^{M+1}\|_{L^2(I_j^m)},$$

because the multiwavelets are orthonormal. Using the notation Δx^m for the element size in level m, we have

$$\|(x - x_j^m)^{M+1}\|_{L^2(I_j^m)} \leq (\Delta x^m)^{M+1} \|1\|_{L^2(I_j^m)} = (\Delta x^m)^{M+1} \sqrt{\Delta x^m} = (\Delta x^m)^{M+3/2}.$$

For the domain $[-1, 1]$, we have $\Delta x^m = 2^{-m+1}$. This means that

$$\|(x - x_j^m)^{M+1}\|_{L^2(I_j^m)} \leq 2^{(-m+1)(M+3/2)},$$

which proves the cancelation property. It should be noticed that this result can be generalized to general grid hierarchies and higher-dimensional problems [6, 10].

The next section contains a discussion of the thresholding technique for one-dimensional multiwavelet expansions.

3.3 Thresholding of the Multiwavelet Coefficients

In this section, the thresholding technique for systems of conservation laws in one dimension is explained, which is based on the cancelation property [8]. This technique is originally used for a multiwavelet-based adaptive strategy in combination with the DG method. However, we are specifically interested in its application for troubled-cell indication.

Following [8], the element I_j^{n-1} is detected as troubled if

$$\max_{\substack{\ell=0,\ldots,k \\ r=1,2,3}} \left(\frac{|d_{\ell j}^{n-1}(r)|}{\max\left\{ \max_{j=0,\ldots,2^n-1} 2^{(n-1)/2}|s_{0j}^n(r)|, 1 \right\}} \right) > \varepsilon_{n-1}\sqrt{2\Delta x}.$$

Here, the value r is related to the conserved quantity in a system of three PDEs. The factor $\sqrt{2\Delta x}$ (with Δx the DG mesh width) occurs because of a scaling difference: the multiwavelets in [8] are scaled with respect to the L^∞-norm, whereas an L^2-norm scaling is used in this paper. The level-dependent threshold value ε_{n-1} is chosen as $\varepsilon_{n-1} = \varepsilon/2$. The parameter ε can be chosen using two different strategies [8]. The first option is to use the *a priori* strategy, which is based on the balance between discretization errors and perturbation errors of adaptive meshes [10]. If the solution contains discontinuities, then the a priori strategy leads to $\varepsilon = C\Delta x^2$. The second option is the *heuristic* approach, which is based on numerous computations for practical applications [8]. This method is more efficient since it is less pessimistic than the a priori strategy. For discontinuous solutions, the heuristic approach uses $\varepsilon = C\Delta x$.

This yields detection of element I_j^{n-1} if

$$\max_{\substack{\ell=0,\ldots,k \\ r=1,2,3}} \left(\frac{|d_{\ell j}^{n-1}(r)|}{\max\left\{ \max_{j=0,\ldots,2^n-1} 2^{(n-1)/2}|s_{0j}^n(r)|, 1 \right\}} \right) > \frac{1}{\sqrt{2}}\Delta x^{\beta+0.5}C,$$

where $\beta = 2$ for the a priori strategy and $\beta = 1$ for the heuristic strategy. Note that the multiwavelet coefficients are scaled by the cell average if this value is greater than 1 in absolute value (to prevent division by zero).

The optimal choice of the parameter C depends on the problem, in particular on the strength of the shock compared to the normal amplitude of the solution. The smaller C is, the more elements are detected. In general, the value $C = 1/(b-a)$ should work for the domain $[a, b]$ [8]. If C is chosen too small, then too many cells are detected as troubled. For the adaptive strategy, this is not really problematic since the approximation is usually more accurate on a finer grid. However, for troubled-cell indication, it is important to detect the correct number of elements.

It should be noticed that this indicator is designed for very fine resolutions (since the strategies use asymptotic arguments). For coarse meshes, smaller values of C should be used, which are difficult to predict a priori.

3.4 Generalized Grids

The algorithm for utilizing Alpert's multiwavelets for a nonuniform grid is given in [7]: the only difference with Alpert's algorithm [1] is that no additional vanishing moments are added. Multiwavelets for one-dimensional irregular meshes have been designed in [12, 13]. It should be noticed that this construction is local, which means that the resulting bases are depending on the level and the position unless there is an affine mapping from the element to a reference element. This leads to slower computations. On the other hand, the use of such multiwavelet space makes it possible to decompose the DG approximation to a multiwavelet expansion exactly. The multiwavelet coefficients will again become small if the underlying function is smooth, and the mesh width between two neighboring elements is not varying too much.

When coupled with a troubled-cell indication variable, it will be necessary to include spatial information of the mesh in the algorithm using the element size. Alternatively, one can use of a window-based technique [3]. A window is a fixed length subsequence of the test sequence, which can be slid through the domain using a sliding step. These issues and resulting numerics are discussed further in [15].

4 Numerical Results

In this section, the different multiwavelet troubled-cell indicators are applied to one-dimensional problems based on the Euler equations of gas dynamics.

The results for the original multiwavelet troubled-cell indicators (both based on a parameter, and based on outlier detection), can be seen in Figs. 1 and 2 (polynomial degree 2, 128 elements for Sod's and Lax's shock tube, and 512 elements for the blast-wave and Shu-Osher problem). The parameter-based technique performs well if a suitable value for the problem-dependent parameter C is chosen. The outlier-detection results are generally better than the original troubled-cell indicator using an optimized parameter: both the weak and the strong shock regions were detected, whereas smooth regions were not selected.

It is also possible to use the thresholding technique for multiwavelet coefficients to detect troubled cells. It turns out that this indicator works very well as long as an appropriate value for C is chosen, and the mesh is taken fine enough. The results for the different test cases are visualized in Fig. 3 using the heuristic strategy (polynomial degree 2, 1024 elements for all models). Here, we take the value $C = 1/(b - a)$ where $[a, b]$ is the domain on which the test problem is defined. Note that this thresholding technique is very accurate. However, many elements should be used to meet the asymptotic properties of the indicator.

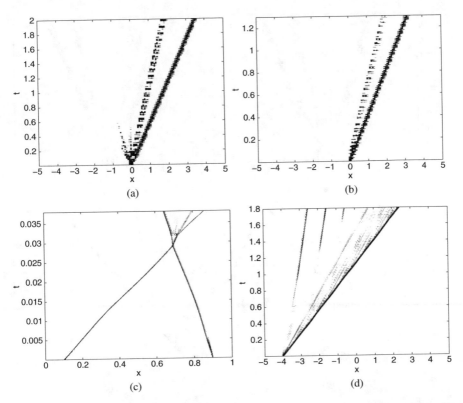

Fig. 1 Time-history plot of detected troubled cells using the parameter-based multiwavelet troubled-cell indicator, polynomial degree 2. (**a**) Sod's shock tube, $C = 0.1$, 128 elements. (**b**) Lax's shock tube, $C = 0.1$, 128 elements. (**c**) Blast-wave problem, $C = 0.05$, 512 elements. (**d**) Shu-Osher, $C = 0.01$, 512 elements

If the number of elements is taken smaller, then C should decrease to detect the correct features. In that case, it is difficult to guess the correct value of C. Another option is to use the a priori strategy for coarser meshes, see Fig. 4 (polynomial degree 2, 128 elements for Sod's and Lax's shock tube, and 512 elements for the blast-wave and Shu-Osher problem). If $C = 1/(b - a)$ is used, then this approach works well for Sod's and Lax's shock tube, but too many elements are detected for the blast-wave and the Shu-Osher problem. Also here, the value of C should be adapted to find the correct results.

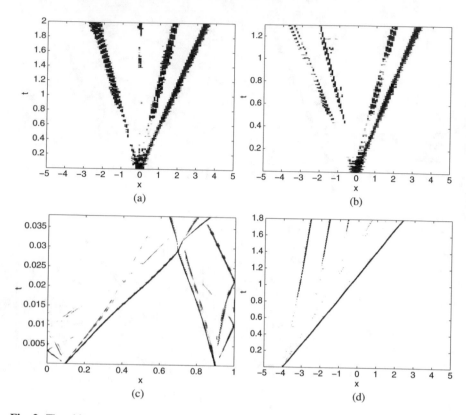

Fig. 2 Time-history plot of detected troubled cells using the outlier-detection multiwavelet troubled-cell indicator, polynomial degree 2. (**a**) Sod's shock tube, 128 elements. (**b**) Lax's shock tube, 128 elements. (**c**) Blast-wave problem, 512 elements. (**d**) Shu-Osher problem, 512 elements

5   Conclusions and Recommendations

In this paper, a new troubled-cell indicator was formed, based on the cancelation property for multiwavelets and the derived thresholding technique. Inspection of this technique reveals that it is very useful to design adaptive meshes [8]. For troubled-cell indication, we found out that detection is very accurate as long as a very fine mesh is used. For coarser meshes, it seems to be more useful to apply a different detection method. Furthermore, it is not straightforward how to choose the parameter C.

More research should be done to see in which way the cancelation property for multiwavelet coefficients can be used for the accurate detection of troubled cells. For example, it could be that this property also relates to the severity of the shocks.

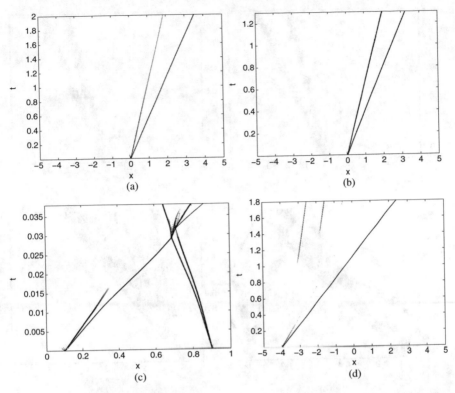

Fig. 3 Thresholding technique with heuristic approach: time-history plot of detected troubled cells, 1024 elements, polynomial degree 2, $C = 1/(b - a)$, with $[a, b]$ the computational domain. (**a**) Sod's shock tube. (**b**) Lax's shock tube. (**c**) Blast-wave problem. (**d**) Shu-Osher problem

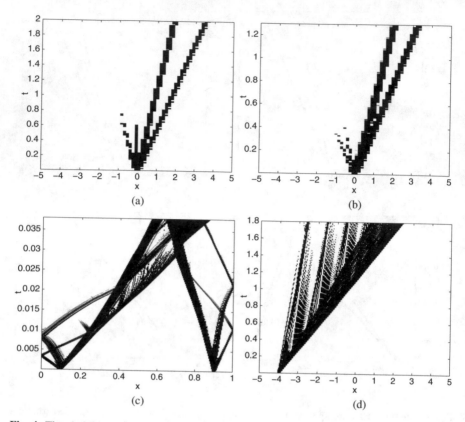

Fig. 4 Thresholding technique with a priori approach on coarser meshes: time-history plot of detected troubled cells, polynomial degree 2, $C = 1/(b-a)$, with $[a, b]$ the computational domain. (**a**) Sod's shock tube, 128 elements. (**b**) Lax's shock tube, 128 elements. (**c**) Blast-wave problem, 512 elements. (**d**) Shu-Osher problem, 512 elements

Acknowledgements The author gratefully wishes to acknowledge the collaboration with Jennifer Ryan and the useful comments provided by Siegfried Müller that helped to shape this work.

References

1. Alpert, B.K.: A class of bases in L^2 for the sparse representation of integral operators. SIAM J. Math. Anal. **24**, 246–262 (1993)
2. Archibald, R.K., Fann, G.I., Shelton, W.A.: Adaptive discontinuous Galerkin methods in multiwavelets bases. Appl. Numer. Math. **61**, 879–890 (2011)
3. Chandola, V.: Anomaly detection for symbolic sequences and time series data. PhD Thesis, University of Minnesota, Minneapolis (2009)
4. Cockburn, B., Shu, C.-W.: TVB Runge-Kutta local projection discontinuous Galerkin finite element method for conservation laws II: general framework. Math. Comput. **52**, 411–435 (1989)
5. Cockburn, B., Lin, S.-Y., Shu, C.-W.: TVB Runge-Kutta local projection discontinuous Galerkin finite element method for conservation laws III: one-dimensional systems. J. Comput. Phys. **84**, 90–113 (1989)
6. Dahmen, W.: Wavelet methods for PDEs—some recent developments. J. Comput. Appl. Math. **128**, 133–185 (2001)
7. Gerhard, N., Müller, S.: Adaptive multiresolution discontinuous Galerkin schemes for conservation laws: multi-dimensional case. Comput. Appl. Math. **35**, 321–349 (2016)
8. Gerhard, N., Iacono, F., May, G., Müller, S., Schäfer, R.: A high-order discontinuous Galerkin discretization with multiwavelet-based grid adaptation for compressible flows. J. Sci. Comput **62**, 25–52 (2015)
9. Hoaglin, D.C., Iglewicz, B., Tukey, J.W.: Performance of some resistant rules for outlier labeling. J. Am. Statist. Assoc. **81**, 991—999 (1986)
10. Hovhannisyan, N., Müller, S., Schäfer, R.: Adaptive multiresolution discontinuous Galerkin schemes for conservation laws. Math. Comput. **83**, 113–151 (2014)
11. Krivodonova, L.: Limiters for high-order discontinuous Galerkin methods. J. Comput. Phys. **226**, 879–896 (2007)
12. Nagel, D.: Effiziente Konstruktion von Multiwavelets auf nicht uniformen dyadischen Gitterhierarchien. MasterÕs thesis from RWTH Aachen University (2015)
13. Pistre, S.: Konstruktion von Multiwavelets auf nicht-uniformen eindimensionalen Gitterhierarchien. BachelorÕs thesis from RWTH Aachen University (2013)
14. Tukey, J.W.: Exploratory Data Analysis. Addison-Wesley, Boston (1977)
15. Vuik, M.J.: The use of multiwavelets and outlier detection for troubled-cell indication in discontinuous Galerkin methods. PhD Thesis from Delft University of Technology (2017)
16. Vuik, M.J., Ryan, J.K.: Multiwavelet troubled-cell indicator for discontinuity detection of discontinuous Galerkin schemes. J. Comput. Phys. **270**, 138–160 (2014)
17. Vuik, M.J., Ryan, J.K.: Multiwavelets and Jumps in DG Approximations. In: Kirby, R.M., Berzins, M., Hesthaven, J.S. (eds.) Spectral and High Order Methods for Partial Differential Equations—ICOSAHOM 2014, pp. 503–511. Springer, Berlin (2015)
18. Vuik, M.J., Ryan, J.K.: Automated parameters for troubled-cell indicators using outlier detection. SIAM J. Sci. Comput. **38**, A84–A104 (2016)

An Anisotropic p-Adaptation Multigrid Scheme for Discontinuous Galerkin Methods

Andrés M. Rueda-Ramírez, Gonzalo Rubio, Esteban Ferrer, and Eusebio Valero

1 Introduction

In recent decades, high-order discontinuous Galerkin (DG) methods have been gaining increasing popularity for high-accuracy solutions of systems of conservation laws, such as the compressible Euler and Navier-Stokes equations [5, 6, 22]. The lack of a continuity constraint on element interfaces makes DG methods robust for describing advection-dominated problems when an appropriate Riemann solver is selected [5, 12, 22].

Multigrid methods speed up the iterative solution of large systems of equations using coarse-grid representations (lower levels). Iterative methods (known as *smoothers* in the multigrid community) are good at eliminating the high frequencies of the error fast; therefore, when applied to coarse-grid representations, they also reduce the low frequencies of the error. They have been broadly used in the high-order community in recent years in the form of p-multigrid [2, 8] (where levels are constructed using different polynomial orders) and hp-multigrid [14, 21] (where both the order and size of the elements are changed). Two types of multigrid methods can be found in the literature: linear and nonlinear multigrid. In our work, we make use of the nonlinear multigrid scheme, also known as the Full Approximation Scheme (FAS), since it enables the estimation of the truncation error of coarse representations, as will be shown. The smoother can be either a time-marching scheme (implicit or explicit), or an iterative method applied to the linearized problem.

A. M. Rueda-Ramírez (✉) · G. Rubio · E. Ferrer · E. Valero
ETSIAE-UPM (School of Aeronautics - Universidad Politécnica de Madrid), Plaza Cardenal Cisneros, Madrid, Spain
e-mail: am.rueda@upm.es; g.rubio@upm.es; esteban.ferrer@upm.es; eusebio.valero@upm.es

© The Author(s) 2020
S. J. Sherwin et al. (eds.), *Spectral and High Order Methods for Partial Differential Equations ICOSAHOM 2018*, Lecture Notes in Computational Science and Engineering 134, https://doi.org/10.1007/978-3-030-39647-3_44

Because of the allowed discontinuities on element interfaces, DG methods are capable of handling non-conforming meshes with hanging nodes and/or different polynomial orders efficiently [7, 13, 15]. It is possible to take advantage of this feature to accelerate the computations through local adaptation strategies. Local adaptation can be performed by subdividing or merging elements (h-adaptation) or by enriching or reducing the polynomial order in certain elements (p-adaptation). The main idea behind these methodologies is to reduce the number of degrees of freedom (NDOF) while maintaining a high accuracy, which translates into shorter computational times and reduced storage requirements. Furthermore, since several 2D and 3D implementations of the DG methods use tensor-product basis functions, it is possible to adapt the polynomial order in each coordinate direction independently. In order to identify the localized regions that need increased or decreased accuracy, an error estimator is commonly used.

There are several approaches to estimate the error and drive an adaptation method. In this work, we focus on truncation error estimates since it has been shown that a reduction of the truncation error controls the numerical accuracy of all functionals [10], hence reducing the truncation error necessarily leads to a more accurate lift and drag. The τ-estimation method [4] is a way to estimate the truncation error locally that has been used to drive mesh adaptation strategies in low-order [9, 20] and high-order methods [10, 17, 18]. The adaptation strategy consists in converging a high order representation (reference mesh) to a specified global residual and then performing a single error estimation followed by a corresponding mesh adaptation process. Rueda-Ramírez et al. [19] developed a new method for estimating the truncation error of anisotropic representations that is cheaper to evaluate than previous implementations, and showed that it produces very accurate extrapolations of the truncation error, which enables the use of coarser reference meshes.

In this work, we employ the anisotropic truncation error estimator developed in [19] and the anisotropic p-adaptation method detailed in [18] to accelerate the computation of the compressible steady viscous flow past a NACA0012 at angle of attack 5°, $Re_\infty = 200$ based on the airfoil chord, and $M_\infty = 0.2$. This particular settings correspond to a steady laminar flow, but the proposed method can be directly used with any steady solution (e.g. RANS). The paper is organized as follows: In Sect. 2, we briefly describe the methods used in this paper. In Sect. 3, we compare the performance of the proposed methods with traditional strategies for solving the flow past a NACA0012 and show the speed-up advantages for different accuracies. Finally, the conclusions are summarized in Sect. 4.

2 Methods

2.1 DG Method

We consider the approximation of systems of conservation laws,

$$\partial_t \mathbf{q} + \nabla \cdot \mathscr{F} = \mathbf{s}, \tag{1}$$

where $\mathbf{q}$ is the vector of conserved variables, $\mathscr{F}$ is the flux dyadic tensor, and $\mathbf{s}$ is a source term. The domain Ω is partitioned in a mesh $\mathcal{T} = \{e\}$ consisting of K non-overlapping elements Ω^e. Multiplying equation (1) by a test function $\mathbf{v}$ and integrating by parts over each subdomain Ω^e yields the weak formulation:

$$\int_{\Omega^e} \partial_t \mathbf{q} \mathbf{v} d\Omega^e - \int_{\Omega^e} \mathscr{F} \cdot \nabla \mathbf{v} d\Omega^e + \int_{\partial \Omega^e} \mathscr{F} \cdot \mathbf{n} \mathbf{v} d\sigma^e = \int_{\Omega^e} \mathbf{s} \mathbf{v} d\Omega^e. \qquad (2)$$

Let $\mathbf{q}, \mathbf{s}, \mathscr{F}$ and $\mathbf{v}$ be approximated by piece-wise polynomial functions defined in the space of L^2 functions: $\mathcal{V}^N = \{\mathbf{v}^N \in L^2(\Omega^e) : \mathbf{v}^N|_{\Omega^e} \in \mathscr{P}^N(\Omega^e) \ \forall \ \Omega^e \in \mathcal{T}\}$, where $\mathscr{P}^N(\Omega^e)$ is the space of polynomials of degree at most N. The functions in $\mathcal{V}^N$ can be represented in each element as a linear combination of basis functions $\phi_i^N \in \mathscr{P}^N(\Omega^e)$ (e.g. $\mathbf{q}^N|_{\Omega^e} = \sum_i \mathbf{Q}_i^N \phi_i^N$), where ϕ_i^N are usually tensor product expansions. After some manipulations, the discontinuous Galerkin finite element discretization system is obtained:

$$[\mathbf{M}]\partial_t \mathbf{Q}^N + \mathbf{F}(\mathbf{Q}^N) = [\mathbf{M}]\mathbf{S}^N, \qquad (3)$$

where $[\mathbf{M}]$ is the mass matrix and $\mathbf{F}$ is a nonlinear operator, which are the assembled global versions of the element-wise mass matrices and nonlinear operators:

$$[\mathbf{M}]_{i,j}^e = \int_{\Omega^e} \phi_i \phi_j d\Omega^e, \qquad (4)$$

$$\mathbf{F}^e(\mathbf{Q})_j = \sum_{i=1}^{\mathrm{NDOF}^e} \left[-\int_{\Omega^e} \mathscr{F}_i^e \cdot \phi_i \nabla \phi_j d\Omega^e \right] + \int_{\partial \Omega^e} \mathscr{F}^{*N} \left(\mathbf{Q}, \mathbf{Q}^-, \mathbf{n}\right) \phi_j d\sigma^e, \qquad (5)$$

where $\mathscr{F}_i^e$ is the ith position of the vector $\mathscr{F}^e$, which contains the value of $\mathscr{F}^e$ for all the degrees of freedom of element e. In the rest of this paper, bold uppercase Roman letters and bold Greek letters are used to note vectors spanning several degrees of freedom, unless specified.

The numerical flux function $\mathscr{F}^*$ allows to uniquely define the flux at the element interfaces and to weakly prescribe the boundary data as a function of the conserved variable on both sides of the boundary/interface and the normal vector. In the present work, we use the scheme by Roe [16] as the advective Riemann solver and the original scheme by Bassi and Rebay [1] (BR1) as the diffusive Riemann solver.

2.2 Full Approximation Scheme p-Multigrid

The Full Approximation Scheme (FAS) is a nonlinear version of the multigrid method that is specially suited to solve systems of nonlinear equations [4]. Departing from Eq. (3) and defining the operator $\mathbf{A}(\mathbf{Q}^N) = [\mathbf{M}]^{-1}\mathbf{F}(\mathbf{Q}^N)$, the steady-state problem of order P yields

$$A(Q^P) = S^P. \tag{6}$$

After β_1 sweeps of a smoother, a non-converged solution $\tilde{Q}^P$ is obtained that has an associated discretization error $\epsilon^P = Q^P - \tilde{Q}^P$. The FAS multigrid procedure consists in obtaining an approximation to the discretization error in a coarse grid of order N and projecting it to the original problem of order P:

$$\epsilon^P = \underline{I}_N^P \epsilon^N = \underline{I}_N^P (Q^N - \underline{I}_P^N \tilde{Q}^P), \tag{7}$$

where $\underline{I}_N^P$ is an L^2 projection operator $N \to P$ and Q^N is the solution to the coarse-grid problem:

$$A^N(Q^N) = S^N, \tag{8}$$

where the source term is defined as

$$S^N = A^N(\underline{I}_P^N \tilde{Q}^P) + \underline{I}_P^N \left(S^P - A^P(\tilde{Q}^P) \right). \tag{9}$$

In practice, several p-multigrid levels are used in V- or W-cycles. The smoothing steps that are performed when coarsening are called pre-smoothing sweeps, and the ones performed when refining back are called post-smoothing sweeps. Furthermore, Q^N is not obtained exactly in the coarse grids, but approximated using an iterative method $\tilde{Q}^N \to Q^N$. In this work, we use a third order low-storage Runge-Kutta (RK3) as the smoother and V-cycles.

2.3 τ-Based p-Adaptation

In this section we show how to drive an anisotropic p-adaptation procedure using the truncation error, which is estimated in the multigrid procedure.

2.3.1 The Anisotropic τ-Estimation Method

The *non-isolated* truncation error of a discretization of order N is defined as

$$\tau^N = \mathcal{R}^N(I^N q) - \mathcal{R}(q), \tag{10}$$

where q is the exact solution to the problem, I^N is a discretizing operator, $\mathcal{R}$ is the continuous partial differentiation operator, and $\mathcal{R}^N$ is the discrete partial differentiation operator. From Eqs. (1) and (3):

$$\mathcal{R}(\mathbf{q}) = \mathbf{s} - \nabla \cdot \mathcal{F}, \tag{11}$$

$$\mathcal{R}^N (\mathbf{I}^N \mathbf{q}) = [\mathbf{M}]\mathbf{S}^N - \mathbf{F}(\mathbf{I}^N \mathbf{q}), \tag{12}$$

where $\mathbf{I}^N$ is an operator that samples the exact solution on the points that correspond to the degrees of freedom of a representation of order N, and therefore Eq. (12) corresponds to the sampled values of $\mathcal{R}^N (\mathbf{I}^N \mathbf{q})$.

Note that in steady cases, $\mathcal{R}(\mathbf{q}) = 0$ holds. Since the exact solution $\mathbf{q}$ is usually not at hand, we utilize the *quasi a-piori* τ-estimation method, which approximates the exact solution with the non-converged solution on a high-order grid $\mathbf{q} \approx \tilde{\mathbf{q}}^P$, where $N < P$. Therefore, the steady *non-isolated* truncation error estimation yields

$$\tau_P^N = \mathcal{R}^N (\mathbf{I}_P^N \tilde{\mathbf{q}}^P) \quad \rightarrow \quad \tau_P^N = \mathcal{R}^N (\mathbf{I}_P^N \tilde{\mathbf{Q}}^P) = [\mathbf{M}]\mathbf{S}^N - \mathbf{F}(\mathbf{I}_P^N \tilde{\mathbf{Q}}^P). \tag{13}$$

On the left side of the arrow is the estimation of the truncation error that lives in the space $\mathcal{V}^N$, and on the right side is the sampled form of the truncation error estimation on the points that correspond to the degrees of freedom. In a DG representation, one can also define the *isolated* truncation error $\hat{\tau}$ as

$$\hat{\tau}_P^N = \hat{\mathcal{R}}^N (\mathbf{I}_P^N \tilde{\mathbf{Q}}^P) = [\mathbf{M}]\mathbf{S}^N - \hat{\mathbf{F}}(\mathbf{I}_P^N \tilde{\mathbf{Q}}^P), \tag{14}$$

where $\hat{\mathbf{F}}$ is the assembled version of the *isolated* nonlinear operator, defined elementwise as

$$\mathbf{F}^e (\mathbf{Q})_j = \sum_{i=1}^{\text{NDOF}^e} \left[-\int_{\Omega^e} \mathcal{F}_i^e \cdot \phi_i \nabla \phi_j \, d\Omega^e \right] + \int_{\partial \Omega^e} \mathcal{F}^N \cdot \mathbf{n} \phi_j \, d\sigma^e. \tag{15}$$

Note that Eq. (15) is (5) without substituting $\mathcal{F}$ by the numerical flux $\mathcal{F}^*$. This change eliminates the influence of the neighboring elements and boundaries on the truncation error of each element. We drop the hat notation in the next statements since they are valid for both the *isolated* and *non-isolated* truncation error.

The τ-estimation method can also be used with anisotropic representations, i.e.

$$\tau_{P_1 P_2}^{N_1 N_2} = \mathcal{R}^{N_1 N_2} (\mathbf{I}_{P_1 P_2}^{N_1 N_2} \tilde{\mathbf{q}}^{P_1 P_2}), \tag{16}$$

where N_i and P_i are the polynomial orders in the direction i of the analyzed representation and the high-order reference solution, respectively, where $N_i < P_i$. Additionally, Rueda-Ramírez et al. [19] showed that the truncation error of an anisotropic representation can be estimated using directional components:

$$\tau^{N_1 N_2} \approx \tau_1^{N_1 N_2} + \tau_2^{N_1 N_2} \approx \tau_{P_1 P_2}^{N_1 P_2} + \tau_{P_1 P_2}^{P_1 N_2}, \tag{17}$$

where the directional components in discrete form are therefore,

$$\tau_1 = \tau_{P_1 P_2}^{N_1 P_2} = [\underline{\mathbf{M}}]S^{N_1 P_2} - [\underline{\mathbf{M}}]A(\underline{\mathbf{I}}_{P_1 P_2}^{N_1 P_2} \tilde{\mathbf{Q}}^{P_1 P_2}), \tag{18}$$

and that these directional components decrease exponentially with the polynomial order in smooth solutions. Consequently, it is possible to use a semi-converged solution $\tilde{\mathbf{q}}^{P_1 P_2}$ to estimate $\tau^{N_1 N_2}$ ($N_i < P_i$) and then extrapolate the directional components τ_i to obtain the values of $\tau^{N_1 N_2}$ for $N_i > P_i$. Figure 1a shows a graphical representation of the truncation error $\tau^{N_1 N_2}$ as estimated with a semi-converged solution of order $P_1 = P_2 = 5$.

2.3.2 The p-Adaptation Multigrid Scheme

It has been shown that the use of FAS p-multigrid methods speeds up the computation of steady-state and unsteady solutions of the compressible Navier-Stokes equations [2, 8]. In addition, Rueda-Ramírez et al. [18] showed that the truncation error of an anisotropic representation can be inexpensively obtained inside an anisotropic p-multigrid cycle that performs the coarsening in one coordinate direction at a time. In fact, the second term of Eq. (18) is naturally computed in an anisotropic multigrid for obtaining the coarse-grid source term (Eq. (9)).

Therefore, we propose a p-adaptation multigrid scheme that makes use of the multigrid as a solver, but also as an error estimator. Every time the error is estimated, an anisotropic p-multigrid strategy is used to generate a truncation error map for each element, like the one in Fig. 1a. Afterwards, the polynomial orders in the different coordinate directions are selected for

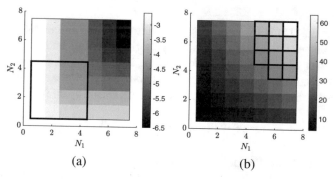

(a) (b)

Fig. 1 (a) Truncation error map for a specific element that shows $\log \left\| \tau_{5,5}^{N_1 N_2} \right\|_\infty$ as a function of N_1 and N_2 (the black box shows the limit between the *estimated* and *extrapolated* maps). (b) Map of degrees of freedom (the black boxes show the polynomial orders that achieve $\left\| \tau_{5,5}^{N_1 N_2} \right\|_\infty < 10^{-5}$)

each element, such that a truncation error threshold τ_{max} is achieved with the minimum NDOF possible, as illustrated in Fig. 1b. In the simulations shown in this paper, the reference representation, $\tilde{q}^P$, is converged to a residual $\tau_{max}/10$ before the p-adaptation stage, so that the truncation error is accurately estimated down to τ_{max}, as was shown necessary by Kompenhans et al. [10].

3 Flow Past a NACA0012 Airfoil

In this section, we compare the performance of the proposed p-adaptation multi-grid scheme with a uniformly adapted p-multigrid method (without local p-adaptation) and a uniformly adapted RK3 method when solving the steady viscous flow past a NACA0012 airfoil at angle of attack 5°, $Re_\infty = 200$ ($L_\infty = L_{chord}$) and $M_\infty = 0.2$. This particular settings correspond to a steady laminar flow, but the proposed method can be directly used with any steady solution (e.g. RANS). An unstructured mesh of 2011 quadrilateral elements is employed (Fig. 2).

In the cases where multigrid is employed, the RK3 scheme is used as the iterative method (smoother), so that additional speed-ups are only due to the methods exposed in Sect. 2. As in [18], a residual-based smoothing strategy is performed. The minimum number of smoothing sweeps is $\beta = 200$ for the coarsest multigrid level ($N = 1$) and $\beta = 50$ for any other level. After every β pre-smoothing sweeps, the residual in the next (coarser) representation is checked. If $\left\| \mathcal{R}^N \right\|_\infty < 1.2 \left\| \mathcal{R}^{N-1} \right\|_\infty$, the pre-smoothing is stopped; otherwise, β additional

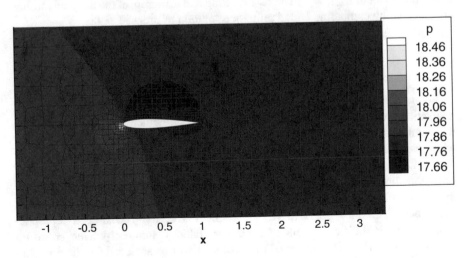

Fig. 2 Pressure contours of the flow past a NACA0012 at angle of attack 5°

sweeps are performed. Similarly, the norm of the residual after the post-smoothing is forced to be at least as low as it was after the pre-smoothing, $\left\|\mathcal{R}_{post}^{N}\right\|_{\infty} \leq \left\|\mathcal{R}_{pre}^{N}\right\|_{\infty}$. If that condition is not fulfilled, additional β sweeps are taken until it is.

The *isolated* truncation error estimate is used to drive the p-adaptation method since it has been shown to provide better results than the *non-isolated* one [17–19]. The conservative form (Eq. (1)) of the compressible Navier-Stokes equations is discretized using the Discontinuous Galerkin Spectral Element Method (DGSEM) [3, 12], which is a nodal (collocation) version of a DG method that uses Gauss points as the solution nodes and quadrature points, obtaining diagonal mass matrices. However, the methods that are exposed here can be applied to any DG scheme with tensor-product basis functions.

In [18] it was explained that, when using the DGSEM in general 3D curved meshes and p-nonconforming representations, the order of the mapping must be at most $M \leq N/2$ for the numerical representation to be free-stream preserving. For this reason, the use of a *conforming algorithm* was proposed, which forces the polynomial orders to be conforming in the first layer of elements on a curved boundary. The use of a *conforming algorithm* is necessary to retain the well-known $M \leq N$ condition of the DGSEM [11]. In this work, we use the *conforming algorithm* on the airfoil surface since it showed to produce better results, although its use is not imperative as the considered test case is 2D.

For the uniformly adapted cases, the polynomial order is varied between $N = 2$ and $N = 7$. For the cases with local p-adaptation, a single-stage anisotropic p-adaptation procedure is performed, and the minimum polynomial order after adaptation is set to $N_{min} = 1$, whereas the maximum polynomial order after adaptation is set to $N_{max} = 7$. The relative drag and lift errors of the adapted meshes are assessed by comparing with a reference solution of order $N = 8$:

$$e_{drag}^{N=8} = \frac{|C_d - C_d^{N=8}|}{C_d^{N=8}}, \ e_{lift}^{N=8} = \frac{|C_d - C_l^{N=8}|}{C_l^{N=8}}. \tag{19}$$

Figure 3 shows a comparison between the errors obtained using the $\hat{\tau}$-based adaptation procedure and the ones using uniform p-refinement. As can be observed, the number of degrees of freedom is substantially reduced for the same accuracy when using the $\hat{\tau}$-based p-adaptation. This reduction translates into a reduction of the CPU-times. It is interesting to point out that, as the *isolated* truncation error threshold $\hat{\tau}_{max}$ is decreased, the polynomial orders of the mesh tend to the maximum specified polynomial order, $N_{max} = 7$. Consequently, the lift and drag coefficients also tends to $C_l^{N=7}$. Using Fig. 3, it is possible to compute a speed-up for different levels of accuracy. Table 1 summarizes the speed-up calculations for the maximum level of accuracy that was achieved for the drag and lift coefficients.

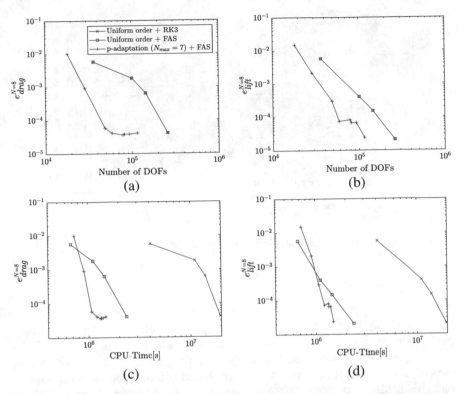

Fig. 3 Relative error in the drag and lift coefficients for different methods for the flow past the NACA0012 airfoil. The blue lines represent uniform refinement, and the red lines represent the $\hat{\tau}$-based p-adaptation procedure with $N_{max} = 7$. (a) Drag error vs. DOFs; (b) lift error vs. DOFs; (c) drag error vs. CPU-time; (d) lift error vs. CPU-time

Table 1 Computation times and speed-up for the different methods after converging until $\|\mathbf{r}\|_\infty < 10^{-9}$

Method	Drag coefficient ($e_{drag} \leq \times 4.1 \times 10^{-5}$)			Lift coefficient ($e_{lift} \leq 2.4 \times 10^{-5}$)		
	CPU-time [s]	Time [%]	Speed-up	CPU-time [s]	Time [%]	Speed-up
RK3	1.95×10^7	100.00%	1.00	1.95×10^7	100.00%	1.00
FAS	2.36×10^6	12.10%	8.26	2.36×10^6	12.10%	8.26
FAS + p-adaptation	1.21×10^6	6.20%	16.13	1.48×10^6	7.58%	13.19

Figure 4 shows the distribution of polynomial orders after the single-stage adaptation procedure for a threshold of $\tau_{max} = 5 \times 10^{-4}$, which has related errors of $e_{drag}^{N=8} = 4.10 \times 10^{-5}$ and $e_{lift}^{N=8} = 7.31 \times 10^{-5}$. As can be observed, the elements that are enriched are mainly the ones on the boundary layer (specially leading and trailing edge), and the zones of the wake where the element size changes significantly.

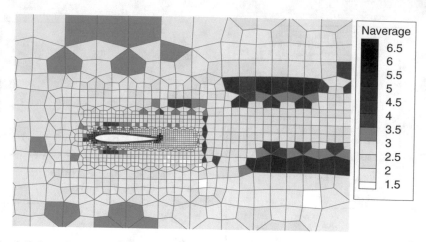

Fig. 4 Polynomial order distribution after the anisotropic p-adaptation. $N_{average} = (N_1 + N_2)/2$

4 Conclusions

In this work, we have applied recently developed error estimators and anisotropic p-adaptation methods in conjunction with multigrid solving strategies for solving the compressible Navier-Stokes equations. In particular, we have shown that the coupling of anisotropic truncation error-based p-adaptation methods with p-multigrid schemes can speed up the computation of steady-state solutions of PDEs. The achieved speed-up depends on the desired accuracy, being this method optimal when high accuracy is required (low errors). In particular, a speed-up of 16.13 was achieved for the computation of the steady compressible viscous flow past a NACA0012 airfoil at angle of attack 5° with respect to the uniformly adapted representation without multigrid.

Acknowledgements This project has received funding from the European Union's Horizon 2020 Research and Innovation Program under the Marie Skłodowska-Curie grant agreement No 675008 for the SSeMID project. The authors acknowledge the computer resources and technical assistance provided by the *Centro de Supercomputación y Visualización de Madrid* (CeSViMa).

References

1. Bassi, F., Rebay, S.: A high-order accurate discontinuous finite element method for the numerical solution of the compressible Navier-Stokes equations. J. Comput. Phys. **131**, 267–279 (1997)
2. Bassi, F., Ghidoni, A., Rebay, S., Tesini, P.F.: High-order accurate p-multigrid discontinuous Galerkin solution of the Euler equations. Int. J. Numer. Methods Fluids **60**, 847–865 (2009)
3. Black, K.: A conservative spectral element method for the approximation of compressible fluid flow. Kybernetika **35**, 133–146 (1999)

4. Brandt, A., Livne, O.E.: Multigrid Techniques: 1984 Guide with Applications to Fluid Dynamics, Revised Edition. SIAM, Philadelphia (2011)
5. Cockburn, B., Shu, C.-W.: The local discontinuous Galerkin method for time-dependent convection-diffusion systems. SIAM J. Numer. Anal. **35**, 2440–2463 (1998)
6. Ferrer, E.: An interior penalty stabilised incompressible discontinuous Galerkin–Fourier solver for implicit large eddy simulations. J. Comput. Phys. **348**, 754–775 (2017)
7. Ferrer, E., Willden, R.H.: A high order Discontinuous Galerkin–Fourier incompressible 3D Navier-Stokes solver with rotating sliding meshes. J. Comput. Phys. **231**, 7037–7056 (2012)
8. Fidkowski, K.J., Oliver, T.A., Lu, J., Darmofal, D.L.: p-Multigrid solution of high-order discontinuous Galerkin discretizations of the compressible Navier-Stokes equations. J. Comput. Phys. **207**, 92–113 (2005)
9. Fraysse, F., Rubio, G., De Vicente, J., Valero, E.: Quasi-a priori mesh adaptation and extrapolation to higher order using τ-estimation. Aerosp. Sci. Technol. **38**, 76–87 (2014)
10. Kompenhans, M., Rubio, G., Ferrer, E., Valero, E.: Adaptation strategies for high order discontinuous Galerkin methods based on Tau-estimation. J. Comput. Phys. **306**, 216–236 (2016)
11. Kopriva, D.A.: Metric identities and the discontinuous spectral element method on curvilinear meshes. J. Sci. Comput. **26**, 301–327 (2006)
12. Kopriva, D.: Implementing Spectral Methods for Partial Differential Equations: Algorithms for Scientists and Engineers. Springer, Berlin (2009)
13. Kopriva, D.A., Woodruff, S.L., Hussaini, M.Y.: Computation of electromagnetic scattering with a non-conforming discontinuous spectral element method. Int. J. Numer. Methods Eng. **53**, 105–122 (2002)
14. Mitchell, W.F., Division, C.S.: The hp -multigrid method applied to hp -adaptive refinement of triangular grids. Numer. Linear Algebra Appl. **17**, 211–228 (2010)
15. Rivière, B.: Discontinuous Galerkin Methods for Solving Elliptic and Parabolic Equations Theory and Implementation. SIAM, Philadelphia (2008)
16. Roe, P.L.: Approximate Riemann solvers, parameter vectors, and difference schemes. J. Comput. Phys. **43**, 357–372 (1981)
17. Rubio, G., Fraysse, F., Kopriva, D.A., Valero, E.: Quasi-a priori truncation error estimation in the DGSEM. J. Sci. Comput. **64**, 425–455 (2015)
18. Rueda-Ramírez, A.M., Manzanero, J., Ferrer, E., Rubio, G., Valero, E.: A p-multigrid strategy with anisotropic p-adaptation based on truncation errors for high-order discontinuous Galerkin methods. J. Comput. Phys. **378**, 209–233 (2019)
19. Rueda-Ramírez, A.M., Rubio, G., Ferrer, E., Valero, E.: Truncation error estimation in the p-anisotropic discontinuous Galerkin spectral element method. J. Sci. Comput. **78**(1), 433–466 (2018)
20. Syrakos, A., Efthimiou, G., Bartzis, J.G., Goulas, A.: Numerical experiments on the efficiency of local grid refinement based on truncation error estimates. J. Comput. Phys. **231**, 6725–6753 (2012)
21. Wang, L., Mavriplis, D.: Adjoint-based h-p adaptive discontinuous Galerkin methods for the compressible Euler equations, J. Comput. Phys. **228**, 7643–7661 (2009)
22. Wang, Z., Fidkowski, K., Abgrall, R., Bassi, F., Caraeni, D., Cary, A., Deconinck, H., Hartmann, R., Hillewaert, K., Huynh, H., Kroll, N., May, G., Persson, P.-O., van Leer, B., Visbal, M.: High-order CFD methods: current status and perspective. Int. J. Numer. Methods Fluids **72**, 811–845 (2013)

A Spectral Element Reduced Basis Method for Navier–Stokes Equations with Geometric Variations

Martin W. Hess, Annalisa Quaini, and Gianluigi Rozza

1 Introduction and Motivation

Spectral element methods (SEM) use high-order polynomial ansatz functions to solve partial differential equations (PDEs) in all fields of science and engineering, see, e.g., [4–7, 12, 16] and references therein for an overview. Typically, an exponential error decay under p-refinement is observed, which can provide an enhanced accuracy over standard finite element methods at the same computational cost. In the following, we assume that the discretization error is much smaller than the model reduction error, small enough not to interfere with our results. In general, this needs to be established with the use of suitable error estimation and adaptivity techniques.

We consider the flow through a channel with a narrowing of variable height. A reduced order model (ROM) is computed from a few high-order SEM solves, which accurately approximates the high-order solutions for the parameter range of interest, i.e., the different narrowing heights under consideration. Since the parametric variations are affine, a mapping to a reference domain is applied without further interpolation techniques. The focus of this work is to show how to use simulations arising from the SEM solver Nektar++ [3] in a ROM context. In particular, the multilevel static condensation of the high-order solver is not applied, but the ROM projection works with the system matrices in local coordinates. See [12] for further details. This is in contrast to our previous work [8], since numerical

M. W. Hess (✉) · G. Rozza
SISSA mathLab, International School for Advanced Studies, Trieste, Italy
e-mail: mhess@sissa.it; gianluigi.rozza@sissa.it

A. Quaini
Department of Mathematics, University of Houston, Houston, TX, USA
e-mail: quaini@math.uh.edu

© The Author(s) 2020
S. J. Sherwin et al. (eds.), *Spectral and High Order Methods for Partial Differential Equations ICOSAHOM 2018*, Lecture Notes in Computational Science and Engineering 134, https://doi.org/10.1007/978-3-030-39647-3_45

experiments have shown that the multilevel static condensation is inefficient in a ROM context. Additionally, we consider affine geometry variations. With SEM as discretization method, we use global approximation functions for the high-order as well as reduced-order methods. The ROM techniques described in this paper are implemented in open-source project ITHACA-SEM.[1]

The outline of the paper is as follows. In Sect. 2, the model problem is defined and the geometric variations are introduced. Section 3 provides details on the spectral element discretization, while Sect. 4 describes the model reduction approach and shows the affine mapping to the reference domain. Numerical results are given in Sect. 5, while Sect. 6 summarizes the work and points out future perspectives.

2 Problem Formulation

Let $\Omega \in \mathbb{R}^2$ be the computational domain. Incompressible, viscous fluid motion in spatial domain Ω over a time interval $(0, T)$ is governed by the incompressible *Navier-Stokes* equations with vector-valued velocity $\mathbf{u}$, scalar-valued pressure p, kinematic viscosity ν and a body forcing $\mathbf{f}$:

$$\frac{\partial \mathbf{u}}{\partial t} + \mathbf{u} \cdot \nabla \mathbf{u} = -\nabla p + \nu \Delta \mathbf{u} + \mathbf{f}, \tag{1}$$

$$\nabla \cdot \mathbf{u} = 0. \tag{2}$$

Boundary and initial conditions are prescribed as

$$\mathbf{u} = \mathbf{d} \quad \text{on } \Gamma_D \times (0, T), \tag{3}$$

$$\nabla \mathbf{u} \cdot \mathbf{n} = \mathbf{g} \quad \text{on } \Gamma_N \times (0, T), \tag{4}$$

$$\mathbf{u} = \mathbf{u}_0 \quad \text{in } \Omega \times 0, \tag{5}$$

with $\mathbf{d}$, $\mathbf{g}$ and $\mathbf{u}_0$ given and $\partial \Omega = \Gamma_D \cup \Gamma_N$, $\Gamma_D \cap \Gamma_N = \emptyset$. The *Reynolds* number Re, which characterizes the flow [11], depends on ν, a characteristic velocity U, and a characteristic length L:

$$Re = \frac{UL}{\nu}. \tag{6}$$

We are interested in computing the steady states, i.e., solutions where $\frac{\partial \mathbf{u}}{\partial t}$ vanishes. The high-order simulations are obtained through time-advancement, while the ROM solutions are obtained with a fixed-point iteration.

---

[1] https://github.com/mathLab/ITHACA-SEM.

2.1 Oseen-Iteration

The *Oseen*-iteration is a secant modulus fixed-point iteration, which in general exhibits a linear rate of convergence [2]. Given a current iterate (or initial condition) $\mathbf{u}^k$, the next iterate $\mathbf{u}^{k+1}$ is found by solving linear system:

$$-\nu\Delta\mathbf{u}^{k+1} + (\mathbf{u}^k \cdot \nabla)\mathbf{u}^{k+1} + \nabla p = \mathbf{f} \text{ in } \Omega,$$

$$\nabla \cdot \mathbf{u}^{k+1} = 0 \text{ in } \Omega,$$

$$\mathbf{u}^{k+1} = \mathbf{d} \quad \text{on } \Gamma_D,$$

$$\nabla\mathbf{u}^{k+1} \cdot \mathbf{n} = \mathbf{g} \quad \text{on } \Gamma_N.$$

Iterations are typical stopped when the relative difference between iterates falls below a predefined tolerance in a suitable norm, like the $L^2(\Omega)$ or $H_0^1(\Omega)$ norm.

2.2 Model Description

We consider the reference computational domain shown in Fig. 1, which is decomposed into 36 triangular spectral elements. The spectral element expansion uses modal Legendre polynomials of the Koornwinder-Dubiner type of order $p = 11$ for the velocity. Details on the discretization method can be found in chapter 3.2 of [12]. The pressure *ansatz* space is chosen of order $p - 2$ to fulfill the inf-sup stability condition [1, 20]. A parabolic inflow profile is prescribed at the inlet (i.e., $x = 0$) with horizontal velocity component $u_x(0, y) = y(3 - y)$ for $y \in [0, 3]$. At the outlet (i.e., $x = 8$) we impose a stress-free boundary condition, everywhere else we prescribe a no-slip condition.

The height of the narrowing in the reference configuration is $\mu = 1$, from $y = 1$ to $y = 2$. See Fig. 1. Parameter μ is considered variable in the interval $\mu \in [0.1, 2.9]$. The narrowing is shrunken or expanded as to maintain the geometry

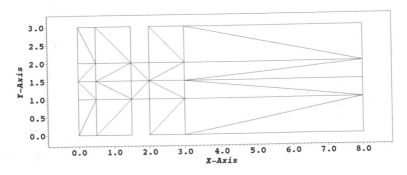

Fig. 1 Reference computational domain for the channel flow, divided into 36 triangles

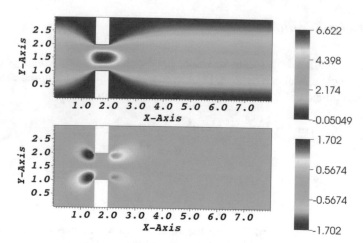

Fig. 2 Full order, steady-state solution for $\mu = 1$: velocity in x-direction (top) and y-direction (bottom)

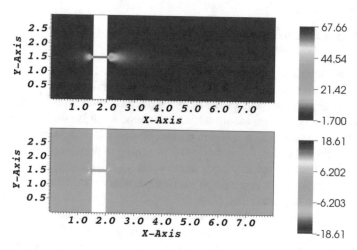

Fig. 3 Full order, steady-state solution for $\mu = 0.1$: velocity in x-direction (top) and y-direction (bottom)

symmetric about line $y = 1.5$. Figures 2, 3, and 4 show the velocity components close to the steady state for $\mu = 1, 0.1, 2.9$, respectively.

The viscosity is kept constant to $\nu = 1$. For these simulations, the *Reynolds* number (6) is between 5 and 10, with maximum velocity in the narrowing as characteristic velocity U and the height of the narrowing characteristic length L. For larger *Reynolds* numbers (about 30), a supercritical pitchfork bifurcation occurs giving rise to the so-called *Coanda* effect [8, 9, 22], which is not subject of the current study. Our model is similar to the model considered in [17, 18], i.e. an

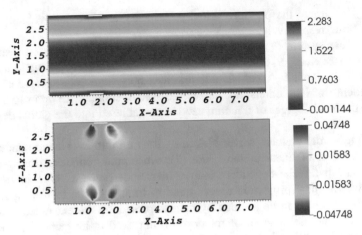

Fig. 4 Full order, steady-state solution for $\mu = 2.9$: velocity in x-direction (top) and y-direction (bottom)

expansion channel with an inflow profile of varying height. However, in [18] the computational domain itself does not change.

3 Spectral Element Full Order Discretization

The *Navier-Stokes* problem is discretized with the spectral element method. The spectral/hp element software framework used is Nektar++ in version 4.4.0.[2] The discretized system of size N_δ to solve at each step of the *Oseen*-iteration for fixed μ can be written as

$$
\begin{bmatrix}
A & -D_{bnd}^T & B \\
-D_{bnd} & 0 & -D_{int} \\
\tilde{B}^T & -D_{int}^T & C
\end{bmatrix}
\begin{bmatrix}
\mathbf{v}_{bnd} \\
\mathbf{p} \\
\mathbf{v}_{int}
\end{bmatrix}
=
\begin{bmatrix}
\mathbf{f}_{bnd} \\
\mathbf{0} \\
\mathbf{f}_{int}
\end{bmatrix},
\tag{7}
$$

where $\mathbf{v}_{bnd}$ and $\mathbf{v}_{int}$ denote velocity degrees of freedom on the boundary and in the interior of the domain, respectively, while $\mathbf{p}$ denotes the pressure degrees of freedom. The forcing terms on the boundary and interior are denoted by $\mathbf{f}_{bnd}$ and $\mathbf{f}_{int}$, respectively. The matrix A assembles the boundary-boundary coupling, B the boundary-interior coupling, $\tilde{B}$ the interior-boundary coupling, and C assembles the interior-interior coupling of elemental velocity *ansatz* functions. In the case of a *Stokes* system, it holds that $B = \tilde{B}^T$, but this is not the case for the *Oseen* equation because of the linearized convective term. The matrices D_{bnd}

---

[2] See www.nektar.info.

and D_{int} assemble the pressure-velocity boundary and pressure-velocity interior contributions, respectively.

The linear system (7) is assembled in local degrees of freedom, resulting in block matrices A, B, $\tilde{B}$, C, D_{bnd} and D_{int}, each block corresponding to a spectral element. This allows for an efficient matrix assembly since each spectral element is independent from the others, but makes the system singular. In order to solve the system, the local degrees of freedom need to be gathered into the global degrees of freedom [12].

The high-order element solver Nektar++ uses a multilevel static condensation for the solution of linear systems like (7). Since static condensation introduces intermediate parameter-dependent matrix inversions (such as C^{-1} in this case) several intermediate projection spaces need to be introduced to use model order reduction [8]. This can be avoided by instead projecting the expanded system (7) directly. The internal degrees of freedom do not need to be gathered, since they are the same in local and global coordinates. Only *ansatz* functions extending over multiple spectral elements need to be gathered.

Next, we will take the boundary-boundary coupling across element interfaces into account. Let M denote the rectangular matrix which gathers the local boundary degrees of freedom into global boundary degrees of freedom. Multiplication of the first row of (7) by $M^T M$ will then set the boundary-boundary coupling in local degrees of freedom:

$$
\begin{bmatrix} M^T M A & -M^T M D_{bnd}^T & M^T M B \\ -D_{bnd} & 0 & -D_{int} \\ \tilde{B}^T & -D_{int}^T & C \end{bmatrix} \begin{bmatrix} \mathbf{v}_{bnd} \\ \mathbf{p} \\ \mathbf{v}_{int} \end{bmatrix} = \begin{bmatrix} M^T M \mathbf{f}_{bnd} \\ 0 \\ \mathbf{f}_{int} \end{bmatrix}. \tag{8}
$$

The action of the matrix in (8) on the degrees of freedom on the Dirichlet boundary is computed and added to the right hand side. Such degrees of freedom are then removed from (8). The resulting system can then be used in a projection-based ROM context [13], of high-order dimension $N_\delta \times N_\delta$ and depending on the parameter μ:

$$
\mathcal{A}(\mu)\mathbf{x}(\mu) = \mathbf{f}. \tag{9}
$$

4 Reduced Order Model

The reduced order model (ROM) computes accurate approximations to the high-order solutions in the parameter range of interest, while greatly reducing the overall computational time. This is achieved by two ingredients. First, a few high-order solutions are computed and the most significant proper orthogonal decomposition (POD) modes are obtained [13]. These POD modes define the reduced order ansatz space of dimension N, in which the system is solved. Second, to reduce the computational time, an offline-online computational procedure is used. See Sect. 4.1.

The POD computes a singular value decomposition of the snapshot solutions to 99.99% of the most dominant modes [10], which define the projection matrix $U \in \mathbb{R}^{N_\delta \times N}$ used to project system (9):

$$U^T \mathcal{A}(\mu) U \mathbf{x}_N(\mu) = U^T \mathbf{f}. \tag{10}$$

The low order solution $\mathbf{x}_N(\mu)$ then approximates the high order solution as $\mathbf{x}(\mu) \approx U \mathbf{x}_N(\mu)$.

4.1 Offline-Online Decomposition

The offline-online decomposition [10] enables the computational speed-up of the ROM approach in many-query scenarios. It relies on an affine parameter dependency, such that all computations depending on the high-order model size can be moved into a parameter-independent offline phase, while having a fast input-output evaluation online.

In the example under consideration here, the parameter dependency is already affine and a mapping to the reference domain can be established without using an approximation technique such as the empirical interpolation method. Thus, there exists an affine expansion of the system matrix $\mathcal{A}(\mu)$ in the parameter μ as

$$\mathcal{A}(\mu) = \sum_{i=1}^{Q} \Theta_i(\mu) \mathcal{A}_i. \tag{11}$$

The coefficients $\Theta_i(\mu)$ are computed from the mapping $\mathbf{x} = T_k(\mu)\hat{\mathbf{x}} + \mathbf{g}_k$, $T_k \in \mathbb{R}^{2 \times 2}$, $\mathbf{g}_k \in \mathbb{R}^2$, which maps the deformed subdomain $\hat{\Omega}_k$ to the reference subdomain Ω_k. See also [19, 21]. Figure 5 shows the reference subdomains Ω_k for the problem under consideration.

For each subdomain $\hat{\Omega}_k$ the elemental basis function evaluations are transformed to the reference domain. For each velocity basis function $\mathbf{u} = (u_1, u_2)$, $\mathbf{v} = (v_1, v_2)$, $\mathbf{w} = (w_1, w_2)$ and each (scalar) pressure basis function ψ, we can write the transformation with summation convention as:

$$\int_{\hat{\Omega}_k} \frac{\partial \hat{\mathbf{u}}}{\partial \hat{x}_i} \hat{v}_{ij} \frac{\partial \hat{\mathbf{v}}}{\partial \hat{x}_j} d\hat{\Omega}_k = \int_{\Omega_k} \frac{\partial \mathbf{u}}{\partial x_i} v_{ij} \frac{\partial \mathbf{v}}{\partial x_j} d\Omega_k,$$

$$\int_{\hat{\Omega}_k} \hat{\psi} \nabla \cdot \hat{\mathbf{u}} d\hat{\Omega}_k = \int_{\Omega_k} \psi \chi_{ij} \frac{\partial u_j}{\partial x_i} d\Omega_k,$$

$$\int_{\hat{\Omega}_k} (\hat{\mathbf{u}} \cdot \nabla)\hat{\mathbf{v}} \cdot \hat{\mathbf{w}} d\hat{\Omega}_k = \int_{\Omega_k} u_i \pi_{ij} \frac{\partial v_j}{\partial x_i} \mathbf{w} d\Omega_k,$$

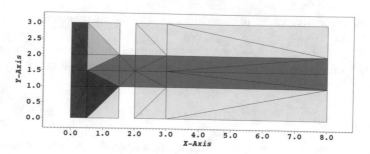

Fig. 5 Reference computational domain with subdomains Ω_1 (green), Ω_2 (yellow), Ω_3 (blue), Ω_4 (grey) and Ω_5 (brown)

with

$$v_{ij} = T_{ii'}\hat{v}_{i'j'}T_{jj'}\det(T)^{-1},$$

$$\chi_{ij} = \pi_{ij} = T_{ij}\det(T)^{-1}.$$

The subdomain Ω_5 (see Fig. 5) is kept constant, so that no interpolation of the inflow profile is necessary. To achieve fast reduced order solves, the offline-online decomposition expands the system matrix as in (11) and computes the parameter independent projections offline, which are stored as small-sized matrices of the order $N \times N$. Since in an *Oseen*-iteration each matrix is dependent on the previous iterate, the submatrices corresponding to each basis function are assembled and then formed online using the reduced basis coordinate representation of the current iterate. This is the same procedure used for the assembly of the nonlinear term in the *Navier-Stokes* case [13].

5 Numerical Results

The accuracy of the ROM is assessed using 40 snapshots sampled uniformly over the parameter domain [0.1, 2.9] for the POD and 40 randomly chosen parameter locations to test the accuracy. Figure 6 (left) shows the decay of the energy of the POD modes. To reach the typical threshold of 99.99% on the POD energy, it takes 9 POD modes as RB ansatz functions. Figure 6 (right) shows the relative $L^2(\Omega)$ approximation error of the reduced order model with respect to the full order model up to 6 digits of accuracy, evaluated at the 40 randomly chosen verification parameter locations. With 9 POD modes the maximum approximation error is less than 0.7% and the mean approximation error is less than 0.5%.

While the full-order solves were computed with Nektar++, the reduced-order computations were done in ITHACA-SEM with a separate python code. To assess the computational gain, the time for a fixed point iteration step using the full-

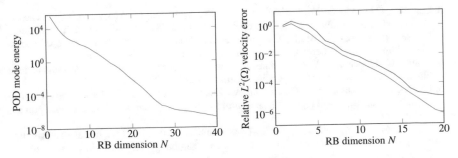

Fig. 6 Left: Decay of POD mode energy. Right: Maximum (red) and mean (blue) relative $L^2(\Omega)$ error for the velocity over increasing reduced basis dimension

order system is compared to the time for a fixed point iteration step of the ROM with dimension 20, both done in python. The ROM online phase reduces the computational time by a factor of over 100. The offline time is dominated by computing the snapshots and the trilinear forms used to project the advection terms. See [13] for detailed explanations.

6 Conclusion and Outlook

We showed that the POD reduced basis technique generates accurate reduced order models for SEM discretized models under parametric variation of the geometry. The potential of a high-order spectral element method with a reduced basis ROM is the subject of current investigations. See also [6]. Since each spectral element comprises a block in the system matrix in local coordinates, a variant of the reduced basis element method (RBEM) [14, 15] can be successfully applied in the future.

Acknowledgements This work was supported by European Union Funding for Research and Innovation through the European Research Council (project H2020 ERC CoG 2015 AROMA-CFD project 681447, P.I. Prof. G. Rozza). This work was also partially supported by NSF through grant DMS-1620384 (Prof. A. Quaini).

References

1. Boffi, D., Brezzi F., Fortin, M.: Mixed Finite Element Methods and Applications. Springer Series in Computational Mathematics. Springer, Berlin (2013)
2. Burger, M.: Numerical Methods for Incompressible Flow. Lecture Notes. UCLA, Los Angeles (2010)

3. Cantwell, C.D., Moxey, D., Comerford, A., Bolis, A., Rocco, G., Mengaldo, G., de Grazia, D., Yakovlev, S., Lombard, J.-E., Ekelschot, D., Jordi, B., Xu, H., Mohamied, Y., Eskilsson, C., Nelson, B., Vos, P., Biotto, C., Kirby, R.M., Sherwin, S.J.: Nektar++: an open-source spectral/hp element framework. Comput. Phys. Commun. **192**, 205–219 (2015)
4. Canuto, C., Hussaini, M.Y., Quarteroni, A., Zhang, Th.A.: Spectral Methods Fundamentals in Single Domains. Scientific Computation. Springer, Berlin (2006)
5. Canuto, C., Hussaini, M.Y., Quarteroni, A., Zhang, Th.A.: Spectral Methods Evolution to Complex Geometries and Applications to Fluid Dynamics. Scientific Computation. Springer, Berlin (2007)
6. Fick, L., Maday, Y., Patera A., Taddei T.: A stabilized POD model for turbulent flows over a range of Reynolds numbers: optimal parameter sampling and constrained projection. J. Comput. Phys. **371**, 214–243 (2018)
7. Herrero, H., Maday, Y., Pla, F.: RB (Reduced Basis) for RB (Rayleigh–Bénard). Comput. Meth. Appl. Mech. Eng. **261–262**, 132–141 (2013)
8. Hess, M.W., Rozza, G.: A spectral element reduced basis method in parametric CFD. In: Numerical Mathematics and Advanced Applications ENUMATH 2017. Springer, Berlin (2018, in press). E-print arXiv:1712.06432
9. Hess, M.W., Alla, A., Quaini, A., Rozza, G., Gunzburger, M.: A localized reduced-order modeling approach for PDEs with bifurcating solutions. In: Computer Methods in Applied Mechanics and Engineering (CMAME) (2019, accepted for publication). E-print. arXiv:1807.08851
10. Hesthaven, J.S., Rozza, G., Stamm, B.: Certified Reduced Basis Methods for Parametrized Partial Differential Equations. SpringerBriefs in Mathematics. Springer, Berlin (2016)
11. Holmes, P., Lumley, J., Berkooz, G.: Turbulence, Coherent Structures, Dynamical Systems and Symmetry. Cambridge University Press, Cambridge (1996)
12. Karniadakis, G., Sherwin, S.: Spectral/hp Element Methods for Computational Fluid Dynamics, 2nd edn. Oxford University Press, Oxford (2005)
13. Lassila, T., Manzoni, A., Quarteroni, A., Rozza, G.: Model order reduction in fluid dynamics: Challenges and perspectives. In: Quarteroni, A., Rozza, G. (eds.) Reduced Order Methods for Modelling and Computational Reduction. MS&A Modeling, Simulation and Applications, vol. 9, pp. 235–273. Springer International Publishing, Cham (2014)
14. Lovgren, A.E., Maday, Y, Ronquist, E.M.: A reduced basis element method for the steady Stokes problem. ESAIM: Math. Model. Numer. Anal. **40**(3), 529–552 (2006)
15. Maday, Y., Ronquist, E.M.: A reduced-basis element method. Comptes Rendus Math. **335**(2), 195–200 (2002)
16. Patera, A.T.: A spectral element method for fluid dynamics; laminar flow in a channel expansion. J. Comput. Phys. **54**(3), 468–488 (1984)
17. Pitton, G., Rozza, G.: On the application of reduced basis methods to bifurcation problems in incompressible fluid dynamics. J. Sci. Comput. **73**, 157–177 (2017)
18. Pitton, G., Quaini, A., Rozza, G.: Computational reduction strategies for the detection of steady bifurcations in incompressible fluid-dynamics: applications to *Coanda* effect in cardiology. J. Comput. Phys. **344**, 534–557 (2017)
19. Quarteroni, A., Rozza, G.: Numerical solution of parametrized Navier-Stokes equations by reduced basis methods. Num. Meth. Part. Diff. Eq. **23**(4), 923–948 (2007)
20. Quarteroni, A., Valli, A.: Numerical Approximation of Partial Differential Equations. Springer, Berlin (1994)
21. Rozza, G.: Real-time reduced basis solutions for Navier-Stokes equations: optimization of parametrized bypass configurations. In: ECCOMAS CFD 2006 Proceedings on CD, vol. 676, pp. 1–16 (2006)
22. Wille, R., Fernholz, H.: Report on the first European mechanics colloquium, on the *Coanda* effect. J. Fluid Mech. **23**(4), 801–819 (1965)

Iterative Spectral Mollification and Conjugation for Successive Edge Detection

Robert E. Tuzun and Jae-Hun Jung

1 Introduction

Detection of edges is a fundamental problem in a variety of applications, including image processing and the numerical solution of differential equations. In applications such as magnetic resonance imaging (MRI), it is required to construct images from Fourier data. Let $\{\hat{f}_k | k = 0, \pm 1, \pm 2, \cdots \}$ be the set of Fourier coefficients of $f(x) \in L_2[-\pi, \pi]$ given by

$$\hat{f}_k = \frac{1}{2\pi} \int_{-\pi}^{\pi} f(x) e^{-ikx} dx,$$

and let f_N be the Fourier partial sum $f_N = \sum_{k=-N}^{N} \hat{f}_k e^{ikx}$. When the underlying function is smooth and periodic, the Fourier reconstruction f_N is accurate to spectral accuracy, but when edges are present, the reconstruction is plagued by the Gibbs phenomenon, also known as the Gibbs ringing in MRI applications.

Various methods have been proposed to address these issues and those methods consist of edge detection followed by reconstruction. Thus, the determination of edge locations is critical. Fourier concentration method has emerged over the past

R. E. Tuzun
Department of Mathematics, University at Buffalo, The State University of New York, Buffalo, NY, USA
e-mail: retuzun@buffalo.edu

J.-H. Jung (✉)
Department of Mathematics, University at Buffalo, The State University of New York, Buffalo, NY, USA

Department of Data Science, Ajou University, Suwon, South Korea
e-mail: jaehun@buffalo.edu

© The Author(s) 2020
S. J. Sherwin et al. (eds.), *Spectral and High Order Methods for Partial Differential Equations ICOSAHOM 2018*, Lecture Notes in Computational Science and Engineering 134, https://doi.org/10.1007/978-3-030-39647-3_46

decade as a robust method for edge detection in a variety of circumstances and applications [5, 6]. Essentially, a certain Fourier partial sum converges to the jump function as the number of Fourier coefficients increases and this convergence can be accelerated by what is known as concentration factors (functions). Use of different types of concentration factors tends to impart trade-offs between oscillations near jump discontinuities and significant non-zero concentration away from them [2]. Several methods have been devised to address this issue, as well as to treat special circumstances such as incomplete Fourier data and the presence of noise [1, 4, 13, 15].

Thanks to the convergence property of the Fourier concentration to the jump function, the concentration method detects edges with large concentrations. Where the function is smooth, the concentration vanishes as the jump function vanishes as $N \to \infty$. In practice, the concentration method is designed to detect edges with magnitudes larger than some given threshold, with the value of the used threshold being problem dependent. The value of the threshold cannot be arbitrarily small; otherwise, too many false edges can be detected. If the magnitude of weak edges is much smaller than other edges, those edges are considered insignificant, but for some cases weak edges are more important than strong edges. For example, it was shown that in the segmentation of MRI of the knee, the cartilage is better characterized by weak edges rather than strong edges for the separation from the tibia and femur [11].

This note shows that an iterative approach based on the successive conjugation and adaptive mollification can detect all edges without any prior threshold. This approach is similar to the iterative method in the context of the radial basis function method [3, 9, 10]. The iterative method is as follows: at each iteration step, all previously found edges are smoothed by a local mollification and new corresponding Fourier coefficients are computed. By applying conjugation and mollification successively, one can distinguish real edges from fake edges. This approach is useful and effective particularly for problems where the weak jump can significantly affect the global solution of differential equations or images where the interesting structure is represented by the weak edges [11].

In Sect. 2, a brief explanation of the Fourier concentration method is given. In Sect. 3, the proposed iterative method is explained based on the adaptive filtering method. The stopping criteria is also explained. Numerical examples with remarks are given. In Sect. 4, a brief concluding remark is provided.

2 Edge Detection Using Fourier Concentration Method

Let $[f](x) = f(x^+) - f(x^-)$ denote the jump function of $f(x) \in L_2[-\pi, \pi]$, where the superscripts $+$ and $-$ denote the limits taken from the right and left, respectively. Given a finite set of Fourier coefficients, $\{\hat{f}_k\}_{|k| \leq N}$, the Fourier concentration method, developed in [5, 6], computes the concentration as a sum of the form

$$S_N^\sigma[f](x) = i \sum_{|k| \leq N} sgn(k) \hat{f}_k \sigma\left(\frac{|k|}{N}\right) e^{ikx} \tag{1}$$

where the $\sigma(\cdot)$ are known as concentration factors and $sgn(k)$ is the sign function. Given certain admissibility conditions [6], the sum converges to the jump function:

$$S_N^\sigma[f](x) = [f](x) + \begin{cases} O\left(\frac{\log N}{N}\right), & d(x) \leq \frac{\log N}{N} \\ O\left(\frac{\log N}{(Nd(x))^s}\right), & d(x) \gg N^{-1} \end{cases} \tag{2}$$

where $d(x)$ denotes the distance to the nearest edge and s depends on the concentration factor. Here we note that Eq. (2) shows that the concentration function $S_N^\sigma[f](x)$ recovers the jump function of $f(x)$ as $N \to \infty$ and the convergence may be slow. Equation (2) also implies that the absolute maximum value of the concentration function $S_N^\sigma[f](x)$ converges to the maximum jump. Accordingly we observe that strong jumps are relatively easier to detect than weak jumps. The common types of concentration factors satisfying the admissibility conditions are polynomial concentrations

$$\sigma(\eta) = p\eta^p, \quad p \geq 1, \tag{3}$$

where p is a positive integer and $\eta = |k|/N$ and exponential concentration functions

$$\sigma(\eta) = C\eta e^{1/(\alpha\eta(\eta-1))}, \tag{4}$$

where $\alpha > 0$ is an order and C is a normalization constant. Cutoffs for edge detection, $\tau \in (0, 1]$, are with respect to the normalized concentration

$$\hat{S}(y) = |S_N^\sigma[f](y)| / \max_y \{|S_N^\sigma[f](y)|\}$$

and the edge set, E, is defined as

$$E = \{x | \hat{S}(x) \geq \tau, x \in [-\pi, \pi]\}. \tag{5}$$

Several approaches have been developed for improving the concentration method. We refer readers some to [1, 2, 4, 6, 7, 12, 13, 15, 16]. All these methods are basically utilizing the edge map. Figure 1 is by the Fourier concentration method for $f_1(x)$

$$f_1(x) = \begin{cases} 3 & \frac{\pi}{4} \leq x \leq \frac{\pi}{2} \\ -2 & \pi \leq x \leq \frac{5}{4}\pi \\ m_w & \frac{3}{2}\pi \leq x \leq \frac{7}{4}\pi \\ 0 & \text{otherwise} \end{cases} \tag{6}$$

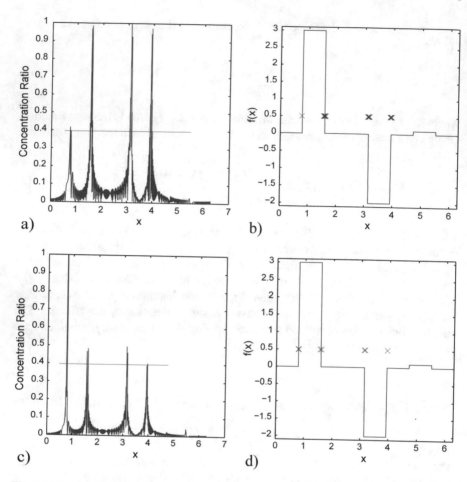

Fig. 1 Edge detection for $f_1(x)$ with $\tau = 0.4$. (**a** and **b**) with the polynomial concentration with $p = 5$. (**c** and **d**) with the exponential concentration with $p = 12$ for $\sigma = \exp\left[-\epsilon_M\left(1 - \left|\frac{k}{N}\right|\right)^p\right]$ and $\epsilon_M = 64$. (**a** and **c**) $\hat{S}(y)$. (**b** and **d**) the detected edges marked by red cross symbols. $N = 128$. The red cross in (**b** and **d**) indicates the edge locations found

where $m_w = 0.1$. That is, the magnitude of the strong jump is 30 times the weakest. As clearly shown in the figure, it is hard to detect the weak edge by looking at the normalized Fourier concentration $\hat{S}(y)$, (a) and (c) in Fig. 1, although the weak edges are clearly visible in (b) and (d) in Fig. 1.

3 Iterative Concentration Method

As clearly seen in Fig. 1, the Fourier concentration method may fail to detect the weak edge when the concentration of the weak edge is too small compared to the strong edge. To find all edges, we propose to apply the Fourier concentration method iteratively based on the local mollification using the local adaptive filtering method.

3.1 Local Adaptive Mollification

The local adaptive mollification is a key step for the iterative algorithm. Consider a smooth function $\phi \in C_0^\infty[0, 2\pi]$ which is compactly supported such that

$$\phi_\epsilon(x) = \frac{1}{\epsilon} \phi \left(\frac{x}{\epsilon} \right), \tag{7}$$

where $\lim_{\epsilon \to 0^+} \phi_\epsilon(x) = \delta(x)$. Here $\delta(x)$ is the Dirac delta function. And further $\int_{-\infty}^{\infty} \phi(x)dx = 1$. With these properties, the limit property is given by

$$\lim_{\epsilon \to 0} (\phi_\epsilon * f)(x) = f(0),$$

where $(*)$ operation denotes convolution. The parameter ϵ is free and it localizes the convolution and is known as the localization factor. The parameter ϵ is a fixed value for every x. Thus a global smoothing occurs everywhere including both the nonsmooth and smooth areas. However, we only want to apply the mollification locally to minimize the Gibbs oscillations near the jump. In order to achieve this, we use a two-parameter family of the spectral mollifier introduced by Gottlieb and Tadmor [8]. Consider the convolution of the Fourier partial sum $f_N(x)$ and the mollifier ϕ. Then by the definition of $f_N(x)$ and ϕ we have

$$(\phi_\epsilon * f_N(x)) = \frac{1}{2\pi} \int_0^{2\pi} \phi_\epsilon(x - y) f_N(y)dy$$

$$= \frac{1}{2\pi} \int_0^{2\pi} \phi_\epsilon(x - y)(D_N(y - z) * f(z))(y)dzdy,$$

where D_N is the Dirichlet kernel of degree N. The idea proposed in [8] is that one changes the degree of the Dirichlet kernel with the localization parameter ϵ so that the two-parameter family of the new mollifier is defined by

$$\phi_{p,\epsilon}(x) = \frac{1}{\epsilon} \rho \left(\frac{x}{\epsilon} \right) D_p \left(\frac{x}{\epsilon} \right), \tag{8}$$

where D_p is the Dirichlet kernel of degree p. Then for all s, the error is given by [8]

$$|\phi_{p,\epsilon} * f_N(x) - f(x)| \leq C||\rho^{(s)}||_\infty \left[N \left(\frac{1+p}{N\epsilon} \right)^{s+1} + p \left(\frac{2}{p} \right)^s ||f^s||_{L_{loc}^\infty} \right],$$
(9)

where $|| \cdot ||_{L_{loc}^\infty} = \sup_{(x-\epsilon\pi, x+\epsilon\pi)} | \cdot |$. The first term in the right hand side of the above inequality is the truncation error and the second term is the regularization error. As we see, the optimization of the error is determined by how the localization parameter ϵ and the degree p are balanced. In [8] those parameters were chosen such that $\epsilon = \frac{d}{\pi}$, where d is the distance to the nearest jump from the current position. The order of the Dirichlet kernel is chosen such that spectral convergence is achieved, say, $p \approx \sqrt{N}$. Here we note that a modification of the two-parameter mollifier for the enhancement of the convergence was proposed in [14], which was designed to reduce the Gibbs oscillations while it provides a sharp reconstruction up to the edge. Note that the adaptive mollifier was used to sharpen the concentration map $\hat{S}$ in [2].

Our proposed iterative method is that once the edge is identified, the edge region is first localized using the value of ϵ so that f_N in the region away from the detected edge is not affected by the mollification. This helps the next available edge to be preserved through the mollification of f_N if existent. Thus as in [8], the localization factor is a function of the distance from the edge, d, i.e. $\epsilon = \epsilon(d)$. Then we adaptively mollify f_N so that a heavy mollification using p is applied to reduce the Gibbs oscillations near the edge. The limit property of p is given as $p \to 0$ if $d \to 0$ and $p \to \infty$ if $d \to 2\pi$. In this work, we use the local adaptive filtering for the mollification.

3.2 Almost Automatic Stopping of the Iteration

To see the proposed method stops almost automatically, consider $f(x) = x, x \in [-\pi, \pi]$ with the Fourier coefficients

$$\hat{f}_k = (-1)^k \frac{i}{k}, \quad k \neq 0$$

and $\hat{f}_0 = 0$. There are two edges ($x = \pm\pi$) and conjugation and local adaptive filtering have the most effect at $\pm\pi$. Therefore, by considering the local behavior near $\pm\pi$, we assume a constant order of filtering p and of conjugation q, with functional forms of $\exp(-\epsilon_M(1 - |k/N|)^p)$ and $\exp(-\epsilon_M|k/N|^q)$, respectively. By letting $\phi_{p,\epsilon}$ and C_N^σ be the corresponding kernels and letting S and F denote conjugation and filtering,

$$C_N^\sigma * (\phi_{p,\epsilon} * f_N) \approx \phi_{p,\epsilon} * (C_N^\sigma * f_N)$$

and after some simplification, we have

$$S[F[f_N]] = C_N^\sigma * (\phi_{p,\epsilon} * f_N) \approx \sum_{k=-N, k\neq 0}^{N} \frac{(-1)^{k+1}}{k} \exp\left(-\epsilon\left(1 - \frac{|k|}{N}\right)^p\right)$$

$$\times \exp(-\epsilon|k/N|^q)e^{ikx}.$$

This has Fourier coefficients

$$\widehat{S}[F[f_N]] \approx \frac{(-1)^{k+1}}{k} \exp\left(-\epsilon\left(1 - \frac{|k|}{N}\right)^p\right) \exp(-\epsilon|k/N|^q).$$

From the sharp localization and heavy filtering near $\pm\pi$, we choose $p, q \longrightarrow 0$. Then setting $y = |k/N|$ yields

$$|\widehat{S}[F[f_N]]| \approx \frac{1}{k} \exp[-\epsilon(y^q) + (1 - y)^p], \quad k \neq 0$$

which approaches 0 exponentially. Thus after all the edges are found through iteration, the concentration decays exponentially small. Thus if the stopping criteria η below is chosen small enough, e.g. $\eta \sim 10^{-10}$, the stopping of the iteration is guaranteed

$$|S[F[f_N]]| \leq \eta. \tag{10}$$

3.3 Numerical Examples

We consider the case that the magnitude of the weak edge is highly small for the function $f_1(x)$ in Eq. (6)

$$m_w = 0.01.$$

Figure 2 shows how the iteration method finds edges. The order of the finding the edges is from left to right (see red arrows). As shown in the figure, the iterative method finds all edges even with $m_w = 0.01$. It is interesting to observe that the weakest edges are found in the 3rd and 7th iteration steps before all the strong edges are found. Now we consider even smaller value of m_w

$$m_w = 0.001.$$

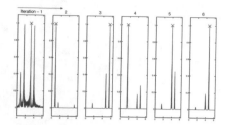 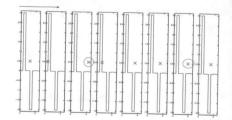

Fig. 2 $f_1(x)$ with $m_w = 0.01$ and successive edge detection. Left: $\hat{S}(y)$. Right: edges found in each iteration marked by red cross symbols with the weak edges circled in green. Note that the weak edges are almost invisible in the right figure of $f_1(x)$. Each figure in the left shows $\hat{S}(y)$ after each iteration from left to right. Each figure in the right shows the actual function and detected edges, from left to right

Figure 3 shows how the iteration method finds all edges. Figure 3 shows similar result as in Fig. 2. As shown in the figure, the method is highly accurate and finds all edges including the highly weak edges.

As an application to the solution of PDEs, namely the shock-density wave interaction equation, we consider finding shocks in the density profile at $t = 2$ with the total number of grid points $N = 300$ computed with the WENO-Z method used in [9]. The left two figures of Fig. 4 show the edges (shocks) found by the Fourier concentration method while the right figure shows the edges (shocks) found by the iterative method. As shown in the figure, the iterative method find all the physical shocks accurately while the Fourier concentration method misses some of shocks.

For two-dimensional examples, we consider a Shepp-Logan image with a faint box added to comprise additional weak edges, and a brain image. To detect edges in two dimensions, edges are detected slicewise in the x and y directions. The x and y coordinates have a range of $[-\pi, \pi]$. For a $2N_x+1 \times 2N_y+1$ image, slices of $f(x, y)$ are taken at evenly spaced x and y with $\Delta x = 2\pi/(2N_x + 1)$ and $\Delta y = 2\pi/(2N_y + 1)$, with $-\pi$ included and π excluded. Within each slice, Fourier coefficients are computed by partial Fourier expansion and the iterative method is applied to find strong and weak edges. Calculation parameters for the two-dimensional calculations were similar to those for the one-dimensional calculations. An edge with a concentration magnitude at or above a fraction $\tau = 0.1$ of the maximum magnitude concentration was considered strong. To detect strong edges, trigonometric concentration factors with $\alpha = \pi$ were used. Figure 5 shows the edges found by the proposed method for the Shepp-Logan image. As in the figure, the weak edges (square box with magnitude of 0.01) are successfully found by the method.

Remarks First, the proposed method is affected by noise as the original Fourier concentration. Consider Eq. (6). Let $\hat{f}_k(m_w = 0)$ be the Fourier coefficients with $m_w = 0$. Then $\hat{f}_k(m_w) = m_w/8$ for $k = 0$ and $\hat{f}_k(m_w) = \frac{m_w}{2\pi k}\left[\sin(7k\pi/4) - \sin(3k\pi/2)\right]$ for $k \neq 0$. The weak edge translates $\hat{f}_k$. Thus we expect that unless SNR is high enough, $|\hat{f}_k(m_w = 0) - \hat{f}_k(m_w)|$ becomes easily smaller than the noise as m_w decays. Figure 6 shows the concentration with $m_w = 0$ (left), the concentration with $SNR = 20$ (middle) and with $SNR = 10$. As in the figure, the weak edges are

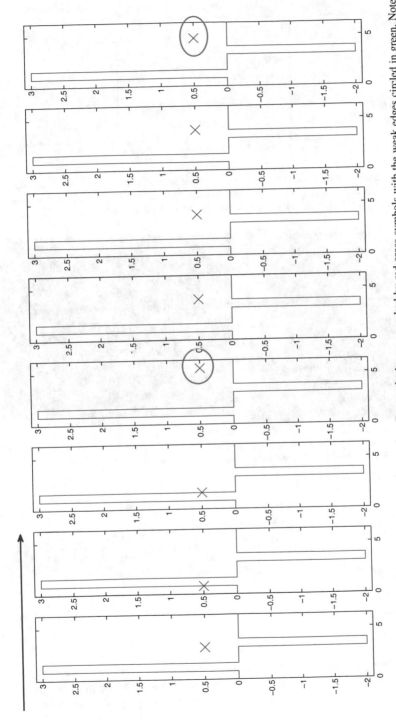

Fig. 3 $f_1(x)$ with $m_w = 0.001$ and successive edge detection. Detected edges are marked by red cross symbols with the weak edges circled in green. Note that the edges in $f_1(x)$ are almost invisible in the figure

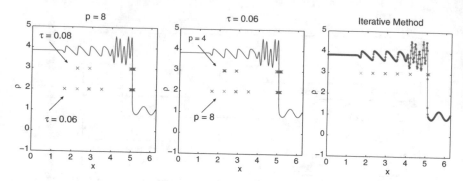

Fig. 4 The density ρ in y-axis versus the x-coordinate in x-axis at $t = 1$ for the shock-density wave interaction and shocks (with cross symbols) found with different values of τ and p. Left two figures: the Fourier concentration method. Right: the iterative method

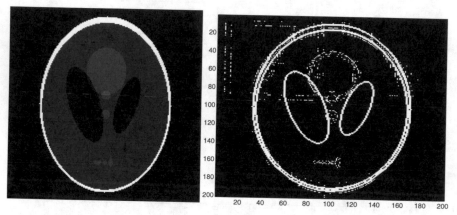

Fig. 5 Original image (left) and edges (right) detected for concentration followed by iterative method on the Shepp-Logan image with a weak square edge of magnitude 0.01 added

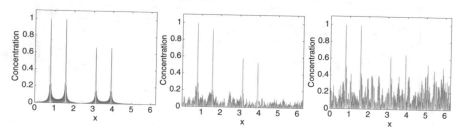

Fig. 6 Concentration in y-axis versus x. Left: without noise. Middle: $SNR=20$. Right: $SNR=10$

easily indistinguishable as SNR decreases. As the main objective of this research is finding the weak edges, a proper noise reduction suitable for the proposed method should be investigated in our future research. Second, the proposed method is to find

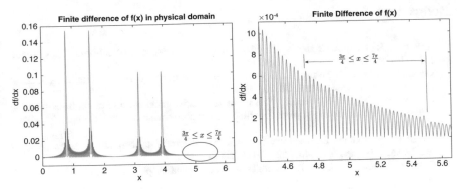

Fig. 7 Left: edges with finite difference in the physical domain. Right: $\frac{3\pi}{2} \le x \le \frac{7\pi}{4}$

the edges by $\hat{f}_k$. Figure 7 shows the edge detection in the physical domain, with the forward difference, generated from the Fourier data (the left figure). The right figure shows the plot in $x \in \frac{3\pi}{2} \le x \le \frac{7\pi}{4}$ where the weak edges exist. As in the figures, the weak edges are still hard to distinguish in the physical domain. Once the strong edges are removed in the Fourier domain and switching back and forth from the Fourier to physical domains, the weak edges are eventually found with the proposed method.

Summary The following is the summary of the proposed iterative concentration method. The procedure stops eventually with a non-zero value of $\eta > 0$ in Eq. (10).

- Step 1: Find edge locations x_o using the Fourier concentration method.
- Step 2: Apply the local filter near x_o and find the new set of Fourier coefficients.
- Step 3: Find a new edge location y_o where the normalized concentration $\hat{S}$ by $\{\hat{f}_k\}$ from Step 2 has the maximum.
- Step 4: Repeat Steps 2 and 3 until all the edges are found (the iteration stops once all edges are found.)

4 Conclusion

We showed that the iterative approach of the Fourier concentration method can detect all edges, which is not the case if the weak edges are too small. We showed that the proposed method is able to detect weak edges 3000 times weaker than the strongest edge, as long as the weak edges are well-separated from the stronger edges without noise and that the proposed method find all weak edges in a PDE application, namely the WENO calculation for the shock-density wave interaction. The iterative method also shows that it stops almost automatically after all the edges are found. Thus the proposed method is accurate and efficient.

References

1. Archibald, R., Gelb, A.: Reducing the effects of noise in image reconstruction. J. Sci. Comput. **17**, 167–179 (2002)
2. Cochran, D., Gelb, A., Wang, Y.: Edge detection from truncated Fourier data using spectral mollifiers. Adv. Comput. Math. **38**, 737–762 (2013)
3. Don, W.-S., Wang, B.-S., Gao, Z.: Fast iterative adaptive multi-quadric radial basis function method for edges detection of piecewise functions–I: uniform mesh. J. Sci. Comput. **75**, 1016–1039 (2018)
4. Engelberg, S., Tadmor, E.: Recovery of edges from spectral data with noise—a new perspective. SIAM J. Numer. Anal. **46**, 2620–2635 (2008)
5. Gelb, A., Tadmor, E.: Detection of edges in spectral data. Appl. Comput. Harmon. Anal. **7**, 101–135 (1999)
6. Gelb, A., Tadmor, E.: Detection of edges in spectral data II: nonlinear enhancement. SIAM J. Numer. Anal. **38**, 1389–1408 (2000)
7. Gelb, A., Tadmor, E.: Adaptive edge detectors for piecewise smooth data based on the minmod limiter. J. Sci. Comput. **28**, 279–306 (2006)
8. Gottlieb, D., Tadmor, E.: Recovering pointwise values of discontinuous data within spectral accuracy. In: Murman, E.M., Abarbanel, S.S. (eds.). Progress and Supercomputing in Computational Fluid Dynamics. Proceedings of a 1984 U.S.-Israel Workshop, Progress in Scientific Computing, vol. 6, pp. 357–375. Birkhauser, Boston (1985)
9. Jung, J.-H., Durante, V.R.: An iterative adaptive multi quadric radial basis function method for the detection of local jump discontinuities. Appl. Numer. Math. **59**, 1449–1466 (2009)
10. Jung, J.-H., Gottlieb, S., Kim, S.: Iterative adaptive RBF methods for detection of edges in two dimensional functions. Appl. Numer. Math. **61**, 77–91 (2011)
11. Pang, J., Miller, E., Driban, J., Tassinari, A., McAlindon, T.: A curve evolution method for identifying weak edges with applications to the segmentation of magnetic resonance images of the knee. In: 2011 IEEE International Symposium on Biomedical Imaging: From Nano to Macro, pp. 1410–1415 (2011)
12. Petersen, A., Gelb, A., Eubank, R.: Hypothesis testing for Fourier based edge detection methods. J. Sci. Comput. **51**, 608–630 (2012)
13. Stefan, W., Viswanathan, A., Gelb, A., Renaut, R.: Sparsity enforcing edge detection method for blurred and noisy Fourier data. J. Sci. Comput. **50**, 536–556 (2012)
14. Tadmor, E., Tanner, J.: Adaptive mollifiers for high resolution recovery of piecewise smooth data from its spectral information. Found. Comput. Math. **2**(2), 155–189 (2002)
15. Tadmor, E., Zou, J.: Three novel edge detection methods for incomplete and noisy spectral data. J. Fourier Anal. Appl. **14**, 744–763 (2008)
16. Viswanathan, A., Gelb, A., Cochran, D.: Iterative design of concentration factors for jump detection. J. Sci. Comput. **51**, 631–649 (2012)

Small Trees for High Order Whitney Elements

Ana Alonso Rodríguez and Francesca Rapetti

1 Introduction

We aim at determining in a constructive way, for the high order case, the finite element solutions of $\operatorname{grad}\phi = \mathbf{E}$, $\operatorname{curl}\mathbf{A} = \mathbf{B}$, $\operatorname{div}\mathbf{D} = \rho$, namely, of the equations linking the electric field $\mathbf{E}$, the magnetic induction $\mathbf{B}$, and the electric charge density ρ, to their potentials ϕ, $\mathbf{A}$ and $\mathbf{D}$, respectively. Stating the necessary and sufficient conditions for assuring that a function defined in a bounded set $\Omega \subset \mathbb{R}^3$ is the gradient of a scalar potential, the curl of a vector potential or the divergence of a vector field is one of the most classical problem of vector analysis (see for example [3, 6, 8]). We aim at providing an explicit and efficient procedure to construct a finite element solution. For example, div-free fields, $\mathbf{W}$, are implicitly characterized in terms of a vector $\mathbf{w}$ of degrees of freedom of $\mathbf{W}$ by the algebraic constraint $\mathrm{D}\mathbf{w} = \mathbf{0}$, with D the matrix of the div operator between finite elements spaces. The same fields, in the case of a domain with connected boundary, are explicitly defined by $\mathbf{w} = \mathrm{R}\,\mathbf{a}$, with no constraint on $\mathbf{a}$, where R is the matrix of the curl operator between finite elements spaces and $\mathbf{a}$ collects the degrees of freedom of the vector potentials $\mathbf{A}$. Similarly, one can wish to compute a vector potential $\mathbf{a}$ such that $\mathrm{R}\,\mathbf{a} = \mathbf{b}$, for a given field $\mathbf{b}$ verifying $\mathrm{D}\mathbf{b} = \mathbf{0}$. As explained in [5], these bases can be constructed by the help of "trees" and "co-trees", which are at the core of this contribution. The case $r = 0$ is largely treated in the literature for different types of topological domains (see for example [2]). In these pages, we develop the

A. Alonso Rodríguez (✉)
Dipartimento di Matematica, Università degli Studi di Trento, Trento, Italy
e-mail: ana.alonso@unitn.it

F. Rapetti
Département de Mathématiques, Université Côte d'Azur, Nice, France
e-mail: francesca.rapetti@univ-cotedazur.fr

© The Author(s) 2020
S. J. Sherwin et al. (eds.), *Spectral and High Order Methods for Partial Differential Equations ICOSAHOM 2018*, Lecture Notes in Computational Science and Engineering 134, https://doi.org/10.1007/978-3-030-39647-3_47

tree and co-tree approaches for $r > 0$ when fields in the high order Whitney spaces are represented on the basis of their weights on small simplices [7, 9, 10]. With this choice of degrees of freedom, the tree and co-tree concepts extend from $r = 0$ to $r > 0$ straightforwardly.

2 Basic Concepts

Let $\Omega \subset \mathbb{R}^3$ be a bounded polyhedral domain with Lipschitz boundary $\partial\Omega$ and $\mathcal{M}$ a simplicial mesh of $\bar{\Omega}$. We denote by $|A|$ the cardinality of the set A. For $0 \le k \le 3$, let $\Delta_k(T)$ (resp. $\Delta_k(\mathcal{M})$) be the set of k-simplices of a mesh tetrahedron T (resp. of the mesh $\mathcal{M}$). Note that $\Delta_k(\mathcal{M}) = \cup_{T \in \mathcal{M}} \Delta_k(T)$. If $\Delta_0(\mathcal{M}) = \{\mathbf{v}_i\}_i$, with $i = 1, \ldots, N_v$, being $N_v = |\Delta_0(\mathcal{M})|$, then each k-simplex $S \in \Delta_k(\mathcal{M})$ has associated an increasing map $m_S : \{0, \ldots, k\} \to \{1, \ldots, N_v\}$. This map induces an (inner) orientation on S (i.e., a way to run along S if $k = 1$, through S if $k = 2$, in S if $k = 3$).

If we assign to each $S \in \Delta_k(\mathcal{M})$ a real number c_S we can define the k-chain $c = \sum_{S \in \Delta_0(\mathcal{M})} c_S\, S$, i.e. a formal weighted sum of k-simplices S in $\mathcal{M}$. One can add k-chains, namely $(c + \tilde{c}) = \sum_S (c_S + \tilde{c}_S)\, S$, and multiply a k-chain by a scalar p, namely $p\, c = \sum_S (p\, c_S)\, S$. The set of all k-chains in $\mathcal{M}$, here denoted $\mathcal{C}_k(\mathcal{M})$, is a vector space, in one-to-one correspondence with the set of real vectors $c = (c_S)_{S \in \Delta_k(\mathcal{M})}$. Each k-simplex $S \in \Delta_k(\mathcal{M})$, can be associated with the elementary k-chain c with entries $c_S = 1$ and $c_{\tilde{S}} = 0$ for $\tilde{S} \ne S$. In the following we will use the same symbol S to denote the oriented k-simplex and the associated elementary k-chain.

The boundary operator ∂ takes a k-simplex S and returns the sum of all its $(k-1)$-faces f with coefficient 1 or -1 depending of whether the orientation of the $(k-1)$-face f matches or not with the orientation induced by that of the simplex S on f. Since the boundary operator is a linear mapping from $\mathcal{C}_k(\mathcal{M})$ to $\mathcal{C}_{k-1}(\mathcal{M})$, it can be represented by a matrix ∂ of dimension $|\Delta_{k-1}(\mathcal{M})| \times |\Delta_k(\mathcal{M})|$, which is rather sparse, gathering the coefficients 0, -1, or $+1$. Note that in three dimensions, there are three nontrivial boundary operators acting, respectively, on edges, triangles and tetrahedra: ∂_1 represented by the matrix $\mathsf{G}^\top$, ∂_2 represented by $\mathsf{R}^\top$, and ∂_3 represented by $\mathsf{D}^\top$. To fully specify ∂, we need to specify the boundary of each simplex S. By definition, we have

$$\partial_1 e = \sum_{n \in \Delta_0(\mathcal{M})} \mathsf{G}_{e,n}\, n, \qquad \partial_2 f = \sum_{e \in \Delta_1(\mathcal{M})} \mathsf{R}_{f,e}\, e, \qquad \partial_3 T = \sum_{f \in \Delta_2(\mathcal{M})} \mathsf{D}_{T,f}\, f,$$

for any $e \in \Delta_1(\mathcal{M})$, any $f \in \Delta_2(\mathcal{M})$ and any $T \in \Delta_3(\mathcal{M})$. For $e = [\mathbf{v}_0, \mathbf{v}_1]$, $f = [\mathbf{v}_0, \mathbf{v}_1, \mathbf{v}_2]$ and $T = [\mathbf{v}_0, \mathbf{v}_1, \mathbf{v}_2, \mathbf{v}_3]$, we have, respectively,

$$\partial_1[\mathbf{v}_0, \mathbf{v}_1] = \mathbf{v}_0 - \mathbf{v}_1, \qquad \partial_2[\mathbf{v}_0, \mathbf{v}_1, \mathbf{v}_2] = [\mathbf{v}_0, \mathbf{v}_1] - [\mathbf{v}_0, \mathbf{v}_2] + [\mathbf{v}_1, \mathbf{v}_2],$$
$$\partial_3[\mathbf{v}_0, \mathbf{v}_1, \mathbf{v}_2, \mathbf{v}_3] = [\mathbf{v}_0, \mathbf{v}_1, \mathbf{v}_2] - [\mathbf{v}_0, \mathbf{v}_1, \mathbf{v}_3] + [\mathbf{v}_0, \mathbf{v}_2, \mathbf{v}_3] - [\mathbf{v}_1, \mathbf{v}_2, \mathbf{v}_3].$$

The subscript is removed when there is no ambiguity, since the operator needed for a particular operation is indicated from the type of the operand (e.g., ∂_3 when ∂ applies to tetrahedra). The notion of boundary can be extended to k-chains by linearity, $\partial c = \partial(\sum_{S \in \Delta_k(\mathcal{M})} \mathbf{c}_S S) = \sum_{S \in \Delta_k(\mathcal{M})} \mathbf{c}_S \partial S$.

We say that a k-chain c is closed if $\partial_k c = 0$. Non-trivial closed k-chains are called k-cycles and constitute the subspace $Z_k(\mathcal{M}) = \ker(\partial_k; C_k(\mathcal{M}))$. A k-chain c is a boundary if it exists a $(k+1)$-chain γ such that $c = \partial_{k+1}\gamma$. The k-boundaries constitute the subspace $B_k(\mathcal{M}) = \partial_{k+1}C_{k+1}(\mathcal{M})$. From the property $\partial\partial = 0$, we know that boundaries are cycles but not all cycles are boundaries, and we have $B_k(\mathcal{M}) \subset Z_k(\mathcal{M})$. The quotient space $\mathcal{H}_k(\mathcal{M}) = [Z_k(\mathcal{M})/B_k(\mathcal{M})]$ is the homology spaces of order k of the mesh $\mathcal{M}$, and the Betti's number $b_k = \mathrm{rank}\,[\mathcal{H}_k(\mathcal{M})]$. The presence of curl-free fields (resp. div-free fields) that are not the gradient of a scalar field (resp. the curl of a vector field) is indicated from the fact that $b_1 \neq 0$ (resp. $b_2 \neq 0$). We recall that Betti's numbers are topological invariants (i.e., they depend on the domain Ω up to a homeomorphism) and do not depend on the mesh $\mathcal{M}$ on $\bar{\Omega}$ that is used to compute them (see [12] and an application in [11]).

For the high order case, we need to introduce some concepts of relative homology. Let $\mathcal{K}_k(\mathcal{M})$ be subspaces of $C_k(\mathcal{M})$ with $\partial_k\mathcal{K}_k(\mathcal{M}) \subset \mathcal{K}_{k-1}(\mathcal{M})$. We thus say that $c \in C_k(\mathcal{M})$ is closed [modulo $\mathcal{K}_k(\mathcal{M})$] if $\partial c \in \mathcal{K}_k(\mathcal{M})$. A $(k-1)$-chain c bounds [modulo $\mathcal{K}_k(\mathcal{M})$] if there exists a k-chain γ such that $c - \partial\gamma \in \mathcal{K}_{k-1}(\mathcal{M})$. We thus talk about relative homology groups.

A k-cochain w (over the mesh $\mathcal{M}$) is a linear mapping from $C_k(\mathcal{M})$ to $\mathbb{R}$. They are discrete analogues to differential forms. For $k > 0$, the exterior derivative of the $(k-1)$-form w is the k-form dw such that $\int_s dw = \int_{\partial s} w$ for all $s \in C_k(\mathcal{M})$. With this simple equation relating the evaluation of dw on a simplex s to the evaluation of w on the boundary of this simplex, the exterior derivative is readily defined. We can naturally extend the notion of evaluation of a differential form w on an arbitrary chain by linearity: $\int_{\sum_i \mathbf{c}_i s_i} w = \sum_i \mathbf{c}_i \int_{s_i} w$. Thus

$$\int_{\sum_i \mathbf{c}_i s_i} dw = \int_{\partial(\sum_i \mathbf{c}_i s_i)} w = \int_{\sum_i \mathbf{c}_i \partial s_i} w = \sum_i \mathbf{c}_i \int_{\partial s_i} w.$$

The operator d is the dual of the boundary operator ∂. As a corollary of the boundary operator property $\partial\partial = 0$, we have that $dd = 0$. Since we used arrays of dimension $|\Delta_k(\mathcal{M})|$ to represent a k-cochain, the operator d can be represented by a matrix $\mathbf{d}$ of dimension $|\Delta_k(\mathcal{M})| \times |\Delta_{k-1}(\mathcal{M})|$, $1 \leq k \leq 3$. Again, we have one matrix for the exterior derivative operator for each simplex dimension. When a metric is introduced on the ambient affine space, the exterior derivative operator d stands for grad, curl, div, according to the value of k from 1 to 3, and it is represented by, respectively, G, R, D, the connectivity matrices of the mesh $\mathcal{M}$.

3 Small Simplices, Weights and Potentials

We introduce the multi-index $\boldsymbol{\alpha} = (\alpha_0, \ldots, \alpha_s)$ of $s + 1$ integers $\alpha_i \geq 0$ and weight $|\boldsymbol{\alpha}| = \sum_{i=1}^{s} \alpha_i$. The set of multi-indices $\boldsymbol{\alpha}$ with $s + 1$ components and weight r is denoted $\mathcal{I}(s + 1, r)$. We denote by $\mathbf{v}_i$ the (Cartesian) coordinates of the node n_i in $\mathbb{R}^3$. Given a multi-index $\boldsymbol{\alpha} \in \mathcal{I}(4, r)$, and a k-subsimplex S of T, the small simplex $\{\boldsymbol{\alpha}, S\}$ is the k-simplex that belongs to the small tetrahedron with barycenter at the point of coordinates $\sum_{i=0}^{3}[(\frac{1}{4} + \alpha_i)\mathbf{v}_{\sigma_T^0(i)}]/(r + 1)$, which is parallel and $1/(r + 1)$-homothetic to the (big) sub-simplex S of T. The notation $\{\boldsymbol{\alpha}, S\}$ was first defined in [9]. The set of small tetrahedra of order $r + 1 > 1$ can be visualized starting from the principal lattice $L_{r+1}(T)$ in the simplex $T = \{n_{\sigma_T^0(0)} \, n_{\sigma_T^0(1)} \, n_{\sigma_T^0(2)} \, n_{\sigma_T^0(3)}\}$ defined as

$$L_{r+1}(T) = \left\{ \mathbf{x} \in T \, : \, \lambda_{\sigma_T^0(i)}(\mathbf{x}) \in \{0, \frac{1}{r + 1}, \frac{2}{r + 1}, \ldots, \frac{r}{r + 1}, 1\}, \, 0 \leq i \leq 3 \right\}.$$

and connecting its points by edges parallel to those of T. (See, e.g., Fig. 1.)

We denote by $\Lambda^k(\Omega)$ the space of all smooth differential k-forms on Ω. The completion of $\Lambda^k(\Omega)$ in the corresponding norm defines the Hilbert space $L^2\Lambda^k(\Omega)$. Let $\mathcal{P}_{r+1}^-\Lambda^k(T)$ be the space of so-called trimmed polynomial k-forms of degree $r + 1$ on T, with $r \geq 0$, (as in [7]), and we define

$$\mathcal{P}_{r+1}^-\Lambda^k(\mathcal{M}) = \{\omega \in H\Lambda^k(\Omega) \, : \, \omega_{|T} \in \mathcal{P}_{r+1}^-\Lambda^k(T), \, T \in \mathcal{M}\}$$

where $H\Lambda^k(\Omega) = \{\omega \in \Lambda^k(\Omega) \, : \, d\omega \in \Lambda^k(\Omega)\}$ is a Hilbert space (see [4]).

Definition 1 The weights of a polynomial k-form $u \in \mathcal{P}_{r+1}^-\Lambda^k(T)$, with $0 \leq k \leq 3$ and $r \geq 0$, are the scalar quantities

$$\int_{\{\alpha, S\}} u, \tag{1}$$

on the small simplices $\{\boldsymbol{\alpha}, S\}$ with $\boldsymbol{\alpha} \in \mathcal{I}(4, r)$ and $S \in \Delta_k(T)$.

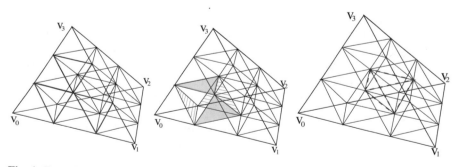

Fig. 1 From the principal lattice of degree $r + 1 = 3$ in a tetrahedron T, we define a decomposition of T into 10 small tetrahedra, 4 octahedra O and 1 reversed tetrahedron. Each face on ∂T is decomposed into 6 small faces and 3 reversed triangles, in solid red line (Left)

We now list some remarkable properties of the small simplices which are useful in the tree construction.

Property 1 The weights (1) of a Whitney k-form $u \in \mathcal{P}^-_{r+1}\Lambda^k(T)$ on all the small simplex $\{\alpha, S\}$ of T are unisolvent, as stated in [7, Proposition 3.14]. The small simplices can thus support the degrees of freedom for fields $u \in \mathcal{P}^-_{r+1}\Lambda^k(T)$, with $0 \leq k \leq 3$ and $r \geq 0$. Since the result on unisolvence holds true also by replacing T with $F \in \Delta_{n-1}(T)$ then $\mathrm{Tr}_F u \in \mathcal{P}^-_{r+1}\Lambda^k(F)$ is uniquely determined by the weights on small simplices in F. It thus follows that a locally defined u, with $u_{|T} \in \mathcal{P}^-_{r+1}\Lambda^k(T)$ and single-valued weights, is in $H\Lambda^k(\Omega)$. We thus can use the weights on the small simplices $\{\alpha, S\}$ as degrees of freedom for the fields in the finite element space $\mathcal{P}^-_{r+1}\Lambda^k(\mathcal{M})$ being aware that their number is greater than the dimension of the space.

Property 2 The weights given in Definition 1 have a meaning as cochains and this relates directly the matrix describing the exterior derivative with the matrix of the boundary operator. The key point is the Stokes' theorem $\int_C du = \int_{\partial C} u$, where u is a $(k-1)$-form and C a k-chain. More precisely, if $u \in \mathcal{P}^-_{r+1}\Lambda^k(\mathcal{M})$ then $z = du \in \mathcal{P}^-_{r+1}\Lambda^{k+1}(\mathcal{M})$ and

$$\int_{\{\alpha,S\}} z = \int_{\{\alpha,S\}} du = \int_{\partial\{\alpha,S\}} u = \sum_{\{\beta,F\}} B_{\{\alpha,S\},\{\beta,F\}} \int_{\{\beta,F\}} u$$

being B the boundary matrix with as many rows as small simplices of dimension k and as many columns as small simplices of dimension $k - 1$. The small simplices $\{\alpha, S\}$ inherit the orientation of the simplex S so the coefficient $B_{\{\alpha,S\},\{\beta,F\}}$ is equal to the coefficient $B_{S,F}$ of the boundary of the simplex S if $\beta = \alpha$. This is straightforward if $\dim(F) > 0$ and when $\dim(F) = 0$, providing that small nodes in T are given in the notation $\{\alpha, n\}$ according to their position in the small simplices when fragmented (see Fig. 1 in [1]).

Property 3 The generated $\binom{r+2}{2}$ small faces on each face F of T, pave F together with the $\binom{r+1}{2}$ reversed triangles, denoted by ∇, contained in F. Similarly, the generated $\binom{r+3}{3}$ small tetrahedra contained in T pave T together with the $\binom{r+2}{3}$ octahedra, denoted by O, and the $\binom{r+1}{3}$ reversed tetrahedra, denoted by $\bot$, contained in T, as shown in Fig. 1. Reversed octahedra and reversed tetrahedra are examples of "holes" in T (see [9, 10]).

Property 4 Since homology is preserved by homotopy, in [10, Section 3.4], it is discussed the fact that the relative homology (i.e., the homology [modulo the holes' boundaries]), of the complex of small simplices is the same of the homology of $\mathcal{M}$. This property is fundamental to build the tree for high order potentials when working with small simplices. The homology [modulo the holes' boundaries] can be translated in matrix notation, by showing that the boundary matrices associated with the small simplices, "modified" and "completed" (in a sense that we explain in the next section) by the relations [10, Proposition 3.5] are incidence matrices of a graph. To apply the theory presented in [10, Section 3.4] in a tetrahedron $T \in \Delta_3(\mathcal{M})$, we need to introduce, for $r > 0$, two sets $\mathcal{K}_1$ and $\mathcal{K}_2$ of chains generated by the small simplices that belong to the boundary of some hole in T as follows:

- $\mathcal{K}_1$ are the chains generated by the boundary of the $\binom{r+1}{2}$ reversed triangle $\nabla \subset F$ and that for each $F \in \Delta_2(T)$, and the boundary of the three faces out of four on the boundary $\partial \perp$ of each of the $\binom{r}{2}$ reversed tetrahedra $\perp$ in T;
- $\mathcal{K}_2$ are the chains generated by 4 out of 8 faces of the $\binom{r+2}{3}$ octahedra O in T. The involved faces are the small faces belonging to the boundary ∂O privated of $\partial O \cap (\Delta_2(T) \cup \partial \perp)$.

The two sets $\mathcal{K}_1$ and $\mathcal{K}_2$ satisfy the property $\partial \mathcal{K}_2 \subset \mathcal{K}_1$, decisive to conclude that the relative homology [modulo the holes' boundaries] of the complex of the small simplices is the same as the homology of the original mesh $\mathcal{M}$ [10].

4 Trees and Graphs

As stated in [12], a directed graph $\mathcal{G}$ consists of two sets $\mathcal{N}$ and $\mathcal{A}$ of nodes and arcs, respectively, subjected to certain incidence relations, collected in the all-vertex incidence matrix $\mathrm{M}^{\mathcal{G}} \in \mathbb{Z}^{|\mathcal{N}| \times |\mathcal{A}|}$ as follows:

$$\mathrm{M}^{\mathcal{G}}_{n,a} = \begin{cases} -1, & \text{if } a \text{ starts from } n, \\ +1, & \text{if } a \text{ ends in } n, \\ 0, & \text{if } a \text{ does not contain } n. \end{cases}$$

An incidence matrix M of the graph $\mathcal{G}$ is any sub-matrix of $\mathrm{M}^{\mathcal{G}}$ with $|\mathcal{N}| - 1$ rows and $|\mathcal{A}|$ columns. The node that corresponds to the row of $\mathrm{M}^{\mathcal{G}}$ that is not in M will be indicated as the reference node of $\mathcal{G}$. A graph $\mathcal{G}$ is connected if there is a path between any two of its nodes. A tree $\mathcal{T}$ of a graph $\mathcal{G}$ is a connected acyclic subgraph of $\mathcal{G}$. A spanning tree $\mathcal{T}_s$ is a tree of $\mathcal{G}$ visiting all its nodes. Any connected graph $\mathcal{G}$ admits a spanning tree $\mathcal{T}_s$. We have now to particularize these notions for small simplices. In each tetrahedron T of the oriented mesh $\mathcal{M}$, we consider the small mesh associated with $L_{r+1}(T)$ composed only of small tetrahedra, for a given r uniform all over the mesh $\mathcal{M}$. The union of the small meshes for all $T \in \Delta_3(\mathcal{M})$ is denoted $\mathcal{M}_{all}$.

A (Primal) Small Tree for the Gradient Problem

For $r = 0$, the graph $\mathcal{G}^1$ has $\mathcal{N} = \Delta_0(\mathcal{M})$ and $\mathcal{A} = \Delta_1(\mathcal{M})$. The boundary matrix $\mathrm{G}^\top$ is the all-vertex incidence matrix of the graph $\mathcal{G}^1$. Extracting a spanning 1-tree $\mathcal{T}_s^1$ from $\mathcal{G}^1$ is equivalent to finding in $\mathrm{G}^\top$, minus one row, a submatrix of maximal rank (see [11] for a suitable and easy way of constructing $\mathcal{T}$). For $r > 0$, we have

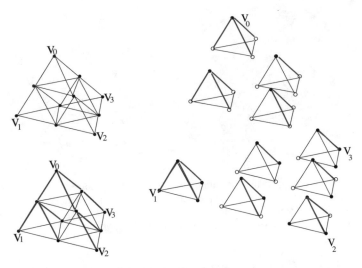

Fig. 2 (Left) The graph $\mathcal{G}^1$ and a spanning tree in thick line, for $r = 1$. (Right) A spanning tree for $r = 2$ in a fragmented layout

to consider the new graph $\mathcal{G}^1$ with $\mathcal{N} = \Delta_0(\mathcal{M}_{all})$ and $\mathcal{A} = \Delta_1(\mathcal{M}_{all})$. Let $\mathrm{G}_{all}^{\mathrm{T}}$ be the all-vertex incidence matrix of this new graph $\mathcal{G}^1$. Note that $\mathrm{G}_{all}^{\mathrm{T}}$ results from the boundary operator ∂_1 on the elementary 1-chains from $\mathcal{M}_{all}$. Extracting a spanning 1-tree $\mathcal{T}_s^1$ from $\mathcal{G}^1$ is equivalent to finding in $\mathrm{G}_{all}^{\mathrm{T}}$, minus one row, a submatrix of maximal rank. Example of spanning 1-tree $\mathcal{T}_s^1$ for $r+1 = 2$ in the right part of Fig. 2 and for $r + 1 = 3$ in Fig. 5 (fragmented visualization). Note that we can repeat this construction in the two-dimensional case.

A (Dual) Small Tree for the Divergence Problem

For $r = 0$, the graph $\mathcal{G}^2$ is built on $\mathcal{M}^*$, the so-called dual mesh of $\mathcal{M}$, as follows. Let us note that an internal face $F \in \Delta_2(\mathcal{M})$ connects two adjacent tetrahedra $T_1, T_2 \in \Delta_3(\mathcal{M})$ whereas a boundary face $F_b \in \Delta_2(\mathcal{M})$ connects a tetrahedron $T_b \in \Delta_3(\mathcal{M})$ and the boundary $\partial\Omega$. We can construct the following connected (dual) graph $\mathcal{G}^2$: the set of nodes, $\mathcal{N}$, contains the barycenter of any tetrahedron $T \in \Delta_3(\mathcal{M})$ together with one additional exterior node representing $\partial\Omega$; the set of arcs, $\mathcal{A}$, contains any face $F \in \Delta_2(\mathcal{M})$. For $r = 0$, the matrix D associated with the boundary operator ∂_3, acting on $C_3(\mathcal{M})$, is an incidence matrix of the (dual) graph $\mathcal{G}^2$, with reference node the one corresponding to $\partial\Omega$. Extracting a spanning tree $\mathcal{T}_s^2$ from $\mathcal{G}^2$ is equivalent to finding in D a submatrix of maximal rank.

For $r > 0$, let $\mathcal{R}_2$ be the set of small faces chosen as follows: one small face for each octahedron O contained in $\mathcal{K}_2$ (see the right side of Fig. 3 for the dashed small face in $\mathcal{R}_2$ when $r+1 = 2$). To construct the graph $\mathcal{G}^2$ for $r > 0$ we need to consider $\mathcal{M}_{all}^*$, the dual mesh associated to $\mathcal{M}_{all}$, where nodes are the small tetrahedra and the arcs the small faces, apart from the ones in $\mathcal{R}_2$. To understand this, we can reason

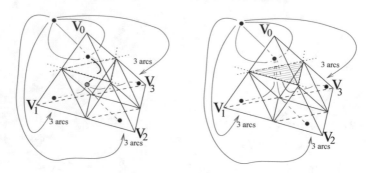

Fig. 3 The (dual) graph $\mathcal{G}_*^2$ associated with the small mesh $\mathcal{M}_{all}$ defined in a tetrahedron T for $r = 1$: the black dots are the nodes, and curved lines the arcs (Left). The (dual) graph $\mathcal{G}^2$ obtained from $\mathcal{G}_*^2$ by merging the nodes corresponding to barycenter of $t_0 = \{(1, 0, 0, 0), T\}$ and of O, thus eliminating the arc associated with the shaded small face f_u^O (Right)

as follows. For $r > 0$, we have one arc connecting two small tetrahedra, say $t_\diamond$, $t_\circ$, when

- either $t_\diamond$, $t_\circ$ share the *same* small face f, i.e. $\partial t_\diamond \cap \partial t_\circ = f$;
- or $t_\diamond$, $t_\circ$ have a small face on the boundary of the *same* octahedron O, i.e. $f_\diamond = \partial t_\diamond \cap \partial O$ and $f_\circ = \partial t_\circ \cap \partial O$ for the *same* octahedron O.

See an example of graph $\mathcal{G}^2$ for $\mathcal{M}_{all}$ (here $\mathcal{M} = \{T\}$) in the left part of Fig. 3 for $r + 1 = 2$, where the node associated with the octahedron O is not a node in the graph, but stands to indicate that the four small tetrahedra are connected one to the other by one arc because they all have one small face on ∂O. Naming t_k the small tetra with a vertex in $\mathbf{v}_k$, $k = 0, 3$, and numbering first the 3×4 faces on $t_k \cap \partial T$, called f_i^k for $i = 1, 2, 3$, second those on ∂O (where $f_u^O, f_\ell^O, f_d^O, f_r^O$ are the small faces up, left, down, right of ∂O), we have

$$
D_{tmp} = \begin{array}{c} \partial T \\ t_0 \\ t_1 \\ t_2 \\ t_3 \\ O \end{array} \left(\begin{array}{cccccccccccccccc} -1 & -1 & -1 & -1 & -1 & -1 & -1 & -1 & -1 & -1 & -1 & -1 & & & & \\ 1 & 1 & 1 & & & & & & & & & & -1 & & & \\ & & & 1 & 1 & 1 & & & & & & & & -1 & & \\ & & & & & & 1 & 1 & 1 & & & & & & -1 & \\ & & & & & & & & & 1 & 1 & 1 & & & & -1 \\ & & & & & & & & & & & & 1 & 1 & 1 & 1 \end{array} \right)
$$
$$
\quad\quad\quad f_1^0 \; f_2^0 \; f_3^0 \; f_1^1 \; f_2^1 \; f_3^1 \; f_1^2 \; f_2^2 \; f_3^2 \; f_1^3 \; f_2^3 \; f_3^3 \; f_u^O \; f_\ell^O \; f_d^O \; f_r^O
$$

Since the octahedron O is not part of the small mesh $\mathcal{M}_{all}$, we have to imagine that its node collapses with the node of one of its neighbouring small tetrahedron, say t_0 with a vertex in $\mathbf{v}_0$, and thus that the corresponding arc (i.e. the small face $f_u^O = \partial t_0 \cap \partial O$, the dashed one in the right part of Fig. 3) is eliminated. From a

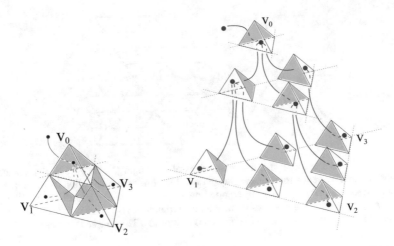

Fig. 4 Example of spanning tree in the (dual) graph $\mathcal{G}^2$, namely a selection of acyclic paths made of arcs, visiting all the nodes of $\mathcal{G}^2$ ($r = 1$, Left and $r = 2$, Right)

matrix point of view, D is obtained by adding the line "O" in D_{tmp} to the line "t_0", and eliminating f_u^O, namely

$$
D = \begin{array}{c} t_0 \\ t_1 \\ t_2 \\ t_3 \end{array}
\begin{pmatrix}
\mathbf{1} & \mathbf{1} & \mathbf{1} & & & & & & & & & & \mathbf{1} & \mathbf{1} & \mathbf{1} \\
& & & 1 & 1 & 1 & & & & & & & & \mathbf{-1} & \\
& & & & & & 1 & 1 & 1 & & & & & & \mathbf{-1} \\
& & & & & & & & & 1 & 1 & 1 & & & \mathbf{-1}
\end{pmatrix}
$$
$$
\quad f_1^0 \; f_2^0 \; f_3^0 \; f_1^1 \; f_2^1 \; f_3^1 \; f_1^2 \; f_2^2 \; f_3^2 \; f_1^3 \; f_2^3 \; f_3^3 \; f_\ell^O \; f_d^O \; f_r^O
$$

(in bold font, the submatrix of maximal rank in D for the spanning tree $\mathcal{T}_s^2$ illustrated in Fig. 4, left part for $r + 1 = 2$). To repeat this construction in the two-dimensional case, when T is a triangle, we have to consider the mesh $\mathcal{M}_{all}$ of small triangles in T and the role of the core octahedra O is played by the reversed triangles $\nabla \in T$. The set $\mathcal{R}_2$ is replaced by $\mathcal{R}_1$, composed of one small edge for each reversed triangle $\nabla \in \mathcal{K}_1$. In two dimensions we do not have reversed tetrahedra, therefore no reversed triangles $\nabla_\perp$.

The construction of the spanning tree in $\mathcal{M}_{all}$ can be done by assembling that of the geometrical mesh $\mathcal{M}$, namely a spanning tree for the Whitney forms of lower degree (blue lines in Fig. 5 (Right)), together with local contributions, one from each element (green lines in Fig. 5 (Right)). Each local contribution results from one fixed on a *reference* element which is *mapped* on the current element (respecting the orientation). In Fig. 5 (Left), in green/red thick line we have marked the small edges of a spanning tree in the graph $\mathcal{G}^1$, for $r = 3$, in the reference triangle. The red ones belong to the spanning tree in the reference triangle, but they are in general omitted in the spanning tree of $\mathcal{M}_{all}$, (indeed, they appear only if they are covered by the

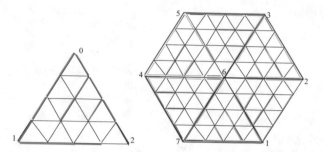

Fig. 5 (Left) In thick colored line, the small edges of the graph $\mathcal{G}^1$, for $r = 3$, that compose a spanning tree in a *reference* triangle. (Right) In thick blue line the contribution of the branches of a spanning tree in a (2D) toy mesh $\mathcal{M}$ reported on $\mathcal{M}_{all}$. In green, the contribution of the small branches *mapped* from the green ones in the reference triangle. It is not necessary to report the red ones since they are either covered by the blue ones or omitted. The co-tree is in black

blue tree). The small co-tree is in black. A similar construction can be repeated in 3D (both for $k = 1$ and $k = 2$) and it reflects the decomposition given, for instance, in [13] (Sect. 5).

References

1. Alonso Rodríguez, A., Rapetti, F.: The discrete relations between fields and potentials with high order Whitney forms. In: Radu, F.A., et al. (eds.) European Conference on Numerical Mathematics and Advanced Applications. Enumath 2017. Lecture Notes in Computational Science and Engineering LNCSE, vol. 126. Springer, Berlin (2018)
2. Alonso Rodríguez, A., Valli, A.: Finite element potentials. Appl. Numer. Math. **95**, 2–14 (2015)
3. Amrouche, C., Bernardi, C., Dauge, M., Girault, V.: Vector potentials in three-dimensional nonsmooth domains. Math. Methods Appl. Sci. **21**, 823–864 (1998)
4. Arnold, D., Falk, R., Winther, R.: Finite element exterior calculus, homological techniques, and applications. Acta Numer. **15**, 1–155 (2006)
5. Bossavit, A.: Computational Electromagnetism: Variational Formulations, Complementarity, Edge Elements. Academic, New York (1998)
6. Cantarella, J., DeTurck, D., Gluck, H.: Vector calculus and the topology of domains in 3-space. Am. Math. Mon. **109**, 409–442 (2002)
7. Christiansen, S.H., Rapetti, F.: On high order finite element spaces of differential forms. Math. Comput. **85/298**, 517–548 (2016)
8. Girault, V., Raviart, P.A.: Finite Element Methods for Navier-Stokes Equations: Theory and Applications. Springer, New York (1986)
9. Rapetti, F., Bossavit, A.: Geometrical localization of the degrees of freedom for Whitney elements of higher order. IEE Proc. Science, Meas. Technol. **1/1**, 63–66 (2007); Special Issue on "Computational Electromagnetism"
10. Rapetti, F., Bossavit, A.: Whitney forms of higher degree. SIAM Numer. Anal. **47/3**, 2369–2386 (2009)
11. Rapetti, F., Dubois, F., Bossavit, A.: Discrete vector potentials for non-simply connected three-dimensional domains. SIAM Numer. Anal. **41/4**, 1505–1527 (2003)

12. Stillwell, J.: Classical Topology and Combinatorial Group Theory. Springer, New-York (1993)
13. Zaglmayr, S.: High Order Finite Element Methods for Electromagnetic Field Computation. Ph.D. Thesis, Johannes Kepler Universität Linz, 2006

Non-conforming Elements in Nek5000: Pressure Preconditioning and Parallel Performance

A. Peplinski, N. Offermans, P. F. Fischer, and P. Schlatter

1 Introduction

One of the most important concerns when solving numerically partial differential equations is finding the optimal grid on which the solution will be computed. Unfortunately in most cases it is not an easy task that could be determined in advance without deep understanding of the studied problem. That is why self-adapting algorithms like e.g. adaptive mesh refinement (AMR) have received much attention in past decades and became an important part of many packages for numerical modelling of fluid dynamics e.g. [9, 18]. The goal of AMR is to control the computational error during the simulation by placing higher resolution grids where it is needed. This makes the numerical modelling more robust, and gives the possibility to increase the accuracy of numerical simulations at minimal computational cost. The drawback is, however, increased solver complexity, and it that can have negative effects on the parallel code performance, in particular related to load balancing.

There are number of different AMR schemes, and in the context of the spectral element method (SEM) [16], in which the discretisation is based on a decomposition of the computational domain into a number of non-overlapping, high-order sub-domains called elements, we can distinguish three different categories: The mesh adaptation in this case can mean adjusting the (local) size of an element (*r*-

A. Peplinski (✉) · N. Offermans · P. Schlatter
Linné FLOW Centre and Swedish e-Science Research Centre (SeRC), KTH Mechanics, Stockholm, Sweden
e-mail: adam@mech.kth.se; nof@mech.kth.se; pschlatt@mech.kth.se

P. F. Fischer
CS and MechSE Depts., University of Illinois at Urbana–Champaign, Champaign, IL, USA
e-mail: fischerp@illinois.edu

© The Author(s) 2020
S. J. Sherwin et al. (eds.), *Spectral and High Order Methods for Partial Differential Equations ICOSAHOM 2018*, Lecture Notes in Computational Science and Engineering 134, https://doi.org/10.1007/978-3-030-39647-3_48

refinement), changing the polynomial order in a particular element (*p*-refinement), or splitting the element into smaller ones (*h*-refinement). In this work we concentrate on an *h*-refinement framework and its implementation in *Nek5000* [8], which is a highly parallel and efficient SEM solver for the incompressible Navier–Stokes equations. In its established version, Nek5000 only supports conformal elements at constant polynomial order throughout the domain.

The present work was started within EU project CRESTA, where the non-conforming solver for advection-diffusion problem was developed and the basic AMR tasks were implemented using existing external libraries. As *h*-refinement affects the element connectivity resulting in non-conforming meshes, a special grid manager is required to perform local refinement/coarsening and to build globally consistent meshes. For this task the *p4est* library [1] has been chosen, as it is designed to manipulate domains composed of multiple, non-overlapping logical cubic sub-domains, which can be represented by a recursive tree structure. This library provides element connectivity information for the dual graph, which is later manipulated by ParMETIS [10] producing a new element-to-processor mapping. The final step of grid refinement/coarsening and redistribution is performed within the non-conforming version of *Nek5000*, which utilises the so-called conforming-space/nonconforming-mesh approach based on the previous work of Fischer et al. [7, 11]. As the solver complexity grows special care has been taken to develop efficient tools that can be used within AMR framework. A more detailed description of them and the related scaling tests can be found in [17].

The goal of ExaFLOW is to extend results of CRESTA to the full incompressible Navier–Stokes equations focusing on proper adaptation of the pressure preconditioners for nonconforming SEM. Defining a robust parallel preconditioning strategy has received much attention in past decades, as the linear sub-problem associated with the divergence-free constraint (pressure-Poisson equation) can become very ill-conditioned. In the context of SEM two possible approaches based on the additive overlapping Schwarz method [4, 6] and the hybrid Schwarz-multigrid method [5, 12] were proposed and implemented in *Nek5000*, leading to a significant reduction of pressure iterations.

In the present paper, we discuss the modifications necessary to adapt *Nek5000* for the *h*-type AMR framework. The article is organised as follows. A short description of SEM and pressure preconditioners is given in Sects. 2 and 3. The following Sects. 4 and 5 describe the algorithmic modifications and parallel performance of the code. Finally, Sect. 6 provides conclusion and future work.

2 SEM Discretisation of the Navier–Stokes Equations

We review briefly the discretisation of the incompressible Navier–Stokes equations to introduce notation and point out algorithm parts that require modification. The more in-depth derivation can be found in e.g. [4]. The temporal discretisation is based on a semi-implicit scheme in which the nonlinear term is treated explicitly and the remaining unsteady Stokes problem is solved implicitly. To avoid spurious pressure modes our spatial discretisation is based on the $\mathbb{P}_N - \mathbb{P}_{N-2}$ SEM, where velocity and pressure spaces are spanned by Lagrangian interpolants on the Gauss–Lobatto–Legendre (GLL) and Gauss–Legendre (GL) quadrature points, respectively. Note that the basis for velocity is continuous across element interfaces, whereas the basis for pressure is not. Assuming $\mathbf{f}^n$ incorporates all nonlinear and source terms treated explicitly at time t^n, the matrix form of the Stokes problem after applying the Uzawa decoupling reads:

$$\begin{bmatrix} \mathbf{H} & -\frac{\Delta t}{\beta_0}\mathbf{H}\mathbf{B}^{-1}\mathbf{D}^T \\ 0 & \mathbf{E} \end{bmatrix} \begin{pmatrix} \mathbf{u}^n \\ \Delta p \end{pmatrix} = \begin{pmatrix} \mathbf{B}\mathbf{f}^n + \mathbf{D}^T p^{n-1} \\ g \end{pmatrix}, \tag{1}$$

where

$$\mathbf{E} = \frac{\Delta t}{\beta_0}\mathbf{D}\mathbf{B}^{-1}\mathbf{D}^T \tag{2}$$

is the Stokes Schur complement governing the pressure, $\Delta p = p^n - p^{n-1}$ is the pressure update, and g is the inhomogeneity arising from Gaussian elimination. In these equations $\mathbf{H} = -\frac{1}{Re}\mathbf{A} + \frac{\beta_0}{\Delta t}\mathbf{B}$ and $\mathbf{D}$ are the discrete Helmholtz and divergence operators, respectively. β_0, $\mathbf{A}$ and $\mathbf{B}$ denote here a coefficient from time derivative, a discrete Laplacian and a diagonal mass matrix associated with the velocity mesh. Applying the Uzawa decoupling we use the inverse mass matrix $\mathbf{B}^{-1}$ as approximation of the inverse Helmholtz operator $\mathbf{H}^{-1}$, giving rise to a splitting error. Note that for this splitting method the diagonality of the mass matrix $\mathbf{B}$ is crucial to avoid costly matrix inversion.

All operators $\mathbf{H}$, $\mathbf{A}$, $\mathbf{B}$ and $\mathbf{E}$ are symmetric positive definite (SPD) and can be solved with a preconditioned conjugate gradient (PCG) method. Moreover, $\mathbf{E}$ has properties similar to a Poisson operator, and is often referred to as a *consistent Poisson operator*. The systems involving $\mathbf{H}$ and $\mathbf{E}$ are solved iteratively with $\mathbf{E}$ being more challenging, and in the next section we will present the preconditioning strategy for the pressure equation,

$$\mathbf{E}\Delta p = -\mathbf{D}\mathbf{u}. \tag{3}$$

We close this section by shortly presenting the SEM operators. SEM introduces a globally unstructured and locally structured basis by tessellating the domain into K non-overlapping subdomains (deformed quadrilaterals), $\Omega = \bigcup_{k=1}^{K}\Omega_k$, and representing functions in each subdomain in terms of tensor-product polynomials on a reference subdomain $\hat{\Omega} = [-1, 1]^d$. In this approach every function or operator is represented by its local counterparts, which in case of functions takes the form of a sum over the subdomains

$$f(\mathbf{x}) = \sum_{k=1}^{K}\sum_{i} f_i^k h_i(\mathbf{r}).$$

Here, f_i^k and h_i are the nodal values of the function in Ω_k and the base functions in $\hat{\Omega}$, respectively, with i representing the natural ordering of nodes in $\hat{\Omega}$. Combining the coefficients f_i^k one can build global $\underline{f}$ and local $\underline{f}_L$ representations of the function. Each global degree of freedom occurs only once in the global representation, but has multiple copies of faces, edges and vertices related to Ω_k in the local one. To enforce function continuity, the global-to-local mapping is defined as the matrix–vector product $\underline{f}_L = Q\underline{f}$, where Q is a binary operator duplicating the basis coefficients in adjoining subdomains. The action $Q^T \underline{f}_L$ sums multiple contributions to the global degree of freedom from their local values. The assembled global stiffness matrix A takes the form

$$\left(\nabla f, \nabla g\right) = \underline{f}^T A \underline{g} = \underline{f}^T Q^T A_L Q \underline{g},$$

where a block diagonal matrix A_L is the unassembled stiffness matrix with each diagonal block consisting of the local stiffness matrix $A_{ij}^k = \int \frac{dh_i}{dx} \frac{dh_j}{dx} dx$. In practise, the global stiffness matrix is never formed explicitly, and the gather–scatter operator QQ^T is used instead. This operator contains all information about element connectivity.

3 Pressure Preconditioner

An efficient solution of Eq. (3) requires finding an SPD preconditioning matrix M^{-1} which can be inexpensively applied and which reduces the condition number of $M^{-1}E$. Preconditioners based on domain decomposition are a natural choice for SEM as the data is structured within an element but is otherwise unstructured.

An overlapping additive Schwarz preconditioner for Eq. (3) was developed in [4] based on linear finite element discretisation of Poisson operator. It combines solutions of the local Poisson problems in overlapping subdomains $R_k^T \hat{A}_k^{-1} R_k$ with the coarse grid problem $R_0^T \hat{A}_0^{-1} R_0$, which is solved on few degrees of freedom, but covers the entire domain

$$M^{-1} = R_0^T \hat{A}_0^{-1} R_0 + \sum_k R_k^T \hat{A}_k^{-1} R_k.$$

For the local problems restriction and prolongation operators, R_k and R_k^T, are Boolean matrices that transfer data to and from the subdomain, and $\hat{A}_k$ is a local stiffness matrix which can be inverted with e.g. a fast diagonalisation method. Note that action of R_k and R_k^T are similar to the gather–scatter operator QQ^T.

The coarse grid problem corresponds to the Poisson problem solved on the element vertices only, with R_0^T being the linear operator interpolating the coarse grid solution onto the tensor product array of GL points. Unlike in [4, 6], $\hat{A}_0$ is defined using local SEM-based Neumann operators performing the projection of local stiffness matrices A_k evaluated on the GLL quadrature points onto the set of coarse base functions b_i representing the linear finite element base on the GLL grid. The coarse base functions are defined in $\hat{\Omega}$ as a tensor-product of the one-dimensional linear functions. The local contribution to $\hat{A}_0$ is given by $b_i^T A_k b_j$, and the full $\hat{A}_0$ is finally assembled by local-to-global mapping summing contributions to the global degree of freedom from their local counterparts. $\hat{A}_0$ is one of few matrices formed explicitly in *Nek5000*.

On the other hand, the hybrid Schwarz-multigrid preconditioner is based on the multiplicative Schwarz method, which for the two-level scheme takes the form,

$$M^{-1} = R_0^T \hat{A}_0^{-1} R_0 \left[\sum_k R_k^T \hat{A}_k^{-1} R_k \right],$$

and leads to the following two-level multigrid scheme,

$$(i) \ u^1 = \sum_k R_k^T \hat{A}_k^{-1} R_k g,$$

$$(ii) \ r = g - Au^1,$$

$$(iii) \ e = R_0^T \hat{A}_0^{-1} R_0 r,$$

$$(iv) \ u = u^1 + e,$$

where g, r, e and u are right-hand side, residual, coarse-grid error and solution of equation $Au = g$, respectively. This method can be extended to a general multilevel solver performing a full V cycle [5, 12]. Notice that by replacing step $ii)$ with $r = g$ we obtain the additive Schwarz preconditioner.

4   Adaptation for Non-conforming Meshes

The important advantage of SEM in the context of AMR is its spatial decomposition into elements that can easily be split into smaller ones, and use of the local representation of the operators which decouples intra- and inter-element operations. As h-type AMR using the conforming-space/nonconforming-mesh approach leaves the approximation spaces unchanged, most of the tensor-product operations evaluated element-by-element are preserved, limiting the changes in the algorithm.

The inter-element operations are mostly performed by the gather–scatter operator QQ^T which has to be redefined to include spectral interpolation at the non-conforming faces. Following [7] we consider a non-conforming face shared by one low resolution element (parent) and two (in 3D four) high resolution elements (children). We introduce a local parent-to-child interpolation operator J^{cp} which is a spectral interpolation operator with entries

$$\left(J^{cp}\right)_{ij} = h_j(\zeta_i^{cp}),$$

where ζ_j^{cp} represents the mapping of GLL points from the child face to its parent. This operator is locally applied to give the desired nodal values on the child face, after Q copies data form the parent to the children. Building a block-diagonal matrix J_L with local matrices J^{cp} one can redefine scatter $J_L Q$ and gather–scatter $J_L Q Q^T J_L^T$ operators, respectively. For more discussion see Fig. 6 and Sect. 4 in [7].

The next crucial modification is diagonalisation of the global mass matrix $Q^T B_L Q$ (B_L is a block-diagonal built of local mass matrices), whose inverse is required in Eqs. (1) and (2). It is non-diagonal due to the fact that the quadrature points in the elements along the non-conforming faces do not coincide. A diagonal-isation procedure is given in [7] and consists of building the global vector $\tilde{\underline{b}}$

$$\tilde{\underline{b}} := B\hat{\underline{e}} = Q^T J_L^T B_L \hat{\underline{e}}_L,$$

and finally setting the lumped mass matrix $\tilde{B}_{ij} = \delta_{ij}\tilde{\underline{b}}_i$. $\hat{\underline{e}}$ and $\hat{\underline{e}}_L$ denote here the global and local vectors containing all ones.

The additive Schwarz preconditioner requires two significant modifications. The first one is related to the assembly of the coarse grid operator $\hat{A}_0$, which gets more complex for non-conforming meshes. This is due to the fact that the non-conforming mesh introduces hanging vertices located in the middle of faces or edges. These hanging vertices are not global degrees of freedom and cannot be included in $\hat{A}_0$. To remove them from consideration one has to modify the set of local coarse base functions b_i, which are thus dependent on the shape of the refined region as well as the position and orientation of the child face with respect to the parent one. Unlike the conforming case, where all b_i could be represented by a tensor product of two or three linear functions, the non-conforming mesh requires 5 basic components in two and 21 in three dimensions to assemble all the possible shapes of b_i.

The last missing components are the restriction and prolongation operators, R_k and R_k^T, for the local Poisson problem. Taking into account the similarity between these operators with QQ^T and following the previous development we use an operator similar to $J_L QQ^T J_L^T$, replacing J_L with the interpolation operator defined on the GL quadrature points. Although this choice seems to be optimal as it preserves properties of the preconditioner and J_L^T is well defined, our numerical experiments showed a significant increase of pressure iterations in some cases. It was found to be caused by the noise introduced by J_L^T in the Schwarz operator. To reduce this noise we replaced the transposed interpolation operator with the inverse one, getting a significant reduction of iterations. Unfortunately, such a preconditioner is no longer SPD and PCG cannot be used as an iterative solver in this case. The other problem is the definition of J_L^{-1}, as J^{cp} can be inverted for square matrices only, thus excluding p-refinement strategies. To avoid this problem we define a child-to-parent interpolation operator J^{pc} with the entries

$$\left(\mathsf{J}^{pc}\right)_{ij} = \begin{cases} h_j(\zeta_i^{pc}) & \text{if } \zeta_j^p \in \partial\Omega^p \cap \partial\Omega^c \\ 0 & \text{otherwise} \end{cases},$$

where $\partial\Omega^p$ and $\partial\Omega^c$ are the parent and child common faces, ζ_j^p is a parent GLL point at the face $\partial\Omega^p$, and ζ_j^{pc} represents the mapping of ζ_j^p to the child face $\partial\Omega^c$. This operator is locally applied to give the desired nodal values on the child face, before Q^T sums data form the children and the parent. Building block-diagonal matrix J_L^{-1} consisting of local matrices J^{pc} one can redefine the gather–scatter $\mathsf{J}_L\mathsf{Q}\mathsf{Q}^T\mathsf{J}_L^{-1}$ operator such that it is appropriate for the pressure preconditioner.

In a similar way we modify the multiplicative Schwarz method, as it shares a number of features with the additive one. In this case we distinguish between Schwarz (acting at single level) and restriction (connecting different levels) operators and apply $\mathsf{J}_L\mathsf{Q}\mathsf{Q}^T\mathsf{J}_L^{-1}$ and $\mathsf{J}_L\mathsf{Q}\mathsf{Q}^T\mathsf{J}_L^T$ to each of them, respectively. Unlike the additive preconditioner, the hybrid one requires also the redefinition of the diagonal weight matrix that indicates the number of sub-domains sharing a given node, and is used to accommodate for overlapping regions. Its value is important as it reduces the largest eigenvalue of the MA operator and defines the smoothing properties of the additive Schwarz step (see [4] and the references therein). In the conforming case its definition is straightforward, however the non-conforming case is more involved as hanging nodes are not real degrees of freedom. In the current implementation the information about node multiplicity on the non-conforming faces is hidden to the parent element, so the parent element sees only one neighbour instead of two (four in 3D). Although this choice gives a preconditioner that significantly reduces the number of pressure iterations, its performance for the studied cases is slightly worse than the performance of the additive Schwarz preconditioner. This can be caused by a non-optimal value of the weight matrix, or by the fact that the hybrid preconditioner is superior over the additive one for high-aspect ratio elements (that are not present in our adaptive simulations).

5 Parallel Performance

The parallel performance test is based on the one of the ExaFLOW flagship calculations, and consists of the turbulent flow around a NACA4412 wing section with 5° angle of attack, at a Reynolds number based on inflow velocity U_∞ and chord length c of $Re_c = 200{,}000$. It was previously studied in a series of well-resolved large-eddy simulations conducted with the conforming *Nek5000* version, and discussed in detail in [19]. This flow configuration was chosen to illustrate the significant benefit of using AMR, in particular when it comes to the farfield region in the computational domain, but for this article we will only briefly discuss the

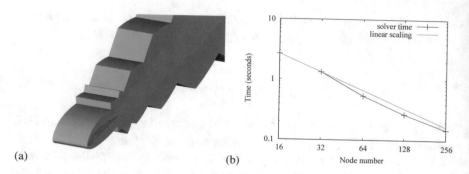

(b)

Fig. 1 (a) Volume visualisation of that part of the domain covered by refinement levels higher than one for the turbulent flow around a wing profile. The wing vicinity and wake region are resolved and a colour indicates different refinement levels. (b) Strong scaling of the non-conforming *Nek5000* solver for the same case performed on Beskow. The plot shows the time per time step as a function of node number. Each node consists of 32 cores

strong scaling results. We omit here a weak scaling test, as *Nek5000* uses iterative solvers and with the current example we cannot provide meaningful data.

The initial coarse and conforming mesh consisted of 2190 elements with polynomial order $N = 7$ and was evolved for 7.2 time units c/U_∞ to evolve the refinement process using spectral error indicators [13, 14], and allowing for 6 refinements levels. The resulting non-conforming grid was built of 224,272 elements with 76.37×10^6 degrees of freedom, resolving the wing surface and the wake, Fig. 1a. This final mesh was used to test the parallel performance of the non-conforming solver using the petascale Cray XC40 system Beskow at PDC (Stockholm). This system consists of 2060 nodes with 32 cores per node and 2.438 PFlops peak performance. We compare our results with the scaling tests of the conforming *Nek5000* presented in Offermans et al. [15]. The most relevant test in this article is pipe flow at $Re_\tau = 360$ (upper-right plot in their Fig. 5), as it is similar in size with the discussed wing case. We should mention here that our goal is not to improve the parallel performance of the conforming code, but rather to retain it despite of a work imbalance introduced by an additional operator in the direct stiffness summation of the non-conforming solver.

To be able to compare to the conforming solver, we focus on the time evolution loop only, excluding code initialisation, finalisation, mesh rebuilding within AMR and I/O operations. The result of the strong scaling test is presented in Fig. 1b showing the time per time step as a function of node count. This plot is almost identical with the reference one in [15]. Both show slight super-linear scaling between 32 and 256 nodes despite growing work imbalance for the non-conforming solver. We also reach the strong scaling limit at around 256 nodes, which for the conforming solver on Beskow was estimated to be between 30,000 and 50,000 degrees of freedom per core [15]. This shows that the parallel performance of the non-conforming and conforming solvers is almost the same and proves the efficiency of our implementation.

The maximum number of the compute nodes used in the test was not set by the parallel properties of the non-conforming *Nek5000*, but by the quality of the domain partitioning provided by *ParMETIS*. Within ExaFLOW we developed a new grid partitioning scheme for *Nek5000* (not discussed in this paper) that takes into account a core distribution among the nodes, and consists of two steps: inter- and intra-node partitioning. Although this two-level partitioning scheme significantly improves the efficiency of a coarse grid operations for XXT, especially during the setup phase, it relies on the quality of an inter-node partitioning. If the first step gives subdomains with disjoint graphs, the second step cannot be performed. We found that the probability of getting disjoint graphs increases with decreasing number of elements per node, virtually prohibiting the runs with less than 1000 elements per node. However, this limit can differ between simulations. We note however that in the standard production use of the solver this limitation is not critical, as according to [15] it is usually close to the strong scaling limit of conforming *Nek5000*.

6   Conclusions

Within the ExaFLOW project we developed a fully functional SEM-based h-type adaptive mesh refinement (AMR) solver for the incompressible Navier–Stokes equations. This allows for much larger flow cases to be run at reduced cost, as the high resolution grid is placed only in those region where it is needed. At the same time the simulation quality is improved, as the computational error can be controlled during the run.

We have optimised for non-conforming meshes the pressure preconditioners based on the additive overlapping Schwarz and hybrid Schwarz-multigrid methods. To achieve this we modified the base functions for the assembly of a coarse-grid operator to remove hanging nodes, and redefined the direct stiffness summation operator to include spectral interpolation at the non-conforming faces and edges. We introduced two operators $J_L QQ^T J_L^T$ and $J_L QQ^T J_L^{-1}$ for the different steps in the pressure calculation. The last crucial modification was the diagonalisation of the global mass matrix.

Using real flow cases we show our AMR implementation to be correct and efficient. An important success is the fact that parallel performance of the conforming and non-conforming solvers is very similar, despite the increased complexity of the non-conforming one.

In the future we are going to investigate other definitions of the weight matrix for the hybrid Schwarz-multigrid method, and to test different pressure preconditioners based on the restricted additive Schwarz method [2, 3]. We are going as well to work on the quality of the graph partition, as the two-level partitioning would not accept disjoint graphs on the node's subdomain.

Acknowledgements We would like to thank Niclas Jansson for sharing his expertise on adaptive mesh refinement. The work presented in this publication was supported by a European Commission Horizon 2020 project grant entitled "ExaFLOW: Enabling Exascale Fluid Dynamics Simulation" (grant reference 671571). Computer time was provided by Swedish National Infrastructure for Computing (SNIC) at PDC (KTH Stockholm) and by ExaFLOW at HLRS Stuttgart.

References

1. Burstedde, C., Wilcox, L., Ghattas, O.: `p4est`: scalable algorithms for parallel adaptive mesh refinement on forests of octrees. SIAM J. Sci. Comput. **33**(3), 1103–1133 (2011)
2. Cai, X.C., Sarkis, M.: A restricted additive Schwarz preconditioner for general sparse linear systems. SIAM J. Sci. Comput. **21**(2), 792–797 (1999)
3. Efstathiou, E., Gander, M.J.: Why restricted additive Schwarz converges faster than additive Schwarz. BIT Numer. Math. **43**(5), 945–959 (2003)
4. Fischer, P.F.: An overlapping Schwarz method for spectral element solution of the incompressible Navier–Stokes equations. J. Comput. Phys. **133**, 84–101 (1997)
5. Fischer, P.F., Lottes, J.W.: Hybrid Schwarz-Multigrid Methods for the Spectral Element Method: Extensions to Navier-Stokes, pp. 35–49. Springer, Berlin (2005)
6. Fischer, P., Miller, N., Tufo, H.: An overlapping Schwarz method for spectral element simulation of three-dimensional incompressible flow. In: Parallel Solution of Partial Differential Equations, pp. 159–180. Springer, New York (2000)
7. Fischer, P.F., Kruse, G.W., Loth, F.: Spectral element methods for transitional flows in complex geometries. J. Sci. Comput. **17**(1–4), 81–98 (2002)
8. Fischer, P.F., Lottes, J.W., Kerkemeier, S.G.: Nek5000 Web page (2008). http://nek5000.mcs.anl.gov
9. Fryxell, B., Olson, K., Ricker, P., Timmes, F.X., Zingale, M., Lamb, D.Q., MacNeice, P., Rosner, R., Truran, J.W., Tufo, H.: FLASH: an adaptive mesh hydrodynamics code for modeling astrophysical thermonuclear flashes. Astrophys. J. Suppl. Ser. **131**, 273–334 (2000)
10. Karypis, G., Kumar, V.: A fast and high quality multilevel scheme for partitioning irregular graphs. SIAM J. Sci. Comput. **20**(1), 359–392 (1998)
11. Kruse, G.W.: Parallel nonconforming spectral element solution of the incompressible Navier–Stokes equations in three dimensions. Ph.D. Thesis, Brown University, Providence, 1997. UMI Order No. GAX97-38573
12. Lottes, J.W., Fischer, P.F.: Hybrid multigrid/Schwarz algorithms for the spectral element method. J. Sci. Comput. **24**(1), 45–78 (2005)
13. Mavriplis, C.: A posteriori error estimators for adaptive spectral element techniques. In: Wesseling, P. (ed.) Proceedings of the Eighth GAMM-Conference on Numerical Methods in Fluid Mechanics. Notes on Numerical Fluid Mechanics, pp. 333–342 (1990)
14. Offermans, N.: Towards adaptive mesh refinement in Nek5000. Licentiate Thesis, KTH, Mechanics, 2017
15. Offermans, N., Marin, O., Schanen, M., Gong, J., Fischer, P., Schlatter, P.: On the strong scaling of the spectral element solver Nek5000 on petascale systems. In: Proceedings of the Exascale Applications and Software Conference 2016, Stockholm, April 26–29, 2016 (2016)
16. Patera, A.T.: A spectral element method for fluid dynamics: Laminar flow in a channel expansion. J. Comput. Phys. **54**(3), 468–488 (1984)
17. Peplinski, A., Fischer, P.F., Schlatter, P.: Parallel performance of h-type adaptive mesh refinement for Nek5000. In: Proceedings of the Exascale Applications and Software Conference 2016, Stockholm, April 26–29, 2016, pp. 4:1–4:9 (2016)

18. Rosenberg, D., Fournier, A., Fischer, P., Pouquet, A.: Geophysical-astrophysical spectral-element adaptive refinement (GASpAR): object-oriented h-adaptive fluid dynamics simulation. J. Comput. Phys. **215**(1), 59–80 (2006)
19. Vinuesa, R., Negi, P.S., Atzori, M., Hanifi, A., Henningson, D.S., Schlatter, P.: Turbulent boundary layers around wing sections up to $Re_c = 1,000,000$. Int. J. Heat Fluid Flow **72**, 86–99 (2018)

Sparse Approximation of Multivariate Functions from Small Datasets Via Weighted Orthogonal Matching Pursuit

Ben Adcock and Simone Brugiapaglia

1 Introduction

In recent years, a new class of approximation strategies based on compressive sensing (CS) has been shown to be able to substantially lessen the curse of dimensionality in the context of approximation of multivariate functions from pointwise data, with applications to the uncertainty quantification of partial differential equations with random inputs. Based on random sampling from orthogonal polynomial systems and on weighted ℓ^1 minimization, these techniques are able to accurately recover a sparse approximation to a function of interest from a small-sized datasets of pointwise samples. In this paper, we show the potential of weighted greedy techniques as an alternative to convex minimization programs based on weighted ℓ^1 minimization in this context.

The contribution of this paper is twofold. First, we propose a weighted orthogonal matching pursuit (WOMP) algorithm based on a rigorous derivation of the corresponding greedy index selection strategy. Second, we numerically show that WOMP is a promising alternative to convex recovery programs based on weighted ℓ^1 minimization, thanks to its ability to compute sparse approximations with an accuracy comparable to those computed via weighted ℓ^1 minimization, but with a considerably lower computational cost when the target sparsity level (and, hence, the number of WOMP iterations) is small enough. It is also worth observing here that WOMP computes approximations that are exactly sparse, as opposed to approaches based on weighted ℓ^1 minimization, which provide compressible approximations in general.

B. Adcock · S. Brugiapaglia (✉)
Simon Fraser University, Burnaby, BC, Canada
e-mail: ben_adcock@sfu.ca; simone_brugiapaglia@sfu.ca

© The Author(s) 2020
S. J. Sherwin et al. (eds.), *Spectral and High Order Methods for Partial Differential Equations ICOSAHOM 2018*, Lecture Notes in Computational Science and Engineering 134, https://doi.org/10.1007/978-3-030-39647-3_49

Brief Literature Review Various approaches for multivariate function approximation based on CS with applications to uncertainty quantification can be found in [1, 3–6, 11–13, 17]. An overview of greedy methods for sparse recovery in CS and, in particular of OMP, can be found in [7, Chapter 3.2]. For a general review on greedy algorithms, we refer the reader to [15] and references therein. Some numerical experiments on a weighted variant of OMP have been performed in the context of CS methods for uncertainty quantification in [4]. Weighted variants of OMP have also been considered in [10, 16], but the weighted procedure is tailored for specific signal processing applications and the term "weighted" does not refer to the weighted sparsity setting of [14] employed here. To the authors' knowledge, the weighted variant of OMP considered in this paper seems to have been proposed here for the first time.

Organization of the Paper In Sect. 2 we describe the setting of sparse multivariate function approximation in orthonormal systems via random sampling and weighted ℓ^1 minimization. Then, in Sect. 3 we formally derive a strategy for the greedy selection in the weighted sparsity setting and present the WOMP algorithm. Finally, we numerically show the effectiveness of the proposed technique in Sect. 4 and give our conclusions in Sect. 5.

2 Sparse Multivariate Function Approximation

We start by briefly introducing the framework of sparse multivariate function approximation from pointwise samples and refer the reader to [3] for further details.
     Our aim is to approximate a function defined over a high-dimensional domain

$$f : D \to \mathbb{C}, \quad \text{with } D = (-1, 1)^d,$$

where $d \gg 1$, from a dataset of pointwise samples $f(t_1), \ldots, f(t_m)$. Let ν be a probability measure on D and let $\{\phi_j\}_{j \in \mathbb{N}_0^d}$ be an orthonormal basis for the Hilbert space $L_\nu^2(D)$. In this paper, we will consider $\{\phi_j\}_{j \in \mathbb{N}_0^d}$ to be a tensorized family of Legendre or Chebyshev orthogonal polynomials, with ν being the uniform or the Chebyshev measure on D, respectively. Assuming that $f \in L_\nu^2(D) \cap L^\infty(D)$, we consider the series expansion

$$f = \sum_{j \in \mathbb{N}_0^d} x_j \phi_j.$$

Then, we choose a finite set of multi-indices $\Lambda \subseteq \mathbb{N}_0^d$ with $|\Lambda| = N$ and obtain the truncated series expansion

$$f_\Lambda = \sum_{j \in \Lambda} x_j \phi_j.$$

In practice, a convenient choice for Λ is the hyperbolic cross of order s, i.e.

$$\Lambda := \left\{ j \in \mathbb{N}_0^d : \prod_{k=1}^d (j_k + 1) \leq s \right\},$$

due to the moderate growth of N with respect to d. Now, assuming we collect $m \ll N$ pointwise samples independently distributed according to ν, namely,

$$f(t_1), \ldots, f(t_m), \quad \text{with} \quad t_1, \ldots, t_m \overset{\text{i.i.d.}}{\sim} \nu,$$

the approximation problem can be recasted as a linear system

$$A x_\Lambda = y + e, \tag{1}$$

with $x_\Lambda = (x_j)_{j \in \Lambda} \in \mathbb{C}^N$, and where the sensing matrix $A \in \mathbb{C}^{m \times N}$ and the measurement vector $y \in \mathbb{C}^m$ are defined as

$$A_{ij} := \frac{1}{\sqrt{m}} \phi_j(t_i), \quad y_i := \frac{1}{\sqrt{m}} f(t_i), \quad \forall i \in [m], \forall j \in [N], \tag{2}$$

with $[k] := \{1, \ldots, k\}$ for every $k \subset \mathbb{N}$. The vector $e \in \mathbb{C}^m$ accounts for the truncation error introduced by Λ and satisfies $\|e\|_2 \leq \eta$, where $\eta > 0$ is an a priori upper bound to the truncation $L^\infty(D)$-error, namely $\|f - f_\Lambda\|_{L^\infty(D)} \leq \eta$. A sparse approximation to the vector can be then computed by means of weighted ℓ^1 minimization.

Given weights $w \in \mathbb{R}^N$ with $w > 0$ (where the inequality is read componentwise), recall that the weighted ℓ^1 norm of a vector $z \in \mathbb{C}^N$ is defined as $\|z\|_{1,w} := \sum_{j \in [N]} |z_j| w_j$. We can compute an approximation $\hat{x}_\Lambda$ to x_Λ by solving the weighted quadratically-constrained basis pursuit (WQCBP) program

$$\hat{x}_\Lambda \in \arg\min_{z \in \mathbb{C}^N} \|z\|_{1,w}, \quad \text{s.t.} \quad \|Az - y\|_2 \leq \eta, \tag{3}$$

where the weights $w \in \mathbb{R}^N$ are defined as

$$w_j = \|\phi_j\|_{L^\infty(D)}. \tag{4}$$

The effectiveness of this particular choice of w is supported by theoretical results and it has been validated from the numerical viewpoint (see [1, 3]). The resulting approximation $\hat{f}_\Lambda$ to f is finally defined as

$$\hat{f}_\Lambda := \sum_{j \in \Lambda} (\hat{x}_\Lambda)_j \phi_j.$$

In this setting, stable and robust recovery guarantees in high probability can be shown for the approximation errors $\|f - f_\Lambda\|_{L^2_\nu(D)}$ and $\|f - f_\Lambda\|_{L^\infty_\nu(D)}$ under a sufficient condition on the number of samples of the form $m \gtrsim s^\gamma \cdot \text{polylog}(s, d)$, with $\gamma = 2$ or $\gamma = \log(3)/\log(2)$ for tensorized Legendre or Chebyshev polynomials, respectively, hence lessening the curse of dimensionality to a substantial extent (see [3] and references therein). We also note in passing that decoders such as the weighted LASSO or the weighted square-root LASSO can be considered as alternatives to (3) for weighted ℓ^1 minimization (see [2]).

3 Weighted Orthogonal Matching Pursuit

In this paper, we consider greedy sparse recovery strategies to find sparse approximate solutions to (1), as alternatives to the WQCBP optimization program (3). With this aim, we propose a variation of the OMP algorithm to the weighted setting.

Before introducing weighted OMP (WOMP) in Algorithm 1, let us recall the rationale behind the greedy index selection rule of OMP (corresponding to Algorithm 1 with $\lambda = 0$ and $w = 1$). For a detailed introduction to OMP, we refer the reader to [7, Section 3.2]. Given a support set $S \subseteq [N]$, OMP solves the least-squares problem

$$\min_{z \in \mathbb{C}^N} G_0(z) \text{ s.t. } \text{supp}(z) \subseteq S,$$

where $G_0(z) := \|y - Az\|_2^2$. In OMP, the support S is iteratively enlarged by one index at the time. Namely, we consider the update $S \cup \{j\}$, where the index $j \in [N]$ is selected in a greedy fashion. In particular, assuming that A has ℓ^2-normalized columns, it is possible to show that (see [7, Lemma 3.3])

$$\min_{t \in \mathbb{C}} G_0(x + te_j) = G_0(x) - |(A^*(y - Ax))_j|^2. \tag{5}$$

This leads to the greedy index selection rule operated by OMP, which prescribes the selection of an index $j \in [N]$ that maximizes the quantity $|(A^*(y - Ax))_j|^2$. We will use this simple intuition to extend OMP to the weighted case by replacing the function G_0 with a suitable function G_λ that takes into account the data-fidelity term and the weighted sparsity prior at the same time.

Let us recall that, given a set of weights $w \in \mathbb{R}^N$ with $w > 0$, the weighted ℓ^0 norm of a vector $z \in \mathbb{C}^N$ is defined as the quantity (see [14])[1]

$$\|z\|_{0,w} := \sum_{j \in \text{supp}(z)} w_j^2.$$

---

[1] The term "norm" here is an abuse of language, but we will stick to it due to its popularity.

Notice that when $w = 1$, then $\| \cdot \|_{0,w} = \| \cdot \|_0$ is the standard ℓ^0 norm. Given $\lambda \geq 0$, we define the function

$$G_\lambda(z) := \|y - Az\|_2^2 + \lambda\|z\|_{0,w}. \tag{6}$$

The tradeoff between the data-fidelity constraint and the weighted sparsity prior is balanced via the choice of the regularization parameter λ. Applying the same rationale employed in OMP for the greedy index selection and replacing G_0 with G_λ leads to Algorithm 1, which corresponds to OMP when $\lambda = 0$ and $w = 1$.

Algorithm 1 Weighted orthogonal matching pursuit (WOMP)

Inputs:

- $A \in \mathbb{C}^{m \times N}$: sampling matrix, with ℓ^2-normalized columns;
- $y \in \mathbb{C}^m$: vector of samples;
- $w \in \mathbb{R}^N$: weights;
- $\lambda \geq 0$: regularization parameter;
- $K \in \mathbb{N}$: number of iterations.

Procedure:

1. Let $\hat{x}_0 = 0$ and $S_0 = \emptyset$;
2. For $k = 1, \ldots, K$:

   a. Find $j_k \in \arg \max_{j \in [N]} \Delta_\lambda(x_{k-1}, S_{k-1}, j)$, with Δ_λ as in (7);
   b. Define $S_k = S_{k-1} \cup \{j_k\}$;
   c. Compute $\hat{x}_k \in \arg \min_{v \in \mathbb{C}^N} \|Av - y\|_2$ s.t. $\text{supp}(v) \subseteq S_k$.

Output:

- $\hat{x}_K \in \mathbb{C}^N$: approximate solution to $Az = y$.

Remark 1 The ℓ^2-normalization of the columns of A is a necessary condition to apply Algorithm 1. If A does not satisfy this hypothesis, is suffices to apply WOMP to the normalized system $\widetilde{A}z = y$, where $\widetilde{A} = AM^{-1}$ and M is the matrix containing the ℓ^2 norms of the columns of A on the main diagonal and zeroes elsewhere. The approximate solution $\hat{x}_K$ to $\widetilde{A}z = y$ computed via WOMP is then rescaled as $M\hat{x}_K$, which approximately solves $Az = y$.

The following proposition justifies the weighted variant of OMP considered in Algorithm 1. In order to minimize G_λ as much as possible, at each iteration, WOMP selects the index j that maximizes the quantity $\Delta_\lambda(x, S, j)$ defined in (7). The following proposition makes the role of the quantity $\Delta_\lambda(x, S, j)$ transparent, generalizing relation (5) to the weighted case, under suitable conditions on A and x that are verified at each iteration of Algorithm 1.

Proposition 1 *Let $\lambda \geq 0$, $S \subseteq [N]$, $A \in \mathbb{C}^{m \times N}$ with ℓ^2-normalized columns, and $x \in \mathbb{C}^N$ satisfying*

$$x \in \arg\min_{z \in \mathbb{C}^N} \|y - Az\|_2 \ s.t. \ \mathrm{supp}(z) \subseteq S.$$

Then, for every $j \in [N]$, the following holds:

$$\min_{t \in \mathbb{C}} G_\lambda(x + te_j) = G_\lambda(x) - \Delta_\lambda(x, S, j),$$

where G_λ is defined as in (6), $\Delta_\lambda : \mathbb{C}^N \times 2^{[N]} \times [N] \to \mathbb{R}$ is defined by

$$\Delta_\lambda(x, S, j) := \begin{cases} \max\left\{|(A^*(y - Ax))_j|^2 - \lambda w_j^2, \ 0\right\} & \text{if } j \notin S \\ \max\left\{\lambda w_j^2 - |x_j|^2, \ 0\right\} & \text{if } j \in S \text{ and } x_j \neq 0 \\ 0 & \text{if } j \in S \text{ and } x_j = 0. \end{cases} \tag{7}$$

Proof Throughout the proof, we will denote the residual as $r := y - Ax$. Let us first assume $j \notin S$. In this case, we compute

$$G_\lambda(x + te_j) = \|y - A(x + te_j)\|_2^2 + \lambda\|x + te_j\|_{0,w}$$

$$= \|r\|_2^2 + \underbrace{|t|^2 - 2\,\mathrm{Re}(\bar{t}(A^*r)_j) + \lambda(1 - \delta_{t,0})w_j^2}_{=:h(t)} + \lambda\|x\|_{0,w},$$

where $\delta_{x,y}$ is the Kronecker delta function. In particular, we have

$$h(t) = \begin{cases} 0 & \text{if } t = 0 \\ |t|^2 - 2\,\mathrm{Re}(\bar{t}(A^*r)_j) + \lambda w_j^2 & \text{if } t \in \mathbb{C} \setminus \{0\}. \end{cases}$$

Now, if $(A^*r)_j = 0$, then $h(t)$ is minimized for $t = 0$ and $\min_{t \in \mathbb{C}} G(x + te_j) = G(x)$. On the other hand, if $(A^*r)_j \neq 0$, by arguing similarly to [7, Lemma 3.3], we see that

$$\min_{t \in \mathbb{C} \setminus \{0\}} h(t) = -|(A^*r)_j|^2 + \lambda w_j^2,$$

where the minimum is realized for some $t \in \mathbb{C}$ with $|t| = |(A^*r)_j| \neq 0$. In summary,

$$\min_{t \in \mathbb{C}} h(t) = \min\left\{-|(A^*r)_j|^2 + \lambda w_j^2, 0\right\} = -\max\left\{|(A^*r)_j|^2 - \lambda w_j^2, 0\right\},$$

which concludes the case $j \notin S$.

Now, assume $j \in S$. Since the vector $x_S = x|_S \in \mathbb{C}^{|S|}$ is a least-squares solution to $A_S z = y$, it satisfies $A_S^*(y - A_S x_S) = 0$ and, in particular, $(A^*r)_j = 0$. (Here, $A_S \in \mathbb{C}^{m \times |S|}$ denotes the submatrix of A corresponding to the columns in S). Therefore, arguing similarly as before, we have

$$G(x + te_j) = \|r\|_2^2 + \underbrace{|t|^2 + \lambda(1 - \delta_{t,-x_j})w_j^2}_{=:\ell(t)} + \lambda\|x - x_j e_j\|_{0,w}.$$

Considering only the terms depending on t, it is not difficult to see that

$$\min_{t \in \mathbb{C}} \ell(t) = \min\{|x_j|^2, \lambda w_j^2\}.$$

As a consequence, for every $j \in S$, we obtain

$$\min_{t \in \mathbb{C}} G(x + te_j) = \|r\|_2^2 + \lambda\|x - x_j e_j\|_{0,w} + \min\{|x_j|^2, \lambda w_j^2\}$$

$$= G(x) + \min\{|x_j|^2, \lambda w_j^2\} - \lambda(1 - \delta_{x_j,0})w_j^2.$$

The results above combined with simple algebraic manipulations lead to the desired result. □

4 Numerical Results

In this section, we show the effectiveness of WOMP (Algorithm 1) in the sparse multivariate function approximation setting described in Sect. 2. In particular, we choose the weights w as in (4). We consider the function

$$f(t) = \ln\left(d + 1 + \sum_{k=1}^{d} t_k\right), \quad \text{with } d = 10. \tag{8}$$

We let $\{\phi_j\}_{j \in \mathbb{N}_0^d}$ be the Legendre and Chebyshev bases and ν be the respective orthogonality measure. In Figs. 1 and 2 we show the relative $L_\nu^2(D)$-error of the approximate solution $\hat{x}_K$ computed via WOMP as a function of iteration K, for different values of the regularization parameter λ in order to solve the linear system $Az = y$, where A and y are defined by (2) and where the ℓ^2-normalization of the columns of A is taken into account according to Remark 1. We consider $\lambda = 0$ (corresponding to OMP) and $\lambda = 10^{-k}$, with $k = 3, 3.5, 4, 4.5, 5$. Here, Λ is the hyperbolic cross of order $s = 10$, corresponding to $N = |\Lambda| = 571$. Moreover, we consider $m = 60$ and $m = 80$. The results are averaged over 25 runs and the $L_\nu^2(D)$-error is computed with respect to a reference solution approximated via least squares

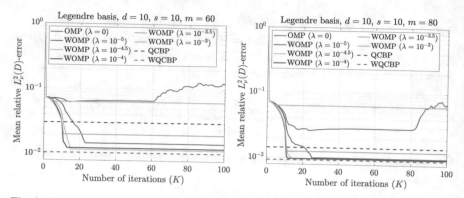

Fig. 1 Plot of the mean relative $L^2_\nu(D)$-error as a function of the number of iterations K of WOMP (Algorithm 1) for different values of the regularization parameter λ for the approximation of the function f defined in (8) and using Legendre polynomials. The accuracy of WOMP is compared with those of QCBP and WQCBP

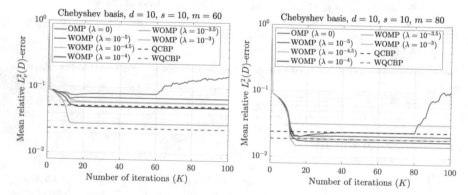

Fig. 2 The same experiment as in Fig. 1, with Chebyshev polynomials

and using $20N = 11,420$ random i.i.d. samples according to ν. We compare the WOMP accuracy with the accuracy obtained via the QCBP program (3) with $\eta = 0$ and WQCBP with tolerance parameter $\eta = 10^{-8}$. To solve these two programs we use CVX Version 1.2, a package for specifying and solving convex programs [8, 9]. In CVX, we use the solver 'mosek' and we set CVX precision to 'high'.

Figures 1 and 2 show the benefits of using weights as compared to the unweighted OMP approach, when the parameter λ is tuned appropriately. A good choice of λ for the setting considered here seem to be between $10^{-4.5}$ and $10^{-3.5}$. We also observe that WOMP is able to reach similar level of accuracy as WQCBP. An interesting feature of WOMP with respect to OMP is its better stability. We observe than after the m-th iteration, the OMP accuracy starts getting substantially worse. This can be explained by the fact that when K approaches N, OMP tends to destroy sparsity by fitting the data too much. This phenomenon is not observed in WOMP, thanks to its

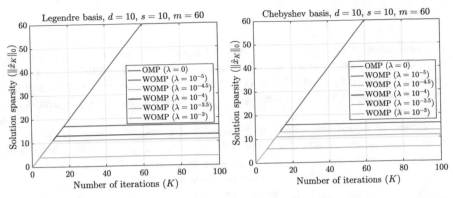

Fig. 3 Plot of the support size of $\hat{x}_K$ as a function of the number of iterations K for WOMP in the same setting as in Figs. 1 and 2, with Legendre (left) and Chebyshev (right) polynomials. The larger the regularization parameter λ, the sparser solution (in the left plot, the curves relative to $\lambda = 10^{-4.5}$ and $\lambda = 10^{-4}$ overlap. In the right plot, the same happens for $\lambda = 10^{-4}$ and $\lambda = 10^{-3.5}$)

Table 1 Comparison of the computing times for WQCBP and $K = 25$ iterations of WOMP

Basis	m	QCBP	WQCBP	OMP	WOMP with λ as below				
					10^{-5}	$10^{-4.5}$	10^{-4}	$10^{-3.5}$	10^{-3}
Legendre	60	1.9e−01	2.0e−01	1.6e−02	1.3e−02	1.2e−02	1.3e−02	1.2e−02	1.2e−02
Legendre	80	2.1e−01	2.1e−01	1.7e−02	1.5e−02	1.3e−02	1.4e−02	1.4e−02	1.3e−02
Chebyshev	60	1.9e−01	1.9e−01	1.5e−02	1.3e−02	1.2e−02	1.2e−02	1.2e−02	1.2e−02
Chebyshev	80	2.1e−01	2.1e−01	1.7e−02	1.5e−02	1.3e−02	1.4e−02	1.4e−02	1.4e−02

ability to keep the support of $\hat{x}_k$ small via the explicit enforcement of the weighted sparsity prior (see Fig. 3).

We show the better computational efficiency of WOMP with respect to the convex minimization programs QCBP and WQCBP solved via CVX by tracking the runtimes for the different approaches. In Table 1 we show the running times for the different recovery strategies. The running times for WOMP are referred to $K = 25$ iterations, sufficient to reach the best accuracy for every value of λ as shown in Figs. 1 and 2. Moreover, the computational times for WOMP take into account the ℓ^2-normalization of the columns of A (see Remark 1). WOMP consistently outperforms convex minimization, being more than ten times faster in all cases. We note that in this comparison a key role is played by the parameter K or, equivalently, by the sparsity of the solution. Indeed, in this case, considering a larger value of K would result is a slower performance of WOMP, but it would not improve the accuracy of the WOMP solution (see Figs. 1 and 2).

5 Conclusions

We have considered a greedy recovery strategy for high-dimensional function approximation from a small set of pointwise samples. In particular, we have proposed a generalization of the OMP algorithm to the setting of weighted sparsity (Algorithm 1). The corresponding greedy selection strategy is derived in Proposition 1.

Numerical experiments show that WOMP is an effective strategy for high-dimensional approximation, able to reach the same accuracy level of WQCBP while being considerably faster when the target sparsity level is small enough. A key role is played by the regularization parameter λ, which may be difficult to tune due to its sensitivity to the parameters of the problem (m, s, and d), and on the polynomial basis employed. In other applications, where explicit formulas for the weights as (4) are not available, there might also be a nontrivial interplay between λ and w. In summary, despite the promising nature of the numerical experiments illustrated in this paper, a more extensive numerical investigation is needed in order to study the sensitivity of WOMP with respect to λ. Moreover, a theoretical analysis of the WOMP approach might highlight practical recipe for the choice of this parameter, similarly to [2]. This type of analysis may also help identifying the sparsity regime where WOMP outperforms weighted ℓ^1 minimization, which, in turn, could be formulated in terms of suitable assumptions on the regularity of f. These questions are beyond the scope of this paper and will be object of future work.

Acknowledgements The authors acknowledge the support of the Natural Sciences and Engineering Research Council of Canada through grant number 611675, and of the Pacific Institute for the Mathematical Sciences (PIMS) Collaborative Research Group "High-Dimensional Data Analysis". S.B. also acknowledges the support of the PIMS Postdoctoral Training Centre in Stochastics.

References

1. Adcock, B.: Infinite-dimensional compressed sensing and function interpolation. Found. Comput. Math. **18**(3), 661–701 (2018)
2. Adcock, B., Bao, A., Brugiapaglia, S.: Correcting for unknown errors in sparse high-dimensional function approximation (2017). Preprint. arXiv:1711.07622
3. Adcock, B., Brugiapaglia, S., Webster, C.G.: Compressed sensing approaches for polynomial approximation of high-dimensional functions. In: Boche, H., Caire, G., Calderbank, R., März, M., Kutyniok, G., Mathar R. (eds.) Compressed Sensing and Its Applications: Second International MATHEON Conference 2015, pp. 93–124. Springer International Publishing, Cham (2017)
4. Bouchot, J.-L., Rauhut, H., Schwab C.: Multi-level compressed sensing Petrov-Galerkin discretization of high-dimensional parametric PDEs (2017). Preprint. arXiv:1701.01671
5. Chkifa, A., Dexter, N., Tran, H., Webster, C.G.: Polynomial approximation via compressed sensing of high-dimensional functions on lower sets. Math. Comp. **87**(311), 1415–1450 (2018)
6. Doostan, A., Owhadi, H.: A non-adapted sparse approximation of PDEs with stochastic inputs. J. Comput. Phys. **230**(8), 3015–3034 (2011)

7. Foucart, S., Rauhut, H.: A Mathematical Introduction to Compressive Sensing. Birkhäuser Basel (2013)
8. Grant, M., Boyd, S.: Graph implementations for nonsmooth convex programs. In: Blondel, V., Boyd, S., Kimura, H. (eds.) Recent Advances in Learning and Control. Lecture Notes in Control and Information Sciences, pp. 95–110. Springer, Berlin (2008)
9. Grant, M., Boyd, S.: CVX: Matlab software for disciplined convex programming, version 2.1. http://cvxr.com/cvx (2014)
10. Li, G.Z., Wang, D.Q., Zhang, Z.K., Li, Z.Y.: A weighted OMP algorithm for compressive UWB channel estimation. In: Applied Mechanics and Materials, vol. 392, pp. 852–856. Trans Tech Publications, Zurich (2013)
11. Mathelin, L., Gallivan, K.A.: A compressed sensing approach for partial differential equations with random input data. Commun. Comput. Phys. **12**(4), 919–954 (2012)
12. Peng, J., Hampton, J., Doostan, A.: A weighted ℓ^1-minimization approach for sparse polynomial chaos expansions. J. Comput. Phys. **267**, 92–111 (2014)
13. Rauhut, H., Schwab, C.: Compressive sensing Petrov-Galerkin approximation of high-dimensional parametric operator equations. Math. Comp. **86**(304), 661–700 (2017)
14. Rauhut, H., Ward, R.: Interpolation via weighted ℓ_1 minimization. Appl. Comput. Harmon. Anal. **40**(2), 321–351 (2016)
15. Temlyakov, V.N.: Greedy approximation. Acta Numer. **17**, 235–409 (2008)
16. Xiao-chuan, W., Wei-bo, D., Ying-ning, D.: A weighted OMP algorithm for Doppler super-resolution. In: 2013 Proceedings of the International Symposium on Antennas & Propagation (ISAP), vol. 2, pp. 1064–1067. IEEE, Piscataway (2013)
17. Yang, X., Karniadakis, G.E.: Reweighted ℓ_1 minimization method for stochastic elliptic differential equations. J. Comput. Phys. **248**, 87–108 (2013)

On the Convergence Rate
of Hermite-Fejér Interpolation

Shuhuang Xiang and Guo He

1 Introduction

For an arbitrarily given system of points

$$\{x_1^{(n)}, x_2^{(n)}, \ldots, x_n^{(n)}\}_{n=1}^{\infty}, \tag{1}$$

Faber [3] in 1914 showed that there exists a continuous function $f(x)$ in $[-1, 1]$ for which the Lagrange interpolation sequence $L_n[f]$ $(n = 1, 2, \ldots)$ is not uniformly convergent to f in $[-1, 1]$, where $\omega_n(x) = (x - x_1^{(n)})(x - x_2^{(n)}) \cdots (x - x_n^{(n)})$

$$L_n[f](x) = \sum_{k=1}^{n} f(x_k^{(n)}) \ell_k^{(n)}(x), \quad \ell_k^{(n)}(x) = \frac{\omega_n(x)}{\omega_n'(x_k^{(n)})(x - x_k^{(n)})}. \tag{2}$$

Whereas, based on the Chebyshev pointsystem

$$x_k^{(n)} = \cos\left(\frac{2k-1}{2n}\pi\right), \quad k = 1, 2, \ldots, n, \quad n = 1, 2, \ldots, \tag{3}$$

S. Xiang
Department of Mathematics, Central South University, Changsha, China
e-mail: xiangsh@mail.csu.edu.cn

G. He (✉)
Department of Mathematics, Jinan University, Guangzhou, Guangdong, China

© The Author(s) 2020
S. J. Sherwin et al. (eds.), *Spectral and High Order Methods for Partial Differential Equations ICOSAHOM 2018*, Lecture Notes in Computational Science and Engineering 134, https://doi.org/10.1007/978-3-030-39647-3_50

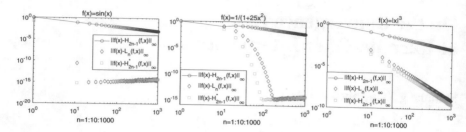

Fig. 1 $\|H_{2n-1}(f, x) - f(x)\|_\infty$, $\|L_n(f, x) - f(x)\|_\infty$ and $\|H^*_{2n-1}(f, x) - f(x)\|_\infty$ at $x = -1$: 0.001 : 1 by using Chebyshev pointsystem (3) for $f(x) = \sin(x)$, $f(x) = \frac{1}{1+25x^2}$ and $f(x) = |x|^3$, respectively

Fejér [4] in 1916 proved that if $f \in C[-1, 1]$, then there is a unique polynomial $H_{2n-1}(f, x)$ of degree at most $2n - 1$ such that $\lim_{n \to \infty} \|H_{2n-1}(f) - f\|_\infty = 0$, where $H_{2n-1}(f, x)$ is determined by

$$H_{2n-1}(f, x_k^{(n)}) = f(x_k^{(n)}), \quad H'_{2n-1}(f, x_k^{(n)}) = 0, \quad k = 1, 2, \ldots, n. \quad (4)$$

This polynomial is known as the Hermite-Fejér interpolation polynomial.

It is of particular notice that the above Hermite-Fejér interpolation polynomial converges much slower compared with the corresponding Lagrange interpolation polynomial at the Chebyshev pointsystem (3) (see Fig. 1).

To get fast convergence, the following Hermite-Fejér interpolation of $f(x)$ at nodes (1) is considered [6, 7]:

$$H^*_{2n-1}(f, x) = \sum_{k=1}^{n} f(x_k^{(n)}) h_k^{(n)}(x) + \sum_{k=1}^{n} f'(x_k^{(n)}) b_k^{(n)}(x), \quad (5)$$

where $h_k^{(n)}(x) = v_k^{(n)}(x) \left(\ell_k^{(n)}(x)\right)^2$, $b_k^{(n)}(x) = (x - x_k^{(n)}) \left(\ell_k^{(n)}(x)\right)^2$ and $v_k^{(n)}(x) = 1 - (x - x_k^{(n)}) \frac{\omega_n''(x_k^{(n)})}{\omega_n'(x_k^{(n)})}$.

Fejér [5] and Grünwald [7] also showed that the convergence of the Hermite-Fejér interpolation of $f(x)$ also depends on the choice of the nodes. The pointsystem (1) is called normal if for all n

$$v_k^{(n)}(x) \geq 0, \quad k = 1, 2, \ldots, n, \quad x \in [-1, 1], \quad (6)$$

while the pointsystem (1) is called strongly normal if for all n

$$v_k^{(n)}(x) \geq c > 0, \quad k = 1, 2, \ldots, n, \quad x \in [-1, 1] \quad (7)$$

for some positive constant c.

Fejér [5] (also see Szegö [12, pp 339]) showed that for the zeros of Jacobi polynomial $P_n^{(\alpha,\beta)}(x)$ of degree n $(\alpha > -1, \beta > -1)$

$$v_k^{(n)}(x) \geq \min\{-\alpha, -\beta\} \quad \text{for } -1 < \alpha \leq 0, -1 < \beta \leq 0, k = 1, 2, \ldots, n \text{ and } x \in [-1, 1].$$

For (strongly) normal pointsystems, Grünwald [7] showed that for every $f \in C^1(-1, 1)$, $\lim_{n \to \infty} \|H_{2n-1}^*(f) - f\|_\infty = 0$ if $\{x_k^{(n)}\}$ is strongly normal satisfying (7) and $\{f'(x_k^{(n)})\}$ satisfies

$$|f'(x_k^{(n)})| < n^{c-\delta} \quad \text{for some given positive number } \delta, \quad k = 1, 2, \ldots, \quad n = 1, 2, \ldots,$$

while $\lim_{n \to \infty} \|H_{2n-1}^*(f) - f\|_\infty = 0$ in $[-1 + \epsilon, 1 - \epsilon]$ for each fixed $0 < \epsilon < 1$ if $\{x_k^{(n)}\}$ is normal and $\{f'(x_k^{(n)})\}$ is uniformly bounded for $n = 1, 2, \ldots$[1]

Moreover, Szabados [11] showed the convergence of the Hermite-Fejér interpolation (5) at the Chebyshev pointsystem (3) satisfies

$$\|f - H_{2n-1}^*(f)\|_\infty = O(1)\|f - p^*\|_{C^1[-1,1]} \tag{8}$$

where p^* is the best approximation polynomial of f with degree at most $2n - 1$ and $\|f - p^*\|_{C^1[-1,1]} = \max_{0 \leq j \leq 1} \|f^{(j)} - p^{*(j)}\|_\infty$.

Hermite-Fejér interpolation has plenty of use in computer geometry aided geometric design with boundary conditions including derivative information. The convergence rate under the infinity norm has been extensively studied in [5–7, 11, 14]. The efficient algorithm on the fast implementation of Hermite-Fejér interpolation at zeros of Jacobi polynomial can be found in [17].

In this paper, the following convergence rates of Hermite-Fejér interpolation $H_{2n-1}^*(f, x)$ at Gauss-Jacobi pointsystems are considered.

- If f is analytic in $\mathcal{E}_\rho$ with $|f(z)| \leq M$, then

$$\|f(x) - H_{2n-1}^*(f, x)\|_\infty = \begin{cases} O\left(\dfrac{4\tau_n M[2n\rho^2 + (1 - 2n)\rho]}{(\rho - 1)^2 \rho^{2n}}\right), & \gamma \leq 0, \\[4mm] O\left(\dfrac{n^{2+2\gamma}[2n\rho^2 + (1 - 2n)\rho]}{(\rho - 1)^2 \rho^{2n}}\right), & \gamma > 0 \end{cases}, \quad \gamma = \max\{\alpha, \beta\} \tag{9}$$

---

[1] In fact, Grünwald in [7] considered more general cases with any vector $\{d_k^{(n)}\}$ instead of $\{f'(x_k^{(n)})\}$.

where

$$\tau_n = \begin{cases} O(n^{-1.5-\min\{\alpha,\beta\}}\log n), & \text{if } -1 < \min\{\alpha,\beta\} \le \gamma \le -\frac{1}{2} \\ O(n^{2\gamma-\min\{\alpha,\beta\}-\frac{1}{2}}), & \text{if } -1 < \min\{\alpha,\beta\} \le -\frac{1}{2} < \gamma \le 0 \\ O(n^{2\gamma}), & \text{if } -\frac{1}{2} < \min\{\alpha,\beta\} \le \gamma \end{cases} \quad (10)$$

- If $f(x)$ has an absolutely continuous $(r-1)$st derivative $f^{(r-1)}$ on $[-1, 1]$ for an integer $r \ge 3$, and a rth derivative $f^{(r)}$ of bounded variation $V_r = \text{Var}(f^{(r)}) < \infty$, then

$$\|f(x) - H^*_{2n-1}(f, x)\|_\infty = \begin{cases} O\left(n^{-r}\log n\right), & \gamma \le -\frac{1}{2}, \\ O\left(n^{2\gamma-r+1}\right), & \gamma > -\frac{1}{2}, \end{cases} \quad (11)$$

while if $f(x)$ is differentiable and $f'(x)$ is bounded on $[-1, 1]$, then

$$\|f(x) - H^*_{2n-1}(f, x)\|_\infty = \begin{cases} O\left(n^{-1}\log n\right), & \gamma \le -\frac{1}{2}, \\ O\left(n^{2\gamma}\right), & \gamma > -\frac{1}{2}. \end{cases}$$

Comparing these results with

$$f(x) - H_{2n-1}(f, x) = \begin{cases} O\left(n^{-1}\log n\right), & \text{if } \gamma \le -\frac{1}{2} \\ O(n^{2\gamma}), & \text{if } \gamma > -\frac{1}{2}, \end{cases} \quad \text{(Vértesi [14])},$$

which is sharp and attainable (see Fig. 2), we see that $H^*_{2n-1}(f, x)$ converges much faster than $H_{2n-1}(f, x)$ for analytic functions or functions of higher regularities (see Fig. 1). Particularly, $H_{2n-1}(f, x)$ diverges at Gauss-Jacobi pointsystems with $\gamma \ge 0$, whereas, $H^*_{2n-1}(f, x)$ converges for functions analytic in the Bernstein ellipse or of finite limited regularity.

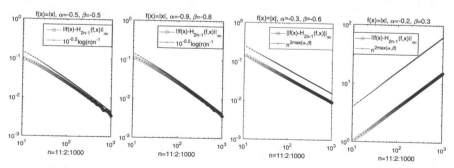

Fig. 2 $\|H_{2n-1}(f, x) - f(x)\|_\infty$ at $x = -1 : 0.001 : 1$ by using Gauss-Jacobi pointsystem for $f(x) = |x|$ with different α and β, respectively

For simplicity, in the following we abbreviate $x_k^{(n)}$ as x_k, $\ell_k^{(n)}(x)$ as $\ell_k(x)$, $h_k^{(n)}(x)$ as $h_k(x)$, and $b_k^{(n)}(x)$ as $b_k(x)$. $A \sim B$ denotes there exist two positive constants c_1 and c_2 such that $c_1 \le |A|/|B| \le c_2$.

2 Main Results

Suppose $f(x)$ satisfies a Dini-Lipschitz condition on $[-1, 1]$, then it has the following absolutely and uniformly convergent Chebyshev series expansion

$$f(x) = \sum_{j=0}^{\infty}{}' c_j T_j(x), \quad c_j = \frac{2}{\pi} \int_{-1}^{1} \frac{f(x) T_j(x)}{\sqrt{1-x^2}} dx, \quad j = 0, 1, \dots. \tag{12}$$

where the prime denotes summation whose first term is halved, $T_j(x) = \cos(j \cos^{-1} x)$ denotes the Chebyshev polynomial of degree j.

Lemma 1

(i) *(Bernstein [2]) If f is analytic with $|f(z)| \le M$ in the region bounded by the ellipse $\mathcal{E}_\rho$ with foci ± 1 and major and minor semiaxis lengths summing to $\rho > 1$, then for each $j \ge 0$,*

$$|c_j| \le \frac{2M}{\rho^j}. \tag{13}$$

(ii) *(Trefethen [13]) For an integer $r \ge 1$, if $f(x)$ has an absolutely continuous $(r-1)$st derivative $f^{(r-1)}$ on $[-1, 1]$ and a rth derivative $f^{(r)}$ of bounded variation $V_r = \mathrm{Var}(f^{(r)}) < \infty$, then for each $j \ge r+1$,*

$$|c_j| \le \frac{2 V_r}{\pi j (j-1) \cdots (j-r)}. \tag{14}$$

Suppose $-1 < x_n < x_{n-1} < \cdots < x_1 < 1$ in decreasing order are the roots of $P_n^{(\alpha, \beta)}(x)$ $(\alpha, \beta > -1)$, and $\{w_j\}_{j=1}^{n}$ are the corresponding weights in the Gauss-Jacobi quadrature.

Lemma 2 *For $j = 1, 2, \dots, n$, it follows*

$$(x - x_j)\ell_j(x) = \sigma_n (-1)^j \frac{\sqrt{(1 - x_j^2) w_j}}{2^{(\alpha+\beta+1)/2}} \sqrt{\frac{n! \Gamma(n + \alpha + \beta + 1)}{\Gamma(n + \alpha + 1)\Gamma(n + \beta + 1)}} P_n^{(\alpha, \beta)}(x), \tag{15}$$

where $\sigma_n = +1$ for even n and $\sigma_n = -1$ for odd n.

Proof Let $z_n = \int_{-1}^{1}(1-x)^\alpha(1+x)^\beta[P_n^{(\alpha,\beta)}(x)]^2 dx$ and K_n the leading coefficient of $P_n^{(\alpha,\beta)}(x)$. From Abramowitz and Stegun [1], we have

$$z_n = \frac{2^{\alpha+\beta+1}}{2n+\alpha+\beta+1} \cdot \frac{\Gamma(n+\alpha+1)\Gamma(n+\beta+1)}{n!\Gamma(n+\alpha+\beta+1)}, \quad K_n = \frac{1}{2^n} \frac{\Gamma(2n+\alpha+\beta+1)}{n!\Gamma(n+\alpha+\beta+1)}.$$

Furthermore, by Szegö [12, (15.3.1)] (also see Wang et al. [15]), we obtain

$$(x-x_j)\ell_j(x) = \frac{1}{\omega_n'(x_j)}\omega_n(x) = \sigma_n(-1)^j\sqrt{\frac{K_n^2 2n(1-x_j^2)w_j}{2n(2n+\alpha+\beta+1)z_n}}\omega_n(x)$$

$$= \sigma_n(-1)^j\sqrt{\frac{(1-x_j^2)w_j}{z_n(2n+\alpha+\beta+1)}}P_n^{(\alpha,\beta)}(x),$$

which implies the desired result (15). $\qquad\square$

Lemma 3 *For $j = 1, 2, \ldots, n$, it follows*

$$(1-x_j^2)w_j = O\left(n^{-1}\right). \tag{16}$$

Proof From $w_j = O\left(\frac{2^{\alpha+\beta+1}\pi}{n}\left(\sin\frac{\theta_j}{2}\right)^{2\alpha+1}\left(\cos\frac{\theta_j}{2}\right)^{2\beta+1}\right)$ Szegö [12, (15.3.10)], we see for $x_j = \cos\theta_j$ that $(1-x_j^2)w_j = O\left(\frac{2^{\alpha+\beta+3}\pi}{n}\left(\sin\frac{\theta_j}{2}\right)^{2\alpha+3}\left(\cos\frac{\theta_j}{2}\right)^{2\beta+3}\right)$, which derives the desired result. $\qquad\square$

Lemma 4 ([10, 16]) *For $t \in [-1, 1]$, let x_m be the root of the Jacobi polynomial $P_n^{(\alpha,\beta)}$ which is closest to t. Then for $k = 1, 2, \ldots, n$, we have*

$$\ell_k(t) = \begin{cases} O\left(|k-m|^{-1} + |k-m|^{\gamma-\frac{1}{2}}\right), & k \neq m \\ O(1) & k = m \end{cases}, \quad \gamma = \max\{\alpha, \beta\}. \tag{17}$$

Lemma 5 (Szegö [12, Theorem 8.1.2]) *Let α, β be real but not necessarily greater than -1 and $x_k = \cos\theta_k$. Then for each fixed k, it follows*

$$\lim_{n\to\infty} n\theta_k = j_k, \tag{18}$$

where j_k is the kth positive zero of Bessel function J_α.

Lemma 6 *For $k = 1, 2, \ldots, n$, it follows*

$$v_k(x) = 1 - (x-x_k)\frac{\omega_n''(x_k)}{\omega_n'(x_k)} = O(n^2). \tag{19}$$

Proof Note that $P_n^{(\alpha,\beta)}(x)$ satisfies the second order linear homogeneous Sturm-Liouville differential equation [12, (4.2.1)]

$$(1 - x^2)y'' + (\beta - \alpha - (\alpha + \beta + 2)x)y' + n(n + \alpha + \beta + 1)y = 0.$$

By $\omega_n(x) = \frac{P_n^{(\alpha,\beta)}(x)}{K_n}$, we get

$$\frac{\omega_n''(x_j)}{\omega_n'(x_j)} = -\frac{\beta - \alpha - (\alpha + \beta + 2)x_j}{1 - x_j^2} \quad ([12, (14.5.1)]). \tag{20}$$

In addition, by Lemma 5 with $x_j = \cos\theta_j$, we see that $\theta_1 \sim \frac{1}{n}$. Similarly, by $P_n^{(\alpha,\beta)}(-x) = (-1)^n P_n^{(\beta,\alpha)}(x)$ we have $\theta_n \sim \frac{1}{n}$. These together yield

$$\frac{1}{1 - x_1^2} = O(n^2), \quad \frac{1}{1 - x_n^2} = O(n^2), \quad \frac{1}{1 - x_j^2} \leq \max\left(\frac{1}{1 - x_1^2}, \frac{1}{1 - x_n^2}\right) = O(n^2)$$

and then by (20) it deduces the desired result. $\qquad\square$

Theorem 1 *Suppose $\{x_j\}_{j=1}^n$ are the roots of $P_n^{(\alpha,\beta)}(x)$ with $\alpha, \beta > -1$, then the Hermite-Fejér interpolation (5) for f analytic in $\mathcal{E}_\rho$ with $|f(z)| < M$ at $\{x_j\}_{j=1}^n$ has the convergence rate (9).*

Proof Since the Chebyshev series expansion of $f(x)$ is uniformly convergent under the assumptions, and the error of Hermite-Fejér interpolation (5) on Chebyshev polynomials satisfies $|E(T_j, x)| = |T_j(x) - H_{2n-1}^*(T_j, x)| = 0$ for $j = 0, 1, \ldots, 2n - 1$, then it yields

$$|E(f, x)| = |f(x) - H_{2n-1}^*(f, x)| = \left|\sum_{j=0}^\infty c_j E(T_j, x)\right| \leq \sum_{j=2n}^\infty |c_j| |E(T_j, x)|. \tag{21}$$

Furthermore, $|E(T_j, x)| = |T_j(x) - \sum_{i=1}^n T_j(x_i)h_i(x) - \sum_{i=1}^n T_j'(x_i)b_i(x)|$. In the following, we will focus on estimates of $|E(T_j, x)|$ for $j \geq 2n$.

In the case $\gamma \leq 0$: Notice that the pointsystem is normal which implies $h_i(x) \geq 0$ for all $i = 1, 2, \ldots, n$ and for all $x \in [-1, 1]$,

$$1 \equiv \sum_{i=1}^n h_i(x) = \sum_{i=1}^n v_i(x)\ell_i^2(x).$$

Then we have

$$\left|\sum_{i=1}^n T_j(x_i)h_i(x)\right| \leq \sum_{i=1}^n h_i(x) = 1, \quad j = 0, 1, \ldots. \tag{22}$$

Additionally, by Lemma 2, it obtains for $j = 2n, 2n + 1, \ldots$ that

$$
\begin{aligned}
&\left| \sum_{i=1}^{n} T_j'(x_i) b_i(x) \right| \\
&= j \left| \sum_{i=1}^{n} U_{j-1}(x_i)(x - x_i)\ell_i^2(x) \right| \\
&= \frac{j}{2^{(\alpha+\beta+1)/2}} \sqrt{\frac{n!\Gamma(n+\alpha+\beta+1)}{\Gamma(n+\alpha+1)\Gamma(n+\beta+1)}} \left| P_n^{(\alpha,\beta)}(x) \sum_{i=1}^{n} U_{j-1}(x_i)\sqrt{(1 - x_i^2)w_i}\,\ell_i(x) \right| \\
&= \frac{j}{2^{(\alpha+\beta+1)/2}} \sqrt{\frac{n!\Gamma(n+\alpha+\beta+1)}{\Gamma(n+\alpha+1)\Gamma(n+\beta+1)}} \left| P_n^{(\alpha,\beta)}(x) \sum_{i=1}^{n} \sin((j-1)\arccos(x_i))\sqrt{w_i}\,\ell_i(x) \right| \\
&= j\, O\left(|P_n^{(\alpha,\beta)}(x)| \sqrt{\|\{w_i\}_{i=1}^{n}\|_\infty}\, \Lambda_n \right)
\end{aligned}
$$

(U_{j-1} is the second kind of Chebyshev polynomial of degree $j - 1$) since $\sqrt{\frac{n!\Gamma(n+\alpha+\beta+1)}{\Gamma(n+\alpha+1)\Gamma(n+\beta+1)}}$ is uniformly bounded in n for $\alpha, \beta > -1$ due to

$$
\frac{(n + 1)!\Gamma(n + \alpha + \beta + 2)}{\Gamma(n + \alpha + 2)\Gamma(n + \beta + 2)} = \left(1 - \frac{\alpha\beta}{(n + 1)^2 + (\alpha + \beta)(n + 1) + \alpha\beta} \right)
$$
$$
\times \frac{n!\Gamma(n + \alpha + \beta + 1)}{\Gamma(n + \alpha + 1)\Gamma(n + \beta + 1)},
$$

which implies $\frac{n!\Gamma(n+\alpha+\beta+1)}{\Gamma(n+\alpha+1)\Gamma(n+\beta+1)}$ is uniformly bounded in n and then $\sqrt{\frac{n!\Gamma(n+\alpha+\beta+1)}{\Gamma(n+\alpha+1)\Gamma(n+\beta+1)}}$ is uniformly bounded. Here $\Lambda_n = \max_{x \in [-1,1]} \sum_{i=1}^{n} |\ell_i(x)|$ is the Lebesgue constant. Then from

$$
P_n^{(\alpha,\beta)}(x) = \begin{cases} O(n^{-\frac{1}{2}}), & \text{if } \max\{\alpha, \beta\} \le -\frac{1}{2} \\ O(n^{\max\{\alpha,\beta\}}), & \text{if } \max\{\alpha, \beta\} > -\frac{1}{2} \end{cases},
$$

$$
w_i = \begin{cases} O(n^{-2-2\min\{\alpha,\beta\}}), & \text{if } \min\{\alpha, \beta\} \le -\frac{1}{2} \\ O(n^{-1}), & \text{if } \min\{\alpha, \beta\} > -\frac{1}{2} \end{cases}
$$

(see Szegö [12, pp 168, 354]) and

$$
\Lambda_n = \begin{cases} O(\log n), & \text{if } \max\{\alpha, \beta\} \le -\frac{1}{2} \\ O(n^{\max\{\alpha,\beta\}+\frac{1}{2}}), & \text{if } \max\{\alpha, \beta\} > -\frac{1}{2} \end{cases} \quad ([12, \text{pp } 338]),
$$

we have

$$
\left| \sum_{i=1}^{n} T_j'(x_i) b_i(x) \right| = j\tau_n. \tag{23}
$$

Then by (22) and (23), we find $|E(T_j, x)| \leq 2 + j\tau_n < 2j\tau_n$ for $j \geq 2n$, and consequently

$$|E(f, x)| = |f(x) - H^*_{2n-1}(f, x)| \leq \sum_{j=2n}^{\infty} |c_j||E(T_j, x)| = 2\tau_n \sum_{j=2n}^{\infty} j|c_j|,$$

which, directly following [18], leads to the desired result.

In the case $\gamma > 0$: From $|E(T_j, x)| = |T_j(x) - \sum_{i=1}^{n} T_j(x_i)h_i(x) - \sum_{i=1}^{n} T'_j(x_i)b_i(x)|$, by Lemmas 3 and 6 we obtain

$$\sum_{i=1}^{n} |v_i(x)||\ell_i^2(x) = O\left(n^2 \int_1^n t^{2\gamma-1} dt\right) = O(n^{2+2\gamma}),$$

and

$$T_j(x) - \sum_{i=1}^{n} T_j(x_i)h_i(x) = T_j(x) - \sum_{i=1}^{n} T_j(x_i)v_i(x)\ell_i^2(x) = O\left(n^{2+2\gamma}\right).$$

These together with

$$\begin{aligned}
&|\sum_{i=1}^{n} T'_j(x_i)b_i(x)| \\
= &\frac{j}{2^{(\alpha+\beta+1)/2}} \sqrt{\frac{n!\Gamma(n+\alpha+\beta+1)}{\Gamma(n+\alpha+1)\Gamma(n+\beta+1)}} |P_n^{(\alpha,\beta)}(x) \sum_{i=1}^{n} \sin((j-1)\arccos(x_i))\sqrt{w_i}\ell_i(x)| \\
= &j\tau_n
\end{aligned}$$

and then $|E(T_j, x)| = O\left(j^{2+2\gamma}\right)$ for $j \geq 2n$, similar to the above proof in the case of $\gamma \leq 0$, implies the desired result.   □

From the definition of τ_n, we see that when $\alpha = \beta = -\frac{1}{2}$ the convergence order on n is the lowest. In addition, if f is of limited regularity, we have

Lemma 7 (Vértesi [14]) *Suppose* $\{x_j\}_{j=1}^{n}$ *are the roots of* $P_n^{(\alpha,\beta)}(x)$, *for every continuous function* $f(x)$ *we have*

$$|H_{2n-1}(f, x) - f(x)| = O(1) \sum_{j=1}^{n} \left[w\left(f; \frac{j\sqrt{1-x^2}}{n}\right) + w\left(f; \frac{j^2|x|}{n^2}\right) \right] j^{2\bar\gamma-1}, \tag{24}$$

where $w(f; t) = w(t)$ *is the modulus of continuity of* $f(x)$, *and* $\bar\gamma = \max\left(\alpha, \beta, -\frac{1}{2}\right)$.

Theorem 2 *Suppose* $\{x_j\}_{j=1}^{n}$ *are the roots of* $P_n^{(\alpha,\beta)}(x)$ $(\alpha, \beta > -1)$, *and* $f(x)$ *has an absolutely continuous* $(r-1)$*st derivative* $f^{(r-1)}$ *on* $[-1, 1]$ *for some* $r \geq 3$, and a rth derivative $f^{(r)}$ of bounded variation $V_r < \infty$, then the Hermite-Fejér interpolation (5) at $\{x_j\}_{j=1}^n$ has the convergence rate (11).

Proof Consider the special functional $L(g) = E_n(g, x)$, where $E_n(g, x)$ is defined for $\forall g \in C^1([-1, 1])$ by

$$E_n(g, x) = g(x) - \sum_{j=1}^n g(x_j)v_j(x)\ell_j^2(x) - \sum_{j=1}^n g'(x_j)(x - x_j)\ell_j^2(x). \quad (25)$$

By the Peano kernel theorem for $n \geq r$ (see Peano [9] or Kowalewski [8]), $E_n(f, x)$ can be represented as

$$E_n(f, x) = \int_{-1}^1 f^{(r)}(t)K_r(t)dt \quad (26)$$

with $K_r(t) = \frac{1}{(r-1)!}L\left((x - t)_+^{r-1}\right)$ for $r = 3, 4, \cdots$, that is

$$K_r(t) = \frac{1}{(r-1)!}(x - t)_+^{r-1} - \frac{1}{(r-1)!}\sum_{j=1}^n (x_j - t)_+^{r-1}v_j(x)\ell_j^2(x)$$

$$-\frac{1}{(r-1)!}\sum_{j=1}^n (x_j - t)_+^{r-2}(x - x_j)\ell_j^2(x),$$

where

$$(x - t)_+^{k-1} = \begin{cases} (x - t)^{k-1}, & x \geq t; \\ 0, & x < t. \end{cases} \quad (k \geq 2), \qquad (x - t)_+^0 = \begin{cases} 1, & x \geq t; \\ 0, & x < t. \end{cases} \quad (k = 1).$$

Moreover, noting that

$$\frac{1}{(k-2)!}(x - u)_+^{k-2} = \int_u^1 \frac{1}{(k-3)!}(x - t)_+^{k-3}(t)dt, \quad k = 3, 4, \cdots,$$

we get the following identity

$$K_{s-1}(u) = \int_u^1 K_{s-2}(t)dt, \quad s = 4, 5, \cdots,$$

where $K_2(t)$ is defined by

$$K_2(t) = (x - t)_+^1 - \sum_{j=1}^n (x_j - t)_+^1 v_j(x)\ell_j^2(x) - \sum_{j=1}^n (x_j - t)_+^0 (x - x_j)\ell_j^2(x).$$

In addition, it can be easily verified that $K_s(-1) = K_s(1) = 0$ for $s = 2, 3, \ldots$.

Since $f^{(r)}$ is of bounded variation, directly applying the similar skills of Theorem 2 and Lemma 4 in [16], we get

$$\|E_n(f, x)\|_\infty \le V_r \|K_{r+1}\|_\infty, \tag{27}$$

and

$$\|K_{s+1}\|_\infty \le \frac{\pi}{2n - s} \sup_{-1 \le t \le 1} |K_s(t)|, \quad \text{for } s = 2, 3, \cdots, \tag{28}$$

respectively. Then from (27) and (28), we can obtain that

$$\|E_n(f, x)\|_\infty \le \frac{\pi^{r-1} V_r}{(2n - 2)(2n - 3) \cdots (2n - r))} \|K_2\|_\infty. \tag{29}$$

In addition, by Lemma 7, we have

$$\left\| (x - t)_+^1 - \sum_{j=1}^n (x_j - t)_+^1 v_j(x) \ell_j^2(x) \right\|_\infty = \begin{cases} O\left(\frac{\log n}{n}\right), & \gamma \le -\frac{1}{2} \\ O\left(n^{2\gamma}\right), & \gamma > -\frac{1}{2}, \end{cases} \tag{30}$$

while by Lemmas 2–3, we get

$$\left| \sum_{j=1}^n (x_j - t)_+^0 (x - x_j) \ell_j^2(x) \right| \le \sum_{j=1}^n |(x - x_j) \ell_j^2(x)| = \begin{cases} O\left(\frac{\log n}{n}\right), & \gamma \le -\frac{1}{2} \\ O\left(n^{2\gamma}\right), & \gamma > -\frac{1}{2}. \end{cases} \tag{31}$$

Together (30) and (31), we can obtain the desired results by using

$$K_2(t) = \begin{cases} O\left(\frac{\log n}{n}\right), & \gamma \le -\frac{1}{2} \\ O\left(n^{2\gamma}\right), & \gamma > -\frac{1}{2}. \end{cases}$$

Finally, We use a function of analytic $f(x) = \frac{1}{1+25x^2}$ and a function of limited regularity $f(x) = |x|^5$ to show that the convergence rate of $\|f(x) - H_{2n-1}^*(f, x)\|_\infty$ is dependent on α and β in Fig. 3.

634   S. Xiang and G. He

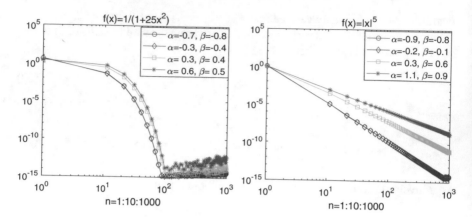

Fig. 3 $\|H^*_{2n-1}(f, x) - f(x)\|_\infty$ at $x = -1 : 0.001 : 1$ by using Gauss-Jacobi pointsystem for $f(x) = \frac{1}{1+25x^2}$ and $f(x) = |x|^5$ with different α and β, respectively

Acknowledgements This author "Shuhuang Xiang" was supported partly by NSF of China (No. 11771454). This author "Guo He" was supported partly by NSF of China (No. 11901242), the Fundamental Research Funds for the Central Universities (No. 21618333), and the Opening Project at the Sun Yat-sen University (No. 2018010).

References

1. Abramowitz, M., Stegun, I.A.: Handbook of Mathematical Functions. National Bureau of Standards, Washington (1964)
2. Bernstein, S.: Sur l'ordre de la meilleure approximation des fonctions continues par les polynômes de degré donné. Mem. Cl. Sci. Acad. Roy. Belg. **4**, 1–103 (1912)
3. Faber, G.: Über die interpolatorische Darstellung stetiger Funktionen. Jahresber. Deut. Math. Verein. **23**, 192–210 (1914)
4. Fejér, L.: Über Interpolation, Nachrichten der Gesellschaft der Wissenschaften zu Göttingen Mathematisch-physikalische Klasse, 66–91 (1916)
5. Fejér, L.: Lagrangesche interpolation und die zugehörigen konjugierten Punkte. Math. Ann. **106**, 1–55 (1932)
6. Fejér, L.: Bestimmung derjenigen Abszissen eines Intervalles, für welche die Quadratsumme der Grundfunktionen der Lagrangeschen Interpolation im Intervalle ein Möglichst kleines Maximum Besitzt. Ann. della Sc. Norm. Super. di Pisa **1**, 263–276 (1932)
7. Grünwald, G.: On the theory of interpolation. Acta Math. **75**, 219–245 (1942)
8. Kowalewski, G.: Interpolation und Genäherte Quadratur. Teubner-Verlag, Leipzig (1932)
9. Peano, G.: Resto nelle formule di quadrature, espresso con un integrale definito. Rom. Acc. L. Rend. **22**, 562–569 (1913)
10. Sun, X.: Lagrange interpolation of functions of generalized bounded variation. Acta Math. Hungar. **53**, 75–84 (1989)
11. Szabados, J.: On the order of magnitude of fundamental polynomials of Hermite interpolation. Acta Math. Hungar. **61**, 357–368 (1993)
12. Szegö, G.: Orthogonal Polynomials, vol. 23. Colloquium Publications, American Mathematical Society, Providence (1939)

13. Trefethen, L.N.: Approximation Theory and Approximation Practice. SIAM, Philadelphia (2012)
14. Vértesi, P.: Notes on the Hermite-Fejér interpolation based on the Jacobi abscissas. Acta Math. Acad. Sci. Hung. **24**, 233–239 (1973)
15. Wang, H., Huybrechs, D., Vandewalle, S.: Explicit barycentric weights for polynomial interpolation in the roots or extrema of classical orthogonal polynomials. Math. Comput. **290**, 2893–2914 (2012)
16. Xiang, S: On interpolation approximation: convergence rates for interpolation for functions of limited regularity. SIAM J. Numer. Anal. **54**, 2081–2113 (2016)
17. Xiang, S., He, G.: The fast implementation of higher order Hermite-Fejér interpolation. SIAM J. Sci. Comput. **37**, A1727–A1751 (2015)
18. Xiang, S., Chen, X., Wang, H: Error bounds for approximation in Chebyshev points. Numer. Math. **116**, 463–491 (2010)

Fifth-Order Finite-Volume WENO on Cylindrical Grids

Mohammad Afzal Shadab, Xing Ji, and Kun Xu

1 Introduction

The conventional WENO scheme is specifically designed for the reconstruction in Cartesian coordinates on uniform grids [1]. The employment of Cartesian-based reconstruction scheme on a cylindrical grid suffers from a number of drawbacks [2, 3], e.g., in the original PPM paper, reconstruction was performed in volume coordinates (than the linear ones) so that algorithm for a Cartesian mesh can be used on a curvilinear mesh. However, the resulting interface states became first-order accurate even for smooth flows [2]. Another example can be the volume average assignment to the geometrical cell center of finite-volume than the centroid [2]. A breakthrough in the field of high order reconstruction in cylindrical coordinates is the application of the Vandermonde-like linear systems of equations with spatially

M. A. Shadab (✉)
Department of Mechanical and Aerospace Engineering, Hong Kong University of Science and Technology, Kowloon City, Hong Kong

X. Ji
Department of Mathematics, Hong Kong University of Science and Technology, Kowloon City, Hong Kong
e-mail: xjiad@connect.ust.hk

K. Xu
Department of Mathematics, Hong Kong University of Science and Technology, Kowloon City, Hong Kong

Department of Mechanical and Aerospace Engineering, Hong Kong University of Science and Technology, Kowloon City, Hong Kong
e-mail: makxu@ust.hk

© The Author(s) 2020
S. J. Sherwin et al. (eds.), *Spectral and High Order Methods for Partial Differential Equations ICOSAHOM 2018*, Lecture Notes in Computational Science and Engineering 134, https://doi.org/10.1007/978-3-030-39647-3_51

varying coefficients [2]. It is reintroduced in the present work to build a basis for the derivation of the high order WENO schemes.

The motivation for the present work is to develop a fifth-order finite-volume WENO reconstruction scheme in the efficient dimension-by-dimension framework, specifically aimed at regularly-spaced and irregularly-spaced grids in cylindrical coordinates.

2 Finite-Volume Discretization in Curvilinear Coordinates

2.1 Evaluation of the Linear Weights

A non-uniform grid spacing with zone width $\Delta \xi_i = \xi_{i+\frac{1}{2}} - \xi_{i-\frac{1}{2}}$ is considered having $\xi \in (x_1, x_2, x_3)$ as the coordinate along the reconstruction direction and $\xi_{i+\frac{1}{2}}$ denoting the location of the cell interface between zones i and $i + 1$. Let $\bar{Q}_i$ be the cell average of conserved quantity Q inside zone i at some given time, which can be expressed in form of Eq. (1).

$$\bar{Q}_i = \frac{1}{\Delta \mathcal{V}_i} \int_{\xi_{i-\frac{1}{2}}}^{\xi_{i+\frac{1}{2}}} Q_i(\xi) \frac{\partial \mathcal{V}}{\partial \xi} d\xi \quad \& \quad \Delta \mathcal{V}_i = \int_{\xi_{i-\frac{1}{2}}}^{\xi_{i+\frac{1}{2}}} \frac{\partial \mathcal{V}}{\partial \xi} d\xi \tag{1}$$

where the local cell volume $\Delta \mathcal{V}_i$ of ith cell in the direction of reconstruction given in Eq. (1) and $\frac{\partial \mathcal{V}}{\partial \xi}$ is the one-dimensional Jacobian. Now, our aim is to find a pth order accurate approximation to the actual solution by constructing a $(p - 1)$th order polynomial distribution, as given in Eq. (2).

$$Q_i(\xi) = a_{i,0} + a_{i,1}(\xi - \xi_i^c) + a_{i,2}(\xi - \xi_i^c)^2 + \ldots + a_{i,p-1}(\xi - \xi_i^c)^{p-1} \tag{2}$$

where $a_{i,n}$ corresponds to a vector of the coefficients which needs to be determined and ξ_i^c can be taken as the cell centroid. However, the final values at the interface are independent of the particular choice of ξ_i^c and one may as well set $\xi_i^c = 0$ [2]. Unlike the cell center, the centroid is not equidistant from the cell interfaces in the case of cylindrical-radial coordinates, and the cell averaged values are assigned at the centroid [2]. Further, the method has to be locally conservative, i.e., the polynomial $Q_i(\xi)$ must fit the neighboring cell averages, satisfying Eq. (3).

$$\int_{\xi_{i+s-\frac{1}{2}}}^{\xi_{i+s+\frac{1}{2}}} Q_i(\xi) \frac{\partial \mathcal{V}}{\partial \xi} d\xi = \Delta \mathcal{V}_{i+s} \bar{Q}_{i+s} \quad \text{for} \quad -i_L \leq s \leq i_R \tag{3}$$

where the stencil includes i_L cells to the left and i_R cells to the right of the ith zone such that $i_L + i_R + 1 = p$. Implementing Eqs. (1)–(2) in Eq. (3) along with a simple mathematical manipulation leads to Eq. (4), which is the fundamental equation for reconstruction in cylindrical coordinates. For the detailed derivation, kindly refer to [3].

$$
\begin{pmatrix}
\beta_{i-i_L,0} & \cdots & \beta_{i-i_L,p-1} \\
\vdots & \ddots & \vdots \\
\beta_{i+i_R,0} & \cdots & \beta_{i+i_R,p-1}
\end{pmatrix}^{T}
\begin{pmatrix}
w^{\pm}_{i,-i_L} \\
\vdots \\
w^{\pm}_{i,i_R}
\end{pmatrix}
=
\begin{pmatrix}
1 \\
\vdots \\
(\xi_{i\pm\frac{1}{2}} - \xi_i^c)^{p-1}
\end{pmatrix}
\tag{4}
$$

where '$\pm$' represents the positive and negative weights i.e. weights for reconstructing right ($+$) and left ($-$) interface values respectively. Also, the grid dependent linear weights ($w^{\pm}_{i,s}$) satisfy the normalization condition [2].

2.2 Optimal Weights

For the case of fifth-order WENO interpolation, the third order interpolated variables are optimally weighed in order to achieve fifth-order accurate interpolated values as given in Eq. (5) for the case of $p = 3$ [1].

$$
q_{i,0}^{(2p-1)\pm} = \sum_{l=0}^{p-1} C^{\pm}_{i,l} q_{i,l}^{p\pm}
\tag{5}
$$

where $C^{\pm}_{i,l}$ is the optimal weight for the positive/negative cases on the ith finite-volume. So, Eq. (4) is used again to evaluate the weights for the fifth-order ($2p-1 = 5$) interpolation ($i_L = 2, i_R = 2$).

Linear and optimal weights are independent of the mesh size for standard regularly-spaced grid cases. They can be evaluated and stored (at a nominal cost) independently before the actual computation. Also, they conform to the original WENO-JS [1] for the limiting case ($R \to \infty$). The weights required for source term and flux integration in one or more dimensions are given in [3].

2.3 Smoothness Indicators and the Nonlinear Weights

The mathematical definition of the smoothness indicator is given in Eq. (6) [1].

$$
IS_{i,l} = \sum_{m=1}^{p-1} \int_{\xi_{j-\frac{1}{2}}}^{\xi_{j+\frac{1}{2}}} \left(\frac{d^m}{d\xi^m} Q_{i,l}(\xi) \right)^2 \Delta\xi_i^{2m-1} d\xi, \quad l = 0, \ldots, p-1
\tag{6}
$$

To evaluate the value of $IS_{i,l}$, a third order polynomial interpolation on ith cell is required using positive and negative reconstructed values by stencil S_l, as given in Eq. (2). Finally, evaluating the values of the coefficient a's and substituting their values in smoothness indicator formula (6) yields the grid-independent fundamental relation (7). The nonlinear weight $(\omega_{i,l}^{\pm})$ for the WENO-C interpolation is defined in Eq. (8) [1], where ϵ is chosen to be 10^{-6} [1, 3].

$$IS_{i,l} = 4(39\bar{Q}_i^2 - 39\bar{Q}_i(q_{i,l}^- + q_{i,l}^+) + 10((q_{i,l}^-)^2 + (q_{i,l}^+)^2) + 19q_{i,l}^- q_{i,l}^+) \tag{7}$$

$$\omega_{i,l}^{\pm} = \frac{\alpha_{i,l}^{\pm}}{\sum_{l=0}^{p-1} \alpha_{i,l}^{\pm}} \quad \& \quad \alpha_{i,l}^{\pm} = \frac{C_{i,l}^{\pm}}{(\epsilon + IS_{i,l})^2} \qquad l = 0, 1, 2 \tag{8}$$

The final interpolated interface values are evaluated from Eq. (9).

$$q_i^{(2p-1)\pm} = \sum_{l=0}^{p-1} \omega_{i,l}^{p\pm} q_{i,l}^{p\pm} \tag{9}$$

3 Stability Analysis of WENO-C for Hyperbolic Conservation Laws

For WENO-C to be practically useful, it is crucial that it enables a stable discretization for hyperbolic conservation laws when coupled with a proper time-integration scheme. In this section, we analyze WENO-C scheme for model problems involving smooth flow in 1D cylindrical-radial coordinates, based on a modified von Neumann stability analysis [4]. We consider scalar advection equation (10) in 1D cylindrical-radial coordinates.

$$\frac{\partial Q}{\partial t} + \frac{1}{(\partial \mathcal{V}/\partial \xi)} \frac{\partial}{\partial \xi} \left(\left(\frac{\partial \mathcal{V}}{\partial \xi} \right) Q v \right) = 0 \qquad \xi \in [0, \infty], \quad t > 0 \tag{10}$$

where Q is the conserved variable, $(\partial \mathcal{V}/\partial \xi) = \xi$ is the one-dimensional Jacobian in cylindrical-radial coordinates. Boundary conditions are not considered in the present approach to reduce the complexity of the analysis. Assuming a uniform grid with $\xi_i = i\Delta\xi$ and $\xi_{i+1} - \xi_i = \Delta\xi \forall i$ and $(i \pm 1/2)$ denotes the boundaries of the finite-volume i. In the finite-volume framework, Eq. (10) transforms into the conservative scheme given in Eq. (11).

$$\frac{\partial \bar{Q}_i}{\partial t} = -\frac{1}{\Delta \mathcal{V}_i} (\hat{F}_{i+1/2} - \hat{F}_{i-1/2}) \tag{11}$$

where numerical flux $\hat{F}_{i+1/2}$ is the Lax-Friedrich flux, and $\bar{Q}_i$ and $\mathcal{V}_i$ are given in Eq. (1). For this particular problem, let $v = 1$ in Eq. (10). Therefore, only the values on the left side of the interface are considered. Based on the von Neumann stability analysis, the semi-discrete solution can be expressed as a discrete Fourier series. By the superposition principle, only one term in the series can be used for analysis, as illustrated in Eq. (12).

$$\bar{Q}_i(t) = \hat{Q}_k(t)e^{ji\theta_k}, \quad \text{where} \quad j = \sqrt{-1} \tag{12}$$

By substituting Eq. (12) in Eq. (11), we can separate the spatial operator L, as given in Eq. (13).

$$L = -\frac{(\hat{F}_{i+1/2} - \hat{F}_{i-1/2})}{\Delta \mathcal{V}_i} = -\frac{[Q(\partial \mathcal{V}/\partial \xi)]^-_{i+1/2} - [Q(\partial \mathcal{V}/\partial \xi)]^-_{i-1/2}}{\Delta \mathcal{V}_i} = -\frac{z(\theta_k)\bar{Q}_i}{\Delta \xi} \tag{13}$$

where the complex function $z(\theta_k)$ is the Fourier symbol. By substituting the values of $Q^-_{i-1/2}$ and $Q^-_{i+1/2}$ using fifth-order positive weights of cells $(i-1)$ and i respectively for a smooth solution, the value of $z(\theta_k)$ for WENO-C can be evaluated using Eq. (14).

$$z(\theta_k) = \frac{m+1}{i^{(m+1)} - (i-1)^{(m+1)}} \sum_{l=-2}^{+2} \left[w^+_{i,l} i^m e^{jl\theta_k} - w^+_{(i-1),l}(i-1)^m e^{j(l-1)\theta_k} \right] \tag{14}$$

where $m = 1$ for cylindrical-radial coordinates. Using the same approach as given in [4], we can plot the spatial spectrum $\{S : -z(\theta_k)$ for $\theta_k \in [0, 2\pi]\}$ and the stability domain S_t for TVD-RK order 3. The maximum stable CFL number of this scheme can be computed by finding the largest rescaling parameter $\tilde{\sigma}$, so that the rescaled spectrum still lies in the stability domain.

It can be observed from Fig. 1 that the spatial spectrums S of WENO-C differs initially with the index numbers i due to the geometrical variation of the finite-volume. However, the spectrums are the same for high index numbers (i), similar to WENO-JS, as the fifth-order interpolation weights converge. Some regions $(i = 1, 2)$ require boundary conditions and thus, are not considered in the present analysis. The values of CFL number for cylindrical-radial coordinates lie in between 1.45 and 1.52. As a final remark, it can be concluded that the proposed scheme is A-stable with third or higher order of RK method with an appropriate value of CFL number for this case.

4   Numerical Tests

In this section, several tests on Euler equations are performed to analyze the performance of the WENO-C reconstruction scheme. Tests are performed on a gamma law gas ($\gamma = 1.4$) in cylindrical coordinates to investigate the essentially non-oscillatory property of WENO-C for discontinuous flows and the convex combination property for smooth flows. For first-order and second-order (MUSCL)

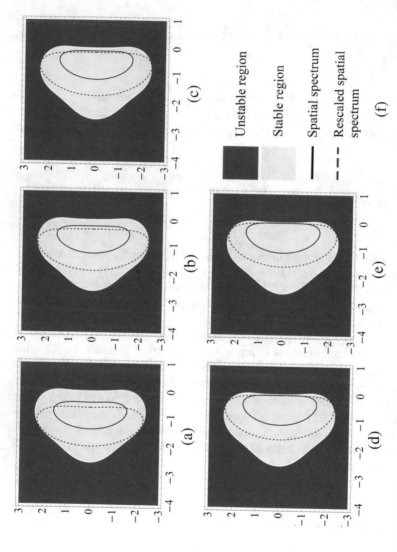

Fig. 1 Rescaled spectrums (with maximum stable CFL number $\tilde{\sigma}$) and stability domains of fifth-order WENO-C in cylindrical-radial coordinates in a complex plane for different cell index numbers i. (**a**) $i = 3$, $\tilde{\sigma} = 1.45$. (**b**) $i = 5$, $\tilde{\sigma} = 1.50$. (**c**) $i = 10$, $\tilde{\sigma} = 1.46$. (**d**) $i = 50$, $\tilde{\sigma} = 1.52$. (**e**) $i = 100$, $\tilde{\sigma} = 1.45$. (**f**) Legend

spatial reconstructions, Euler time marching and Maccormack (predictor-corrector) schemes are respectively employed. For WENO-C, time marching is done with TVD-RK order 3 for 1D cases and RK order 5 for the 2D case.

4.1 Acoustic Wave Propagation

A smooth problem involving a nonlinear system of 1D gas dynamical equations is solved to test fifth-order accuracy of the spatial discretization scheme [3]. The Euler equations in cylindrical-radial coordinates can be written in the form of Eq. (15).

$$
\frac{\partial}{\partial t} \begin{pmatrix} \rho \\ \rho u \\ E \end{pmatrix} + \frac{1}{R} \frac{\partial}{\partial R} \begin{pmatrix} \rho u R \\ (\rho u^2 + p)R \\ (E + p)u R \end{pmatrix} = \begin{pmatrix} 0 \\ p/R \\ 0 \end{pmatrix} \tag{15}
$$

where ρ is the mass density, u is the radial velocity, p is the pressure, and E is the total energy. Equation (16) serves as the adiabatic equation of state.

$$
E = \frac{p}{\gamma - 1} + \frac{1}{2} \rho u^2 \tag{16}
$$

The initial conditions are provided in Eq. (17) with the perturbation given in Eq. (18). The interface flux is evaluated with Rusanov scheme [3].

$$
\rho(R, 0) = 1 + \varepsilon f(R), \quad u(R, 0) = 0, \quad p(R, 0) = 1/\gamma + \varepsilon f(R) \tag{17}
$$

$$
f(R) = \begin{cases} \frac{sin^4(5\pi R)}{R} & \text{if } 0.4 \le R \le 0.6 \\ 0 & \text{otherwise} \end{cases} \tag{18}
$$

A sufficiently small perturbation with $\varepsilon = 10^{-4}$ yields a smooth solution. The interface flux is evaluated using Rusanov scheme with a CFL number of 0.3.

The initial perturbation splits into two acoustic waves traveling in opposite directions. The final time ($t = 0.3$) is set such that the waves remain in the domain and the problem is free from the boundary effects. The computational domain of unity length is uniformly divided into N different zones i.e. $N = 16, 32, 64, 128, 256$. Although an exact solution known up to $O(\varepsilon^2)$ is known, the solution on the finest mesh $N = 1024$ is taken as the reference. Figure 2 illustrate the spatial variation of density at $t = 0.3$ inside the domain. From Table 1, it clear that the scheme approaches the desired fifth-order accuracy.

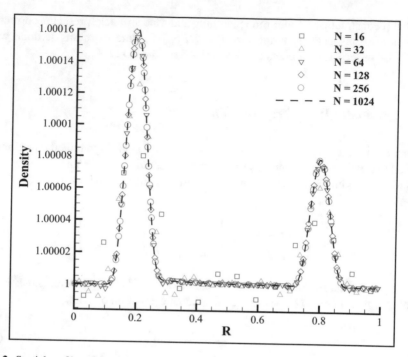

Fig. 2 Spatial profiles of density at $t = 0.3$ for acoustic wave propagation test in cylindrical-radial coordinates

Table 1 L_1 norm errors and order of convergence table for acoustic wave propagation test

N	$\epsilon(\rho)$	O_{L_1}
16	1.01E−05	−
32	4.91E−06	1.036
64	6.74E−07	2.865
128	3.24E−08	4.380
256	1.27E−09	4.670

4.2 Sedov Explosion Test

Sedov explosion test is performed to investigate code's ability to deal with strong shocks and non-planar symmetry [3]. The problem involves a self-similar evolution of a cylindrical blastwave in a uniform grid ($N = 100$) from a localized initial pressure perturbation (delta-function) in an otherwise homogeneous medium. Governing equations are given in Eq. (15) and the fluxes are evaluated with Rusanov scheme and GKS [5]. For the code initialization, dimensionless energy $\epsilon = 1$

is deposited into a small region of radius $\delta = 3\Delta R$. Inside this region, the dimensionless pressure P'_0 is given by Eq. (19).

$$P'_0 = \frac{3(\gamma - 1)\epsilon}{(m + 2)\pi \delta^{(m+1)}} \qquad (19)$$

where $m = 1$ for cylindrical geometry. Reflecting boundary condition is employed at the center ($R = 0$), whereas boundary condition at $R = 1$ is not required for this problem. The initial velocity and density inside the domain are 0 and 1 respectively and the initial pressure everywhere except the kernel is 10^{-5}. As the source term is very stiff, the CFL number is set to be 0.1. The final time is $t = 0.05$.

Figure 3 shows that the peak for WENO-C is higher for density and is closest to the analytical value, similar to fifth-order finite difference version [3], but MUSCL has higher offset peaks for pressure and velocity. GKS performs slightly better than RS, as the peaks are slightly higher for all the cases.

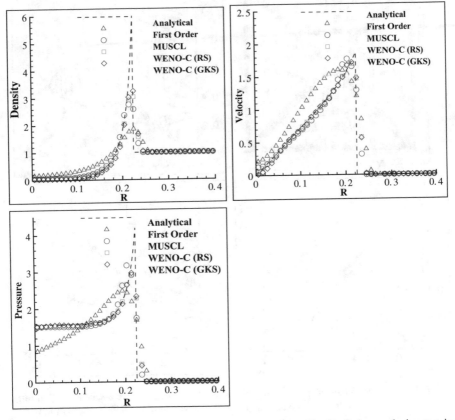

Fig. 3 Variation of density, velocity, and pressure with the radius for Sedov explosion test in cylindrical-radial coordinates. Domain is restricted to $R = 0.4$ for the sake of clarity

4.3 Modified 2D Riemann Problem in (R − z) Coordinates

The final test for the present scheme involves a modified 2D Riemann problem in cylindrical $(R - z)$ coordinates, as illustrated in Fig. 4 (top left). The problem involves 2 contact discontinuities and 2 shocks as the initial condition, resulting in the formation of a self-similar structure propagating towards the low density-low pressure region (region 3). The governing equations in cylindrical $(R - z)$ coordinates are provided in Eq. (20).

The computations are performed until $t = 0.2$ with a CFL number of 0.5 on a domain $(R, z)=[0,1]\times[0,1]$ divided into 500×500 zones. The boundary conditions are symmetry at the center (except for the antisymmetric radial velocity) and outflow elsewhere. HLL Riemann solver is used for flux evaluations. Rich small-scale structures in the contact-contact region (region 1) can be observed from

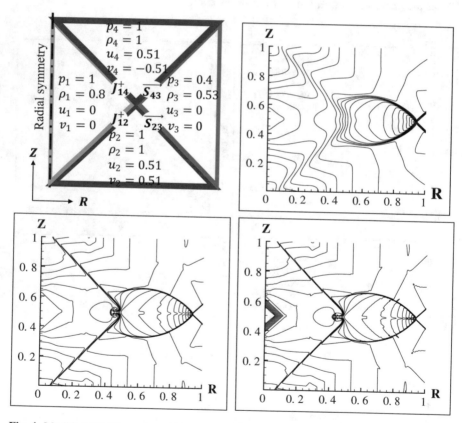

Fig. 4 Modified 2D Riemann problem in cylindrical $(r - z)$ coordinates: schematic (top left), density contours at $t = 0.2$ with first-order (top right), second-order MUSCL (bottom left), and WENO-C (bottom right) reconstruction schemes

Fig. 4 for WENO-C reconstruction, when compared with first and second-order MUSCL reconstruction. Structures are highly smeared for the case of first-order reconstruction.

$$\frac{\partial}{\partial t}\begin{pmatrix} \rho \\ \rho v_R \\ \rho v_z \\ \rho e \end{pmatrix} + \frac{1}{R}\frac{\partial}{\partial R}\begin{pmatrix} \rho v_R R \\ (\rho v_R^2 + p)R \\ \rho v_R v_z R \\ (\rho e + p)v_R R \end{pmatrix} + \frac{\partial}{\partial z}\begin{pmatrix} \rho v_z \\ \rho v_R v_z \\ \rho v_z^2 + p \\ (\rho e + p)v_z \end{pmatrix} = \begin{pmatrix} 0 \\ p/R \\ 0 \\ 0 \end{pmatrix} \quad (20)$$

5 Conclusions

The fifth-order finite-volume WENO-C reconstruction scheme is proposed for structured grids in cylindrical coordinates to achieve high order spatial accuracy along with ENO transition. A grid independent smoothness indicator is derived for this scheme. For uniform grids, the analytical values in cylindrical-radial coordinates for the limiting case ($R \to \infty$) conform to WENO-JS. Linear stability analysis of the present scheme is performed using a scalar advection equation in radial coordinates. Several tests involving smooth and discontinuous flows are performed, which testify for the fifth-order accuracy and ENO property of the scheme.

References

1. Jiang, G.-S., Shu, C.-W.: Efficient implementation of weighted ENO schemes. J. Comput. Phys. **126**(1), 202–228 (1996)
2. Mignone, A.: High-order conservative reconstruction schemes for finite volume methods in cylindrical and spherical coordinates. J. Comput. Phys. **270**, 784–814 (2014)
3. Shadab, M.A., Balsara, D., Shyy, W., Xu, K.: Fifth order finite volume WENO in general orthogonally-curvilinear coordinates. Comput. Fluids. **190**, 398–424 (2019)
4. Liu, H., Jiao, X.: WLS-ENO: Weighted-least-squares based essentially non-oscillatory schemes for finite volume methods on unstructured meshes. J. Comput. Phys. **314**, 749–773 (2016)
5. Xu, K.: A gas-kinetic BGK scheme for the Navier–Stokes equations and its connection with artificial dissipation and Godunov method. J. Comput. Phys. **171**(1), 289–335 (2001)

Editorial Policy

1. Volumes in the following three categories will be published in LNCSE:

i)   Research monographs
ii) Tutorials
iii) Conference proceedings

Those considering a book which might be suitable for the series are strongly advised to contact the publisher or the series editors at an early stage.

2. Categories i) and ii). Tutorials are lecture notes typically arising via summer schools or similar events, which are used to teach graduate students. These categories will be emphasized by Lecture Notes in Computational Science and Engineering. **Submissions by interdisciplinary teams of authors are encouraged**. The goal is to report new developments – quickly, informally, and in a way that will make them accessible to non-specialists. In the evaluation of submissions timeliness of the work is an important criterion. Texts should be well-rounded, well-written and reasonably self-contained. In most cases the work will contain results of others as well as those of the author(s). In each case the author(s) should provide sufficient motivation, examples, and applications. In this respect, Ph.D. theses will usually be deemed unsuitable for the Lecture Notes series. Proposals for volumes in these categories should be submitted either to one of the series editors or to Springer-Verlag, Heidelberg, and will be refereed. A provisional judgement on the acceptability of a project can be based on partial information about the work: a detailed outline describing the contents of each chapter, the estimated length, a bibliography, and one or two sample chapters – or a first draft. A final decision whether to accept will rest on an evaluation of the completed work which should include

- at least 100 pages of text;
- a table of contents;
- an informative introduction perhaps with some historical remarks which should be accessible to readers unfamiliar with the topic treated;
- a subject index.

3. Category iii). Conference proceedings will be considered for publication provided that they are both of exceptional interest and devoted to a single topic. One (or more) expert participants will act as the scientific editor(s) of the volume. They select the papers which are suitable for inclusion and have them individually refereed as for a journal. Papers not closely related to the central topic are to be excluded. Organizers should contact the Editor for CSE at Springer at the planning stage, see *Addresses* below.

In exceptional cases some other multi-author-volumes may be considered in this category.

4. Only works in English will be considered. For evaluation purposes, manuscripts may be submitted in print or electronic form, in the latter case, preferably as pdf- or zipped ps-files. Authors are requested to use the LaTeX style files available from Springer at http://www.springer.com/gp/authors-editors/book-authors-editors/manuscript-preparation/5636 (Click on LaTeX Template → monographs or contributed books).

For categories ii) and iii) we strongly recommend that all contributions in a volume be written in the same LaTeX version, preferably LaTeX2e. Electronic material can be included if appropriate. Please contact the publisher.

Careful preparation of the manuscripts will help keep production time short besides ensuring satisfactory appearance of the finished book in print and online.

5. The following terms and conditions hold. Categories i), ii) and iii):

Authors receive 50 free copies of their book. No royalty is paid.
Volume editors receive a total of 50 free copies of their volume to be shared with authors, but no royalties.

Authors and volume editors are entitled to a discount of 40 % on the price of Springer books purchased for their personal use, if ordering directly from Springer.

6. Springer secures the copyright for each volume.

Addresses:

Timothy J. Barth
NASA Ames Research
Center NAS Division
Moffett Field, CA 94035, USA
barth@nas.nasa.gov

Michael Griebel
Institut für Numerische Simulation
der Universität Bonn
Wegelerstr. 6
53115 Bonn, Germany
griebel@ins.uni-bonn.de

David E. Keyes
Mathematical and Computer Sciences
and Engineering
King Abdullah University of Science
and Technology
P. O. Box 55455
Jeddah 21534, Saudi Arabia
david.keyes@kaust.edu.sa

and

Department of Applied Physics
and Applied Mathematics
Columbia University
500 W. 120th Street
New York, NY 10027, USA
kd2112@columbia.edu

Risto M. Nieminen
Department of Applied Physics
Aalto University School of Science
and Technology
00076 Aalto, Finland
risto.nieminen@aalto.fi

Dirk Roose
Department of Computer Science
Katholieke Universiteit Leuven
Celestijnenlaan 200A
3001 Leuven-Heverlee, Belgium
dirk.roose@cs.kuleuven.be

Tamar Schlick
Department of Chemistry
and Courant Institute
of Mathematical Sciences
New York University
251 Mercer Street
New York, NY 10012, USA
schlick@nyu.edu

Editor for Computational Science
and Engineering at Springer:

Martin Peters
Springer-Verlag
Mathematics Editorial IV
Tiergartenstrasse 17
69121 Heidelberg, Germany
martin.peters@springer.com

Lecture Notes in Computational Science and Engineering

1. D. Funaro, *Spectral Elements for Transport-Dominated Equations.*

2. H.P. Langtangen, *Computational Partial Differential Equations.* Numerical Methods and Diffpack Programming.

3. W. Hackbusch, G. Wittum (eds.), *Multigrid Methods V.*

4. P. Deuflhard, J. Hermans, B. Leimkuhler, A.E. Mark, S. Reich, R.D. Skeel (eds.), *Computational Molecular Dynamics: Challenges, Methods, Ideas.*

5. D. Kröner, M. Ohlberger, C. Rohde (eds.), *An Introduction to Recent Developments in Theory and Numerics for Conservation Laws.*

6. S. Turek, *Efficient Solvers for Incompressible Flow Problems.* An Algorithmic and Computational Approach.

7. R. von Schwenn, *Multi Body System SIMulation.* Numerical Methods, Algorithms, and Software.

8. H.-J. Bungartz, F. Durst, C. Zenger (eds.), *High Performance Scientific and Engineering Computing.*

9. T.J. Barth, H. Deconinck (eds.), *High-Order Methods for Computational Physics.*

10. H.P. Langtangen, A.M. Bruaset, E. Quak (eds.), *Advances in Software Tools for Scientific Computing.*

11. B. Cockburn, G.E. Karniadakis, C.-W. Shu (eds.), *Discontinuous Galerkin Methods.* Theory, Computation and Applications.

12. U. van Rienen, *Numerical Methods in Computational Electrodynamics.* Linear Systems in Practical Applications.

13. B. Engquist, L. Johnsson, M. Hammill, F. Short (eds.), *Simulation and Visualization on the Grid.*

14. E. Dick, K. Riemslagh, J. Vierendeels (eds.), *Multigrid Methods VI.*

15. A. Frommer, T. Lippert, B. Medeke, K. Schilling (eds.), *Numerical Challenges in Lattice Quantum Chromodynamics.*

16. J. Lang, *Adaptive Multilevel Solution of Nonlinear Parabolic PDE Systems.* Theory, Algorithm, and Applications.

17. B.I. Wohlmuth, *Discretization Methods and Iterative Solvers Based on Domain Decomposition.*

18. U. van Rienen, M. Günther, D. Hecht (eds.), *Scientific Computing in Electrical Engineering.*

19. I. Babuška, P.G. Ciarlet, T. Miyoshi (eds.), *Mathematical Modeling and Numerical Simulation in Continuum Mechanics.*

20. T.J. Barth, T. Chan, R. Haimes (eds.), *Multiscale and Multiresolution Methods.* Theory and Applications.

21. M. Breuer, F. Durst, C. Zenger (eds.), *High Performance Scientific and Engineering Computing.*

22. K. Urban, *Wavelets in Numerical Simulation.* Problem Adapted Construction and Applications.

23. L.F. Pavarino, A. Toselli (eds.), *Recent Developments in Domain Decomposition Methods.*

24. T. Schlick, H.H. Gan (eds.), *Computational Methods for Macromolecules: Challenges and Applications*.

25. T.J. Barth, H. Deconinck (eds.), *Error Estimation and Adaptive Discretization Methods in Computational Fluid Dynamics*.

26. M. Griebel, M.A. Schweitzer (eds.), *Meshfree Methods for Partial Differential Equations*.

27. S. Müller, *Adaptive Multiscale Schemes for Conservation Laws*.

28. C. Carstensen, S. Funken, W. Hackbusch, R.H.W. Hoppe, P. Monk (eds.), *Computational Electromagnetics*.

29. M.A. Schweitzer, *A Parallel Multilevel Partition of Unity Method for Elliptic Partial Differential Equations*.

30. T. Biegler, O. Ghattas, M. Heinkenschloss, B. van Bloemen Waanders (eds.), *Large-Scale PDE-Constrained Optimization*.

31. M. Ainsworth, P. Davies, D. Duncan, P. Martin, B. Rynne (eds.), *Topics in Computational Wave Propagation*. Direct and Inverse Problems.

32. H. Emmerich, B. Nestler, M. Schreckenberg (eds.), *Interface and Transport Dynamics*. Computational Modelling.

33. H.P. Langtangen, A. Tveito (eds.), *Advanced Topics in Computational Partial Differential Equations*. Numerical Methods and Diffpack Programming.

34. V John, *Large Eddy Simulation of Turbulent Incompressible Flows*. Analytical and Numerical Results for a Class of LES Models.

35. E. Bänsch (ed.), *Challenges in Scientific Computing- CISC 2002*.

36. B.N. Khoromskij, G. Wittum, *Numerical Solution of Elliptic Differential Equations by Reduction to the Interface*.

37. A. Iske, *Multiresolution Methods in Scattered Data Modelling*.

38. S.-I. Niculescu, K. Gu (eds.), *Advances in Time-Delay Systems*.

39. S. Attinger, P. Koumoutsakos (eds.), *Multiscale Modelling and Simulation*.

40. R. Kornhuber, R. Hoppe, J. Périaux, O. Pironneau, O. Wildlund, J. Xu (eds.), *Domain Decomposition Methods in Science and Engineering*.

41. T. Plewa, T. Linde, V.G. Weirs (eds.), *Adaptive Mesh Refinement – Theory and Applications*.

42. A. Schmidt, K.G. Siebert, *Design of Adaptive Finite Element Software*. The Finite Element Toolbox ALBERTA.

43. M. Griebel, M.A. Schweitzer (eds.), *Meshfree Methods for Partial Differential Equations II*.

44. B. Engquist, P. Lötstedt, O. Runborg (eds.), *Multiscale Methods in Science and Engineering*.

45. P. Benner, V. Mehrmann, D.C. Sorensen (eds.), *Dimension Reduction of Large-Scale Systems*.

46. D. Kressner, *Numerical Methods for General and Structured Eigenvalue Problems*.

47. A. Boriçi, A. Frommer, B. Joó, A. Kennedy, B. Pendleton (eds.), *QCD and Numerical Analysis III*.

48. F. Graziani (ed.), *Computational Methods in Transport*.

49. B. Leimkuhler, C. Chipot, R. Elber, A. Laaksonen, A. Mark, T. Schlick, C. Schütte, R. Skeel (eds.), *New Algorithms for Macromolecular Simulation*.

50. M. Bücker, G. Corliss, P. Hovland, U. Naumann, B. Norris (eds.), *Automatic Differentiation: Applications, Theory, and Implementations.*

51. A.M. Bruaset, A. Tveito (eds.), *Numerical Solution of Partial Differential Equations on Parallel Computers.*

52. K.H. Hoffmann, A. Meyer (eds.), *Parallel Algorithms and Cluster Computing.*

53. H.-J. Bungartz, M. Schäfer (eds.), *Fluid-Structure Interaction.*

54. J. Behrens, *Adaptive Atmospheric Modeling.*

55. O. Widlund, D. Keyes (eds.), *Domain Decomposition Methods in Science and Engineering XVI.*

56. S. Kassinos, C. Langer, G. Iaccarino, P. Moin (eds.), *Complex Effects in Large Eddy Simulations.*

57. M. Griebel, M.A Schweitzer (eds.), *Meshfree Methods for Partial Differential Equations III.*

58. A.N. Gorban, B. Kégl, D.C. Wunsch, A. Zinovyev (eds.), *Principal Manifolds for Data Visualization and Dimension Reduction.*

59. H. Ammari (ed.), *Modeling and Computations in Electromagnetics: A Volume Dedicated to Jean-Claude Nédélec.*

60. U. Langer, M. Discacciati, D. Keyes, O. Widlund, W. Zulehner (eds.), *Domain Decomposition Methods in Science and Engineering XVII.*

61. T. Mathew, *Domain Decomposition Methods for the Numerical Solution of Partial Differential Equations.*

62. F. Graziani (ed.), *Computational Methods in Transport: Verification and Validation.*

63. M. Behrendorf, *Hierarchical Matrices. A Means to Efficiently Solve Elliptic Boundary Value Problems.*

64. C.H. Bischof, H.M. Bücker, P. Hovland, U. Naumann, J. Utke (eds.), *Advances in Automatic Differentiation.*

65. M. Griebel, M.A. Schweitzer (eds.), *Meshfree Methods for Partial Differential Equations IV.*

66. B. Engquist, P. Lötstedt, O. Runborg (eds.), *Multiscale Modeling and Simulation in Science.*

67. I.H. Tuncer, Ü. Gülcat, D.R. Emerson, K. Matsuno (eds.), *Parallel Computational Fluid Dynamics 2007.*

68. S. Yip, T. Diaz de la Rubia (eds.), *Scientific Modeling and Simulations.*

69. A. Hegarty, N. Kopteva, E. O'Riordan, M. Stynes (eds.), *BAIL 2008 – Boundary and Interior Layers.*

70. M. Bercovier, M.J. Gander, R. Kornhuber, O. Widlund (eds.), *Domain Decomposition Methods in Science and Engineering XVIII.*

71. B. Koren, C. Vuik (eds.), *Advanced Computational Methods in Science and Engineering.*

72. M. Peters (ed.), *Computational Fluid Dynamics for Sport Simulation.*

73. H.-J. Bungartz, M. Mehl, M. Schäfer (eds.), *Fluid Structure Interaction II - Modelling, Simulation, Optimization.*

74. D. Tromeur-Dervout, G. Brenner, D.R. Emerson, J. Erhel (eds.), *Parallel Computational Fluid Dynamics 2008.*

75. A.N. Gorban, D. Roose (eds.), *Coping with Complexity: Model Reduction and Data Analysis.*

76. J.S. Hesthaven, E.M. Rønquist (eds.), *Spectral and High Order Methods for Partial Differential Equations.*

77. M. Holtz, *Sparse Grid Quadrature in High Dimensions with Applications in Finance and Insurance.*

78. Y. Huang, R. Kornhuber, O.Widlund, J. Xu (eds.), *Domain Decomposition Methods in Science and Engineering XIX.*

79. M. Griebel, M.A. Schweitzer (eds.), *Meshfree Methods for Partial Differential Equations V.*

80. P.H. Lauritzen, C. Jablonowski, M.A. Taylor, R.D. Nair (eds.), *Numerical Techniques for Global Atmospheric Models.*

81. C. Clavero, J.L. Gracia, F.J. Lisbona (eds.), *BAIL 2010 – Boundary and Interior Layers, Computational and Asymptotic Methods.*

82. B. Engquist, O. Runborg, Y.R. Tsai (eds.), *Numerical Analysis and Multiscale Computations.*

83. I.G. Graham, T.Y. Hou, O. Lakkis, R. Scheichl (eds.), *Numerical Analysis of Multiscale Problems.*

84. A. Logg, K.-A. Mardal, G. Wells (eds.), *Automated Solution of Differential Equations by the Finite Element Method.*

85. J. Blowey, M. Jensen (eds.), *Frontiers in Numerical Analysis - Durham 2010.*

86. O. Kolditz, U.-J. Gorke, H. Shao, W. Wang (eds.), *Thermo-Hydro-Mechanical-Chemical Processes in Fractured Porous Media- Benchmarks and Examples.*

87. S. Forth, P. Hovland, E. Phipps, J. Utke, A. Walther (eds.), *Recent Advances in Algorithmic Differentiation.*

88. J. Garcke, M. Griebel (eds.), *Sparse Grids and Applications.*

89. M. Griebel, M.A. Schweitzer (eds.), *Meshfree Methods for Partial Differential Equations VI.*

90. C. Pechstein, *Finite and Boundary Element Tearing and Interconnecting Solvers for Multiscale Problems.*

91. R. Bank, M. Holst, O. Widlund, J. Xu (eds.), *Domain Decomposition Methods in Science and Engineering XX.*

92. H. Bijl, D. Lucor, S. Mishra, C. Schwab (eds.), *Uncertainty Quantification in Computational Fluid Dynamics.*

93. M. Bader, H.-J. Bungartz, T. Weinzierl (eds.), *Advanced Computing.*

94. M. Ehrhardt, T. Koprucki (eds.), *Advanced Mathematical Models and Numerical Techniques for Multi-Band Effective Mass Approximations.*

95. M. Azaïez, H. El Fekih, J.S. Hesthaven (eds.), *Spectral and High Order Methods for Partial Differential Equations ICOSAHOM 2012.*

96. F. Graziani, M.P. Desjarlais, R. Redmer, S.B. Trickey (eds.), *Frontiers and Challenges in Warm Dense Matter.*

97. J. Garcke, D. Pflüger (eds.), *Sparse Grids and Applications – Munich 2012.*

98. J. Erhel, M. Gander, L. Halpern, G. Pichot, T. Sassi, O. Widlund (eds.), *Domain Decomposition Methods in Science and Engineering XXI.*

99. R. Abgrall, H. Beaugendre, P.M. Congedo, C. Dobrzynski, V. Perrier, M. Ricchiuto (eds.), *High Order Nonlinear Numerical Methods for Evolutionary PDEs - HONOM 2013.*

100. M. Griebel, M.A. Schweitzer (eds.), *Meshfree Methods for Partial Differential Equations VII.*

101. R. Hoppe (ed.), *Optimization with PDE Constraints - OPTPDE 2014.*

102. S. Dahlke, W. Dahmen, M. Griebel, W. Hackbusch, K. Ritter, R. Schneider, C. Schwab, H. Yserentant (eds.), *Extraction of Quantifiable Information from Complex Systems.*

103. A. Abdulle, S. Deparis, D. Kressner, F. Nobile, M. Picasso (eds.), *Numerical Mathematics and Advanced Applications - ENUMATH 2013.*

104. T. Dickopf, M.J. Gander, L. Halpern, R. Krause, L.F. Pavarino (eds.), *Domain Decomposition Methods in Science and Engineering XXII.*

105. M. Mehl, M. Bischoff, M. Schäfer (eds.), *Recent Trends in Computational Engineering - CE2014.* Optimization, Uncertainty, Parallel Algorithms, Coupled and Complex Problems.

106. R.M. Kirby, M. Berzins, J.S. Hesthaven (eds.), *Spectral and High Order Methods for Partial Differential Equations - ICOSAHOM'14.*

107. B. Jüttler, B. Simeon (eds.), *Isogeometric Analysis and Applications 2014.*

108. P. Knobloch (ed.), *Boundary and Interior Layers, Computational and Asymptotic Methods – BAIL 2014.*

109. J. Garcke, D. Pflüger (eds.), *Sparse Grids and Applications – Stuttgart 2014.*

110. H. P. Langtangen, *Finite Difference Computing with Exponential Decay Models.*

111. A. Tveito, G.T. Lines, *Computing Characterizations of Drugs for Ion Channels and Receptors Using Markov Models.*

112. B. Karazösen, M. Manguoğlu, M. Tezer-Sezgin, S. Göktepe, Ö. Uğur (eds.), *Numerical Mathematics and Advanced Applications - ENUMATH 2015.*

113. H.-J. Bungartz, P. Neumann, W.E. Nagel (eds.), *Software for Exascale Computing - SPPEXA 2013–2015.*

114. G.R. Barrenechea, F. Brezzi, A. Cangiani, E.H. Georgoulis (eds.), *Building Bridges: Connections and Challenges in Modern Approaches to Numerical Partial Differential Equations.*

115. M. Griebel, M.A. Schweitzer (eds.), *Meshfree Methods for Partial Differential Equations VIII.*

116. C.-O. Lee, X.-C. Cai, D.E. Keyes, H.H. Kim, A. Klawonn, E.-J. Park, O.B. Widlund (eds.), *Domain Decomposition Methods in Science and Engineering XXIII.*

117. T. Sakurai, S.-L. Zhang, T. Imamura, Y. Yamamoto, Y. Kuramashi, T. Hoshi (eds.), *Eigenvalue Problems: Algorithms, Software and Applications in Petascale Computing.* EPASA 2015, Tsukuba, Japan, September 2015.

118. T. Richter (ed.), *Fluid-structure Interactions.* Models, Analysis and Finite Elements.

119. M.L. Bittencourt, N.A. Dumont, J.S. Hesthaven (eds.), *Spectral and High Order Methods for Partial Differential Equations ICOSAHOM2016.* Selected Papers from the ICOSAHOM Conference, June 27-July 1, 2016, Rio de Janeiro, Brazil.

120. Z. Huang, M. Stynes, Z. Zhang (eds.), *Boundary and Interior Layers, Computational and Asymptotic Methods BAIL 2016.*

121. S.P.A. Bordas, E.N. Burman, M.G. Larson, M.A. Olshanskii (eds.), *Geometrically Unfitted Finite Element Methods and Applications.* Proceedings of the UCL Workshop 2016.

122. A. Gerisch, R. Penta, J. Lang (eds.), *Multiscale Models in Mechano and Tumor Biology*. Modeling, Homogenization, and Applications.

123. J. Garcke, D. Pflüger, C.G. Webster, G. Zhang (eds.), *Sparse Grids and Applications - Miami 2016*.

124. M. Schäfer, M. Behr, M. Mehl, B. Wohlmuth (eds.), *Recent Advances in Computational Engineering*. Proceedings of the 4th International Conference on Computational Engineering (ICCE 2017) in Darmstadt.

125. P.E. Bjørstad, S.C. Brenner, L. Halpern, R. Kornhuber, H.H. Kim, T. Rahman, O.B. Widlund (eds.), *Domain Decomposition Methods in Science and Engineering XXIV*. 24th International Conference on Domain Decomposition Methods, Svalbard, Norway, February 6–10, 2017.

126. F.A. Radu, K. Kumar, I. Berre, J.M. Nordbotten, I.S. Pop (eds.), *Numerical Mathematics and Advanced Applications – ENUMATH 2017*.

127. X. Roca, A. Loseille (eds.), *27th International Meshing Roundtable*.

128. Th. Apel, U. Langer, A. Meyer, O. Steinbach (eds.), *Advanced Finite Element Methods with Applications*. Selected Papers from the 30th Chemnitz Finite Element Symposium 2017.

129. M. Griebel, M.A. Schweitzer (eds.), *Meshfree Methods for Partial Differential Equations IX*.

130. S. Weißer, *BEM-based Finite Element Approaches on Polytopal Meshes*.

131. V.A. Garanzha, L. Kamenski, H. Si (eds.), *Numerical Geometry, Grid Generation and Scientific Computing*. Proceedings of the 9th International Conference, NUMGRIG2018/Voronoi 150, Celebrating the 150th Anniversary of G. F. Voronoi, Moscow, Russia, December 2018.

For further information on these books please have a look at our mathematics catalogue at the following URL: www.springer.com/series/3527

Monographs in Computational Science and Engineering

1. J. Sundnes, G.T. Lines, X. Cai, B.F. Nielsen, K.-A. Mardal, A. Tveito, *Computing the Electrical Activity in the Heart.*

For further information on this book, please have a look at our mathematics catalogue at the following URL: www.springer.com/series/7417

Texts in Computational Science and Engineering

1. H. P. Langtangen, *Computational Partial Differential Equations.* Numerical Methods and Diffpack Programming. 2nd Edition

2. A. Quarteroni, F. Saleri, P. Gervasio, *Scientific Computing with MATLAB and Octave.* 4th Edition

3. H. P. Langtangen, *Python Scripting for Computational Science.* 3rd Edition

4. H. Gardner, G. Manduchi, *Design Patterns for e-Science.*

5. M. Griebel, S. Knapek, G. Zumbusch, *Numerical Simulation in Molecular Dynamics.*

6. H. P. Langtangen, *A Primer on Scientific Programming with Python.* 5th Edition

7. A. Tveito, H. P. Langtangen, B. F. Nielsen, X. Cai, *Elements of Scientific Computing.*

8. B. Gustafsson, *Fundamentals of Scientific Computing.*

9. M. Bader, *Space-Filling Curves.*

10. M. Larson, F. Bengzon, *The Finite Element Method: Theory, Implementation and Applications.*

11. W. Gander, M. Gander, F. Kwok, *Scientific Computing: An Introduction using Maple and MATLAB.*

12. P. Deuflhard, S. Röblitz, *A Guide to Numerical Modelling in Systems Biology.*

13. M. H. Holmes, *Introduction to Scientific Computing and Data Analysis.*

14. S. Linge, H. P. Langtangen, *Programming for Computations - A Gentle Introduction to Numerical Simulations with MATLAB /Octave.*

15. S. Linge, H. P. Langtangen, *Programming for Computations - A Gentle Introduction to Numerical Simulations with Python.*

16. H.P. Langtangen, S. Linge, *Finite Difference Computing with PDEs - A* Modern Software Approach.

17. B. Gustafsson, *Scientific Computing from a Historical Perspective.*

18. J. A. Trangenstein, *Scientific Computing.* Volume I - Linear and Nonlinear Equations.

19. J. A. Trangenstein, *Scientific Computing*. Volume II - Eigenvalues and Optimization.

20. J. A. Trangenstein, *Scientific Computing*. Volume III - Approximation and Integration.

For further information on these books please have a look at our mathematics catalogue at the following URL: www.springer.com/series/5151

Printed in the United States
by Baker & Taylor Publisher Services